Conversion from U.S. customary units to SI units

To convert from	To	Multiply by: Accurate†	Multiply by: Common
Foot (ft)	Metre (m)	3.048 000 E − 01*	0.305
Horsepower (hp)	Watt (W)	7.456 999 E + 02	746
Inch (in)	Metre (m)	2.540 000 E − 02*	0.025 4
Mile, U.S. statute (mi)	Metre (m)	1.609 344 E + 03*	1610
Pound force (lb)	Newton (N)	4.448 222 E + 00	4.45
Pound mass (lbm)	Kilogram (kg)	4.535 924 E − 01	0.454
Poundal (lbm · ft/s^2)	Newton (N)	1.382 550 E − 01	0.138
Pound-foot (lb · ft)	Newton-metre (N · m)	1.355 818 E + 00	1.35
	Joule (J)	1.355 818 E + 00	1.35
Pound-foot/second (lb · ft/s)	Watt (W)	1.355 818 E + 00	1.35
Pound-inch (lb · in)	Newton-metre (N · m)	1.128 182 E − 01	0.113
	Joule (J)	1.128 182 E − 01	0.113
Pound-inch/second (lb · in/s)	Watt (W)	1.128 182 E − 01	0.113
Pound/foot^2 (lb/ft^2)	Pascal (Pa)	4.788 026 E + 01	47.9
Pound/inch^2 (lb/in^2), (psi)	Pascal (Pa)	6.894 757 E + 03	6890
Revolutions/minute (rpm)	Radian/second (rad/s)	1.047 198 E − 01	0.105
Slug	Kilogram (kg)	1.459 390 E + 01	14.6
Ton, short (2000 lbm)	Kilogram (kg)	9.071 847 E + 02	907

† An asterisk indicates that the conversion factor is exact.

THEORY OF MACHINES AND MECHANISMS

McGraw-Hill Series in Mechanical Engineering

Consulting Editor
Jack P. Holman, *Southern Methodist University*

Barron: *Cryogenic Systems*
Eckert: *Introduction to Heat and Mass Transfer*
Eckert and Drake: *Analysis of Heat and Mass Transfer*
Eckert and Drake: *Heat and Mass Transfer*
Ham, Crane, and Rogers: *Mechanics of Machinery*
Hartenberg and Denavit: *Kinematic Synthesis of Linkages*
Hinze: *Turbulence*
Hutton: *Applied Mechanical Vibrations*
Jacobsen and Ayre: *Engineering Vibrations*
Juvinall: *Engineering Considerations of Stress, Strain, and Strength*
Kays and Crawford: *Convective Heat and Mass Transfer*
Lichty: *Combustion Engine Processes*
Martin: *Kinematics and Dynamics of Machines*
Phelan: *Dynamics of Machinery*
Phelan: *Fundamentals of Mechanical Design*
Raven: *Automatic Control Engineering*
Schenck: *Theories of Engineering Experimentation*
Schlichting: *Boundary-Layer Theory*
Shigley: *Dynamic Analysis of Machines*
Shigley: *Kinematic Analysis of Mechanisms*
Shigley: *Mechanical Engineering Design*
Shigley: *Simulation of Mechanical Systems*
Shigley and Uicker: *Theory of Machines and Mechanisms*
Stoecker: *Refrigeration and Air Conditioning*

THEORY OF MACHINES AND MECHANISMS

Joseph Edward Shigley

Professor Emeritus of Mechanical Engineering
The University of Michigan

John Joseph Uicker, Jr.

Professor of Mechanical Engineering
University of Wisconsin, Madison

McGraw-Hill Book Company

New York St. Louis San Francisco Auckland Bogotá Hamburg
Johannesburg London Madrid Mexico Montreal New Delhi Panama
Paris São Paulo Singapore Sydney Tokyo Toronto

THEORY OF MACHINES AND MECHANISMS

890 DODO 898765

This book was set in Times Roman.
The editors were Julienne V. Brown and James W. Bradley;
the production supervisor was Phil Galea.
R. R. Donnelley & Sons Company was printer and binder.

Library of Congress Cataloging in Publication Data

Shigley, Joseph Edward.
Theory of machines and mechanisms.

(McGraw-Hill series in mechanical engineering)
Includes index.
1. Mechanical engineering. I. Uicker, John Joseph, joint author. II. Title.
TJ145.S54 621.8 79-19705
ISBN 0-07-056884-7

CONTENTS

PREFACE

This book is intended to cover that field of engineering theory, analysis, design, and practice which is generally described as mechanisms and kinematics and dynamics of machines. While the book is written primarily for students of engineering, there is much material that will be of value to practicing engineers. After all, a good engineer knows that he or she must remain a student throughout his or her professional career.

The continued tremendous growth of knowledge of kinematics and dynamics of machinery in the past decade has resulted in a strengthening of the engineering curricula in many schools by the substitution of these subjects for weaker ones, and has produced a need for a textbook that satisfies the requirements of these new course structures. Much of this new knowledge exists in a large variety of technical papers, each couched in its own singular language and nomenclature, and each requiring additional background for comprehension. These individual contributions can be used to strengthen the engineering course structure by first providing the necessary foundation and establishing a common notation and nomenclature. These new developments can then be integrated into the existing body of knowledge so as to provide a logical, a modern, and a comprehensive treatise. That, briefly, is the purpose of this book.

So as to develop a broad and basic comprehension, all the methods of analysis and development common to the literature of the field are employed. We have used graphical methods of analysis and synthesis extensively throughout the book because we are firmly of the opinion that graphical computation is so basic and teachable. In addition, it is usually the fastest method of checking the results of machine computation. Conventional vector analysis and the Chace method of vector analysis are used too, because of their compactness, because they are employed in so much research literature, and because they are easy to program for computer analysis. We have also, for the same reasons, employed Raven's method, especially in the basic chapters. Finally, complex-number methods, both polar and rectangular, as well as algebraic methods, are used freely throughout the book.

With certain exceptions we have endeavored to use English units and SI units in about equal proportions in this book. The International System of Units (SI) is presented and used in this book in accordance with the rules and recommendations as given in the National Bureau of Standards Special Publication 330, revised August 1977.†

One of the dilemmas that all writers on the subject of this book have faced is how to distinguish between the motion of two different points on the same moving body and the motion of two different points on two moving bodies. This dilemma always arises with the coincident-point problem where both kinds of motion occur. In the past it has been customary to describe both of these motions as "relative motion"; but since there are two kinds, the student has difficulty in distinguishing between them. We believe we have solved this problem by the introduction of the terms *motion difference* and *apparent motion.* Thus, for example, the book contains the terms *velocity difference* and *apparent velocity* instead of the term "relative velocity," which will not be found. This approach is introduced beginning with the concepts of position and displacement, used extensively in the chapter on velocity, and brought to its fulfillment with the coincident-point problem in the chapter on acceleration where the Coriolis component arises.

The extensive use of machine methods of computation, especially by practicing engineers, has made it necessary that we include a chapter on numerical methods. The home and office computers such as the programmable calculators and the microcomputers are so useful for solving entire cycles of motion that their use is already quite extensive. And the computer-aided design methods with graphic-display terminals used in conjunction with large-capacity computers are proving to be of great value in the solution of many complex problems of analysis and synthesis of mechanisms and machines. In this as well as other chapters of the book where computer methods of analysis are discussed, we have taken great care to avoid the presentation of specific programs and computer languages. Programming is a highly individual effort and most people prefer to write their own program using a computer language of their choice. For these reasons we have presented the program steps required to solve many analytical problems that occur frequently, and have added suggestions that we feel will be helpful. Such an approach will not become obsolete as computers and the languages used with them undergo the expected changes.

† In the past, the Americans and the English have differed on the correct spelling of the words metre and kilogram: the English used metre and kilogramme; the usual American spellings were meter and kilogram. When SI units were first introduced into the United States, the spellings metre and kilogram were recommended with the hope of securing worldwide uniformity in the English spelling of the names of SI units. But the Americans have resisted the new spelling so strongly that the authorities have now sanctioned the use of the spelling meter for the unit of length. However, we have chosen to use the spelling metre for the unit of length in this book because of the international character of the book.

The methods of cam design needed to deliver a specified motion, and the kinematic and dynamic performance of cam systems, are treated extensively using graphical, analytical, and machine-computation methods. A new set of cam-design charts is presented which greatly shortens the kinematic-design time needed. And the methods of dynamic analysis used make it easy, for example, to choose a follower retaining spring so as to avoid follower jump or lift-off, and to compute the contact and camshaft bearing forces.

The kinematic and dynamic analysis of gears and gear trains is treated thoroughly. Especially useful for the analysis of planetary trains are the twelve variations of Lévai and the Lévai notation, which are included.

The research literature concerned with the design or synthesis of linkages to accomplish specific purposes is so great that it would take one person many months to digest it all. We believe that Chapter 10, Synthesis of Linkages, contains enough of the techniques to permit one to solve most of the synthesis problems that arise in engineering. Both graphical and analytical methods are used. Position and path synthesis of the slider-crank and crank-rocker mechanisms are discussed extensively.

The chapter on spatial mechanisms contains all the material needed for a thorough introduction to the subject and its problems. Three-dimensional problems are, in fact, a natural and obvious extension for the reader and not a special case. Both graphical and analytical methods are used in the kinematic analyses of position, velocity, and acceleration of this class of mechanisms.

The two chapters on static and dynamic analysis of forces in machine systems define the terminology and methods that are used in the remaining chapters of the book. Graphical, vectorial, and machine methods of computation are used in about equal proportions. These chapters include material on the concept of mass moment of inertia and its experimental measurement. While most readers will have had a previous introduction to the concept of moment of inertia, teaching experience has shown that it is important to reemphasize this topic in the study of dynamics.

It is also important to include material on the dynamics of reciprocating engines in a dynamics of machinery study. The engine mechanism is a simple and excellent example of the need for force analysis of bearings and sliders, of the need for the balancing of machine systems and components, and of the need for the use of flywheels in machinery.

The study of balancing begins with explanations of the causes and effects of rotating unbalance together with a brief discussion of balancing machines. The two-plane field balancing problem for large rotors is analyzed in detail because this is such a good problem for solution by the programmable calculator. The balancing of single- and multicylinder engines is explained using the imaginary-mass or imaginary-rotor approach. The volume of literature concerned with the balancing of linkages, such as the four-bar mechanism, is so large that it is difficult to make a completely satisfactory selection. We have chosen to present the Berkof-Lowen method of balancing of linkages because it is quite general, it is comprehensive, it can be applied to any linkage system, and because it utilizes the fundamentals already discussed in the book. The problem of force balancing of complete machines as well as the balancing of the shaking moment is also discussed in the balancing chapter.

We are very proud to acknowledge the assistance of Professors George N. Sandor of the University of Florida, Sanjay G. Dhande of the University of Florida, Dennis A. Guenther of the Ohio State University, Glenn C. Tolle of Texas A & M University, Robert A. Lucas of Lehigh University, Edward N. Stevensen, Jr., of the University of Hartford, and Robert J. Williams of Pennsylvania State University in planning the book and reviewing and advising us on the preliminary manuscript and outline. Their critical analyses and careful comments have helped us greatly in organizing the methods and the content of the book.

The final manuscript was reviewed in detail by Professors Robert W. Adamson of California Polytechnic State University, Ferdinand Freudenstein of Columbia University, and Edward N. Stevensen, Jr., of the University of Hartford. We appreciate very much the time and effort spent by these people in aiding us to put the finishing touches on the manuscript.

Finally, we wish to acknowledge our indebtedness to our editor, Julienne V. Brown. Her enthusiasm and her willingness to go the second mile in helping us to work out the thorniest of problems is sincerely appreciated.

Joseph Edward Shigley
John Joseph Uicker, Jr.

CHAPTER

ONE

THE GEOMETRY OF MOTION

1-1 INTRODUCTION

The theory of mechanisms and machines is an applied science which is used to understand the relationships between the geometry and motions of the parts of a machine or mechanism and the forces which produce these motions. The subject, and therefore this book, naturally divides itself into three parts. Chapters 1 to 5 are concerned with kinematics, the analysis of the motions of machine parts. This lays the groundwork for Chaps. 6 to 11, where we study methods of design of mechanisms and machine components. Finally, in Chaps. 12 to 17, we take up the study of kinetics, the time-varying forces in machines and the resulting dynamic phenomena which must be considered in their design.

As shown in Fig. 1-1, the design of a modern machine is often very complex. In the design of a new engine, for example, the automotive engineer must deal with many interrelated questions. What is the relationship between the motion of the piston and the motion of the crankshaft? What will be the sliding velocities and the loads at the lubricated surfaces, and what lubricants are available for the purpose? How much heat will be generated, and how will the engine be cooled? What are the synchronization and control requirements, and how will they be met? What will be the cost to the consumer, both for initial purchase and for continued operation and maintenance? What materials and manufacturing methods will be used? What will be the fuel economy, noise, and exhaust emissions; will they meet legal requirements? Although all these and many other important questions must be answered before the design can be completed, obviously not all can be addressed in a book of this

Figure 1-1 A Figee floating crane with lemniscate boom configuration (*B.V. Machinefabriek Figee, Haarlem, Holland*).

size. Just as people with diverse skills must be brought together to produce an adequate design, so too many branches of science must be brought to bear. This book brings together material which falls into the science of mechanics as it relates to the design of mechanisms and machines.

1-2 ANALYSIS AND SYNTHESIS

There are two completely different aspects of the study of mechanical systems, *design* and *analysis*. The concept embodied in the word "design" might be more properly termed *synthesis*, the process of contriving a scheme or a method of accomplishing a given purpose. Design is the process of prescribing the sizes, shapes, material compositions, and arrangements of parts so that the resulting machine will perform the prescribed task.

Although there are many phases in the design process which can be approached in a well-ordered, scientific manner, the overall process is by its very nature as much of an art as a science. It calls for imagination, intuition, creativity, judgment, and experience. The role of science in the design process is merely to provide tools to be used by the designers as they practice their art.

It is in the process of evaluating the various interacting alternatives that designers find need for a large collection of mathematical and scientific tools. These tools, when properly applied, can provide more accurate and more reliable information for use in judging a design than one can achieve through intuition or estimation. Thus they can be of tremendous help in deciding between alternatives. However, scientific tools cannot make decisions for designers; they have every right to exert their imagination and creative abilities, even to the extent of overruling the mathematical predictions.

Probably the largest collection of scientific methods at the designer's disposal fall into the category called *analysis*. These are the techniques which allow the designer to critically examine an already existing or proposed design in order to judge its suitability for the task. Thus analysis, in itself, is not a creative science but one of evaluation and rating of things already conceived.

We should always bear in mind that although most of our effort may be spent on analysis, the real goal is synthesis, the design of a machine or system. Analysis is simply a tool. It is, however, a vital tool and will inevitably be used as one step in the design process.

1-3 THE SCIENCE OF MECHANICS

That branch of scientific analysis which deals with motions, time, and forces is called *mechanics* and is made up of two parts, statics and dynamics. *Statics* deals with the analysis of stationary systems, i.e., those in which time is not a factor, and *dynamics* deals with systems which change with time.

As shown in Fig. 1-2, dynamics is also made up of two major disciplines, first recognized as separate entities by Euler in 1775:†

> The investigation of the motion of a rigid body may be conveniently separated into two parts, the one geometrical, the other mechanical. In the first part, the transference of the body from a given position to any other position must be investigated without respect to the causes of the motion, and must be represented by analytical formulae, which will define the position of each point of the body. This investigation will therefore be referable solely to geometry, or rather to stereotomy.
>
> It is clear that by the separation of this part of the question from the other, which belongs properly to Mechanics, the determination of the motion from dynamical principles will be made much easier than if the two parts were undertaken conjointly.

These two aspects of dynamics were later recognized as the distinct sciences of *kinematics* (from the Greek word *kinema*, meaning motion) and *kinetics*, and deal with motion and the forces producing it, respectively.

† *Novi comment. Acad. Petrop.*, vol. 20, 1775; also in "Theoria motus corporum," 1790. The translation is by Willis, "Principles of Mechanism," 2nd ed., p. viii, 1870.

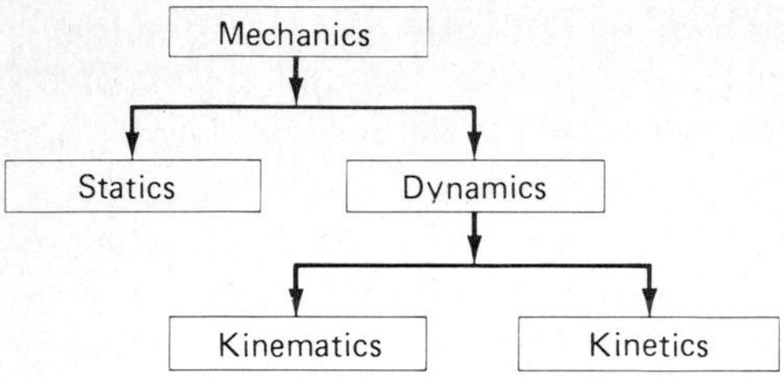

Figure 1-2

The initial problem in the design of a mechanical system is therefore understanding its kinematics. *Kinematics* is the study of motion, quite apart from the forces which produce that motion. More particularly, kinematics is the study of position, displacement, rotation, speed, velocity, and acceleration. The study, say, of planetary or orbital motion is also a problem in kinematics, but in this book we shall concentrate our attention on kinematic problems which arise in the design of mechanical systems. Thus, the kinematics of machines and mechanisms is the focus of the next several chapters of this book. Statics and kinetics, however, are also vital parts of a complete design analysis, and they are covered as well in later chapters.

It should be carefully noted in the above quotation that Euler based his separation of dynamics into kinematics and kinetics on the assumption that they should deal with *rigid* bodies. It is this very important assumption that allows the two to be treated separately. For flexible bodies, the shapes of the bodies themselves, and therefore their motions, depend on the forces exerted on them. In this situation, the study of force and motion must take place simultaneously, thus significantly increasing the complexity of the analysis.

Fortunately, although all real machine parts are flexible to some degree, machines are usually designed from relatively rigid materials, keeping part deflections to a minimum. Therefore, it is common practice to *assume* that deflections are negligible and parts are rigid when analyzing a machine's kinematic performance, and then, after the dynamic analysis when loads are known, to design the parts so that this assumption is justified.

1-4 TERMINOLOGY, DEFINITIONS, AND ASSUMPTIONS

Reuleaux† defines a *machine*‡ as a "*combination of resistant bodies so arranged that by their means the mechanical forces of nature can be com-*

† Much of the material of this section is based on definitions originally set down by F. Reuleaux (1829–1905), a German kinematician whose work marked the beginning of a systematic treatment of kinematics. For additional reading see A. B. W. Kennedy, "*Reuleaux' Kinematics of Machinery*," Macmillan, London, 1876; republished by Dover, New York, 1963.

‡ There appears to be no agreement at all on the proper definition of a machine. In a footnote Reuleaux gives 17 definitions, and his translator gives 7 more and discusses the whole problem in detail.

pelled to do work accompanied by certain determinate motions." He also defines a *mechanism* as an "*assemblage of resistant bodies, connected by movable joints, to form a closed kinematic chain with one link fixed and having the purpose of transforming motion.*"

Some light can be shed on these definitions by contrasting them with the term *structure.* A structure is also a combination of resistant (rigid) bodies connected by joints, but its purpose is not to do work or to transform motion. A structure (such as a truss) is intended to be rigid. It can perhaps be moved from place to place and is movable in this sense of the word; however, it has no *internal* mobility, no *relative motions* between its various members, whereas both machines and mechanisms do. Indeed the whole purpose of a machine or mechanism is to utilize these relative internal motions in transmitting power or transforming motion.

A machine is an arrangement of parts for doing work, a device for applying power or changing its direction. It differs from a mechanism in its purpose. In a machine, terms such as force, torque, work, and power describe the predominant concepts. In a mechanism, though it may transmit power or force, the predominant idea in the mind of the designer is one of achieving a desired motion. There is a direct analogy between the terms structure, mechanism, and machine, and the three branches of mechanics shown in Fig. 1-2. The term "structure" is to statics as the term "mechanism" is to kinematics as the term "machine" is to kinetics.

We shall use the word *link* to designate a machine part or a component of a mechanism. As discussed in the previous section, a link is assumed to be completely rigid. Machine components which do not fit this assumption of rigidity, such as springs, usually have no effect on the kinematics of a device but do play a role in supplying forces. Such members are not called links; they are usually ignored during kinematic analysis, and their force effects are introduced during dynamic analysis. Sometimes, as with a belt or chain, a machine member may possess one-way rigidity; such a member would be considered a link when in tension but not under compression.

The links of a mechanism must be connected together in some manner in order to transmit motion from the *driver,* or input link, to the *follower,* or output link. These connections, joints between the links, are called *kinematic pairs* (or just *pairs*) because each joint consists of a pair of mating surfaces, two *elements,* one mating surface or element being a part of each of the joined links. Thus we can also define a link as the *rigid connection between two or more elements of different kinematic pairs.*

Stated explicitly, the assumption of rigidity is that there can be no relative motion (change in distance) between two arbitrarily chosen points on the same link. In particular, the relative positions of pairing elements on any given link do not change. In other words, the purpose of a link is to hold a constant spatial relationship between the elements of its pairs.

As a result of the assumption of rigidity, many of the intricate details of the actual part shapes are unimportant when studying the kinematics of a machine or mechanism. For this reason it is common practice to draw highly simplified schematic diagrams, which contain important features of the shape of each link, such as the relative locations of pair elements, but which completely subdue the real geometry of the manufactured parts. The slider-crank mechanism of the internal combustion engine, for example, can be simplified to the schematic diagram shown in Fig. 1-4*b* for purposes of analysis. Such simplified schematics are a great help since they eliminate confusing factors which do not affect the analysis; such diagrams are used extensively throughout this text. However, these schematics also have the drawback of bearing little resemblance to physical hardware. As a result they may give the impression that they represent only academic constructs rather than real machinery. We should always bear in mind that these simplified diagrams are intended to carry only the minimum necessary information so as not to confuse the issue with all the unimportant detail (for kinematic purposes) or complexity of the true machine parts.

When several links are movably connected together by joints, they are said to form a *kinematic chain*. Links containing only two pair-element connections are called *binary* links; those having three are called *ternary* links, and so on. If every link in the chain is connected to at least two other links, the chain forms one or more closed loops and is called a *closed* kinematic chain; if not, the chain is referred to as *open*. When no distinction is made the chain is assumed closed. If the chain consists entirely of binary links, it is *simple-closed; compound-closed* chains, however, include other than binary links and thus form more than a single closed loop.

Recalling Reuleaux' definition of a mechanism, we see that it is necessary to have a closed kinematic chain *with one link fixed*. When we say that one link is fixed, we mean that it is chosen as a frame of reference for all other links; i.e., that the motions of all other points on the linkage will be measured with respect to this link, thought of as being fixed. This link in a practical machine usually takes the form of a stationary platform or base (or a housing rigidly attached to such a base) and is called the *frame* or *base link*. The question of whether this reference frame is truly stationary (in the sense of being an inertial reference frame) is immaterial in the study of kinematics but becomes important in the investigation of kinetics, where forces are to be considered. In either case, once a frame member is designated (and other conditions are met), the kinematic chain becomes a mechanism and as the driver is moved through various positions, called *phases*, all other links have well-defined motions with respect to the chosen frame of reference. We use the term *kinematic chain* to specify a particular arrangement of links and joints when it is not clear which link is to be treated as the frame. When the frame link is specified, the kinematic chain is called a *mechanism*.

In order for a mechanism to be useful, the motions between links cannot be completely arbitrary; they too must be constrained to produce the *proper* relative motions, those chosen by the designer for the particular task to be performed. These desired relative motions are obtained by a proper choice of the number of links and the kinds of joints used to connect them.

Thus we are led to the concept that, in addition to the distances between successive joints, the nature of the joints themselves and the relative motions which they permit are essential in determining the kinematics of a mechanism. For this reason it is important to look more closely at the nature of joints in general terms, and in particular at several of the more common types.

The controlling factor which determines the relative motions allowed by a given joint is the shapes of the mating surfaces or elements. Each type of joint has its own characteristic shapes for the elements, and each allows a given type of motion, which is determined by the possible ways in which these elemental surfaces can move with respect to each other. For example, the pin joint in Fig. 1-3*a* has cylindric elements and, assuming that the links cannot slide axially, these surfaces only permit relative rotational motion. Thus a pin joint allows the two connected links to experience relative rotation about the pin center. So too, other joints each have their own characteristic element shapes and relative motions. These shapes restrict the totally arbitrary motion of two unconnected links to some prescribed type of relative motion and form the constraining conditions or *constraints* on the mechanism's motion.

It should be pointed out that the element shapes may often be subtly disguised and difficult to recognize. For example, a pin joint might include a needle bearing so that two mating surfaces, as such, are not distinguishable.

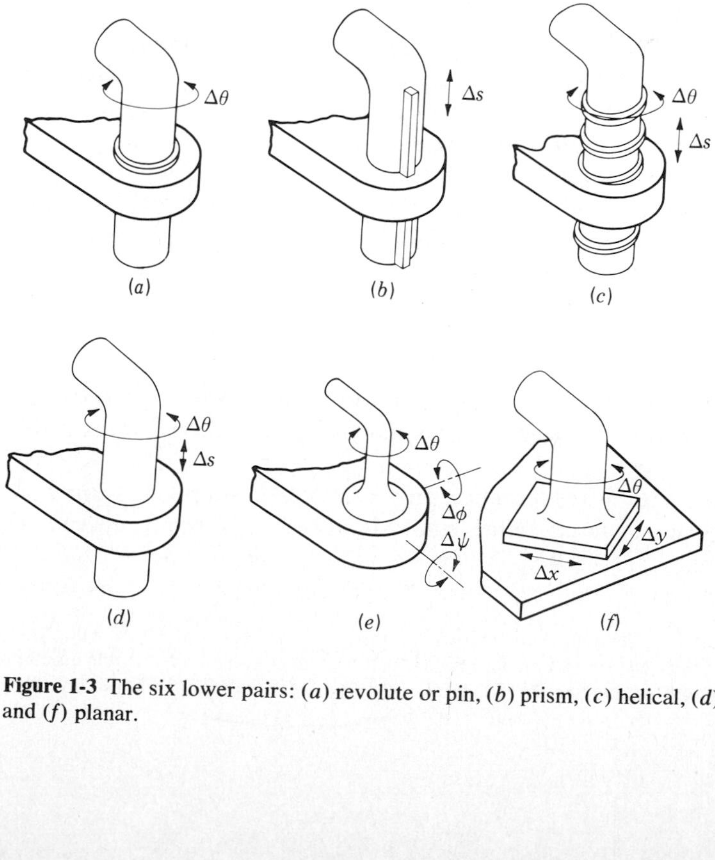

Figure 1-3 The six lower pairs: (*a*) revolute or pin, (*b*) prism, (*c*) helical, (*d*) cylindric, (*e*) spheric, and (*f*) planar.

Nevertheless, if the motions of the individual rollers are not of interest, the motions allowed by the joints are equivalent and the pairs are of the same generic type. Thus the criterion for distinguishing different pair types is the relative motions which they permit and not necessarily the shapes of the elements, though these may provide vital clues. The diameter of the pin used (or other dimension data) is also of no more importance than the exact sizes and shapes of the connected links. As stated previously, the kinematic function of a link is to hold a fixed geometrical relationship between the pair elements. In a similar way, the only kinematic function of a joint or pair is to determine the relative motion between the connected links. All other features are determined for other reasons and are unimportant in the study of kinematics.

When a kinematic problem is formulated, it is necessary to recognize the type of relative motion permitted in each of the pairs and to assign to it some variable parameter(s) for measuring or calculating the motion. There will be as many of these parameters as there are degrees of freedom of the joint in question, and they are referred to as the *pair variables*. Thus the pair variable of a pinned joint will be a single angle measured between reference lines fixed in the adjacent links, while a spheric pair will have three pair variables (all angles) to specify its three-dimensional rotation.

Kinematic pairs were divided by Reuleaux into *higher pairs* and *lower pairs*, the latter category consisting of six prescribed types to be discussed next. He distinguished between the categories by noting that the lower pairs, such as the pin joint, have surface contact between the pair elements, while higher pairs, such as the connection between a cam and its follower, have line or point contact between the elemental surfaces. However, as noted in the case of a needle bearing, this criterion may be misleading. We should rather look for distinguishing features in the relative motion(s) which the joint allows.

The six lower pairs are illustrated in Fig. 1-3. Table 1-1 lists the names of the lower pairs and the symbols employed by Hartenberg and Denavit† for them, together with the number of degrees of freedom and the pair variables for each of the six.

The *turning pair* or *revolute* (Fig. 1-3*a*) permits only relative rotation and hence has one degree of freedom. This pair is often referred to as a pin joint.

The *prismatic pair* (Fig. 1-3*b*) permits only a relative sliding motion and therefore is often called a sliding joint. It also has a single degree of freedom.

† R. S. Hartenberg and J. Denavit, "Kinematic Synthesis of Linkages," McGraw-Hill, New York, 1964. This book is a classic on kinematics and the title is misleading; a considerable amount of material on the history of kinematics, kinematic theory, and kinematic analysis is also included.

Table 1-1 The lower pairs

Pair	Symbol	Pair variable	Degrees of freedom	Relative motion
Revolute	R	$\Delta\theta$	1	Circular
Prism	P	Δs	1	Linear
Screw	S	$\Delta\theta$ or ΔS	1	Helical
Cylinder	C	$\Delta\theta$ and Δs	2	Cylindric
Sphere	G	$\Delta\theta, \Delta\phi, \Delta\psi$	3	Spheric
Flat	F	$\Delta x, \Delta y, \Delta\theta$	3	Planar

The *screw pair* or *helical pair* (Fig. 1-3*c*) has only one degree of freedom because the sliding and rotational motions are related by the helix angle of the thread. Thus the pair variable may be chosen as either Δs or $\Delta\theta$ but not both. Note that the screw pair reverts to a revolute if the helix angle is made zero and to a prismatic pair if the helix angle is made 90°.

The *cylindric pair* (Fig. 1-3*d*) permits both angular rotation and an independent sliding motion. Thus the cylindric pair has two degrees of freedom.

The *globular* or *spheric pair* (Fig. 1-3*e*) is a ball-and-socket joint. It has three degrees of freedom, a rotation about each of the coordinate axes.

The *flat pair* or *planar pair* (Fig. 1-3*f*) is seldom, if ever, found in mechanisms in its undisguised form. It has three degrees of freedom.

All other joint types are called higher pairs. Examples are mating gear teeth, a wheel rolling on a rail, a ball rolling on a flat surface, and a cam contacting its roller follower. Since there are an infinite number of higher pairs, a systematic accounting of them is not a realistic objective. We shall treat each as a separate situation as it arises.

Among the higher pairs there is a subcategory known as *wrapping pairs*. Examples are the connection between a belt and pulley, between a chain and sprocket, or between a rope and a drum. In each case one of the links has one-way rigidity.

In the treatment of various joint types, whether lower or higher pairs, there is another important limiting assumption. Throughout the book, we will assume that the actual joint, as manufactured, can be reasonably represented by a mathematical abstraction having perfect geometry. That is, when a real machine joint is assumed to be a spheric pair, for example, it is also assumed that there is no "play" or clearance between the joint elements and that any deviations in the spherical geometry of the elements is negligible. When a pin joint is treated as a revolute, it is assumed that no axial motion can take place; if it is necessary to study the small axial motions resulting from clearances between the real elements, the joint must be treated as cylindric, thus allowing for the axial motion.

The term "mechanism," as defined earlier, can refer to a wide variety of devices, including both higher and lower pairs. There is a more restrictive term, however, which refers to those mechanisms having only lower pairs; such a mechanism is called a *linkage.* A linkage, then, is connected only by the lower pairs of Fig. 1-3.

1-5 PLANAR, SPHERICAL, AND SPATIAL MECHANISMS

Mechanisms may be categorized in several different ways to emphasize their similarities and differences. One such grouping divides mechanisms into *planar, spherical,* and *spatial* categories. All three groups have many things in common; the criterion which distinguishes the groups, however, is to be found in the characteristics of the motions of the links.

A *planar mechanism* is one in which all particles describe plane curves in space and all these curves lie in parallel planes; i.e., the loci of all points are plane curves parallel to a single common plane. This characteristic makes it possible to represent the locus of any chosen point of a planar mechanism in its true size and shape on a single drawing or figure. The motion transformation of any such mechanism is called *coplanar.* The plane four-bar linkage, the plate cam and follower, and the slider-crank mechanism are familiar examples of planar mechanisms. The vast majority of mechanisms in use today are planar.

Planar mechanisms utilizing only lower pairs are called *planar linkages;* they may include only revolute and prismatic pairs. Although a planar pair might theoretically be included, this would impose no constraint and thus be equivalent to an opening in the kinematic chain. Planar motion also requires that all revolute axes be normal to the plane of motion, and that all prismatic pair axes be parallel to the plane.

A *spherical mechanism* is one in which each link has some point which remains stationary as the linkage moves and in which the stationary points of all links lie at a common location; i.e., the locus of each point is a curve contained in a spherical surface, and the spherical surfaces defined by several arbitrarily chosen points are all *concentric.* The motions of all particles can therefore be completely described by their radial projections, or "shadows," on the surface of a sphere with properly chosen center. Hooke's universal joint is perhaps the most familiar example of a spherical mechanism.

Spherical linkages are constituted entirely of revolute pairs. A spheric pair would produce no additional constraints and would thus be equivalent to an opening in the chain, while all other lower pairs have nonspheric motion. In spheric linkages, the axes of all revolute pairs must intersect at a point.

Spatial mechanisms, on the other hand, include *no* restrictions on the relative motions of the particles. The motion transformation is not necessarily coplanar, nor must it be concentric. A spatial mechanism may have particles with loci of double curvature. Any linkage which contains a screw pair, for example, is a spatial mechanism, since the relative motion within a screw pair is helical.

Thus, the overwhelmingly large category of planar mechanisms and the category of spherical mechanisms are only special cases, or subsets, of the all-inclusive category spatial mechanisms. They occur as a consequence of special geometry in the particular orientations of their pair axes.

If planar and spherical mechanisms are only special cases of spatial mechanisms, why is it desirable to identify them separately? Because of the particular geometric conditions which identify these types, many simplifications are possible in their design and analysis. As pointed out earlier, it is possible to observe the motions of all particles of a planar mechanism in true size and shape from a single direction. In other words, all motions can be represented graphically in a single view. Thus, graphical techniques are well suited to their solution. Since spatial mechanisms do not all have this fortunate geometry, visualization becomes more difficult and more powerful techniques must be developed for their analysis.

Since the vast majority of mechanisms in use today are planar, one might question the need of the more complicated mathematical techniques used for spatial mechanisms. There are a number of reasons why more powerful methods are of value even though the simpler graphical techniques have been mastered.

1. They provide new, alternative methods which will solve the problems in a different way. Thus they provide a means of checking results. Certain problems by their nature may also be more amenable to one method than another.
2. Methods which are analytical in nature are better suited to solution by calculator or digital computer than graphical techniques.
3. Even though the majority of useful mechanisms are planar and well suited to graphical solution, the few remaining must also be analyzed, and techniques should be known for analyzing them.
4. One reason that planar linkages are so common is that good methods of analysis for the more general spatial linkages have not been available until quite recently. Without methods for their analysis, their design and use has not been common, even though they may be inherently better suited in certain applications.
5. We will discover that spatial linkages are much more common in practice than their formal description indicates.

Consider a four-bar linkage. It has four links connected by four pins whose axes are parallel. This "parallelism" is a mathematical hypothesis; it is not a reality. The axes as produced in a shop—in any shop, no matter how good—will only be approximately parallel. If they are far out of parallel, there will be binding in no uncertain terms, and the mechanism will only move because the "rigid" links flex and twist, producing loads in the bearings. If the axes are nearly parallel, the mechanism operates because of the looseness of the running fits of the bearings or flexibility of the links. A common way of compensating for small nonparallelism is to connect the links with self-aligning bearings, actually spherical joints allowing three-dimensional rotation. Such a "planar" linkage is thus a low-grade spatial linkage.

1-6 MOBILITY

One of the first concerns in either the design or the analysis of a mechanism is the number of degrees of freedom, also called the *mobility* of the device. The mobility† of a mechanism is the number of input parameters (usually pair variables) which must be independently controlled in order to bring the device into a particular position. Ignoring for the moment certain exceptions to be mentioned later, it is possible to determine the mobility of a mechanism directly from a count of the number of links and the number and types of joints which it includes.

To develop this relationship, consider that before they are connected together, each link of a planar mechanism has three degrees of freedom when moving relative to the fixed link. Not counting the fixed link, therefore, an n-link planar mechanism has $3(n-1)$ degrees of freedom before any of the joints are connected. Connecting a joint which has one degree of freedom, such as a revolute pair, has the effect of providing two constraints between the connected links. If a two degree-of-freedom pair is connected, it provides one constraint. When the constraints for all joints are subtracted from the total freedoms of the unconnected links, we find the resulting mobility of the connected mechanism. When we use j_1 to denote the number of single-degree-of-freedom pairs and j_2 for the number of two-degree-of-freedom pairs, the resulting mobility m of a planar n-link mechanism is given by

$$m = 3(n-1) - 2j_1 - j_2 \tag{1-1}$$

Written in this form, Eq. (1-1) is called the *Kutzbach criterion* for the mobility of a planar mechanism. Its application is shown for several simple cases in Fig. 1-4.

If the Kutzbach criterion yields $m > 0$, the mechanism has m degrees of freedom. If $m = 1$, the mechanism can be driven by a single input motion. If $m = 2$, then two separate input motions are necessary to produce constrained motion for the mechanism; such a case is shown in Fig. 1-4*d*.

If the Kutzbach criterion yields $m = 0$, as in Fig. 1-4*a*, motion is im-

† The German literature distinguishes between *movability* and *mobility*. Movability includes the six degrees of freedom of the device as a whole, as though the ground link were not fixed, and thus applies to a kinematic chain. Mobility neglects these and considers only the internal relative motions, thus applying to a mechanism. The English literature seldom recognizes this distinction, and the terms are used somewhat interchangeably.

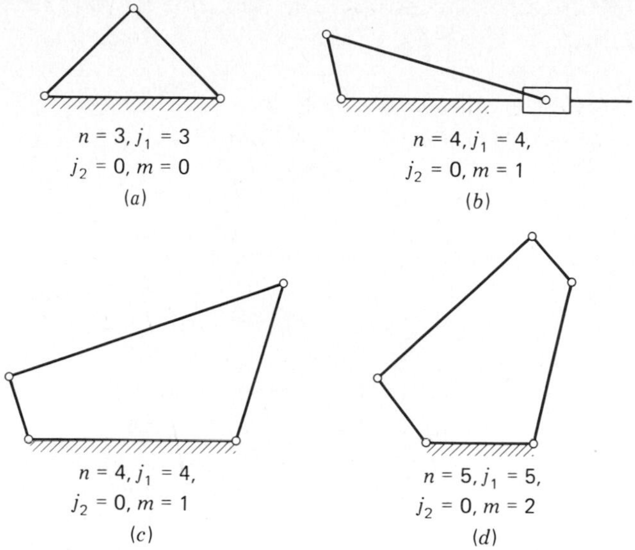

Figure 1-4 Applications of the Kutzbach mobility criterion.

possible and the mechanism forms a structure. If the criterion gives $m = -1$ or less, then there are redundant constraints in the chain and it forms a statically indeterminate structure. Examples are shown in Fig. 1-5. Note in these examples that when three links are joined by a single pin, two joints must be counted; such a connection is treated as two separate but concentric pairs.

Figure 1-6 shows examples of Kutzbach's criterion applied to mechanisms with two-degree-of-freedom joints. Particular attention should be paid to the contact (pair) between the wheel and the fixed link in Fig. 1-6*b*. Here it was assumed that slipping is possible between the links. If this contact included gear teeth or if friction was high enough to prevent slipping, the joint would be counted as a one-degree-of-freedom pair, since only one relative motion would be possible between the links.

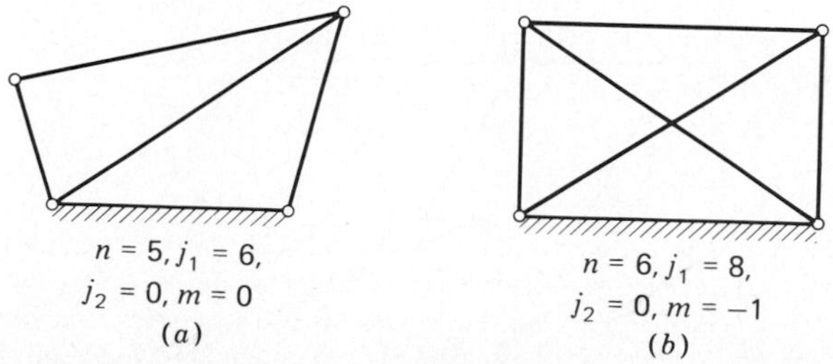

Figure 1-5 Applications of the Kutzbach criterion to structures.

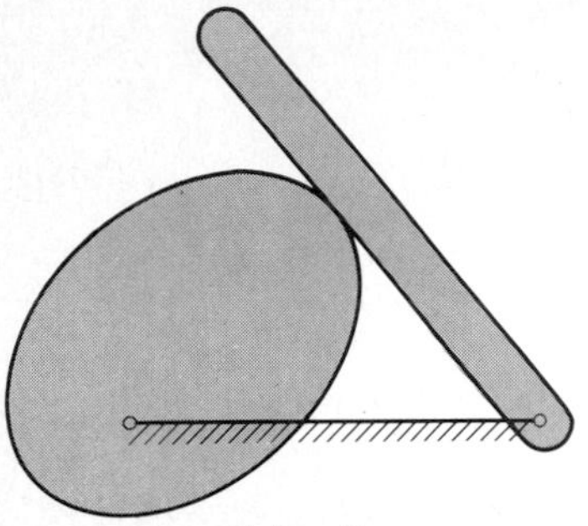

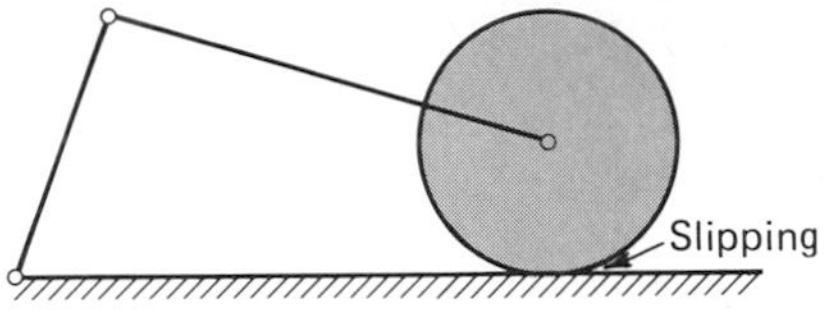

Figure 1-6

Sometimes the Kutzbach criterion will give an incorrect result. Notice that Fig. 1-7*a* represents a structure and that the criterion properly predicts $m = 0$. However, if link 5 is arranged as in Fig. 1-7*b*, the result is a double-parallelogram linkage with a mobility of 1 even though Eq. (1-1) indicates that it is a structure. The actual mobility of 1 results only if the parallelogram geometry is achieved. Since in the development of the Kutzbach criterion no consideration was given to the lengths of the links or other dimensional properties, it is not surprising that exceptions to the criterion can be found for particular cases with equal link lengths, parallel links, or other special geometric features.

Even though the criterion has exceptions, it remains useful because it is so easily applied. To avoid exceptions it would be necessary to include all the dimensional properties of the mechanism. The resulting criterion would be very complex and would be useless at the early stages of design when dimensions may not be known.

An earlier mobility criterion named after Grübler applies to mechanisms with only single-degree-of-freedom joints where the overall mobility of the mechanism is unity. Putting $j_2 = 0$ and $m = 1$ into Eq. (1-1), we find *Grübler's*

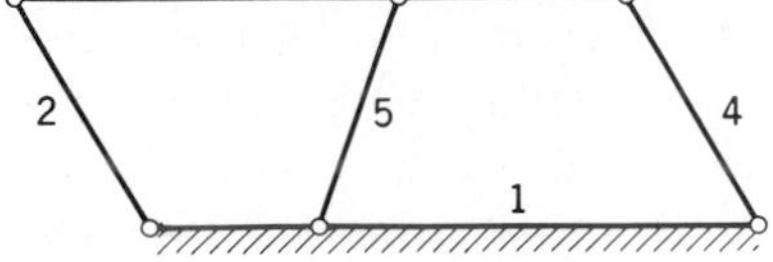

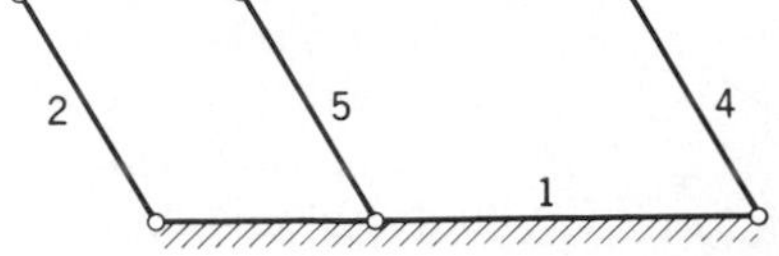

Figure 1-7

criterion for planar mechanisms with constrained motion

$$3n - 2j_1 - 4 = 0 \tag{1-2}$$

From this we can see, for example, that a planar mechanism with a mobility of 1 and only single-degree-of-freedom joints cannot have an odd number of links. Also, we can find the simplest possible mechanism of this type; by assuming all binary links we find $n = j_1 = 4$. This shows why the four-bar linkage, Fig. 1-4*c*, and the slider-crank mechanism, Fig. 1-4*b*, are so common in application.

Both the Kutzbach criterion, Eq. (1-1), and the Grübler criterion, Eq. (1-2), were derived for the case of planar mechanisms. If similar criteria are developed for spatial mechanisms, we must recall that each unconnected link has six degrees of freedom and each revolute pair, for example, provides five constraints. Similar arguments then lead to the three-dimensional form of the Kutzbach criterion,

$$m = 6(n - 1) - 5j_1 - 4j_2 - 3j_3 - 2j_4 - j_5 \tag{1-3}$$

and the Grübler criterion

$$6n - 5j_1 - 7 = 0 \tag{1-4}$$

The simplest form of a spatial mechanism† with all single-freedom pairs and a mobility of 1 is therefore, $n = j_1 = 7$.

1-7 KINEMATIC INVERSION

In Sec. 1-4 we noted that every mechanism has a fixed link called the frame. Until a frame link has been chosen, a connected set of links is called a kinematic chain. When different links are chosen as the frame for a given kinematic chain, the *relative* motions between the various links are not altered but their *absolute* motions (those measured with respect to the frame link) may be changed drastically. The process of choosing different links of a chain for the frame is known as *kinematic inversion*.

In an n-link kinematic chain, choosing each link in turn as the frame yields n distinct kinematic inversions of the chain, n different mechanisms. As an example, the four-link slider-crank chain of Fig. 1-8 has four different inversions.

Figure 1-8*a* shows the basic slider-crank mechanism, as found in most internal combustion engines today. Link 4, the piston, is driven by the expanding gases and forms the input; link 2, the crank, is the driven output. The frame is the cylinder block, link 1. By reversing the roles of the input and output, this same mechanism can be used as a compressor.

† Note that all planar mechanisms are exceptions to the spatial-mobility criteria. They have special geometric characteristics in that all revolute axes are parallel and perpendicular to the plane of motion and all prism axes lie in the plane of motion.

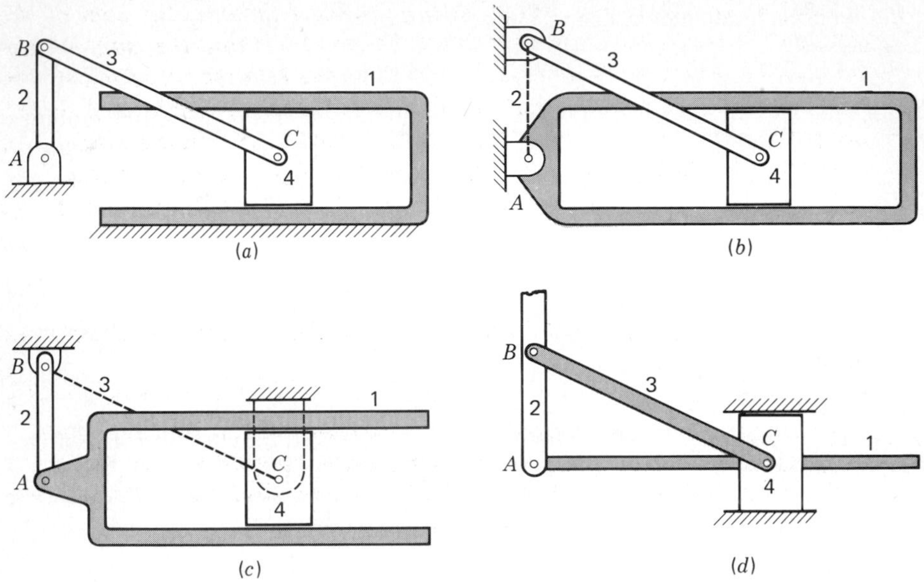

Figure 1-8 Four inversions of the slider-crank mechanism.

Figure 1-8*b* shows the same kinematic chain; however, it is now inverted and link 2 is stationary. Link 1, formerly the frame, now rotates about the revolute at *A*. This inversion of the slider-crank mechanism was used as the basis of the rotary engine found in early aircraft.

Another inversion of the same slider-crank chain is shown in Fig. 1-8*c*; it has link 3, formerly the connecting rod, as the frame link. This mechanism was used to drive the wheels of early steam locomotives, link 2 being a wheel.

The fourth and final inversion of the slider-crank chain has the piston, link 4, stationary. Although it is not found in engines, by rotating the figure 90° clockwise this mechanism can be recognized as part of a garden water pump. It will be noted in the figure that the prismatic pair connecting links 1 and 4 is also inverted; i.e. the "inside" and "outside" elements of the pair have been reversed.

1-8 GRASHOF'S LAW

A very important consideration when designing a mechanism to be driven by a motor, obviously, is to ensure that the input crank can make a complete revolution. Mechanisms in which no link makes a complete revolution would not be useful in such applications. For the four-bar linkage, there is a very simple test of whether this is the case.

Grashof's law states that *for a planar four-bar linkage, the sum of the shortest and longest link lengths cannot be greater than the sum of the remaining two link lengths if there is to be continuous relative rotation between two members.* This is illustrated in Fig. 1-9, where the longest link has length l, the shortest link has length s, and the other two links have lengths p and q. In this notation, Grashof's law states that one of the links, in particular the shortest link, will rotate continuously relative to the other three links if and only if

$$s + l \leq p + q \tag{1-5}$$

If this inequality is not satisfied, no link will make a complete revolution relative to another.

Attention is called to the fact that nothing in Grashof's law specifies the order in which the links are connected or which link of the four-bar chain is fixed. We are free, therefore, to fix any of the four links. When we do so, we create the four inversions of the four-bar linkage shown in Fig. 1-9. All these

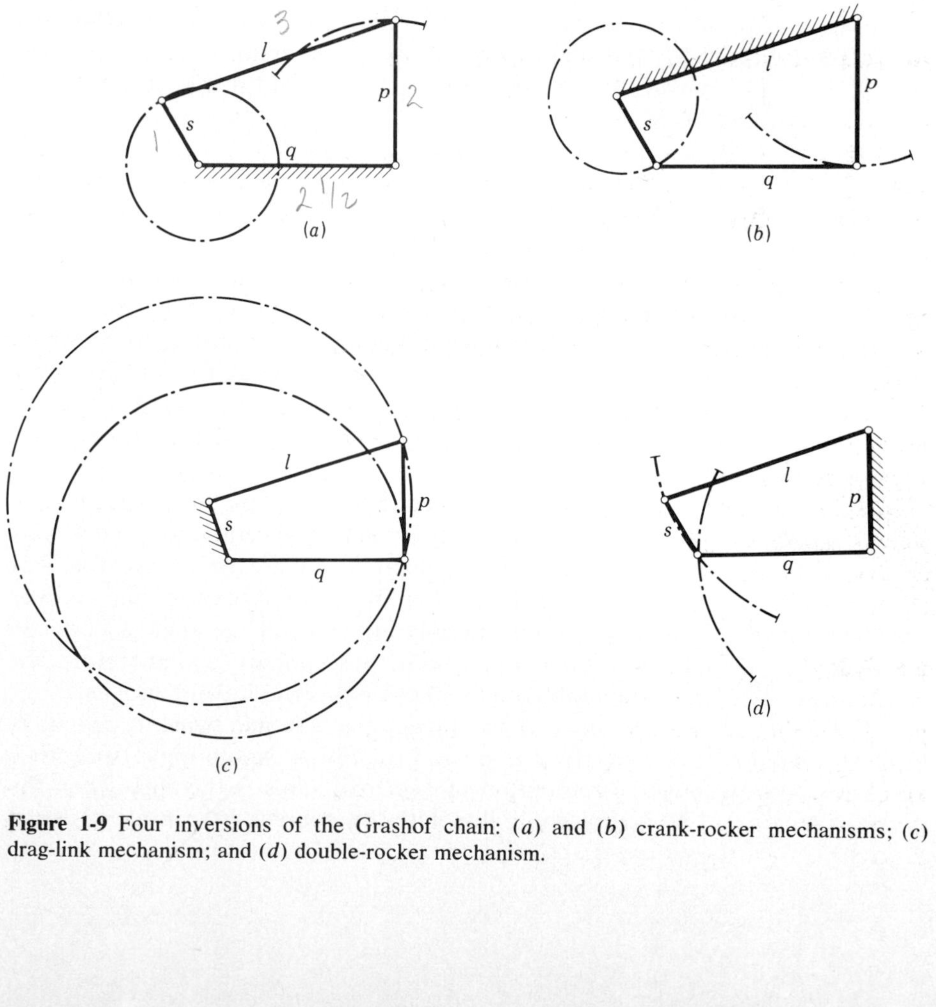

Figure 1-9 Four inversions of the Grashof chain: (*a*) and (*b*) crank-rocker mechanisms; (*c*) drag-link mechanism; and (*d*) double-rocker mechanism.

fit Grashof's law, and in each the link s makes a complete revolution relative to the other links. The different inversions are distinguished by the location of the link s relative to the fixed link.

If the shortest link s is adjacent to the fixed link, as shown in Fig. 1-9*a* and *b*, we obtain what is called a *crank-rocker* linkage. Link s is, of course, the crank since it is able to rotate continuously, and link p, which can only oscillate between limits, is the rocker.

The *drag-link* mechanism, also called the *double-crank* linkage, is obtained by fixing the shortest link s as the frame. In this inversion, shown in Fig. 1-9*c*, both links adjacent to s can rotate continuously, and both are properly described as cranks; the shorter of the two is generally used as the input. Although this is a very common mechanism, you will find it an interesting challenge to devise a practical working model which can operate through the full cycle.

By fixing the link opposite to s we obtain the fourth inversion, the *double-rocker* mechanism of Fig. 1-9*d*. Note that although link s is able to make a complete revolution, neither link adjacent to the frame can do so; both must oscillate between limits and are therefore rockers.

In each of these inversions, the shortest link s is adjacent to the longest link l. However, exactly the same types of linkage inversions will occur if the longest link l is opposite the shortest link s; you should demonstrate this to your own satisfaction.

1-9 MECHANICAL ADVANTAGE

Because of the widespread use of the four-bar linkage, a few remarks are in order here which will help to judge the quality of such a linkage for its intended application. Consider the four-bar linkage shown in Fig. 1-10. Since, according to Grashof's law, this particular linkage is of the crank-rocker variety, it is likely that link 2 is the driver and link 4 is the follower. Link 1 is the frame and link 3 is called the *coupler* since it couples the motions of the input and output cranks.

The *mechanical advantage* of a linkage is the ratio of the output torque exerted by the driven link to the necessary input torque required at the driver. In Sec. 3-16 we will prove that the mechanical advantage of the four-bar linkage is directly proportional to the sine of the angle γ between the coupler and the follower and inversely proportional to the sine of the angle β between the coupler and the driver. Of course, both these angles, and therefore the mechanical advantage, are continuously changing as the linkage moves.

When the sine of the angle β becomes zero, the mechanical advantage becomes infinite; thus, at such a position, only a small input torque is necessary to overcome a large output torque load. This is the case when the driver AB of Fig. 1-10 is directly in line with the coupler BC; it occurs when

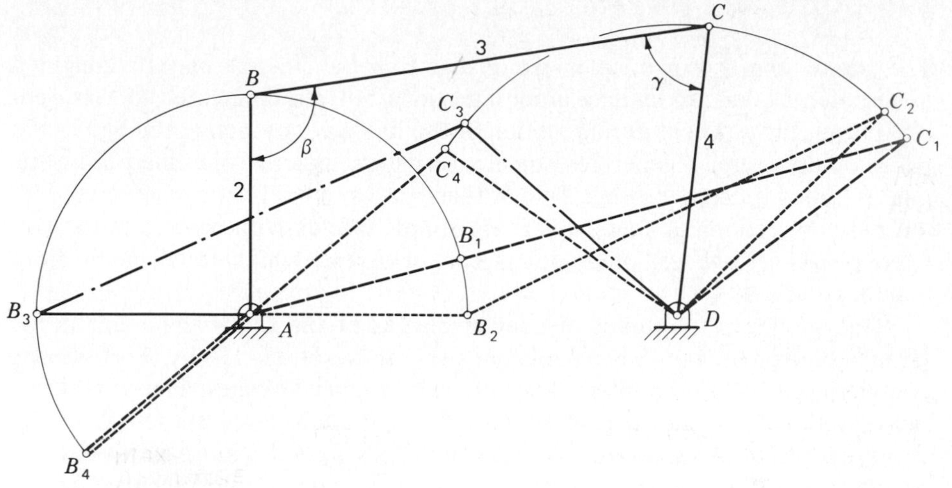

Figure 1-10

the crank is in position AB_1 and again when the crank is in position AB_4. Note that these also define the extreme positions of travel of the rocker DC_1 and DC_4. When the four-bar linkage is in either of these positions, the mechanical advantage is infinite and the linkage is said to be in a *toggle* position.

The angle γ between the coupler and the follower is called the *transmission angle.* As this angle becomes small, the mechanical advantage decreases and even a small amount of friction will cause the mechanism to lock or jam. A common rule of thumb is that a four-bar linkage should not be used in the region where the transmission angle is less than, say, 45 or 50°. The extreme values of the transmission angle occur when the crank AB lies along the line of the frame AD. In Fig. 1-10 the transmission angle is minimum when the crank is in position AB_2 and maximum when the crank has position AB_3. Because of the ease with which it can be visually inspected, the transmission angle has become a commonly accepted measure of the quality of the design of a four-bar linkage.

Note that the definitions of mechanical advantage, toggle, and transmission angle depend on the choice of the driver and driven links. If, in the same figure, link 4 is used as the driver and link 2 as the follower, the roles of β and γ are reversed. In this case the linkage has no toggle position, and its mechanical advantage becomes zero when link 2 is in position AB_1 or AB_4 since the transmission angle is then zero.

These and other methods of rating the suitability of the four-bar or other linkages are discussed more thoroughly in Sec. 3-16.

1-10 COUPLER CURVES

The connecting rod or coupler of a planar four-bar linkage may be imagined as an infinite plane extending in all directions but pin-connected to the input and output links. Then, during motion of the linkage, any point attached to the plane of the coupler generates some path with respect to the fixed link; this path is called a *coupler curve.* Two of these paths, namely, those generated by the pin connections of the coupler, are simple circles with centers at the two fixed pivots. However, other points can be found which trace much more complex curves.

One of the best sources of coupler curves of the four-bar linkage is the Hrones-Nelson atlas.† This book consists of a set of 11- by 17-in charts containing over 7000 coupler curves of crank-rocker linkages. Figure 1-11 is a reproduction of a typical page of this atlas. In each case, the length of the crank is unity and the lengths of other links are varied from page to page to produce the different combinations. On each page a number of different coupler points are chosen and their coupler curves are shown. This atlas of coupler curves is invaluable to the designer who needs a linkage to generate a curve with specified characteristics.

The algebraic equation of a coupler curve is, in general, of sixth order; thus it is possible to find coupler curves with a wide variety of shapes and many interesting features. Some coupler curves have sections which are nearly straight line segments (see Sec. 1-11); others have circular arc sections; still others have one or more cusps or cross over themselves like a figure eight. Therefore it is often not necessary to use a mechanism with a large number of links to obtain a fairly complex motion.

Yet the complexity of the coupler-curve equation is also a hindrance; it means that hand calculation methods can become very cumbersome. Thus, over the years, many mechanisms have been designed by strictly intuitive procedures and proved with cardboard models, without the use of kinematic principles or procedures. Until quite recently, those techniques which did offer a rational approach have been graphical, again avoiding tedious computations. Finally, with the availability of digital computers, and particularly with the growth of computer graphics, useful design methods are now emerging which can deal directly with the complicated calculations required without burdening the designer with the computational drudgery (see Sec. 5-5 for details of these design methods).

One of the more curious and interesting facts about the coupler-curve equation is that the same curve can always be generated by three different linkages. These are called *cognate linkages*, and the theory is developed in Sec. 10-11.

† J. A. Hrones and G. L. Nelson, "Analysis of the Four-Bar Linkage," M.I.T.-Wiley, New York, 1951.

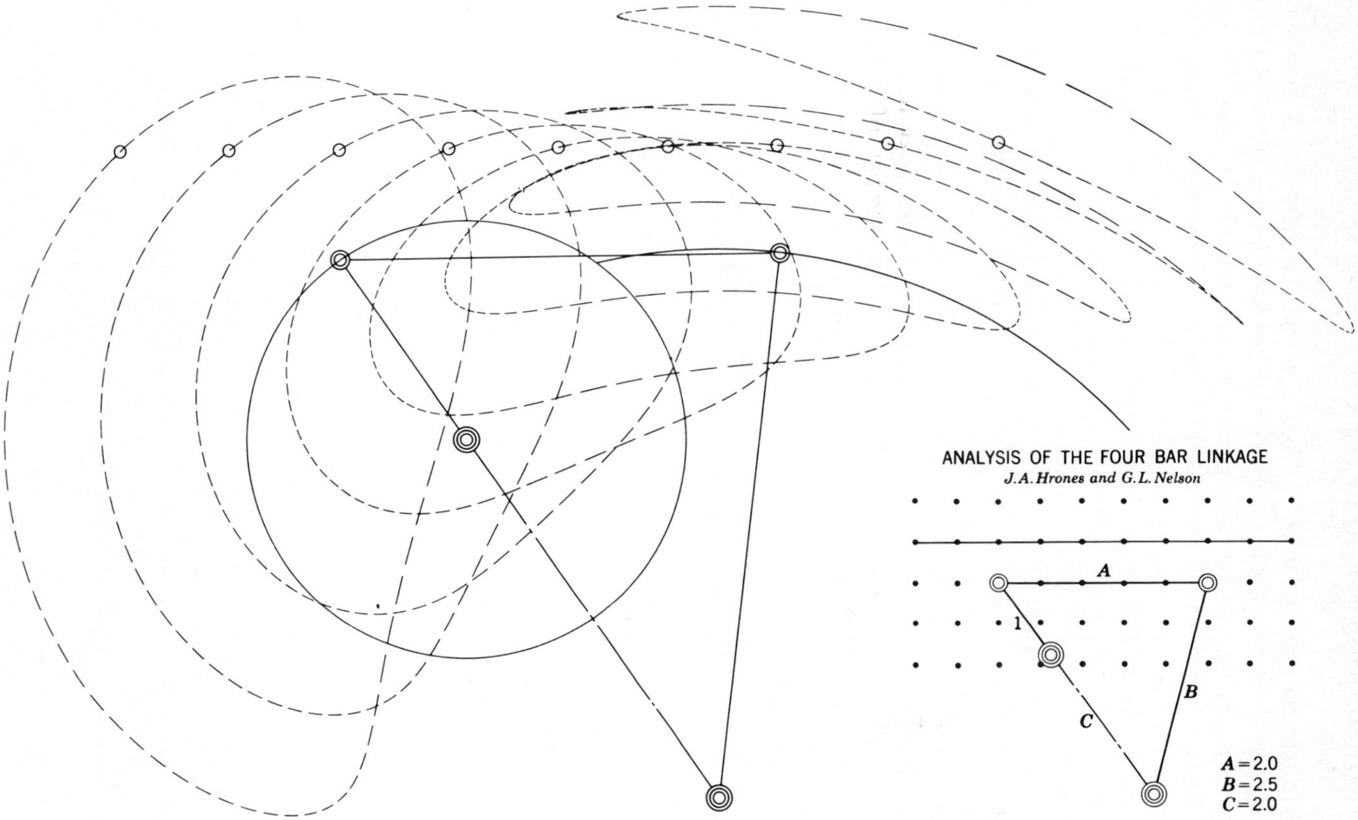

Figure 1-11 A reproduction of one of the Hrones-Nelson pages. (*Reproduced by permission of the publishers, The Technology Press, M.I.T., Cambridge, Mass., and John Wiley & Sons, Inc., New York.*)

1-11 STRAIGHT-LINE MECHANISMS

In the late seventeenth century, before the development of the milling machine, it was extremely difficult to machine straight, flat surfaces. For this reason, good prismatic pairs without backlash were not easy to make. During that era, much thought was given to the problem of attaining a straight-line motion as a part of the coupler curve of a linkage having only revolute connections. Probably the best-known result of this search is the straight-line mechanism development by Watt for guiding the piston of early steam engines. Figure 1-12*a* shows *Watt's linkage* to be a four-bar linkage developing an approximate straight line as a part of its coupler curve. Although it does not generate an exact straight line, a good approximation is achieved over a considerable distance of travel.

Another four-bar linkage in which the tracing point *P* generates an approximate straight-line coupler curve segment is *Roberts' mechanism* (Fig. 1-12*b*). The dashed lines in the figure indicate that the linkage is defined by forming three congruent isosceles triangles; thus $BC = AD/2$.

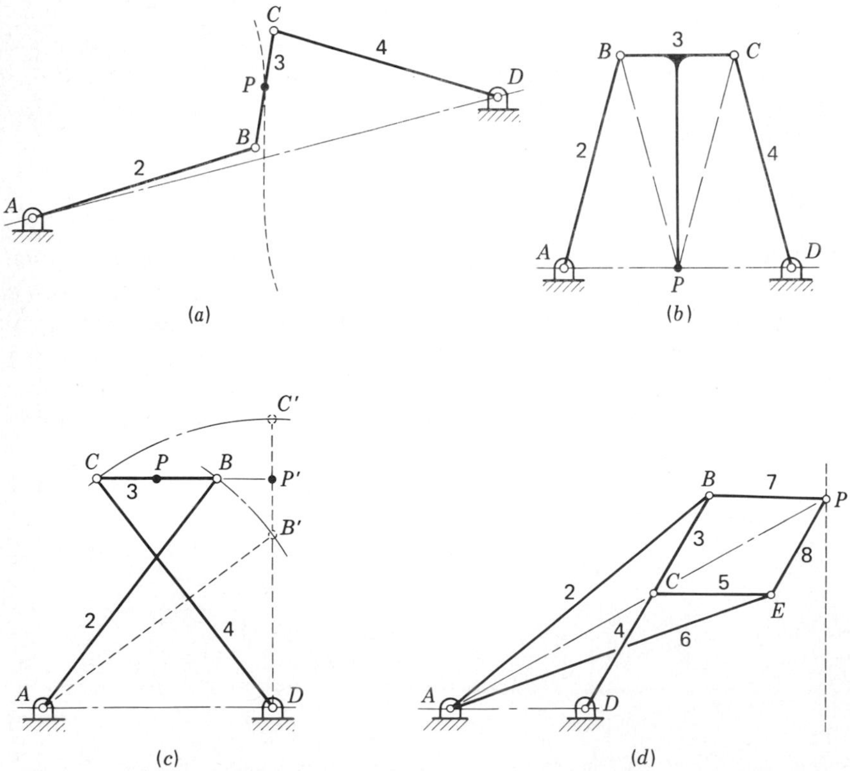

Figure 1-12 Straight-line mechanisms: (*a*) Watt's linkage; (*b*) Roberts' mechanism; (*c*) Chebychev linkage; and (*d*) Peaucillier inversor.

The tracing point P of the *Chebychev linkage* in Fig. 1-12*c* also generates an approximate straight line. The linkage is formed by creating a 3-4-5 triangle with link 4 in the vertical position as shown by the dashed lines; thus $DB' = 3$, $AD = 4$, and $AB' = 5$. Since $AB = DC$, $DC' = 5$ and the tracing point P' is the midpoint of link BC. Note that $DP'C$ also forms a 3-4-5 triangle and hence that P and P' are two points on a straight line parallel to AD.

Yet another mechanism which generates a straight-line segment is the *Peaucillier inversor* shown in Fig. 1-12*d*. The conditions describing its geometry are that $BC = BP = EC = EP$ and $AB = AE$ such that, by symmetry, points A, C, and P always lie on a straight line passing through A. Under these conditions $AC \cdot AP = k$, a constant, and the curves generated by C and P are said to be *inverses* of each other. If we place the other fixed pivot D such that $AD = CD$, then point C must trace a circular arc and point P will follow an *exact* straight line. Another interesting property is that if AD is not equal to CD, point P can be made to trace a true circular arc of very large radius.

Hunt, Fink, and Nayar† give the dimensions of a class of four-bar linkages which generate a symmetrical triangular path with two of the sides approximating straight lines.

Hartenberg and Denavit‡ and Hall§ illustrate most of the classical straight-line generators. Tesar and Vidosic¶ have investigated approximate-straight-line mechanisms in great detail and have developed considerable design information on this class of mechanisms.

1-12 QUICK-RETURN MECHANISMS

In many applications, mechanisms are used to perform repetitive operations such as pushing parts along an assembly line, clamping parts together while they are welded, or folding cardboard boxes in an automated packaging machine. In such applications it is often desirable to use a constant-speed motor; this led us to a discussion of Grashof's law in Sec. 1-8. In addition, however, we should also give some consideration to the power and timing requirements.

In these repetitive operations there is usually a part of the cycle when the mechanism is under load, called the *advance* or *working stroke*, and a part of the cycle, called the *return stroke*, when the mechanism is not working but

† K. H. Hunt, N. Fink, and J. Nayar, Linkage Geneva Mechanisms: A Design Study in Mechanism Geometry, *Proc. Inst. Mech. Eng.*, Vol. 174, no. 21, pp. 643–668, 1960; see also J. Hirschhorn, "Kinematics and Dynamics of Plane Mechanisms," McGraw-Hill, New York, 1964, pp. 349–353.

‡ Op. cit.

§ A. S. Hall, Jr., "Kinematics and Linkage Design," Prentice-Hall, Englewood Cliffs, N.J., 1961.

¶D. Tesar and J. P. Vidosic, Analysis of Approximate Four-Bar Straight-Line Mechanisms, *J. Eng. Ind.*, Vol. 87, no. 3, 1965.

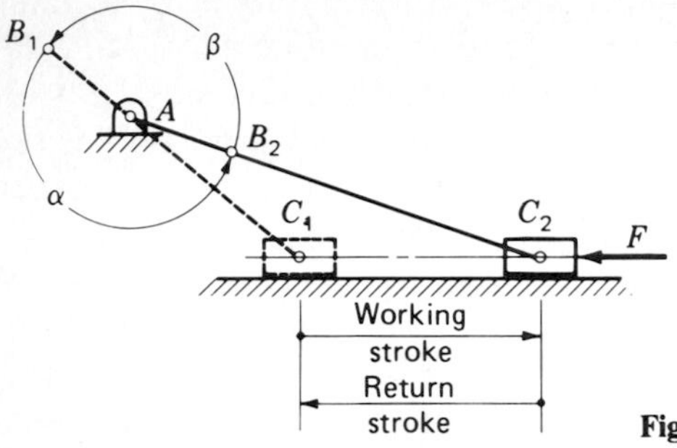

Figure 1-13 Offset slider-crank mechanism.

simply returning to repeat the operation. In the offset slider-crank mechanism of Fig. 1-13, for example, work may be required to overcome the load F while the piston moves to the right from C_1 to C_2 but not during its return to position C_1 since the load may have been removed. In such situations, in order to keep the power requirements of the motor to a minimum and to avoid wasting valuable time, it is desirable to design the mechanism so that the piston will move much faster through the return stroke than it does during the working stroke, i.e., to use a higher fraction of the cycle time for doing work than for returning.

A measure of the suitability of a mechanism from this viewpoint, called *advance- to return-time ratio*, is defined by the formula

$$Q = \frac{\text{time of advance stroke}}{\text{time of return stroke}} \qquad (a)$$

A mechanism for which the value of Q is high is more desirable for such repetitive operations than one in which Q is lower. Certainly, any such operation would use a mechanism for which Q is greater than unity. Because of this, mechanisms with Q greater than unity are called *quick-return* mechanisms.

Assuming that the driving motor operates at constant speed, it is easy to find the time ratio. As shown in Fig. 1-13, the first thing is to determine the two crank positions AB_1 and AB_2, which mark the beginning and end of the working stroke. Next, noticing the direction of rotation of the crank, we can measure the crank angle α traveled through during the advance stroke and the remaining crank angle β of the return stroke. Then, if the period of the motor is τ, the time of the advance stroke is

$$\text{Time of advance stroke} = \frac{\alpha}{2\pi}\tau \qquad (b)$$

and of the return stroke is

$$\text{Time of return stroke} = \frac{\beta}{2\pi}\tau \qquad (c)$$

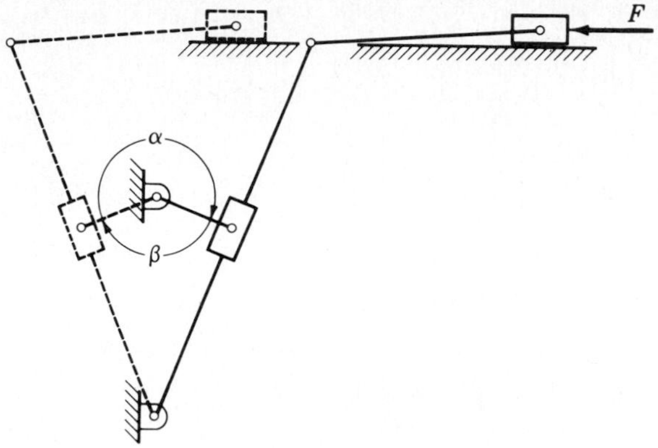

Figure 1-14 Whitworth quick-return mechanism.

Finally, combining Eqs. (*a*), (*b*), and (*c*), we get the following simple equation for time ratio:

$$Q = \frac{\alpha}{\beta} \tag{1-6}$$

Notice that the time ratio of a quick-return mechanism does not depend on the amount of work being done or even on the speed of the driving motor. It is a kinematic property of the mechanism itself and can be found strictly from the geometry of the device.

We also notice, however, that there is a proper and an improper direction of rotation for such a device. If the motor were reversed in the example of Fig. 1-13, the roles of α and β would also reverse and the time ratio would be less than 1. Thus the motor must rotate counterclockwise for this mechanism to have the quick-return property.

Many other mechanisms can be found with quick-return characteristics. Another example is the *Whitworth mechanism*, also called the *crank-shaper mechanism*, shown in Fig. 1-14. Although the determination of the angles α and β is different for each mechanism, Eq. (1-6) applies to all.

PROBLEMS

1-1 Sketch at least six different examples of the use of a planar four-bar linkage in practice. They can be found in the workshop, in domestic appliances, on vehicles, on agricultural machines, etc.

1-2 The link lengths of a planar four-bar linkage are 1, 3, 5, and 5 in. Assemble them in all possible combinations and sketch the four inversions of each. Do these linkages satisfy Grashof's law? Describe each inversion by name, e.g., a crank-rocker mechanism or a drag-link mechanism.

1-3 A crank-rocker linkage has a 100-mm frame, a 25-mm crank, a 90-mm coupler, and a 75-mm rocker. Draw the linkage and find the maximum and minimum values of the transmission angle. Locate both toggle positions and record the corresponding crank angles and transmission angles.

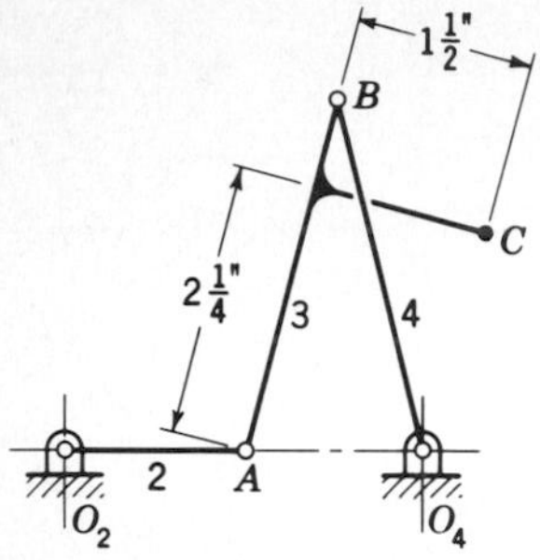

Problem 1-4

1-4 In the figure, point C is attached to the coupler; plot its complete path.

1-5 Find the mobility of each mechanism shown in the figure.

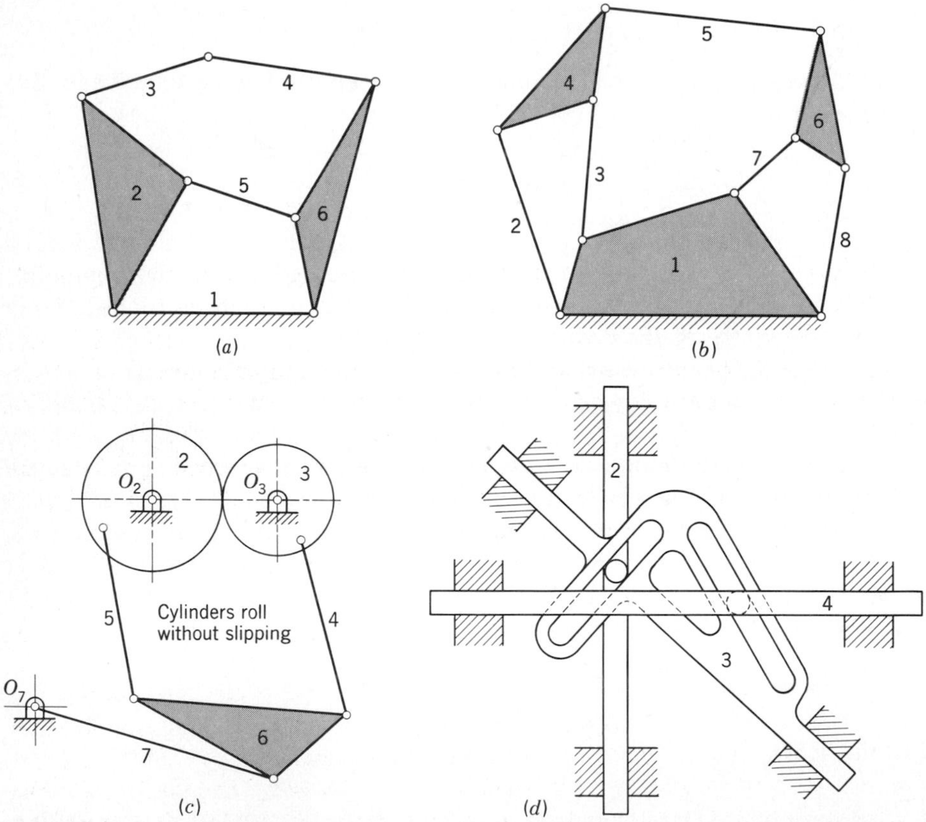

Problem 1-5

1-6 Use the mobility criterion to find a planar mechanism containing a moving quaternary link. How many variations of this mechanism can you find?

1-7 Find the time ratio of the linkage of Prob. 1-3.

1-8 Devise a practical working model of the drag-link mechanism.

1-9 Plot the complete coupler curve of Roberts' mechanism of Fig. 1-12b. Use $AB = CD = AD =$ 2.5 in and $BC = 1.25$ in.

CHAPTER

TWO

POSITION AND DISPLACEMENT

In analyzing motion, the first and most basic problem encountered is that of defining and dealing with the concepts of position and displacement. Since motion can be thought of as a time series of displacements between successive positions, it is important to understand exactly the meaning of the term *position*; rules or conventions must be established to make the definition precise.

Although many of the concepts in this chapter may appear intuitive and almost trivial, many subtleties are explained here which are required for an understanding of the next several chapters.

2-1 COORDINATE SYSTEMS

In speaking of the position of a particle or point, we are really answering the question: Where is the point or what is its location? We are speaking of something which exists in nature and posing the question of how to express this (in words or symbols or numbers) in such a way that the meaning is clear. We soon discover that position cannot be defined on a truly absolute basis. We must define the position of a point in terms of some agreed-upon frame of reference, some reference coordinate system.

As shown in Fig. 2-1*a*, once we have agreed upon the xyz coordinate system as the frame of reference, we can say that point P is located x units along the x axis, y units along the y axis, and z units along the z axis *from the origin O*. In this very statement we see that three vitally important parts of the

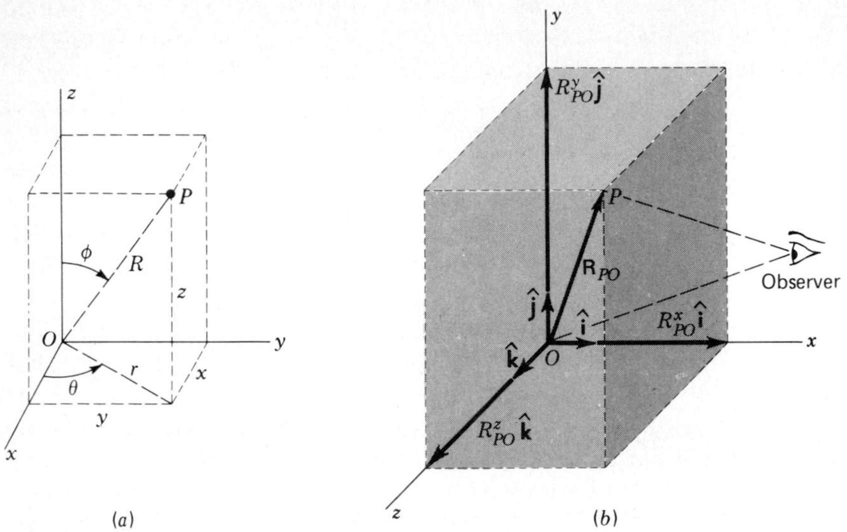

Figure 2-1 (*a*) A right-handed three-dimensional coordinate system; (*b*) position of a point.

definition depend on the existence of the reference coordinate system:

1. The *origin* of coordinates O provides an agreed-upon location from which to measure the location of point P.
2. The *coordinate axes* provide agreed-upon *directions* along which the measurements are to be made; they also provide known lines and planes for the definition and measurement of angles.
3. The unit distance along any of the axes provides a *scale* for quantifying distances.

These observations are not restricted to the cartesian coordinates (x, y, z) of point P. All three properties of the coordinate system are also necessary in defining the cylindrical coordinates (r, θ, z), spherical coordinates (R, θ, ϕ), or any other coordinates of point P. The same properties would also be required if point P were restricted to remain in a single plane and a two-dimensional coordinate system were used. No matter how defined, the concept of the position of a point cannot be related without the definition of a reference coordinate system.

2-2 POSITION OF A POINT

The physical process involved in observing the position of a point, as shown in Fig. 2-1*b*, implies that the observer is actually keeping track of the *relative* location of *two* points, P and O, by looking at both, performing a mental comparison, and recognizing that point P has a certain location *relative to point* O. In this determination two properties are noted, the distance from O to P (based on the unit distance or grid size of the reference coordinate system) and the relative angular orientation of the line OP in the coordinate system. These two properties, magnitude and direction, are precisely those required for a vector. Therefore we define the *position of a point* as the *vector from the origin of a specified reference coordinate system to the point.* We choose the symbol $\mathbf{R}_{PO}$ to denote the vector position of point P relative to the point O.

The reference coordinate system is therefore in a special way related to a particular observer's concept of what he sees. What is the relationship? What properties must this coordinate system have to ensure that position measurements made with respect to it will truly represent his observations? The key to this relationship is that the coordinate system is *stationary* with respect to this observer. In other words, the observer thinks of himself as remaining stationary in his chosen reference coordinate system. If he moves, either through a distance or by turning, his coordinate system moves with him. In this way it is assured that objects that appear stationary *with respect to him*, i.e., as seen through his eyes, do not change their positions within his coordinate system and their position vectors remain constant. Points which he observes as moving have changing position vectors.

Notice that there has been no mention of the observer's actual *location* within the frame of reference. He may be anywhere within the coordinate system. It is not necessary to know his position since the positions of observed points are found relative to the origin of coordinates rather than relative to his own location.

Often it is convenient to express the position vector in terms of its components along the coordinate axes

$$\mathbf{R}_{PO} = R^x_{PO}\hat{\mathbf{i}} + R^y_{PO}\hat{\mathbf{j}} + R^z_{PO}\hat{\mathbf{k}} \tag{2-1}$$

where superscripts are used to denote the direction of each component. As in the remainder of this book, $\hat{\mathbf{i}}$, $\hat{\mathbf{j}}$, and $\hat{\mathbf{k}}$ are used to designate unit vectors in the x, y, and z axis directions, respectively. While vectors are denoted throughout the book by boldface symbols, the scalar magnitude of a vector is signified by the same symbol in italic, without boldface. The magnitude of the position vector, for example, is

$$R_{PO} = |\mathbf{R}_{PO}| = \sqrt{\mathbf{R}_{PO} \cdot \mathbf{R}_{PO}} = \sqrt{(R^x_{PO})^2 + (R^y_{PO})^2 + (R^z_{PO})^2} \tag{2-2}$$

The *unit vector* in the direction of $\mathbf{R}_{PO}$ is denoted by the same boldface symbol with a caret

$$\hat{\mathbf{R}}_{PO} = \frac{\mathbf{R}_{PO}}{R_{PO}} \tag{2-3}$$

The direction of $\mathbf{R}_{PO}$ can be expressed, among other ways, by the direction cosines

$$\cos\alpha = \frac{R_{PO}^x}{R_{PO}} \qquad \cos\beta = \frac{R_{PO}^y}{R_{PO}} \qquad \cos\gamma = \frac{R_{PO}^z}{R_{PO}} \tag{2-4}$$

where the angles α, β, and γ are, respectively, the angles measured from the positive coordinate axes to the vector $\mathbf{R}_{PO}$.

One means of expressing the motion of a point or particle is to define its components along the reference axes as functions of some parameter such as time

$$R_{PO}^x = R_{PO}^x(t) \qquad R_{PO}^y = R_{PO}^y(t) \qquad R_{PO}^z = R_{PO}^z(t) \tag{2-5}$$

If these relations are known, the position vector $\mathbf{R}_{PO}$ can be found for any time t. This is the general case for the motion of a particle and is illustrated in the following example.

Example 2-1 Describe the motion of a particle P whose position changes with time according to the equations $R_{PO}^x = a\cos 2\pi t$, $R_{PO}^y = a\sin 2\pi t$, and $R_{PO}^z = bt$.

SOLUTION Substitution of values for t from 0 to 2 gives values as shown in the following table:

t	R_{PO}^x	R_{PO}^y	R_{PO}^z
0	a	0	0
$\frac{1}{4}$	0	a	$b/4$
$\frac{1}{2}$	$-a$	0	$b/2$
$\frac{3}{4}$	0	$-a$	$3b/4$
1	a	0	b
$\frac{5}{4}$	0	a	$5b/4$
$\frac{3}{2}$	$-a$	0	$3b/2$
$\frac{7}{4}$	0	$-a$	$7b/4$
2	a	0	$2b$

As shown in Fig. 2-2, the point moves with *helical motion* with radius a around the z axis and with a lead of b. Note that if $b = 0$, $R_{PO}^z(t) = 0$, the moving particle is confined to the xy plane and the motion is a circle with its center at the origin.

We have used the words particle and point interchangeably. When we use the word *point*, we have in mind something which has no dimensions, something with zero length, zero width, and zero thickness. When the word *particle* is used, we have in mind something whose dimensions are small and unimportant, i.e., a tiny material body whose dimensions are negligible, a body small enough for its dimensions to have no effect on the analysis to be performed.

The successive positions of a moving point define a line or curve. This curve has no thickness since the point has no dimensions. However, the curve does have length since the point occupies different positions as time changes.

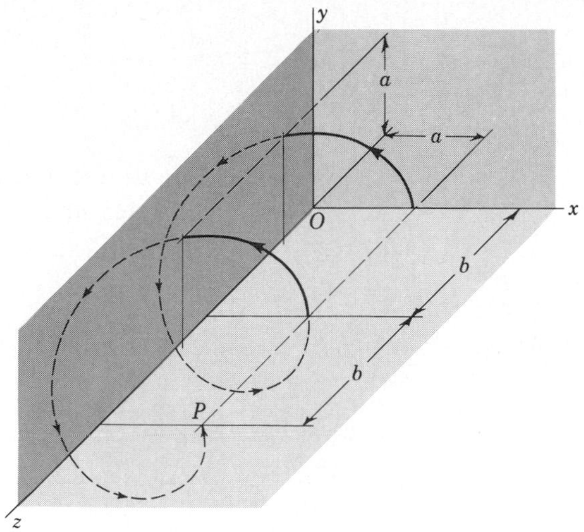

Figure 2-2 Helical motion of a particle.

This curve, representing the successive positions of the point, is called the *path* or *locus* of the moving point in the reference coordinate system.

If three coordinates are necessary to describe the path of a moving point, the point is said to have *spatial motion.* If the path can be described by only two coordinates, i.e., if the coordinate axes can be chosen such that one coordinate is always zero or constant, the path is contained in a single plane and the point is said to have *planar motion.* Sometimes it happens that the path of a point can be described by a single coordinate. This means that two of its spatial position coordinates can be taken as zero or constant. In this case the point moves in a straight line and is said to have *rectilinear motion.*

In each of the three cases described, it is assumed that the coordinate system is chosen so as to obtain the least number of coordinates necessary to describe the motion of the point. Thus the description of rectilinear motion requires one coordinate, a point whose path is a *plane curve* requires two coordinates, and a point whose locus is a *space curve,* sometimes called a *skew curve,* requires three position coordinates.

2-3 THE POSITION DIFFERENCE BETWEEN TWO POINTS

We now investigate the relationship between the position vectors of two different points. The situation is shown in Fig. 2-3*a*. In the last section we found that an observer fixed in the *xyz* coordinate system would observe the positions of points P and Q by comparing each with the position of the origin. The positions of the two points are defined by the vectors $\mathbf{R}_{PO}$ and $\mathbf{R}_{QO}$.

Inspection of the figure shows that they are related by a third vector $\mathbf{R}_{PQ}$,

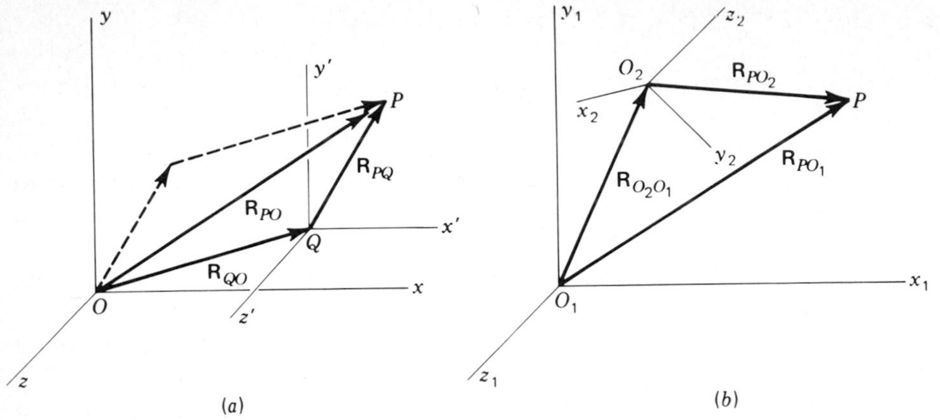

Figure 2-3 (*a*) Position difference between two points, *P* and *Q*. (*b*) Apparent position of a point *P*.

the *position difference* between points P and Q. The figure shows this relationship to be

$$\mathbf{R}_{PQ} = \mathbf{R}_{PO} - \mathbf{R}_{QO} \tag{2-6}$$

The physical interpretation is now slightly different from that of the position vector itself. The observer is no longer comparing the position of point P with the position of the origin. He is now comparing it with the position of point Q. In other words, he is observing the position of point P as though in another temporary coordinate system $x'y'z'$ with origin at Q and axes directed parallel† to those of his basic reference frame xyz. One can use either point of view for the interpretation; both should be understood since both will be used in future developments.

Having now generalized our concept of relative position to include the position difference between any two points, we reflect again on the above discussion of the position vector itself. We notice that it is merely the special case where we agree to measure using the origin of coordinates as the second point. Thus, to be consistent in notation, we have denoted the position vector of a single point P by the dual subscripted symbol $\mathbf{R}_{PO}$. However, in the interest of brevity, we will henceforth agree that when the second subscript is not given explicitly, it is understood to be the origin of the observer's coordinate system

$$\mathbf{R}_P = \mathbf{R}_{PO} \tag{2-7}$$

† That these coordinate systems have parallel axes is a convenience rather than a necessary condition. This concept will be maintained throughout the book, however, since it causes no loss of generality and simplifies the visualization when coordinate systems are in motion.

2-4 THE APPARENT POSITION OF A POINT

Up to now, in discussing the position vector, our point of view has been entirely that of a single observer in a single coordinate system. However, it is often desirable to make observations in a secondary coordinate system, i.e., as seen by a second observer in a different coordinate system, and then to convert this information into the basic coordinate system. Such a situation is illustrated in Fig. 2-3*b*.

If two observers, one using the reference frame $x_1y_1z_1$ and the other using $x_2y_2z_2$, were both asked to give the location of a particle at P, they would report different results. The observer in coordinate system $x_1y_1z_1$ would observe the vector $\mathbf{R}_{PO_1}$, while the second observer, using the $x_2y_2z_2$ coordinate system, would report the position vector $\mathbf{R}_{PO_2}$. We note from Fig. 2-3*b* that these vectors are related by

$$\mathbf{R}_{PO_1} = \mathbf{R}_{O_2O_1} + \mathbf{R}_{PO_2} \tag{2-8}$$

The difference in the positions of the two origins is not the only incompatability between the two observations of the position of point P. Since the two coordinate systems are not aligned,† the two observers would be using different reference lines for their measurements of direction; the first observer would report components measured along the $x_1y_1z_1$ axes while the second would measure in the $x_2y_2z_2$ directions.

A third and very important distinction between these two observations becomes clear when we consider that the two coordinate systems may be moving with respect to each other. Whereas point P may appear stationary with respect to one observer, it may be in motion with respect to the other; i.e., the position vector $\mathbf{R}_{PO_1}$ may appear constant to observer 1 while $\mathbf{R}_{PO_2}$ appears to vary as seen by observer 2.

When any of these conditions exist, it will be convenient to add an additional subscript to our notation which will distinguish which observer is being considered. When we are considering the position of P as seen by the observer using coordinate system $x_1y_1z_1$, we will denote this by the symbol $\mathbf{R}_{PO_1/1}$ or, since O_1 is the origin for this observer,‡ by $\mathbf{R}_{P/1}$. The observations made by the second observer, done in coordinate system $x_2y_2z_2$, will be denoted as $\mathbf{R}_{PO_2/2}$ or $\mathbf{R}_{P/2}$. With this extension of the notation, Eq. (2-8) becomes

$$\mathbf{R}_{P/1} = \mathbf{R}_{O_2/1} + \mathbf{R}_{P/2} \tag{2-9}$$

We refer to $\mathbf{R}_{P/2}$ as the *apparent position of point P to an observer in coordinate system 2,* and we note that it is by no means equal to the apparent position vector $\mathbf{R}_{P/1}$ seen by observer 1.

† Note that the condition that the coordinate systems have parallel axes was assumed for the position difference vector, Fig. 2-3*a*, but is not assumed for the apparent position vector.

‡Note that $\mathbf{R}_{PO_2/1}$ cannot be abbreviated as $\mathbf{R}_{P/1}$ since O_2 is not the origin used by observer 1.

We have now made note of certain intrinsic differences between $\mathbf{R}_{P/1}$ and $\mathbf{R}_{P/2}$ and found Eq. (2-9) to relate them. However, there is no reason why components of either vector *must* be taken along the natural axes of the observer's coordinate system. As with all vectors, components can be found along any convenient set of axes.

In applying the apparent-position equation (2-9) we must use a single consistent set of axes during the numerical evaluation. Though the observer in coordinate system 2 would find it most natural to measure the components of $\mathbf{R}_{P/2}$ along the $x_2y_2z_2$ axes, these components must be transformed into the equivalent components in the $x_1y_1z_1$ system before the addition is actually performed

$$\begin{aligned}\mathbf{R}_{P/1} &= \mathbf{R}_{O_2/1} + \mathbf{R}_{P/2}\\ &= R^{x_1}_{O_2/1}\hat{\mathbf{i}}_1 + R^{y_1}_{O_2/1}\hat{\mathbf{j}}_1 + R^{z_1}_{O_2/1}\hat{\mathbf{k}}_1 + R^{x_1}_{P/2}\hat{\mathbf{i}}_1 + R^{y_1}_{P/2}\hat{\mathbf{j}}_1 + R^{z_1}_{P/2}\hat{\mathbf{k}}_1\\ &= (R^{x_1}_{O_2/1} + R^{x_1}_{P/2})\hat{\mathbf{i}}_1 + (R^{y_1}_{O_2/1} + R^{y_1}_{P/2})\hat{\mathbf{j}}_1 + (R^{z_1}_{O_2/1} + R^{z_1}_{P/2})\hat{\mathbf{k}}_1\\ &= R^{x_1}_{P/1}\hat{\mathbf{i}}_1 + R^{y_1}_{P/1}\hat{\mathbf{j}}_1 + R^{z_1}_{P/1}\hat{\mathbf{k}}_1\end{aligned}$$

The addition can be performed equally well if all vector components are transformed into the $x_2y_2z_2$ system or, for that matter, to any other consistent set of directions. However, they cannot be added algebraically when they have been evaluated along inconsistent axes. The additional subscript in the apparent position vector therefore does not specify a set of directions to be used in the evaluation of components; it merely states the coordinate system in which the vector is defined, the coordinate system in which the observer is stationary.

2-5 ABSOLUTE POSITION OF A POINT

We now turn to the meaning of *absolute position.* We saw in Sec. 2-2 that every position vector is defined relative to a second point, the origin of the observer's coordinate reference frame. It is a special case of the position-difference vector, studied in Sec. 2-3, where the reference point is the origin of coordinates.

In Sec. 2-4 we noted that it may be convenient in certain problems to consider the apparent positions of a single point as viewed by more than one observer using different coordinate systems. When a particular problem leads us to consider multiple coordinate systems, however, the application will lead us to single out one of the coordinate systems as primary or most basic. Most often this is the coordinate system in which the final result is to be expressed, and this coordinate system is usually considered stationary. It is referred to as the *absolute* coordinate system. The *absolute position* of a point is defined as its apparent position as seen by an observer in the absolute coordinate system.

Which coordinate system is designated as absolute (most basic) is an arbitrary decision and unimportant in the study of kinematics. Whether the absolute coordinate system is truly stationary is also a moot point since, as we have seen, all position (and motion) information is measured relative to something else; nothing is truly absolute in the strict sense. When analyzing the kinematics of an automobile suspension, for example, it may be convenient to choose an "absolute" coordinate system attached to the frame of the car and to study the motion of the suspension relative to this. It is then unimportant whether the car is moving or not; motions of the suspension relative to the frame would be defined as absolute.

It is a common convention to number the absolute coordinate system 1 and to use other numbers for other moving coordinate systems. Since we adopt this convention throughout this book, absolute-position vectors are those apparent-position vectors viewed by an observer in coordinate system 1 and carry symbols of the form $\mathbf{R}_{P/1}$. In the interest of brevity and to reduce the complexity, we will agree that when the coordinate system number is not shown explicitly it is assumed to be 1; thus $\mathbf{R}_{P/1}$ can be abbreviated as $\mathbf{R}_P$. Similarly, the apparent-position equation (2-9) can be written† as

$$\mathbf{R}_P = \mathbf{R}_{O_2} + \mathbf{R}_{P/2} \tag{2-10}$$

2-6 THE LOOP-CLOSURE EQUATION

Our discussion of the position-difference and apparent-position vectors has been quite abstract so far, the intent being to develop a rigorous foundation for the analysis of motion in mechanical systems. Certainly precision is not without merit since it is this rigor which permits science to predict a correct result in spite of the personal prejudices and emotions of the analyst. However, tedious developments are not interesting unless they lead to applications in problems of real life. Although there are yet many fundamental principles to be discovered, it might be well at this point to show the relationship between the relative-position vectors discussed above and some of the typical linkages met in real machines.

One of the most common and most useful of all mechanisms is the four-bar linkage. One example is the clamping device shown in Fig. 2-4. A brief study of the assembly drawing shows that as the handle of the clamp is lifted, the clamping bar swings away from the clamping surface, opening the clamp. As the handle is pressed, the clamping bar swings down and the clamp closes again. If we wish to design such a clamp accurately, however, things are not quite so simple. It may be desirable, for example, for the clamp to open at a given rate for a certain rate of lift of the handle. Such relationships

† Reviewing Secs. 2-1 through 2-3 will show that the position-difference vector $\mathbf{R}_{PQ}$ was treated entirely from the absolute coordinate system and is an abbreviation of the notation $\mathbf{R}_{PQ/1}$. We will have no need to treat the completely general case $\mathbf{R}_{PQ/2}$, the apparent-position-difference vector.

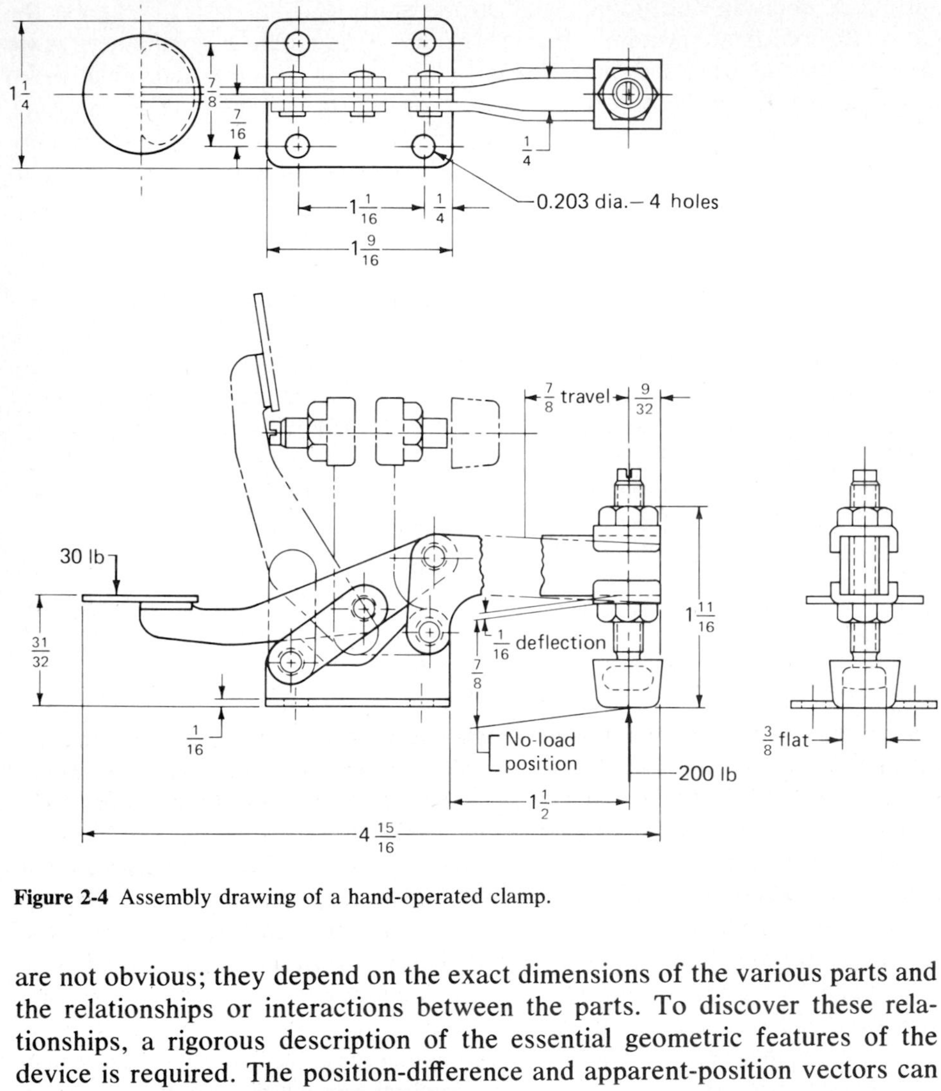

Figure 2-4 Assembly drawing of a hand-operated clamp.

are not obvious; they depend on the exact dimensions of the various parts and the relationships or interactions between the parts. To discover these relationships, a rigorous description of the essential geometric features of the device is required. The position-difference and apparent-position vectors can be used to provide such a description.

Figure 2-5 shows the detail drawings of the individual links of the disassembled clamp. Although not shown, the detail drawings would be completely dimensioned, thus fixing once and for all the complete geometry of each link. The assumption that each is a *rigid* link ensures that the position of any point on any one of the links can be determined precisely relative to any other point on the same link by simply identifying the proper points and scaling the appropriate detail drawing.

The features which are lost in the detail drawings, however, are the interrelationships between the individual parts, i.e., the *constraints* which ensure that each link will move relative to its neighbors in the prescribed

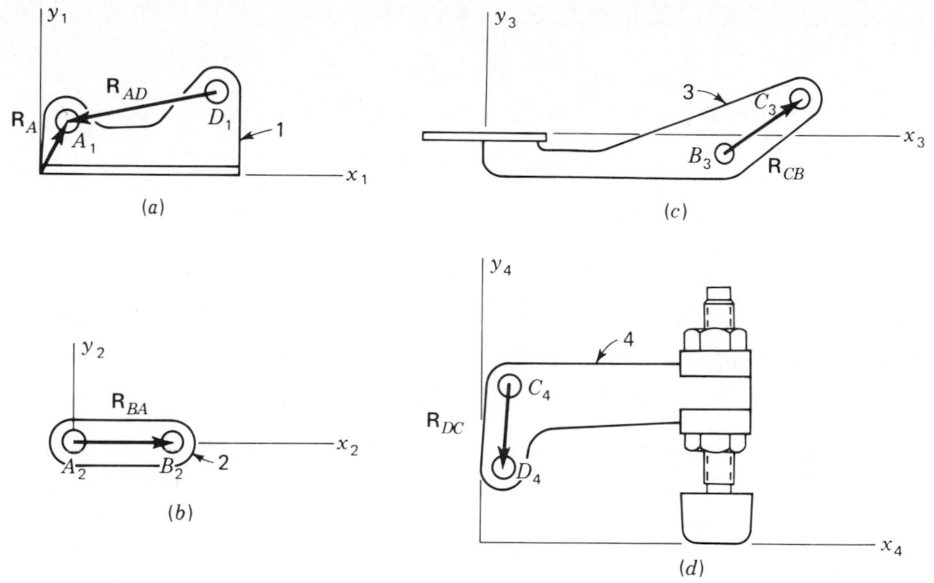

Figure 2-5 Detail drawings of the clamp of Fig. 2-4: (*a*) frame link, (*b*) connecting link, (*c*) handle, (*d*) clamping bar.

fashion. These constraints are, of course, provided by the four pinned joints. Anticipating that they will be of importance in any description of the linkages, we label these pin centers *A*, *B*, *C*, and *D*, and identify the appropriate points on link 1 as A_1 and D_1, those on link 2 as A_2 and B_2, etc. As shown in Fig. 2-5, we also pick a different coordinate system rigidly attached to each link.

Since it is necessary to relate the relative positions of the successive joint centers, we define the position difference vectors $\mathbf{R}_{AD}$ on link 1, $\mathbf{R}_{BA}$ on link 2, $\mathbf{R}_{CB}$ on link 3, and $\mathbf{R}_{DC}$ on link 4. We note again that each of these vectors appears constant to an observer fixed in the coordinate system of that particular link; the magnitudes of these vectors are obtainable from the constant dimensions of the links.

A vector equation can also be written to describe the constraints provided by each of the revolute (pinned) joints. Notice that no matter which position or which observer is chosen, the two points describing each pin center, for example A_1 and A_2, remain coincident. Thus

$$\mathbf{R}_{A_2A_1} = \mathbf{R}_{B_3B_2} = \mathbf{R}_{C_4C_3} = \mathbf{R}_{D_1D_4} = 0 \tag{2-11}$$

Let us now develop vector equations for the absolute position of each of the pin centers. Since link 1 is the frame, absolute positions are those defined relative to an observer in coordinate system 1. Point A_1 is, of course, at the position described by $\mathbf{R}_A$. Next, we mathematically connect link 2 to link 1 by writing

$$\mathbf{R}_{A_2} = \mathbf{R}_{A_1} + \overset{0}{\cancel{\mathbf{R}_{A_2A_1}}} = \mathbf{R}_A \tag{a}$$

Transferring to the other end of link 2, we attach link 3

$$\mathbf{R}_B = \mathbf{R}_A + \mathbf{R}_{BA} \tag{b}$$

Connecting joints C and D in the same manner, we obtain

$$\mathbf{R}_C = \mathbf{R}_B + \mathbf{R}_{CB} = \mathbf{R}_A + \mathbf{R}_{BA} + \mathbf{R}_{CB} \tag{c}$$

$$\mathbf{R}_D = \mathbf{R}_C + \mathbf{R}_{DC} = \mathbf{R}_A + \mathbf{R}_{BA} + \mathbf{R}_{CB} + \mathbf{R}_{DC} \tag{d}$$

Finally, we transfer back across link 1 to point A

$$\mathbf{R}_A = \mathbf{R}_D + \mathbf{R}_{AD} = \mathbf{R}_A + \mathbf{R}_{BA} + \mathbf{R}_{CB} + \mathbf{R}_{DC} + \mathbf{R}_{AD} \tag{e}$$

and from this we obtain

$$\mathbf{R}_{BA} + \mathbf{R}_{CB} + \mathbf{R}_{DC} + \mathbf{R}_{AD} = 0 \tag{2-12}$$

This important equation is called the *loop-closure equation* for the clamp. As shown in Fig. 2-6, it expresses the fact that the mechanism forms a closed loop and therefore the polygon formed by the position difference vectors through successive links and joints must remain closed as the mechanism moves. The constant lengths of these vectors ensure that the joint centers remain at constant distances apart, the requirement for rigid links. The relative rotations between successive vectors indicate the motions within the pinned joints, while the rotation of each individual position-difference vector shows the rotational motion of a particular link. Thus the loop-closure equation holds within it all the important constraints which determine how this particular clamp operates. It forms a mathematical description, or *model*, of the linkage, and many of the later developments in this book are based on this model as a starting point.

Of course, the form of the loop-closure equation depends on the type of linkage. This is illustrated by another example, the *Geneva mechanism* or *Maltese cross*, shown in Fig. 2-7. One early application of this mechanism was to prevent overwinding a watch. Today it finds wide use as an indexing device, e.g., in a milling machine with automatic tool changer.

Although the frame of the mechanism, link 1, is not shown in the figure, it

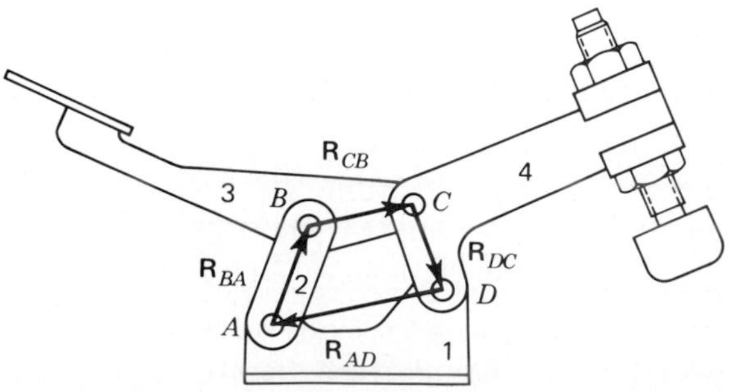

Figure 2-6 The loop-closure equation.

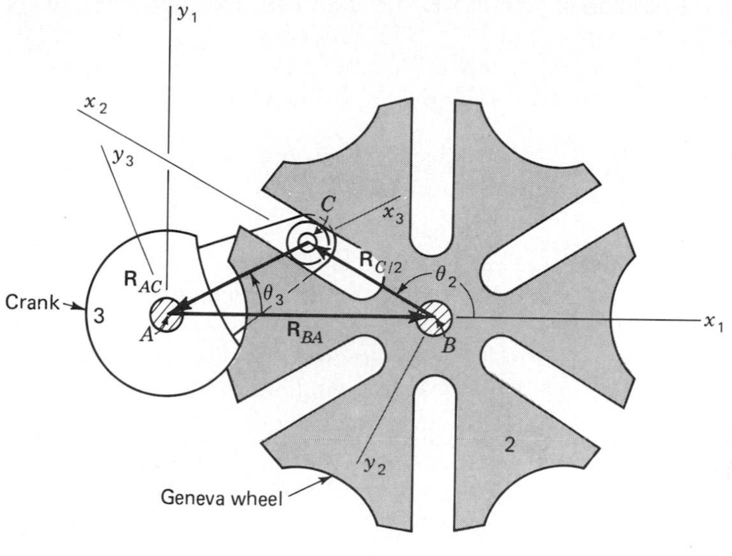

Figure 2-7 The Geneva mechanism or Maltese cross.

is an important part of the mechanism since it holds the two shafts with centers A and B a constant distance apart. Thus we define the vector $\mathbf{R}_{BA}$ to show this dimension. The left crank, link 3, is attached to a shaft, usually rotated at constant speed, and carries a roller at C, running in the slot of the Geneva wheel. The vector $\mathbf{R}_{AC}$ has a constant magnitude equal to the crank length, the distance from the center of the roller C to the shaft center A. The rotation of this vector relative to link 1 will later be used to describe the angular speed of the crank. The x_2 axis is aligned along one slot of the wheel; thus the roller is constrained to ride along this slot, the vector $\mathbf{R}_{C/2}$ has the same rotation as the wheel, link 2. Also, its changing length $\Delta\mathbf{R}_{C/2}$ shows the relative sliding motion taking place between the roller on link 3 and the slot on link 2.

From the figure we see that the loop-closure equation for this mechanism is

$$\mathbf{R}_{BA} + \mathbf{R}_{C/2} + \mathbf{R}_{AC} = 0 \tag{2-13}$$

Notice that the term $\mathbf{R}_{C/2}$ is equivalent to $\mathbf{R}_{CB}$ since point B is the origin of coordinate system 2.

This form of the loop-closure equation is a valid mathematical model only as long as roller C remains in the slot along x_2. However, this condition does not hold throughout the entire cycle of motion. Once the roller leaves the slot, the motion is controlled by the two mating circular arcs on links 2 and 3. A new form of the loop-closure equation is required for this part of the cycle.

Mechanisms can, of course, be connected together forming a multiple-loop kinematic chain. In such a case more than one loop-closure equation is required to model the system completely. The procedures for obtaining the equations, however, are identical to those illustrated in the above examples.

2-7 GRAPHICAL POSITION ANALYSIS OF PLANAR MECHANISMS

When the paths of the moving points in a mechanism lie in a single plane or in parallel planes, it is called a *planar mechanism*. Since a substantial portion of the investigations in this book deal with planar mechanisms, the development of special methods suited to such problems is justified. As we will see in the next section, the nature of the loop-closure equation often leads to the solution of simultaneous nonlinear equations when approached analytically and can become quite cumbersome. Yet, particularly for planar mechanisms, the solution is usually straightforward when approached graphically.

First let us briefly review the process of vector addition. Any two known vectors **A** and **B** can be added graphically as shown in Fig. 2-8*a*. After a scale is chosen, the vectors are drawn tip to tail in either order and their sum **C** is identified

$$\mathbf{C} = \mathbf{A} + \mathbf{B} = \mathbf{B} + \mathbf{A} \tag{2-14}$$

Notice that the magnitudes and the directions of both vectors **A** and **B** are

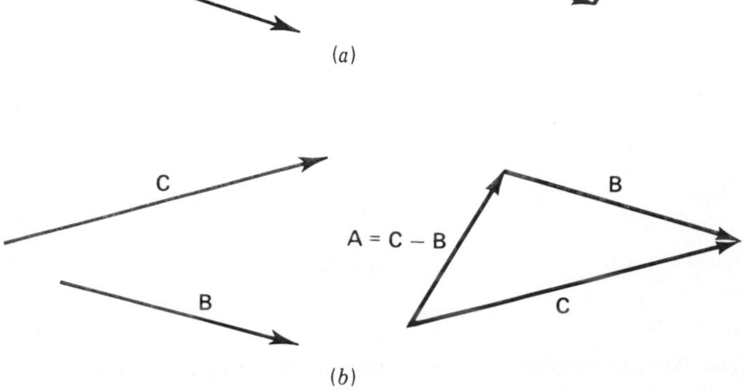

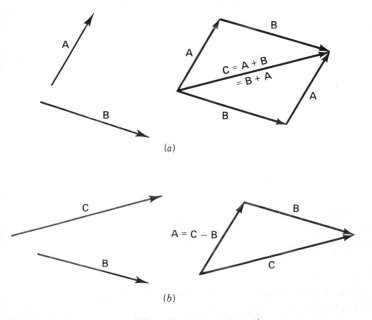

Figure 2-8 (*a*) Vector addition. (*b*) Vector subtraction.

used in performing the addition and that both the magnitude and direction of the sum **C** are found as a result.

The operation of graphical vector subtraction is illustrated in Fig. 2-8*b*, where the vectors are drawn tip to tip in solving the equation

$$\mathbf{A} = \mathbf{C} - \mathbf{B} \tag{2-15}$$

These graphical vector operations should be carefully studied and understood since they are used extensively throughout the book.

A three-dimensional vector equation

$$\mathbf{C} = \mathbf{D} + \mathbf{E} + \mathbf{B} \qquad (a)$$

can be divided into components along any convenient axes, leading to the three scalar equations

$$C^x = D^x + E^x + B^x \qquad C^y = D^y + E^y + B^y \qquad C^z = D^z + E^z + B^z \qquad (b)$$

Since they are components of the same vector equation, these three scalar equations must be consistent. If the three are also linearly independent, they can be solved simultaneously for three unknowns, which may be three magnitudes, three directions, or any combination of three magnitudes and directions. For some combinations, however, the problem is highly nonlinear and quite difficult to solve. Therefore, we shall delay consideration of the three-dimensional problem until Chap. 11, where it is needed.

A two-dimensional vector equation can be solved for two unknowns: two magnitudes, two directions, or one magnitude and one direction. Sometimes it is desirable to indicate the known (√) and unknown (o) quantities above each vector in an equation like this:

$$\overset{o\surd}{\mathbf{C}} = \overset{\surd\surd}{\mathbf{D}} + \overset{\surd\surd}{\mathbf{E}} + \overset{\surd o}{\mathbf{B}} \qquad (c)$$

where the first symbol (√ or o) above each vector indicates its magnitude and the second indicates its direction. Another equivalent form is

$$\overset{0\,\surd}{C\hat{\mathbf{C}}} = \overset{\surd\surd}{D\hat{\mathbf{D}}} + \overset{\surd\surd}{E\hat{\mathbf{E}}} + \overset{\surd 0}{B\hat{\mathbf{B}}} \qquad (d)$$

Either of these equations clearly identifies the unknowns and indicates whether a solution can be achieved. In Eq. (*c*) the vectors **D** and **E** are completely defined and can be replaced by their sum

$$\mathbf{A} = \mathbf{D} + \mathbf{E} \qquad (e)$$

giving

$$\mathbf{C} = \mathbf{A} + \mathbf{B} \tag{2-16}$$

In like manner any plane vector equation, if it is solvable, can be reduced to a three-term equation with two unknowns.

Depending on the forms of the two unknowns, four distinct cases occur.

Chace†,‡ classifies them according to the unknowns; the cases and corresponding unknowns are:

Case 1 Magnitude and direction of the same vector, for example, C, $\hat{\mathbf{C}}$.
Case 2*a* Magnitudes of two different vectors, for example, A, B.
Case 2*b* Magnitude of one vector and direction of another, for example, A, $\hat{\mathbf{B}}$.
Case 2*c* Directions of two different vectors, for example, $\hat{\mathbf{A}}$, $\hat{\mathbf{B}}$.

We will illustrate the solutions of these four cases graphically in this section and analytically in the next.

In case 1 the two unknowns are the magnitude and direction of the same vector. This case can be solved by straightforward graphical addition or subtraction of the remaining vectors, which are completely defined. This was illustrated in Fig. 2-8.

For case 2*a*, two magnitudes, say A and B, are to be found

$$\overset{\surd\surd}{\mathbf{C}} = \overset{\circ\surd}{\mathbf{A}} + \overset{\circ\surd}{\mathbf{B}} \qquad (2\text{-}17)$$

The solution is shown in Fig. 2-9. The steps are as follows:

1. Choose a coordinate system and scale factor and draw vector $\mathbf{C}$.
2. Construct a line through the origin of $\mathbf{C}$ parallel to $\hat{\mathbf{A}}$.
3. Construct another line through the terminus of $\mathbf{C}$ parallel to $\hat{\mathbf{B}}$.
4. The intersection of these two lines defines both magnitudes, A and B, which may be either positive or negative.

† Milton A. Chace, Development and Application of Vector Mathematics for Kinematic Analysis of Three-Dimensional Mechanisms, Ph.D. thesis, The University of Michigan, Ann Arbor, Mich. 1964, p. 19.

‡ See Table 11-1 for all cases.

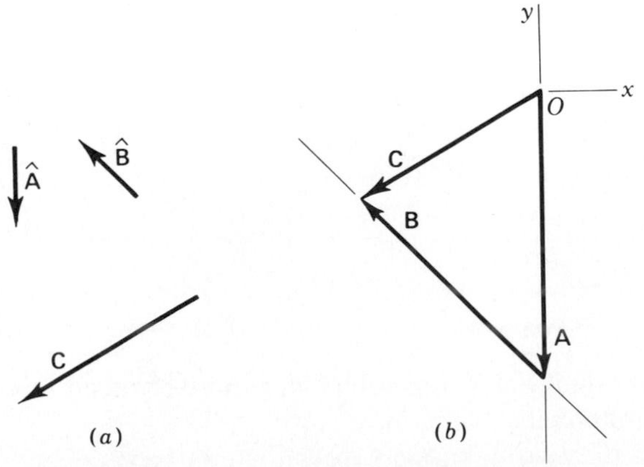

Figure 2-9 Graphical solution of case 2*a*: (*a*) given: $\mathbf{C}$, $\hat{\mathbf{A}}$, and $\hat{\mathbf{B}}$; (*b*) solution for A and B.

Note that case $2a$ has a unique solution unless the lines are collinear; if the lines are parallel but distinct, the magnitudes A and B are both infinite.

For case $2b$, a magnitude and a direction from different vectors, say A and $\hat{\mathbf{B}}$ are to be found

$$\overset{\checkmark\checkmark}{\mathbf{C}} = \overset{\circ\checkmark}{\mathbf{A}} + \overset{\checkmark\circ}{\mathbf{B}} \tag{2-18}$$

The solution, shown in Fig. 2-10, is obtained as follows:

1. Choose a coordinate system and scale factor and draw vector $\mathbf{C}$.
2. Construct a line through the origin of $\mathbf{C}$ parallel to $\hat{\mathbf{A}}$.
3. Adjust a compass to the scaled magnitude B and construct a circular arc with center at the terminus of $\mathbf{C}$.
4. The two intersections of the line and arc define the two sets of solutions A, $\hat{\mathbf{B}}$ and A', $\hat{\mathbf{B}}'$.

Finally, for case $2c$, the directions of two vectors, $\hat{\mathbf{A}}$ and $\hat{\mathbf{B}}$, are to be found

$$\overset{\checkmark\checkmark}{\mathbf{C}} = \overset{\checkmark\circ}{\mathbf{A}} + \overset{\checkmark\circ}{\mathbf{B}} \tag{2-19}$$

The steps in the solution are illustrated in Fig. 2-11.

1. Choose a coordinate system and a scale factor and draw vector $\mathbf{C}$.
2. Construct a circular arc of radius A centered at the origin of $\mathbf{C}$.
3. Construct a circular arc of radius B centered at the terminus of $\mathbf{C}$.
4. The two intersections of these arcs define the two sets of solutions, $\hat{\mathbf{A}}$, $\hat{\mathbf{B}}$ and $\hat{\mathbf{A}}'$, $\hat{\mathbf{B}}'$. Note that a real solution is possible only if $A + B \geq C$.

Let us now apply these procedures to the solution of the loop-closure

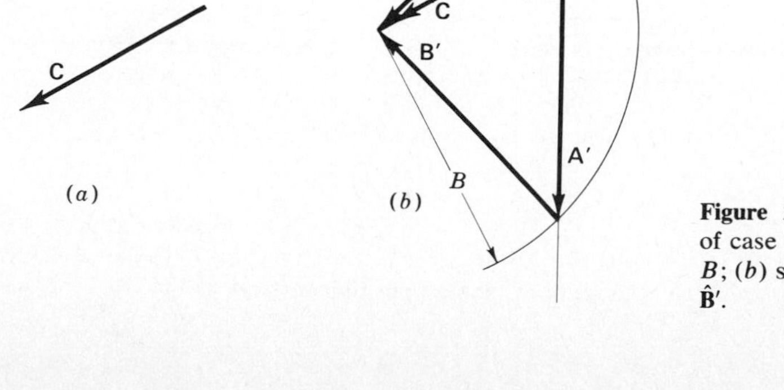

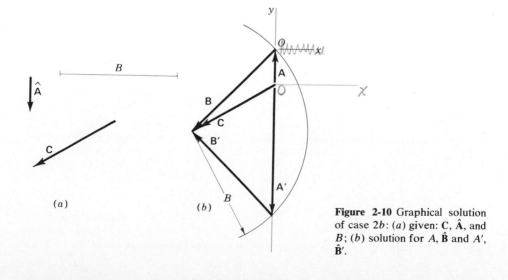

Figure 2-10 Graphical solution of case $2b$: (a) given: $\mathbf{C}$, $\hat{\mathbf{A}}$, and B; (b) solution for A, $\hat{\mathbf{B}}$ and A', $\hat{\mathbf{B}}'$.

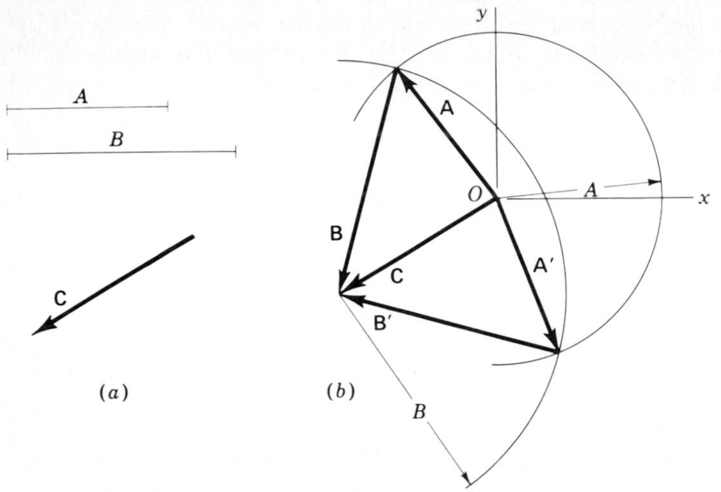

Figure 2-11 Graphical solution of case 2c: (a) given: **C**, A, and B; (b) solution for $\hat{\mathbf{A}}$, $\hat{\mathbf{B}}$ and $\hat{\mathbf{A}}'$, $\hat{\mathbf{B}}'$.

equation. To illustrate, let us consider the slider-crank mechanism shown in Fig. 2-12a. Here link 2 is a crank, constrained to rotate about the fixed pivot A. Link 3 is the connecting rod and link 4 the slider. The loop-closure equation, found by the approach of Sec. 2-6, is

$$\mathbf{R}_C = \mathbf{R}_{BA} + \mathbf{R}_{CB} \tag{f}$$

The problem of position analysis is to determine the values of all position variables (the positions of all points and joints) given the dimensions of each link and the value(s) of the independent variable(s), those chosen to represent the degree(s) of freedom of the mechanism. In the slider-crank mechanism,

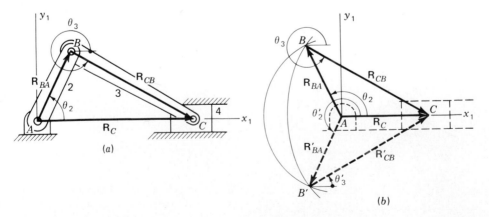

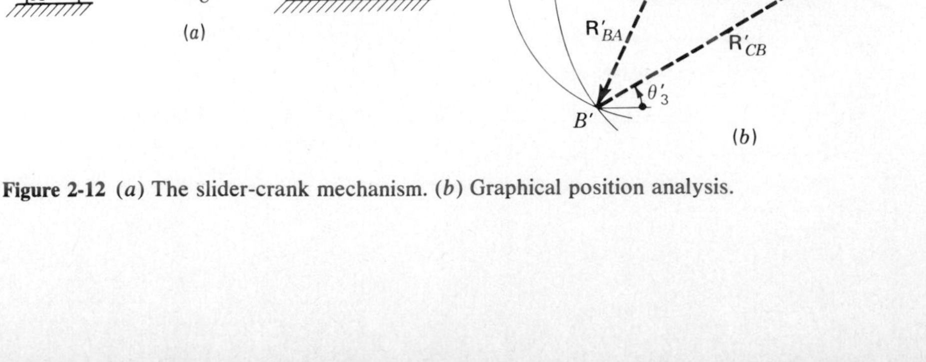

Figure 2-12 (a) The slider-crank mechanism. (b) Graphical position analysis.

upon moving the slider to some known location R_C we must find the unknown angles θ_2 and θ_3, the directions of $\hat{\mathbf{R}}_{BA}$ and $\hat{\mathbf{R}}_{CB}$. After identifying the known dimensions of the links,

$$\overset{\surd\surd}{\mathbf{R}_C} = \overset{\surd\circ}{\mathbf{R}_{BA}} + \overset{\surd\circ}{\mathbf{R}_{CB}} \tag{g}$$

we recognize that this is case 2c of the loop-closure equation. The graphical solution procedure explained earlier is applied in Fig. 2-12b. We note that two possible solutions are found, θ_2, θ_3 and θ_2', θ_3', and correspond to two different configurations of the linkage, i.e., two ways to assemble the links, both of which are consistent with the given position of the slider. Both solutions are equally valid roots to the loop-closure equation and it is necessary to choose between them according to the application.

As another example consider the four-bar linkage shown in Fig. 2-13. Here we may wish to find the position of the coupler point P which corresponds to some given crank angle θ_2. The loop-closure equation is

$$\overset{\surd\surd}{\mathbf{R}_{BA}} + \overset{\surd\circ}{\mathbf{R}_{CB}} = \overset{\surd\surd}{\mathbf{R}_{DA}} + \overset{\surd\circ}{\mathbf{R}_{CD}} \tag{h}$$

and the position of point P is given by the position-difference equation

$$\overset{\circ\circ}{\mathbf{R}_P} = \overset{\surd\surd}{\mathbf{R}_{BA}} + \overset{\surd\circ}{\mathbf{R}_{PB}} \tag{i}$$

Although it appears that this equation has three unknowns, it can be reduced to two once the loop-closure equation (h) is solved by noticing the constant angular relationship between $\mathbf{R}_{PB}$ and $\mathbf{R}_{CB}$,

$$\theta_5 = \theta_3 + \alpha \tag{j}$$

The graphical solution for this problem is started by combining the two known terms of Eq. (h), thus locating the positions of points B and D as shown in Fig. 2-14

$$\mathbf{S} = \overset{\surd\surd}{\mathbf{R}_{DA}} - \overset{\surd\surd}{\mathbf{R}_{BA}} = \overset{\surd\circ}{\mathbf{R}_{CB}} - \overset{\surd\circ}{\mathbf{R}_{CD}} \tag{k}$$

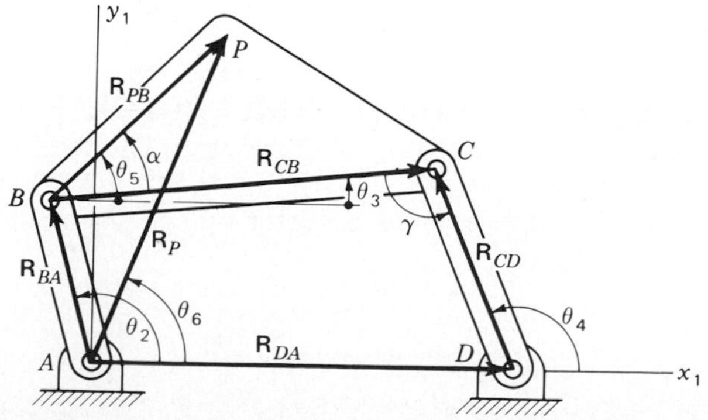

Figure 2-13 Four-bar linkage.

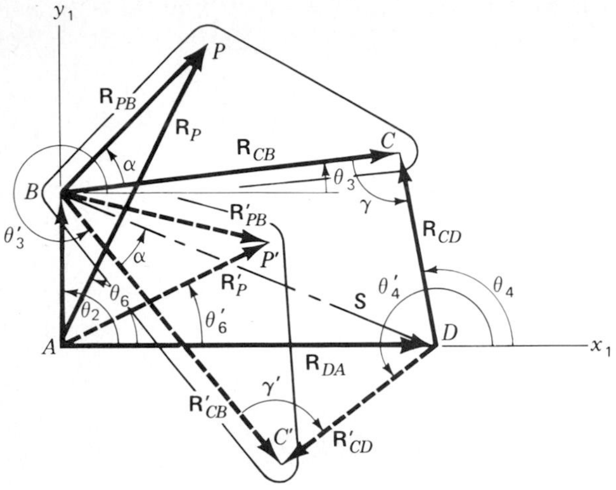

Figure 2-14 Graphical position analysis of the four-bar linkage.

The solution procedure for case 2*c*, two unknown directions, is then used to find the location of point *C*. Two possible solutions θ_3, θ_4 and θ_3', θ_4' are found.

Next we apply Eq. (*j*) to determine the two possible directions of $\hat{\mathbf{R}}_{PB}$. Equation (*i*) can then be solved by the procedures for case 1. Two solutions are finally achieved for the position of point *P*, $\mathbf{R}_P$ and $\mathbf{R}_P'$. Both are valid solutions to Eqs. (*h*) to (*j*) although the position $\mathbf{R}_P'$ could not be achieved physically from the configuration shown in Fig. 2-13 without disassembling the mechanism.

From the slider-crank and four-bar linkage examples it is clear that graphical position analysis requires precisely the same constructions which would be chosen naturally in drafting a scale drawing of the mechanism at the position being considered. For this reason the procedure seems trivial and not truly worthy of the title analysis. Yet this is highly misleading. As we will see in the coming sections, the position analysis of a mechanism is a nonlinear algebraic problem when approached by analytical or computer methods. It is, in fact, the most difficult problem in kinematic analysis, and this is the primary reason graphical solution techniques have retained their attraction for planar-mechanism analysis.

2-8 COMPLEX-ALGEBRA SOLUTIONS OF PLANAR VECTOR EQUATIONS

In planar problems it is often desirable to express a vector by specifying its magnitude and direction in *polar notation*

$$\mathbf{R} = R\underline{/\theta} \tag{2-20}$$

In Fig. 2-15*a* the two-dimensional vector

$$\mathbf{R} = R^x\hat{\mathbf{i}} + R^y\hat{\mathbf{j}} \tag{2-21}$$

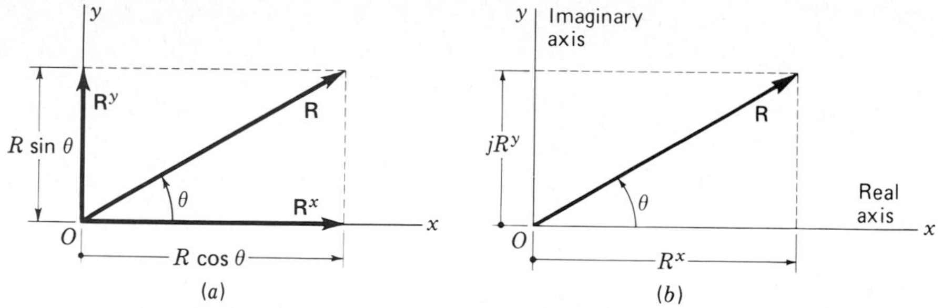

Figure 2-15 Correlation of planar vectors and complex numbers.

has two rectangular components of magnitudes

$$R^x = R \cos \theta \qquad R^y = R \sin \theta \tag{2-22}$$

with

$$R = \sqrt{(R^x)^2 + (R^y)^2} \qquad \theta = \tan^{-1} \frac{R^y}{R^x} \tag{2-23}$$

Note that we have made the arbitrary choice here of accepting the positive square root for the magnitude R when calculating from the components of **R**. Therefore we must be careful to interpret the signs of R^x and R^y individually when deciding upon the quadrant of θ. Note that *θ is defined as the angle from the positive x axis to the positive end of vector* **R**, *measured about the origin of the vector, and is positive when measured counterclockwise.*

Example 2-2 Express the vectors $\mathbf{A} = 10\underline{/30^\circ}$ and $\mathbf{B} = 8\underline{/-15^\circ}$ in rectangular notation† and find their sum.

SOLUTION The vectors are shown in Fig. 2-16 and are

$$\mathbf{A} = 10 \cos 30^\circ\, \hat{\mathbf{i}} + 10 \sin 30^\circ\, \hat{\mathbf{j}} = 8.66\hat{\mathbf{i}} + 5.00\hat{\mathbf{j}}$$
$$\mathbf{B} = 8 \cos(-15^\circ)\, \hat{\mathbf{i}} + 8 \sin(-15^\circ)\, \hat{\mathbf{j}} = 7.73\hat{\mathbf{i}} - 2.07\hat{\mathbf{j}}$$
$$\mathbf{C} = \mathbf{A} + \mathbf{B} = (8.66 + 7.73)\hat{\mathbf{i}} + (5.00 - 2.07)\hat{\mathbf{j}}$$
$$= 16.39\hat{\mathbf{i}} + 2.93\hat{\mathbf{j}}$$

The magnitude of the resultant is found from Eq. (2-23)

$$C = \sqrt{16.39^2 + 2.93^2} = 16.6$$

as is the angle

$$\theta = \tan^{-1} \frac{2.93}{16.39} = 10.1^\circ$$

The final result in planar notation is

$$\mathbf{C} = 16.6\underline{/10.1^\circ} \qquad \textit{Ans.}$$

† Many calculators are equipped to perform polar-rectangular and rectangular-polar conversions directly.

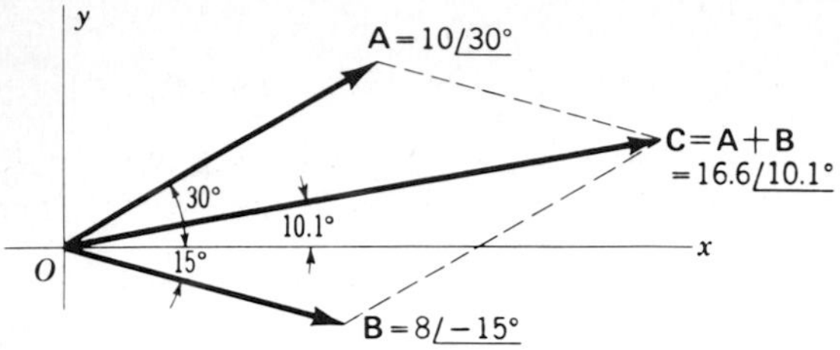

Figure 2-16 Example 2-2.

Another way of treating two-dimensional vector problems analytically makes use of complex algebra. Although complex numbers are not vectors, they can be used to represent vectors in a plane by choosing an origin and real and imaginary axes. In two-dimensional kinematics problems, these axes can conveniently be chosen coincident with the x_1y_1 axes of the absolute coordinate system.

As shown in Fig. 2-15*b*, the location of any point in the plane can be specified either by its absolute-position vector or by its corresponding real and imaginary coordinates

$$\mathbf{R} = R^x + jR^y \tag{2-24}$$

where the operator j is defined as the unit imaginary number

$$j = \sqrt{-1} \tag{2-25}$$

The real usefulness of complex numbers in planar analysis stems from the ease with which they can be switched to polar form. Employing complex rectangular notation for the vector **R**, we can write

$$\mathbf{R} = R\underline{/\theta} = R\cos\theta + jR\sin\theta \tag{2-26}$$

But using the well-known Euler equation from trigonometry,

$$e^{\pm j\theta} = \cos\theta \pm j\sin\theta \tag{2-27}$$

we can also write **R** in *complex polar* form as

$$\mathbf{R} = Re^{j\theta} \tag{2-28}$$

where the magnitude and direction of the vector appear explicitly. As we will see in the next two chapters, expression of a vector in this form is especially useful when differentiation is required.

Some familiarity with useful manipulation techniques for vectors written in complex polar forms can be gained by solving the four cases of the loop-closure equation again. Writing Eq. (2-16) in complex polar form, we have

$$Ce^{j\theta_C} = Ae^{j\theta_A} + Be^{j\theta_B} \tag{2-29}$$

In case 1 the two unknowns are C and θ_C. We begin the solution by separating the real and imaginary parts of the equation. By substituting Euler's equation (2-27) we obtain

$$C(\cos\theta_C + j\sin\theta_C) = A(\cos\theta_A + j\sin\theta_A) + B(\cos\theta_B + j\sin\theta_B) \qquad (a)$$

On equating the real terms and the imaginary terms separately we obtain two real equations corresponding to the horizontal and vertical components of the two-dimensional vector equation

$$C\cos\theta_C = A\cos\theta_A + B\cos\theta_B \qquad (b)$$

$$C\sin\theta_C = A\sin\theta_A + B\sin\theta_B \qquad (c)$$

By squaring and adding these two equations θ_C is eliminated and a solution is found for C

$$C = \sqrt{A^2 + B^2 + 2AB\cos(\theta_B - \theta_A)} \qquad (2\text{-}30)$$

The positive square root was chosen arbitrarily; the negative square root would yield a negative solution for C with θ_C differing by 180°. The angle θ_C is found from

$$\theta_C = \tan^{-1}\frac{A\sin\theta_A + B\sin\theta_B}{A\cos\theta_A + B\cos\theta_B} \qquad (2\text{-}31)$$

where the signs of the numerator and denominator must be considered separately in determining the proper quadrant† of θ_C. Only a single solution is found for case 1, as previously illustrated in Fig. 2-8.

For case 2*a* the two unknowns of Eq. (2-29) are the two magnitudes A and B. The graphical solution to this case was shown in Fig. 2-9. One convenient way of solving in complex polar form is to first divide Eq. (2-29) by $e^{j\theta_A}$

$$Ce^{j(\theta_C - \theta_A)} = A + Be^{j(\theta_B - \theta_A)} \qquad (d)$$

Comparing this equation with Fig. 2-17, we see that division by the complex polar form of a *unit* vector $e^{j\theta_A}$ has the effect of rotating the real and imaginary axes by the angle θ_A such that the real axis lies along the vector **A**. We can now use Euler's equation (2-27) to separate the real and imaginary components

$$C\cos(\theta_C - \theta_A) = A + B\cos(\theta_B - \theta_A) \qquad (e)$$

$$C\sin(\theta_C - \theta_A) = B\sin(\theta_B - \theta_A) \qquad (f)$$

and we note that the vector **A**, now real, has been eliminated from one equation. The solution for B is easily found:

$$B = C\frac{\sin(\theta_C - \theta_A)}{\sin(\theta_B - \theta_A)} \qquad (2\text{-}32)$$

† Calculators of different brands vary somewhat in the treatment of the units and quadrant of angles. Each of us must become familiar with the characteristics of his or her own calculator.

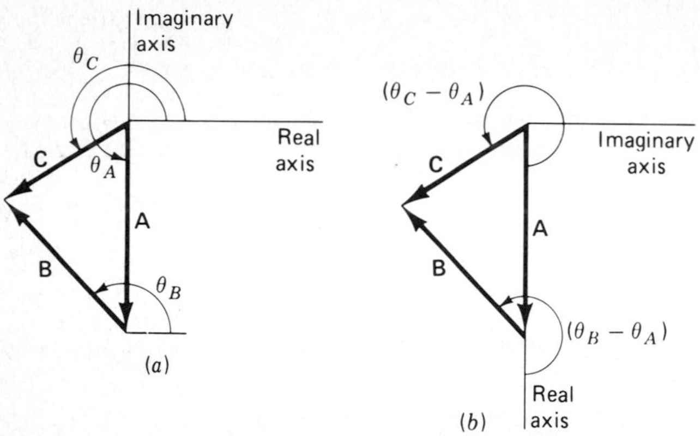

Figure 2-17 Rotation of axes by division of complex polar equation by $e^{j\theta_A}$. (*a*) Original axes; (*b*) rotated axes.

The solution for the other unknown magnitude A is found in completely analogous fashion. Dividing Eq. (2-29) by $e^{j\theta_B}$ aligns the real axis along vector **B**. The equation is then separated into real and imaginary parts and yields

$$A = C\frac{\sin(\theta_C - \theta_B)}{\sin(\theta_A - \theta_B)} \tag{2-33}$$

As before, case 2*a* yields a unique solution.

The graphical solution to case 2*b* was shown in Fig. 2-10. The two unknowns are A and θ_B. We begin by aligning the real axis along vector **A** and separating real and imaginary parts as in case 2*a*. The solutions are then obtained directly from Eqs. (*e*) and (*f*)

$$\theta_B = \theta_A + \sin^{-1}\frac{C\sin(\theta_C - \theta_A)}{B} \tag{2-34}$$

$$A = C\cos(\theta_C - \theta_A) - B\cos(\theta_B - \theta_A) \tag{2-35}$$

We note that the arc sine term is double-valued and therefore case 2*b* has two distinct solutions, A, θ_B and A', θ'_B.

Case 2*c* has the two angles θ_A and θ_B as unknowns. The graphical solution was shown in Fig. 2-11. In this case we align the real axis along vector **C**

$$C = Ae^{j(\theta_A - \theta_C)} + Be^{j(\theta_B - \theta_C)} \tag{g}$$

Using Euler's equation to separate components and then rearranging terms, we obtain

$$A\cos(\theta_A - \theta_C) = C - B\cos(\theta_B - \theta_C) \tag{h}$$

$$A\sin(\theta_A - \theta_C) = -B\sin(\theta_B - \theta_C) \tag{i}$$

Squaring both equations and adding, results in

$$A^2 = C^2 + B^2 - 2BC \cos(\theta_B - \theta_C) \tag{j}$$

We recognize this as the law of cosines for the vector triangle. It can be solved for θ_B

$$\theta_B = \theta_C \mp \cos^{-1} \frac{C^2 + B^2 - A^2}{2CB} \tag{2-36}$$

Putting C on the other side of Eq. (h) before squaring and adding results in another form of the law of cosines, from which

$$\theta_A = \theta_C \pm \cos^{-1} \frac{C^2 + A^2 - B^2}{2CA} \tag{2-37}$$

The plus-or-minus signs in these two equations are a reminder that the arc cosines are each double-valued and therefore θ_B and θ_A each have two solutions. These two pairs of angles can be paired naturally together as θ_A, θ_B and θ'_A, θ'_B under the restriction of Eq. (i) above. Thus case 2c has two distinct solutions, as shown in Fig. 2-11.

2-9 THE CHACE SOLUTIONS TO PLANAR VECTOR EQUATIONS

As we saw in the last section, the algebra involved in solving even simple planar vector equations can become cumbersome. Chace was the first to take advantage of the brevity of vector notation in obtaining explicit closed-form solutions to both two- and three-dimensional vector equations.† In this section we will study his solutions for planar equations in terms of the four cases of the loop-closure equation.

We again recall Eq. (2-16), the typical planar vector equation. In terms of magnitudes and unit vectors it can be written

$$C\hat{\mathbf{C}} = A\hat{\mathbf{A}} + B\hat{\mathbf{B}} \tag{2-38}$$

and it may contain two unknowns consisting of two magnitudes, two directions, or one magnitude and one direction.

Case 1 is the situation where the magnitude and direction of the same vector, say C and $\hat{\mathbf{C}}$, form the two unknowns. The method of solution for this case was shown in Example 2-2. The general form of the solution is

$$\mathbf{C} = (\mathbf{A} \cdot \hat{\mathbf{i}} + \mathbf{B} \cdot \hat{\mathbf{i}})\hat{\mathbf{i}} + (\mathbf{A} \cdot \hat{\mathbf{j}} + \mathbf{B} \cdot \hat{\mathbf{j}})\hat{\mathbf{j}} \tag{2-39}$$

In case 2a the magnitudes of two different vectors, say A and B, are unknown. The Chace approach for this case consists in eliminating one of the unknowns by taking the dot product of every vector with a new vector chosen

†M. A. Chace, Vector Analysis of Linkages, *J. Eng. Ind.*, ser. B, vol. 55, no. 3, pp. 289-297, August 1963.

so that one of the unknowns is eliminated. We can eliminate the vector **B** by taking the dot product of every term of the equation with $\hat{\mathbf{B}} \times \hat{\mathbf{k}}$.

$$\mathbf{C} \cdot (\hat{\mathbf{B}} \times \hat{\mathbf{k}}) = A\hat{\mathbf{A}} \cdot (\hat{\mathbf{B}} \times \hat{\mathbf{k}}) + B\hat{\mathbf{B}} \cdot (\hat{\mathbf{B}} \times \hat{\mathbf{k}}) \tag{a}$$

Thus, since $\hat{\mathbf{B}} \times \hat{\mathbf{k}}$ is perpendicular to $\hat{\mathbf{B}}$, $\hat{\mathbf{B}} \cdot (\hat{\mathbf{B}} \times \hat{\mathbf{k}}) = 0$; hence

$$A = \frac{\mathbf{C} \cdot (\hat{\mathbf{B}} \times \hat{\mathbf{k}})}{\hat{\mathbf{A}} \cdot (\hat{\mathbf{B}} \times \hat{\mathbf{k}})} \tag{2-40}$$

In a similar manner, we obtain the unknown magnitude B

$$B = \frac{\mathbf{C} \cdot (\hat{\mathbf{A}} \times \hat{\mathbf{k}})}{\hat{\mathbf{B}} \cdot (\hat{\mathbf{A}} \times \hat{\mathbf{k}})} \tag{2-41}$$

For case 2*b*, the unknowns are the magnitude of one vector and the direction of another, say A and $\hat{\mathbf{B}}$. We begin the solution by eliminating A from Eq. (2-38)

$$\mathbf{C} \cdot (\hat{\mathbf{A}} \times \hat{\mathbf{k}}) = B\hat{\mathbf{B}} \cdot (\hat{\mathbf{A}} \times \hat{\mathbf{k}}) \tag{b}$$

Now, from the definition of the dot product of two vectors

$$\mathbf{P} \cdot \mathbf{Q} = PQ \cos \phi$$

we note that

$$B\hat{\mathbf{B}} \cdot (\hat{\mathbf{A}} \times \hat{\mathbf{k}}) = B \cos \phi \tag{c}$$

where ϕ is the angle between the vectors $\hat{\mathbf{B}}$ and $(\hat{\mathbf{A}} \times \hat{\mathbf{k}})$. Thus

$$\cos \phi = \hat{\mathbf{B}} \cdot (\hat{\mathbf{A}} \times \hat{\mathbf{k}}) \tag{d}$$

The vectors $\hat{\mathbf{A}}$ and $\hat{\mathbf{A}} \times \hat{\mathbf{k}}$ are perpendicular to each other; hence we are free to choose another coordinate system $\hat{\boldsymbol{\lambda}}\hat{\boldsymbol{\mu}}$ having the directions $\hat{\boldsymbol{\lambda}} = \hat{\mathbf{A}} \times \hat{\mathbf{k}}$ and $\hat{\boldsymbol{\mu}} = \hat{\mathbf{A}}$. In this reference system, the unknown unit vector $\hat{\mathbf{B}}$ can be written as

$$\hat{\mathbf{B}} = \cos \phi \, (\hat{\mathbf{A}} \times \hat{\mathbf{k}}) + \sin \phi \, \hat{\mathbf{A}} \tag{e}$$

If we now substitute Eq. (*d*) into (*b*) and solve for $\cos \phi$, we obtain

$$\cos \phi = \frac{\mathbf{C} \cdot (\hat{\mathbf{A}} \times \hat{\mathbf{k}})}{B} \tag{f}$$

Then

$$\sin \phi = \pm\sqrt{1 - \cos^2 \phi} = \pm \frac{1}{B} \sqrt{B^2 - [\mathbf{C} \cdot (\hat{\mathbf{A}} \times \hat{\mathbf{k}})]^2} \tag{g}$$

Substituting Eqs. (*f*) and (*g*) into (*e*) and multiplying both sides by the known magnitude B gives

$$\mathbf{B} = [\mathbf{C} \cdot (\hat{\mathbf{A}} \times \hat{\mathbf{k}})](\hat{\mathbf{A}} \times \hat{\mathbf{k}}) \pm \sqrt{B^2 - [\mathbf{C} \cdot (\hat{\mathbf{A}} \times \hat{\mathbf{k}})]^2} \, \hat{\mathbf{A}} \tag{2-42}$$

To obtain the vector **A** we may wish to use Eq. (2-38) directly and perform the vector subtraction. Alternatively, if we substitute Eq. (2-42) and rearrange, we obtain

$$\mathbf{A} = \mathbf{C} - [\mathbf{C} \cdot (\hat{\mathbf{A}} \times \hat{\mathbf{k}})](\hat{\mathbf{A}} \times \hat{\mathbf{k}}) \pm \sqrt{B^2 - [\mathbf{C} \cdot (\hat{\mathbf{A}} \times \hat{\mathbf{k}})]^2} \, \hat{\mathbf{A}} \tag{h}$$

The first two terms of this equation can be simplified as shown in Fig. 2-18*a*. The $\hat{\mathbf{A}} \times \hat{\mathbf{k}}$ direction is located 90° clockwise from the $\hat{\mathbf{A}}$ direction. The magnitude $\mathbf{C} \cdot (\hat{\mathbf{A}} \times \hat{\mathbf{k}})$ is the projection of $\mathbf{C}$ in the $\hat{\mathbf{A}} \times \hat{\mathbf{k}}$ direction. Therefore, when $[\mathbf{C} \cdot (\hat{\mathbf{A}} \times \hat{\mathbf{k}})](\hat{\mathbf{A}} \times \hat{\mathbf{k}})$ is subtracted from $\mathbf{C}$, the result is a vector of magnitude $\mathbf{C} \cdot \hat{\mathbf{A}}$ in the $\hat{\mathbf{A}}$ direction. With this substitution, Eq. (*h*) becomes

$$\mathbf{A} = \left[\mathbf{C} \cdot \hat{\mathbf{A}} \mp \sqrt{B^2 - [\mathbf{C} \cdot (\hat{\mathbf{A}} \times \hat{\mathbf{k}})]^2}\right] \hat{\mathbf{A}} \tag{2-43}$$

Finally, in case 2*c* the unknowns are the directions of two different vectors, say $\hat{\mathbf{A}}$ and $\hat{\mathbf{B}}$. This case is illustrated in Fig. 2-18*b*, where the vector $\mathbf{C}$ and the two magnitudes A and B are given. The problem is solved by finding the points of intersection of two circles of radii A and B. We begin by defining a new coordinate system $\hat{\boldsymbol{\lambda}}\hat{\boldsymbol{\mu}}$ whose axes are directed so that $\hat{\boldsymbol{\lambda}} = \hat{\mathbf{C}} \times \hat{\mathbf{k}}$ and $\hat{\boldsymbol{\mu}} = \hat{\mathbf{C}}$, as shown in the figure. If the coordinates of one of the points of intersection in the $\hat{\boldsymbol{\lambda}}\hat{\boldsymbol{\mu}}$ coordinate system are designated as u and v, then

$$\mathbf{A} = u\hat{\boldsymbol{\lambda}} + v\hat{\boldsymbol{\mu}} \qquad \text{and} \qquad \mathbf{B} = -u\hat{\boldsymbol{\lambda}} + (C - v)\hat{\boldsymbol{\mu}} \tag{i}$$

The equation of the circle of radius A is

$$u^2 + v^2 = A^2 \tag{j}$$

The circle of radius B has the equation

$$u^2 + (v - C)^2 = B^2$$

or

$$u^2 + v^2 - 2Cv + C^2 = B^2 \tag{k}$$

Subtracting Eq. (*k*) from (*j*) and solving for v yields

$$v = \frac{A^2 - B^2 + C^2}{2C} \tag{l}$$

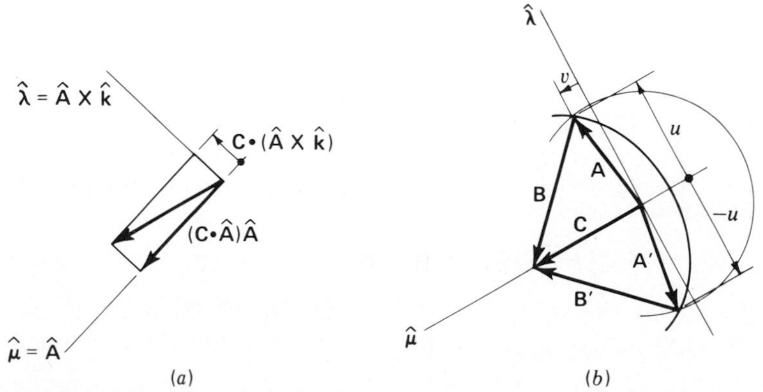

Figure 2-18

Substituting this into Eq. (*j*) and solving for *u* gives

$$u = \pm\sqrt{A^2 - \left(\frac{A^2 - B^2 + C^2}{2C}\right)^2} \tag{m}$$

The final step is to substitute these values of u and v into Eqs. (*i*) and to replace $\hat{\boldsymbol{\lambda}}$ and $\hat{\boldsymbol{\mu}}$ according to their definitions. The results are

$$\mathbf{A} = \pm\sqrt{A^2 - \left(\frac{A^2 - B^2 + C^2}{2C}\right)^2}\,(\hat{\mathbf{C}} \times \hat{\mathbf{k}}) + \frac{A^2 - B^2 + C^2}{2C}\hat{\mathbf{C}} \tag{2-44}$$

$$\mathbf{B} = \mp\sqrt{A^2 - \left(\frac{A^2 - B^2 + C^2}{2C}\right)^2}\,(\hat{\mathbf{C}} \times \hat{\mathbf{k}}) + \frac{B^2 - A^2 + C^2}{2C}\hat{\mathbf{C}} \tag{2-45}$$

2-10 ALGEBRAIC POSITION ANALYSIS OF PLANAR LINKAGES

This section illustrates several algebraic methods of approach to the position analysis of planar mechanisms. The three main advantages of algebraic methods over the graphical approach of Sec. 2-7 are (1) the increased accuracy which can be achieved, (2) the fact that they are suitable for computer or calculator evaluation, and (3) the fact that once the form of the solution has been found, it can be evaluated for any set of link dimensions or for different positions without starting over again. The main disadvantage, as we will see, is that the nature of the equations may lead to tedious algebraic manipulations in finding the form of the solution.

Let us return to the analysis of the slider-crank mechanism of Fig. 2-12, which was solved graphically in Sec. 2-7. One common way of formulating this problem algebraically is to notice from the figure that the vertical position of point B can be related to the length and angle of link 2 or of link 3. Thus

$$R_{BA} \sin\theta_2 = -R_{CB} \sin\theta_3 \tag{a}$$

so that

$$\sin\theta_3 = -\frac{R_{BA}}{R_{CB}} \sin\theta_2 \tag{b}$$

Also from the geometry of Fig. 2-12*a* we see that

$$R_C = R_{BA} \cos\theta_2 + R_{CB} \cos\theta_3 \tag{c}$$

which can be rearranged to read

$$R_C - R_{BA} \cos\theta_2 = R_{CB} \cos\theta_3 \tag{d}$$

Next, squaring and adding Eqs. (*a*) and (*d*), we eliminate the unknown θ_3

$$R_C^2 - 2R_C R_{BA} \cos\theta_2 + R_{BA}^2 = R_{CB}^2 \tag{e}$$

This equation can be solved for the unknown angle θ_2 as a function of the slider position R_C:

$$\theta_2 = \cos^{-1} \frac{R_C^2 + R_{BA}^2 - R_{CB}^2}{2R_C R_{BA}} \tag{2-46}$$

Substituting this solution into Eq. (d) gives an equation which can be solved for the other unknown angle θ_3

$$\theta_3 = \cos^{-1} \frac{R_C^2 + R_{CB}^2 - R_{BA}^2}{2R_C R_{CB}} \tag{2-47}$$

Although transcendental, these are closed-form solutions that can be quickly evaluated for any set of dimensional parameters at any position R_C of the slider.

In the more usual applications of the slider-crank mechanism, the angle of the crank θ_2 is given and the angle of the connecting rod θ_3 and the position of the slider R_C are to be found. This problem can be solved by recalling that since

$$\cos \theta_3 = \pm\sqrt{1 - \sin^2 \theta_3}$$

we have, from Eq. (b),

$$\cos \theta_3 = \frac{1}{R_{CB}} \sqrt{R_{CB}^2 - R_{BA}^2 \sin^2 \theta_2} \tag{2-48}$$

where the positive square root was chosen to correspond to Fig. 2-12a; the negative square root designates a different assembly of the links with the piston to the left of point A. From Eqs. (c) and (2-48), the position of point C is

$$R_C = R_{BA} \cos \theta_2 + \sqrt{R_{CB}^2 - R_{BA}^2 \sin^2 \theta_2} \tag{2-49}$$

We may ask ourselves in starting the algebraic analysis how we will recognize the "proper" equations from the figure; how we will know where to look or when we have enough equations. One of the advantages of the complex algebra approach of Sec. 2-8 is that it guides us in the development of these initial equations. Referring again to Fig. 2-12a, we can write the loop-closure equation in complex polar form

$$R_C = R_{BA}e^{j\theta_2} + R_{CB}e^{j\theta_3} \tag{f}$$

where x_1 is taken as the real axis. Using Euler's formula (2-27), we can separate the real and imaginary terms of the above equation. The two equations resulting are precisely those derived from the figure as Eqs. (c) and (a).

Whether these equations are found from the figure directly or by the use of the complex polar loop-closure equation, the solution process can proceed as above, using whatever manipulations are required to solve these equations simultaneously. With the complex-algebra approach, however, it is often possible to recognize the original loop-closure equation as one of the four standard cases and thus to write down the solution immediately from those derived in Sec. 2-8. Equations (2-46) and (2-47), for example, result directly by the form of Eq. (*f*) as case 2*c* and substituting the proper symbols into the standard solution, Eqs. (2-36) and (2-37). Similarly, Eqs. (2-48) and (2-49) are examples of case 2*b* and could be found directly from Eqs. (2-34) and (2-35).

To solve the same problem by using the Chace approach, we start by writing the loop-closure equation from Fig. 2-12*a*

$$R_C\hat{\mathbf{R}}_C = R_{BA}\hat{\mathbf{R}}_{BA} + R_{CB}\hat{\mathbf{R}}_{CB} \qquad (g)$$

With θ_2 given, the unknowns in this equation are the magnitude R_C and the direction $\hat{\mathbf{R}}_{CB}$. The solution corresponds to case 2*b* and is found by making appropriate substitutions in Eqs. (2-42) and (2-43)

$$\mathbf{R}_{CB} = -[\mathbf{R}_{BA}\cdot(\hat{\mathbf{R}}_C\times\hat{\mathbf{k}})](\hat{\mathbf{R}}_C\times\hat{\mathbf{k}}) + \sqrt{R_{CB}^2 - [\mathbf{R}_{BA}\cdot(\hat{\mathbf{R}}_C\times\hat{\mathbf{k}})]^2}\,\hat{\mathbf{R}}_C \qquad (2\text{-}50)$$

$$\mathbf{R}_C = [\mathbf{R}_{BA}\cdot\hat{\mathbf{R}}_C + \sqrt{R_{CB}^2 - [\mathbf{R}_{BA}\cdot(\hat{\mathbf{R}}_C\times\hat{\mathbf{k}})]^2}]\hat{\mathbf{R}}_C \qquad (2\text{-}51)$$

Example 2-3 Use the Chace equations to find the position of the slider of Fig. 2-12 with $R_{BA} = 25$ mm, $R_{CB} = 75$ mm, and $\theta_2 = 150°$.

SOLUTION Putting the given information in vector form we have

$$\mathbf{R}_{BA} = 25(\cos 150)\hat{\mathbf{i}} + 25(\sin 150)\hat{\mathbf{j}} = -21.7\hat{\mathbf{i}} + 12.5\hat{\mathbf{j}}$$

$$R_{CB} = 75 \qquad \hat{\mathbf{R}}_C = \hat{\mathbf{i}}$$

Note that $\hat{\mathbf{R}}_C \times \hat{\mathbf{k}} = -\hat{\mathbf{j}}$. Then substituting into Eq. (2-51) gives

$$\mathbf{R}_C = \left\{(-21.7\hat{\mathbf{i}} + 12.5\hat{\mathbf{j}})\cdot\hat{\mathbf{i}} + \sqrt{(75)^2 - [(-21.7\hat{\mathbf{i}} + 12.5\hat{\mathbf{j}})\cdot\hat{\mathbf{j}}]^2}\right\}\hat{\mathbf{i}}$$

$$= 50.2\hat{\mathbf{i}} \text{ mm} \qquad \textit{Ans.}$$

The analysis of the four-bar linkage is a classic problem with solution dating back over a century. The graphical solution was shown in Figs. 2-13 and 2-14. The same problem is presented here to illustrate the algebraic solution techniques further. The notation used here is defined in Fig. 2-19.

Note from the figure that s is the diagonal distance *BD*. The law of cosines can be written for the triangle *BAD* and again for the triangle *BCD*. In terms of the link lengths and angles defined in the figure, we have

$$s = \sqrt{r_1^2 + r_2^2 - 2r_1r_2\cos\theta_2} \qquad (h)$$

$$\gamma = \pm\cos^{-1}\frac{r_3^2 + r_4^2 - s^2}{2r_3r_4} \qquad (i)$$

where the plus-or-minus signs refer to the two solutions for the transmission angle γ and γ', respectively. The law of cosines can be written again for the same two triangles to find the angles ϕ and ψ

$$\phi = \cos^{-1}\frac{r_1^2 + s^2 - r_2^2}{2r_1s} \qquad (j)$$

$$\psi = \cos^{-1}\frac{r_4^2 + s^2 - r_3^2}{2r_4s} \qquad (k)$$

where it is noted from the figure that the magnitudes of ϕ and ψ are both less than 180° and that ψ is always positive while $\sin \phi$ is of the same sign as $\sin \theta_2$. From these we find the unknown angles θ_3 and θ_4

$$\theta_4 = 180° - \phi \mp \psi \tag{2-52}$$

$$\theta_3 = \theta_4 - \gamma \tag{2-53}$$

where the minus or plus sign again signifies the two closures, θ_4 and θ_4', respectively.

To solve the same problem using the Chace approach, we first form the vector

$$\mathbf{s} = \mathbf{r}_1 - \mathbf{r}_2 \tag{l}$$

The triangle *BCD* then gives the vector equation

$$\mathbf{s} = r_3\hat{\mathbf{r}}_3 - r_4\hat{\mathbf{r}}_4 \tag{m}$$

where the two directions $\hat{\mathbf{r}}_3$ and $\hat{\mathbf{r}}_4$ are unknown. This is case 2*c*, and the solutions are given by Eqs. (2-44) and (2-45). Substituting gives

$$\mathbf{r}_3 = \pm\sqrt{r_3^2 - \left(\frac{r_3^2 - r_4^2 + s^2}{2s}\right)^2}(\hat{\mathbf{s}} \times \hat{\mathbf{k}}) + \frac{r_3^2 - r_4^2 + s^2}{2s}\hat{\mathbf{s}} \tag{2-54}$$

$$\hat{\mathbf{r}}_4 = \pm\sqrt{r_3^2 - \left(\frac{r_3^2 - r_4^2 + s^2}{2s}\right)^2}(\hat{\mathbf{s}} \times \hat{\mathbf{k}}) + \frac{r_4^2 - r_3^2 + s^2}{2s}\hat{\mathbf{s}} \tag{2-55}$$

The upper set of signs gives the solution for the crossed linkage; the lower set therefore applies to the open linkage of Fig. 2-19.

Solving the same problem using complex algebra, we could adapt the standard solution of case 2*c* as done above. However, it will illustrate some useful manipulation techniques if we do not. We start by writing the loop-closure equation in complex polar form. Using the notation of Fig. 2-19, we have

$$r_2e^{j\theta_2} + r_3e^{j\theta_3} = r_1 + r_4e^{j\theta_4} \tag{n}$$

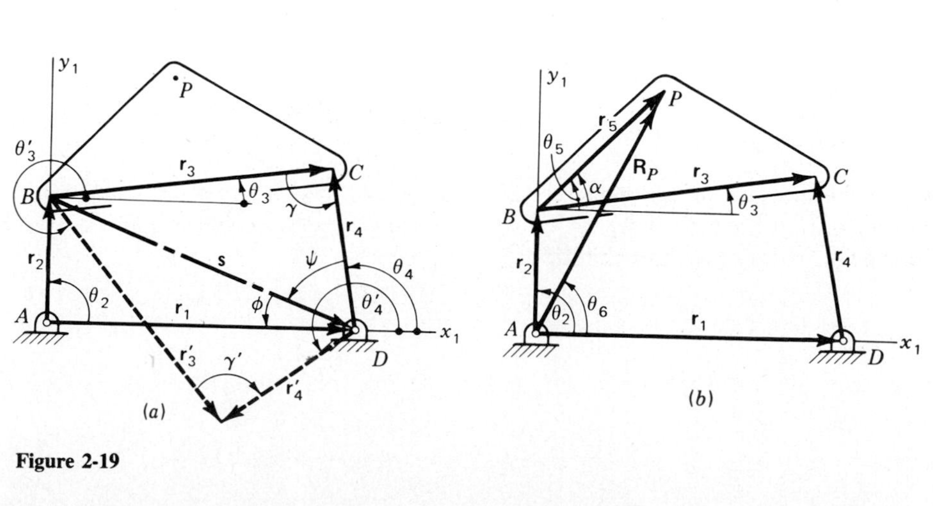

Figure 2-19

where x_1 is chosen as the real axis. Using Euler's formula, we separate the real and imaginary parts of the equation

$$r_2 \cos \theta_2 + r_3 \cos \theta_3 = r_1 + r_4 \cos \theta_4 \tag{o}$$

$$r_2 \sin \theta_2 + r_3 \sin \theta_3 = r_4 \sin \theta_4 \tag{p}$$

where angles θ_3 and θ_4 are the two unknowns. Next, we rearrange these equations to isolate the θ_3 terms

$$r_3 \cos \theta_3 = r_4 \cos \theta_4 - r_2 \cos \theta_2 + r_1$$

$$r_3 \sin \theta_3 = r_4 \sin \theta_4 - r_2 \sin \theta_2$$

and square and add the two equations

$$r_3^2 = r_4^2 + r_2^2 + r_1^2 + 2r_1r_4 \cos \theta_4 - 2r_1r_2 \cos \theta_2 - 2r_2r_4 \cos (\theta_4 - \theta_2) \tag{q}$$

thus eliminating the unknown θ_3.

We can combine a number of the known quantities of this equation and reduce its complexity by noting from the figure that

$$s^x = r_1 - r_2 \cos \theta_2 \tag{r}$$

$$s^y = -r_2 \sin \theta_2 \tag{s}$$

$$\gamma = \cos^{-1} \frac{r_3^2 + r_4^2 - r_1^2 - r_2^2 + 2r_1r_2 \cos \theta_2}{2r_3r_4} \tag{2-56}$$

where this last equation is equivalent to Eqs. (h) and (i) above. After making these substitutions and rearranging, Eq. (q) reduces to

$$s^x \cos \theta_4 + s^y \sin \theta_4 - r_3 \cos \gamma + r_4 = 0 \tag{t}$$

When dealing with both sine and cosine of the same unknown angle in a single equation, it is sometimes helpful to substitute the half-angle identities from trigonometry

$$\cos \eta = \frac{1 - \tan^2 (\eta/2)}{1 + \tan^2 (\eta/2)} \qquad \sin \eta = \frac{2 \tan (\eta/2)}{1 + \tan^2 (\eta/2)} \tag{2-57}$$

Substituting these into Eq. (t), clearing fractions, and rearranging terms, we obtain a quadratic equation

$$(r_4 - r_3 \cos \gamma - s^x) \tan^2 \frac{\theta_4}{2} + 2s^y \tan \frac{\theta_4}{2} + (r_4 - r_3 \cos \gamma + s^x) = 0 \tag{u}$$

from which we obtain two solutions

$$\tan \frac{\theta_4}{2} = \frac{-s^y \mp \sqrt{(s^y)^2 - r_4^2 + 2r_3r_4 \cos \gamma - r_3^2 \cos^2 \gamma + (s^x)^2}}{r_4 - r_3 \cos \gamma - s^x} \tag{v}$$

When we substitute from Eqs. (r), (s) and (2-56), this reduces to

$$\tan \frac{\theta_4}{2} = \frac{-s^y \mp r_3\sqrt{1 - \cos^2 \gamma}}{r_4 - r_3 \cos \gamma - s^x} \tag{w}$$

Therefore
$$\theta_4 = 2\tan^{-1}\frac{r_2 \sin\theta_2 \mp r_3 \sin\gamma}{r_4 - r_1 + r_2\cos\theta_2 - r_3\cos\gamma} \tag{2-58}$$

The solution for the other unknown, the angle θ_3, can be found by a completely analogous procedure. Isolating the θ_4 terms of Eqs. (o) and (p) before squaring and adding eliminates θ_4 and leaves a quadratic equation which can be solved for θ_3. The solution is

$$\theta_3 = 2\tan^{-1}\frac{-r_2 \sin\theta_2 \pm r_4 \sin\gamma}{r_3 + r_1 - r_2\cos\theta_2 - r_4\cos\gamma} \tag{2-59}$$

Having solved the basic four-bar linkage, we now seek an expression for the position of the coupler point P. From Fig. 2-19, in complex polar notation we write

$$\mathbf{R}_P = R_P e^{j\theta_6} = r_2 e^{j\theta_2} + r_5 e^{j(\theta_3+\alpha)} \tag{2-60}$$

We recognize this as case 1 since R_P and θ_6 are the two unknowns. The solutions can be found directly by applying Eqs. (2-30) and (2-31)

$$R_P = \sqrt{r_2^2 + r_5^2 + 2r_2r_5\cos(\theta_3 + \alpha - \theta_2)} \tag{2-61}$$

$$\theta_6 = \tan^{-1}\frac{r_2\sin\theta_2 + r_5\sin(\theta_3+\alpha)}{r_2\cos\theta_2 + r_5\cos(\theta_3+\alpha)} \tag{2-62}$$

Note that both these equations give double values coming from the double values for θ_3 and corresponding to the two closures of the linkage.

Example 2-4 Calculate and plot the coupler curve of a four-bar linkage with the following proportions: $r_1 = 200$ mm, $r_2 = 100$ mm, $r_3 = 250$ mm, $r_4 = 300$ mm, $r_5 = 150$ mm, and $\alpha = -45°$. Notation is as defined in Fig. 2-19.

SOLUTION For each crank angle θ_2, the transmission angle γ is evaluated from Eq. (2-56). Next, Eq. (2-59) is applied to give θ_3. Finally, the coupler point position is calculated from Eqs. (2-61) and (2-62). The solutions for the first several crank angles are given in Table 2-1. The full coupler curve is shown in Fig. 2-20. Note that only one of the two solutions is calculated and plotted.

Table 2-1 Calculation of the coupler curve for Example 2-4

θ_2, deg	γ, deg	θ_3, deg	R_P, mm	θ_6, deg	R_P^x, mm	R_P^y, mm
0.0	18.2	110.5	212	40.1	162	136
10.0	18.9	99.4	232	36.9	186	139
20.0	21.0	87.8	245	33.7	204	136
30.0	23.9	77.5	250	31.5	213	131
40.0	27.4	69.2	248	30.5	213	126
50.0	31.3	62.9	241	30.7	207	123
60.0	35.2	58.4	230	31.8	196	121
70.0	39.2	55.2	218	33.5	182	120
80.0	43.1	53.0	204	35.8	166	119
90.0	46.9	51.8	199	38.4	149	118

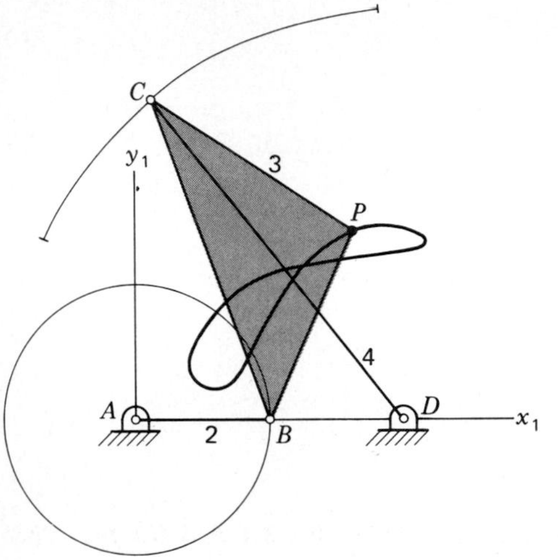

Figure 2-20 Plot of coupler curve of Example 2-4.

Before leaving the subject of the four-bar linkage, let us consider again Eq. (2-56), which defines the transmission angle. As the crank angle θ_2 is varied, the extremes of the transmission angle γ can be found by differentiating Eq. (2-56) with respect to θ_2 and setting the result equal to zero. This shows that the extremes occur at $\theta_2 = 0$ and $\theta_2 = 180°$ and are given by

$$\frac{r_3^2 + r_4^2 - (r_1 + r_2)^2}{2r_3r_4} \leq \cos\gamma \leq \frac{r_3^2 + r_4^2 - (r_1 - r_2)^2}{2r_3r_4} \tag{2-63}$$

The above, of course, assumes that the input crank is capable of making a complete rotation. If it is not a Grashof chain (Sec. 1-8) or is not of the crank-rocker or double-crank type, the crank will be limited to a range of values for θ_2. The calculations will then cause trouble outside of this range; the magnitude of the argument of the arc cosine of Eq. (2-56) will be greater than unity and a real solution will not be found for γ. The limits on this range are given by

$$\frac{r_1^2 + r_2^2 - (r_3 + r_4)^2}{2r_1r_2} \leq \cos\theta_2 \leq \frac{r_1^2 + r_2^2 - (r_3 - r_4)^2}{2r_1r_2} \tag{2-64}$$

2-11 DISPLACEMENT OF A MOVING POINT

We have concerned ourselves so far with only a single instantaneous position of a point, but since we wish to study motion, we must be concerned with the relationship between a succession of positions.

In Fig. 2-21 a particle, originally at point P, is moving along the path shown and, some time later, arrives at the position P'. The *displacement* of the point $\Delta\mathbf{R}_P$ during the time interval is defined as the *net change in position*

$$\Delta\mathbf{R}_P = \mathbf{R}'_P - \mathbf{R}_P \tag{2-65}$$

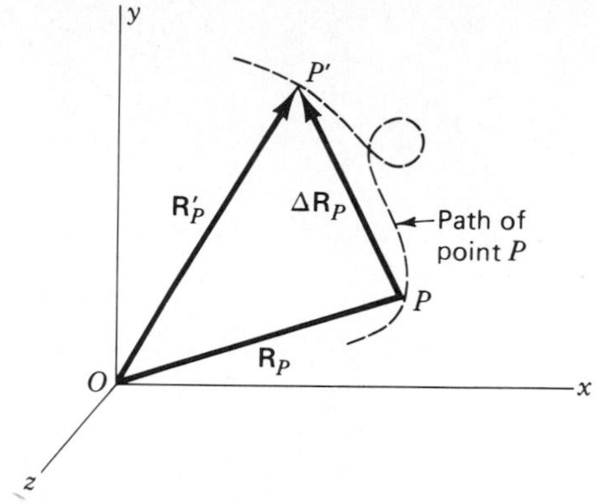

Figure 2-21 Displacement of a moving point.

Displacement is a vector quantity having the magnitude and direction of the vector from point P to point P'.

It is important to note that the displacement $\mathbf{\Delta R}_P$ is the *net* change in position and does not depend on the particular path taken between points P and P'. Its magnitude is *not* necessarily equal to the length of the path (the distance traveled), and direction is *not* necessarily along the tangent to the path, although both these are true when the displacement is infinitesimally small. Knowledge of the path actually traveled from P to P' is not even necessary to find the displacement vector as long as the initial and final positions are known.

2-12 DISPLACEMENT DIFFERENCE BETWEEN TWO POINTS

In this section we consider the difference in the displacements of two moving points. In particular we are concerned with the case where the two moving points are both particles of the same rigid body. The situation is shown in Fig. 2-22, where rigid body 2 moves from an initial position defined by $x_2y_2z_2$ to a later position defined by $x_2'y_2'z_2'$.

From Eq. (2-6), the position difference between the two points P and Q of body 2 at the initial instant is

$$\mathbf{R}_{PQ} = \mathbf{R}_P - \mathbf{R}_Q \tag{a}$$

After the displacement of body 2, the two points are located at P' and Q'. At that time the position difference is

$$\mathbf{R}'_{PQ} = \mathbf{R}'_P - \mathbf{R}'_Q \tag{b}$$

During the time interval of the movement the two points have undergone individual displacements of $\mathbf{\Delta R}_P$ and $\mathbf{\Delta R}_Q$, respectively.

As the name implies, the *displacement difference* between the two points is defined as the net difference between their respective displacements and

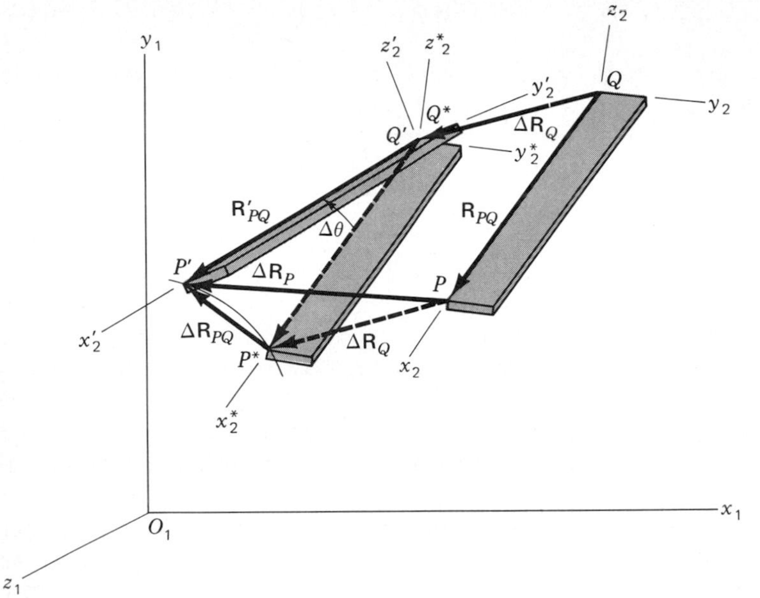

Figure 2-22 Displacement difference between two points on the same rigid body.

given the symbol $\mathbf{\Delta R}_{PQ}$

$$\mathbf{\Delta R}_{PQ} = \mathbf{\Delta R}_P - \mathbf{\Delta R}_Q \tag{2-66}$$

Note that this equation corresponds to the vector triangle PP^*P' in Fig. 2-22. As stated in the previous section, the displacement depends only on the *net* change in position and not on the path by which it was achieved. Thus, no matter how the body containing points P and Q was *actually* displaced, we are free to visualize the path as we choose. Equation (2-66) leads us to visualize the displacement as having taken place in two stages. First, the body translates (slides without rotation) from $x_2y_2z_2$ to $x_2^*y_2^*z_2^*$; during this movement all particles, including P and Q, have the same displacement $\mathbf{\Delta R}_Q$. Next we visualize the body as rotating about point Q' through the angle $\Delta\theta$ to the final position $x_2'y_2'z_2'$.

By manipulating Eq. (2-66) we can obtain a different interpretation

$$\begin{aligned}\mathbf{\Delta R}_{PQ} &= (\mathbf{R}'_P - \mathbf{R}_P) - (\mathbf{R}'_Q - \mathbf{R}_Q)\\ &= (\mathbf{R}'_P - \mathbf{R}'_Q) - (\mathbf{R}_P - \mathbf{R}_Q)\end{aligned} \tag{c}$$

and then, from Eqs. (a) and (b),

$$\mathbf{\Delta R}_{PQ} = \mathbf{R}'_{PQ} - \mathbf{R}_{PQ} \tag{2-67}$$

This equation corresponds to the vector triangle $Q'P^*P'$ in Fig. 2-22 and shows that the displacement difference, defined as the difference between two displacements, is equal to the net change between the position-difference vectors.

In either interpretation we are illustrating *Euler's theorem*, which states *that any displacement of a rigid body is equivalent to the sum of a net translation of one point (Q) and a net rotation of the body about that point.* We also see that only the rotation contributes to the displacement difference between two points on the same rigid body; i.e., *there is no difference between the displacements of any two points of the same rigid body as the result of a translation.* (See Sec. 2-13 for the definition of the term translation.)

In view of the above discussion, we can visualize the displacement difference $\mathbf{\Delta R}_{PQ}$ as the displacement which would be seen for point P by a moving observer who travels along, always staying coincident with point Q *but not turning* with the moving body, i.e., always using the absolute-coordinate axes $x_1y_1z_1$ to measure direction. It is important to understand the difference between the interpretation of an observer moving with point Q but not rotating and the case of the observer *on* the moving body. To an observer on body 2 both points P and Q would appear stationary; neither would be seen to have a displacement since they do not move relative to the observer, and the displacement difference seen by such an observer would be zero.

2-13 ROTATION AND TRANSLATION

Using the concept of displacement difference between two points of the same rigid body, we are now able to define translation and rotation.

Translation is defined as a state of motion of a body for which the displacement difference between any two points P and Q of the body is zero or, from the displacement-difference equation (2-66),

$$\begin{aligned} \mathbf{\Delta R}_{PQ} &= \mathbf{\Delta R}_P - \mathbf{\Delta R}_Q = 0 \\ \mathbf{\Delta R}_P &= \mathbf{\Delta R}_Q \end{aligned} \tag{2-68}$$

which states that *the displacements of any two points of the body are equal.* Rotation is a state of motion of the body for which different points of the body exhibit different displacements.

Figure 2-23*a* illustrates a situation where the body has moved along a curved path from position x_2y_2 to position $x_2'y_2'$. In spite of the fact that the point paths are curved,† $\mathbf{\Delta R}_P$ is still equal to $\mathbf{\Delta R}_Q$ and the body has undergone a translation. Note that in translation the point paths described by any two points on the body are identical and there is no change of angular orientation between the moving coordinate system and the observer's coordinate system; that is, $\Delta\theta_2 = \theta_2' - \theta_2 = 0$.

† Translation in which the point paths are not straight lines is sometimes referred to as *curvilinear translation.*

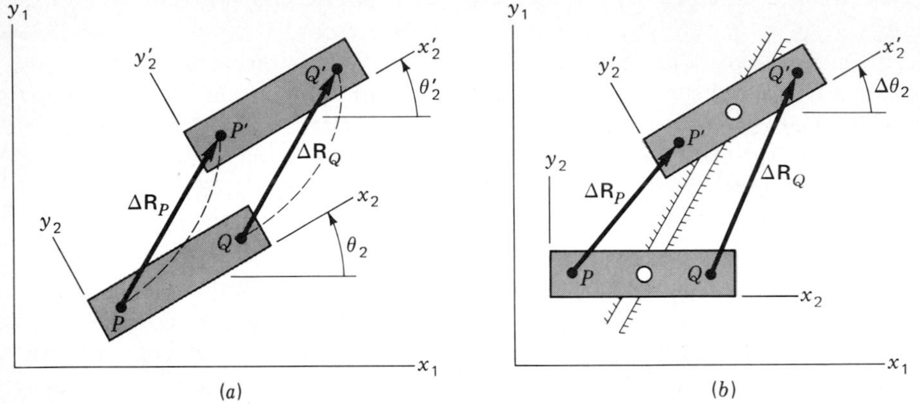

Figure 2-23 (*a*) Translation: $\mathbf{\Delta R}_P = \mathbf{\Delta R}_Q$, $\Delta\theta_2 = 0$; (*b*) rotation: $\mathbf{\Delta R}_P \neq \mathbf{\Delta R}_Q$, $\Delta\theta_2 \neq 0$.

In Fig. 2-23*b* the center point of the moving body is *constrained* to move along a straight-line path. Yet, as it does so, the body rotates so that $\Delta\theta_2 = \theta_2' - \theta_2 \neq 0$ and the displacements $\mathbf{\Delta R}_P$ and $\mathbf{\Delta R}_Q$ are not equal. Even though there is no obvious point on the body about which it has rotated, the coordinate system $x_2'y_2'$ has changed angular orientation relative to x_1y_1 and the body is said to have undergone a rotation. Note that the point paths described by P and Q are not equal.

We see from these two examples that rotation or translation of a body cannot be defined from the motion of a single point. These are characteristic motions of a body or a coordinate system. It is improper to speak of "rotation of a point" since there is no meaning for angular orientation of a point. It is also improper to associate the terms rotation and translation with the rectilinear or curvilinear characteristics of a single point path. Although it does not matter which points of the body are chosen, the motion of two or more points must be compared before meaningful definitions exist for these terms.

2-14 APPARENT DISPLACEMENT

We have already observed that the displacement of a moving point does not depend on the particular path traveled. However, since displacement is computed from the position vectors of the endpoints of the path, a knowledge of the coordinate system of the observer is essential.

Consider a particle P_3 moving along a known path in a moving coordinate system $x_2y_2z_2$, which, in turn, is moving with respect to the absolute reference frame $x_1y_1z_1$, as shown in Fig. 2-24. Let us also define another point P_2 which is rigidly attached to the moving body 2, i.e., stationary with respect to coordinate system $x_2y_2z_2$, and which is initially coincident with point P_3.

As seen by an absolute observer (in coordinate system $x_1y_1z_1$), the particle

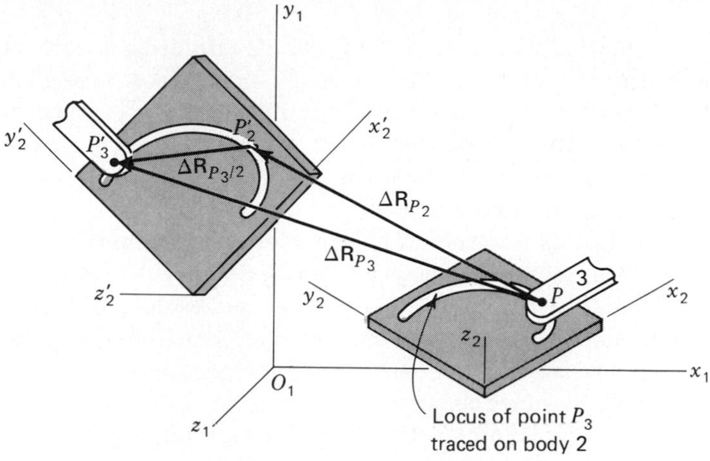

Figure 2-24 Apparent displacement of a point.

P_3, after some time interval, appears to have moved to a new location P_3' with the displacement $\mathbf{\Delta R}_{P_3}$. The point P_2, being part of body 2, moves differently from P_3; it achieves the new location P_2' with the displacement $\mathbf{\Delta R}_{P_2}$.

As seen by an observer in the moving coordinate system $x_2y_2z_2$, however, the situation seems quite different. This observer sees only the *apparent displacement* $\mathbf{\Delta R}_{P_3/2}$ of the particle P_3 as it moves along the path in his coordinate system. Since the path is fixed in his coordinate system, he does not sense its movement and therefore does not observe the same displacement for P_3 as the absolute observer. The point P_2 appears stationary to this observer, and therefore $\mathbf{\Delta R}_{P_2/2} = 0$.

From the vector triangle in Fig. 2-24 we see that the two observers' sensings are related by the *apparent-displacement equation*

$$\mathbf{\Delta R}_{P_3} = \mathbf{\Delta R}_{P_2} + \mathbf{\Delta R}_{P_3/2} \tag{2-69}$$

We can take this equation as the definition of the apparent-displacement vector, although it is important also to understand the physical concepts involved. Notice that the apparent-displacement vector relates the absolute displacements of two *coincident points* which are particles of *different moving bodies*. Notice also that there is no restriction on the actual location of the observer moving with coordinate system 2, only that he be fixed in that coordinate system so that he senses no displacement for point P_2.

One primary use of the apparent displacement is to determine an absolute displacement. It is not uncommon in machines to find a point such as P_3 which is constrained to move along a known slot or path or guideway defined by the shape of another moving link 2. In such cases it may be much more convenient to measure or calculate $\mathbf{\Delta R}_{P_2}$ and $\mathbf{\Delta R}_{P_3/2}$ and to use Eq. (2-69) than to measure the absolute displacement $\mathbf{\Delta R}_{P_3}$ directly.

2-15 ABSOLUTE DISPLACEMENT

In reflecting on the definition and concept of the apparent-displacement vector, we conclude that the absolute displacement of a moving point $\Delta\mathbf{R}_{P_3/1}$ is the special case of an apparent displacement where the observer is fixed in the absolute coordinate system. As explained for the position vector, the notation is often abbreviated to read $\Delta\mathbf{R}_{P_3}$ or just $\Delta\mathbf{R}_P$ and an absolute observer is implied whenever not noted explicitly.

Perhaps a better physical understanding of apparent displacement can be achieved by relating it to absolute displacement. Imagine an automobile P_3 traveling along a roadway and under observation by an absolute observer some distance off to one side. Consider how this observer visually senses the motion of the car. Although he may not be conscious of all of the following steps, the contention here is that the observer first imagines a point P_1, coincident with P_3, which he defines in his mind as stationary; he may relate to a fixed point of the roadway or a nearby tree, for example. He then compares his later observations of the car P_3 with those of P_1 to sense displacement. Notice that he does not compare with his own location but with the initially coincident point P_1. In this instance the apparent-displacement equation becomes an identity

$$\Delta\mathbf{R}_{P_3} = \overset{0}{\cancel{\Delta\mathbf{R}_{P_1}}} + \Delta\mathbf{R}_{P_3/1}$$

PROBLEMS†

2-1 Describe and sketch the locus of a point A which moves according to the equations $R_A^x = at\cos 2\pi t$, $R_A^y = at\sin 2\pi t$, $R_A^z = 0$.

2-2 Find the position difference from point P to point Q on the curve $y = x^2 + x - 16$, where $R_P^x = 2$ and $R_Q^x = 4$.

2-3 The path of a moving point is defined by the equation $y = 2x^2 - 28$. Find the position difference from point P to point Q if $R_P^x = 4$ and $R_Q^x = -3$.

2-4 The path of a moving point P is defined by the equation $y = 60 - x^3/3$. What is the displacement of the point if its motion begins when $R_P^x = 0$ and ends when $R_P^x = 3$?

2-5 If point A moves on the locus of Prob. 2-1, find its displacement from $t = 2$ to $t = 2.5$.

2-6 The position of a point is given by the equation $\mathbf{R} = 100e^{j2\pi t}$. What is the path of the point? Determine the displacement of the point from $t = 0.10$ to $t = 0.40$.

2-7 The equation $\mathbf{R} = (t^2 + 4)e^{-j\pi t/10}$ defines the position of a point. In which direction is the position vector rotating? Where is the point located when $t = 0$? What is the next value t can be if the direction of the position vector is to be the same as it is when $t = 0$? What is the displacement from the first position of the point to the second?

2-8 The location of a point is defined by the equation $\mathbf{R} = (4t + 2)e^{j\pi t^2/30}$ where t is time in seconds. Motion of the point is initiated when $t = 0$. What is the displacement during the first 3 s? Find the change in angular orientation of the position vector during the same time interval.

† When assigning problems, the instructor may wish to specify the method of solution to be used since a variety of approaches have been presented in the text.

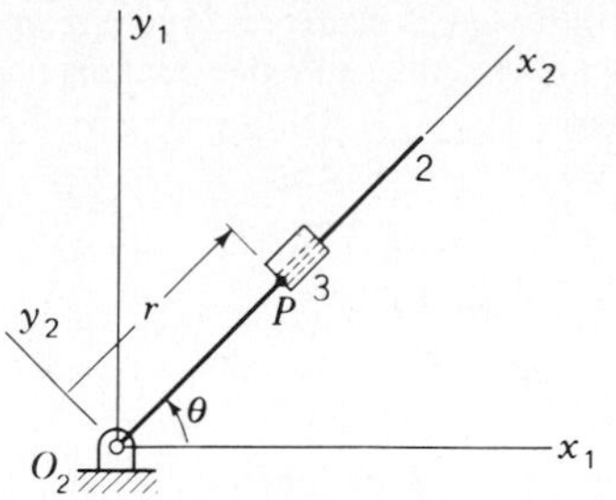

Problem 2-9

2-9 Link 2 in the figure rotates according to the equation $\theta = \pi t/4$. Block 3 slides outward on link 2 according to the equation $r = t^2 + 2$. What is the absolute displacement $\Delta \mathbf{R}_{P_3}$ from $t = 1$ to $t = 2$? What is the apparent displacement $\Delta \mathbf{R}_{P_{3/2}}$?

2-10 A wheel with center at O rolls without slipping so that its center is displaced 10 in to the right. What is the displacement of point P on the periphery during this interval?

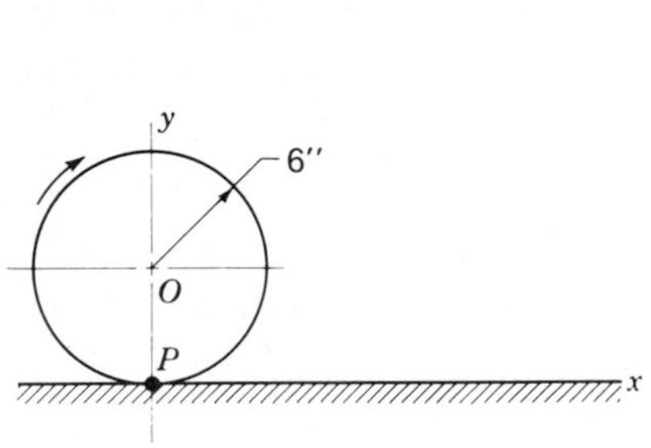

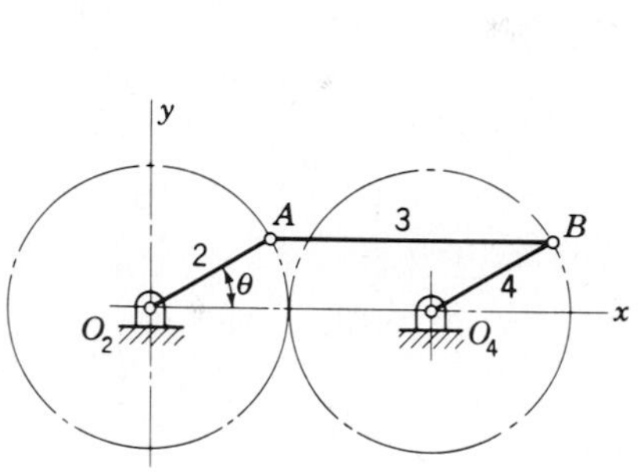

Problem 2-10 Rolling wheel and **Problem 2-11** $R_{AO_2} = R_{BO_4} = 3$ in, $R_{BA} = R_{O_4O_2} = 6$ in.

2-11 A point Q moves from A to B along link 3 while link 2 rotates from $\theta_2 = 30°$ to $\theta_2' = 120°$. Find the absolute displacement of Q.

2-12 The linkage shown is driven by moving the sliding block 2. Write the loop-closure equation. Solve analytically for the position of sliding block 4. Check the result graphically for the position where $\phi = -45°$.

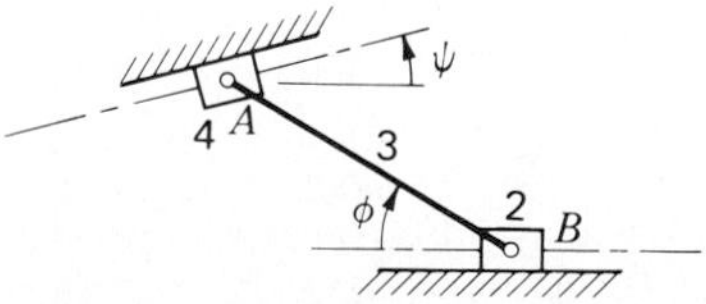

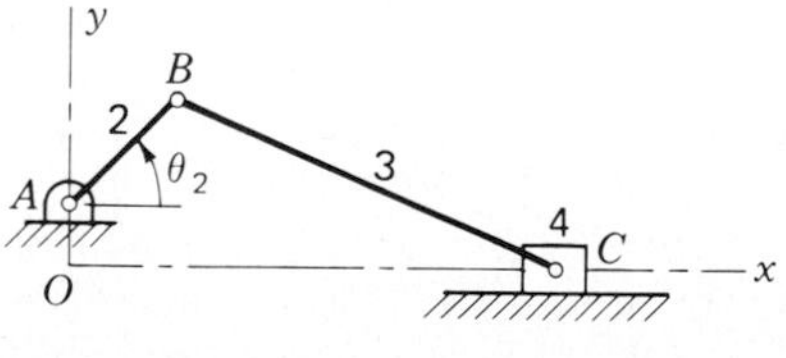

Problem 2-12 $R_{AB} = 200$ mm, $\psi = 15°$; **Problem 2-13** $R_{AO} = 1$ in, $R_{BA} = 2.5$ in, $R_{CB} = 7$ in.

2-13 The offset slider-crank mechanism is driven by rotating crank 2. Write the loop-closure equation. Solve for the position of the slider 4 as a function of θ_2.

2-14 Write a calculator program to find the sum of any number of two-dimensional vectors expressed in mixed rectangular or polar forms. The result should be obtainable in either form with the magnitude and angle of the polar form having only positive values.

2-15 Write a computer program to plot the coupler curve of any crank-rocker or double-crank form of the four-bar linkage. The program should accept four link lengths and either rectangular or polar coordinates of the coupler point relative to the coupler.

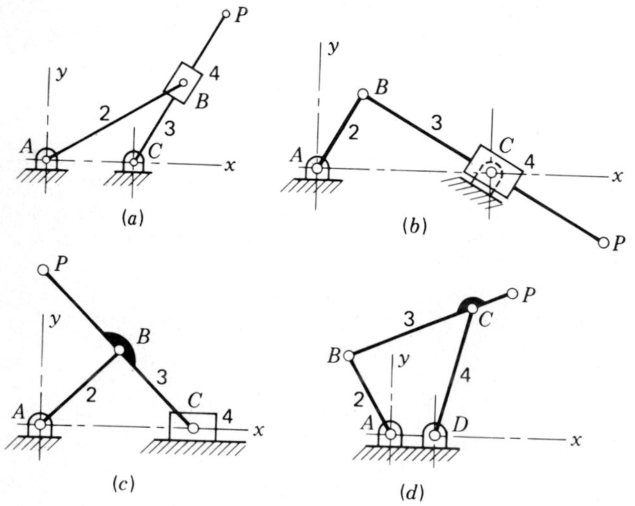

Problem 2-16 (*a*) $R_{CA} = 2$ in, $R_{BA} = 3.5$ in, $R_{PC} = 4$ in, (*b*) $R_{CA} = 40$ mm, $R_{BA} = 20$ mm, $R_{PB} = 65$ mm; (*c*) $R_{BA} = R_{CB} = R_{PB} = 25$ mm; (*d*) $R_{DA} = 1$ in, $R_{BA} = 2$ in, $R_{CB} = R_{DC} = 3$ in, $R_{PB} = 4$ in.

2-16 For each linkage shown in the figure find the path of point P: (*a*) inverted slider-crank mechanism; (*b*) second inversion of the slider-crank mechanism; (*c*) straight-line mechanism; (*d*) drag-link mechanism.

CHAPTER

THREE

VELOCITY

3-1 DEFINITION OF VELOCITY

In Fig. 3-1 a moving point is first observed at location P defined by the absolute-position vector $\mathbf{R}_P$. After a short time interval Δt, its location is observed to have changed to P', defined by $\mathbf{R}'_P$. From Eq. (2-65) we recall that the displacement during this time interval is defined as

$$\mathbf{\Delta R}_P = \mathbf{R}'_P - \mathbf{R}_P$$

The *average velocity* of the point during the time interval Δt is $\mathbf{\Delta R}_P/\Delta t$. Its *instantaneous velocity* (hereinafter simply called *velocity*) is defined by the limit of this ratio for an infinitesimally small time interval and is given by

$$\mathbf{V}_P = \lim_{\Delta t \to 0} \frac{\mathbf{\Delta R}_P}{\Delta t} = \frac{d\mathbf{R}_P}{dt} \tag{3-1}$$

Since $\mathbf{\Delta R}_P$ is a vector, there are *two* convergences in taking this limit, the *magnitude* and the *direction.* Therefore the velocity of a point is a vector quantity equal to the time rate of change of its position. Like the position and displacement vectors, the velocity vector is defined for a specific point; "velocity" should not be applied to a line, coordinate system, volume, or other collection of points, since the velocity of each point may differ.

We recall that the position vectors $\mathbf{R}_P$ and $\mathbf{R}'_P$ depend upon the location and orientation of the observer's coordinate system for their definitions. The displacement vector $\mathbf{\Delta R}_P$ and the velocity vector $\mathbf{V}_P$, on the other hand, are independent of the initial location of the coordinate system or the observer's location within the coordinate system. However, the velocity vector $\mathbf{V}_P$ does

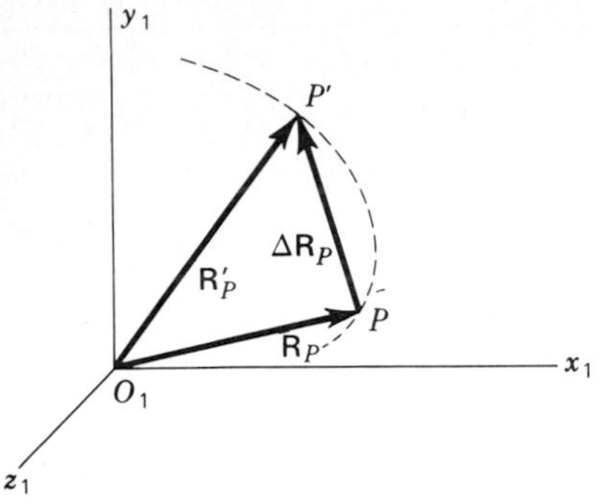

Figure 3-1 Displacement of a moving particle.

depend critically on the motion, if any, of the observer or the coordinate system during the time interval; it is for this reason that the observer is assumed to be stationary within the coordinate system. If the coordinate system involved is the absolute coordinate system, the velocity is referred to as an *absolute velocity* and is denoted by $\mathbf{V}_{P/1}$ or simply $\mathbf{V}_P$. This is consistent with the notation used for absolute displacement.

3-2 ROTATION OF A RIGID BODY

When a rigid body translates, as we saw in Sec. 2-13, the motion of any single particle is equal to the motion of every other particle of the same body. When the body rotates, however, two arbitrarily chosen particles P and Q do not undergo the same motion and a coordinate system attached to the body does not remain parallel to its initial orientation; i.e., the body undergoes some angular displacement $\Delta\theta$.

Angular displacements were not treated in detail in Chap. 2 because, in general, they cannot be treated as vectors. The reason is that they do not obey the usual laws of vector addition; if several gross angular displacements in three dimensions are undergone in succession, the result depends on the order in which they take place.

To illustrate, consider the rectangle $ABCO$ in Fig. 3-2a. The rectangular body is first rotated by $-90°$ about the y axis and then rotated by $+90°$ about the x axis. The final position of the body is seen to be in the yz plane. In Fig. 3-2b the body occupies the same starting position and is again rotated about the same axes, through the same angles, and in the same directions; however, the first rotation is about the x axis and the second about the y axis. The order of the rotations is reversed, and the final position of the rectangle is now seen to be in the xz plane rather than the yz plane, as it was before. Since this characteristic does not correspond to the commutative law of vector addition, three-dimensional angular displacements cannot be treated as vectors.

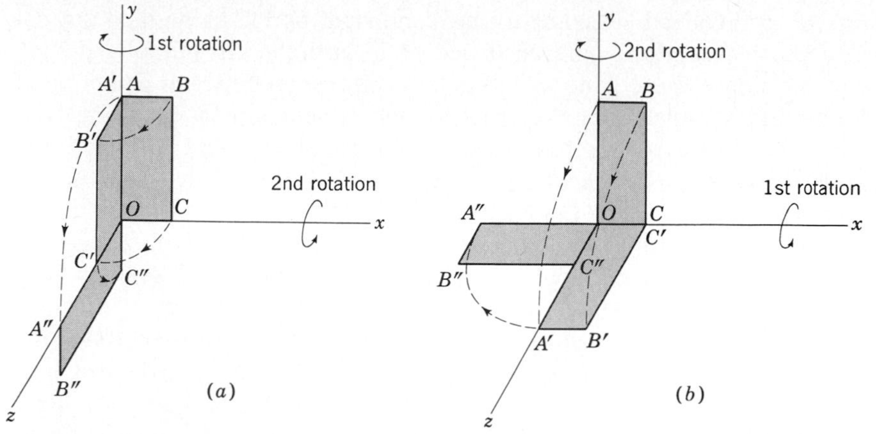

Figure 3-2 Angular displacements cannot be added vectorially because the result depends upon the order in which they are added.

Angular displacements which occur about the same axis or parallel axes, on the other hand, do follow the commutative law. Also, infinitesimally small angular displacements are commutative. To avoid confusion we will treat all finite angular displacements as scalar quantities. However, we will have occasion to treat infinitesimal angular displacements as vectors.

In Fig. 3-3 we recall the definition of the displacement difference between

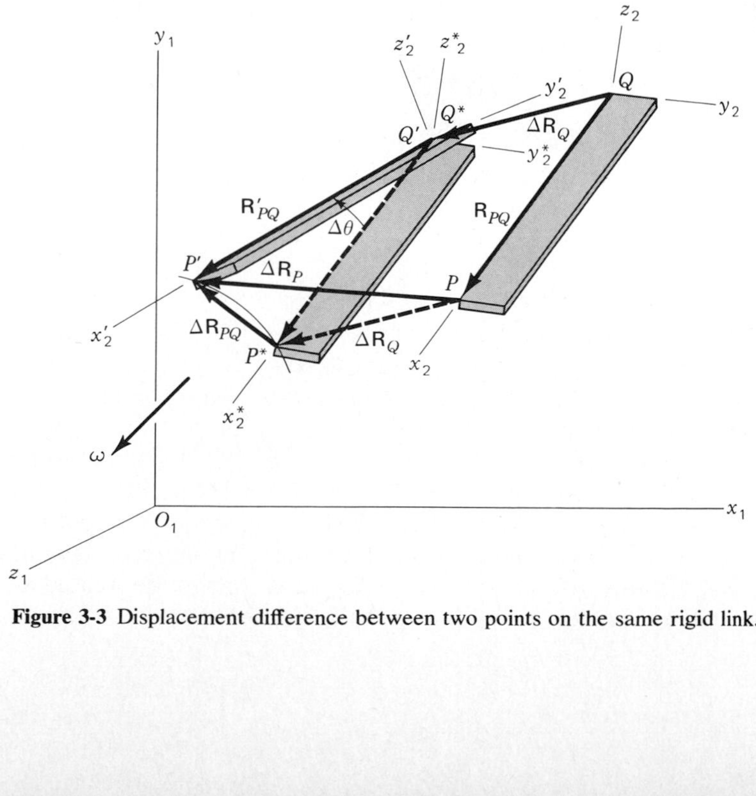

Figure 3-3 Displacement difference between two points on the same rigid link.

two points, P and Q, both attached to the same rigid body. As pointed out in Sec. 2-12, the displacement-difference vector is entirely attributable to the rotation of the body; there is no displacement difference between points in a body undergoing a translation. We reached this conclusion by picturing the displacement as occurring in two steps. First the body was assumed to translate through the displacement $\mathbf{\Delta R}_Q$ to the position $x_2^* y_2^* z_2^*$. Next the body was rotated about point Q^* to the position $x_2' y_2' z_2'$.

Another way to picture the displacement difference $\mathbf{\Delta R}_{PQ}$ is to conceive of a moving coordinate system whose origin travels along with point Q but whose axes remain parallel to the absolute axes $x_1 y_1 z_1$. Note that this coordinate system does not rotate. An observer in this moving coordinate system observes no motion for point Q since it remains at the origin of his coordinate system. For the displacement of point P he will observe the displacement difference vector $\mathbf{\Delta R}_{PQ}$. It seems to such an observer that point Q remains fixed and that the body rotates about this fixed point as shown in Fig. 3-4.

No matter whether the observer is in the ground coordinate system or in the moving coordinate system described, the body appears to rotate through some total angle $\Delta\theta$ in its displacement from $x_2 y_2 z_2$ to $x_2' y_2' z_2'$. If we take the point of view of the fixed observer, the location of the axis of rotation is not obvious. As seen by the translating observer, the axis passes through the apparently stationary point Q; all points in the body appear to travel in circular paths about this axis, and any line in the body whose direction is normal to this axis appears to undergo an identical angular displacement of $\Delta\theta$.

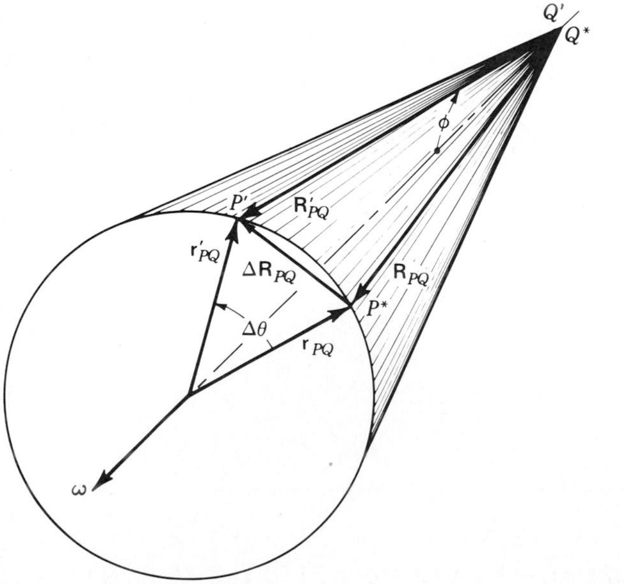

Figure 3-4 Displacement difference $\mathbf{\Delta R}_{PQ}$ as seen by a translating observer.

The *angular velocity* of a rotating body is now defined as a vector quantity $\boldsymbol{\omega}$ having a direction along the instantaneous axis of rotation. The magnitude of the angular-velocity vector is defined as the time rate of change of the angular orientation of any line in the body whose direction is normal to the axis of rotation. If we designate the angular displacement of any of these lines as $\Delta\theta$ and the time interval as Δt, the magnitude of the angular velocity vector $\boldsymbol{\omega}$ is

$$\omega = \lim_{\Delta t \to 0} \frac{\Delta\theta}{\Delta t} = \frac{d\theta}{dt} \tag{3-2}$$

Since we have agreed to treat counterclockwise rotations as positive, the sense of the $\boldsymbol{\omega}$ vector along the axis of rotation is in accordance with the right-hand rule.

3-3 VELOCITY DIFFERENCE BETWEEN POINTS OF THE SAME RIGID BODY

Figure 3-5*a* shows another view of the same rigid-body displacement pictured in Fig. 3-3. This is the view seen by an observer in the absolute coordinate system looking directly along the axis of rotation of the moving body, from the tip end of the $\boldsymbol{\omega}$ vector. In this view the angular displacement $\Delta\theta$ is observed in true size, and *all* lines in the body rotate through this same angle

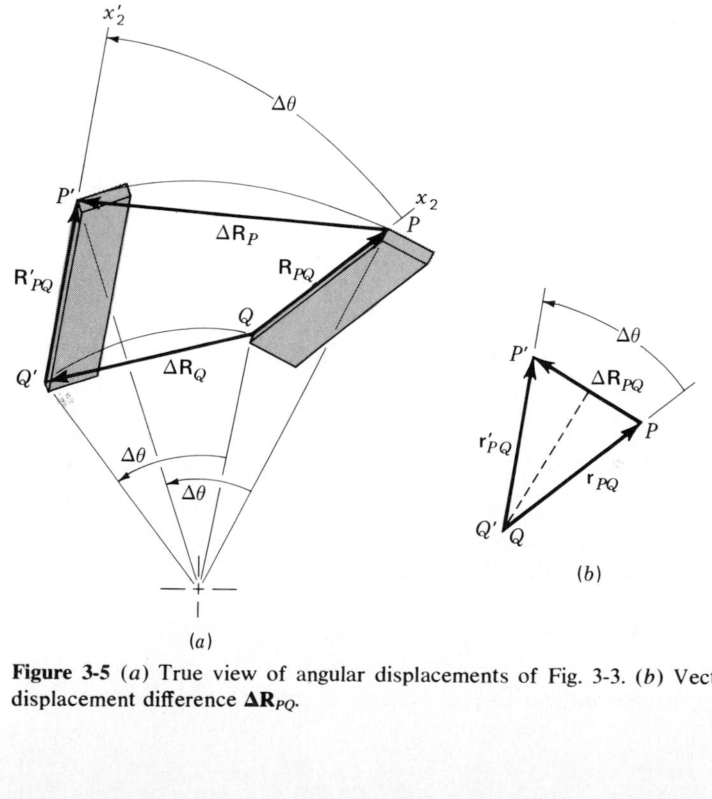

Figure 3-5 (*a*) True view of angular displacements of Fig. 3-3. (*b*) Vector subtraction to form displacement difference $\Delta\mathbf{R}_{PQ}$.

during the displacement. The displacement vectors and thc position difference vectors shown are not necessarily seen in true size; they may appear foreshortened under this viewing angle.

Figure 3-5*b* shows the same rigid-body rotation from the same viewing angle but this time from the point of view of the translating observer. Thus this figure corresponds to the base of the cone shown in Fig. 3-4. We note that the two vectors labeled $\mathbf{r}_{PQ}$ and $\mathbf{r}'_{PQ}$ are the foreshortened views of $\mathbf{R}_{PQ}$ and $\mathbf{R}'_{PQ}$ and from Fig. 3-4 we observe that their magnitudes are

$$r_{PQ} = r'_{PQ} = R_{PQ} \sin \phi \tag{a}$$

where ϕ is the constant angle from the angular velocity vector $\boldsymbol{\omega}$ to the rotating position-difference vector $\mathbf{R}_{PQ}$ as it traverses the cone.

Looking again at Fig. 3-5*b*, we see that it can also be interpreted as a scale drawing corresponding to Eq. (2-67). It shows that the displacement-difference vector $\Delta\mathbf{R}_{PQ}$ is equal to the vector change in the absolute position difference $\mathbf{R}_{PQ}$ produced during the displacement

$$\Delta\mathbf{R}_{PQ} = \mathbf{R}'_{PQ} - \mathbf{R}_{PQ} \tag{b}$$

We are now ready to calculate the magnitude of the displacement-difference vector $\Delta\mathbf{R}_{PQ}$. In Fig. 3-5*b*, where it appears in true size, we construct its perpendicular bisector, from which we see that

$$\Delta R_{PQ} = 2r_{PQ} \sin \frac{\Delta\theta}{2} \tag{c}$$

and, from Eq. (*a*)

$$\Delta R_{PQ} = 2(R_{PQ} \sin \phi) \sin \frac{\Delta\theta}{2} \tag{d}$$

If we now limit ourselves to small motions, the sine of the angular displacement term can be approximated by the angle itself

$$\Delta R_{PQ} \approx 2(R_{PQ} \sin \phi) \frac{\Delta\theta}{2} = \Delta\theta\, R_{PQ} \sin \phi \tag{e}$$

Dividing by the small time increment Δt, noting that the magnitude R_{PQ} and the angle ϕ are constant during the interval, and taking the limit, we get

$$\lim_{\Delta t \to 0} \frac{\Delta R_{PQ}}{\Delta t} = \lim_{\Delta t \to 0} \left(\frac{\Delta\theta}{\Delta t}\right) R_{PQ} \sin \phi = \omega R_{PQ} \sin \phi \tag{f}$$

On recalling the definition of ϕ as the angle between the $\boldsymbol{\omega}$ and $\mathbf{R}_{PQ}$ vectors, we can restore the vector attributes of the above equation by recognizing it as the form of a cross product. Thus

$$\lim_{\Delta t \to 0} \frac{\Delta\mathbf{R}_{PQ}}{\Delta t} = \frac{d\,\mathbf{R}_{PQ}}{dt} = \boldsymbol{\omega} \times \mathbf{R}_{PQ} \tag{g}$$

This form is so important and so useful that it is given its own name and symbol; it is called the *velocity-difference vector* and denoted $\mathbf{V}_{PQ}$

$$\mathbf{V}_{PQ} = \frac{d\mathbf{R}_{PQ}}{dt} = \boldsymbol{\omega} \times \mathbf{R}_{PQ} \tag{3-3}$$

Let us now recall the displacement-difference equation (2-66)

$$\Delta\mathbf{R}_P = \Delta\mathbf{R}_Q + \Delta\mathbf{R}_{PQ} \tag{h}$$

Dividing this equation by Δt and taking the limit gives

$$\lim_{\Delta t \to 0} \frac{\Delta\mathbf{R}_P}{\Delta t} = \lim_{\Delta t \to 0} \frac{\Delta\mathbf{R}_Q}{\Delta t} + \lim_{\Delta t \to 0} \frac{\Delta\mathbf{R}_{PQ}}{\Delta t} \tag{i}$$

which, by Eqs. (3-1) and (3-3), becomes

$$\mathbf{V}_P = \mathbf{V}_Q + \mathbf{V}_{PQ} \tag{3-4}$$

This extremely important equation is called the *velocity-difference equation*; together with Eq. (3-3) it forms one of the primary bases of all velocity-analysis techniques. Equation (3-4) can be written for any two points with no restriction. However, as will be recognized by reviewing the above derivation, Eq. (3-3) should not be applied to any arbitrary pair of points. *This form is valid only if the two points are both attached to the same rigid body.* This restriction can perhaps be better remembered if all subscripts are written explicitly,

$$\mathbf{V}_{P_2Q_2} = \boldsymbol{\omega}_2 \times \mathbf{R}_{P_2Q_2} \tag{j}$$

but, in the interest of brevity, the link-number subscripts are often suppressed. Note that the link-number subscripts are the same throughout Eq. (*j*). If a mistaken attempt is made to apply Eq. (3-3) when points P and Q are not part of the same link, the error should be discovered since it will not be clear which $\boldsymbol{\omega}$ vector should be used.

3-4. GRAPHICAL VELOCITY ANALYSIS; VELOCITY POLYGONS

One major approach to velocity analysis is graphical. As seen in graphical position analysis, it is primarily of use in two-dimensional problems when only a single position requires solution. Its major advantages are that a solution can be achieved quickly and that visualization of, and insight into, the problem are enhanced by the graphical approach.

As a first example of graphical velocity analysis, let us consider the two-dimensional motion of the unconstrained link shown in Fig. 3-6*a*. Suppose that we know the velocities of points A and B and wish to determine the velocity of point C and the angular velocity of the link. We assume that a scale diagram, Fig. 3-6*a*, has already been drawn of the link at the instant considered, i.e., that a position analysis has been completed and that position difference vectors can be measured from the diagram.

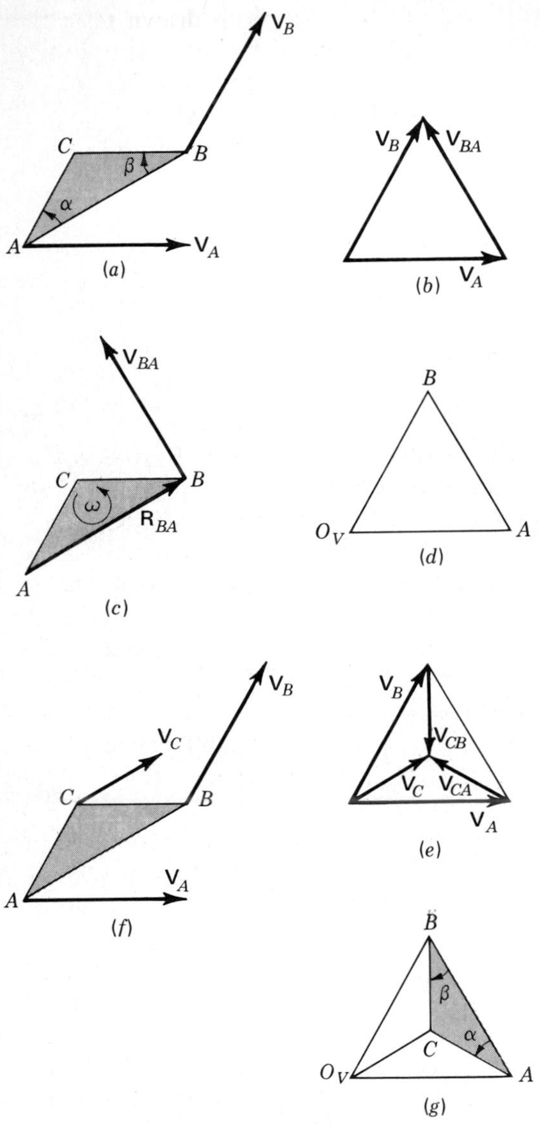

Figure 3-6

Next we consider the velocity-difference equation (3-4) relating points A and B

$$\overset{\surd\surd}{\mathbf{V}}_B = \overset{\surd\surd}{\mathbf{V}}_A + \overset{\circ\circ}{\mathbf{V}}_{BA} \tag{a}$$

where the two unknowns are the magnitude and direction of the velocity-difference vector $\mathbf{V}_{BA}$, as indicated above this symbol in the equation. Figure

3-6*b* shows the graphical solution to the equation. After choosing a scale to represent velocity vectors, the vectors $\mathbf{V}_A$ and $\mathbf{V}_B$ are both drawn to scale, starting from a common origin and in the given directions. The vector spanning the termini of $\mathbf{V}_A$ and $\mathbf{V}_B$ is the velocity-difference vector $\mathbf{V}_{BA}$ and is correct, within graphical accuracy, in both magnitude and direction.

The angular velocity $\boldsymbol{\omega}$ for the link can now be found from Eq. (3-3)

$$\mathbf{V}_{BA} = \boldsymbol{\omega} \times \mathbf{R}_{BA} \tag{b}$$

Since the link is in planar motion, the $\boldsymbol{\omega}$ vector lies perpendicular to the plane of motion, i.e., perpendicular to the vectors $\mathbf{V}_{BA}$ and $\mathbf{R}_{BA}$. Therefore, considering the magnitudes in the above equation,

$$V_{BA} = \omega R_{BA}$$

or

$$\omega = V_{BA}/R_{BA} \tag{c}$$

The numerical magnitude ω can therefore be found by scaling V_{BA} from Fig. 3-6*b* and R_{BA} from Fig. 3-6*a*, being careful to apply the scale factors for units properly; it is usual practice to evaluate ω in units of radians per second.

The magnitude ω is not a complete solution for the angular-velocity vector; the direction must also be determined. As observed above, the $\boldsymbol{\omega}$ vector is perpendicular to the plane of the link itself because the motion is planar. However, this does not say whether $\boldsymbol{\omega}$ is directed into or out of the plane of the figure. This is determined as shown in Fig. 3-6*c*. Taking the point of view of a translating observer, moving with point *A* but not rotating, we can picture the link as rotating about point *A*. The velocity difference $\mathbf{V}_{BA}$ is the only velocity seen by such an observer. Therefore, interpreting $\mathbf{V}_{BA}$ to indicate the direction of rotation of point *B* about point *A*, we find the direction of $\boldsymbol{\omega}$, counterclockwise in this example. Although not strict vector notation, it is common practice in two-dimensional problems to indicate the final solution in the form $\boldsymbol{\omega} = 15$ rad/s ccw, which indicates both magnitude and direction.

The practice of constructing vector diagrams in thick black lines, such as Fig. 3-6*b*, makes them easy to read, but when the diagram is the graphical solution of an equation, it is not very accurate. For this reason it is customary to construct the graphical solution with thin sharp lines, made with a hard drawing pencil, as shown in Fig. 3-6*d*. The solution is started by choosing a scale and a point labeled O_V to represent zero velocity. Absolute velocities such as $\mathbf{V}_A$ and $\mathbf{V}_B$ are constructed with their origins at O_V and their termini are labeled as points *A* and *B*. The line *from A to B* then represents the velocity difference $\mathbf{V}_{BA}$. It will be seen as we continue that these labels at the vertices are sufficient to determine the precise notation of all velocity differences represented by lines in the diagram. Notice, for example, that $\mathbf{V}_{AB}$ is represented by the vector *from* point *B to* point *A*. With the labeling convention, no arrowheads or additional notation are necessary and do not clutter the diagram. Such a diagram is called a *velocity polygon* and, as we will see, adds considerable convenience to the graphical solution technique.

A danger of this convention, however, is that the analyst will begin to think of the technique as a series of graphical "tricks" and lose sight of the fact that each line drawn can and should be fully justified by a corresponding vector equation. The graphics are merely a convenient solution technique, not a substitute for a sound theoretical basis.

Returning to Fig. 3-6*c*, it may have appeared as coincidence that the vector $\mathbf{V}_{BA}$ was perpendicular to $\mathbf{R}_{BA}$. Looking back to Eq. (*b*), however, we see that it was a necessary outcome, resulting from the cross product with the $\boldsymbol{\omega}$ vector. We will take advantage of this property in the next step.

Now that $\boldsymbol{\omega}$ has been found, let us determine the absolute velocity of point *C*. We can relate this by velocity-difference equations to the absolute velocities of both points *A* and *B*

$$\overset{\circ\circ}{\mathbf{V}}_C = \overset{\checkmark\checkmark}{\mathbf{V}}_A + \overset{\circ\checkmark}{\mathbf{V}}_{CA} = \overset{\checkmark\checkmark}{\mathbf{V}}_B + \overset{\circ\checkmark}{\mathbf{V}}_{CB} \qquad (d)$$

Since points *A*, *B*, and *C* are all on the same rigid link, each of the velocity difference vectors, $\mathbf{V}_{CA}$ and $\mathbf{V}_{CB}$, is of the form $\boldsymbol{\omega} \times \mathbf{R}$, using $\mathbf{R}_{CA}$ and $\mathbf{R}_{CB}$, respectively. As a result $\mathbf{V}_{CA}$ is perpendicular to $\mathbf{R}_{CA}$ and $\mathbf{V}_{CB}$ is perpendicular to $\mathbf{R}_{CB}$. The directions of these two terms are therefore indicated as known in Eq. (*d*).

Since $\boldsymbol{\omega}$ has already been determined, it is easy to calculate the magnitudes of $\mathbf{V}_{CA}$ and $\mathbf{V}_{CB}$ by using a formula like Eq. (*c*); however, we will assume that this is not done. Instead, we form the graphical solution to Eq. (*d*). This equation states that a vector which is perpendicular to $\mathbf{R}_{CA}$ must be added to $\mathbf{V}_A$ and that the result will equal the sum of $\mathbf{V}_B$ and a vector perpendicular to $\mathbf{R}_{CB}$. The solution is illustrated in Fig. 3-6*e*. In practice the solution is continued on the same diagram as Fig. 3-6*d* and results in Fig. 3-6*g*. A line perpendicular to $\mathbf{R}_{CA}$ (representing $\mathbf{V}_{CA}$) is drawn starting at point *A* (representing addition to $\mathbf{V}_A$); similarly, a line is drawn perpendicular to $\mathbf{R}_{CB}$ starting at point *B*. The point of intersection of these two lines is labeled *C* and represents the solution to Eq. (*d*). The line from O_V to point *C* now represents the absolute velocity $\mathbf{V}_C$. This velocity can be transferred back to the link and interpreted as $\mathbf{V}_C$ in both magnitude and direction as shown in Fig. 3-6*f*.

In seeing the shading and the labeled angles α and β in Fig. 3-6*g* and *a*, we are led to investigate whether the two triangles labeled *ABC* in each of these figures are similar in shape, as they appear to be. In reviewing the construction steps we see that, indeed, they are since the velocity-difference vectors $\mathbf{V}_{BA}$, $\mathbf{V}_{CA}$, and $\mathbf{V}_{CB}$ are perpendicular to the respective position-difference vectors, $\mathbf{R}_{BA}$, $\mathbf{R}_{CA}$, and $\mathbf{R}_{CB}$. This property would be true no matter what the shape of the moving link; a similarly shaped figure would appear in the velocity polygon. Its sides are always scaled up or down by a factor equal to the angular velocity of the link, and it is always rotated by 90° in the direction of the angular velocity. The properties result from the fact that each velocity-difference vector between two points on the link is of the form of a cross product of the same $\boldsymbol{\omega}$ vector with the corresponding position-difference vector. This similarly shaped figure in the velocity polygon is commonly referred to as the *velocity image* of the link, and any moving link will have a corresponding velocity image in the velocity polygon.

If the concept of the velocity image had been known initially, the solution process could have been speeded up considerably. Once the solution has progressed to the state of Fig. 3-6*d*, the velocity-image points A and B are known. One can use these two points as the base of a triangle similar to the link shape and label the image point C directly, without writing Eq. (*d*). Care must be taken not to allow the triangle to be flipped over between the position diagram and the velocity image, but the solution can proceed quickly, accurately, and naturally, resulting in Fig. 3-6*g*. Here again the caution is repeated that all steps in the solution are based on strictly derived vector equations and are not tricks. It is very wise to continue to write the corresponding vector equations until one is thoroughly familiar with the procedure.

To increase familiarity with graphical velocity-analysis techniques, we analyze two typical example problems.

Example 3-1 The four-bar linkage shown to scale in Fig. 3-7*a* with all necessary dimensions is driven by crank 2 at a constant angular velocity of $\boldsymbol{\omega}_2 = 900$ rpm ccw. Find the instantaneous velocities of points E and F and the angular velocities of links 3 and 4 at the position shown.

SOLUTION To obtain a graphical solution we first calculate the angular velocity of link 2 in radians per second. This is

$$\omega_2 = \left(900\,\frac{\text{rev}}{\text{min}}\right)\left(2\pi\,\frac{\text{rad}}{\text{rev}}\right)\left(\frac{1\text{ min}}{60\text{ s}}\right) = 94.2\text{ rad/s ccw} \tag{1}$$

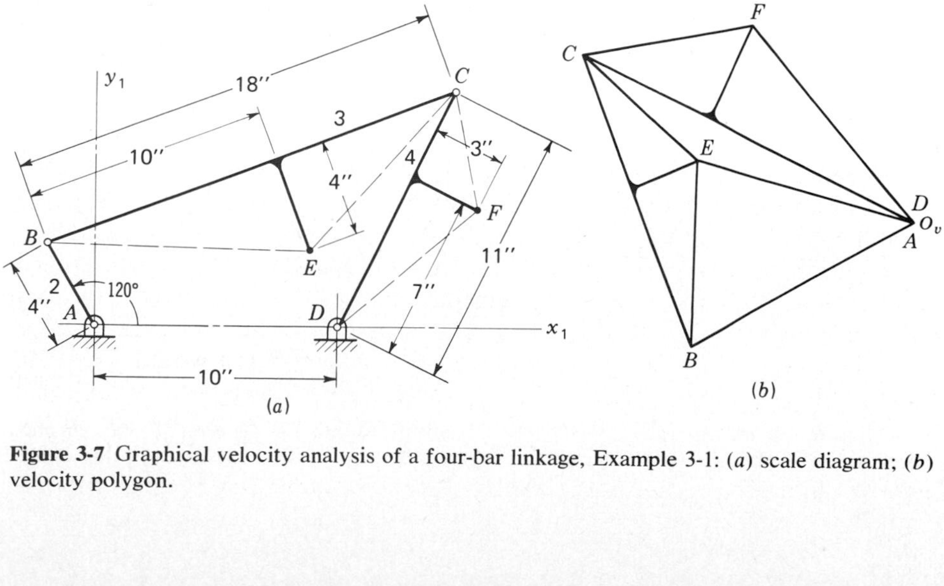

Figure 3-7 Graphical velocity analysis of a four-bar linkage, Example 3-1: (*a*) scale diagram; (*b*) velocity polygon.

Then we notice that the point A remains fixed and calculate the velocity of point B

$$\mathbf{V}_B = \cancel{\mathbf{V}_A}^{0} + \mathbf{V}_{BA} = \boldsymbol{\omega}_2 \times \mathbf{R}_{BA}$$

$$V_B = (94.2\ \text{rad/s})(\tfrac{4}{12}\ \text{ft}) = 31.4\ \text{ft/s} \tag{2}$$

We note that the form $\boldsymbol{\omega} \times \mathbf{R}$ was used for the velocity difference, not for the absolute velocity $\mathbf{V}_B$ directly. In Fig. 3-7*b* we choose the point O_V and a velocity scale factor. We note that the image point A is coincident with O_V and construct the line AB perpendicular to $\mathbf{R}_{BA}$ and toward the left because of the counterclockwise direction of $\boldsymbol{\omega}_2$; this line represents $\mathbf{V}_{BA}$.

If we attempt at this time to write an equation directly for the velocity of point E, we find by counting unknowns that it cannot be solved yet. Therefore, we next write two equations for the velocity of point C. Since the velocities of points C_3 and C_4 must be equal (links 3 and 4 are pinned together at C),

$$\overset{\circ\circ}{\mathbf{V}}_C = \overset{\surd\surd}{\mathbf{V}}_B + \overset{\circ\surd}{\mathbf{V}}_{CB} = \cancel{\mathbf{V}_D}^{0} + \overset{\circ\surd}{\mathbf{V}}_{CD} \tag{3}$$

We now construct two lines in the velocity polygon; the line BC is drawn from B perpendicular to $\mathbf{R}_{CB}$ and the line DC is drawn from D (coincident with O_V since $\mathbf{V}_D = 0$) perpendicular to $\mathbf{R}_{CD}$. We label the point of intersection as point C. When the lengths of these lines are scaled, we find that $V_{CB} = 38.4\ \text{ft/s}$ and $V_C = V_{CD} = 45.6\ \text{ft/s}$. The angular velocities of links 3 and 4 can now be found:

$$\omega_3 = \frac{V_{CB}}{R_{CB}} = \frac{38.4\ \text{ft/s}}{18/12\ \text{ft}} = 25.6\ \text{rad/s ccw} \qquad \textit{Ans.} \tag{4}$$

$$\omega_4 = \frac{V_{CD}}{R_{CD}} = \frac{45.5\ \text{ft/s}}{11/12\ \text{ft}} = 49.6\ \text{rad/s ccw} \qquad \textit{Ans.} \tag{5}$$

where the directions of $\boldsymbol{\omega}_3$ and $\boldsymbol{\omega}_4$ were found by the technique illustrated in Fig. 3-6*c*.

There are now several methods of finding $\mathbf{V}_E$. In one method we measure R_{EB} from the scale drawing of Fig. 3-7*a*, and then, since points B and E are both attached to link 3, we can calculate†

$$V_{EB} = \omega_3 R_{EB} = (25.6\ \text{rad/s})\left(\frac{10.8}{12}\ \text{ft}\right) = 23.0\ \text{ft/s} \tag{6}$$

We can now construct the line BE in the velocity polygon, drawn to the proper scale and perpendicular to $\mathbf{R}_{EB}$, thus solving‡ the velocity-difference equation

$$\overset{\circ\circ}{\mathbf{V}}_E = \overset{\surd\surd}{\mathbf{V}}_B + \overset{\surd\surd}{\mathbf{V}}_{EB} \tag{7}$$

The result is

$$V_E = 27.6\ \text{ft/s} \qquad \textit{Ans.}$$

as scaled from the velocity polygon.

Alternatively, V_E can be found from

$$\overset{\circ\circ}{\mathbf{V}}_E = \overset{\surd\surd}{\mathbf{V}}_C + \overset{\surd\surd}{\mathbf{V}}_{EC} \tag{8}$$

† There is *no* restriction in our derivation which requires that R_{EB} lie along the material part of link 3 in order to use Eq. (6).

‡ Note that numerical values should *not* be substituted into Eq. (7) directly; this equation requires vector addition, *not scalar*, and this is precisely the purpose of constructing the velocity polygon.

by an identical procedure to that used for Eq. (7). This solution would produce the triangle O_VEC in the velocity polygon.

Suppose we wish to find $\mathbf{V}_E$ without the intermediate step of calculating ω_3. In this case we write Eqs. (7) and (8) simultaneously,

$$\overset{\circ\circ}{\mathbf{V}}_E = \overset{\surd\surd}{\mathbf{V}}_B + \overset{\circ\surd}{\mathbf{V}}_{EB} = \overset{\surd\surd}{\mathbf{V}}_C + \overset{\circ\surd}{\mathbf{V}}_{EC} \tag{9}$$

Drawing lines EB (perpendicular to $\mathbf{R}_{EB}$) and EC (perpendicular to $\mathbf{R}_{EC}$) in the velocity polygon, we find their intersection and so solve Eq. (9).

Perhaps the easiest method of solving for $\mathbf{V}_E$, however, is to take advantage of the concept of the velocity image of link 3. Recognizing that the velocity-image points B and C have already been found, we can construct the triangle BEC in the velocity polygon, similar in shape to the triangle BEC in the scale diagram of link 3. This locates point E in the velocity polygon and thus gives a solution for $\mathbf{V}_E$.

The velocity $\mathbf{V}_F$ can also be found by any of the above methods using points C, D, and F of link 4. The result is

$$V_F = 31.8 \text{ ft/s} \qquad \textit{Ans.}$$

Example 3-2 The offset slider-crank mechanism of Fig. 3-8*a* is driven by slider 4 at a speed of $V_C = 10$ m/s to the left at the phase shown. Determine the instantaneous velocity of point D and the angular velocities of links 2 and 3.

SOLUTION The velocity scale and pole O_V are chosen and $\mathbf{V}_C$ is drawn, thus locating point C as shown in Fig. 3-8*b*. Simultaneous equations are then written for the velocity of point B,

$$\overset{\circ\circ}{\mathbf{V}}_B = \overset{\surd\surd}{\mathbf{V}}_C + \overset{\circ\surd}{\mathbf{V}}_{BC} = \cancelto{0}{\mathbf{V}_A} + \overset{\circ\surd}{\mathbf{V}}_{BA} \tag{10}$$

and solved for the location of point B in the velocity polygon.

Having found points B and C, we can construct the velocity image of link 3 as shown to locate point D; we then scale the line O_VD, which gives

$$V_D = 12.0 \text{ m/s} \qquad \textit{Ans.}$$

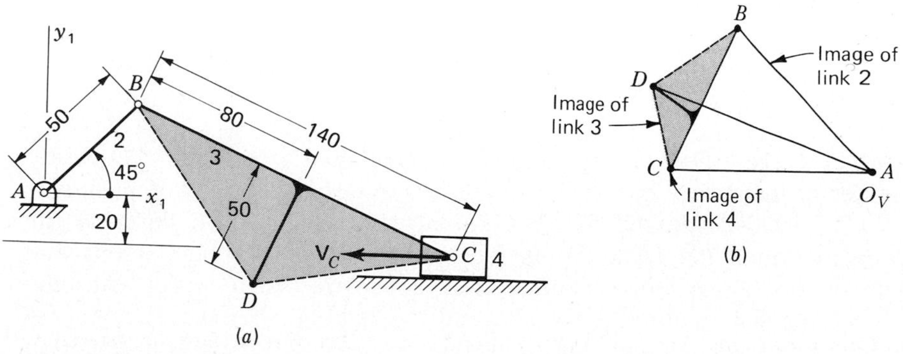

Figure 3-8 Example 3-2: (*a*) scale diagram of offset slider-crank mechanism (dimensions in millimetres); (*b*) velocity polygon.

The angular velocities of links 2 and 3 are

$$\omega_2 = \frac{V_{BA}}{R_{BA}} = \frac{10.0 \text{ m/s}}{0.05 \text{ m}} = 200 \text{ rad/s cw} \qquad \textit{Ans.} \tag{11}$$

$$\omega_3 = \frac{V_{BC}}{R_{BC}} = \frac{7.5 \text{ m/s}}{0.14 \text{ m}} = 53.6 \text{ rad/s cw} \qquad \textit{Ans.} \tag{12}$$

In this second example problem, Fig. 3-8*b*, the velocity image of each link is indicated in this polygon. If the analysis of any problem is carried through completely, there will be a velocity image for each link of the mechanism. The following points are true in general and can be verified in the above examples:

1. The velocity image of each link is a scale reproduction of the shape of the link in the velocity polygon.
2. The velocity image of each link is rotated 90° in the direction of the angular velocity of the link.
3. The letters identifying the vertices of each link are the same as those in the velocity polygon and progress around the velocity image in the same order and in the same angular direction as around the link.
4. The ratio of the size of the velocity image of a link to the size of the link itself is equal to the magnitude of the angular velocity of the link. In general, it is *not* the same for different links in the same mechanism.
5. The velocity of all points on a translating link are equal, and the angular velocity is zero. Therefore, the velocity image of a link which is translating shrinks to a single point in the velocity polygon.
6. The point O_V in the velocity polygon is the image of all points with zero absolute velocity; it is the velocity image of the fixed link.
7. The absolute velocity of any point on any link is represented by the line from O_V to the image of the point. The velocity difference vector between any two points, say P and Q, is represented by the line from image point P to image point Q.

3-5 THE APPARENT VELOCITY OF A POINT IN A MOVING COORDINATE SYSTEM

In analyzing the velocities of various machine components we frequently encounter problems in which it is convenient to describe how a point moves relative to another moving link but not at all convenient to describe the absolute motion of the point. An example of this occurs when a rotating link contains a slot along which another link is constrained to slide. With the motion of the link containing the slot and the relative sliding motion taking place in the slot as known quantities, we may wish to find the absolute motion of the sliding member. It was for problems of this type that the apparent-displacement vector was defined in Sec. 2-14, and we now wish to extend this concept to velocity.

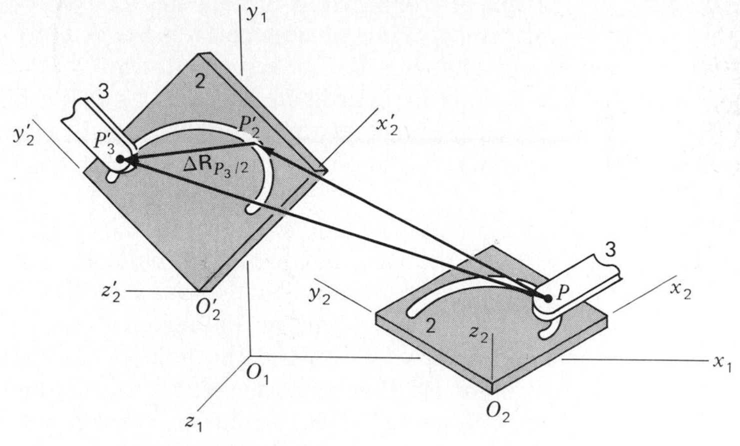

Figure 3-9 Apparent displacement.

In Fig. 3-9 we recall the definition of the apparent-displacement vector. A rigid link having some general motion carries a coordinate system $x_2y_2z_2$ attached to it. At a certain time t the coordinate system lies at $x_2y_2z_2$, and after some short time interval Δt it moves to its new location $x'_2y'_2z'_2$. All points of link 2 move with the coordinate system.

Also, during the same time interval, another point P_3 of another link 3 is constrained in some manner to move along a known path relative to link 2. In Fig. 3-9 this constraint is depicted as a slot carrying a pin from link 3; the center of the pin is the point P_3. Although pictured in this way, the constraint may occur in a variety of different forms. The only *assumption here is that the path which the moving point P_3 traces in coordinate system $x_2y_2z_2$, that is, the locus of the tip of the apparent-position vector $\mathbf{R}_{P_3/2}$, is known.*

Recalling the apparent-displacement equation (2-69),

$$\Delta\mathbf{R}_{P_3} = \Delta\mathbf{R}_{P_2} + \Delta\mathbf{R}_{P_3/2}$$

we divide by Δt and take the limit

$$\lim_{\Delta t \to 0} \frac{\Delta\mathbf{R}_{P_3}}{\Delta t} = \lim_{\Delta t \to 0} \frac{\Delta\mathbf{R}_{P_2}}{\Delta t} + \lim_{\Delta t \to 0} \frac{\Delta\mathbf{R}_{P_3/2}}{\Delta t}$$

We now define the *apparent-velocity* vector as

$$\mathbf{V}_{P_3/2} = \lim_{\Delta t \to 0} \frac{\Delta\mathbf{R}_{P_3/2}}{\Delta t} = \frac{d\mathbf{R}_{P_3/2}}{dt} \tag{3-5}$$

and, in the limit, the above equation becomes

$$\mathbf{V}_{P_3} = \mathbf{V}_{P_2} + \mathbf{V}_{P_3/2} \tag{3-6}$$

called the *apparent-velocity equation.*

We note from its definition, Eq. (3-5), that the apparent velocity resembles the absolute velocity except that it comes from the apparent displacement rather than the absolute displacement. Thus, in concept, it is the velocity of the moving point P_3 *as it would appear to an observer attached to the moving link* 2 and making observations in coordinate system $x_2y_2z_2$. This concept accounts for its name. We also note that the absolute velocity is a special case of the apparent velocity where the observer happens to be fixed to the $x_1y_1z_1$ coordinate system.

We can get further insight into the nature of the apparent-velocity vector by studying Fig. 3-10. This figure shows the view of the moving point P_3 as it would be seen by the moving observer. To him, the path traced on link 2 appears stationary and the moving point moves along this path from P_3 to P_3'. Working in this coordinate system, suppose we locate the point C as the center of curvature of the path at point P_2. For small distances from P_2 the path follows the circular arc P_3P_3' with center C and radius of curvature $\boldsymbol{\rho}$. We now define the unit-vector extension of $\boldsymbol{\rho}$, labeled $\hat{\boldsymbol{\rho}}$, and we define the unit vector tangent to the path $\hat{\boldsymbol{\tau}}$ with positive sense in the direction of motion. We note that these are at right angles to each other and complete a right-hand cartesian coordinate system by defining the normal unit vector

$$\hat{\boldsymbol{\nu}} = \hat{\boldsymbol{\rho}} \times \hat{\boldsymbol{\tau}} \tag{3-7}$$

This coordinate system moves with its origin tracking the motion of point P_3. However, it rotates with the radius-of-curvature vector (through the angle $\Delta\phi$) as the motion progresses, *not* the same rotation as either link 2 or link 3.

We now define the scalar Δs as the distance along the curve from P_3 to P_3' and note that $\Delta\mathbf{R}_{P_3/2}$ is the chord of the same arc. However, for very short Δt,

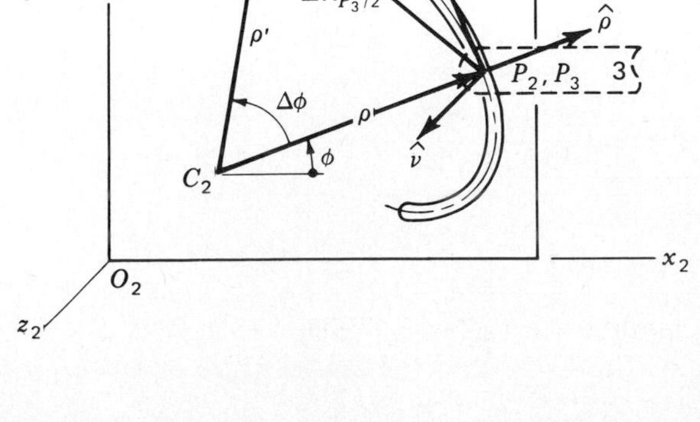

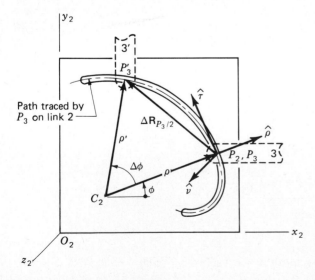

Figure 3-10 Apparent displacement of point P_3 as seen by an observer on link 2.

the magnitude of the chord and the arc distance approach equality. Therefore,

$$\lim_{\Delta s \to 0} \frac{\Delta \mathbf{R}_{P_3/2}}{\Delta s} = \frac{d\mathbf{R}_{P_3/2}}{ds} = \hat{\boldsymbol{\tau}} \tag{3-8}$$

Here both $\Delta \mathbf{R}_{P_3/2}$ and Δs are considered to be functions of time. Therefore, from Eq. (3-5)

$$\mathbf{V}_{P_3/2} = \lim_{\Delta t \to 0} \left(\frac{\Delta \mathbf{R}_{P_3/2}}{\Delta s} \frac{\Delta s}{\Delta t} \right) = \frac{d\mathbf{R}_{P_3/2}}{ds} \frac{ds}{dt} = \frac{ds}{dt} \hat{\boldsymbol{\tau}} \tag{3-9}$$

There are two important conclusions from this result: the magnitude of the apparent velocity is equal to the *speed* with which the point P_3 progresses along the path and the apparent-velocity vector is *always tangent to the path traced* by the point in the coordinate system of the observer. The first of these two results is seldom useful in the solution of problems, although it is an important concept. The second result is extremely useful since the apparent path traced can often be visualized from the nature of the constraints and thus the direction of the apparent-velocity vector becomes known. Note that only the tangent to the path needs to be determined; the radius of curvature $\boldsymbol{\rho}$ is not needed until we attempt acceleration analysis in the next chapter.

Example 3-3 Shown in Fig. 3-11*a* is an inversion of the slider-crank mechanism. Link 2, the crank, is driven at an angular velocity of 36 rad/s cw. Link 3 slides on link 4 and is pivoted to the crank at *A*. Find the angular velocity of link 4.

SOLUTION We first calculate the velocity of point *A*

$$\mathbf{V}_A = \overset{0}{\cancel{\mathbf{V}_E}} + \mathbf{V}_{AE} = \boldsymbol{\omega}_2 \times \mathbf{R}_{AE}$$
$$V_A = (36\ \text{rad/s})(\tfrac{3}{12}\ \text{ft}) = 9\ \text{ft/s} \tag{1}$$

and we plot this from the pole O_V to locate point *A* in the velocity polygon, Fig. 3-11*b*.

Next we distinguish two different points B_3 and B_4 at the location of sliding. Point B_3 is part of link 3, and B_4 is attached to link 4, but at the instant shown the two are coincident. Note that, as seen by an observer on link 4, point B_3 seems to slide along link 4, thus

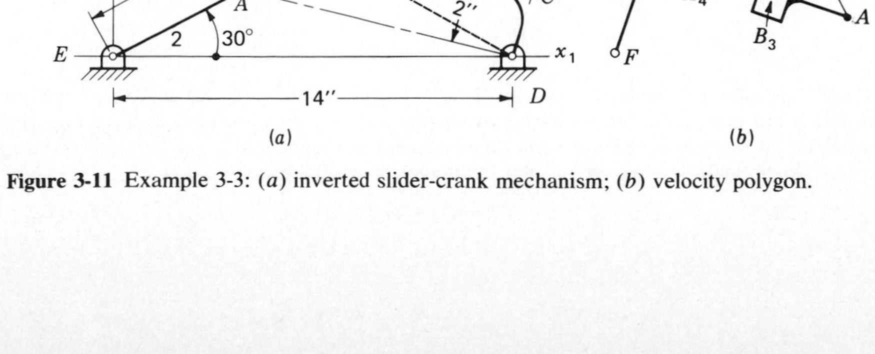

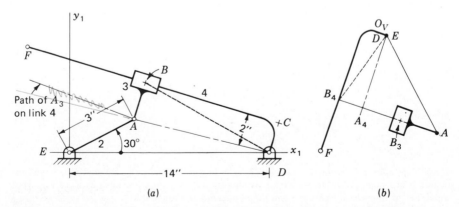

Figure 3-11 Example 3-3: (*a*) inverted slider-crank mechanism; (*b*) velocity polygon.

defining a straight-line path along the line CF. Thus we can write the apparent-velocity equation

$$\mathbf{V}_{B_3} = \mathbf{V}_{B_4} + \mathbf{V}_{B_3/4} \tag{2}$$

When point B_3 is related to A and point B_4 to D by velocity differences, expansion of Eq. (2) gives

$$\overset{\surd\surd}{\mathbf{V}_A} + \overset{\circ\surd}{\mathbf{V}_{B_3A}} = \overset{0}{\cancel{\mathbf{V}_D}} + \overset{\circ\surd}{\mathbf{V}_{B_4D}} + \overset{\circ\surd}{\mathbf{V}_{B_3/4}} \tag{3}$$

where $\mathbf{V}_{B_3A}$ is perpendicular to $\mathbf{R}_{BA}$, $\mathbf{V}_{B_4D}$ is perpendicular to $\mathbf{R}_{BD}$ (shown dashed), and $\mathbf{V}_{B_3/4}$ has a direction defined by the tangent to the path of sliding at B.

Although Eq. (3) appears to have three unknowns, if we note that $\mathbf{V}_{B_3A}$ and $\mathbf{V}_{B_3/4}$ have identical directions, the equation can be rearranged as

$$\overset{\surd\surd}{\mathbf{V}_A} + (\overset{\circ\surd}{\mathbf{V}_{B_3A} - \mathbf{V}_{B_3/4}}) = \overset{\circ\surd}{\mathbf{V}_{B_4D}} \tag{4}$$

and the difference shown in parentheses can be treated as a single vector of known direction. The equation is now reduced to two unknowns and can be solved graphically to locate point B_4 in the velocity polygon.

The magnitude R_{BD} can be computed or measured from the diagram, and V_{B_4D} can be scaled from the velocity polygon (the dashed line from O_V to B_4). Therefore,

$$\omega_4 = \frac{V_{B_4D}}{R_{BD}} = \frac{7.3\ \text{ft/s}}{11.6/12\ \text{ft}} = 7.55\ \text{rad/s ccw} \qquad \textit{Ans.} \tag{5}$$

Although the problem as stated is now completed, the velocity polygon has been extended to show the images of links 2, 3, and 4. In doing so it was necessary to note that since links 3 and 4 always remain perpendicular to each other, they must rotate at the same rate. Thus $\boldsymbol{\omega}_3 = \boldsymbol{\omega}_4$. This allowed the calculation of $\mathbf{V}_{BA} = \boldsymbol{\omega}_3 \times \mathbf{R}_{BA}$ and the plotting of the velocity-image point B_3. We also note that the velocity images of links 3 and 4 are of comparable size since $\boldsymbol{\omega}_3 = \boldsymbol{\omega}_4$. However, they have quite a different scale than the velocity image of link 2, the line O_VA, since $\boldsymbol{\omega}_2$ is a larger angular velocity.

Another approach to the same problem avoids the need to combine terms as in Eq. (4). If we consider an observer riding on link 4 and ask what he would see for the path of point A in his coordinate system, we find that this path is a straight line parallel to the line CF, as indicated in Fig. 3-11a. Let us now define one point of this path as A_4. At the instant shown the point A_4 is located coincident with points A_2 and A_3. However, A_4 *does not move with the pin; it is attached to link 4 and rotates with the path around the fixed point D.* Since we can identify the path traced by A_2 and A_3 on link 4, we can write the apparent-velocity equation

$$\mathbf{V}_{A_2} = \mathbf{V}_{A_4} + \mathbf{V}_{A_2/4} \tag{6}$$

and, since point A_4 is a part of link 4,

$$\mathbf{V}_{A_4} = \overset{0}{\cancel{\mathbf{V}_D}} + \mathbf{V}_{A_4D} \tag{7}$$

Substituting Eq. (6) into (7) gives

$$\overset{\surd\surd}{\mathbf{V}_{A_2}} = \overset{\circ\surd}{\mathbf{V}_{A_4D}} + \overset{\circ\surd}{\mathbf{V}_{A_2/4}} \tag{8}$$

where $\mathbf{V}_{A_4D}$ is perpendicular to $\mathbf{R}_{AD}$ and $\mathbf{V}_{A_2/4}$ is tangent to the path. Solving this equation locates image point A_4 in the velocity polygon and allows a solution for $\omega_4 = V_{A_4D}/R_{AD}$. The remainder of the velocity polygon can then be found as shown above.

It would indicate a wrong concept to attempt to use the equation

$$\mathbf{V}_{A_4} = \mathbf{V}_{A_2} + \mathbf{V}_{A_4/2}$$

rather than Eq. (6) since *the path* traced by point A_4 in a coordinate system attached to link 2 *is not known*†.

Another insight into the nature and use of the apparent-velocity equation is provided by the following example.

Example 3-4 As shown in Fig. 3-12, an airplane traveling at a speed of 300 km/h is turning a circle of radius 5 km with center at *C*. As it does so, the pilot sees a rocket 30 km away traveling on a straight course at 2000 km/h. What is the velocity of the rocket as seen by the pilot of the plane?

SOLUTION Since the plane is on a circular course, the point C_2, attached to the coordinate system of the plane but coincident with *C*, has no motion. Therefore the angular velocity of the plane is

$$\omega_2 = \frac{V_{PC}}{R_{PC}} = \frac{V_P}{R_{PC}} = \frac{300\text{ km/h}}{5\text{ km}} = 60\text{ rad/h ccw} \tag{9}$$

The question asked obviously requires the calculation of the apparent velocity $\mathbf{V}_{R_3/2}$, but this can be applied only *between coincident points*. Therefore, we define another point R_2, attached to the rotating coordinate system of the plane but located coincident with the rocket R_3 at the instant shown. As part of the plane the velocity of this point is

$$\mathbf{V}_{R_2} = \mathbf{V}_P + \boldsymbol{\omega}_2 \times \mathbf{R}_{RP} = 300\,\frac{\text{km}}{\text{h}} + \left(60\,\frac{\text{rad}}{\text{h}}\right)(30\text{ km}) = 2100\text{ km/h} \tag{10}$$

where the values are added algebraically because the vectors are parallel. The apparent velocity can now be calculated

$$\mathbf{V}_{R_3/2} = \mathbf{V}_{R_3} - \mathbf{V}_{R_2}$$

$$V_{R_3/2} = 2000\text{ km/h} - 2100\text{ km/h} = -100\text{ km/h} \qquad \textit{Ans.} \tag{11}$$

† Although the use of this equation would suggest a faulty understanding, it still yields a correct solution. If the corresponding path were found, it would be tangent to the path used at point *A*. Since the *tangents to the two paths are the same even though the paths are not*, the solution would yield an accurate result. This is *not* true in acceleration analysis, Chap. 4; therefore, the concept should be studied and this "backward" use should be avoided.

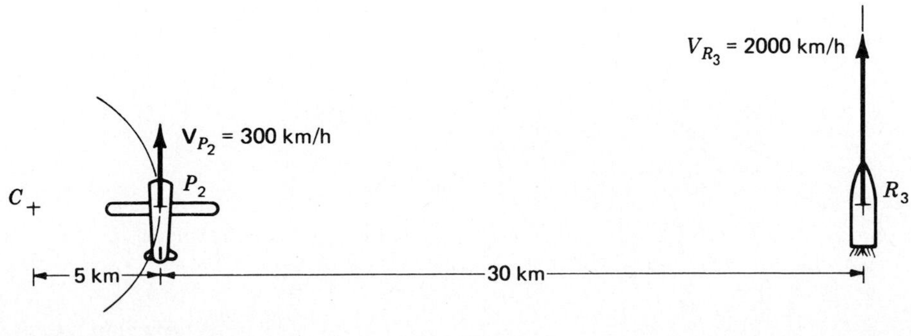

Figure 3-12 Example 3-4.

Thus, as seen by the pilot of the plane, the rocket appears to be *backing up* at a speed of 100 km/h. This result becomes better understood as we consider the motion of point R_2. This point is treated as being *attached to the plane* and therefore seems stationary to the pilot. Yet, in the absolute coordinate system, this point is traveling faster than the rocket; the rocket is not keeping up with this point and therefore appears to the pilot to be backing up.

3-6 APPARENT ANGULAR VELOCITY

When two rigid bodies rotate with different angular velocities, the vector difference between the two is defined as the *apparent angular velocity*. Thus

$$\boldsymbol{\omega}_{3/2} = \boldsymbol{\omega}_3 - \boldsymbol{\omega}_2 \tag{3-10}$$

which can also be written

$$\boldsymbol{\omega}_3 = \boldsymbol{\omega}_2 + \boldsymbol{\omega}_{3/2} \tag{3-11}$$

It will be seen that $\boldsymbol{\omega}_{3/2}$ is the angular velocity of body 3 as it would appear to an observer attached to, and rotating with, body 2. Compare this equation with Eq. (3-6) for the apparent velocity of a point.

3-7 DIRECT CONTACT AND ROLLING CONTACT

Two elements of a mechanism which are in direct contact with each other have relative motion which may or may not involve sliding between the links at the point of direct contact. In the cam-and-follower system shown in Fig. 3-13*a*, the cam, link 2, drives the follower, link 3, by direct contact. We see that if slip were not possible between links 2 and 3 at point *P*, the triangle *PAB* would form a truss; therefore, sliding as well as rotation must take place between the links.

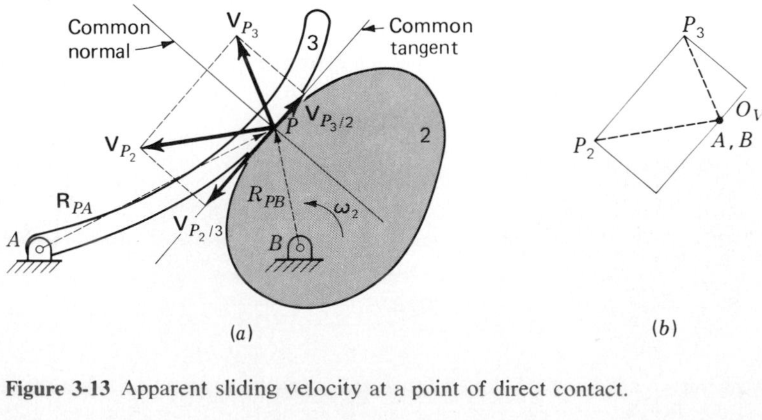

Figure 3-13 Apparent sliding velocity at a point of direct contact.

Let us distinguish between the two points P_2 attached to link 2 and P_3 attached to link 3. They are coincident points, both located at P at the instant shown; therefore we can write the apparent-velocity equation

$$\mathbf{V}_{P_3/2} = \mathbf{V}_{P_3} - \mathbf{V}_{P_2} \tag{3-12}$$

If the two absolute velocities $\mathbf{V}_{P_3}$ and $\mathbf{V}_{P_2}$ were both known, they could be subtracted to find $\mathbf{V}_{P_3/2}$. Components could then be taken along directions defined by the common normal and common tangent to the surfaces at the point of direct contact. The components of $\mathbf{V}_{P_3}$ and $\mathbf{V}_{P_2}$ along the common normal must be equal, and this component of $\mathbf{V}_{P_3/2}$ must be zero. Otherwise, either the two links would separate or they would interfere, both contrary to our basic assumption that contact persists. The total apparent velocity $\mathbf{V}_{P_3/2}$ ***must therefore lie along the common tangent*** and is the velocity of the relative sliding motion within the direct-contact interface. The velocity polygon for this sytem is shown in Fig. 3-13*b*.

It is possible in other mechanisms for there to be direct contact between links without slip between the links. The cam-follower system of Fig. 3-14, for example, might have high friction between the roller, link 3, and the cam surface, link 2, and restrain the wheel to roll against the cam without slip. Henceforth we will restrict the use of the term *rolling contact* to situations where *no slip* takes place. By "no slip" we imply that the apparent slipping velocity of Eq. (3-12) is zero

$$\mathbf{V}_{P_3/2} = 0 \tag{3-13a}$$

This equation is sometimes referred to as the *rolling-contact condition* for velocity. By Eq. (3-12) it can also be written as

$$\mathbf{V}_{P_3} = \mathbf{V}_{P_2} \tag{3-13b}$$

which says that ***the absolute velocities of two points in rolling contact are equal.***

The graphical solution of the problem of Fig. 3-14 is also shown in the figure. Given $\boldsymbol{\omega}_2$, the velocity difference $\mathbf{V}_{P_2B}$ can be calculated and plotted,

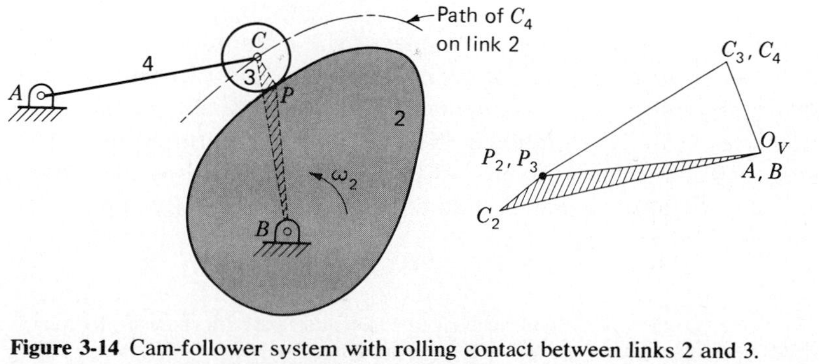

Figure 3-14 Cam-follower system with rolling contact between links 2 and 3.

thus locating point P_2 in the velocity polygon. Using Eq. (3-13), the rolling-contact condition, we also label this point P_3. Next, writing simultaneous equations for $\mathbf{V}_C$, using $\mathbf{V}_{CP_3}$ and $\mathbf{V}_{CA}$, we can find the velocity-image point C. Then $\boldsymbol{\omega}_3$ and $\boldsymbol{\omega}_4$ can be found from $\mathbf{V}_{CP}$ and $\mathbf{V}_{CA}$, respectively.

Another approach to the solution of the same problem involves defining a fictitious point C_2 which is located instantaneously coincident with points C_3 and C_4 but which is understood to be attached to, and move with, link 2, as shown by the shaded triangle BPC. When the velocity-image concept is used for link 2, the velocity-image point C_2 can be located. Noticing that point C_4 (and C_3) traces a known path on link 2, we can write and solve the apparent-velocity equation involving $\mathbf{V}_{C_4/2}$, thus obtaining the velocity $\mathbf{V}_{C_4}$ (and $\boldsymbol{\omega}_4$, if desired) without dealing with the point of direct contact. This second approach would be necessary if we had not assumed rolling contact (no slip) at P.

3-8 VELOCITY ANALYSIS USING COMPLEX ALGEBRA

We recall from Sec. 2-8 that complex algebra provides an alternative formulation for two-dimensional kinematics problems. As we saw, the complex-algebra formulation provides the advantage of increased accuracy, and it is amenable to solution by digital computer at a large number of positions once the program is written. On the other hand, the solution of the loop-closure equation for its unknown position variables is a nonlinear problem and can lead to tedious algebraic manipulations. Fortunately, as we will see, the extension of the complex-algebra approach to velocity analysis leads to a set of *linear* equations, and solution is quite straightforward.

Recalling the complex polar form of a two-dimensional vector from Eq. (2-28),

$$\mathbf{R} = Re^{j\theta}$$

we find the general form of its time derivative

$$\dot{\mathbf{R}} = \frac{d\mathbf{R}}{dt} = \dot{R}e^{j\theta} + j\dot{\theta}Re^{j\theta} \tag{3-14}$$

where $\dot{R}$ and $\dot{\theta}$ denote the time rates of change of the magnitude and angle of $\mathbf{R}$, respectively. We will see in the following examples that the first term of this equation usually represents an apparent velocity and the second term often represents a velocity difference. The methods illustrated in these examples were developed by Raven. Although the original work† gives methods applicable to both planar and spatial mechanisms, only the planar aspects are shown here.

† F. H. Raven, Velocity and Acceleration Analysis of Plane and Space Mechanisms by Means of Independent-Position Equations, *J. Appl. Mech.*, *ASME Trans.*, ser. E, vol. 80, pp. 1–6, 1958.

To illustrate Raven's approach let us analyze the inversion of the slider-crank mechanism shown in Fig. 3-15*a*. We will consider link 2, the driver, to have a known angular position θ_2 and a known angular velocity ω_2 at the instant considered. We wish to derive expressions for the angular velocity of link 4 and the absolute velocity of point *P*.

To simplify notation in this example we will use the symbolism shown in Fig. 3-15*b* for the position-difference vectors; thus, $\mathbf{R}_{AB}$ is denoted $\mathbf{r}_1$, $\mathbf{R}_{C_2A}$ is denoted $\mathbf{r}_2$, and $\mathbf{R}_{C_4B}$ is denoted $\mathbf{r}_4$. In terms of these symbols, the loop-closure equation is

$$\mathbf{r}_1 + \mathbf{r}_2 = \mathbf{r}_4 \tag{a}$$

where $\mathbf{r}_1$ has constant magnitude and direction.† The vector $\mathbf{r}_2$ has constant magnitude and its direction θ_2 varies, but is the input angle. We assume θ_2 is known or, more precisely, that other unknowns will be solved for as functions of θ_2. The vector $\mathbf{r}_4$ has unknown magnitude and direction.

Recognizing this as case 1 (Sec. 2-8), we obtain the position solution from Eqs. (2-30) and (2-31).

$$r_4 = \sqrt{r_1^2 + r_2^2 - 2r_1r_2 \cos \theta_2} \tag{b}$$

$$\theta_4 = \tan^{-1} \frac{r_2 \sin \theta_2}{r_2 \cos \theta_2 - r_1} \tag{c}$$

The velocity solution is initiated by differentiating the loop-closure equation (*a*) with respect to time. Applying the general form, Eq. (3-14), to each term of this equation in turn, and keeping in mind that r_1, θ_1, and r_2 are

† Note particularly that the angle of $\mathbf{r}_1$ is $\theta_1 = 180°$, not zero.

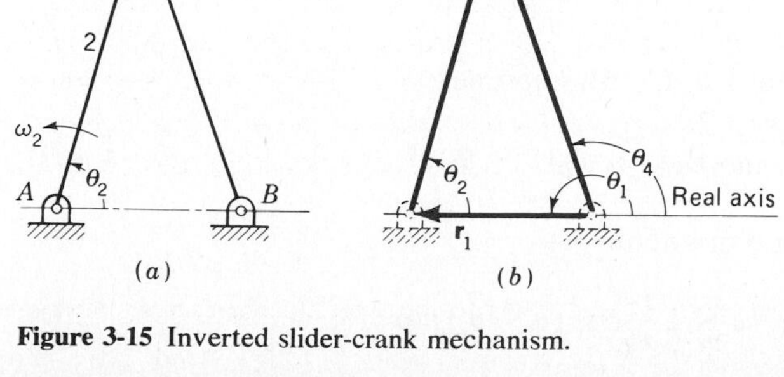

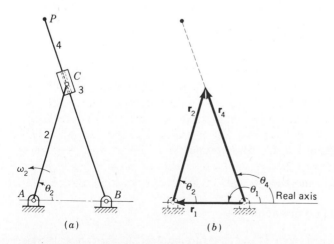

Figure 3-15 Inverted slider-crank mechanism.

constants, we obtain

$$j\dot{\theta}_2 r_2 e^{j\theta_2} = \dot{r}_4 e^{j\theta_4} + j\dot{\theta}_4 r_4 e^{j\theta_4} \tag{d}$$

Since $\dot{\theta}_2$ and $\dot{\theta}_4$ are the same as ω_2 and ω_4, respectively, and since we recognize that

$$\dot{\theta}_2 r_2 = V_{C_2} \qquad \dot{r}_4 = V_{C_2/4} \qquad \dot{\theta}_4 r_4 = V_{C_4}$$

we see that Eq. (*d*) is, in fact, the complex polar form of the apparent-velocity equation

$$\mathbf{V}_{C_2} = \mathbf{V}_{C_4} + \mathbf{V}_{C_2/4}$$

(This is pointed out for comparison only and is not a necessary step in the solution process.)

The velocity solution is performed by using Euler's formula to separate Eq. (*d*) into its real and imaginary components. This gives

$$-\omega_2 r_2 \sin\theta_2 = \dot{r}_4 \cos\theta_4 - \omega_4 r_4 \sin\theta_4 \tag{e}$$

$$\omega_2 r_2 \cos\theta_2 = \dot{r}_4 \sin\theta_4 + \omega_4 r_4 \cos\theta_4 \tag{f}$$

Solving these two equations simultaneously for the two unknowns $\dot{r}_4$ and ω_4 yields

$$\dot{r}_4 = \omega_2 r_2 \sin(\theta_4 - \theta_2) \tag{3-15}$$

$$\omega_4 = \omega_2 \frac{r_2}{r_4} \cos(\theta_4 - \theta_2) \tag{3-16}$$

Although the variables r_4 and θ_4 could be substituted from Eqs. (*b*) and (*c*) to reduce these results to functions of θ_2 and ω_2 alone, the above forms are considered sufficient since in writing a computer program numeric values are normally found first for r_4 and θ_4 while performing the position analysis, and these numeric values can then be used in finding $\dot{r}_4$ and ω_4 at each phase of angle θ_2.

To find the velocity of point P we write

$$\mathbf{R}_P = R_{PB} e^{j\theta_4} \tag{g}$$

and use Eq. (3-14) to differentiate with respect to time, remembering that R_{PB} is a constant length. This yields

$$\mathbf{V}_P = j\omega_4 R_{PB} e^{j\theta_4} \tag{h}$$

which, upon substituting from Eq. (3-16), becomes

$$\mathbf{V}_P = j\omega_2 R_{PB} \frac{r_2}{r_4} \cos(\theta_4 - \theta_2) e^{j\theta_4} \tag{3-17}$$

The horizontal and vertical components are

$$V_P^x = -\omega_2 R_{PB} \frac{r_2}{r_4} \cos(\theta_4 - \theta_2) \sin\theta_4 \tag{i}$$

$$V_P^y = \omega_2 R_{PB} \frac{r_2}{r_4} \cos(\theta_4 - \theta_2) \cos \theta_4 \tag{j}$$

As another illustration of Raven's approach, consider the following example problem.

Example 3-5 Develop an equation for the relationship between the angular velocities of the input and output cranks of a four-bar linkage.

SOLUTION We recall the loop-closure equation from Sec. 2-10, Eq. (*n*),

$$R_{BA}e^{j\theta_2} + R_{CB}e^{j\theta_3} = R_{DA} + R_{CD}e^{j\theta_4} \tag{1}$$

Remembering that all lengths remain constant, we use Eq. (3-14) to take the time derivative. This gives

$$j\dot{\theta}_2 R_{BA}e^{j\theta_2} + j\dot{\theta}_3 R_{CB}e^{j\theta_3} = j\dot{\theta}_4 R_{CD}e^{j\theta_4} \tag{2}$$

Equating the real and imaginary parts and rearranging terms yields

$$\omega_3 R_{CB} \sin \theta_3 - \omega_4 R_{CD} \sin \theta_4 = -\omega_2 R_{BA} \sin \theta_2 \tag{3}$$

$$\omega_3 R_{CB} \cos \theta_3 - \omega_4 R_{CD} \cos \theta_4 = -\omega_2 R_{BA} \cos \theta_2 \tag{4}$$

Finally, these two simultaneous equations are solved for ω_3 and ω_4.

$$\omega_3 = \frac{R_{BA} \sin(\theta_2 - \theta_4)}{R_{CB} \sin(\theta_4 - \theta_3)} \omega_2 \tag{3-18}$$

$$\omega_4 = \frac{R_{BA} \sin(\theta_2 - \theta_3)}{R_{CD} \sin(\theta_4 - \theta_3)} \omega_2 \qquad \textit{Ans.} \tag{3-19}$$

Since solutions are known for θ_3 and θ_4 from Eqs. (2-59) and (2-58), this equation for ω_4 can be numerically evaluated and is considered a complete solution.

Note that in both the above problems the simultaneous equations which were solved were *linear* equations. This was not a coincidence but is true in all velocity solutions. It results from the fact that the general equation (3-14) is linear in the velocity variables. When real and imaginary components are taken, the *coefficients* may become complicated, but the equations remain linear with respect to the velocity unknowns. Therefore, their solution is straightforward.

Another symptom of the linearity of velocity relationships is seen by recalling that in the graphical velocity solutions of previous sections it was possible to pick an arbitrary scale factor for a velocity polygon. If the input speed of a mechanism is doubled, the scale factor of the velocity polygon could be doubled and the same polygon would be valid. This is an indication of linear equations.

It is also worth noting that both Eqs. (3-18) and (3-19) include $\sin(\theta_4 - \theta_3)$ in their denominators. In general, any velocity-analysis problem will have similar denominators in the solution for each of the velocity unknowns; these denominators are the determinant of the matrix of coefficients of the unknowns of the linear equations, as will be recognized by recalling Cramer's rule. In the four-bar linkage it can be seen from Fig. 2-13 that $\theta_4 - \theta_3$ is the transmission angle. When the transmission angle becomes small, the ratio of the output to input velocity becomes very large and difficulty results.

3-9 VELOCITY ANALYSIS USING VECTOR ALGEBRA

The Chace approach to position analysis was discussed in Sec. 2-9. It will be shown here how this approach is applied to the velocity analysis of linkages. The method is illustrated by again solving the inverted slider-crank mechanism of Fig. 3-15.

The procedure begins by writing the loop-closure equation

$$\mathbf{r}_1 + \mathbf{r}_2 = \mathbf{r}_4 \tag{a}$$

The velocity relationships are found by differentiating this equation with respect to time. The derivative of a typical term becomes

$$\dot{\mathbf{R}} = \frac{d}{dt}(R\hat{\mathbf{R}}) = \dot{R}\hat{\mathbf{R}} + R\dot{\hat{\mathbf{R}}} \tag{b}$$

However, since $\hat{\mathbf{R}}$ is of constant length, and since it usually rotates with one of the links, $\dot{\hat{\mathbf{R}}}$ can be expressed as

$$\dot{\hat{\mathbf{R}}} = \boldsymbol{\omega} \times \hat{\mathbf{R}} = \omega(\hat{\mathbf{k}} \times \hat{\mathbf{R}}) \tag{3-20}$$

from which Eq. (*b*) becomes

$$\dot{\mathbf{R}} = \dot{R}\hat{\mathbf{R}} + \omega R(\hat{\mathbf{k}} \times \hat{\mathbf{R}}) \tag{3-21}$$

Using this general form and recognizing that the magnitudes r_1 and r_2 and the direction $\hat{\mathbf{r}}_1$ are constant, we can take the time derivative of the loop-closure equation (*a*). This yields

$$\omega_2 r_2(\hat{\mathbf{k}} \times \hat{\mathbf{r}}_2) = \dot{r}_4\hat{\mathbf{r}}_4 + \omega_4 r_4(\hat{\mathbf{k}} \times \hat{\mathbf{r}}_4) \tag{c}$$

Since it is assumed that r_4 and $\hat{\mathbf{r}}_4$ would be known from a previous position analysis, perhaps following the Chace approach of Sec. 2-9, and since ω_2 is a known driving speed, the only two unknowns of this equation are the velocities $\dot{r}_4$ and ω_4.

Rather than taking components of Eq. (*c*) in the horizontal and vertical directions, which would lead to two simultaneous equations in two unknowns, Chace's approach leads to the elimination of one unknown by careful choice of the directions along which components are taken. In Eq. (*c*), for example, we note that the unit vector $\hat{\mathbf{r}}_4$ is perpendicular to $\hat{\mathbf{k}} \times \hat{\mathbf{r}}_4$, and therefore

$$\hat{\mathbf{r}}_4 \cdot (\hat{\mathbf{k}} \times \hat{\mathbf{r}}_4) = 0 \tag{d}$$

We take advantage of this fact to eliminate the unknown $\dot{r}_4$. Taking the dot product of each term of Eq. (*c*) with $\hat{\mathbf{k}} \times \hat{\mathbf{r}}_4$, we obtain

$$\omega_2 r_2(\hat{\mathbf{k}} \times \hat{\mathbf{r}}_2) \cdot (\hat{\mathbf{k}} \times \hat{\mathbf{r}}_4) = \omega_4 r_4$$

from which we solve for ω_4

$$\omega_4 = \omega_2 \frac{r_2}{r_4} (\hat{\mathbf{k}} \times \hat{\mathbf{r}}_2) \cdot (\hat{\mathbf{k}} \times \hat{\mathbf{r}}_4) \tag{e}$$

Similarly, we can take the dot product of Eq. (*c*) with the unit vector $\hat{\mathbf{r}}_4$ and thus eliminate ω_4. This gives

$$\dot{r}_4 = \omega_2 r_2 (\hat{\mathbf{k}} \times \hat{\mathbf{r}}_2) \cdot \hat{\mathbf{r}}_4 \tag{f}$$

That these solutions are indeed the same as those obtained by Raven's method can be shown quite simply. From Eq. (*e*) we can write

$$(\hat{\mathbf{k}} \times \hat{\mathbf{r}}_2) = \begin{vmatrix} \hat{\mathbf{i}} & \hat{\mathbf{j}} & \hat{\mathbf{k}} \\ 0 & 0 & 1 \\ \cos\theta_2 & \sin\theta_2 & 0 \end{vmatrix} = -\sin\theta_2\, \hat{\mathbf{i}} + \cos\theta_2\, \hat{\mathbf{j}}$$

and, similarly,

$$(\hat{\mathbf{k}} \times \hat{\mathbf{r}}_4) = -\sin\theta_4\, \hat{\mathbf{i}} + \cos\theta_4\, \hat{\mathbf{j}}$$

Then

$$\begin{aligned} (\hat{\mathbf{k}} \times \hat{\mathbf{r}}_2) \cdot (\hat{\mathbf{k}} \times \hat{\mathbf{r}}_4) &= (-\sin\theta_2\, \hat{\mathbf{i}} + \cos\theta_2\, \hat{\mathbf{j}}) \cdot (-\sin\theta_4\, \hat{\mathbf{i}} + \cos\theta_4\, \hat{\mathbf{j}}) \\ &= \sin\theta_2 \sin\theta_4 + \cos\theta_2 \cos\theta_4 \\ &= \cos(\theta_4 - \theta_2) \end{aligned} \tag{g}$$

and, in a similar manner,

$$(\hat{\mathbf{k}} \times \hat{\mathbf{r}}_2) \cdot \hat{\mathbf{r}}_4 = \sin(\theta_4 - \theta_2) \tag{h}$$

When the terms of Eqs. (*g*) and (*h*) are substituted into Eqs. (*e*) and (*f*), the results match identically with those obtained from Raven's approach, Eqs. (3-15) and (3-16).

3-10 INSTANTANEOUS CENTER OF VELOCITY

One of the more interesting concepts in kinematics is that of an instantaneous velocity axis for rigid bodies which move relative to one another. In particular, we shall find that an axis exists which is common to both bodies and about which either body can be considered as rotating with respect to the other.

Since our study of these axes will be restricted to planar motions,† each axis is perpendicular to the plane of the motion. We shall refer to them as *instant centers* or *poles*. These instant centers are regarded as a pair of

† For three-dimensional motion, this axis is referred to as the *instantaneous screw* axis. The classic work covering its properties is R. S. Ball, "A Treatise on the Theory of Screws," Cambridge University Press, Cambridge, 1900.

coincident points, one attached to each body, about which one body has an apparent rotation relative to the other. This property is true only *instantaneously*, and a new pair of coincident points will become the instant center at the next instant. It is not correct, therefore, to speak of an instant center as the center of rotation, since it is generally not located at the center of curvature of the apparent point path which a point of one body generates with respect to the coordinate system of the other. Even with this restriction, however, we will find that instant centers contribute substantially to understanding the kinematics of planar motion.

The instantaneous center of velocity is defined as *the instantaneous location of a pair of coincident points of two different rigid bodies for which the absolute velocities of the two points are equal.* It may also be defined as the location of a pair of coincident points of two different rigid bodies for which *the apparent velocity of one of the points is zero as seen by an observer on the other body.*

Let us consider a rigid body 2 having some general motion relative to the x_1y_1 plane; the motion might be translation, rotation, or a combination of both. As shown in Fig. 3-16*a*, suppose that some point A of the body has a known velocity $\mathbf{V}_A$ and that the body has a known angular velocity $\boldsymbol{\omega}_2$. With these two quantities known, the velocity of any other point of the body can be found from the velocity-difference equation. Suppose we define a point P, for example, whose position difference $\mathbf{R}_{PA}$ from point A is chosen to be

$$\mathbf{R}_{PA} = \frac{\boldsymbol{\omega}_2 \times \mathbf{V}_A}{\omega_2^2} \tag{3-22}$$

Because of the cross product we see that the point P is located on the perpendicular to $\mathbf{V}_A$ and the vector $\mathbf{R}_{PA}$ is rotated from the direction of $\mathbf{V}_A$ in the direction of $\boldsymbol{\omega}_2$, as shown in Fig. 3-16*b*. The length of $\mathbf{R}_{PA}$ can be calculated from the above equation, and point P can be located. We see that

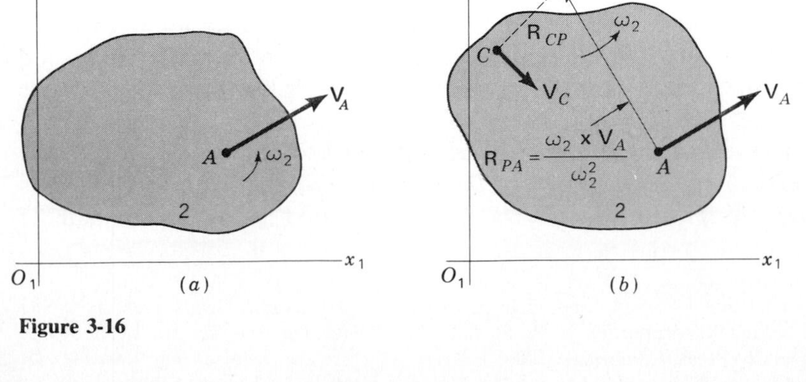

Figure 3-16

its velocity is

$$\mathbf{V}_P = \mathbf{V}_A + \mathbf{V}_{PA} = \mathbf{V}_A + \boldsymbol{\omega}_2 \times \mathbf{R}_{PA} = \mathbf{V}_A + \boldsymbol{\omega}_2 \times \frac{\boldsymbol{\omega}_2 \times \mathbf{V}_A}{\omega_2^2}$$

But on replacing this triple product by a vector identity we get

$$\mathbf{V}_P = \mathbf{V}_A + \frac{(\boldsymbol{\omega}_2 \cdot \mathbf{V}_A \overset{0}{\nearrow})\boldsymbol{\omega}_2 - (\boldsymbol{\omega}_2 \cdot \boldsymbol{\omega}_2)\mathbf{V}_A}{\omega_2^2} = \mathbf{V}_A - \mathbf{V}_A = 0 \qquad (a)$$

Since the absolute velocity of the particular point P chosen is zero, the same as the velocity of the coincident point of the fixed link, this point P is the instant center between links 1 and 2.

The velocity of any third point C of the moving body can now be found

$$\mathbf{V}_C = \mathbf{V}_P \overset{0}{\nearrow} + \mathbf{V}_{CP} = \boldsymbol{\omega}_2 \times \mathbf{R}_{CP} \qquad (b)$$

as shown in Fig. 3-16*b*.

The instant center can be located more easily when the absolute velocities of two points are given. In Fig. 3-17*a* suppose that points A and C have known velocities $\mathbf{V}_A$ and $\mathbf{V}_C$. Perpendiculars to $\mathbf{V}_A$ and $\mathbf{V}_C$ intersect at P, the instant center. Figure 3-17*b* shows how to locate the instant center P when the points A, C, and P happen to fall on the same straight line.

The instant center between two bodies, in general, is not a stationary point. It changes its location relative to both bodies as the motion progresses and describes a path or locus on each. These paths of the instant centers, called *centrodes*, will be discussed in Sec. 3-17.

Since we have adopted the convention of numbering the links of a mechanism, it is convenient to designate an instant center by using the numbers of the two links associated with it. Thus P_{32} identifies the instant center between links 3 and 2. This same center could be identified as P_{23} since the order of the numbers has no significance. A mechanism has as many

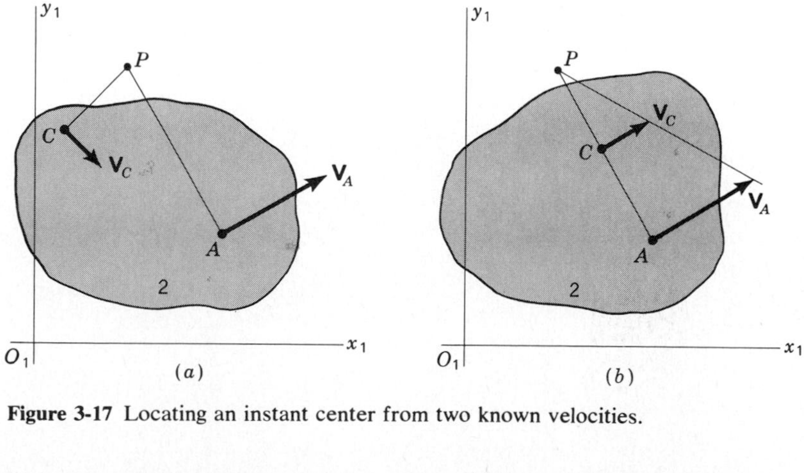

Figure 3-17 Locating an instant center from two known velocities.

instant centers as there are ways of pairing the link numbers. Thus the number of instant centers in an n-link mechanism is

$$N = \frac{n(n-1)}{2} \tag{3-23}$$

3-11 THE ARONHOLD-KENNEDY THEOREM OF THREE CENTERS

According to Eq. (3-23), the number of instant centers in a four-bar linkage is six. As shown in Fig. 3-18*a* we can identify four of them by inspection; we see that the four pins can be identified as instant centers P_{12}, P_{23}, P_{34}, and P_{14}, since each satisfies the definition. Point P_{23}, for example, is a point of link 2 about which link 3 appears to rotate; it is a point of link 3 which has no apparent velocity as seen from link 2; it is a pair of coincident points of links 2 and 3 which have the same absolute velocities.

A good method of keeping track of which instant centers have been found is to space the link numbers around the perimeter of a circle, as shown in Fig. 3-18*b*. Then, as each pole is identified, a line is drawn connecting the corresponding pair of link numbers. Figure 3-18*b* shows that P_{12}, P_{23}, P_{34}, and P_{14} have been found; it also shows missing lines where P_{13} and P_{24} have not been located. Those two cannot be found by applying the definition visually.

After finding as many of the instant centers as possible by inspection, i.e., by locating points which obviously fit the definition, others are located by applying the Aronhold-Kennedy theorem (often just called Kennedy's theorem†) of three centers. This theorem states that *the three instant centers shared by three rigid bodies in relative motion to one another (whether or not connected) all lie on the same straight line.*

† This theorem is named after its two independent discoverers, Aronhold, 1872, and Kennedy, 1886. It is known as the Aronhold theorem in German-speaking countries, and Kennedy's theorem in English-speaking countries.

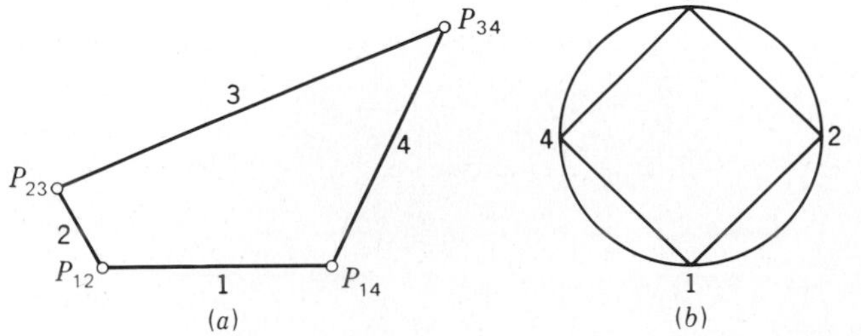

Figure 3-18

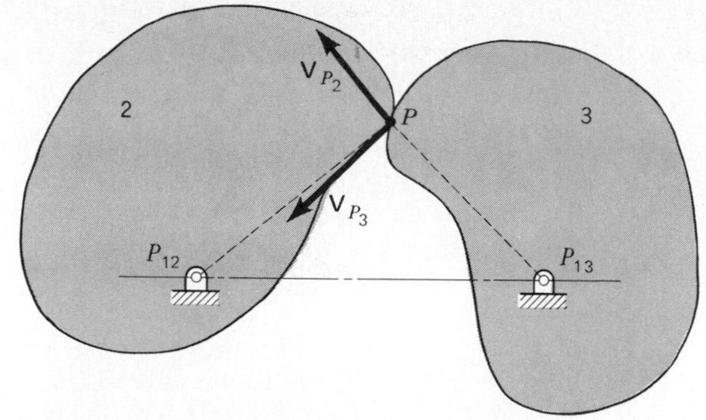

Figure 3-19 Aronhold-Kennedy theorem.

The theorem can be proven by contradiction, as shown in Fig. 3-19. Link 1 is a stationary frame, and instant center P_{12} is located where link 2 is pin-connected to it. Similarly, P_{13} is located at the pin connecting links 1 and 3. The shapes of links 2 and 3 are arbitrary. The Aronhold-Kennedy theorem states that the three instant centers P_{12}, P_{13}, and P_{23} must all lie on the same straight line, the line connecting the two pins. Let us suppose that this were not true; in fact, let us suppose that P_{23} were located at the point labeled P in Fig. 3-19. Then the velocity of P as a point of link 2 would have the direction $\mathbf{V}_{P_2}$, perpendicular to $\mathbf{R}_{PP_{12}}$. But the velocity of P as a point of link 3 would have the direction $\mathbf{V}_{P_3}$, perpendicular to $\mathbf{R}_{PP_{13}}$. The directions are inconsistent with the definition that an instant center must have equal absolute velocities as a part of either link. The point P chosen therefore cannot be the instant center P_{23}. This same contradiction in the directions of $\mathbf{V}_{P_2}$ and $\mathbf{V}_{P_3}$ occurs for any location chosen for point P unless it is chosen on the straight line through P_{12} and P_{13}.

3-12 LOCATING INSTANT CENTERS OF VELOCITY

In the last two sections we have seen several means of locating instant centers of velocity. They can often be located by inspecting the figure of a mechanism and visually seeking out a point which fits the definition, such as a pin-joint center. Also, once some instant centers are found, others can be found from them by using the theorem of three centers. Section 3-10 demonstrated that an instant center between a moving body and the fixed link can be found if the directions of the absolute velocities of two points of the body are known or if the absolute velocity of one point and the angular velocity of the body are known. The purpose of this section is to expand this list of techniques and to present examples.

Consider the cam-follower system of Fig. 3-20. Instant centers P_{12} and P_{13}

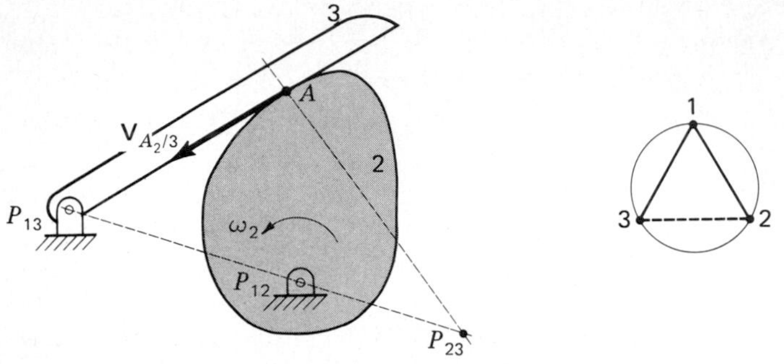

Figure 3-20 Instant centers of a disk cam with flat-faced follower.

can be located by inspection at the two pin centers. However, the remaining instant center, P_{23}, is not as obvious. According to the Aronhold-Kennedy theorem, it must lie on the straight line connecting P_{12} and P_{13}, but where on this line? After some reflection we see that the direction of the apparent velocity $\mathbf{V}_{A_2/3}$ must be along the common tangent to the two moving links at the point of contact and, as seen by an observer on link 3, this velocity must appear as a result of the apparent rotation of body 2 about the instant center P_{23}. Therefore, P_{23} must lie on the perpendicular to $\mathbf{V}_{A_2/3}$. This line now locates P_{23} as shown. The concept illustrated in this example should be remembered since it is often useful in locating the instant centers of mechanisms involving direct contact.

A special case of direct contact, as we have seen before, is rolling contact without slip. Considering the mechanism of Fig. 3-21, we can immediately locate instant centers P_{12}, P_{23}, and P_{34}. If the contact between links 1 and 4 involves any slippage, we can only say that instant center P_{14} is located on the

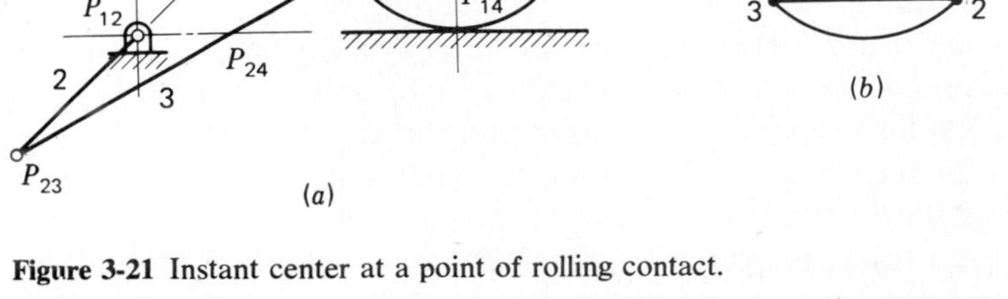

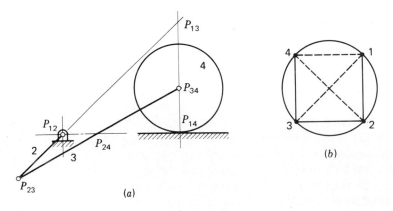

Figure 3-21 Instant center at a point of rolling contact.

vertical line through the point of contact. However, if we also know that there is no slippage, i.e., *if there is rolling contact, the instant center is located at the point of contact.* This is also a general principle, as can be seen by comparing the definition of rolling contact, Eq. (3-13), and the definition of an instant center; they are equivalent.

Another special case of direct contact is evident between links 3 and 4 in Fig. 3-22. In this case there is an apparent (slip) velocity $\mathbf{V}_{A_3/4}$ between points A of links 3 and 4, but *there is no apparent rotation between the links.* Here, as in Fig. 3-20, the instant center P_{34} lies along a common perpendicular to the known line of sliding, but now it is located *infinitely far away*, in the direction defined by this perpendicular line. This infinite distance can be shown by considering the kinematic inversion of the mechanism in which link 4 becomes stationary. Writing Eq. (3-22) for the inverted mechanism, we see that

$$\mathbf{R}_{P_{34}A} = \frac{\boldsymbol{\omega}_{3/4} \times \mathbf{V}_{A_3/4}}{\omega_{3/4}^2} = \frac{\hat{\boldsymbol{\omega}}_{3/4} \times \mathbf{V}_{A_3/4}}{\omega_{3/4}} = \infty \tag{3-24}$$

The direction stated earlier is confirmed by the numerator of this equation. We also see that, *since there is no relative rotation* between links 3 and 4, *the denominator is zero and the distance to P_{34} is infinite.* The other instant centers of Fig. 3-22 are found by inspection or by the Aronhold-Kennedy theorem. Notice in this figure how the line through P_{14} and P_{34} (at infinity) was used in locating P_{13}.

One final example will illustrate the above principles again.

Example 3-6 Locate all instant centers of the mechanism of Fig. 3-23 assuming rolling contact between links 1 and 2.

SOLUTION Instant centers P_{13}, P_{34}, and P_{15} are located by inspection. Also, P_{12} is located at the point of rolling contact. P_{24} may possibly be noticed by the fact that this is the center of the apparent rotation between links 2 and 4; if not, it can be located by drawing perpendicular lines to the directions of the apparent velocities at two of the corners of link 4.

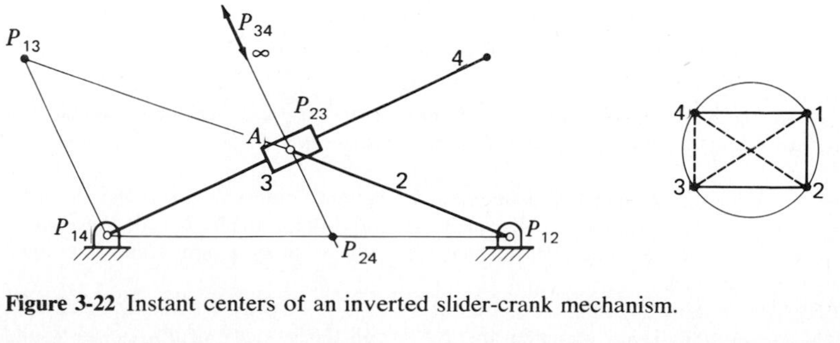

Figure 3-22 Instant centers of an inverted slider-crank mechanism.

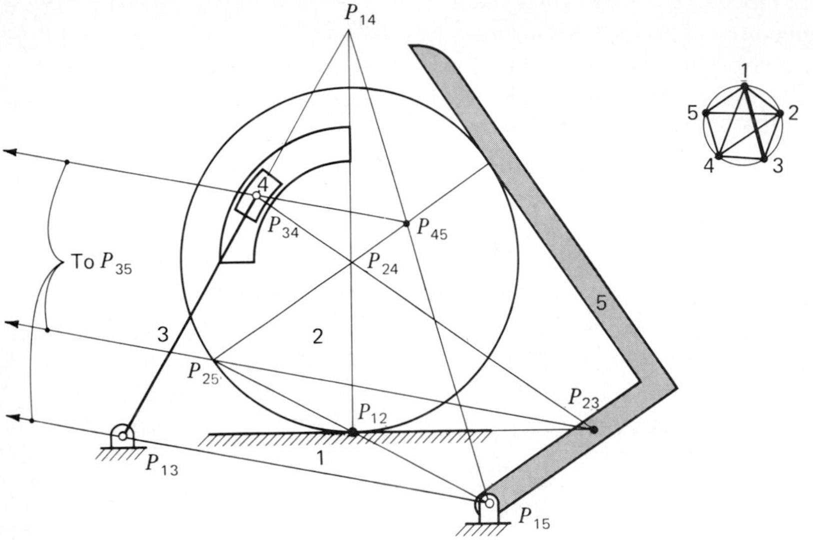

Figure 3-23 Example 3-6.

One line for instant center P_{25} comes from noticing the direction of slipping between links 2 and 5; the other comes from the line $P_{12}P_{15}$. From these, all other instant centers can be found by repeated applications of the theorem of three centers.

It should be noted before closing this section that in all the above examples the locations of all instant centers were found without having to specify the actual operating speed of the mechanism. This is another indication of the linearity of the equations relating velocities, as pointed out in Sec. 3-8. *For any single-degree-of-freedom mechanism, the locations of all instant centers are uniquely determined by the geometry alone and do not depend on the operating speed.*

3-13 VELOCITY ANALYSIS USING INSTANT CENTERS

The properties of instant centers also provide a simple graphical approach for the velocity analysis of planar-motion mechanisms.

Example 3-7 In Fig. 3-24*a* we assume that the angular velocity ω_2 of crank 2 is given, and we wish to find the velocities of *B*, *D*, and *E* at the instant shown.

SOLUTION Consider the straight line defined by the instant centers P_{12}, P_{14}, and P_{24}. This must be a straight line according to the Kennedy-Aronhold theorem and is called the *line of centers.* According to its definition P_{24} is common to both links 2 and 4 and has equal absolute velocities in each.

First consider instant center P_{24} as a point of link 2. The velocity $\mathbf{V}_A$ can be found from ω_2 using the velocity-difference equation about P_{12}, and the velocity of P_{24} can be found

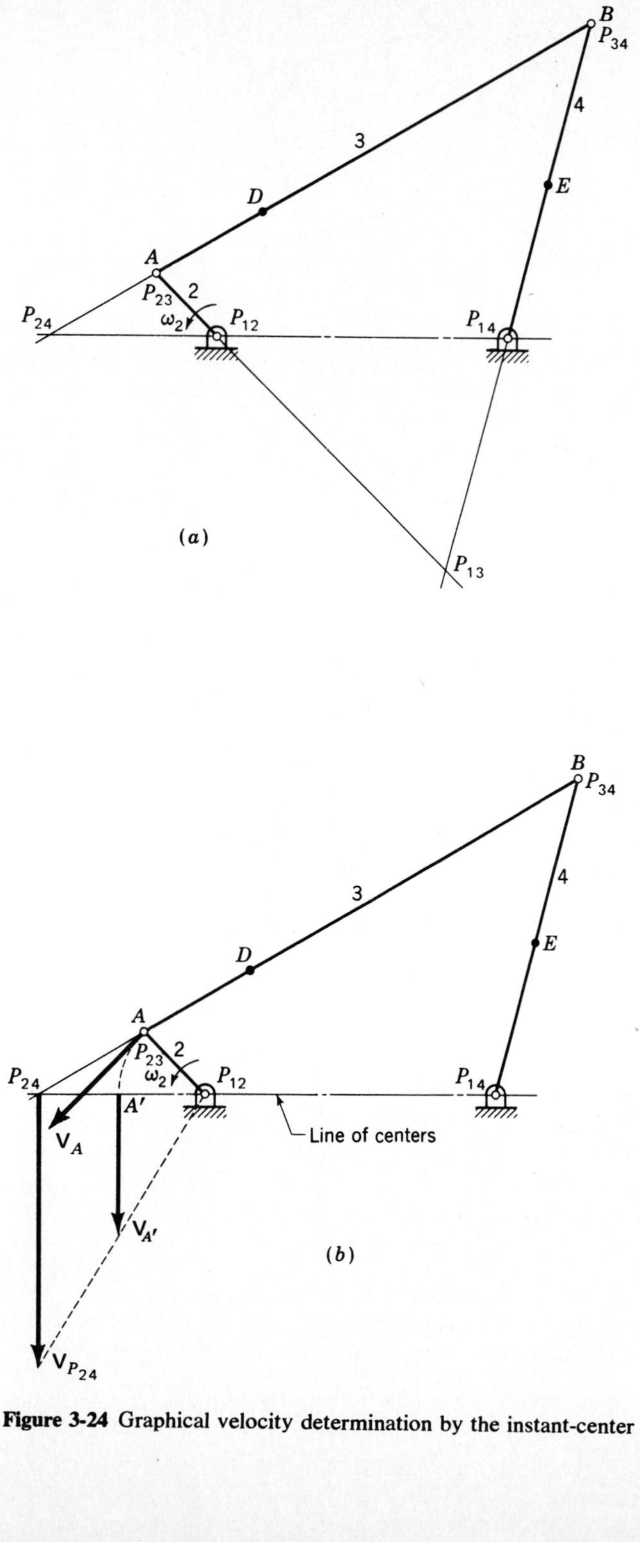

Figure 3-24 Graphical velocity determination by the instant-center method.

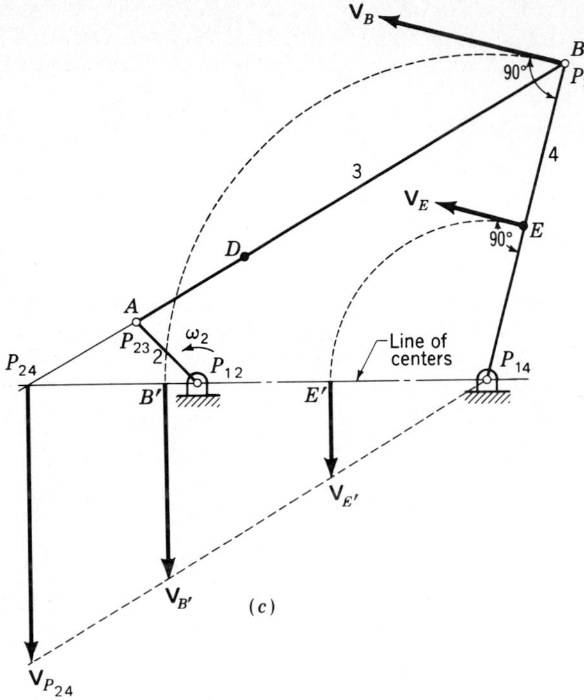

(c)

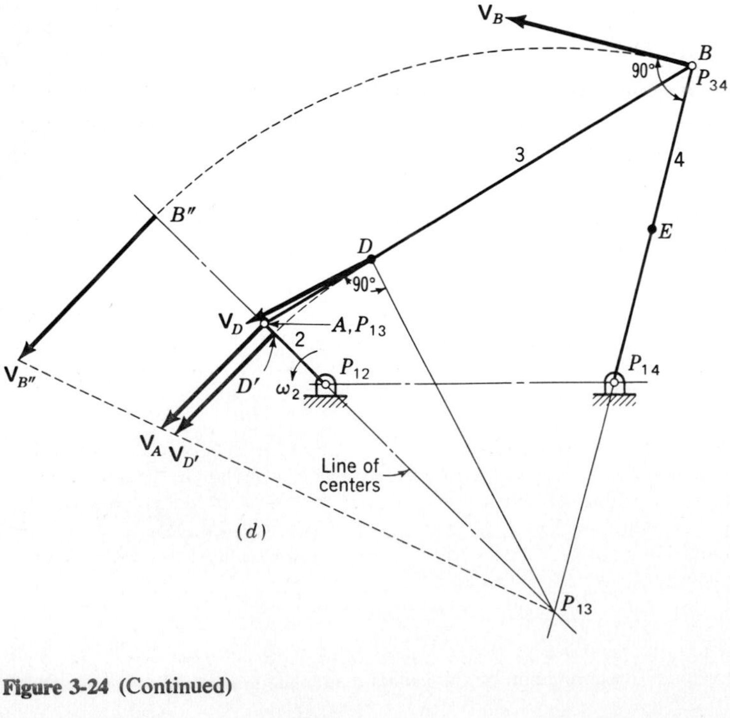

(d)

Figure 3-24 (Continued)

from it; the graphical construction is shown in Fig. 3-24*b*. When point A' of link 2 is located on the line of centers at an equal distance from P_{12}, its absolute velocity $\mathbf{V}_{A'}$ is equal in magnitude to $\mathbf{V}_A$. Now the magnitude of $\mathbf{V}_{P_{24}}$ can be found† by constructing a line from P_{12} through the terminus of $\mathbf{V}_{A'}$ as shown.

Next consider P_{24} as a point of link 4 rotating about P_{14}. Knowing $\mathbf{V}_{P24}$, we can find the velocity of any other point of link 4, such as B'or E' (Fig. 3-24*c*), by using the reverse construction. Since B' and E' were chosen to have the same radii from P_{14} as B and E, their velocities have magnitudes equal to $\mathbf{V}_B$ and $\mathbf{V}_E$, respectively, and these can be laid out in their proper directions as shown in Fig. 3-24*c*.

To obtain $\mathbf{V}_D$ we note that D is in link 3; the known velocity $\boldsymbol{\omega}_2$ (or $\mathbf{V}_A$) is for link 2, and the reference link is link 1. Therefore a new line of centers $P_{12}P_{13}P_{23}$ is chosen, as shown in Fig. 3-24*d*. Using $\boldsymbol{\omega}_2$ and P_{12}, we find the absolute velocity of the common instant center P_{23}. Here this step is trivial since $\mathbf{V}_{P_{23}} = \mathbf{V}_A$. Locating point D' on the new line of centers, we find $\mathbf{V}_D$ as shown and use its magnitude to find the desired velocity $\mathbf{V}_D$. Note that, according to the definition, instant center P_{13} as part of link 3 has zero velocity at this instant. Since B can also be considered a point of link 3, its velocity can be found in a similar manner by finding $\mathbf{V}_{B''}$ as shown.

The *line-of-centers method* of velocity analysis using instant centers can be summarized as follows:

1. Identify the three link numbers associated with the given velocity and the velocity to be found. One of these is usually link 1 since usually absolute velocity information is given and requested.
2. Locate the three instant centers defined by the links of step 1 and draw the line of centers.
3. Find the velocity of the common instant center by treating it as a point of the link whose velocity is given.
4. With the velocity of the common instant center known, consider it as a point of the link whose velocity is to be found. The velocity of any other point in that link can now be found.

Another example will illustrate the procedure and will show how to treat instant centers located at infinity.

Example 3-8 Only some of the links of the device shown in Fig. 3-25 can be seen; others are enclosed in a housing, but it is known that the instant center P_{25} has the location shown. Find the angular velocity of the crank $\boldsymbol{\omega}_2$ which is necessary to produce a velocity $\mathbf{V}_C$ of 10 m/s to the right.

SOLUTION Since we are given $\mathbf{V}_{C5/1}$ and want $\boldsymbol{\omega}_{2/1}$, we need to use the instant centers P_{15}, P_{12}, and P_{25}. After locating P_{25}, P_{56}, and P_{16} by inspection and applying the theorem of three centers, we locate P_{15} at infinity as shown. We now draw the line of centers $P_{12}P_{25}P_{15}$.

Considering P_{25} as a part of link 5, we wish to find its velocity from the given $\mathbf{V}_C$. We have difficulty in locating a point C' on the line of centers at the same radius from P_{15} as C because P_{15} is at infinity. How can we proceed?

† Note that $\mathbf{V}_{P_{24}}$ could have been found directly from its velocity difference from P_{12}. This construction was shown to illustrate the principle of the graphical method.

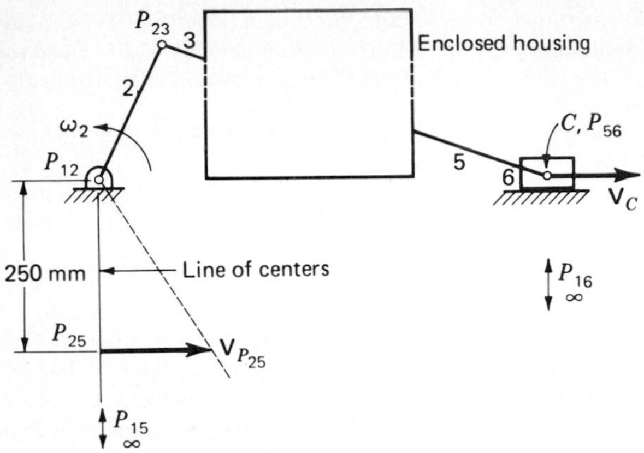

Figure 3-25 Example 3-8.

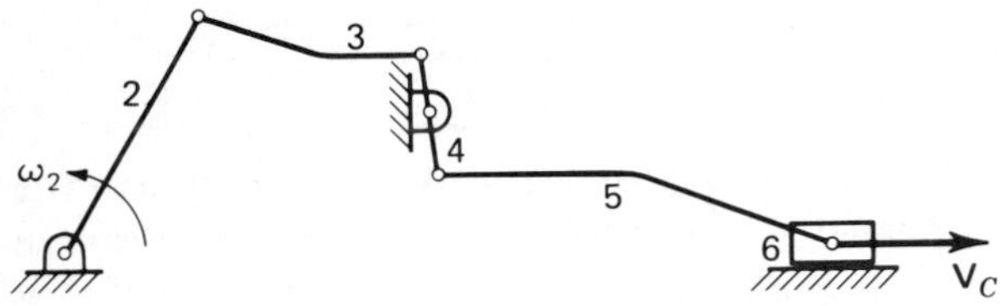

Figure 3-26

Recalling the discussion of Sec. 3-12 and Eq. (3-24), we see that since P_{15} is at infinity, the relative motion between links 5 and 1 is a translation and $\boldsymbol{\omega}_{5/1} = 0$. Since this is true, *every* point of link 5 has the same absolute velocity, including $\mathbf{V}_{P_{25}} = \mathbf{V}_C$. Thus we lay out $\mathbf{V}_{P_{25}}$ in the figure.

Next we treat P_{25} as a point of link 2, rotating about P_{12}, and we solve for $\boldsymbol{\omega}_2$

$$\omega_2 = \frac{V_{P_{25}}}{R_{P_{25}P_{12}}} = \frac{10 \text{ m/s}}{0.25 \text{ m}} = 40 \text{ rad/s ccw} \qquad \textit{Ans.}$$

Noticing the apparent paradox between the directions of $\mathbf{V}_C$ and $\boldsymbol{\omega}_2$, we may speculate on the validity of our solution. This would be resolved, however, by opening the enclosed housing and discovering the linkage shown in Fig. 3-26.

3-14 THE ANGULAR-VELOCITY-RATIO THEOREM

In Fig. 3-27, P_{24} is the instant center common to links 2 and 4. Its absolute velocity $\mathbf{V}_{P_{24}}$ is the same whether P_{24} is considered as a point of link 2 or of link 4. Considering it each way, we can write

$$\mathbf{V}_{P_{24}} = \cancelto{0}{\mathbf{V}_{P_{12}}} + \boldsymbol{\omega}_{2/1} \times \mathbf{R}_{P_{24}P_{12}} = \cancelto{0}{\mathbf{V}_{P_{14}}} + \boldsymbol{\omega}_{4/1} \times \mathbf{R}_{P_{24}P_{14}} \qquad (a)$$

where $\boldsymbol{\omega}_{2/1}$ and $\boldsymbol{\omega}_{4/1}$ are the same as $\boldsymbol{\omega}_2$ and $\boldsymbol{\omega}_4$, respectively, but the additional subscript has been written to emphasize the presence of the third link (the frame).

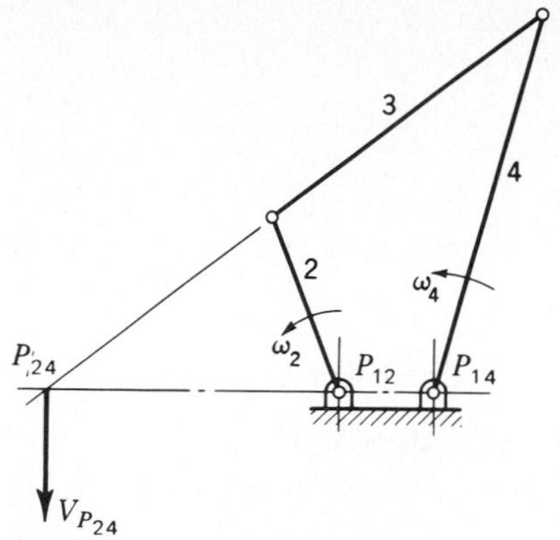

Figure 3-27 The angular-velocity ratio theorem.

Considering magnitudes only, we can rearrange Eq. (*a*) to read

$$\frac{\omega_{4/1}}{\omega_{2/1}} = \frac{R_{P_{24}P_{12}}}{R_{P_{24}P_{14}}} \tag{b}$$

This system illustrates the *angular-velocity-ratio theorem.* The theorem states that *the angular-velocity ratio of any two bodies in planar motion relative to a third body is inversely proportional to the segments into which the common instant center cuts the line of centers.* Written in general notation for the motion of bodies *j* and *k* relative to body *i*, the equation is

$$\frac{\omega_{k/i}}{\omega_{j/i}} = \frac{R_{P_{jk}P_{ij}}}{R_{P_{jk}P_{ik}}} \tag{3-25}$$

Picking an arbitrary positive direction along the line of centers, you should prove for yourself that the angular-velocity ratio is positive when the common instant center falls outside the other two centers and negative when it falls between them.

3-15 FREUDENSTEIN'S THEOREM

In the analysis and design of linkages it is often important to know the phases of the linkage at which the extreme values of the output velocity occur or, more precisely, the phases at which the ratio of the output and input velocities reaches its extremes.

The earliest work in determining extreme values is apparently that of Krause,† who stated that the velocity ratio ω_4/ω_2 of the drag-link mechanism (Fig. 3-28) reaches an extreme value when the connecting rod and follower,

† R. Krause, Die Doppelkurbel und Ihre Geschwindigkeitsgrenzen, *Maschinenbau/Getriebetechnik*, vol. 18, pp. 37–41, 1939; Zur Synthese der Doppelkurbel, *Maschinenbau/Getriebetechnik*, vol. 18, pp. 93–94, 1939.

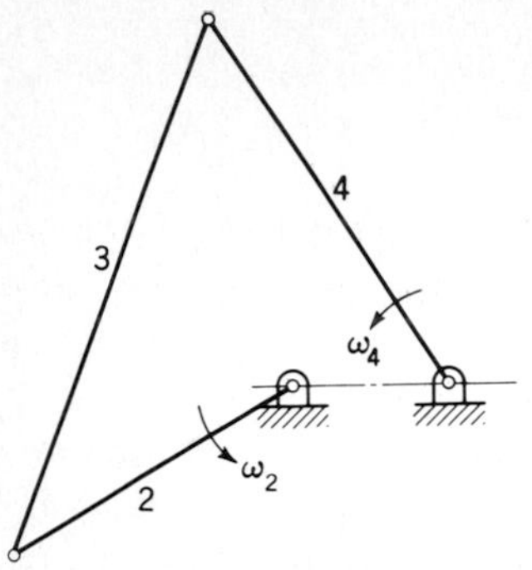

Figure 3-28 The drag-link mechanism.

links 3 and 4, become perpendicular to each other. Rosenauer, however, has shown that this is not strictly true.† Following Krause, Freudenstein developed a simple graphical method for determining the phases of the four-bar linkage at which the extreme values of the velocity do occur.‡

Freudenstein's theorem makes use of the line connecting instant centers P_{13} and P_{24} (Fig. 3-29), called the *collineation axis.* The theorem states that *at an extreme of the output- to input-angular-velocity ratio of a four-bar linkage, the collineation axis is perpendicular to the coupler link.*§

† N. Rosenauer, Synthesis of Drag-Link Mechanisms for Producing Nonuniform Rotational Motion with Prescribed Reduction Ratio Limits, *Aust. J. Appl. Sci.*, vol. 8, pp. 1–6, 1957.

‡ F. Freudenstein, On the Maximum and Minimum Velocities and Accelerations in Four-Link Mechanisms, *Trans. ASME*, vol. 78, pp. 779–787, 1956.

§ A. S. Hall, Jr. contributed a rigorous proof of this theorem in an appendix to Freudenstein's paper.

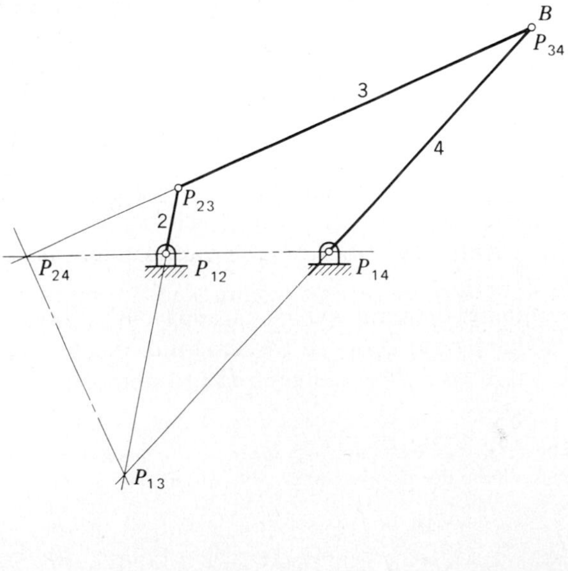

Figure 3-29 The collineation axis.

Using the angular-velocity-ratio theorem, Eq. (3-25), we write

$$\frac{\omega_4}{\omega_2} = \frac{R_{P_{24}P_{12}}}{R_{P_{24}P_{12}} + R_{P_{12}P_{14}}}$$

Since $R_{P_{12}P_{14}}$ is the fixed length of the frame link, the extremes of the velocity ratio occur when $R_{P_{24}P_{12}}$ is either a maximum or a minimum. Such positions may occur on either or both sides of P_{12}. Thus the problem reduces to finding the geometry of the linkage for which $R_{P_{24}P_{12}}$ is an extremum.

During motion of the linkage, P_{24} travels along the line $P_{12}P_{14}$, as seen by the theorem of three centers, but at an extreme value of the velocity ratio P_{24} must instantaneously be at rest (its direction of travel on this line must be reversing). This occurs when the velocity of P_{24}, considered as a point of link 3, is directed along the coupler link. This will be true only when the coupler link is perpendicular to the collineation axis since P_{13} is the instant center of link 3.

An inversion of the theorem (treating link 2 as fixed) states that *an extreme value of the velocity ratio ω_3/ω_2 of a four-bar linkage occurs when the collineation axis is perpendicular to the follower (link 4).*

3-16 INDEXES OF MERIT; MECHANICAL ADVANTAGE

In this section we will study some of the various ratios, angles, and other parameters of mechanisms which tell us whether a mechanism is a good one or a poor one. Many such parameters have been defined by various authors over the years, and there is no common agreement on a single "index of merit" for all mechanisms. Yet the many used have a number of features in common, including the fact that most can be related to the velocity ratios of the mechanism and therefore can be determined solely from the geometry of the mechanism. In addition, most depend on some knowledge of the application of the mechanism, especially of which are the input and output links. It is often desirable in the analysis or synthesis of mechanisms to plot these indexes of merit for a revolution of the input crank and to notice in particular their minimum and maximum values when evaluating the design of the mechanism or its suitability to a given application.

In Sec. 3-14 we learned that the ratio of the angular velocity of the output link to the input link of a mechanism is inversely proportional to the segments into which the common instant center cuts the line of centers. Thus, in the four-bar linkage of Fig. 3-30, if links 2 and 4 are the input and output links, respectively, then

$$\frac{\omega_4}{\omega_2} = \frac{R_{PA}}{R_{PD}}$$

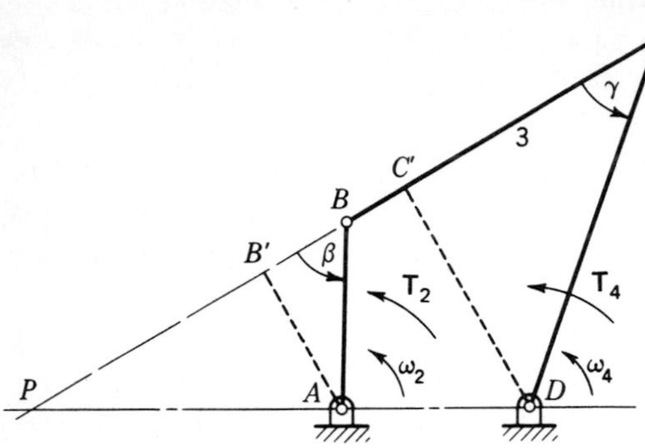

Figure 3-30 Four-bar linkage.

is the equation for the output- to input-velocity ratio. We also learned in Sec. 3-15 that the extremes of this ratio occur when the collineation axis is perpendicular to the coupler, link 3.

If we now assume that the linkage of Fig. 3-30 has no friction or inertia forces during its operation or that these are negligible compared with the input torque T_2, applied to link 2, and the output torque T_4, the resisting load torque on link 4, then we can derive a relation between T_2 and T_4. Since friction and inertia forces are negligible, the input power applied to link 2 is the negative of the power applied to link 4 by the load; hence

$$T_2\omega_2 = -T_4\omega_4 \tag{a}$$

or

$$\frac{T_4}{T_2} = -\frac{\omega_2}{\omega_4} = -\frac{R_{PD}}{R_{PA}} \tag{3-26}$$

The *mechanical advantage* of a mechanism is the instantaneous ratio of the output force (torque) to the input force (torque). Here we see that the mechanical advantage is the negative reciprocal of the velocity ratio. Either can be used as an index of merit in judging a mechanism's ability to transmit force or power.

The mechanism is redrawn in Fig. 3-31 at the position where links 2 and 3 are on the same straight line. At this position, R_{PA} and ω_4 are passing through

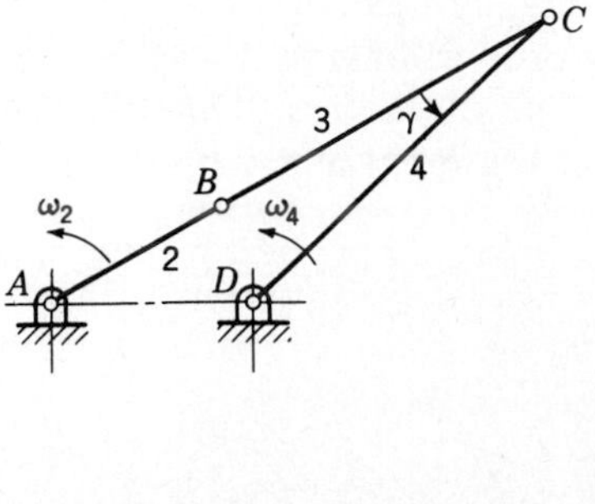

Figure 3-31 Four-bar linkage in toggle.

zero; hence an extreme value of the mechanical advantage (infinity) is obtained. A mechanism in this phase is said to be *in toggle.* Such toggle positions are often used to produce a high mechanical advantage; an example is the clamping mechanism of Fig. 2-6.

Proceeding further, construct $B'A$ and $C'D$ perpendicular to the line PBC in Fig. 3-30. Also let β and γ be the acute angles made by the coupler, or its extension and the output and input angles, respectively. Then, by similar triangles

$$\frac{R_{PD}}{R_{PA}} = \frac{R_{C'D}}{R_{B'A}} = \frac{R_{CD} \sin \gamma}{R_{BA} \sin \beta} \tag{b}$$

Then, using Eq. (3-26), we see that another expression for mechanical advantage is†

$$\frac{T_4}{T_2} = -\frac{\omega_2}{\omega_4} = -\frac{R_{CD} \sin \gamma}{R_{BA} \sin \beta} \tag{3-27}$$

Equation (3-27) shows that the mechanical advantage is infinite whenever the angle β is 0 or 180°, that is, whenever the mechanism is in toggle.

In Sec. 1-9 we defined the angle γ between the coupler and the follower link as the *transmission angle.* This angle is also often used as an index of merit for a four-bar linkage. Equation (3-27) shows that the mechanical advantage diminishes when the transmission angle is much less than a right angle. If the transmission angle becomes too small, the mechanical advantage becomes small and even a very small amount of friction will cause a mechanism to lock or jam. To avoid this, a common rule of thumb is that a four-bar linkage should not be used in a region where the transmission angle is less than, say, 45 or 50°. The best four-bar linkage, based on the quality of its force transmission, will have a transmission angle which deviates from 90° by the smallest amount.

In other mechanisms, e.g., meshing gear teeth or a cam-follower system, the *pressure angle* is used as an index of merit. The pressure angle is defined as the acute angle between the direction of the output force and the direction of the velocity of the point where the output force is applied. Pressure angles will be discussed more thoroughly in Chaps. 6 and 7. In the four-bar linkage, the pressure angle is the complement of the transmission angle.

Another index of merit which has been proposed‡ is the determinant of the coefficients of the simultaneous equations relating the dependent velocities of a mechanism. In Example 3-5, for example, we saw that the

† Compare this result with Eq. (3-19).

‡ J. Denavit et al., Velocity, Acceleration, and Static Force Analysis of Spatial Linkages, *J. Appl. Mech., ASME Trans.*, vol. 87, ser. E, no. 4, pp. 903–910, 1965.

dependent velocities of a four-bar linkage are related by the equations

$$R_{CB}\sin\theta_3\omega_3 - R_{CD}\sin\theta_4\omega_4 = -R_{BA}\sin\theta_2\omega_2$$

$$R_{CB}\cos\theta_3\omega_3 - R_{CD}\cos\theta_4\omega_4 = -R_{BA}\cos\theta_2\omega_2$$

The determinant of the coefficients is

$$\Delta = \begin{vmatrix} R_{CB}\sin\theta_3 & -R_{CD}\sin\theta_4 \\ R_{CB}\cos\theta_3 & -R_{CD}\cos\theta_4 \end{vmatrix} = R_{CB}R_{CD}\sin(\theta_4 - \theta_3)$$

As is obvious from Cramer's rule, the solutions for the dependent velocities, in this case ω_3 and ω_4, must include this determinant in the denominator. This is borne out in the solution of the four-bar linkage, Eqs. (3-18) and (3-19). Although the form of this determinant changes for different mechanisms, such a determinant can always be defined and always appears in the denominators of all dependent velocity solutions.

When this determinant becomes small, the mechanical advantage also becomes small and the usefulness of the mechanism is reduced in such regions. We have not yet seen it, but it is also true that this same determinant also appears in the denominator of the dependent accelerations and all other quantities which require taking derivatives of the loop-closure equation. If this determinant is small, the mechanism will function poorly in all respects—force transmission, motion transformation, sensitivity to manufacturing errors, and so on.

3-17 CENTRODES

We noted in Sec. 3-10 that the location of the instant center of velocity was defined only instantaneously and would change as the mechanism moves. If the locations of the instant centers are found for all possible phases of the mechanism, they describe curves or loci, called *centrodes.*† In Fig. 3-32 the instant center P_{13} is located at the intersection of the extensions of links 2 and 4. As the linkage is moved through all possible positions, P_{13} traces out the curve called the *fixed centrode* on link 1.

† Opinion seems about equally divided on whether these loci should be termed *centrodes* or *polodes*. Generally, those preferring the name *instant center* call them *centrodes* and those who use the word *pole* call them *polodes*. The name *roulettes* has also been applied. The three-dimensional equivalents are ruled surfaces and are referred to as *axodes*.

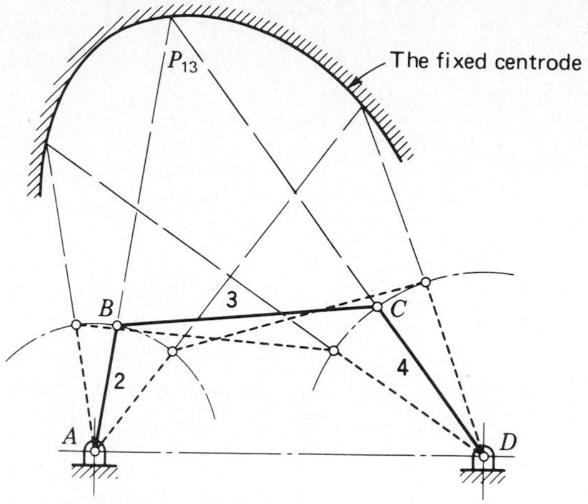

Figure 3-32 The fixed centrode.

Figure 3-33 shows the inversion of the same linkage in which link 3 is fixed and link 1 is movable. When this inversion is moved through all possible positions, P_{13} traces a ***different*** curve on link 3. For the original linkage, with link 1 fixed, this is the curve traced by P_{13} on the coordinate system of the moving link 3; it is called the *moving centrode.*

Figure 3-34 shows the moving centrode, attached to link 3, and the fixed centrode, attached to link 1. It is imagined here that links 1 and 3 have been machined to the actual shapes of the respective centrodes and that links 2 and 4 have been removed entirely. If the moving centrode is now permitted to roll on the fixed centrode without slip, link 3 will have exactly the same motion as it had in the original linkage. This remarkable property, which stems from the

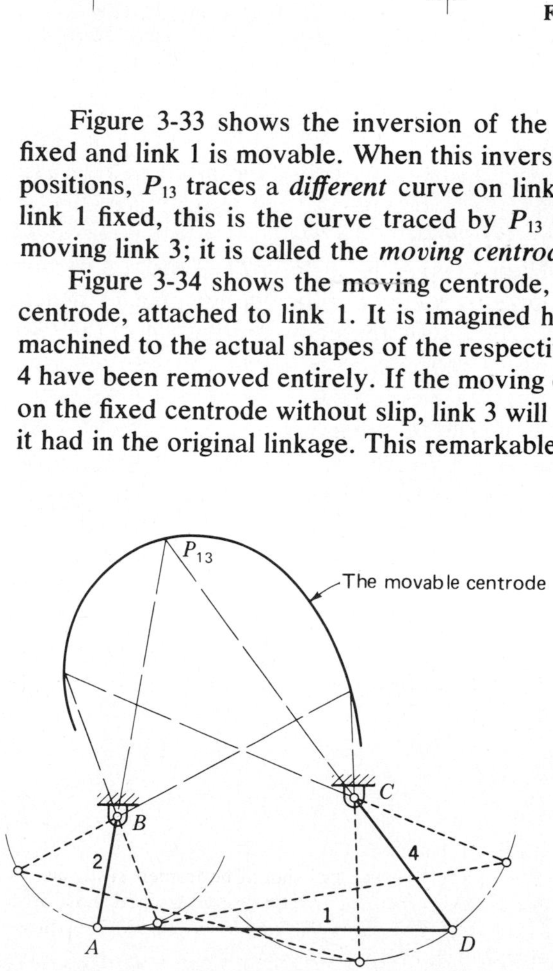

Figure 3-33 The moving centrode.

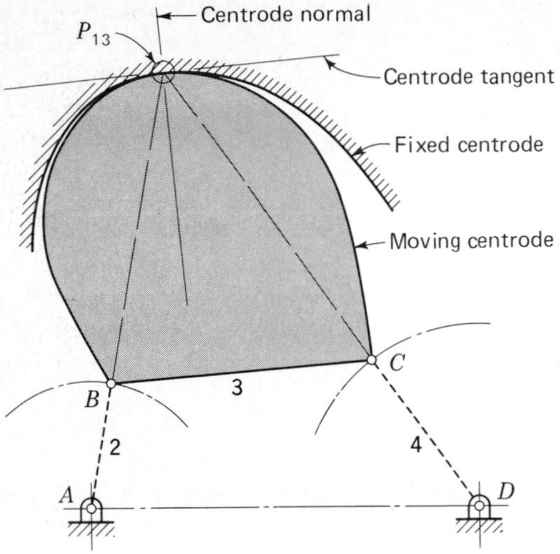

Figure 3-34 Rolling contact between centrodes.

fact that a point of rolling contact is an instant center, turns out to be quite useful in the synthesis of linkages.

We can restate this property as follows: *The plane motion of one rigid body relative to another is completely equivalent to the rolling motion of one centrode on the other.* The instantaneous point of rolling contact is the instant center, as shown in Fig. 3-34. Also shown are the common tangent to the two centrodes and the common normal, called the *centrode tangent* and the *centrode normal*; they are often used as the axes of a coordinate system for developing equations for a coupler curve or other properties of the motion.

The centrodes of Fig. 3-34 were generated by the instant center P_{13} on links 1 and 3. Another set of centrodes, both moving, is generated on links 2

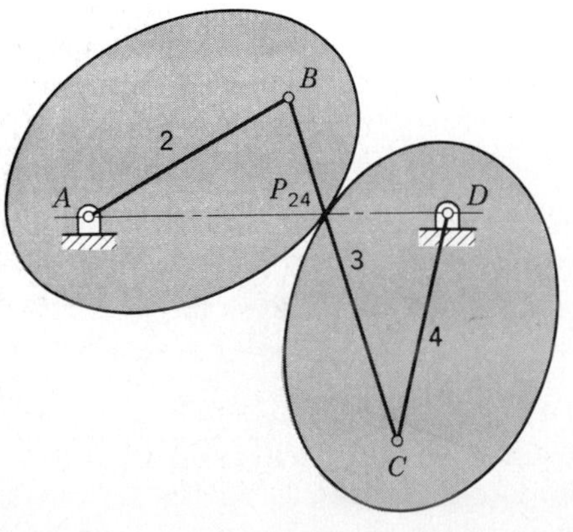

Figure 3-35

and 4 when instant center P_{24} is considered. Figure 3-35 shows these as two ellipses for the case of a crossed double-crank linkage with equal cranks. These two centrodes roll upon each other and describe the identical motion between links 2 and 4 which would result from the operation of the original four-bar linkage. This construction can be used as the basis for the development of a pair of elliptical gears.

PROBLEMS†

3-1 The position vector of a point is given by the equation $\mathbf{R} = 100e^{j\pi t}$, where R is in inches. Find the velocity of the point at $t = 0.40$ s.

3-2 The equation $\mathbf{R} = (t^2 + 4)e^{-j\pi t/10}$ defines the path of a particle. If R is in metres, find the velocity of the particle at $t = 20$ s.

3-3 If automobile A is traveling south at 55 mi/h and automobile B north 60° east at 40 mi/h, what is the velocity difference between B and A? What is the apparent velocity of B to the driver of A?

3-4 In the figure, wheel 2 rotates at 600 rpm and drives wheel 3 without slipping. Find the velocity difference between points B and A.

3-5 Two points A and B, located along the radius of a wheel (see figure), have speeds of 80 and 140 m/s, respectively. The distance between the points is $R_{BA} = 300$ mm.

(*a*) What is the diameter of the wheel?

(*b*) Find $\mathbf{V}_{AB}$, $\mathbf{V}_{BA}$, and the angular velocity of the wheel.

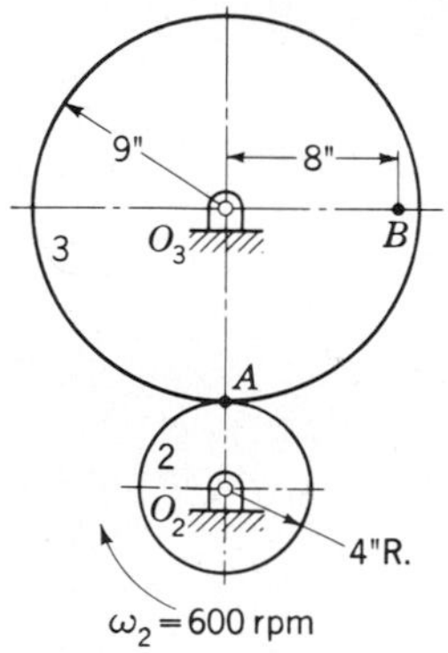

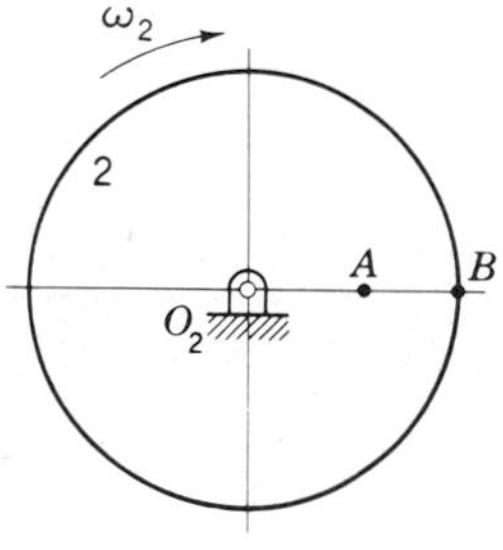

Problems 3-4 and 3-5

3-6 A plane leaves point B and flies east at 350 mi/h. Simultaneously at point A, 200 miles southeast (see figure), a plane leaves and flies northeast at 390 mi/h.

(*a*) How close will the planes come to each other if they fly at the same altitude?

(*b*) If they both leave at 6:00 P.M., at what time will this occur?

3-7 To the data of Prob. 3-6, add a wind of 30 mi/h from the west.

(*a*) If A flies the same heading, what is its new path?

(*b*) What change does the wind make in the results of Prob. 3-6?

† When assigning problems, the instructor may wish to specify the method of solution to be used, since a variety of approaches are presented in the text.

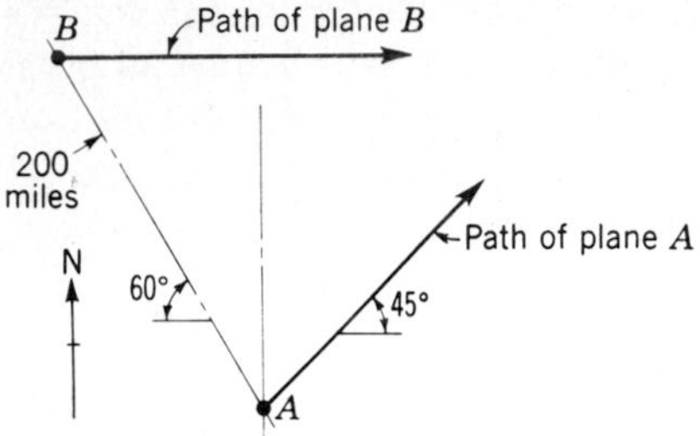

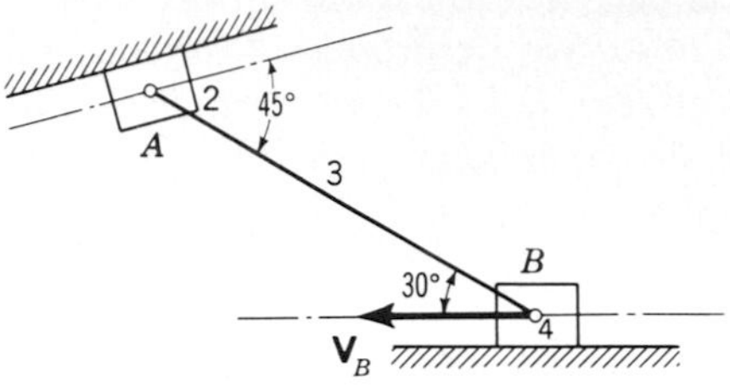

Problems 3-6 and 3-8 $R_{AB} = 400$ mm.

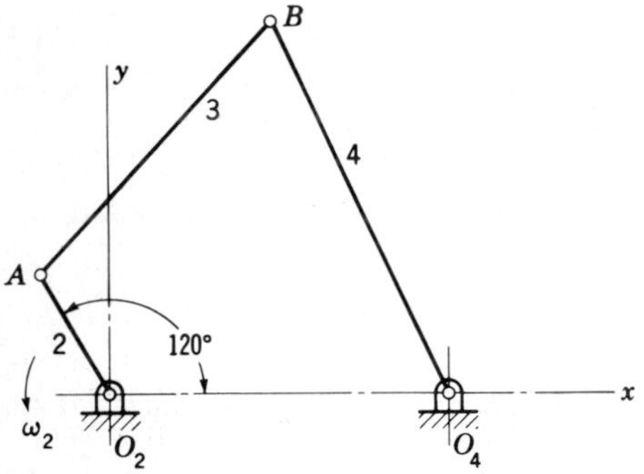

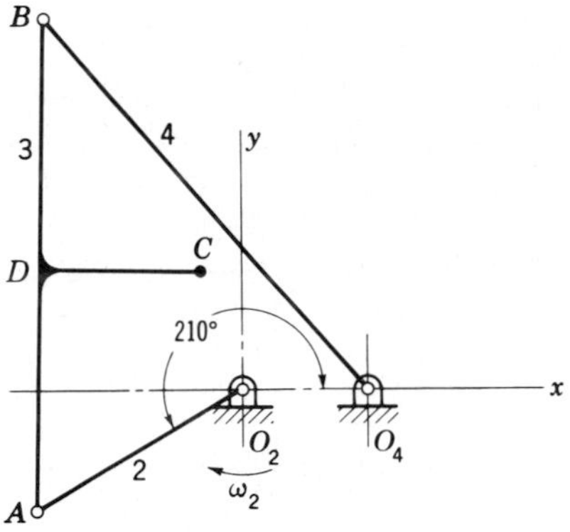

Problem 3-9 $R_{AO_2} = 4$ in, $R_{BA} = 10$ in, $R_{O_4O_2} = 10$ in, $R_{BO_4} = 12$ in. **Problem 3-10** $R_{AO_2} = 150$ mm, $R_{BA} = 300$ mm, $R_{O_4O_2} = 75$ mm, $R_{BO_4} = 300$ mm, $R_{DA} = 150$ mm, $R_{CD} = 100$ mm.

3-8 The velocity of point B of the linkage shown in the figure is 40 m/s. Find the velocity of point A and the angular velocity of link 3.

3-9 The mechanism shown in the figure is driven by link 2 at $\omega_2 = 45$ rad/s ccw. Find the angular velocities of links 3 and 4.

3-10 Crank 2 of the push-link mechanism shown in the figure is driven at $\omega_2 = 60$ rad/s cw. Find the velocities of points B and C and the angular velocities of links 3 and 4.

3-11 Find the velocity of point C on link 4 of the mechanism shown in the figure if crank 2 is driven at $\omega_2 = 48$ rad/s ccw. What is the angular velocity of link 3?

3-12 The figure illustrates a parallel-bar linkage, in which opposite links have equal lengths. For this linkage show that ω_3 is always zero and that $\omega_4 = \omega_2$. How would you describe the motion of link 4 relative to link 2?

3-13 The figure illustrates the antiparallel or crossed-bar linkage. If link 2 is driven at $\omega_2 =$ 1 rad/s ccw, find the velocities of points C and D.

3-14 Find the velocity of point C of the linkage shown in the figure, assuming that link 2 has an angular velocity of 60 rad/s ccw. Also find the angular velocities of links 3 and 4.

3-15 The inversion of the slider-crank mechanism shown is driven by link 2 at $\omega_2 = 60$ rad/s ccw. Find the velocity of point B and the angular velocities of links 3 and 4.

3-16 Find the velocity of the coupler point C and the angular velocities of links 3 and 4 of the mechanism shown if crank 2 has an angular velocity of 30 rad/s cw.

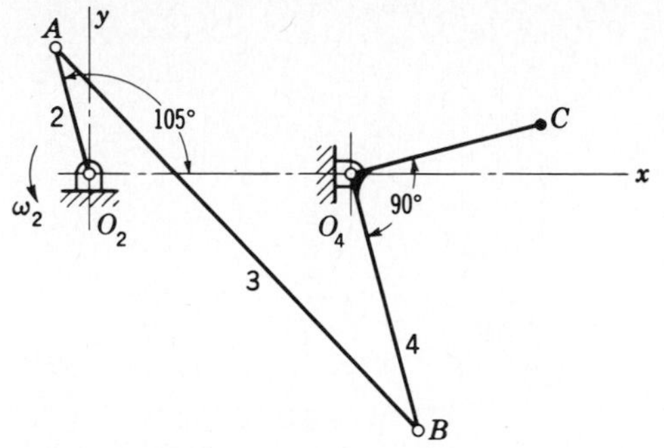

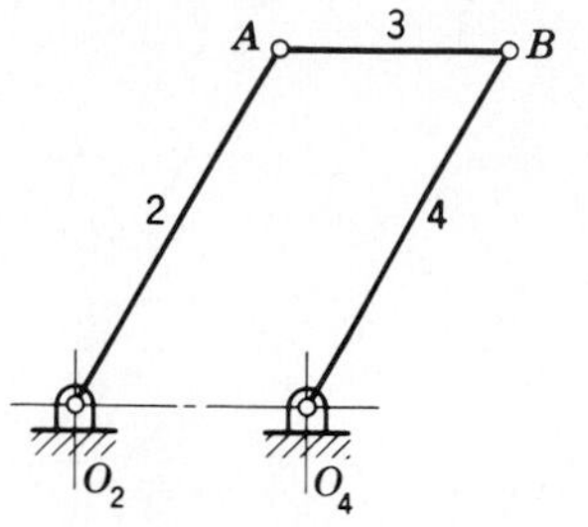

Problem 3-11 $R_{AO_2} = 8$ in, $R_{BA} = 32$ in, $R_{O_4O_2} = 16$ in, $R_{BO_4} = 16$ in, $R_{CO_4} = 12$ in. **Problem 3-12.**

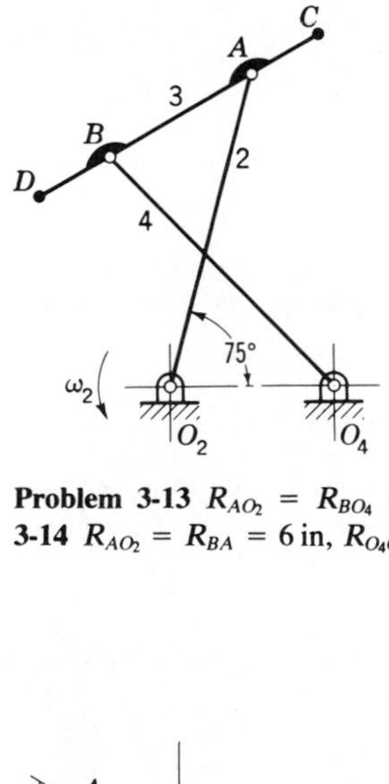

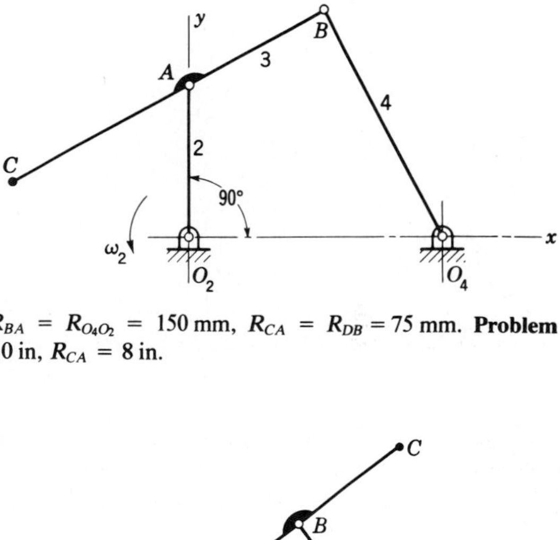

Problem 3-13 $R_{AO_2} = R_{BO_4} = 300$ mm, $R_{BA} = R_{O_4O_2} = 150$ mm, $R_{CA} = R_{DB} = 75$ mm. **Problem 3-14** $R_{AO_2} = R_{BA} = 6$ in, $R_{O_4O_2} = R_{BO_4} = 10$ in, $R_{CA} = 8$ in.

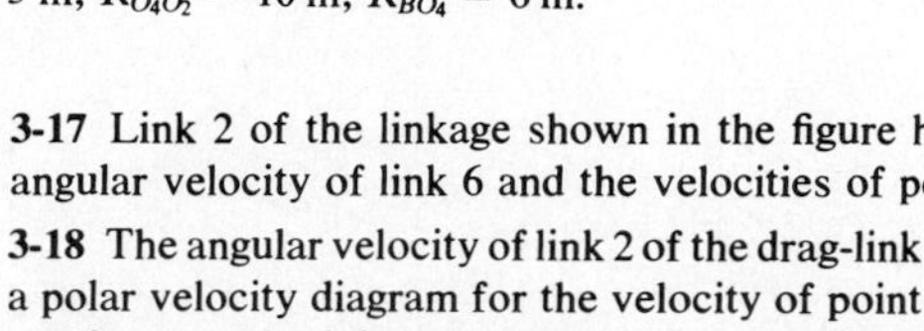

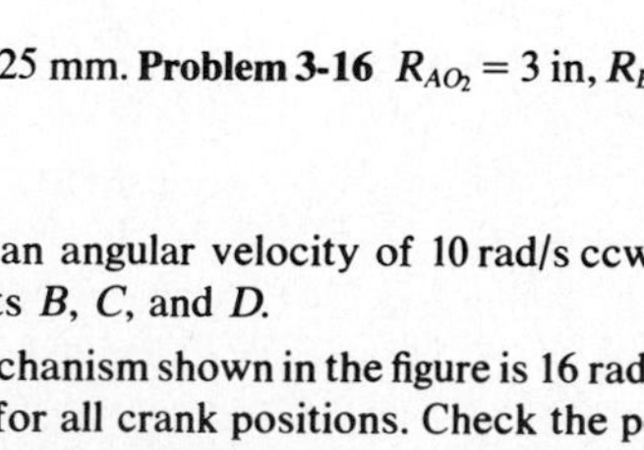

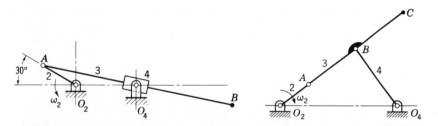

Problem 3-15 $R_{AO_2} = 75$ mm, $R_{BA} = 400$ mm, $R_{O_4O_2} = 125$ mm. **Problem 3-16** $R_{AO_2} = 3$ in, $R_{BA} = R_{CB} = 5$ in, $R_{O_4O_2} = 10$ in, $R_{BO_4} = 6$ in.

3-17 Link 2 of the linkage shown in the figure has an angular velocity of 10 rad/s ccw. Find the angular velocity of link 6 and the velocities of points *B*, *C*, and *D*.

3-18 The angular velocity of link 2 of the drag-link mechanism shown in the figure is 16 rad/s cw. Plot a polar velocity diagram for the velocity of point *B* for all crank positions. Check the positions of maximum and minimum velocities by using Freudenstein's theorem.

3-19 Link 2 of the mechanism shown in the figure is driven at $\omega_2 = 36$ rad/s cw. Find the angular velocity of link 3 and the velocity of point *B*.

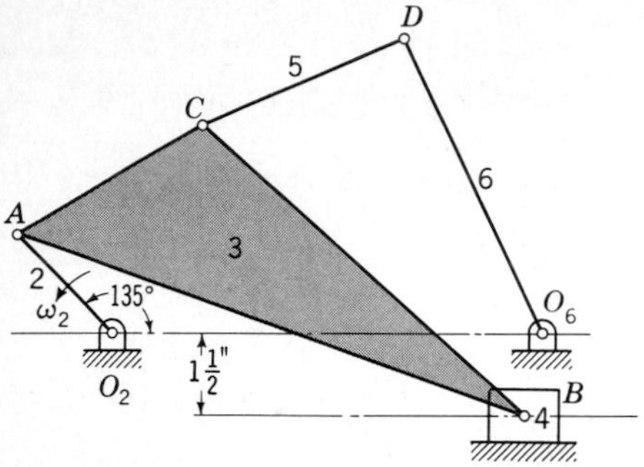

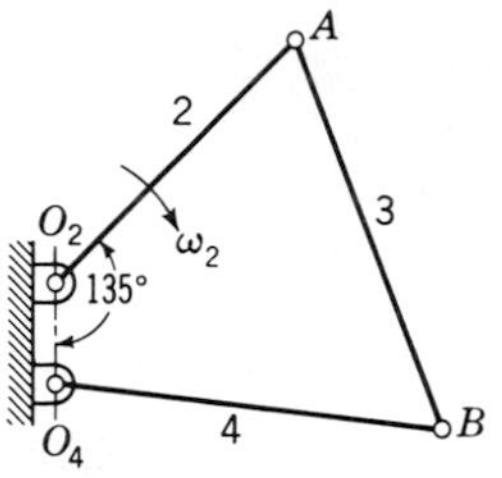

Problem 3-17 $R_{AO_2} = 2.5$ in, $R_{BA} = 10$ in, $R_{CB} = 8$ in, $R_{CA} = R_{DC} = 4$ in, $R_{O_6O_2} = 8$ in, $R_{DO_6} = 6$ in. **Problem 3-18** $R_{AO_2} = 350$ mm, $R_{BA} = 425$ mm, $R_{O_4O_2} = 100$ mm, $R_{BO_4} = 400$ mm.

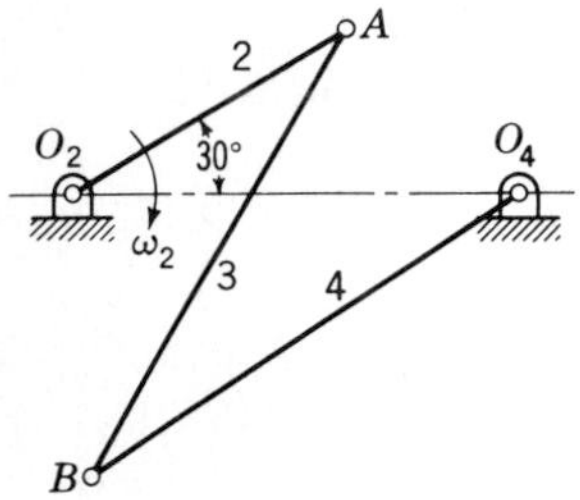

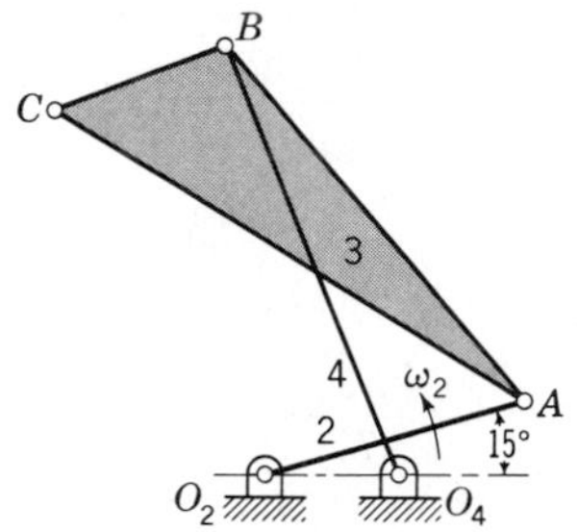

Problem 3-19 $R_{AO_2} = 5$ in, $R_{BA} = R_{BO_4} = 8$ in, $R_{O_4O_2} = 7$ in. **Problem 3-20** $R_{AO_2} = 150$ mm, $R_{BA} = R_{BO_4} = 250$ mm, $R_{O_4O_2} = 75$ mm, $R_{CA} = 300$ mm, $R_{CB} = 100$ mm.

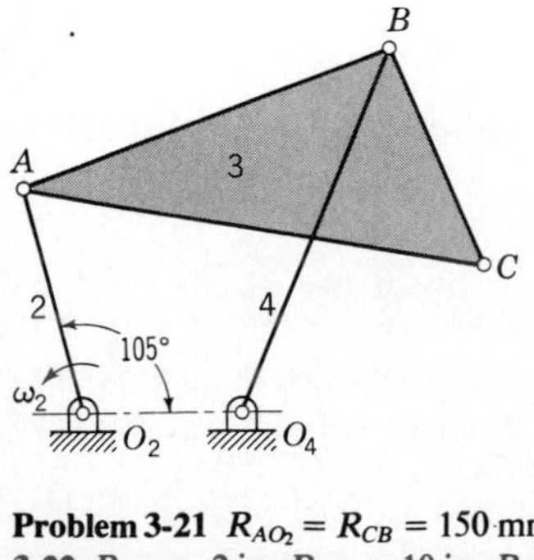

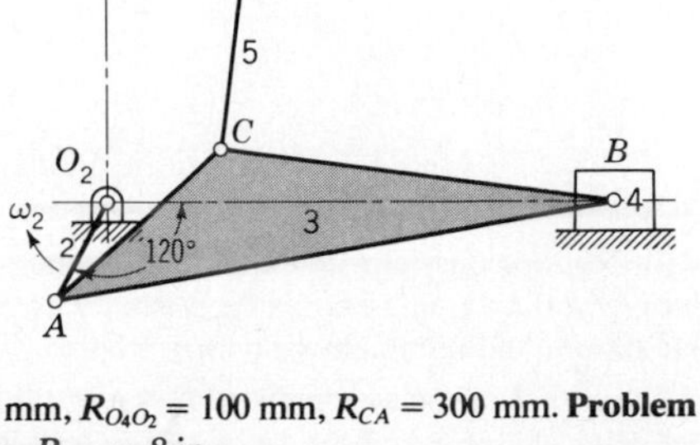

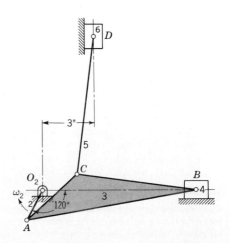

Problem 3-21 $R_{AO_2} = R_{CB} = 150$ mm, $R_{BA} = R_{BO_4} = 250$ mm, $R_{O_4O_2} = 100$ mm, $R_{CA} = 300$ mm. **Problem 3-22** $R_{AO_2} = 2$ in, $R_{BA} = 10$ in, $R_{CA} = 4$ in, $R_{CB} = 7$ in, $R_{DC} = 8$ in.

3-20 Find the velocity of point C and the angular velocity of link 3 of the push-link mechanism shown in the figure. Link 2 is the driver and rotates at 8 rad/s ccw.

3-21 Link 2 of the mechanism shown in the figure has an angular velocity of 56 rad/s ccw. Find $\mathbf{V}_C$.

3-22 Find the velocities of points B, C, and D of the double-slider mechanism shown in the figure if crank 2 rotates at 42 rad/s ccw.

3-23 The figure shows the mechanism used in a two-cylinder 60° V engine consisting, in part, of an articulated connecting rod. Crank 2 rotates at 2000 rpm cw. Find the velocities of points B, C, and D.

3-24 Make a complete velocity analysis of the linkage shown in the figure, given that $\omega_2 =$ 24 rad/s cw. What is the absolute velocity of point B? What is its apparent velocity to an observer moving with link 4?

3-25 Find $\mathbf{V}_B$ for the linkage shown in the figure if $V_A = 1$ ft/s.

3-26 The figure shows a variation of the Scotch-yoke mechanism. It is driven by crank 2 at $\omega_2 = 36$ rad/s ccw. Find the velocity of the crosshead, link 4.

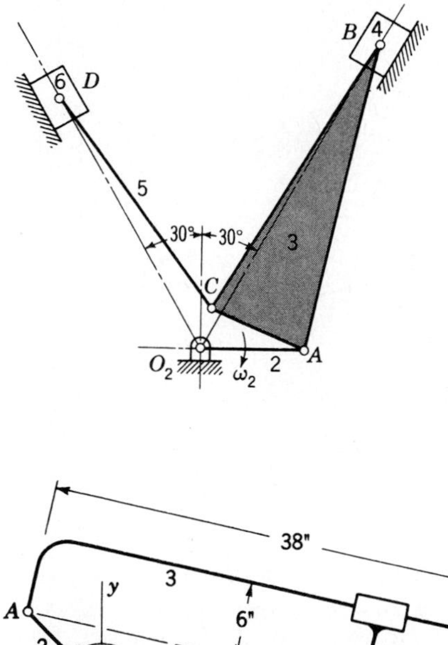

Problem 3-23 $R_{AO_2} = 2$ in, $R_{BA} = R_{CB} = 6$ in, $R_{CA} = 2$ in, $R_{DC} = 5$ in.

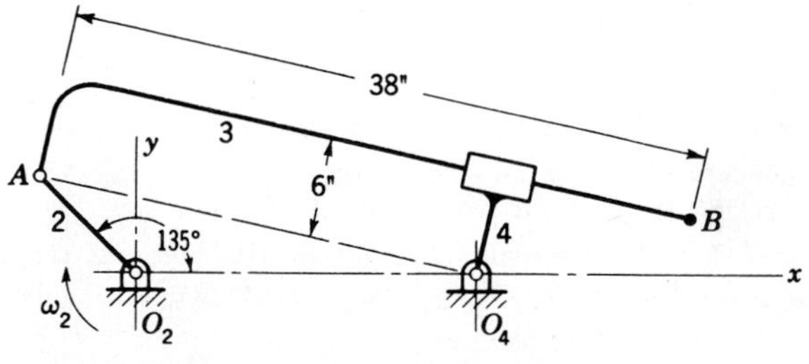

Problem 3-24 $R_{AO_2} = 8$ in, $R_{O_4O_2} = 20$ in.

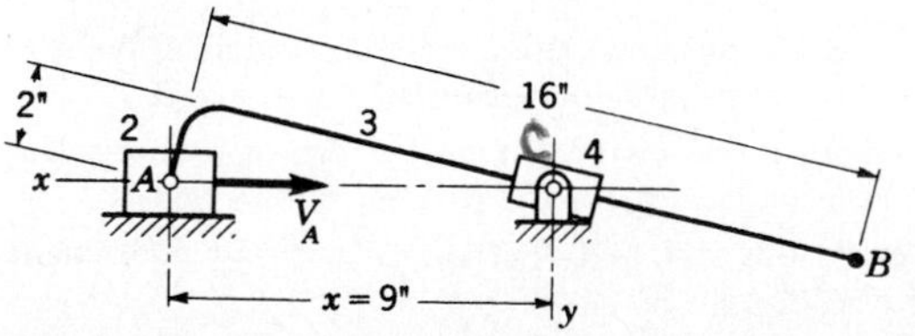

Problem 3-25

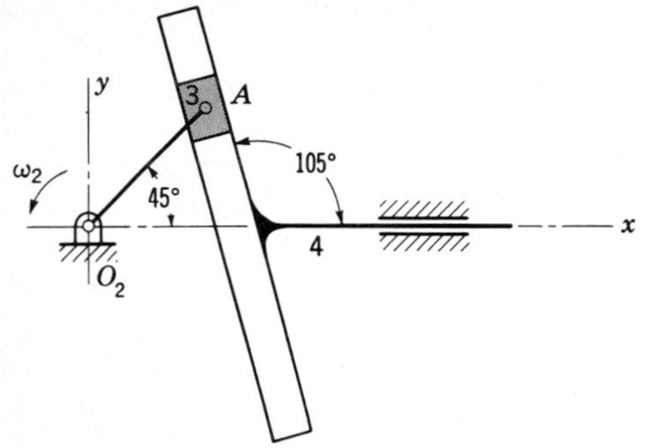

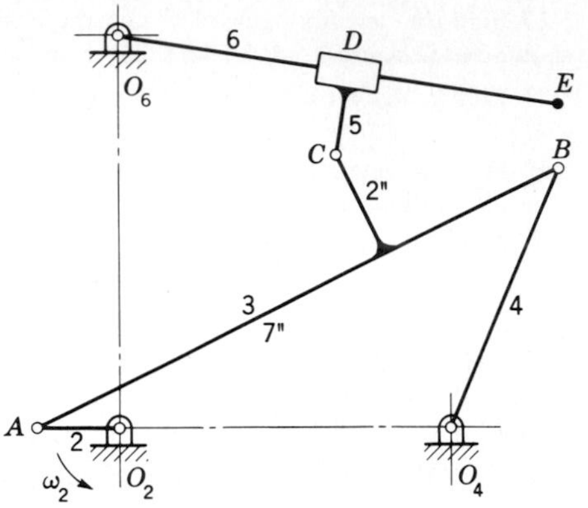

Problem 3-26 $R_{AO_2} = 250$ mm. **Problem 3-27** $R_{AO_2} = R_{DC} = 1.5$ in, $R_{BA} = 10.5$ in, $R_{O_4O_2} = 6$ in, $R_{BO_4} = 5$ in, $R_{O_6O_2} = 7$ in, $R_{EO_6} = 8$ in.

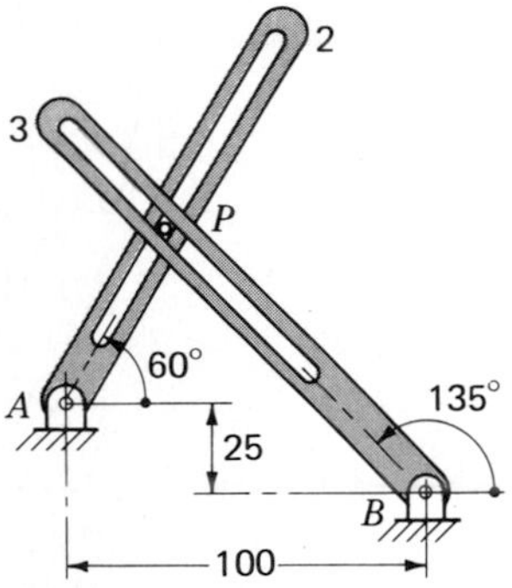

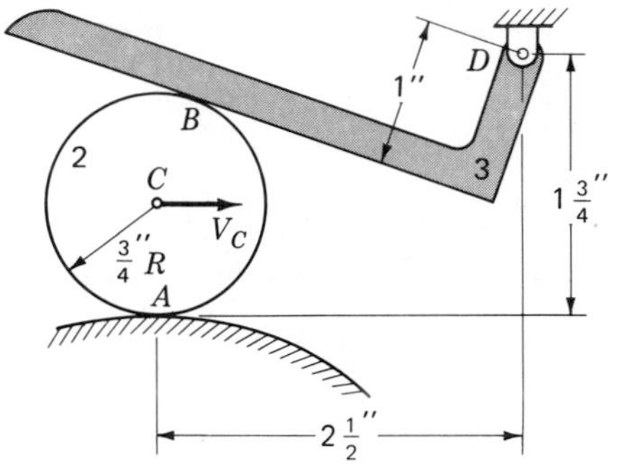

Problem 3-28 Dimensions in millimetres. **Problem 3-29.**

3-27 Make a complete velocity analysis of the linkage shown in the figure for $\omega_2 = 72$ rad/s ccw.

3-28 Slotted links 2 and 3 are driven independently at $\omega_2 = 30$ rad/s cw and $\omega_3 = 20$ rad/s cw, respectively. Find the absolute velocity of the center of the pin P_4 carried in the two slots.

3-29 The mechanism shown is driven such that $V_C = 10$ in/s to the right. Rolling contact is assumed between links 1 and 2, but slip is possible between links 2 and 3. Determine the angular velocity of link 3.

3-30 The circular cam shown is driven at an angular velocity of $\omega_2 = 15$ rad/s cw. There is rolling contact between the cam and the roller, link 3. Find the angular velocity of the oscillating follower, link 4.

3-31 The mechanism shown is driven by link 2 at 10 rad/s ccw. There is rolling contact at point F. Determine the velocity of points E and G and the angular velocities of links 3, 4, 5 and 6.

3-32 The figure shows the schematic diagram for a two-piston pump. It is driven by a circular eccentric, link 2, at $\omega_2 = 25$ rad/s ccw. Find the velocities of the two pistons, links 6 and 7.

3-33 The epicyclic gear train shown is driven by the arm, link 2, at $\omega_2 = 10$ rad/s cw. Determine the angular velocity of the output shaft, attached to gear 3.

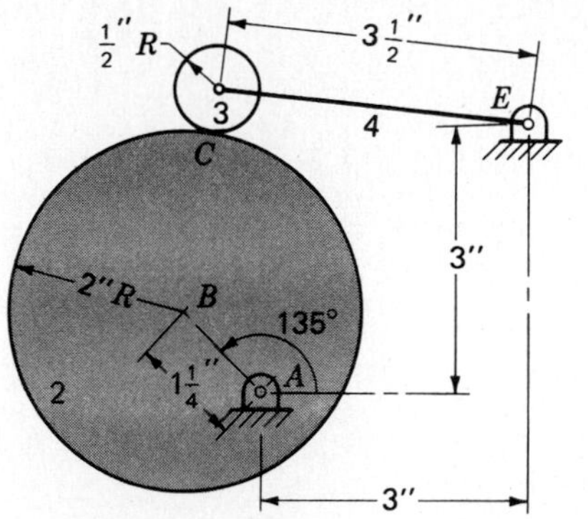

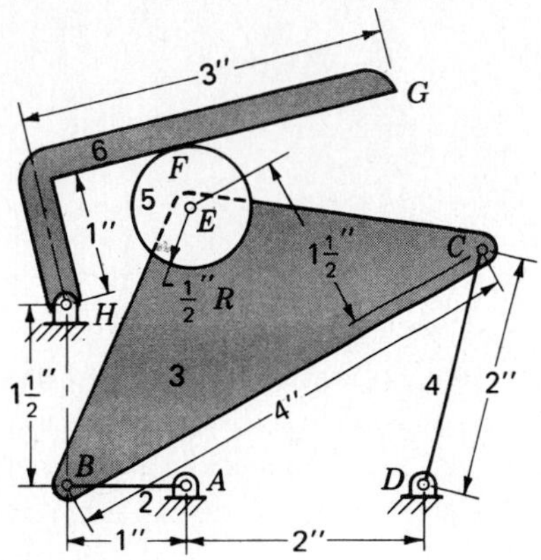

Problems 3-30 and 3-31.

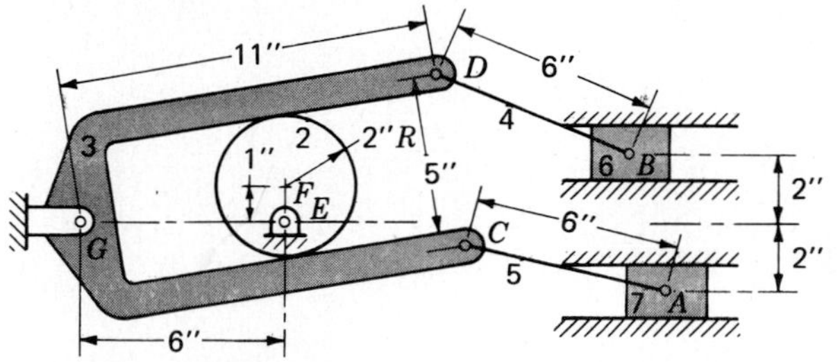

Problem 3-32

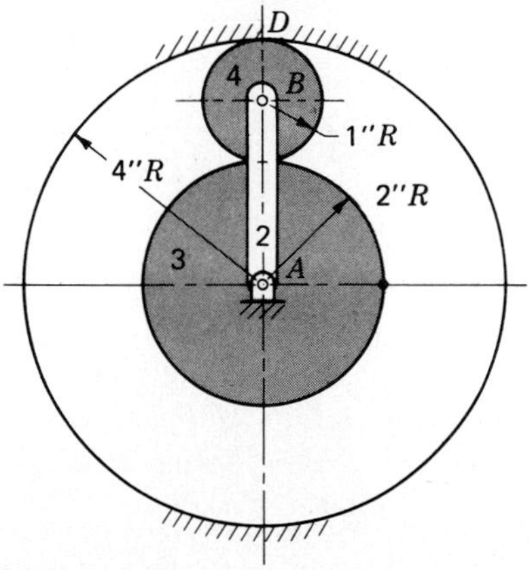

Problem 3-33

3-34 The diagram shows a planar schematic approximation of an automotive front suspension. The *roll center* is the term used by the industry to describe the point about which the auto body seems to rotate relative to the ground. The assumption is made that there is pivoting but no slip between the tires and the road. After making a sketch, use the concepts of instant centers to find a technique to locate the roll center.

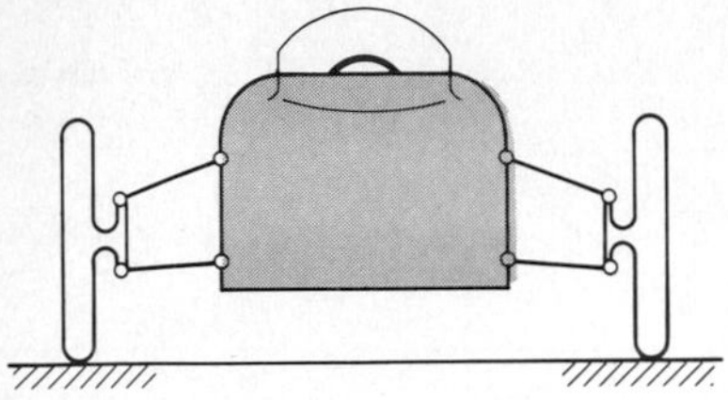

Problem 3-34

3-35 Locate all instant centers for the linkage of Prob. 3-22.

3-36 Find all instant centers for the mechanism of Prob. 3-25.

3-37 Locate all instant centers for the mechanism of Prob. 3-26.

3-38 Locate all instant centers for the mechanism of Prob. 3-27.

3-39 Find all instant centers for the mechanism of Prob. 3-29.

3-40 Find all instant centers for the mechanism of Prob. 3-30.

CHAPTER
FOUR
ACCELERATION

4-1 DEFINITION OF ACCELERATION

In Fig. 4-1*a* a moving point is first observed at location P, where it has a velocity of $\mathbf{V}_P$. After a short time interval Δt, the point is observed to have moved along some path to a new location P', and its velocity has changed to $\mathbf{V}'_P$, which may differ from $\mathbf{V}_P$ in both magnitude and direction. We can evaluate the change in velocity $\mathbf{\Delta V}_P$ as shown in Fig. 4-1*b*

$$\mathbf{\Delta V}_P = \mathbf{V}'_P - \mathbf{V}_P$$

The *average acceleration* of the point P during the time interval is $\mathbf{\Delta V}_P/\Delta t$. The *instantaneous acceleration* (hereafter called simply the *acceleration*) of point P is defined as the time rate of change of its velocity, i.e., the limit of the average acceleration for an infinitesimally small time interval

$$\mathbf{A}_P = \lim_{\Delta t \to 0} \frac{\mathbf{\Delta V}_P}{\Delta t} = \frac{d\mathbf{V}_P}{dt} = \frac{d^2\mathbf{R}_P}{dt^2} \tag{4-1}$$

Since velocity is a vector quantity, $\mathbf{\Delta V}_P$ and the acceleration $\mathbf{A}_P$ are also vector quantities and have both magnitude and direction. Also, like velocity, the acceleration vector is properly defined only for a point; the term should not be applied to a line, coordinate system, volume, or other collection of points since the accelerations of the several points involved may differ.

Like velocity, the acceleration of a moving point will appear differently to different observers. Acceleration does not depend on the actual location of the observer but does depend critically on the observer's motion or, more precisely, on the motion of the observer's coordinate system. If the ac-

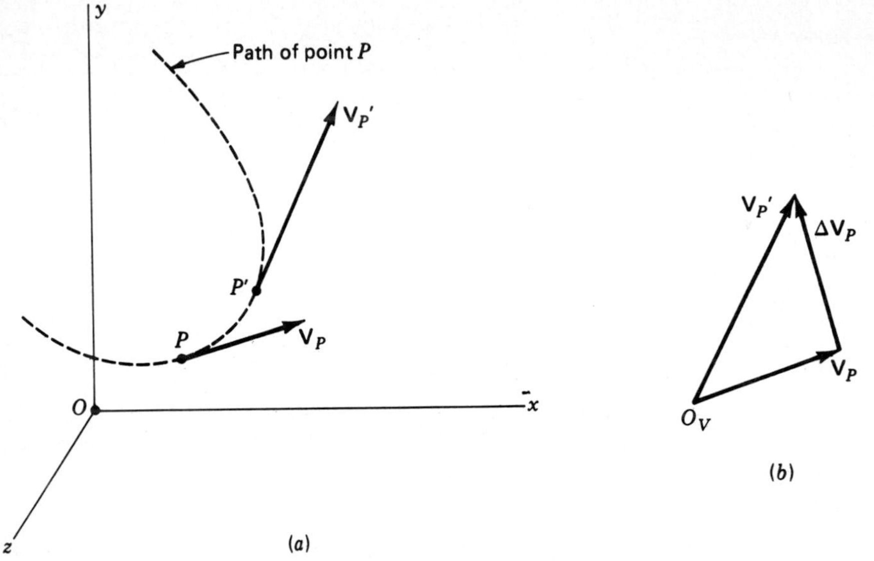

Figure 4-1 Change in velocity of a moving point.

celeration is sensed by an observer in the absolute coordinate system, it is referred to as an *absolute acceleration* and is denoted by the symbol $\mathbf{A}_{P/1}$ or simply $\mathbf{A}_P$, which is consistent with the notation used for position, displacement, and velocity.

4-2 ANGULAR ACCELERATION OF A RIGID BODY

In Fig. 4-2 we consider the motion of a rigid body. Two points of the body P and Q first undergo small displacements during a short time interval Δt and arrive at new locations P' and Q'. Next, during a second short time interval, they undergo further small displacements to reach locations P'' and Q''. We recall (Sec. 3-3) that we used such displacements to derive the velocity-difference vector $\mathbf{V}_{PQ}$ and to define the angular-velocity vector $\boldsymbol{\omega}$ of the moving body. In doing so, we took the point of view of an observer in a moving coordinate system whose origin travels with point Q but whose axes remain parallel to the absolute coordinate axes. Such an observer, we recall, sees only the rotation of the body about point Q and, as we saw in Fig. 3-4, the position-difference vector $\mathbf{R}_{PQ}$ seems to him to describe a cone whose axis defines the direction of $\boldsymbol{\omega}$.

In developing formulas for acceleration we now wish to extend this point of view for two successive time intervals. In Fig. 4-3a we take the view of the same moving but nonrotating observer. During the first time interval, the body

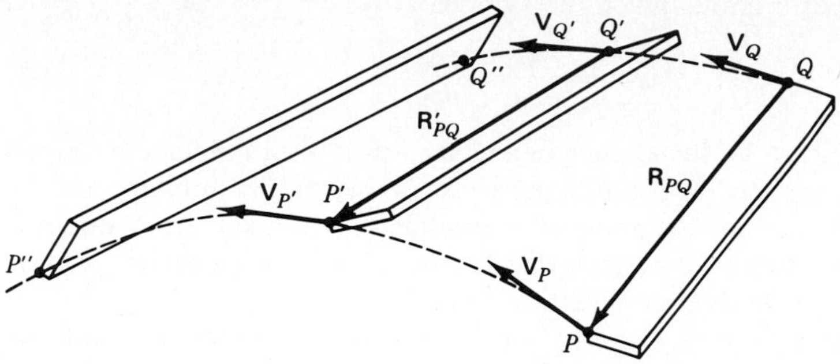

Figure 4-2

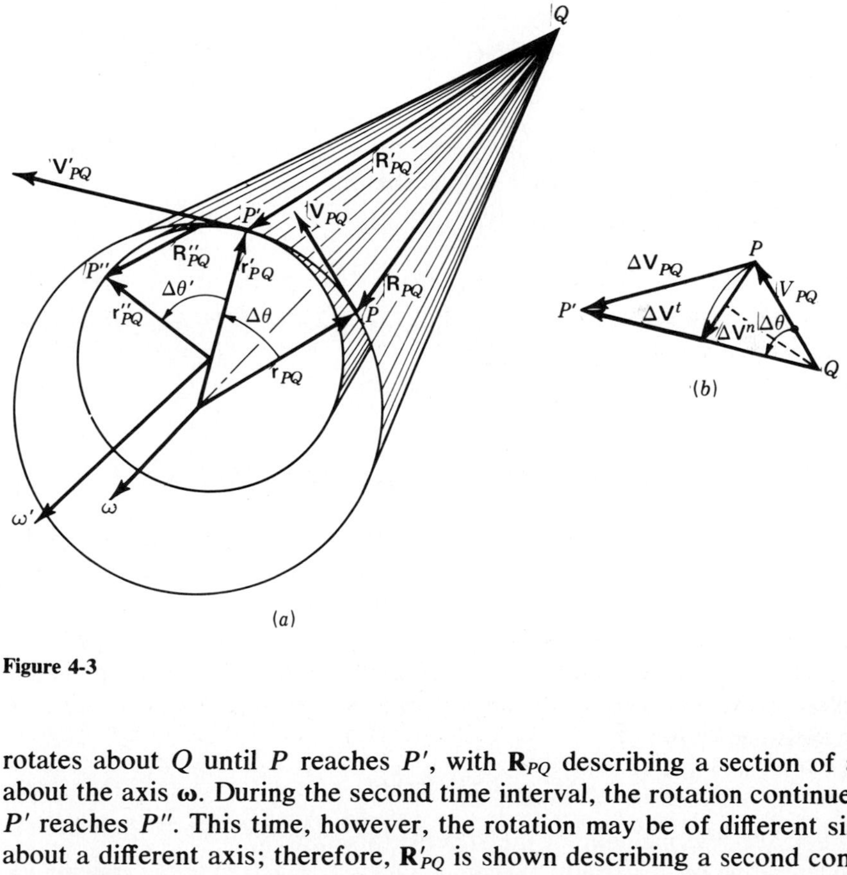

Figure 4-3

rotates about Q until P reaches P', with $\mathbf{R}_{PQ}$ describing a section of a cone about the axis $\boldsymbol{\omega}$. During the second time interval, the rotation continues until P' reaches P''. This time, however, the rotation may be of different size and about a different axis; therefore, $\mathbf{R}'_{PQ}$ is shown describing a second cone with the modified axis $\boldsymbol{\omega}'$. The change in angular velocity of the body is given by

$$\Delta\boldsymbol{\omega} = \boldsymbol{\omega}' - \boldsymbol{\omega}$$

The angular acceleration of the body is defined as the time rate of change of its angular velocity and is given the symbol $\boldsymbol{\alpha}$

$$\boldsymbol{\alpha} = \lim_{\Delta t \to 0} \frac{\Delta \boldsymbol{\omega}}{\Delta t} = \frac{d\boldsymbol{\omega}}{dt} \tag{4-2}$$

As we see from Fig. 4-3*a*, the change in angular velocity can include a change in magnitude (if the speed of rotation increases or decreases) and/or a change in direction (if the axis of the rotation is modified). Like $\Delta\boldsymbol{\omega}$, from which it comes, there is no reason to believe that $\boldsymbol{\alpha}$ has a direction along either $\boldsymbol{\omega}$ or $\boldsymbol{\omega}'$; it may have an entirely new direction.

Like the angular-velocity vector $\boldsymbol{\omega}$, the angular-acceleration vector $\boldsymbol{\alpha}$ applies to the absolute rotation of the entire rigid body and is often subscripted by the number of the coordinate system of the moving body, for example, $\boldsymbol{\alpha}_2$ or $\boldsymbol{\alpha}_{2/1}$.

4-3 ACCELERATION DIFFERENCE BETWEEN POINTS OF A RIGID BODY

Continuing with Fig. 4-3*a*, we can write the velocity-difference equation coming from each of the successive displacements

$$\mathbf{V}_{PQ} = \mathbf{V}_P - \mathbf{V}_Q = \boldsymbol{\omega} \times \mathbf{R}_{PQ} \tag{a}$$

and

$$\mathbf{V}'_{PQ} = \mathbf{V}'_P - \mathbf{V}'_Q = \boldsymbol{\omega}' \times \mathbf{R}'_{PQ} \tag{b}$$

The two velocity-difference vectors are shown tangent to the respective cones at P and P'.

Subtracting Eq. (*a*) from Eq. (*b*), we get

$$\Delta\mathbf{V}_{PQ} = \mathbf{V}'_{PQ} - \mathbf{V}_{PQ} \tag{c}$$

$$= \Delta\mathbf{V}_P - \Delta\mathbf{V}_Q \tag{d}$$

Figure 4-3*b* shows the graphical subtraction of Eq. (*c*) as its outside boundary. It will be noticed that $\mathbf{V}'_{PQ}$ and $\mathbf{V}_{PQ}$ differ in direction by $\Delta\theta$ since, according to Eqs. (*b*) and (*a*), they are perpendicular to the cone radii $\mathbf{r}'_{PQ}$ and $\mathbf{r}_{PQ}$, respectively. The magnitudes V'_{PQ} and V_{PQ} are not necessarily equal.

To aid in the evaluation of $\Delta\mathbf{V}_{PQ}$, we next divide it into two components, $\Delta\mathbf{V}^n$, taken as the chord of a circular arc with center Q and radius $\mathbf{V}_{PQ}$, and $\Delta\mathbf{V}^t$, taken along $\mathbf{V}'_{PQ}$

$$\Delta\mathbf{V}_{PQ} = \Delta\mathbf{V}^n + \Delta\mathbf{V}^t \tag{e}$$

We will discover the meaning of the superscripts shortly.

Concentrating on $\Delta\mathbf{V}^n$ only, we can evaluate its magnitude by drawing its perpendicular bisector through Q. Then

$$\Delta V^n = 2V_{PQ} \sin\frac{\Delta\theta}{2}$$

If we assume that the time interval Δt (and therefore the angular displacement) is small, the sine of the small angle can be approximated by the angle itself

$$\Delta V^n \approx 2V_{PQ}\frac{\Delta\theta}{2} = \Delta\theta\, V_{PQ}$$

We can now divide by Δt and take the limit, thus defining what is called the *normal component* of the *acceleration difference.* It is given the symbol $\mathbf{A}^n_{PQ}$

$$A^n_{PQ} = \lim_{\Delta t\to 0}\frac{\Delta V^n}{\Delta t} = \lim_{\Delta t\to 0}\left(\frac{\Delta\theta}{\Delta t}V_{PQ}\right)$$

Recalling the definition of angular velocity, this becomes

$$A^n_{PQ} = \omega V_{PQ}$$

Also, in the limit, we notice from Fig. 4-3*b* that the chord $\Delta\mathbf{V}^n$ becomes perpendicular to $\mathbf{V}_{PQ}$. Therefore, the vector attributes can be restored to the equation by writing

$$\mathbf{A}^n_{PQ} = \boldsymbol{\omega}\times\mathbf{V}_{PQ}$$

Recalling Eq. (3-3) for the velocity difference vector, we can write this in the form

$$\mathbf{A}^n_{PQ} = \boldsymbol{\omega}\times(\boldsymbol{\omega}\times\mathbf{R}_{PQ}) \tag{4-3}$$

If the body containing the two points P and Q is in planar motion, other useful forms can be found from Eqs. (4-3) and (3-3) for evaluating $\mathbf{A}^n_{PQ}$

$$\mathbf{A}^n_{PQ} = -\omega^2\mathbf{R}_{PQ} \tag{4-4}$$

$$A^n_{PQ} = \frac{V^2_{PQ}}{R_{PQ}} \tag{4-5}$$

We next turn our attention to $\Delta\mathbf{V}^t$, the other term of Eq. (*e*). Because $\Delta\mathbf{V}^n$ is the chord of a circular arc, the magnitude of $\Delta\mathbf{V}^t$ can be evaluated as

$$\Delta V^t = V'_{PQ} - V_{PQ} = |\boldsymbol{\omega}'\times\mathbf{R}'_{PQ}| - |\boldsymbol{\omega}\times\mathbf{R}_{PQ}| = \omega' r'_{PQ} - \omega r_{PQ}$$

Next we divide by Δt and take the limit, thus defining the *tangential component* of the *acceleration difference* $\mathbf{A}^t_{PQ}$

$$A^t_{PQ} = \lim_{\Delta t\to 0}\frac{\Delta V^t}{\Delta t} = \lim_{\Delta t\to 0}\frac{\omega' r'_{PQ} - \omega r_{PQ}}{\Delta t} = \lim_{\Delta t\to 0}\left(\frac{\Delta\omega}{\Delta t}r_{PQ}\right)$$

We see that, in the limit, the directions of $\Delta\mathbf{V}^t$, $\mathbf{V}_{PQ}$, and $\mathbf{V}'_{PQ}$ all approach the tangent to the cone at P. Therefore, we can restore the vector properties of this equation as follows

$$\mathbf{A}^t_{PQ} = \lim_{\Delta t\to 0}\left(\frac{\Delta\boldsymbol{\omega}}{\Delta t}\times\mathbf{R}_{PQ}\right)$$

or, remembering Eq. (4-2),

$$\mathbf{A}^t_{PQ} = \boldsymbol{\alpha}\times\mathbf{R}_{PQ} \tag{4-6}$$

Now, having looked at the separate components, we divide Eq. (*e*) by Δt, take the limit, and define the *acceleration-difference vector* between two points P and Q of a rigid body

$$\mathbf{A}_{PQ} = \lim_{\Delta t \to 0} \frac{\Delta \mathbf{V}_{PQ}}{\Delta t} = \frac{d\mathbf{V}_{PQ}}{dt} = \mathbf{A}^n_{PQ} + \mathbf{A}^t_{PQ} \tag{4-7}$$

When the same limit is formed from Eq. (*d*), we obtain the *acceleration-difference equation*

$$\mathbf{A}_{PQ} = \mathbf{A}_P - \mathbf{A}_Q$$

or

$$\mathbf{A}_P = \mathbf{A}_Q + \mathbf{A}_{PQ} \tag{4-8}$$

This important equation is one of the primary bases for acceleration analysis. It allows the acceleration of a point P to be found from the acceleration of any other point Q *of the same rigid body* and the acceleration difference between the points. According to Eq. (4-7), the acceleration difference consists of two components which can be evaluated from Eqs. (4-3) and (4-6) if the rotational-motion properties $\boldsymbol{\omega}$ and $\boldsymbol{\alpha}$ of the body are known.

The directions of the acceleration-difference components are shown in Fig. 4-4, which again shows the conical motion which would be seen for $\mathbf{R}_{PQ}$ by an observer in a coordinate system translating with point Q. Both components lie in the plane defined by the base of the cone. The n and t superscripts refer to components which are *normal* and *tangent* to the circle of the base of the cone. The normal component $\mathbf{A}^n_{PQ}$ is always toward the

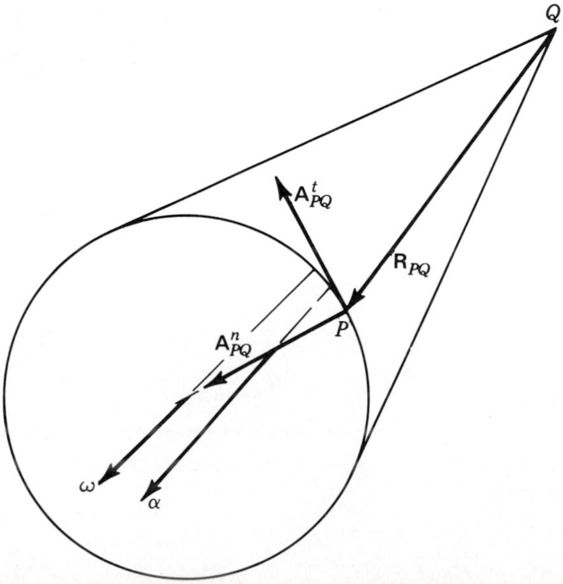

Figure 4-4

center of this circle; the direction of $\mathbf{A}^t_{PQ}$ is always tangent to this circle, but its sense depends on the sense of $\boldsymbol{\alpha}$.

Again we emphasize that $\boldsymbol{\alpha}$ and $\boldsymbol{\omega}$ generally do not have the same direction in three-dimensional space.

The acceleration-difference equation can be solved by means very similar to those employed in Chap. 3 for the velocity-difference equation.

4-4 GRAPHICAL ACCELERATION ANALYSIS; ACCELERATION POLYGONS

As in velocity analysis, the graphical approach provides a powerful and easily used method of analyzing accelerations in two-dimensional mechanisms.

As a first example of graphical acceleration analysis, consider the motion of the unconstrained link shown in Fig. 4-5*a* with velocities as shown in the velocity polygon, Fig. 4-5*b*. Suppose we are given the acceleration of two points, A and B, and wish to determine the acceleration of point C and the angular acceleration of the link (it will be noticed that this is a continuation of Sec. 3-4, Fig. 3-6). In general, it is desirable to draw the figure to scale and solve for all important velocities before beginning an acceleration analysis.

We next consider the acceleration-difference equation (4-8)

$$\overset{\surd\surd}{\mathbf{A}_B} = \overset{\surd\surd}{\mathbf{A}_A} + \overset{\circ\circ}{\mathbf{A}_{BA}} \tag{a}$$

The graphical solution of this equation for $\mathbf{A}_{BA}$ is shown in Fig. 4-5*c*. In obtaining it, it is necessary to choose a scale for the graphical representation of acceleration vectors; a starting point O_A is also chosen. The vectors $\mathbf{A}_A$ and $\mathbf{A}_B$ are plotted to the chosen scale, both originating from O_A and terminating at points A and B since they are absolute accelerations. According to Eq. (*a*), the vector spanning their termini is now identified as the acceleration difference $\mathbf{A}_{BA}$ and, within graphical accuracy, gives a correct representation of both the magnitude and direction.

The direction of $\mathbf{R}_{BA}$ is known from the diagram of the link, Fig. 4-5*a*. Based on this direction, the vector $\mathbf{A}_{BA}$ is now divided into its normal and tangential components

$$\mathbf{A}_{BA} = \mathbf{A}^n_{BA} + \mathbf{A}^t_{BA} \tag{b}$$

These are shown in Fig. 4-5*c* and again on the drawing of the link in Fig. 4-5*d*, where their directions can be better visualized.

The angular acceleration can be found by scaling the magnitude of the tangential component of the acceleration difference and the distance between the points and using Eq. (4-6). For planar motion the $\boldsymbol{\alpha}$ vector is perpendicular to the plane of motion, and its magnitude is given by

$$\alpha = \frac{A^t_{BA}}{R_{BA}} \tag{c}$$

Its sense can be determined visually from Fig. 4-5*d*. Taking the point of view of a nonrotating observer traveling with point *A*, the tangential component $\mathbf{A}^t_{BA}$ can be visualized as indicating a rotation of the link about point *A* in the direction of $\boldsymbol{\alpha}$, in this case clockwise. Notice that if $\mathbf{A}_{AB}$ had been found instead of $\mathbf{A}_{BA}$, the sense of $\mathbf{A}^t_{AB}$ would have been opposite to the sense of $\mathbf{A}^t_{BA}$. However, it would be visualized as indicating a rotation of the link about point *B*; thus the sense of $\boldsymbol{\alpha}$ would still have been found to be clockwise.

Now that $\boldsymbol{\alpha}$ has been found, the absolute acceleration of point *C* can be determined by relating it to points *A* and *B* by acceleration-difference equations

$$\overset{\circ\circ}{\mathbf{A}_C} = \overset{\surd\surd}{\mathbf{A}_A} + \overset{\surd\surd}{\mathbf{A}^n_{CA}} + \overset{\circ\surd}{\mathbf{A}^t_{CA}} = \overset{\surd\surd}{\mathbf{A}_B} + \overset{\surd\surd}{\mathbf{A}^n_{CB}} + \overset{\circ\surd}{\mathbf{A}^t_{CB}} \qquad (d)$$

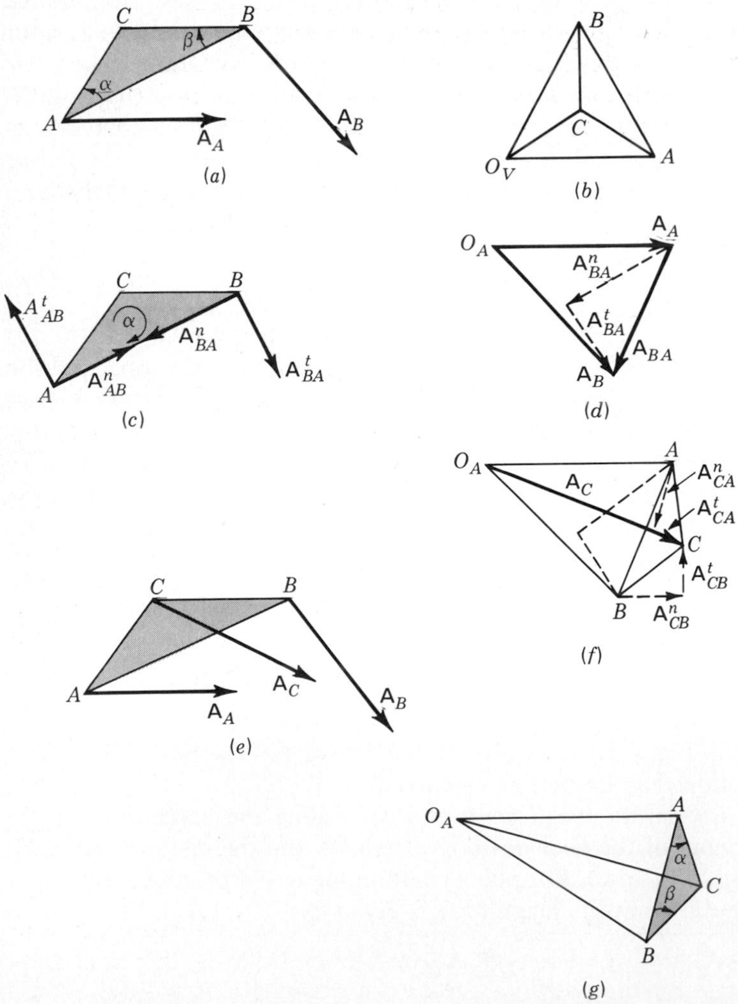

Figure 4-5

Since points A, B, and C are all on the same link, the normal components $\mathbf{A}^n_{CA}$ and $\mathbf{A}^n_{CB}$ are each of the form $-\omega^2\mathbf{R}$ [Eq. (4-4)]. Since $\boldsymbol{\omega}$ is known (or found from $\mathbf{V}_{BA}$), the two magnitudes can be calculated as $\omega^2 R$ using R_{CA} and R_{CB}, respectively. These are then added graphically to $\mathbf{A}_A$ and $\mathbf{A}_B$, as shown in Fig. 4-5*e*. Note that the minus sign of Eq. (4-4) means that $\mathbf{A}^n_{CA}$ is parallel in direction but opposite in sense to $\mathbf{R}_{CA}$, and similarly for $\mathbf{A}^n_{CB}$ and $\mathbf{R}_{CB}$. Continuing to follow Eq. (*d*), we must now add the tangential components $\mathbf{A}^t_{CA}$ and $\mathbf{A}^t_{CB}$, which, according to Eq. (4-6), are perpendicular to $\mathbf{R}_{CA}$ and $\mathbf{R}_{CB}$, respectively. These two lines are drawn as shown in Fig. 4-5*e* and intersect at the point labeled C. We see from Eq. (*d*) that the absolute acceleration of point C is given by the vector from O_A to C in the acceleration polygon. It is shown properly positioned on the diagram of the link in Fig. 4-5*f*.

In the approach taken above, the previously calculated value of $\boldsymbol{\alpha}$ was not used. An alternate approach would have been to use $\boldsymbol{\alpha}$ and Eq. (4-6) to calculate either $\mathbf{A}^t_{CA}$ or $\mathbf{A}^t_{CB}$. Only one of the two equations (*d*) would have been necessary with this approach to locate point C and determine $\mathbf{A}_C$.

Figure 4-5*g* shows the same acceleration polygon with the triangle ABC shaded and the normal and tangential acceleration-difference components suppressed. We see again that the triangle ABC in the acceleration polygon appears similar in shape to the original link shape ABC. We can prove that this is indeed the case by writing equations for the magnitude of each side. Each acceleration-difference vector is made up of a normal and tangential component and the three form a right triangle, as shown in Fig. 4-5*f*. Thus, using the pythagorean theorem, we can find the magnitude of $\mathbf{A}_{BA}$, for example, as follows:

$$A_{BA} = \sqrt{(A^n_{BA})^2 + (A^t_{BA})^2} = \sqrt{(\omega^2 R_{BA})^2 + (\alpha R_{BA})^2} = R_{BA}\sqrt{\omega^4 + \alpha^2} \tag{e}$$

Similarly,

$$A_{CA} = R_{CA}\sqrt{\omega^4 + \alpha^2} \tag{f}$$

and

$$A_{CB} = R_{CB}\sqrt{\omega^4 + \alpha^2} \tag{g}$$

Thus we see that the sides of the triangle ABC in the acceleration polygon are proportional to the sides of the original link ABC, with the proportionality factor depending on the rotational motion of the link. This similarly shaped figure in the acceleration polygon is referred to as the *acceleration image* of the link, and every moving link has a corresponding acceleration image in the acceleration polygon.

As with the velocity image, the concept of the acceleration image can be used to simplify the solution of the above example greatly. Once the acceleration-image points A and B have been located, the acceleration-image triangle ABC can be constructed by proportioning the sides to those of the link or by constructing the angles α and β, as shown in Fig. 4-5*g*. Note that the calculation of the two normal components of Eq. (*d*) is avoided by this approach. Although the angle of rotation of the acceleration image relative to the link itself is not an easily determined value (it depends on the magnitude of $\boldsymbol{\omega}$ and both the magnitude and direction of $\boldsymbol{\alpha}$), the other properties of the velocity image do carry over to acceleration images:

1. The acceleration image of each rigid link is a scale reproduction of the shape of the link in the acceleration polygon.
2. The letters identifying the vertices of each link are the same as those in the acceleration polygon and they progress around the acceleration image in the same order and in the same angular direction as around the link.
3. The ratio of the size of the acceleration image of a link and the size of the link itself depends on the rotational motion of the link. In general, it is *not* the same for different links in a mechanism.
4. The point O_A in the acceleration polygon is the image of all points with zero absolute acceleration. It is the acceleration image of the fixed link.
5. The absolute acceleration of any point on any link is represented by the line from O_A to the image of the point in the acceleration polygon. The acceleration difference between two points, say P and Q, is represented by the line from image point P to image point Q.

As mentioned with regard to graphical velocity analysis, the convenience of the acceleration-image concept can be used to speed up the solution and reduce numerical computations. It may, however, give the illusion of a graphical trick without a sound theoretical foundation. It is a wise practice to continue to write the corresponding velocity- and acceleration-difference equations whenever using the image concept until the underlying principles become thoroughly familiar. To give increased experience with graphical acceleration analysis, two more examples will be presented here.

Example 4-1 The four-bar linkage shown in Fig. 4-6*a* was analyzed for velocities in Example 3-1. Its velocity polygon was shown in Fig. 3-7*b*. Assuming that link 2 is driven at

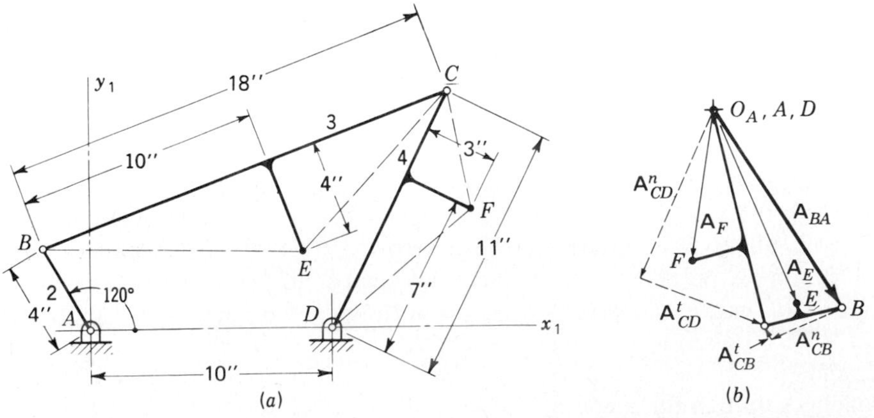

Figure 4-6 Graphical acceleration analysis of a four-bar linkage, Example 4-1: (*a*) scale diagram and (*b*) acceleration polygon.

constant angular velocity, determine the absolute accelerations of points E and F and the angular accelerations of links 3 and 4.

SOLUTION Starting from the fixed pivot point A, we first write the acceleration-difference equation for the acceleration of point B

$$\overset{\circ\circ}{\mathbf{A}}_B = \cancel{\mathbf{A}_A}^{0} + \overset{\checkmark\checkmark}{\mathbf{A}}{}^n_{BA} + \overset{\checkmark\checkmark}{\mathbf{A}}{}^t_{BA} \tag{1}$$

The components of the acceleration-difference equation are calculated from the given angular motion of link 2

$$A^n_{BA} = \omega_2^2 R_{BA} = (94.2\text{ rad/s})^2\left(\frac{4}{12}\text{ ft}\right) = 2958\text{ ft/s}^2$$

$$A^t_{BA} = \alpha_2 R_{BA} = (0\text{ rad/s}^2)\left(\frac{4}{12}\text{ ft}\right) = 0$$

Point O_A and an acceleration scale are chosen and $\mathbf{A}^n_{BA}$ is plotted (opposite in direction to $\mathbf{R}_{BA}$) to locate the acceleration-image point B, as shown in Fig. 4-6b, thus solving Eq. (1).

Next, the acceleration-difference equations are written relating point C to points B and D

$$\mathbf{A}_C = \overset{\checkmark\checkmark}{\mathbf{A}}_B + \overset{\checkmark\checkmark}{\mathbf{A}}{}^n_{CB} + \overset{\circ\checkmark}{\mathbf{A}}{}^t_{CB} = \cancel{\mathbf{A}_D}^{0} + \overset{\checkmark\checkmark}{\mathbf{A}}{}^n_{CD} + \overset{\circ\checkmark}{\mathbf{A}}{}^t_{CD} \tag{2}$$

Using information scaled from the velocity polygon, we calculate the magnitudes of the two acceleration-difference normal components

$$A^n_{CB} = \frac{V^2_{CB}}{R_{CB}} = \frac{(38.4\text{ ft/s})^2}{18/12\text{ ft}} = 938\text{ ft/s}^2$$

$$A^n_{CD} = \frac{V^2_{CD}}{R_{CD}} = \frac{(45.5\text{ ft/s})^2}{11/12\text{ ft}} = 2268\text{ ft/s}^2$$

These two normal components have directions opposite to $\mathbf{R}_{CB}$ and $\mathbf{R}_{CD}$, respectively. As required by Eq. (2), they are added to the acceleration polygon originating from points B and D, respectively, and are shown by the dashed lines of Fig. 4-6b. Perpendicular dashed lines are then drawn through the termini of these two normal components; they represent the addition of the two tangential components $\mathbf{A}^t_{CB}$ and $\mathbf{A}^t_{CD}$, as required, thus completing Eq. (2). Their intersection is labeled as acceleration-image point C.

The angular accelerations of links 3 and 4 are now found from the two tangential components

$$\alpha_3 = \frac{A^t_{CB}}{R_{CB}} = \frac{160\text{ ft/s}^2}{18/12\text{ ft}} = 107\text{ rad/s}^2\text{ ccw} \qquad \textit{Ans.}$$

$$\alpha_4 = \frac{A^t_{CD}}{R_{CD}} = \frac{1670\text{ ft/s}^2}{11/12\text{ ft}} = 1822\text{ rad/s}^2\text{ cw} \qquad \textit{Ans.}$$

where the directions are found by the visualization technique illustrated in the last example, Fig. 4-5c.

The absolute acceleration of point E can now be found by relating it through acceleration-difference equations to points B and C, also on link 3

$$\mathbf{A}_E = \mathbf{A}_B + \mathbf{A}^n_{EB} + \mathbf{A}^t_{EB} = \mathbf{A}_C + \mathbf{A}^n_{EC} + \mathbf{A}^t_{EC} \tag{3}$$

The solution of these equations can follow the same methods used for Eq. (2) if desired. A second approach is to use the value of $\boldsymbol{\alpha}_3$, now known, to calculate one or both of the tangential components. Probably the easiest approach, however, is to form the acceleration-image triangle BCE for line 3, using $\mathbf{A}_{CB}$ and the shape of link 3 as the basis. Any of these approaches leads to the location of the acceleration-image point E shown in the acceleration polygon, Fig. 4-6*b*. The absolute acceleration of point E is then scaled and found to be

$$A_E = 2580 \text{ ft/s}^2 \qquad Ans.$$

Any of these approaches can also be used to find the absolute acceleration of point F. The appropriate acceleration-difference equations, relating it to points C and D on link 4, are

$$\mathbf{A}_F = \mathbf{A}_D + \mathbf{A}^n_{FD} + \mathbf{A}^t_{FD} = \mathbf{A}_C + \mathbf{A}^n_{FC} + \mathbf{A}^t_{FC} \tag{4}$$

Their solution results in the location of image point F, as shown in the acceleration polygon. The result is

$$A_F = 1960 \text{ ft/s}^2 \qquad Ans.$$

On reviewing this example it is clear that the overall strategy for graphical acceleration analysis, the order and number of equations written, follows identically the approach used in graphical velocity analysis. Although there are two components in each acceleration difference and only one for each velocity difference, the normal components can always be calculated from information contained in the velocity polygon; they never contain an unknown. The unknowns of the acceleration-difference equation usually come from the unknown magnitude of the tangential component, which depends on the angular acceleration of a link, and the unknown magnitude or direction of one of the absolute accelerations.

Example 4-2 The velocity analysis of an offset slider-crank mechanism was performed in Example 3-2. The velocity polygon was shown in Fig. 3-8*b*. Assuming that the given velocity of the slider was constant, determine the instantaneous absolute acceleration of point D and the angular accelerations of links 2 and 3.

SOLUTION The scale diagram of the mechanism is shown again in Fig. 4-7*a*. The acceleration polygon is started by choosing a scale and pole O_A, as shown in Fig. 4-7*b*. Since the velocity $\mathbf{V}_C$ is given as constant, its acceleration is zero; thus the acceleration-image point C is identified with O_A.

Acceleration-difference equations are written next for the acceleration of point B

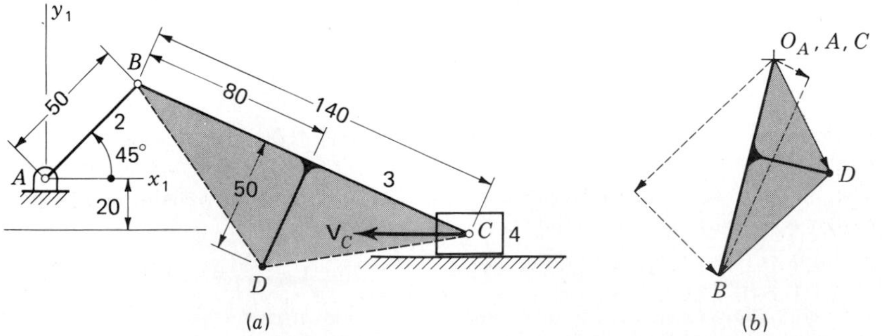

Figure 4-7 Graphical acceleration analysis of an offset slider-crank mechanism, Example 4-2: (*a*) scale diagram (dimensions in millimetres) and (*b*) acceleration polygon.

relating it to two points whose accelerations are known,† points C and A

$$\mathbf{A}_B = \cancel{\mathbf{A}_C}^{0} + \mathbf{A}^n_{BC} + \mathbf{A}^t_{BC} = \cancel{\mathbf{A}_A}^{0} + \mathbf{A}^n_{BA} + \mathbf{A}^t_{BA} \tag{5}$$

The magnitudes of the two normal components can be calculated from information obtained from the position diagram and the velocity polygon

$$A^n_{BC} = \frac{V^2_{BC}}{R_{BC}} = \frac{(7.5\ \mathrm{m/s})^2}{0.14\ \mathrm{m}} = 402\ \mathrm{m/s^2}$$

$$A^n_{BA} = \frac{V^2_{BA}}{R_{BA}} = \frac{(10.0\ \mathrm{m/s})^2}{0.05\ \mathrm{m}} = 2000\ \mathrm{m/s^2}$$

These are drawn parallel to but opposite in direction to $\mathbf{R}_{BC}$ and $\mathbf{R}_{BA}$, respectively; they are added to $\mathbf{A}_C$ and $\mathbf{A}_A$, as shown by dashed lines in Fig. 4-7*b*.

The tangential components of Eq. (5) are now added, drawing them perpendicular to $\mathbf{R}_{BC}$ and $\mathbf{R}_{BA}$, respectively. Their intersection is identified as the acceleration-image point B.

Since image points B and C are now known, the acceleration image of link 3 can be drawn to locate image point D. Taking care that the image is not flipped over, it is shown shaded in the acceleration polygon. The absolute acceleration of point D can now be scaled from O_A to image point D; the result is

$$A_D = 1300\ \mathrm{m/s^2} \qquad \textit{Ans.}$$

The angular accelerations of links 2 and 3 are found from the two tangential components of Eq. (5)

$$\alpha_2 = \frac{A^t_{BA}}{R_{BA}} = \frac{1260\ \mathrm{m/s^2}}{0.05\ \mathrm{m}} = 25\,200\ \mathrm{rad/s^2\ cw} \qquad \textit{Ans.}$$

$$\alpha_3 = \frac{A^t_{BC}}{R_{BC}} = \frac{2300\ \mathrm{m/s^2}}{0.14\ \mathrm{m}} = 16\,400\ \mathrm{rad/s^2\ ccw} \qquad \textit{Ans.}$$

Note that $\boldsymbol{\alpha}_2$ is found to be clockwise even though the motion of the slider is to the left. This example should provide fair warning to those inclined to use intuition in determining the directions of accelerations; they are not easily visualized and should be solved for from basic principles rather than guessed. From Example 3-2 we saw that $\boldsymbol{\omega}_2$ is counterclockwise as expected; the clockwise result for $\boldsymbol{\alpha}_2$ indicates that link 2 is decelerating in its rotational motion.

4-5 THE APPARENT ACCELERATION OF A POINT IN A MOVING COORDINATE SYSTEM

In Sec. 3-5 we found it necessary to develop the apparent-velocity equation for situations where it was convenient to describe the path along which a point moves relative to another moving link but where it was not convenient to describe the absolute motion of the same point. Let us now investigate the acceleration of such a point.

To review, Fig. 4-8 illustrates a point P_3 of link 3 which moves along a known path, the slot, relative to the moving reference frame $x_2y_2z_2$. Point P_2 is

† Although the acceleration-image points C and A are known, it would be a gross error to draw an "acceleration image" of the triangle ABC since these points are not all on the same link.

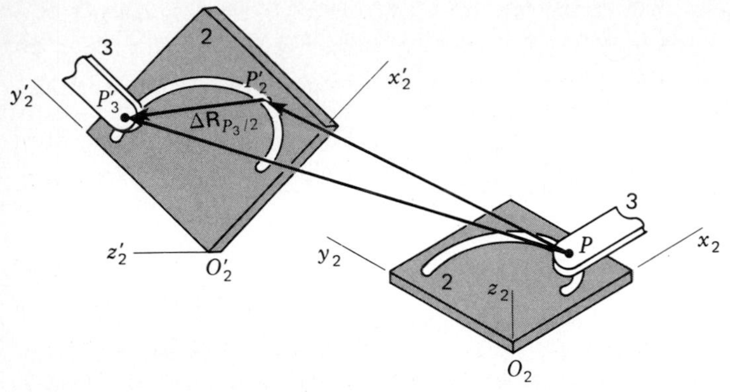

Figure 4-8 Apparent displacement.

fixed to the moving link 2 and is instantaneously coincident with P_3. The problem is to find an equation relating the accelerations of points P_3 and P_2 in terms of meaningful parameters which can be calculated (or measured) in a typical mechanical system.

In Fig. 4-9 we recall how the same situation would be perceived by a moving observer attached to link 2. To him the *path* of P_3, the slot, would appear stationary and the point P_3 would appear to move along its tangent with the apparent velocity $\mathbf{V}_{P_3/2}$.

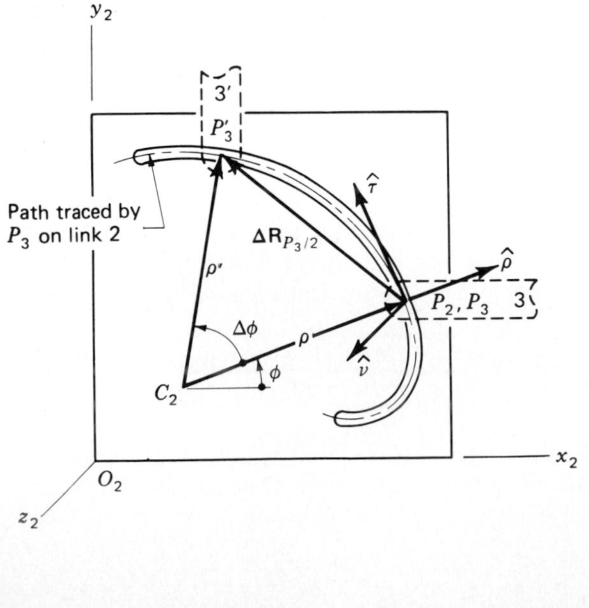

Figure 4-9 Apparent displacement of point P_3 as seen by an observer on link 2.

We recall defining in Sec. 3-5 another moving coordinate system $\hat{\boldsymbol{\rho}}\hat{\boldsymbol{\tau}}\hat{\boldsymbol{\nu}}$, where $\hat{\boldsymbol{\rho}}$ was defined as a unit vector in the direction of the radius of curvature vector of the path, $\hat{\boldsymbol{\tau}}$ was defined as the unit tangent vector to the path at P, and $\hat{\boldsymbol{\nu}}$ was normal to the plane containing $\hat{\boldsymbol{\rho}}$ and $\hat{\boldsymbol{\tau}}$, thus forming a right-hand cartesian coordinate system. After defining s as a scalar arc distance measuring the travel of P_3 along the curved path, we derived Eq. (3-9) for the apparent velocity

$$\mathbf{V}_{P_3/2} = \frac{ds}{dt}\hat{\boldsymbol{\tau}} \tag{a}$$

Consider the rotation of the radius of curvature vector; it sweeps through some small angle $\Delta\phi$ as P_3 travels the small arc distance Δs during a short time interval Δt. The small angle and arc distance are related by

$$\Delta\phi = \frac{\Delta s}{\rho}$$

Dividing this by Δt and taking the limit for infinitesimally small Δt, we find

$$\frac{d\phi}{dt} = \frac{1}{\rho}\frac{ds}{dt} = \frac{V_{P_3/2}}{\rho} \tag{b}$$

This is the angular rate at which the radius-of-curvature vector $\boldsymbol{\rho}$ (and also $\hat{\boldsymbol{\tau}}$) appears to rotate as seen by a moving observer in coordinate system 2 as point P_3 moves along its path. We can give this rotational speed its proper vector properties as an apparent angular velocity by noticing that the axis of this rotation is parallel to $\hat{\boldsymbol{\nu}}$. Thus we define

$$\dot{\boldsymbol{\phi}} = \frac{d\phi}{dt}\hat{\boldsymbol{\nu}} = \frac{V_{P_3/2}}{\rho}\hat{\boldsymbol{\nu}} \tag{c}$$

Next we seek to find the time derivative of the unit vector $\hat{\boldsymbol{\tau}}$ so that we can differentiate Eq. (a). Since $\hat{\boldsymbol{\tau}}$ is a unit vector, its length does not change; however, it does have a derivative due to its change in direction, its rotation. In the absolute coordinate system $\hat{\boldsymbol{\tau}}$ is subject to the rotation $\dot{\boldsymbol{\phi}}$ and also to the angular velocity $\boldsymbol{\omega}$, with which the moving coordinate system 2 is rotating. Therefore,

$$\frac{d\hat{\boldsymbol{\tau}}}{dt} = (\boldsymbol{\omega} + \dot{\boldsymbol{\phi}}) \times \hat{\boldsymbol{\tau}} = \boldsymbol{\omega} \times \hat{\boldsymbol{\tau}} + \dot{\boldsymbol{\phi}} \times \hat{\boldsymbol{\tau}} \tag{d}$$

But, when Eq. (c) is used, this becomes

$$\frac{d\hat{\boldsymbol{\tau}}}{dt} = \boldsymbol{\omega} \times \hat{\boldsymbol{\tau}} + \frac{V_{P_3/2}}{\rho}\hat{\boldsymbol{\nu}} \times \hat{\boldsymbol{\tau}} = \boldsymbol{\omega} \times \hat{\boldsymbol{\tau}} - \frac{V^2_{P_3/2}}{\rho}\hat{\boldsymbol{\rho}} \tag{e}$$

Now, taking the time derivative of Eq. (a), we find that

$$\frac{d\mathbf{V}_{P_3/2}}{dt} = \frac{d^2s}{dt^2}\hat{\boldsymbol{\tau}} + \frac{ds}{dt}\frac{d\hat{\boldsymbol{\tau}}}{dt} = \frac{d^2s}{dt^2}\hat{\boldsymbol{\tau}} + \frac{ds}{dt}\hat{\boldsymbol{\omega}} \times \hat{\boldsymbol{\tau}} - \frac{ds}{dt}\frac{V_{P_3/2}}{\rho}\hat{\boldsymbol{\rho}}$$

and, on using Eq. (a), this reduces to

$$\frac{d\mathbf{V}_{P_3/2}}{dt} = \frac{d^2s}{dt^2}\hat{\boldsymbol{\tau}} + \boldsymbol{\omega} \times \mathbf{V}_{P_3/2} - \frac{V^2_{P_3/2}}{\rho}\hat{\boldsymbol{\rho}} \tag{f}$$

Notice that the three terms in the above equation are *not* defined as the apparent-acceleration components. To be consistent, the term *apparent acceleration* should include only those components *which would be seen by an observer attached to the moving coordinate system.* The above equation is derived in the absolute coordinate system and includes the rotational effect of $\boldsymbol{\omega}$, which would not be sensed by the moving observer. The apparent acceleration, which is given the notation $\mathbf{A}_{P_3/2}$, can easily be found, however, by setting $\boldsymbol{\omega}$ to zero in Eq. (f). This gives the two remaining components

$$\mathbf{A}_{P_3/2} = \mathbf{A}^n_{P_3/2} + \mathbf{A}^t_{P_3/2} \tag{4-9}$$

where

$$\mathbf{A}^n_{P_3/2} = -\frac{V^2_{P_3/2}}{\rho}\hat{\boldsymbol{\rho}} \tag{4-10}$$

is called the *normal component*, indicating that it is always normal to the path and directed toward the center of curvature (the $-\hat{\boldsymbol{\rho}}$ direction), while

$$\mathbf{A}^t_{P_3/2} = \frac{d^2s}{dt^2}\hat{\boldsymbol{\tau}} \tag{4-11}$$

is called the *tangential component*, indicating that it is always tangent to the path (the $\hat{\boldsymbol{\tau}}$ direction).

Next we note that the radius-of-curvature vector $\boldsymbol{\rho}$ rotates due to both $\boldsymbol{\omega}$ and $\dot{\boldsymbol{\phi}}$. Therefore, its derivative is†

$$\frac{d\boldsymbol{\rho}}{dt} = (\boldsymbol{\omega} + \dot{\boldsymbol{\phi}}) \times \boldsymbol{\rho} = \boldsymbol{\omega} \times \boldsymbol{\rho} + \mathbf{V}_{P_3/2} \tag{g}$$

We can now write the position equation from Fig. 4-9

$$\mathbf{R}_{P_3} = \mathbf{R}_{C_2} + \boldsymbol{\rho}$$

and with the help of Eq. (g) its time derivative can be taken‡

$$\mathbf{V}_{P_3} = \mathbf{V}_{C_2} + \boldsymbol{\omega} \times \boldsymbol{\rho} + \mathbf{V}_{P_3/2} \tag{h}$$

Differentiating this equation again with respect to time, we obtain

$$\mathbf{A}_{P_3} = \mathbf{A}_{C_2} + \boldsymbol{\alpha} \times \boldsymbol{\rho} + \boldsymbol{\omega} \times \frac{d\boldsymbol{\rho}}{dt} + \frac{d\mathbf{V}_{P_3/2}}{dt}$$

† Note that the magnitude of $\boldsymbol{\rho}$ is treated as constant in the vicinity of point P due to its definition. It is not truly constant, but at a stationary value; its second derivative is nonzero, but its first derivative is zero at the instant considered.

‡ The first two terms of Eq. (h) are equal to $\mathbf{V}_{P_2}$; thus this is equivalent to the apparent-velocity equation. Note, however, that even though $\boldsymbol{\rho} = \mathbf{R}_{P_2C_2}$, their derivatives are not equal; they do not rotate at the same rate. Thus, some of the terms of the next equation would be missed if the apparent-velocity equation were differentiated instead.

and, with the help of Eqs. (f) and (g), this becomes

$$\mathbf{A}_{P_3} = \mathbf{A}_{C_2} + \boldsymbol{\alpha} \times \boldsymbol{\rho} + \boldsymbol{\omega} \times (\boldsymbol{\omega} \times \boldsymbol{\rho}) + 2\boldsymbol{\omega} \times \mathbf{V}_{P_{3/2}} - \frac{V^2_{P_{3/2}}}{\rho}\hat{\boldsymbol{\rho}} + \frac{d^2 s}{dt^2}\hat{\boldsymbol{\tau}} \tag{i}$$

We recognize the first three terms of this equation as the components of $\mathbf{A}_{P_2}$ and the last two terms as the components of the apparent acceleration $\mathbf{A}_{P_{3/2}}$. We therefore define a symbol for the remaining term

$$\mathbf{A}^c_{P_3P_2} = 2\boldsymbol{\omega}_2 \times \mathbf{V}_{P_{3/2}} \tag{4-12}$$

This term is called the *Coriolis component of acceleration.* We note that it is one term of the apparent-acceleration equation. However, unlike the components of $\mathbf{A}_{P_{3/2}}$, it is not sensed by a moving observer attached to the moving coordinate system 2. Still, it is a necessary term in Eq. (i) and is a part of the difference between $\mathbf{A}_{P_3}$ and $\mathbf{A}_{P_2}$ sensed by an absolute observer.

With the definition, Eq. (i) can be written in the following form, called the *apparent-acceleration equation,*

$$\mathbf{A}_{P_3} = \mathbf{A}_{P_2} + \mathbf{A}^c_{P_3P_2} + \mathbf{A}^n_{P_{3/2}} + \mathbf{A}^t_{P_{3/2}} \tag{4-13}$$

where the definitions of the individual components are given in Eqs. (4-10) to (4-12).

It is extremely important in applications to recognize certain features of this equation: (1) It serves the objectives of this section because it relates the accelerations of two coincident points on different links in a meaningful way. (2) There is only one new unknown among the three new components defined. The normal and Coriolis components can be calculated from Eqs. (4-10) and (4-12) from velocity information; they do not contribute any new unknowns. The tangential component $\mathbf{A}^t_{P_{3/2}}$, however, will almost always have an unknown magnitude in application, since d^2s/dt^2 cannot be found. (3) It is important to notice the dependence of Eq. (4-13) on the ability to recognize in each application the point path which P_3 traces on coordinate system 2. This path is the basis for the directions of the normal and tangential components and is also necessary for determination of ρ for Eq. (4-10).

Finally, a word of warning, *the path described by P_3 on link 2 is not necessarily the same as the path described by P_2 on link* 3. In Fig. 4-9 the path of P_3 on link 2 is clear; it is the curved slot. The path of P_2 on link 3 is not at all clear. As a result there is a natural right and wrong way to write the apparent-acceleration equation for that situation. The equation

$$\mathbf{A}_{P_2} = \mathbf{A}_{P_3} + \mathbf{A}^c_{P_2P_3} + \mathbf{A}^n_{P_{2/3}} + \mathbf{A}^t_{P_{2/3}}$$

is a perfectly valid equation but *useless* since ρ is not known for the normal component. Note also that $\mathbf{A}^c_{P_3P_2}$ makes use of $\boldsymbol{\omega}_2$ while $\mathbf{A}^c_{P_2P_3}$ makes use of $\boldsymbol{\omega}_3$. We must be extremely careful to write the appropriate equation for each application, recognizing which path is known.

Example 4-3 In Fig. 4-10 a block 3 slides outward on link 2 at a uniform rate of 30 m/s, while link 2 is rotating at a constant angular velocity of 50 rad/s ccw. Determine the absolute acceleration of point A of the block.

SOLUTION We first calculate the absolute acceleration of the coincident point A_2, immediately under the block but attached to link 2

$$\mathbf{A}_{A_2} = \cancel{\mathbf{A}_{O_2}}^{\,0} + \mathbf{A}^n_{A_2O_2} + \cancel{\mathbf{A}^t_{A_2O_2}}^{\,0}$$

$$A^n_{A_2O_2} = \omega_2^2 R_{A_2O_2} = (50\ \text{rad/s})^2(500\ \text{mm}) = 1250\ \text{m/s}^2$$

We plot this, determining the acceleration-image point A_2. Next, we recognize that point A_3 is constrained to travel only along the axis of link 2. This provides a path for which we can write the apparent-acceleration equation

$$\mathbf{A}_{A_3} = \mathbf{A}_{A_2} + \mathbf{A}^c_{A_3A_2} + \mathbf{A}^n_{A_3/2} + \mathbf{A}^t_{A_3/2}$$

The terms for this equation are computed as follows and graphically added in the acceleration polygon:

$$A^c_{A_3A_2} = 2\omega_2 V_{A_3/2} = 2(50\ \text{rad/s})(30\ \text{m/s}) = 3000\ \text{m/s}^2$$

$$A^n_{A_3/2} = \frac{V^2_{A_3/2}}{\rho} = \frac{(30\ \text{m/s})^2}{\infty} = 0$$

$$A^t_{A_3/2} = \frac{d^2s}{dt^2} = 0 \qquad \text{uniform rate along path}$$

This locates the acceleration-image point A_3 and the result is

$$A_{A_3} = 3250\ \text{m/s}^2 \qquad \textit{Ans.}$$

Example 4-4 Perform an acceleration analysis of the linkage shown in Fig. 4-11 for the constant input speed of $\omega_2 = 18$ rad/s cw.

SOLUTION A complete velocity analysis is performed first, as shown in the figure. This yields

$$V_A = 12\ \text{ft/s} \qquad V_{B_3A} = 10.1\ \text{ft/s} \qquad V_{B_3/4} = 6.5\ \text{ft/s}$$

$$\omega_3 = \omega_4 = 7.77\ \text{rad/s cw}$$

To solve the accelerations we first find

$$\mathbf{A}_A = \cancel{\mathbf{A}_{O_2}}^{\,0} + \mathbf{A}^n_{AO_2} + \cancel{\mathbf{A}^t_{AO_2}}^{\,0}$$

$$A^n_{AO_2} = \omega_2^2 R_{AO_2} = (18\ \text{rad/s})^2\left(\frac{8}{12}\ \text{ft}\right) = 216\ \text{ft/s}^2$$

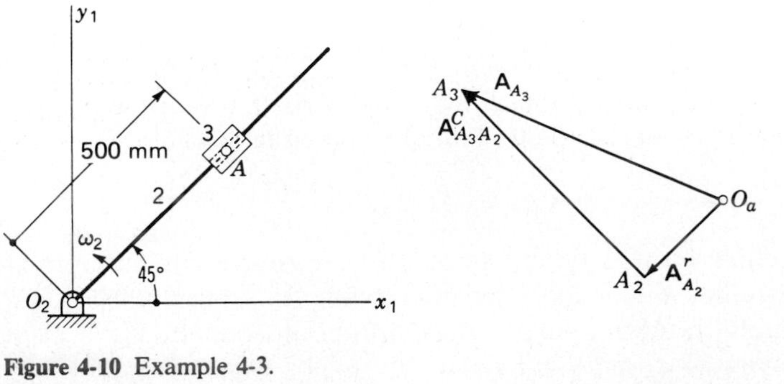

Figure 4-10 Example 4-3.

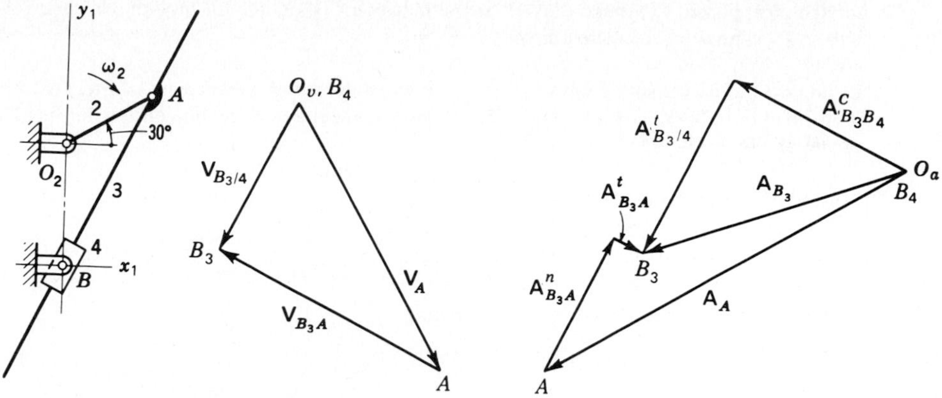

Figure 4-11 Example 4-4: $R_{AO_2} = 8$ in, $R_{BO_2} = 10$ in.

and plot this to locate the acceleration image point A. Next we write the acceleration-difference equation

$$\mathbf{A}_{B_3} = \overset{\surd\surd}{\mathbf{A}_A} + \overset{\surd\surd}{\mathbf{A}^n_{B_3A}} + \overset{o\surd}{\mathbf{A}^t_{B_3A}} \tag{1}$$

$$A^n_{B_3A} = \frac{V^2_{B_3A}}{R_{BA}} = \frac{(10.1\ \text{ft/s})^2}{15.6/12\ \text{ft}} = 78.5\ \text{ft/s}^2$$

The term $\mathbf{A}^n_{B_3A}$ is directed from B toward A and is added to the acceleration polygon as shown. The term $\mathbf{A}^t_{B_3A}$ has unknown magnitude but is perpendicular to $\mathbf{R}_{BA}$.

Since Eq. (1) has three unknowns, it cannot be solved. Therefore we seek a second equation for A_{B_3}. Consider the view of an observer located on link 4; he would see point B_3 moving on a straight-line path along the centerline of the block. Using this path, we can write the apparent-acceleration equation

$$\mathbf{A}_{B_3} = \cancel{\mathbf{A}_{B_4}}^{\,0} + \overset{\surd\surd}{\mathbf{A}^c_{B_3B_4}} + \overset{\surd\surd}{\mathbf{A}^n_{B_3/4}} + \overset{o\surd}{\mathbf{A}^t_{B_3/4}} \tag{2}$$

Since point B_4 is pinned to the ground link, it has zero acceleration. The other components of Eq. (2) are

$$A^c_{B_3B_4} = 2\omega_4 V_{B_3/4} = 2(7.77\ \text{rad/s})(6.5\ \text{ft/s}) = 101\ \text{ft/s}^2$$

$$A^n_{B_3/4} = \frac{V^2_{B_3/4}}{\rho} = \frac{(6.5\ \text{ft/s})^2}{\infty} = 0$$

The Coriolis component is added to the acceleration polygon, originating at point B_4 (O_A) as shown. Finally, $\mathbf{A}^t_{B_3/4}$, whose magnitude is unknown, is graphically added to this in the direction defined by the path tangent. It crosses the unknown line from $\mathbf{A}^t_{B_3A}$, Eq. (1), locating the acceleration-image point B_3. When the polygon is scaled, the results are found to be

$$A^t_{B_3/4} = 103\ \text{ft/s}^2 \qquad A^t_{B_3A} = 16\ \text{ft/s}^2 \qquad A_{B_3} = 144\ \text{ft/s}^2$$

The angular accelerations of links 3 and 4 are

$$\alpha_4 = \alpha_3 = \frac{A^t_{B_3A}}{R_{BA}} = \frac{16\ \text{ft/s}^2}{15.6/12\ \text{ft}} = 12.3\ \text{rad/s}^2\ \text{ccw}$$

We note that in this example the path of B_3 on link 4 and the path of B_4 on link 3 can both be visualized and either could have been used in deciding the approach. However, even though B_4 is pinned to ground (link 1), the path of point B_3 on link 1 is not known. Thus, the term $\mathbf{A}^n_{B_3/1}$ cannot be calculated directly.

Example 4-5 The velocity analysis of the inverted slider-crank mechanism of Fig. 4-12 was performed in Example 3-3 (Fig. 3-11). Determine the angular acceleration of link 4 if link 2 is driven at constant speed.

SOLUTION From Example 3-3 we recall that

$$V_{A_2} = 9\text{ ft/s} \qquad V_{AD} = 7.24\text{ ft/s} \qquad V_{A_2/4} = 5.52\text{ ft/s}$$

$$\omega_2 = 36\text{ rad/s cw} \qquad \omega_3 = \omega_4 = 7.55\text{ rad/s ccw}$$

To analyze for accelerations we start by writing

$$\mathbf{A}_{A_2} = \cancel{\mathbf{A}_E}^{\,0} + \mathbf{A}^n_{AE} + \cancel{\mathbf{A}^t_{AE}}^{\,0}$$

$$A^n_{AE} = \omega_2^2 R_{AE} = (36\text{ rad/s})^2\left(\frac{3}{12}\text{ ft}\right) = 324\text{ ft/s}^2$$

and plot this as shown in the figure.

Next we notice that point A_2 travels along the straight-line path shown relative to an observer on link 4. Knowing this path, we write

$$\overset{\surd\surd}{\mathbf{A}}_{A_2} = \overset{o\surd}{\mathbf{A}}_{A_4} + \overset{\surd\surd}{\mathbf{A}}{}^c_{A_2A_4} + \cancel{\mathbf{A}^n_{A_2/4}}^{\,0} + \overset{o\surd}{\mathbf{A}}{}^t_{A_2/4} \tag{3}$$

where $A^n_{A_2/4} = 0$ since $\rho = \infty$ and

$$\mathbf{A}_{A_4} = \cancel{\mathbf{A}_D}^{\,0} + \overset{\surd\surd}{\mathbf{A}}{}^n_{A_4D} + \overset{o\surd}{\mathbf{A}}{}^t_{A_4D} \tag{4}$$

$$A^n_{A_4D} = \frac{V^2_{A_4D}}{R_{A_4D}} = \frac{(7.24\text{ ft/s})^2}{11.6/12\text{ ft}} = 54.2\text{ ft/s}^2$$

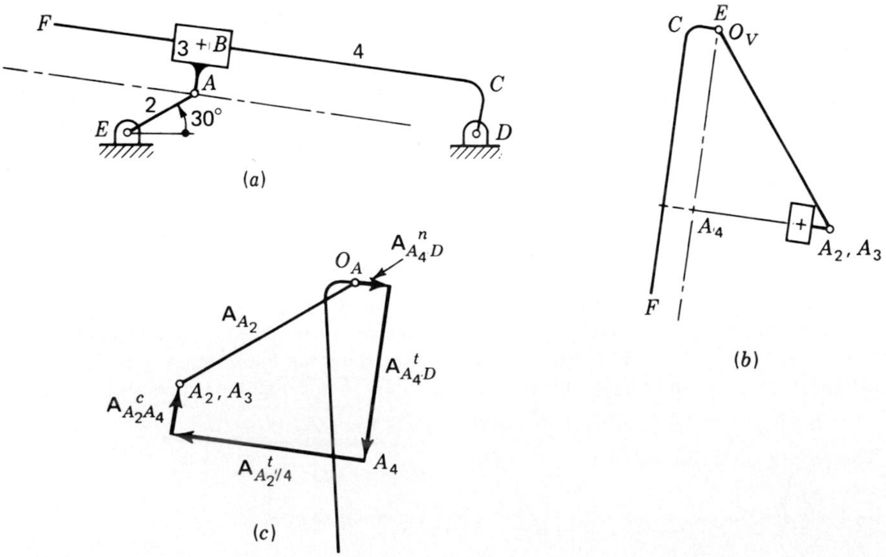

Figure 4-12 Example 4-5.

The term $\mathbf{A}^n_{A_4D}$ is added from O_A followed by a line of unknown length for $\mathbf{A}^t_{A_4D}$. Since the image point A_4 is not yet known, the terms $\mathbf{A}^c_{A_2A_4}$ and $\mathbf{A}^t_{A_2/4}$ cannot be added as directed by Eq. (3). However, these two terms can be transferred to the other side of Eq. (3) and graphically subtracted from image point A_2, thus completing the acceleration polygon. The angular acceleration of link 4 can now be found

$$\alpha_4 = \frac{A^t_{A_4D}}{R_{AD}} = \frac{284 \text{ ft/s}}{11.6/12 \text{ ft}} = 294 \text{ rad/s}^2 \text{ ccw} \qquad \textit{Ans.}$$

This need to subtract vectors is common in acceleration problems involving the Coriolis component and should be studied carefully. Note that the opposite equation involving $\mathbf{A}_{A_4/2}$ cannot be used since ρ and therefore $A^n_{A_4/2}$ would be an additional (third) unknown.

4-6 APPARENT ANGULAR ACCELERATION

Although it is seldom useful, completeness suggests that we should also define the term *apparent angular acceleration.* When two rigid bodies rotate with different angular accelerations, the vector difference between them is defined as the apparent angular acceleration

$$\boldsymbol{\alpha}_{3/2} = \boldsymbol{\alpha}_3 - \boldsymbol{\alpha}_2$$

The apparent-angular-acceleration equation can also be written

$$\boldsymbol{\alpha}_3 = \boldsymbol{\alpha}_2 + \boldsymbol{\alpha}_{3/2} \qquad (4\text{-}14)$$

It will be seen that $\boldsymbol{\alpha}_{3/2}$ is the angular acceleration of body 3 as it would appear to an observer attached to, and rotating with, body 2.

4-7 DIRECT CONTACT AND ROLLING CONTACT

We recall from Sec. 3-7 that the relative motion between two bodies in direct contact can be of two different kinds; there may be an apparent slipping velocity between the bodies, or there may be no such slip. We defined the term rolling contact to imply that no slip is possible and developed the rolling-contact condition, Eq. (3-13), to indicate that the apparent velocity at

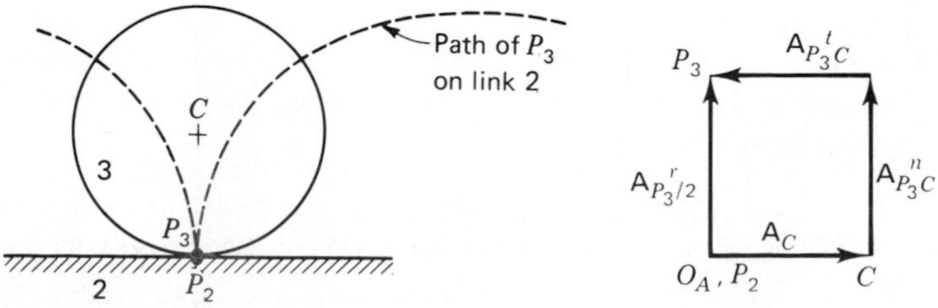

Figure 4-13 Rolling contact.

such a point is zero. Here we intend to investigate the apparent acceleration at a point of rolling contact.

Consider the case of a circular wheel in rolling contact with another straight link, as shown in Fig. 4-13. Although this is admittedly a very simplified case, the arguments made and the conclusions reached are completely general and apply to any rolling-contact situation, no matter what the shapes of the two bodies or whether either is the ground link. To keep this clear in our minds, the ground link has been numbered 2 for this example.

Once the acceleration $\mathbf{A}_C$ of the center point of the wheel is given, the pole O_A can be chosen and the acceleration polygon can be started by plotting $\mathbf{A}_C$. In relating the accelerations of points P_3 and P_2 at the rolling-contact point, however, we are dealing with two coincident points of different bodies. Therefore, we can think of using the apparent-acceleration equation. To do this we must identify a path one of these points traces on the other body. The path† which point P_3 traces on link 2 is shown in the figure. Although the precise shape of the path depends on the shapes of the two contacting links, it will always have a cusp at the point of rolling contact and the tangent to this cusp-shaped path will always be perpendicular to the surfaces in contact.

Since this path is known, we are free to write the apparent-acceleration equation

$$\mathbf{A}_{P_3} = \mathbf{A}_{P_2} + \mathbf{A}^c_{P_3P_2} + \mathbf{A}^n_{P_3/2} + \mathbf{A}^t_{P_3/2}$$

In evaluating the components, we keep in mind the rolling-contact velocity condition, that $\mathbf{V}_{P_3/2} = 0$. Then

$$A^c_{P_3P_2} = 2\omega_2 V_{P_3/2} = 0 \qquad \text{and} \qquad A^n_{P_3/2} = \frac{V^2_{P_3/2}}{\rho} = 0$$

Thus only one component of the apparent acceleration, $\mathbf{A}^t_{P_3/2}$, can be nonzero.

Because of possible confusion in calling this nonzero term a tangential component (tangent to the cusp-shaped path) while its direction is normal to the rolling surfaces, we will adopt a new superscript and refer to the *rolling-contact acceleration* $\mathbf{A}^r_{P_3/2}$. Thus, for rolling contact, the apparent-acceleration equation becomes

$$\mathbf{A}_{P_3} = \mathbf{A}_{P_2} + \mathbf{A}^r_{P_3/2} \qquad (4\text{-}15)$$

and the term $\mathbf{A}^r_{P_3/2}$ is known always to have a direction perpendicular to the surfaces at the point of rolling contact.

To help in understanding the graphical approach to the acceleration analysis of direct-contact and rolling-contact mechanisms, we will contrast the solutions of two very similar examples.

Example 4-6 Given the scale drawing and velocity analysis of the direct-contact circular cam with oscillating flat-faced follower system shown in Fig. 4-14*a*, determine the angular

† This particular curve is called a *cycloid*.

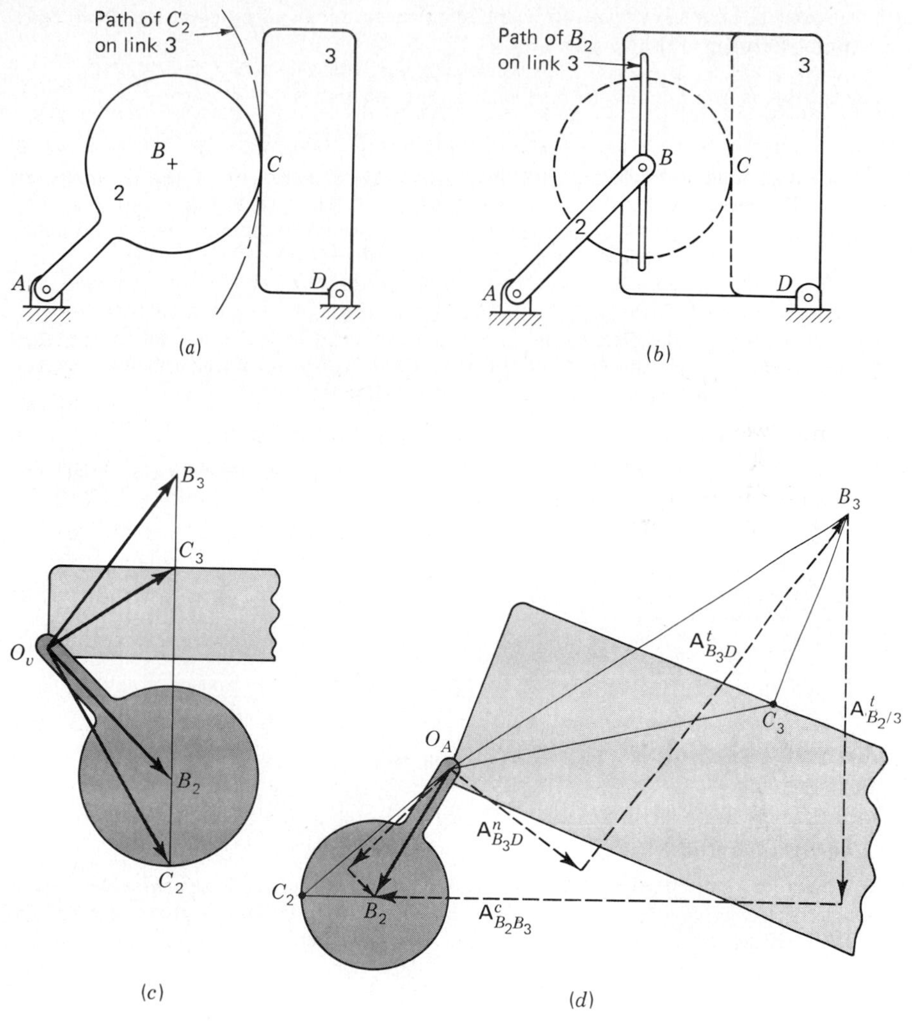

Figure 4-14 Example 4-6.

acceleration of the follower at the instant shown. The angular velocity of the cam is $\omega_2 = 10$ rad/s cw, and its angular acceleration is $\alpha_2 = 25$ rad/s^2 cw.

SOLUTION The velocity polygon of the cam-and-follower system is shown in Fig. 4-14*c*.

The acceleration polygon, Fig. 4-14*d*, is begun by calculating and plotting the acceleration of point B_2 relative to point A

$$\mathbf{A}_{B_2} = \mathbf{A}_A + \mathbf{A}^n_{B_2A} + \mathbf{A}^t_{B_2A} \tag{1}$$

$$A^n_{B_2A} = \frac{V^2_{B_2A}}{R_{B_2A}} = \frac{(30 \text{ in/s})^2}{3 \text{ in}} = 300 \text{ in/s}^2$$

$$A^t_{B_2A} = \alpha_2 R_{B_2A} = (25 \text{ rad/s}^2)(3 \text{ in}) = 75 \text{ in/s}^2$$

These are plotted as shown, and the acceleration-image point C_2 is found by constructing the acceleration image of the triangle AB_2C_2.

Proceeding as in Sec. 3-7, we would next find the velocity of the point C_3. If we attempt this approach to acceleration analysis, the equation is

$$\mathbf{A}_{C_2} = \mathbf{A}_{C_3} + \mathbf{A}^c_{C_2C_3} + \mathbf{A}^n_{C_2/3} + \mathbf{A}^t_{C_2/3} \tag{2}$$

and is based on the path which point C_2 traces on link 3, shown in Fig. 4-14*a*. However, this approach is useless since the radius of curvature of this path is not known and $A^n_{C_2/3}$ therefore cannot be calculated. To avoid this problem, we take a different approach; we look for another pair of coincident points where the curvature of the path is known.

If we consider the path traced by point B_2 on the (extended) link 3, we see that it remains a constant distance from the surface; it is a straight line. Our new approach is better visualized if we consider the mechanism of Fig. 4-14*b* and note that it has motion equivalent to the original. Having thus visualized a slot in the equivalent mechanism for a path, it becomes clear how to proceed; the appropriate equation is

$$\mathbf{A}_{B_2} = \mathbf{A}_{B_3} + \mathbf{A}^c_{B_2B_3} + \mathbf{A}^n_{B_2/3} + \mathbf{A}^t_{B_2/3} \tag{3}$$

where B_3 is a point coincident with B_2 but attached to link 3. Since the path is a straight line,

$$A^n_{B_2/3} = \frac{V^2_{B_2/3}}{\rho} = \frac{(50 \text{ in/s})^2}{\infty} = 0$$

$$A^c_{B_2B_3} = 2\omega_3 V_{B_2/3} = 2(10 \text{ rad/s})(50 \text{ in/s}) = 1000 \text{ in/s}^2$$

The acceleration of B_3 can be found from

$$\mathbf{A}_{B_3} = \cancelto{0}{\mathbf{A}_D} + \mathbf{A}^n_{B_3D} + \mathbf{A}^t_{B_3D} \tag{4}$$

where $$A^n_{B_3D} = \frac{V^2_{B_3D}}{R_{B_3D}} = \frac{(35.4 \text{ in/s})^2}{3.58 \text{ in}} = 350 \text{ in/s}^2$$

Substituting Eq. (4) into Eq. (3) and rearranging terms, we arrive at an equation with only two unknowns

$$\overset{\surd\surd}{\mathbf{A}_{B_2}} - \overset{\surd\surd}{\mathbf{A}^c_{B_2B_3}} - \overset{o\surd}{\mathbf{A}^t_{B_2/3}} = \overset{\surd\surd}{\mathbf{A}^n_{B_3D}} + \overset{o\surd}{\mathbf{A}^t_{B_3D}} \tag{5}$$

This equation is solved graphically as shown in Fig. 4-14*d*. Once the image point B_3 has been found, the image point C_3 is easily found by constructing the acceleration image of the triangle DB_3C_3, all on link 3. Figure 4-14*d* has been extended to show the complete acceleration images of links 2 and 3 to aid visualization and to illustrate once again that there is no obvious relation between the final locations of image points C_2 and C_3, as suggested by Eq. (2).

Finally, the angular acceleration of link 3 can be found as follows:

$$\alpha_3 = \frac{A^t_{B_3D}}{R_{B_3D}} = \frac{938 \text{ in/s}^2}{3.58 \text{ in}} = 262 \text{ rad/s}^2 \text{ cw} \qquad \textit{Ans.} \tag{6}$$

Next let us consider another, very closely related, example problem.

Example 4-7 Given the scale drawing and velocity analysis of the circular roller, rolling without slip against the oscillating flat-faced follower shown in Fig. 4-15*a*, determine the angular accelerations of both the follower and the roller at the instant shown. The angular velocity of link 2 is $\omega_2 = 10$ rad/s cw, and its angular acceleration is $\alpha_2 = 25$ rad/s^2 cw.

SOLUTION The completed velocity polygon is shown in Fig. 4-15*b*. The acceleration analysis proceeds exactly as shown in the previous example. Again it is fruitless to proceed with equations for the accelerations of points C_3 and C_4 initially, and again the equivalent

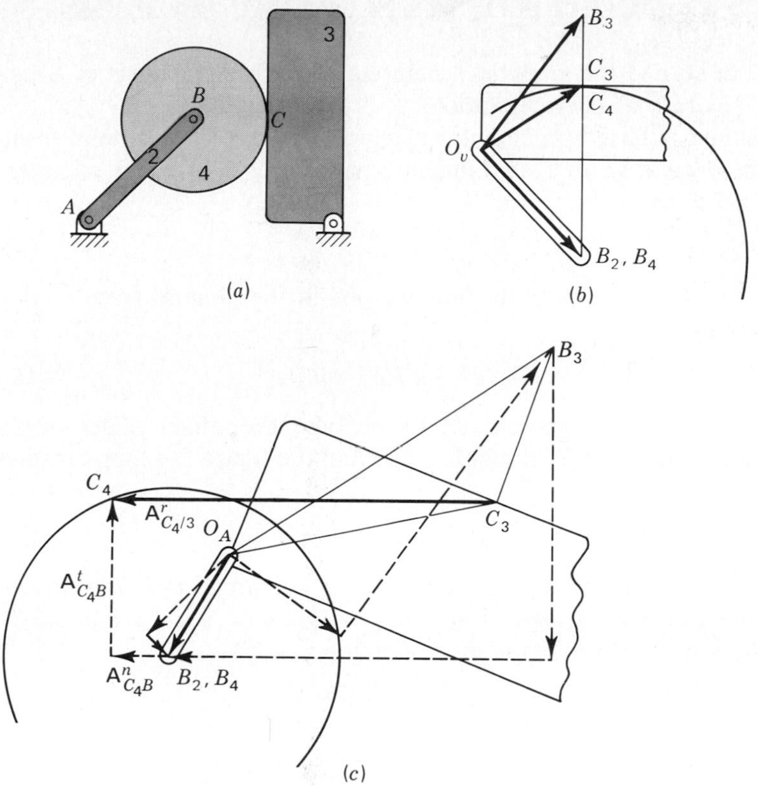

Figure 4-15 Example 4-7.

mechanism of Fig. 4-14*b* must be used. Only after the acceleration of point C_3 has been found (as in the previous example) should the rolling-contact condition be considered.

Then we can relate the acceleration of point C_4 to that of point B_4

$$\mathbf{A}_{C_4} = \mathbf{A}_{B_4} + \mathbf{A}^n_{C_4B_4} + \mathbf{A}^t_{C_4B_4}$$

$$A^n_{C_4B_4} = \frac{V^2_{C_4B_4}}{R_{CB}} = \frac{(14.8 \text{ in/s})^2}{1.50 \text{ in}} = 146 \text{ in/s}^2 \tag{7}$$

We can also write the rolling-contact acceleration equation (4-15) for this situation

$$\mathbf{A}_{C_4} = \mathbf{A}_{C_3} + \mathbf{A}^r_{C_4/3} \tag{8}$$

Remembering that $\mathbf{A}^r_{C_4/3}$ is perpendicular to the surfaces at point C, we can graphically construct the simultaneous solution to Eqs. (7) and (8) shown in Fig. 4-15*c*.

Finally, the acceleration of the roller can be found as follows:

$$\alpha_4 = \frac{A^t_{C_4B_4}}{R_{CB}} = \frac{406 \text{ in/s}^2}{1.50 \text{ in}} = 271 \text{ rad/s}^2 \text{ ccw} \qquad \textit{Ans.} \tag{9}$$

The angular acceleration of link 3 is identical to that found in Example 4-6

$$\alpha_3 = 262 \text{ rad/s}^2 \text{ cw} \qquad \textit{Ans.}$$

4-8 ANALYTICAL METHODS OF ACCELERATION ANALYSIS

In this section we extend the analytical methods of velocity analysis of Secs. 3-8 and 3-9 to the analysis of accelerations.

Raven's method is based on complex algebra. We recall the general form of the first time derivative of a two-dimensional vector written in complex polar form from Eq. (3-14)

$$\dot{\mathbf{R}} = \dot{R}e^{j\theta} + j\dot{\theta}Re^{j\theta} \tag{a}$$

Differentiating again with respect to time, we obtain the general form of the second time derivative

$$\ddot{\mathbf{R}} = \ddot{R}e^{j\theta} + j2\dot{\theta}\dot{R}e^{j\theta} - \dot{\theta}^2Re^{j\theta} + j\ddot{\theta}Re^{j\theta} \tag{4-16}$$

To illustrate Raven's approach, let us analyze the offset slider-crank mechanism of Fig. 4-16. For the symbols defined in the figure the loop-closure equation is

$$\mathbf{r}_1 + \mathbf{r}_4 = \mathbf{r}_2 + \mathbf{r}_3 \tag{b}$$

where r_1, $\theta_1 = -90°$, r_2, r_3, and $\theta_4 = 0$ are constant. The angle θ_2 is the driven input angle and is assumed known. Using the methods of Secs. 2-8 and 3-8, we find the unknown position and velocity variables to be

$$\theta_3 = -\sin^{-1}\frac{r_1 + r_2 \sin\theta_2}{r_3} \tag{4-17}$$

$$r_4 = r_2\cos\theta_2 + r_3\cos\theta_3 \tag{4-18}$$

$$\dot{\theta}_3 = -\frac{r_2\cos\theta_2}{r_3\cos\theta_3}\dot{\theta}_2 \tag{4-19}$$

$$\dot{r}_4 = -r_2\sin\theta_2\,\dot{\theta}_2 - r_3\sin\theta_3\,\dot{\theta}_3 \tag{4-20}$$

The accelerations are found by using the general form, Eq. (4-16), to take the second time derivative of the loop-closure equation. This yields

$$\ddot{r}_4e^{j0} = j\ddot{\theta}_2r_2e^{j\theta_2} - \dot{\theta}_2^2r_2e^{j\theta_2} + j\ddot{\theta}_3r_3e^{j\theta_3} - \dot{\theta}_3^2r_3e^{j\theta_3} \tag{c}$$

Using Euler's formula to separate this complex polar equation into its real and

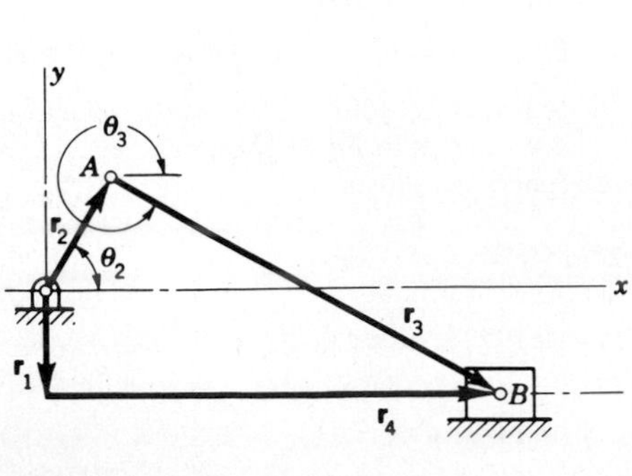

Figure 4-16 Offset slider-crank mechanism.

imaginary components gives

$$\ddot{r}_4 = -\ddot{\theta}_2 r_2 \sin\theta_2 - \dot{\theta}_2^2 r_2 \cos\theta_2 - \ddot{\theta}_3 r_3 \sin\theta_3 - \dot{\theta}_3^2 r_3 \cos\theta_3 \tag{d}$$

$$0 = \ddot{\theta}_2 r_2 \cos\theta_2 - \dot{\theta}_2^2 r_2 \sin\theta_2 + \ddot{\theta}_3 r_3 \cos\theta_3 - \dot{\theta}_3^2 r_3 \sin\theta_3 \tag{e}$$

These two equations can be solved simultaneously for the two acceleration unknowns, $\ddot{\theta}_3$ and $\ddot{r}_4$

$$\ddot{\theta}_3 = \frac{-r_2 \cos\theta_2\, \ddot{\theta}_2 + r_2 \sin\theta_2\, \dot{\theta}_2^2 + r_3 \sin\theta_3\, \dot{\theta}_3^2}{r_3 \cos\theta_3} \tag{4-21}$$

$$\ddot{r}_4 = -r_2 \sin\theta_2\, \ddot{\theta}_2 - r_3 \sin\theta_3\, \ddot{\theta}_3 - r_2 \cos\theta_2\, \dot{\theta}_2^2 - r_3 \cos\theta_3\, \dot{\theta}_3^2 \tag{4-22}$$

The solution is now considered complete, since Eqs. (4-17) to (4-22) can be numerically evaluated (in that order) for each crank angle θ_2 given the dimensions r_1, r_2, and r_3 and the input velocity and acceleration $\dot{\theta}_2$ and $\ddot{\theta}_2$.

In preparation for the study of the dynamics of the internal combustion engine in Chap. 14, however, we should point out that, with substitutions from Eqs. (4-17), (4-19), and (4-21) and considerable manipulation, Eqs. (4-20) and (4-22) can be written as

$$\dot{r}_4 = r_2 \frac{\sin(\theta_3 - \theta_2)}{\cos\theta_3} \dot{\theta}_2 \tag{4-23}$$

$$\ddot{r}_4 = r_2 \frac{\sin(\theta_3 - \theta_2)}{\cos\theta_3} \ddot{\theta}_2 - r_2 \frac{\cos(\theta_3 - \theta_2)}{\cos\theta_3} \dot{\theta}_2^2 - \frac{r_2^2 \cos^2\theta_2}{r_3 \cos^3\theta_3} \dot{\theta}_2^2 \tag{4-24}$$

If we take the case of the radial slider-crank mechanism ($r_1 = 0$) and assume that r_3 is much greater than r_2 ($\cos\theta_3 \approx 1$), the following approximate solutions result

$$\dot{r}_4 \approx -r_2\left(\sin\theta_2 + \frac{r_2}{2r_3}\sin 2\theta_2\right)\dot{\theta}_2 \tag{4-25}$$

$$\ddot{r}_4 \approx -r_2\left(\sin\theta_2 + \frac{r_2}{2r_3}\sin 2\theta_2\right)\ddot{\theta}_2 - r_2\left(\cos\theta_2 + \frac{r_2}{r_3}\sin 2\theta_2\right)\dot{\theta}_2^2 \tag{4-26}$$

As another illustration of Raven's approach, consider the following example.

Example 4-8 Develop an expression for the angular acceleration of the output crank of a four-bar linkage.

SOLUTION Since this a continuation of Example 3-5, the position and velocity solutions are known. The loop-closure equation is taken from that example, and the notation corresponds to Fig. 2-13:

$$R_{BA}e^{j\theta_2} + R_{CB}e^{j\theta_3} = R_{DA} + R_{CD}e^{j\theta_4} \tag{1}$$

Remembering that all lengths are constant, we use Eq. (4-16) to take the second time derivative. This gives

$$-\dot{\theta}_2^2 R_{BA}e^{j\theta_2} + j\ddot{\theta}_2 R_{BA}e^{j\theta_2} - \dot{\theta}_3^2 R_{CB}e^{j\theta_3} + j\ddot{\theta}_3 R_{CB}e^{j\theta_3} = -\dot{\theta}_4^2 R_{CD}e^{j\theta_4} + j\ddot{\theta}_4 R_{CD}e^{j\theta_4} \tag{2}$$

Later manipulation is made easier if this equation is divided by $e^{j\theta_3}$

$$-\dot{\theta}_2^2 R_{BA} e^{j(\theta_2-\theta_3)} + j\ddot{\theta}_2 R_{BA} e^{j(\theta_2-\theta_3)} - \dot{\theta}_3^2 R_{CB} + j\ddot{\theta}_3 R_{CB} = -\dot{\theta}_4^2 R_{CD} e^{j(\theta_4-\theta_3)} + j\ddot{\theta}_4 R_{CD} e^{j(\theta_4-\theta_3)} \quad (3)$$

Because of this rotation of the real axis, the real component of Eq. (3) does not contain the unknown $\ddot{\theta}_3$,

$$-\dot{\theta}_2^2 R_{BA}\cos(\theta_2-\theta_3) - \ddot{\theta}_2 R_{BA}\sin(\theta_2-\theta_3) - \dot{\theta}_3^2 R_{CB} = -\dot{\theta}_4^2 R_{CD}\cos(\theta_4-\theta_3) - \ddot{\theta}_4 R_{CD}\sin(\theta_4-\theta_3) \quad (4)$$

and can be easily solved for $\ddot{\theta}_4$

$$\ddot{\theta}_4 = \frac{R_{BA}\sin(\theta_2-\theta_3)\ddot{\theta}_2 + R_{BA}\cos(\theta_2-\theta_3)\dot{\theta}_2^2 + R_{CB}\dot{\theta}_3^2 - R_{CD}\cos(\theta_4-\theta_3)\dot{\theta}_4^2}{R_{CD}\sin(\theta_4-\theta_3)} \qquad \textit{Ans.} \quad (4\text{-}27)$$

By dividing Eq. (2) by $e^{j\theta_4}$ and taking the real components, a solution can also be found for $\ddot{\theta}_3$

$$\ddot{\theta}_3 = \frac{R_{BA}\sin(\theta_2-\theta_4)\ddot{\theta}_2 + R_{BA}\cos(\theta_2-\theta_4)\dot{\theta}_2^2 + R_{CB}\cos(\theta_4-\theta_3)\dot{\theta}_3^2 - R_{CD}\dot{\theta}_4^2}{R_{CB}\sin(\theta_4-\theta_3)} \quad (4\text{-}28)$$

It will be noted from both the above examples that, as pointed out for the velocity equations in Sec. 3-8, the acceleration equations are also always linear in the unknowns. Therefore, their solution, although perhaps tedious, is also straightforward.

Chace's approach to acceleration analysis involves the derivatives of unit vectors. From Eq. (3-21), the first time derivative of a typical vector $\mathbf{R}$ is

$$\dot{\mathbf{R}} = \dot{R}\hat{\mathbf{R}} + \omega R(\hat{\mathbf{k}} \times \hat{\mathbf{R}}) \quad (f)$$

where $\omega\hat{\mathbf{k}}$ is the angular velocity of the vector $\mathbf{R}$. Differentiating again with respect to time gives

$$\ddot{\mathbf{R}} = \ddot{R}\hat{\mathbf{R}} + \dot{R}\dot{\hat{\mathbf{R}}} + \dot{\omega}R(\hat{\mathbf{k}} \times \hat{\mathbf{R}}) + \omega\dot{R}(\hat{\mathbf{k}} \times \hat{\mathbf{R}}) + \omega R(\hat{\mathbf{k}} \times \dot{\hat{\mathbf{R}}}) \quad (g)$$

But if we identify $\dot{\omega}$ as α, the angular acceleration of the $\mathbf{R}$ vector, and use Eq. (3-20), this reduces to

$$\ddot{\mathbf{R}} = \ddot{R}\hat{\mathbf{R}} + 2\omega\dot{R}(\hat{\mathbf{k}} \times \hat{\mathbf{R}}) - \omega^2 R\hat{\mathbf{R}} + \alpha R(\hat{\mathbf{k}} \times \hat{\mathbf{R}}) \quad (4\text{-}29)$$

This is a general expression for the second time derivative of any two-dimensional† vector.

The Chace approach to acceleration analysis will be illustrated by solving for the accelerations in the inverted slider-crank mechanism of Fig. 3-15. The loop-closure equation is

$$\mathbf{r}_1 + \mathbf{r}_2 = \mathbf{r}_4 \quad (h)$$

Using the general form, Eq. (4-29), and recognizing that r_1, r_2, and $\hat{\mathbf{r}}_1$ are constant, we take the second time derivative of Eq. (h)

$$-\omega_2^2 r_2\hat{\mathbf{r}}_2 + \alpha_2 r_2(\hat{\mathbf{k}} \times \hat{\mathbf{r}}_2) = \ddot{r}_4\hat{\mathbf{r}}_4 + 2\omega_4\dot{r}_4(\hat{\mathbf{k}} \times \hat{\mathbf{r}}_4) - \omega_4^2 r_4\hat{\mathbf{r}}_4 + \alpha_4 r_4(\hat{\mathbf{k}} \times \hat{\mathbf{r}}_4) \quad (i)$$

† The two-dimensional restriction results from the assumption that $\hat{\boldsymbol{\omega}}$ is the same as $\hat{\mathbf{k}}$.

Since the position and velocity solutions are known from Sec. 3-9, and since ω_2 and α_2 are given as input crank conditions, the only two unknowns in this equation are $\ddot{r}_4$ and α_4.

As was done in velocity analysis using Chace's method, we seek to eliminate one of the unknowns by careful choice of the directions along which components are taken. Noting that

$$\hat{\mathbf{r}}_4 \cdot (\hat{\mathbf{k}} \times \hat{\mathbf{r}}_4) = 0 \qquad \text{and} \qquad (\hat{\mathbf{k}} \times \hat{\mathbf{r}}_4) \cdot (\hat{\mathbf{k}} \times \hat{\mathbf{r}}_4) = 1$$

we take the dot product of each term of Eq. (*i*) with $\hat{\mathbf{k}} \times \hat{\mathbf{r}}_4$ in order to eliminate $\ddot{r}_4$

$$-\omega_2^2 r_2 \hat{\mathbf{r}}_2 \cdot (\hat{\mathbf{k}} \times \hat{\mathbf{r}}_4) + \alpha_2 r_2 (\hat{\mathbf{k}} \times \hat{\mathbf{r}}_2) \cdot (\hat{\mathbf{k}} \times \hat{\mathbf{r}}_4) = 2\omega_4 \dot{r}_4 + \alpha_4 r_4 \qquad (j)$$

from which we solve for α_4

$$\alpha_4 = \alpha_2 \frac{r_2}{r_4} (\hat{\mathbf{k}} \times \hat{\mathbf{r}}_2) \cdot (\hat{\mathbf{k}} \times \hat{\mathbf{r}}_4) - \omega_2^2 \frac{r_2}{r_4} \hat{\mathbf{r}}_2 \cdot (\hat{\mathbf{k}} \times \hat{\mathbf{r}}_4) - \frac{2\omega_4}{r_4} \dot{r}_4 \qquad (k)$$

Similarly, we can take the dot product of Eq. (*i*) with $\hat{\mathbf{r}}_4$ and eliminate α_4. This gives

$$\ddot{r}_4 = \alpha_2 r_2 (\hat{\mathbf{k}} \times \hat{\mathbf{r}}_2) \cdot \hat{\mathbf{r}}_4 - \omega_2^2 r_2 \hat{\mathbf{r}}_2 \cdot \hat{\mathbf{r}}_4 + \omega_4^2 r_4 \qquad (l)$$

4-9 THE INSTANT CENTER OF ACCELERATION

While it is of little use in analysis, it is desirable to define the *instant center of acceleration*, or *acceleration pole*, for a planar-motion mechanism, if only to avoid the implication that the instant center of velocity is also the instant center of acceleration. The instant center of acceleration is defined as *the instantaneous location of a pair of coincident points of two different rigid bodies where the absolute accelerations of the two points are equal.* If we consider a fixed and a moving body, the instantaneous center of acceleration is the point of the moving body which has zero absolute acceleration at the instant considered.

In Fig. 4-17*a*, let P be the instant center of acceleration, a point of zero absolute acceleration whose location is unknown. Assume that another point A of the moving plane has a known acceleration $\mathbf{A}_A$ and that $\boldsymbol{\omega}$ and $\boldsymbol{\alpha}$ of the moving plane are known. The acceleration-difference equation can then be written

$$\mathbf{A}_P = \mathbf{A}_A - \omega^2 \mathbf{R}_{PA} + \boldsymbol{\alpha} \times \mathbf{R}_{PA} = 0 \qquad (a)$$

Solving for $\mathbf{A}_A$ gives

$$\mathbf{A}_A = \omega^2 R_{PA} \hat{\mathbf{R}}_{PA} - \alpha R_{PA} (\hat{\mathbf{k}} \times \hat{\mathbf{R}}_{PA}) \qquad (b)$$

Now, since $\hat{\mathbf{R}}_{PA}$ is perpendicular to $\hat{\mathbf{k}} \times \hat{\mathbf{R}}_{PA}$, the two terms on the right of Eq. (*b*) are the rectangular components of $\mathbf{A}_A$, as shown in Fig. 4-17*b*. From this

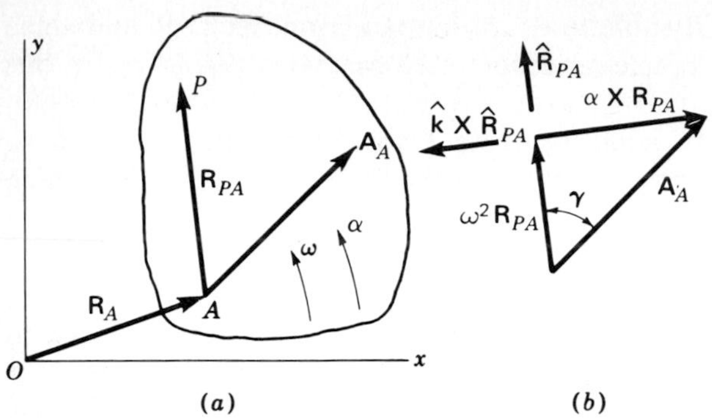

Figure 4-17 The instantaneous center of acceleration.

figure we can solve for the magnitude and direction of $\mathbf{R}_{PA}$

$$\gamma = \tan^{-1} \frac{\alpha}{\omega^2} \tag{4-30}$$

$$R_{PA} = \frac{A_A}{\sqrt{\omega^4 + \alpha^2}} = \frac{A_A \cos \gamma}{\omega^2} \tag{4-31}$$

Equation (4-31) states that the distance R_{PA} from point A to the instant center of acceleration can be found from the magnitude of the acceleration A_A of *any* point of the moving plane. Since the denominator ω^2 is always positive, the angle γ is always acute.

There are many graphical methods of locating the instant center of acceleration.† Here, without proof, we present one method. In Fig. 4-18 we are given points A and B and their absolute accelerations $\mathbf{A}_A$ and $\mathbf{A}_B$. Extend

† N. Rosenauer and A. H. Willis, "Kinematics of Mechanisms," Associated General Publications, Sydney, Australia, 1953, pp. 145–156; republished by Dover, New York, 1967; K. Hain (trans. by T. P. Goodman et al.), "Applied Kinematics," 2d ed., McGraw-Hill, New York, 1967, pp. 149–158.

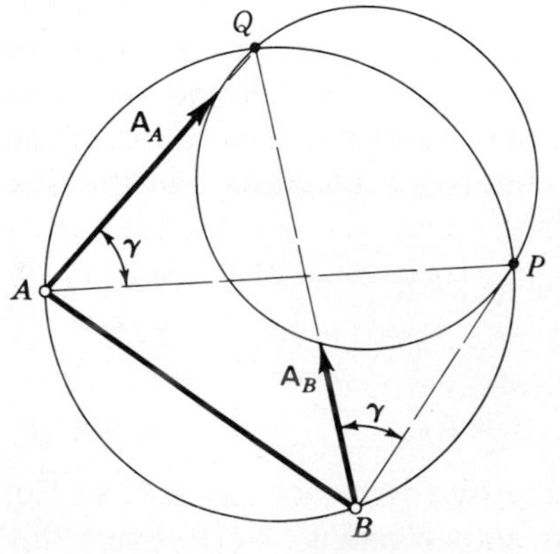

Figure 4-18 The four-circle method of locating the instantaneous center of acceleration P.

$\mathbf{A}_A$ and $\mathbf{A}_B$ until they intersect at Q; then construct a circle through points A, B, and Q. Now draw another circle through the termini of $\mathbf{A}_A$ and $\mathbf{A}_B$ and point Q. The intersection of the two circles locates point P, the instant center of acceleration.

4-10 THE EULER-SAVARY EQUATION†

In Sec. 4-5 we developed the apparent-acceleration equation (4-13). Then, in the examples which followed, we found that it was very important to carefully choose a point whose apparent path was known so that the radius of curvature of the path, needed for the normal component in Eq. (4-10), could be found by inspection. This need to know the radius of curvature of the path often dictates the method of approach to such a problem, as in Fig. 4-6*b*, and sometimes even requires the visualization of an equivalent mechanism. It would be more convenient if an arbitrary point could be chosen and the radius of curvature of its path could be calculated. In planar mechanisms this can be done by the methods presented below.

When two rigid bodies move relative to each other with planar motion, any arbitrarily chosen point A of one describes a path or locus relative to a coordinate system fixed to the other. At any given instant there is a point A', attached to the other body, which is the center of curvature of the locus of A. If we take the kinematic inversion of this motion, A' also describes a locus relative to the body containing A, and it so happens that A is the center of curvature of this locus. Each point therefore acts as the center of curvature of the path traced by the other, and the two points are called *conjugates* of each other. The distance between these two conjugate points is the radius of curvature of either locus.

Figure 4-19 shows two circles with centers at C and C'. Let us think of the circle with center C' as the fixed centrode and the circle with center C as the moving centrode of two bodies experiencing some particular relative planar motion. In actuality the fixed centrode need not be fixed but is attached to the body which contains the path whose curvature is sought. Also it is not necessary that the two centrodes be circles; we are interested only in instantaneous values and, for convenience, we will think of the centrodes as circles matching the curvatures of the two actual centrodes in the region near their point of contact P. As pointed out in Sec. 3-16, when the bodies containing the two centrodes move relative to each other, the centrodes appear to roll against each other without slip. Their point of contact P, of

† The most important and most useful references on this subject are Rosenauer and Willis, op. cit., chap. 4; A. E. R. de Jonge, A Brief Account of Modern Kinematics, *Trans. ASME*, vol. 65, 1943, pp. 663–683; R. S. Hartenberg and J. Denavit, "Kinematic Synthesis of Linkages," McGraw-Hill, New York, 1964, chap. 7; A. S. Hall, Jr., "Kinematics and Linkage Design," Prentice-Hall, Englewood Cliffs, N.J., 1961, chap. 5 (this book is a real classic on the theory of mechanisms and contains many useful examples); Hain, op. cit., chap. 4.

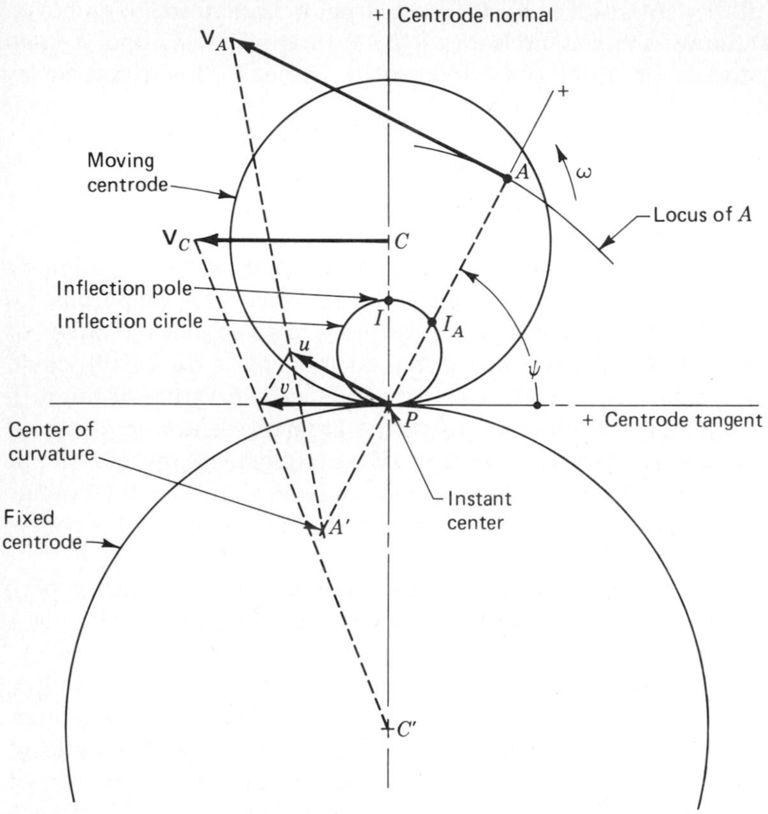

Figure 4-19 The Hartmann construction.

course, is the instant center of velocity. Because of these properties, we can think of the two circular centrodes as actually representing the shapes of the two moving bodies if this helps in visualizing the motion.

If the moving centrode has some angular velocity ω relative to the fixed centrode, the instantaneous velocity† of point C is

$$V_C = \omega R_{CP} \tag{a}$$

Similarly, the arbitrary point A, whose conjugate point A' we wish to find, has a velocity of

$$V_A = \omega R_{AP} \tag{b}$$

As the motion progresses, the point of contact of the two centrodes, and therefore the location of the instant center P, moves along both centrodes

† All velocities used in this section are actually apparent velocities, relative to the coordinate system of the fixed centrode; they are written as absolute velocities to simplify the notation.

with some velocity $\mathbf{v}$. As shown in the figure, $\mathbf{v}$ can be found by connecting a straight line from the terminus of $\mathbf{V}_C$ to the point C'. Alternatively, its size can be found from

$$v = \frac{R_{PC'}}{R_{CC'}} V_C \tag{c}$$

A graphical construction for A', the center of curvature of the locus of point A, is shown in Fig. 4-19 and is called the *Hartmann construction.* First the component $\mathbf{u}$ of the instant center's velocity $\mathbf{v}$ is found as that component parallel to $\mathbf{V}_A$ or perpendicular to $\mathbf{R}_{AP}$. Then the intersection of the line AP and a line connecting the termini of the velocities $\mathbf{V}_A$ and $\mathbf{u}$ gives the location of the conjugate point A'. The radius of curvature $\boldsymbol{\rho}$ of the locus of point A is $\boldsymbol{\rho} = \mathbf{R}_{AA'}$.

An analytical expression for locating point A' would also be desirable and can be derived from the Hartmann construction. The magnitude of the velocity $\mathbf{u}$ is given by

$$u = v \sin \psi \tag{d}$$

where ψ is the angle measured from the centrode tangent to the line of $\mathbf{R}_{AP}$. Then, noticing the similar triangles in Fig. 4-19, we can also write

$$u = \frac{R_{PA'}}{R_{AA'}} V_A \tag{e}$$

Now, equating the expressions of Eqs. (d) and (e) and substituting from Eqs. (a), (b), and (c) gives

$$u = \frac{R_{PC'}R_{CP}}{R_{CC'}} \omega \sin \psi = \frac{R_{PA'}R_{AP}}{R_{AA'}} \omega \tag{f}$$

Dividing through by $\omega \sin \psi$ and inverting, we get

$$\frac{R_{AA'}}{R_{AP}R_{PA'}} \sin \psi = \frac{R_{CC'}}{R_{CP}R_{PC'}} = \frac{\omega}{v} \tag{g}$$

Next, on noticing that $\mathbf{R}_{AA'} = \mathbf{R}_{AP} - \mathbf{R}_{A'P}$ and $\mathbf{R}_{CC'} = \mathbf{R}_{CP} - \mathbf{R}_{C'P}$, we can reduce this equation to the form

$$\left(\frac{1}{R_{AP}} - \frac{1}{R_{A'P}}\right) \sin \psi = \frac{1}{R_{CP}} - \frac{1}{R_{C'P}} \tag{4-32}$$

This important equation is one form of the *Euler-Savary equation.* Once the radii of curvature of the two centrodes R_{CP} and $R_{C'P}$ are known, this equation can be used to determine the positions of the two conjugate points A and A' relative to the instant center P.

Before continuing, a few words on sign conventions are in order. In using the Euler-Savary equation, we may arbitrarily choose a positive sense for the centrode tangent; the positive centrode normal is then 90° counterclockwise from it. This establishes a positive direction for the line CC' which may be used in assigning appropriate signs to R_{CP} and $R_{C'P}$. Similarly, an arbitrary positive direction can be chosen for the line AA'. The angle ψ is then taken as positive counterclockwise from the positive centrode tangent to the positive sense of line AA'. The sense of line AA' also gives the appropriate signs for R_{AP} and $R_{A'P}$ for Eq. (4-32).

There is a major drawback with the above form of the Euler-Savary equation in that the radii of curvature of both centrodes, R_{CP} and $R_{C'P}$, must be found. Usually they are not known any more than the curvature of the locus itself was known. However, this difficulty can be overcome by seeking a new form of the equation.

Let us consider the particular point labeled I in Fig. 4-19. This point is located on the centrode normal at a location defined by

$$\frac{1}{R_{IP}} = \frac{1}{R_{CP}} - \frac{1}{R_{C'P}} \tag{h}$$

If this particular point is chosen for A in Eq. (4-32), we find that its conjugate point I' must be located at infinity. The radius of curvature of the path of point I is infinite, and the locus of I therefore has an inflection point at I. The point I is called the *inflection pole.*

Let us now consider whether there are any other points I_A of the moving body which also have infinite radii of curvature at the instant considered. If so, then for each of these points $R_{I_AP} = \infty$ and, from Eqs. (4-32) and (h),

$$R_{I_AP} = R_{IP} \sin \psi \tag{4-33}$$

This equation defines a circle called the *inflection circle* whose diameter is R_{IP}, as shown in Fig. 4-19. Every point on this circle has its conjugate point at infinity, and each therefore has an infinite radius of curvature at the instant shown.

Now, with the help of Eq. (4-33), the Euler-Savary equation can be written in the form

$$\frac{1}{R_{AP}} - \frac{1}{R_{A'P}} = \frac{1}{R_{I_AP}} \tag{4-34}$$

Also, after some further manipulations, this can be put in the form

$$\rho = R_{AA'} = \frac{R_{AP}^2}{R_{AI_A}} \tag{4-35}$$

Either of these two forms of the Euler-Savary equation, Eqs. (4-34) and (4-35), is more useful in practice than Eq. (4-32), since they do not require knowledge of the curvatures of the two centrodes. They do require finding the inflection circle, but it will be shown how this can be done in the following example.

Example 4-9 Find the inflection circle for the motion of the coupler of the slider-crank linkage of Fig. 4-20 and determine the instantaneous radius of curvature of the path of the coupler point C.

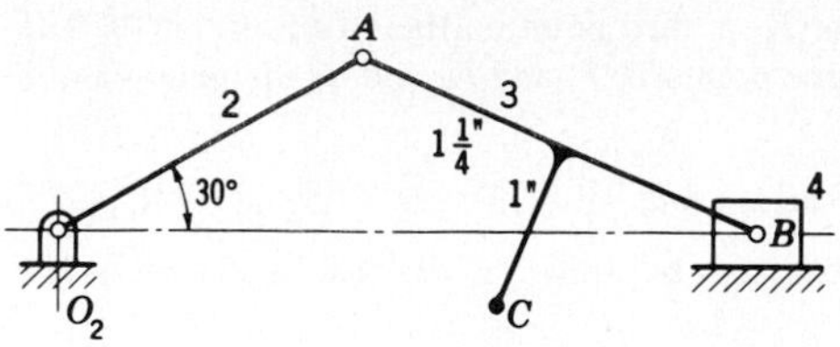

Figure 4-20 Example 4-9. $R_{AO_2} = 2$ in, $R_{BA} =$ 2.5 in.

SOLUTION We begin in Fig. 4-21 by locating the instant center P at the intersection of the line O_2A and a line through B perpendicular to its direction of travel. Points B and P must both lie on the inflection circle by definition; hence we need only one additional point to construct the circle.

The center of curvature of A is, of course, at O_2, which we now call A'. Taking the positive sense of the line AP as being downward and to the left, we measure $R_{AA'} = -2$ in and $R_{AP} = 2.64$ in. Then, substituting into Eq. (4-35), we obtain

$$R_{AI_A} = \frac{R_{AP}^2}{R_{AA'}} = \frac{2.64^2}{-2.00} = -3.48 \text{ in} \tag{1}$$

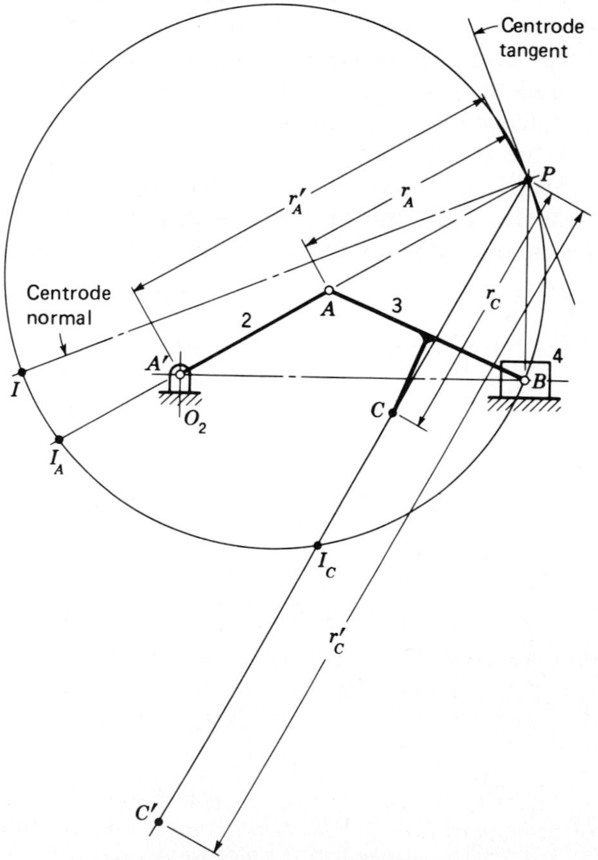

Figure 4-21 Example 4-9.

With this, we lay off 3.48 in from A to locate I_A, a third point on the inflection circle. The circle can now be constructed through the three points B, P, and I_A, and its diameter can be found

$$R_{IP} = 6.28 \text{ in} \qquad \textit{Ans.}$$

The centrode normal and centrode tangent can also be drawn, if desired, as shown in the figure.

Next, drawing the ray R_{CI_C} and taking its positive sense as downward and to the left, we can measure $R_{CP} = 3.1$ in and $R_{CI_C} = -1.75$ in. Substituting these into Eq. (4-35), we can solve for the instantaneous radius of curvature of the path of point C.

$$\rho = R_{CC'} = \frac{R_{CP}^2}{R_{CI_C}} = \frac{3.1^2}{-1.75} = -5.49 \text{ in} \qquad \textit{Ans.} \tag{2}$$

where the negative sign indicates that C' is below C on the line $C'CP$.

4-11 THE BOBILLIER CONSTRUCTIONS

The Hartmann construction, Sec. 4-10, provides one graphical method of finding the conjugate point and the radius of curvature of the path of a moving point, but it requires a knowledge of the curvature of the fixed and moving centrodes. It would be desirable to have *graphical* methods of obtaining the inflection circle and the conjugate of a given point without requiring the curvature of the centrodes. Such graphical solutions are presented in this section and are called the *Bobillier constructions.*

To understand these constructions, consider the inflection circle and the centrode normal N and centrode tangent T shown in Fig. 4-22. Let us select any two points A and B of the moving body which are not on a straight line

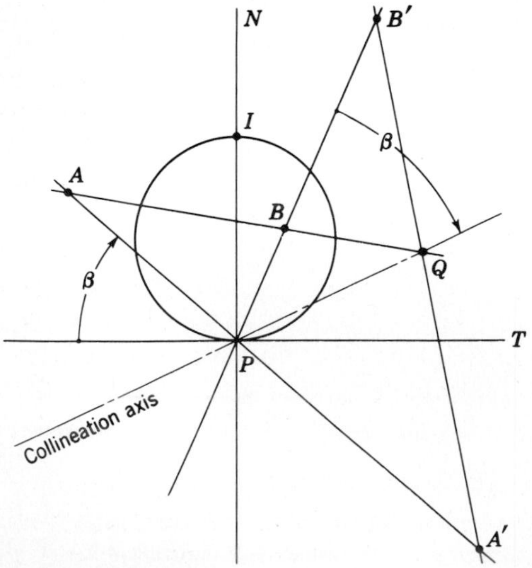

Figure 4-22 The Bobillier theorem.

through P. Now, by using the Euler-Savary equation, we could find the two corresponding conjugate points A' and B'. The intersection of the lines AB and $A'B'$ is labeled Q. Then, the straight line drawn through P and Q is called the *collineation axis*. This axis applies only to the two lines AA' and BB' and so is said to *belong* to these two rays; also the point Q will be located differently on the collineation axis if another set of points A and B are chosen on the same rays. Nevertheless, there is a unique relationship between the collineation axis and the two rays used to define it. This relationship is expressed in *Bobillier's theorem* which states that *the angle from the centrode tangent to one of these rays is the negative of the angle from the collineation axis to the other ray.*

In applying the Euler-Savary equation to a planar mechanism, two pairs of conjugate points can usually be found by inspection, and from these we wish to graphically determine the inflection circle. For example, a four-bar linkage with a crank O_2A and a follower O_4B has A and O_2 as one set of conjugate points and B and O_4 as the other when we are interested in the motion of the coupler relative to the frame. Given these two pairs of conjugate points, how do we use the Bobillier theorem to find the inflection circle?

In Fig. 4-23*a* let A and A' and B and B' represent the known pairs of conjugate points. Rays constructed through each pair intersect at P, the instant center of velocity, giving one point on the inflection circle. Point Q is located next by the intersection of a ray through A and B with a ray through A' and B'. Then the collineation axis can be drawn as the line PQ.

The next step is shown in Fig. 4-23*b*. Drawing a straight line through P parallel to $A'B'$, we identify the point W as the intersection of this line with the line AB. Now, through W we draw a second line parallel to the collineation axis. This line intersects AA' at I_A and BB' at $I_{B'}$, the two additional points on the inflection circle for which we are searching.

We could now construct the circle through the three points I_A, I_B, and P, but there is an easier way. Remembering that a triangle inscribed in a semicircle is a right triangle having the diameter as its hypotenuse, we erect a perpendicular to AP at I_A and another to BP at I_B. The intersection of these two perpendiculars gives point I, the inflection pole, as shown in Fig. 4-23*c*. Since PI is the diameter, the inflection circle, the centrode normal N, and the centrode tangent T can all be easily constructed.

To show that this construction satisfies the Bobillier theorem, note that the arc from P to I_A is inscribed by the angle which I_AP makes with the centrode tangent. But this same arc is also inscribed by the angle PI_BI_A. Therefore these two angles are equal. But the line I_AI_B was originally constructed parallel to the collineation axis. Therefore the line PI_B also makes the same angle β with the collineation axis.

Our final problem is to learn how to use the Bobillier theorem to find the conjugate of another arbitrary point, say C, when the inflection circle is given. In Fig. 4-24 we join C with the instant center P and locate the intersection

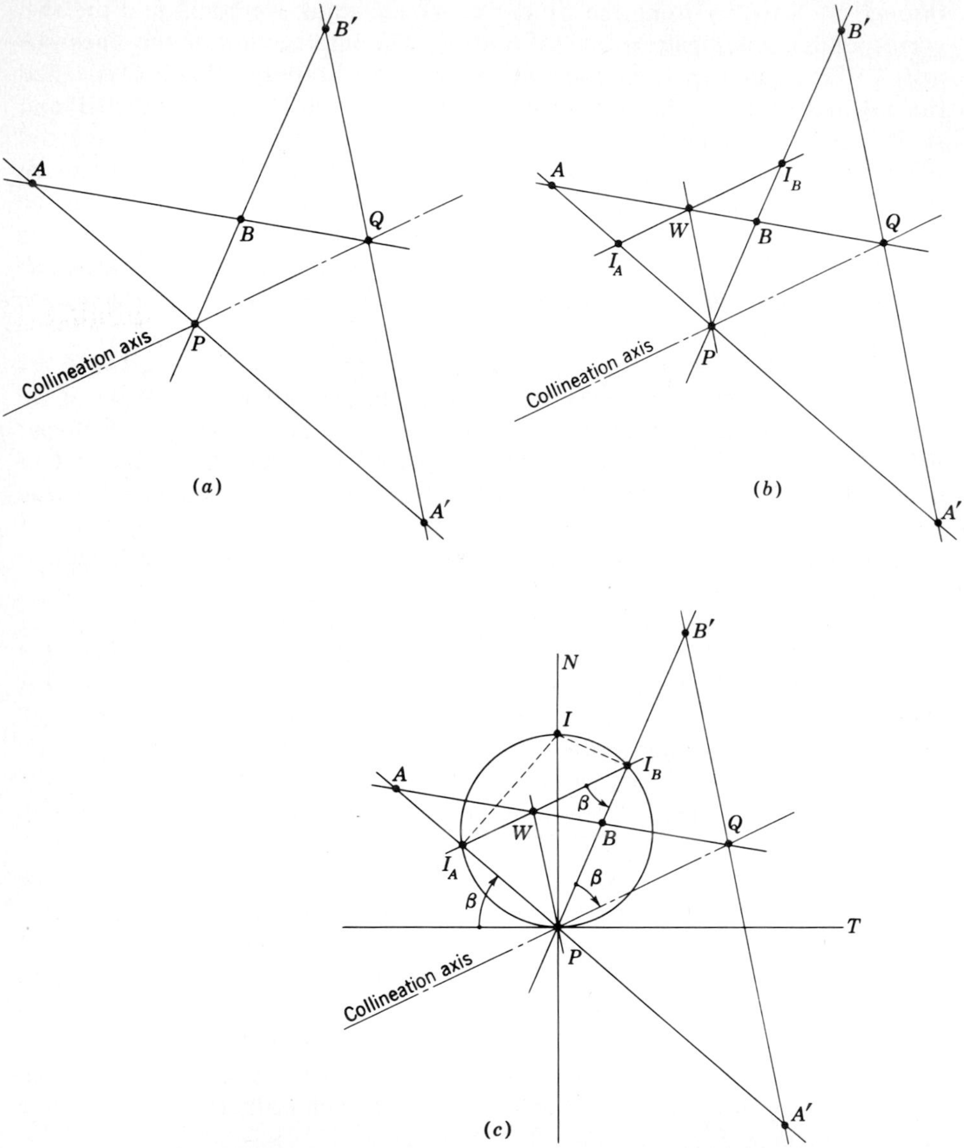

Figure 4-23 The Bobillier construction for locating the inflection circle.

point I_C with the inflection circle. This ray serves as one of the two necessary to locate the collineation axis. For the other we may as well use the centrode normal since I and its conjugate point I', at infinity, are both known. For these two rays the collineation axis is a line through P parallel to the line $I_C I$, as we learned in Fig. 4-22*c*. The balance of the construction is similar to that of Fig. 4-23. Q is located by the intersection of a line through I and C with

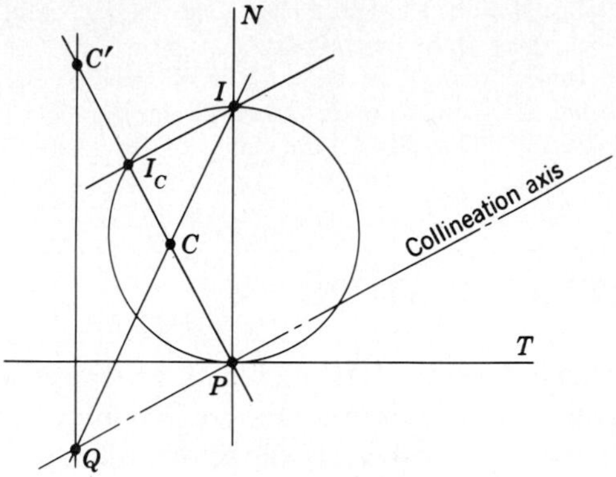

Figure 4-24 The Bobillier construction for locating the conjugate point C'.

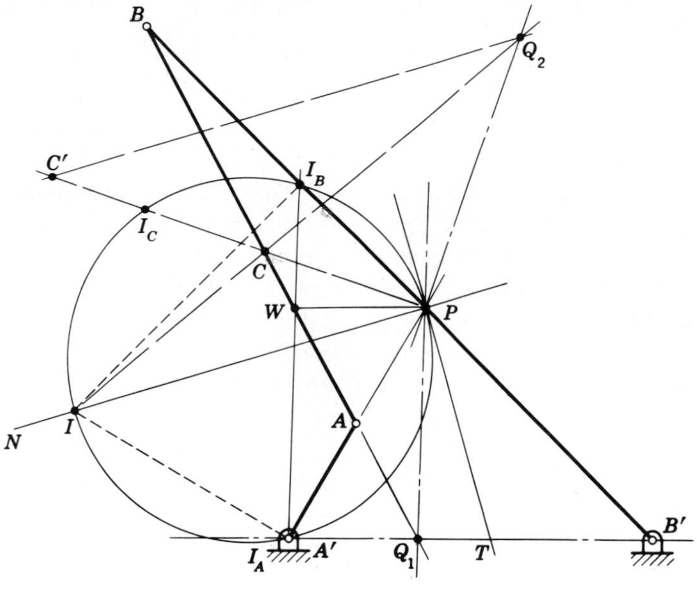

Figure 4-25 Example 4-10.

the collineation axis. Then a line through Q and I' at infinity intersects the ray PC at C', the conjugate point for C.

Example 4-10 Use the Bobillier theorem to find the center of curvature of the coupler curve of point C of the four-bar linkage shown in Fig. 4-25.

SOLUTION Locate the instant center P at the intersection of AA' and BB'; also locate Q_1 at the intersection of AB and $A'B'$. PQ_1 is the first collineation axis. Through P draw a line parallel to $A'B'$ to locate W on AB. Through W draw a line parallel to PQ_1 to locate I_A on AA' and I_B on BB'. Then, through I_A draw a perpendicular to AA' and through I_B draw a perpendicular to BB'. These perpendiculars intersect at the inflection pole I and define the inflection circle, the centrode normal N, and the centrode tangent T.

To obtain the conjugate point of C draw the ray PC and locate I_C on the inflection circle. The second collineation axis PQ_2, belonging to the pair of rays PC and PI, is a line through P parallel to a line (not shown) from I to I_C. The point Q_2 is obtained as the intersection of this collineation axis and a line IC. Now, through Q_2 draw a line parallel to the centrode normal; its intersection with the ray PC yields C', the center of curvature of the path of C.

4-12 THE CUBIC OF STATIONARY CURVATURE

Consider a point on the coupler of a planar four-bar linkage which generates a path relative to the frame whose radius of curvature, at the instant considered, is ρ. For most cases, since the coupler curve is of sixth order, this radius of curvature continuously changes as the point moves. In certain situations, however, the path will have stationary curvature, which means that

$$\frac{d\rho}{ds} = 0 \qquad (a)$$

where s is the distance traveled along the path. The locus of all points on the coupler or moving plane which have stationary curvature at the instant considered is called the *cubic of stationary curvature* or sometimes the *circling-point curve.* It should be noted that stationary curvature does not necessarily mean constant curvature but rather that the continually varying radius of curvature is passing through a maximum or minimum.

Here we will present a fast and simple graphical method of obtaining the cubic of stationary curvature, as described by Hain.† In Fig. 4-26 we have the four-bar linkage $A'ABB'$, with A' and B' the frame pivots. Then A and B have stationary curvature, in fact constant curvature about centers at A' and B'; hence, A and B lie on the cubic.

The first step of the construction is to obtain the centrode normal and centrode tangent. Since the inflection circle is not needed, we locate the collineation axis PQ as shown and draw the centrode tangent T at the angle ψ from the line PB', equal but opposite in direction to the angle ψ from the line PA' to the collineation axis. This construction follows directly from Bobillier's theorem. We can also construct the centrode normal N. At this point it is convenient to reorient the drawing on the board so that the T square or horizontal edge of the drafting machine lies along the centrode normal.

Next construct a line through A perpendicular to PA and another line through B perpendicular to PB. These lines intersect the centrode normal and centrode tangent at A_N, A_T and B_N, B_T, respectively, as shown in Fig. 4-26. Now draw the two rectangles $PA_NA_GA_T$ and $PB_NB_GB_T$; the points A_G and B_G define an auxiliary line G which is used to obtain other points on the cubic.

Now choose any point S_G on the line G. A ray parallel to N locates S_T and another parallel to T locates S_N. Connect S_T with S_N and draw a

† Hain, op. cit., pp. 498–502.

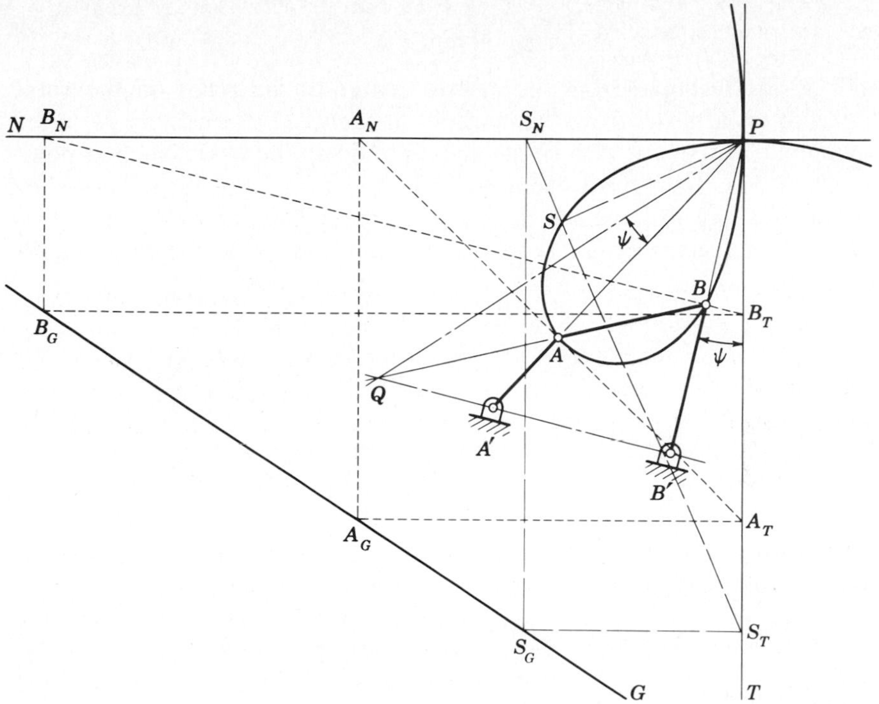

Figure 4-26 The cubic of stationary curvature.

perpendicular to this line through P; this locates point S, another point on the cubic of stationary curvature. This process is now repeated as often as desired by choosing different points on G, and the cubic is drawn as a smooth curve through all the points S obtained.

Note that the cubic of stationary curvature has two tangents at P, the *centrode-normal tangent* and the *centrode-tangent tangent*. The radius of curvature of the cubic at these tangents is obtained as follows. Extend G to intersect T at G_T and N at G_N (not shown). Then, half the distance PG_T is the radius of curvature of the cubic at the centrode-normal tangent, and half the distance PG_N is the radius of curvature of the cubic at the centrode-tangent tangent.

A point with interesting properties occurs at the intersection of the cubic of stationary curvature with the inflection circle; it is called *Ball's point*. A point of the coupler coincident with Ball's point describes a path which is approximately a straight line because it has stationary curvature and is located at an inflection point of its path.

The equation of the cubic of stationary curvature† is

† For a derivation of this equation see Hall, op. cit., p. 98, or Hartenberg and Denavit, op. cit., p. 206.

$$\frac{1}{M \sin \psi} + \frac{1}{N \cos \psi} - \frac{1}{r} = 0 \tag{4-36}$$

where r is the distance from the instant center to the point on the cubic, measured at an angle ψ from the centrode tangent. The constants M and N are obtained by using any two points known to lie on the cubic, such as points A and B of Fig. 4-26. It so happens† that M and N are, respectively, the diameters PG_T and PG_N of the circles centered on the centrode tangent and centrode normal whose radii represent the curvatures of the cubic at the instant center.

PROBLEMS‡

4-1 The position vector of a point is defined by the equation

$$\mathbf{R} = \left(4t - \frac{t^3}{3}\right)\hat{\mathbf{i}} + 10\hat{\mathbf{j}}$$

where R is in inches and t is in seconds. Find the acceleration of the point at $t = 2$ s.

4-2 Find the acceleration at $t = 3$ s of a point which moves according to the equation

$$\mathbf{R} = \left(t^2 - \frac{t^3}{6}\right)\hat{\mathbf{i}} + \frac{t^3}{3}\hat{\mathbf{j}}$$

The units are metres and seconds.

4-3 The path of a point is described by the equation

$$\mathbf{R} = (t^2 + 4)e^{-j\pi t/10}$$

where R is in millimetres and t is in seconds. For $t = 20$ s find the unit tangent vector for the path, the normal and tangential components of the point's absolute acceleration, and the radius of curvature of the path.

4-4 The motion of a point is described by the equations

$$x = 4t \cos \pi t^3 \qquad \text{and} \qquad y = \frac{t^3 \sin 2\pi t}{6}$$

where x and y are in feet and t is in seconds. Find the acceleration of the point at $t = 1.40$ s.

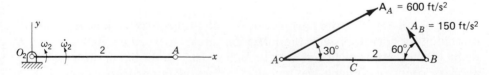

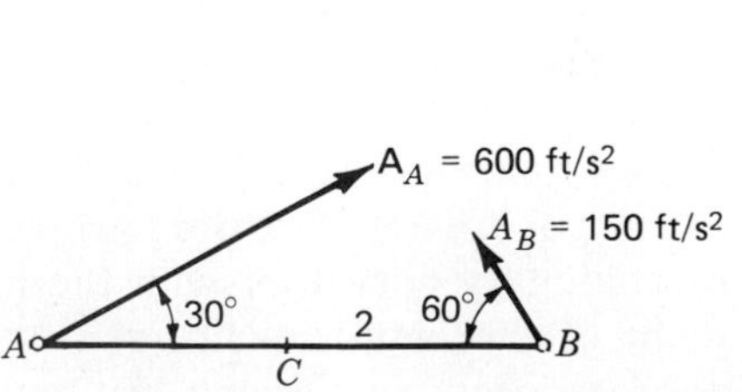

Problem 4-5 $R_{AO_2} = 500$ mm. **Problem 4-6** $R_{BA} = 20$ in.

† D. C. Tao, "Applied Linkage Synthesis," Addison-Wesley, Reading, Mass., 1964, p. 111.

‡ When assigning problems, the instructor may wish to specify the method of solution to be used since a variety of approaches are provided in the text.

4-5 Link 2 in the figure has an angular velocity of $\omega_2 = 120$ rad/s ccw and an angular acceleration of 4800 rad/s^2 ccw at the instant shown. Determine the absolute acceleration of point A.

4-6 Link 2 is rotating clockwise as shown in the figure. Find its angular velocity and acceleration and the acceleration of its midpoint C.

4-7 For the data given in the figure find the velocity and acceleration of points B and C.

4-8 For the straight-line mechanism shown in the figure $\omega_2 = 20$ rad/s cw and $\alpha_2 = 140$ rad/s^2 cw. Determine the velocity and acceleration of point B and the angular acceleration of link 3.

4-9 In the figure, the slider 4 is moving to the left with a constant velocity of 20 m/s^2. Find the angular velocity and acceleration of link 2.

4-10 Solve Prob. 3-8 for the acceleration of point A and the angular acceleration of link 3.

4-11 For Prob. 3-9 find the angular accelerations of links 3 and 4.

4-12 Solve Prob. 3-10 for the acceleration of point C and the angular accelerations of links 3 and 4.

4-13 For Prob. 3-11 find the acceleration of point C and the angular accelerations of links 3 and 4.

4-14 Using the data of Prob. 3-13, solve for the accelerations of points C and D and the angular acceleration of link 4.

4-15 For Prob. 3-14 find the acceleration of point C and the angular acceleration of link 4.

4-16 Solve Prob. 3-16 for the acceleration of point C and the angular acceleration of link 4.

4-17 For Prob. 3-17 find the acceleration of point B and the angular accelerations of links 3 and 6.

4-18 For the data of Prob. 3-18, what angular acceleration must be given to link 2 for the position shown to make the angular acceleration of link 4 zero?

4-19 For the data of Prob. 3-19, what angular acceleration must be given to link 2 for the angular acceleration of link 4 to be 100 rad/s^2 cw at the instant shown?

4-20 Solve Prob. 3-20 for the acceleration of point C and the angular acceleration of link 3.

4-21 For Prob. 3-21 find the acceleration of point C and the angular acceleration of link 3.

4-22 Find the acceleration of points B and D of Prob. 3-22.

4-23 Find the accelerations of points B and D of Prob. 3-23.

4-24 to 4-30 The nomenclature for this group of problems is shown in the figure, and the dimensions and data are given in the accompanying table. For each problem determine θ_3, θ_4, $\boldsymbol{\omega}_3$,

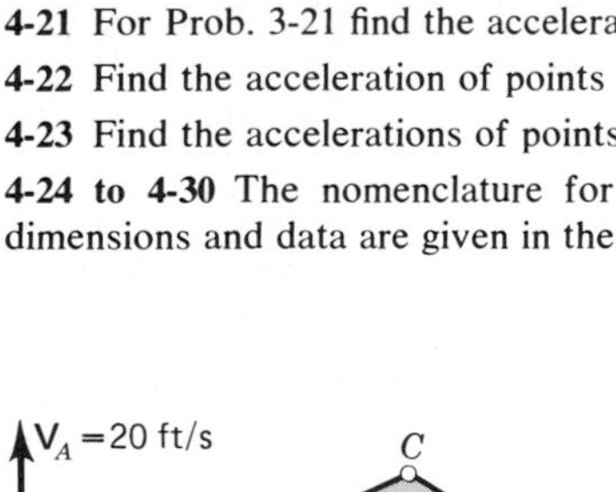

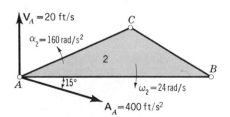

Problem 4-7 $R_{BA} = 16$ in, $R_{CA} = 10$ in, $R_{CB} = 8$ in.

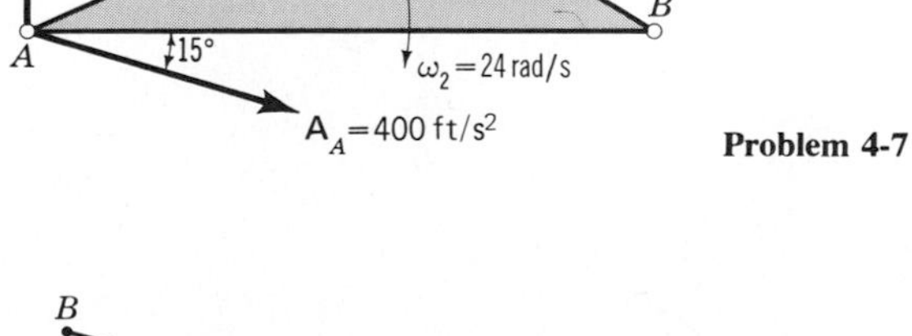

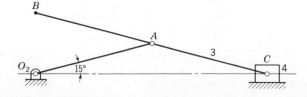

Problems 4-8 and 4-9 $R_{AO_2} = R_{CA} = R_{BA} = 100$ mm.

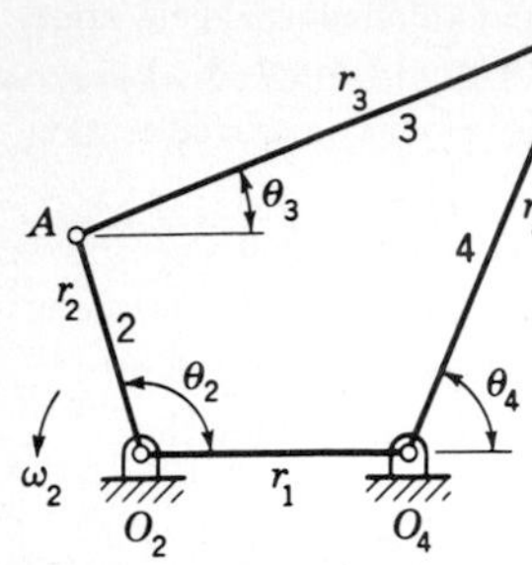

Problems 4-24 to 4-30

ω_4, α_3, and α_4. The angular velocity ω_2 is constant for each problem, and a negative sign is used to indicate the clockwise direction. The dimensions of even-numbered problems are in inches; odd-numbered problems are given in millimetres.

Prob.	r_1	r_2	r_3	r_4	θ_2, deg	ω_2, rad/s
4-24	4	6	9	10	240	1
4-25	100	150	250	250	−45	56
4-26	14	4	14	10	0	10
4-27	250	100	500	400	70	−6
4-28	8	2	10	6	40	12
4-29	400	125	300	300	210	−18
4-30	16	5	12	12	315	−18

4-31 Crank 2 of the system shown has a speed of 60 rpm ccw. Find the velocity and acceleration of point B and the angular velocity and acceleration of link 4.

4-32 The mechanism shown in the figure is a marine steering gear called *Rapson's slide.* O_2B is the tiller, and AC is the actuating rod. If the velocity of AC is 10 in/min to the left, find the angular acceleration of the tiller.

4-33 Determine the acceleration of link 4 of Prob. 3-26.

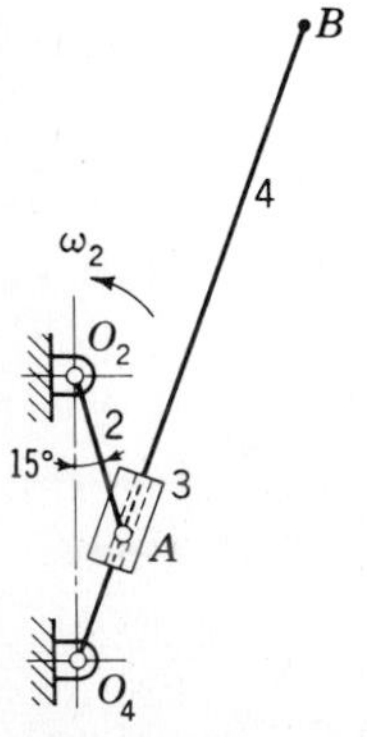

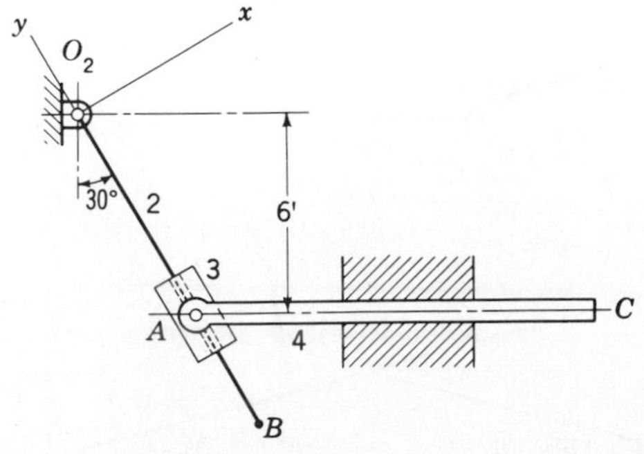

Problem 4-31 $R_{O_4O_2} = 12$ in, $R_{AO_2} = 7$ in, $R_{BO_4} = 28$ in. **Problem 4-32.**

4-34 For Prob. 3-27 find the acceleration of point *E*.

4-35 Find the acceleration of point *B* and the angular acceleration of link 4 of Prob. 3-24.

4-36 For Prob. 3-25 find the acceleration of point *B* and the angular acceleration of link 3.

4-37 Assuming that both links 2 and 3 of Prob. 3-28 are rotating at constant speed, find the acceleration of point P_4.

4-38 Solve Prob. 3-32 for the accelerations of points *A* and *B*.

4-39 For Prob. 3-33, determine the acceleration of point C_4 and the angular acceleration of link 3 if crank 2 is given an angular acceleration of 2 rad/s² ccw.

4-40 Determine the angular accelerations of links 3 and 4 of Prob. 3-30.

4-41 For Prob. 3-31 determine the acceleration of point *G* and the angular accelerations of links 5 and 6.

4-42 Find the inflection circle for motion of the coupler of the double-slider mechanism shown in the figure. Select several points on the centrode normal and find their conjugate points. Plot portions of the paths of these points to demonstrate for yourself that the conjugates are indeed the centers of curvature.

4-43† Find the inflection circle for motion of the coupler relative to the frame of the linkage shown in the figure. Find the center of curvature of the coupler curve of point *C* and generate a portion of the path of *C* to verify your findings.

4-44 For the motion of the coupler relative to the frame, find the inflection circle, the centrode normal, the centrode tangent, and the centers of curvature of points *C* and *D* of the linkage of

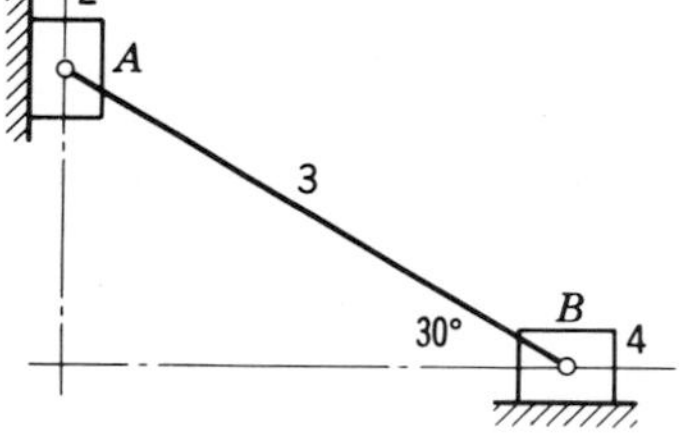

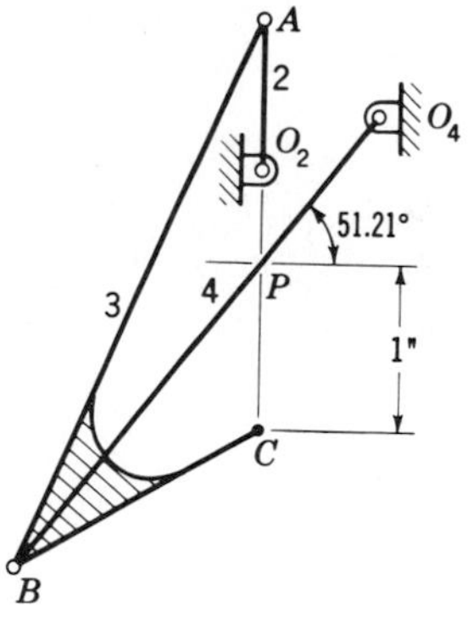

Problem 4-42 R_{BA} = 125 mm. **Problem 4-43** $R_{CA} = 2.5$ in, $R_{AO_2} = 0.9$ in, $R_{BO_4} = 3.5$ in, $R_{PO_4} = 1.17$ in.

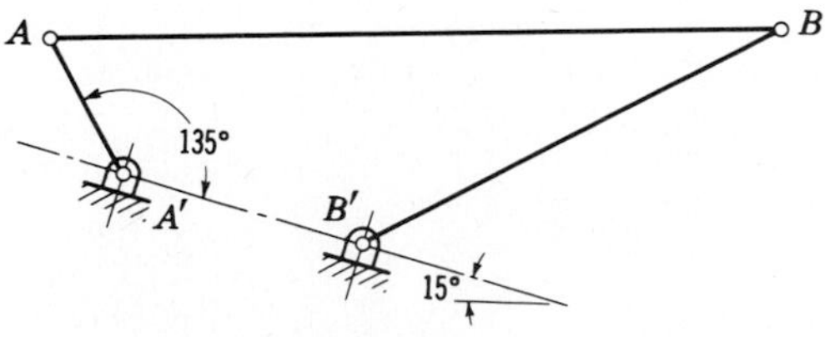

Problem 4-45 $R_{AA'} = 1$ in, $R_{BA} = 5$ in, $R_{B'A'} = 1.75$ in, $R_{BB'} = 3.25$ in.

† This mechanism appears in D. Tesar and J. C. Wolford, Five Point Exact Four-Bar Straight-Line Mechanisms, *Trans. 7th Conf. Mech.*, Penton, Cleveland, Ohio, 1962.

Prob. 3-13. Choose points on the coupler coincident with the instant center and inflection pole and plot their paths.

4-45 On 18- by 24-in paper draw the linkage shown in the figure in full size, placing A' at 6 in from the lower edge and 7 in from the right edge. Better utilization of the paper is obtained by tilting the frame through about 15° as shown.

(*a*) Find the inflection circle.

(*b*) Draw the cubic of stationary curvature.

(*c*) Choose a coupler point C coincident with the cubic and plot a portion of its coupler curve in the vicinity of the cubic.

(*d*) Find the conjugate point C'. Draw a circle through C with center at C' and compare this circle with the actual path of C.

(*e*) Find Ball's point. Locate a point D on the coupler at Ball's point and plot a portion of its path. Compare the result with a straight line.

CHAPTER

FIVE

NUMERICAL METHODS IN KINEMATIC ANALYSIS

5-1 INTRODUCTION

The first four chapters have been concerned with the development of a sound theoretical background for the kinematic analysis of mechanisms. Methods for position, displacement, velocity, and acceleration analysis have been presented, and examples have been shown of how these methods may be applied to the solution of planar problems.

By their very definitions, velocity and acceleration solutions are problems of *vector* analysis. Yet, although a rigorous vector notation has been used in all the preceding developments, a variety of solution techniques have been presented, including graphical solutions, algebraic techniques, vector algebra, and complex-algebra methods. As we have seen, the theoretical foundation for all of these approaches is the same; yet each method of solution has its own characteristic strengths and weaknesses.

Historically, graphical techniques have played the predominant role in the solution of planar kinematics problems. This is readily understood by considering the advantages of the graphical approach: it is quickly and easily performed and offers a good insight into the workings of a particular mechanism because of the ease with which the solution steps can be visualized. It also avoids the tedious algebra inherent in the solution of higher-order or transcendental equations.

Yet, there are also disadvantages to the graphical approach. When working to a reasonable scale, a solution accurate to 1 or 2 percent can be achieved in most problems if sufficient care is taken. However, a high degree of accuracy cannot be expected from a graphical solution. Also, a graphical method is a good choice when analyzing a mechanism at a single position, but it becomes very laborious for multiple positions since each position must be started over as a completely new problem. Often the design of a machine requires finding the *maximum* velocity of a point or the *maximum* force transmitted through a joint throughout its cycle of operation. In these situations, when working graphically, it has been common to solve at only a few positions, to assume, without proof, that the values obtained are representative, and then to use a suitable factor of safety to cover this bold assumption.

Algebraic methods, on the other hand, whether based on complex algebra or vector algebra, do not suffer from either of the above disadvantages. The accuracy of the method is not limited by the algebra but only by the accuracies of the problem *data* and the care employed in the final *numerical evaluation* of the results. Also, once the *algebraic form* of the solution has been derived, it can be *evaluated* as often as desired at different positions of the mechanism with very little effort. The drawbacks of the algebraic approaches are the need for tedious mathematical manipulations which may be required in finding the form of the solution and the chance of a mathematical error since the close link to visualization and physical intuition is reduced.

Although, in balance, the historical preference has been in favor of the graphical approach, this preference was completely upset by the development of the digital computer and, more recently, the pocket electronic calculator. Before these tools emerged, the promise of increased accuracy for the algebraic approaches was somewhat fictitious since the slide rule showed little better accuracy than careful graphical constructions. The accuracy of the calculator or digital computer, on the other hand, far surpasses that required for mechanical design problems and requires no more effort on the part of the designer than careful preparation of the input data.

The second major advantage of the computer is the ability to save and reuse a working program. Thus it becomes worthwhile to perform the tedious mathematical manipulations in finding the form of the solution since this now need be done only once and can then be used for a wide variety of problems with differing dimensions or at different positions. Although the effort is perhaps greater for solving a particular problem at one position, this can be repaid by almost instantaneous solutions of other positions or with changes in link dimensions. This ability to reuse a program, although at first restricted to large digital computers, is now quite readily available also in *programmable* pocket calculators, with magnetic memory strips on which operational programs can be stored for later use.

With the ability to save and reuse them, it has become worthwhile to write quite complex computer or calculator programs, since the effort spent in writing them can be justified by repeated use. Some quite sophisticated programs (see Sec. 5-5) are generally available which allow a wide range of analysis capabilities even on very complex problems yet require only a minimal effort in data preparation by the designer. Indeed such areas as stress analysis have been completely revolutionized by computer-based approaches developed over the past several years. In time, this may happen also in the fields of kinematics of mechanisms or machine dynamics. However, at present, the primary need is for a basic understanding of the underlying principles of how the computer can be used in these areas, since the development and adoption of large general programs is still in its infancy.

The purpose of this chapter is to present a basic understanding of how the electronic calculator or digital computer can be used in solving the kinematic relations of the previous chapters. The basic methods used are the algebraic techniques, including vectors and complex algebra, which have already been treated in some depth. It is not the purpose of this chapter to redevelop the complex-algebra techniques of Raven, for example, but to present guidelines showing how they can be programmed for digital computation. The material for position, displacement, velocity, and acceleration analyses has been presented without mention of computers so far, so that this chapter can deal with these together. Here, we will see the general approach to using a computer in such problems and many hints on programming styles and procedures. Then, as we proceed to later chapters, we will often pause to reflect how their subject matter can be programmed.

This chapter is not intended to be a treatise on numerical analysis, nor is it intended to present the details of any particular computer programming language. The presentation here is intended to be quite general since each of us will find ourselves limited in our choice of computing facilities and available languages. Furthermore, computer technology is still advancing rapidly and any more specific approach would soon be outdated.

5-2 PROGRAMMING FOR AN ELECTRONIC CALCULATOR

In this section we present several examples suitable for solution using a programmable calculator. They are, of course, also of use on larger computers. Readers using a nonprogrammable calculator or one with only limited storage capability would follow the same strategies in solving these examples but would find it necessary to key in the operations anew each time.

In approaching a problem to be solved on a programmable calculator, the first step is to develop a suitable method, called an *algorithm*. It must be remembered that a calculator (or computer) can manipulate only numeric quantities, not algebraic symbols. Therefore it is necessary to develop a *closed-form algebraic solution* to the desired problem completely *before* it can be programmed. A calculator *cannot* be used for this purpose; it becomes useful only when it is time to *evaluate* the numeric answer for a specific set of numeric data.

In developing the algorithm for a problem in kinematics, any of the algebraic methods of the previous chapters can be used. Of course, the calculator cannot be made to read data such as link lengths from a drawing;

therefore, very careful consideration must be given to what minimum set of data will be required from the user. Also care must be taken to see that the solution steps are so ordered that at each step the required data are available either from the user or from a previous calculation step.

Example 5-1 Develop an algorithm suitable for a programmable electronic calculator for finding the sum

$$\mathbf{R} = \mathbf{r}_1 + \mathbf{r}_2 + \cdots + \mathbf{r}_i + \cdots + \mathbf{r}_m + \mathbf{a}_1 \times \mathbf{b}_1 + \mathbf{a}_2 \times \mathbf{b}_2 + \cdots + \mathbf{a}_j \times \mathbf{b}_j + \cdots + \mathbf{a}_n \times \mathbf{b}_n \tag{1}$$

where the input data are to be the cartesian coordinates of the three-dimensional vectors $\mathbf{r}_i$, $\mathbf{a}_j$, and $\mathbf{b}_j$

$$\mathbf{r}_i = r_i^x\hat{\mathbf{i}} + r_i^y\hat{\mathbf{j}} + r_i^z\hat{\mathbf{k}} \tag{2}$$

$$\mathbf{a}_j = a_j^x\hat{\mathbf{i}} + a_j^y\hat{\mathbf{j}} + a_j^z\hat{\mathbf{k}} \tag{3}$$

$$\mathbf{b}_j = b_j^x\hat{\mathbf{i}} + b_j^y\hat{\mathbf{j}} + b_j^z\hat{\mathbf{k}} \tag{4}$$

The final result is to be stored in cartesian coordinate form in memories 1, 2, and 3.

SOLUTION The following algorithm is presented as a series of steps, although it could as easily be shown in flowchart form.

Step 1. Set memories to zero.
Step 2. Accept integer data for m; store $-m$ in memory 4.
Step 3. If memory 4 is zero or positive, go to step 8.
Step 4. Accept data for r_i^x and add to memory 1.
Step 5. Accept data for r_i^y and add to memory 2.
Step 6. Accept data for r_i^z and add to memory 3.
Step 7. Add 1 to memory 4 and return to step 3.
Step 8. Accept integer data for n; store $-n$ in memory 4.
Step 9. If memory 4 is zero or positive, go to step 16.
Step 10. Accept data for a_j^x, a_j^y, and a_j^z; store in memories 5, 6, and 7, respectively.
Step 11. Accept data for b_j^x, b_j^y, and b_j^z; store in memories 8, 9, and 10, respectively.
Step 12. Compute $a_j^y b_j^z - a_j^z b_j^y$ and add result to memory 1.
Step 13. Compute $a_j^z b_j^x - a_j^x b_j^z$ and add result to memory 2.
Step 14. Compute $a_j^x b_j^y - a_j^y b_j^x$ and add result to memory 3.
Step 15. Add 1 to memory 4 and return to step 9.
Step 16. Display successively the contents of memories 1, 2, and 3 as the results for R^x, R^y, and R^z, respectively.

The following data can be used to check your programming. Given $m = 2$, $\mathbf{r}_1 = -4\hat{\mathbf{i}} + 2\hat{\mathbf{j}}$, $\mathbf{r}_2 = 2\hat{\mathbf{i}} - 3\hat{\mathbf{k}}$, $n = 2$, $\mathbf{a}_1 = \hat{\mathbf{i}} - 3\hat{\mathbf{j}}$, $\mathbf{b}_1 = 2\hat{\mathbf{i}} + 2\hat{\mathbf{k}}$, $\mathbf{a}_2 = 4\hat{\mathbf{i}} + 3\hat{\mathbf{j}}$, $\mathbf{b}_2 = \hat{\mathbf{i}}$. The solution vector is $\mathbf{R} = -8\hat{\mathbf{i}}$.

It will be noted from the above example that the care taken in stating each step of an algorithm precisely will greatly reduce the time required for writing a program and will eliminate many potential sources of error. Writing out each step or drawing a flowchart before programming will also aid in later searches for possible errors and in documenting the final program.

As the algorithm is developed, due consideration should also be given to the efficient use of available memories. Most calculator programs will find insufficient memories to be the limiting factor in the complexity of the algorithms which may be used. In the above example it can be seen how memory 4 was used to store both m and n and how each vector accepted as data was used before the next was accepted, rather than accepting and storing all vectors before the computations were begun. Thus the resulting program requires only 10 memories and is not limited as to the numbers of vectors m and n.

When programming is complete, be sure to document your program carefully. Otherwise you may forget the solution procedure when you next want to use the program. The documentation should include, as a minimum, a short description of the method used, any limiting assumptions, a list of the number, order, and form of the required input data, and a description of the number, order, form, and location of the final results. An example problem, with numerical data and solution, is also a recommended part of a well-documented program. Good documentation is probably the most important part of writing a program and yet is often the most neglected. This often results in costly redevelopment of already existing programs, since they have inadequate documentation and are therefore useless when the need for them arises.

Example 5-2 Develop an algorithm for a calculator program to compute the position, velocity, and acceleration of all links of an offset slider-crank mechanism. The dimensions r_1, r_2, and r_3, shown in Fig. 5-1, are to be accepted as data. The solution is to start at the specified crank angle θ_2 and be incremented by the specified angle $\Delta\theta_2$ as often as desired. The angular velocity of the crank, specified by the user, is to be assumed constant.

SOLUTION The development of the position, velocity, and acceleration equations for the offset slider-crank mechanism can be done by Raven's method using complex algebra, as described in detail in Secs. 2-10, 3-8, and 4-8. The final equations are Eqs. (4-17) to (4-22). The algorithm for their evaluation by a programmable calculator might be as follows:

Step 1. Accept numeric data for r_1, r_2, and r_3 and store in memories 1, 2, and 3, respectively.
Step 2. Accept numeric data for θ_2, $\Delta\theta_2$, and $\dot{\theta}_2$ and store in memories 4, 5, and 6, respectively.
Step 3. Compute $r_2 \sin\theta_2$ and $r_2 \cos\theta_2$ and store in memories 7 and 8, respectively.
Step 4. Compute and display $\theta_3 = -\sin^{-1}[(r_1 + r_2 \sin\theta_2)/r_3]$.
Step 5. Compute $r_3 \sin\theta_3$ and $r_3 \cos\theta_3$, and store in memories 9 and 10, respectively.
Step 6. Compute and display $r_4 = r_2 \cos\theta_2 + r_2 \cos\theta_3$.
Step 7. Compute and display $\dot{\theta}_3 = -(\dot{\theta}_2 r_2 \cos\theta_2)/(r_2 \cos\theta_3)$.
Step 8. Compute and display $\dot{r}_4 = -\dot{\theta}_2 r_2 \sin\theta_2 - \dot{\theta}_3 r_3 \sin\theta_3$.

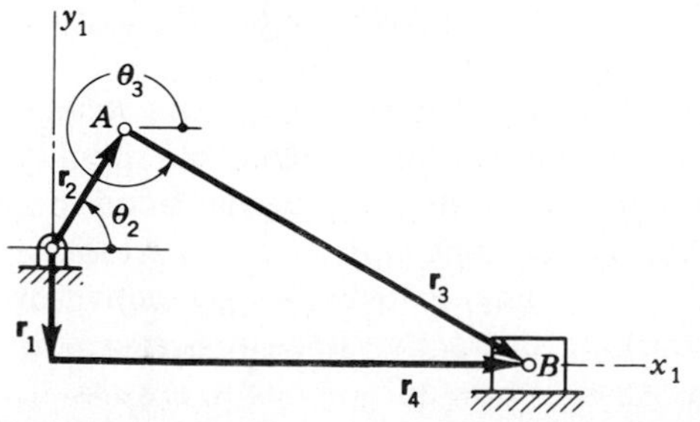

Figure 5-1 Example 5-2. Offset slider-crank mechanism.

Step 9. Compute and display $\ddot{\theta}_3 = (\dot{\theta}_2^2 r_2 \sin\theta_2 + \dot{\theta}_3^2 r_3 \sin\theta_3)/(r_3 \cos\theta_3)$.
Step 10. Compute and display $\ddot{r}_4 = -(\ddot{\theta}_3 r_3 \sin\theta_3 + \dot{\theta}_2^2 r_2 \cos\theta_2 + \dot{\theta}_3^2 r_3 \cos\theta_3)$.
Step 11. Add $\Delta\theta_2$ from memory 5 to θ_2 in memory 4.
Step 12. Return to step 3 and repeat.

As a check on the accuracy of the programming, use $r_1 = 0.150$ m, $r_2 = 0.300$ m, $r_3 = 0.900$ m, $\theta_2 = 0$, $\Delta\theta_2 = 90°$, and $\dot{\theta}_2 = 40$ rad/s. When the position $\theta_2 = 270°$ is reached, the set of rounded results displayed should be $\theta_3 = 9.594°$, $r_4 = 0.887$ m, $\dot{\theta}_3 = 0$, $\dot{r}_4 = 12$ m/s, $\ddot{\theta}_3 = -540.899$ rad/s^2, and $\ddot{r}_4 = 81.135$ m/s^2.

This example raises the issue of units. It is good practice to derive the equations for programs such as this without reference to a particular set of units. Then, any set of units can be used with the program as long as they are consistent. The alternative is to restrict the program to a particular set of units. In either case the choice of units should be made clear in the documentation. In the above example, the program itself is independent of the set of units used. However, since the test data were given in metres, the results had the units metres per second and metres per second squared.

The example also shows a typical case of compromises which arise between efficient use of memories and improved speed of computations. Memories 7 to 10 have been used in steps 3 and 5 to store geometric terms which recur repeatedly in the calculations of steps 4 and 6 to 10. By using these four memories the trigonometric functions are calculated only once each, thus saving considerable time. If sufficient memories were not available, they might be recalculated each time. Further memories were also saved by displaying the results immediately upon calculation rather than storing them.

Another issue which arises is the units to be used for the stored numeric values of angles. Certainly, for ease of use, any input data involving angles, such as θ_2 and $\Delta\theta_2$ in the above example, should be in degrees rather than radians; yet radians are preferable for angular velocities and accelerations. On most calculators, angles can be in either degrees or radians and trigonometric functions will be performed according to the setting of a keyboard option indicating the choice. In these cases, it is preferable to leave all angles in degrees. On some calculators, however, this choice is not available. Also, on digital computers using FORTRAN or BASIC, the trigonometric functions such as SIN, COS, or TAN will assume that angles are given in radians. In these cases it is necessary to convert angular input data into radians through extra program steps and to convert any calculated angles back into degrees for display.

Usually the best method of checking a program is by comparing results with a graphical solution of the same problem. Angles can be quickly measured using a protractor. Whereas a decimal point may be misplaced in a computer program, a vector which is 10 times too long will not be overlooked in a graphical solution. Also, graphical accuracy is usually sufficient for verifying that a computer program is working correctly, since programming errors will usually result in drastic rather than subtle differences in results.

Example 5-3 Develop an algorithm for a calculator program to compute the angular position and velocity of all links of a planar four-bar linkage. The link lengths r_1, r_2, r_3, and r_4 shown in Fig. 5-2, are to be accepted as data, along with the specified starting angle θ_2, the increment angle $\Delta\theta_2$, and the angular velocity $\dot{\theta}_2$ of the input crank.

SOLUTION The development of appropriate equations was done in Secs. 2-10 and 3-8. These are Eqs. (2-58), (2-59), (3-18), and (3-19). The algorithm for their evaluation is as follows:

Step 1. Accept numeric data for r_1, r_2, r_3, and r_4 and store in memories 1, 2, 3, and 4, respectively.
Step 2. Accept numeric data for θ_2, $\Delta\theta_2$, and $\dot{\theta}_2$ and store in memories 5, 6, and 7, respectively.
Step 3. Compute $a = (r_3^2 + r_4^2 - r_1^2 - r_2^2)/2r_3r_4$ and store in memory 8.
Step 4. Compute $b = r_1/r_3r_4$ and store in memory 9.
Step 5. Compute $c = r_2 \sin \theta_2$ and store in memory 10.
Step 6. Compute $d = r_2 \cos \theta_2$ and store in memory 11.
Step 7. Compute and display the transmission angle $\gamma = \cos^{-1}(a + bd)$.
Step 8. Compute $\sin \gamma$ and $\cos \gamma$ and store in memories 12 and 13.
Step 9. Compute and display

$$\theta_3 = 2\tan^{-1}\frac{-c + r_4 \sin\gamma}{d + r_3 - r_1 - r_4\cos\gamma}$$

and store in memory 14.
Step 10. Compute and display

$$\theta_4 = 2\tan^{-1}\frac{c - r_3\sin\gamma}{d + r_4 - r_1 - r_3\cos\gamma}.$$

Step 11. Compute and display

$$\dot{\theta}_3 = \frac{\dot{\theta}_2 r_2 \sin(\theta_4 - \theta_2)}{r_3 \sin\gamma}$$

Step 12. Compute and display

$$\dot{\theta}_4 = \frac{\dot{\theta}_2 r_2 \sin(\theta_3 - \theta_2)}{r_4 \sin\gamma}$$

Step 13. Add $\Delta\theta_2$ from memory 6 to θ_2 in memory 5.
Step 14. Return to step 5 and repeat.

To check your program, use a crank-rocker linkage having the dimensions $r_1 = 10$ in, $r_2 = 4$ in, $r_3 = 10$ in, and $r_4 = 12$ in. At $\theta_2 = 0$, with $\dot{\theta}_2 = 45$ rad/s, the rounded results are $\gamma = 30°$, $\theta_3 = 93.8°$, $\theta_4 = 123.7°$, and $\dot{\theta}_3 = \dot{\theta}_4 = -30$ rad/s.

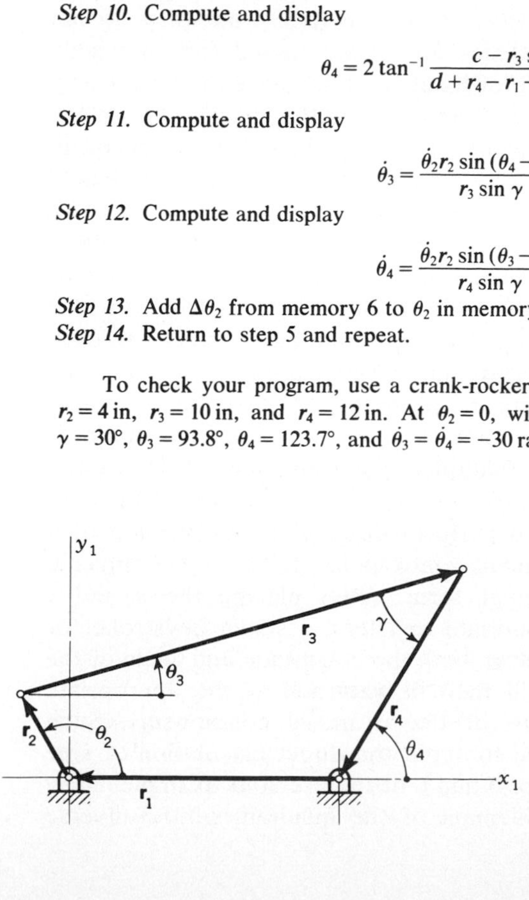

Figure 5-2 Example 5-3.

The algorithm described above gives the solution for the open configuration of a planar four-bar linkage. Depending on the starting situation, the user may wish to solve for the crossed configuration. As shown by Eqs. (2-58) and (2-59), the crossed-configuration case would be obtained by changing the plus and minus signs in the numerators of steps 9 and 10. This does require slight reprogramming as each set of problem data is run, an undesirable situation. An alternative would be to request another piece of input data specifying which configuration was sought and then modifying the algorithm to branch to the appropriate equations. Another alternative would be to calculate and display both configurations for each set of problem data. Each of these alternatives, however, requires more memories and a longer program.

A very closely related issue is the question of whether the original equations were developed precisely enough to permit the solutions for the open and crossed configurations of the linkage to be distinguished. If Eqs. (2-52) and (2-53) had been used rather than Eqs. (2-58) and (2-59), for example, there would be the need to see that the angles ϕ and ψ were both less than 180° and that ψ was positive while $\sin \phi$ was of the same sign as $\sin \theta_2$. These are not easy conditions to program and can be misleading when judging the suitability of an analytical solution. At first glance Eqs. (2-58) and (2-59) seem more complicated than Eqs. (2-52) and (2-53), but they do not require these extra tests. Such subtleties are often missed (or knowingly neglected) when developing equations for hand solution, since they can easily be dealt with by physical insight at the proper stage in the calculations. When writing a program, however, every detail must be defined with precision in such a way that it can be done on a calculator or computer, a highly unintelligent device.

Closely related is the question of the quadrants of angles, particularly when computing the inverse trigonometric functions such as the arc cosine of step 7 or the arc tangents of steps 9 and 10. Such inverse trigonometric functions are, by their mathematical definitions, multiple-valued. Yet each calculator or computer program will return only a single answer, chosen at the option of the manufacturer rather than the programmer. Extreme care must be taken to understand the particular choice of quadrant which will result on a particular calculator when using such multiple-valued functions. They are not always the same for different calculators.

Some calculators are equipped to perform directly the conversion of a two-dimensional vector from rectangular into polar form or the inverse conversion from polar into rectangular form. After placing the x and y components of a vector in the appropriate registers, a single keystroke (or program step) can then be made to return both the magnitude and angle of the vector. What is more, the angle will then be returned in the appropriate quadrant, determined from the signs of the x and y components. Such conversions, if available, can be used to avoid the direct calculation of sine and cosine, say as alternatives to steps 5 and 6 of the previous example. They can also be used in avoiding the dilemma of the quadrant of the inverse trigonometric functions if both the cosine and sine of the angle can be computed.†

5-3 PROGRAMMING THE CHACE EQUATIONS

When analyzing planar mechanisms, it is particularly useful to have a set of programs already written and tested for the solution of the four cases of the planar loop-closure equation involving two unknowns. These four cases were identified and discussed in Chap. 2. Graphical solution procedures were detailed in Sec. 2-7, complex-algebra solutions in Sec. 2-8, and Chace's vector-algebra solutions in Sec. 2-9. In this section we present algorithms for the numerical solution of each of the four cases using Chace's approach on an electronic calculator.

For notation, we assume that the two-dimensional loop-closure equation to be solved has been reduced to three vectors with two unknowns. It has the form

$$C\hat{\mathbf{C}} = A\hat{\mathbf{A}} + B\hat{\mathbf{B}} \tag{5-1}$$

where $\hat{\mathbf{C}}$, for example, is a unit vector along the vector $\mathbf{C}$; it makes an angle θ_C relative to the x axis and has components of $\hat{C}^x$ and $\hat{C}^y$ in the x and y directions.

Since the four cases will often be used together, it is useful to organize their data in similar fashion. Therefore, it is assumed that memories 1 to 12 are reserved for values of C, θ_C, C^x, C^y, A, θ_A, A^x, A^y, B, θ_B, B^x, and B^y, respectively. It is assumed that when the solution of a particular case is needed, known data will be entered into the appropriate memories. The program will then be loaded and run, leaving results in other appropriate memories. The problems of input of data and later display of results are therefore considered to be separate from the solution algorithms.

Example 5-4 Develop an algorithm for the solution of case 1 of Eq. (5-1) with C and θ_C as unknowns. Assume that data for A^x, A^y, B^x, and B^y are already stored in memories 5, 6, 9, and 10, respectively. Values of C^x, C^y, C, and θ_C should be computed and stored in memories 1 to 4, respectively.

SOLUTION The appropriate equation for the solution is Eq. (2-39). The algorithm is:

Step 1. Compute $C^x = A^x + B^x$ and store in memory 1.
Step 2. Compute $C^y = A^y + B^y$ and store in memory 2.
Step 3. Compute $C = \sqrt{(C^x)^2 + (C^y)^2}$ and store in memory 3.

† When writing a computer program in FORTRAN, the same advantage can be gained by use of the ATAN2 function rather than ATAN, ASIN, or ACOS. However, on some computers, ATAN2 (Y, X) will simply divide the first argument by the second and then use ATAN. Since this defeats the entire purpose of the ATAN2 function, it is necessary to write a subprogram ARCT (Y, X) to do what ATAN2 should do.

Step 4. Compute $\theta_C = \tan^{-1}(C^y/C^x)$; use signs of C^x and C^y to give correct quadrant; and store in memory 4.
Step 5. Stop.

To check the program, the following data may be used: store $A^x = 5$, $A^y = -8.661$, $B^x = -20$, $B^y = 0$ in memories 5, 6, 9, and 10, respectively. The rounded results, stored in memories 1 to 4, should be $C^x = -15.000$, $C^y = -8.661$, $C = 17.321$, and $\theta_C = 210.000°$.

Example 5-5 Develop an algorithm for the solution of case 2*a* of Eq. (5-1) with A and B as unknowns. Assume that data for C^x, C^y, θ_A, and θ_B are already stored in memories 1, 2, 8, and 12, respectively.

SOLUTION The solution for case 2*a* is given by Eqs. (2-40) and (2-41). The algorithm is:

Step 1. Compute $\hat{A}^x = \cos\theta_A$ and $\hat{A}^y = \sin\theta_A$ and store in memories 5 and 6.
Step 2. Compute $\hat{B}^x = \cos\theta_B$ and $\hat{B}^y = \sin\theta_B$ and store in memories 9 and 10.
Step 3. Compute $P = \cos(\theta_B - \theta_A)$ and store in memory 13.
Step 4. Compute $A = (C^x\hat{B}^y - C^y\hat{B}^x)/P$ and store in memory 7.
Step 5. Compute $B = (C^y\hat{A}^x - C^x\hat{A}^y)/P$ and store in memory 11.
Step 6. Multiply contents of memories 5 and 6 by A.
Step 7. Multiply contents of memories 9 and 10 by B.
Step 8. Stop.

The following data can be used to check the program: store $C^x = -15$, $C^y = -8.661$, $\theta_A = -60°$, $\theta_B = 180°$ in memories 1, 2, 8, and 12, respectively. The rounded results, stored in memories 5 to 7 and 9 to 11 should be $A^x = 5.000$, $A^y = -8.661$, $A = 10.000$, $B^x = -20.000$, $B^y = 0$, and $B = 20.000$.

Example 5-6 Develop an algorithm for the solution of case 2*b* of Eq. (5-1), with A and θ_B as unknowns. Assume that data for C^x, C^y, θ_A, and B are already stored in memories 1, 2, 8, and 11, respectively.

SOLUTION The solution for case 2*b* is given by Eqs. (2-42) and (2-43). The algorithm is:

Step 1. Compute $\hat{A}^x = \cos\theta_A$ and $\hat{A}^y = \sin\theta_A$ and store in memories 5 and 6.
Step 2. Compute $P = C^x\hat{A}^y - C^y\hat{A}^x$ and store in memory 13.
Step 3. Compute $Q = \sqrt{B^2 - P^2}$ and store in memory 14.
Step 4. Compute $A = C^x\hat{A}^x + C^y\hat{A}^y \mp Q$ and store in memory 7.
Step 5. Compute $B^x = P\hat{A}^y \pm Q\hat{A}^x$ and store in memory 9.
Step 6. Compute $B^y = -P\hat{A}^x \pm Q\hat{A}^y$ and store in memory 10.
Step 7. Multiply contents of memories 5 and 6 by A.
Step 8. Compute $\theta_B = tan^{-1}(B^y/B^x)$; use signs of B^x and B^y to give correct quadrant; and store in memory 12.
Step 9. Stop.

There are two solutions for case 2*b*, as discussed in Sec. 2-7. They appear as multiple signs in steps 4 to 6. It is recommended that two separate programs be written, one called case 2*b* using the upper signs, and one called case 2*b*′ using the lower signs.

The programs can be checked by using the following data: $C^x = -15$, $C^y = -8.661$, $\theta_A = -60°$, and $B = 20$, stored in memories 1, 2, 8, and 11, respectively. Program 2*b* should then give the rounded results $A^x = -5.000$, $A^y = 8.661$, $A = -10.000$, $B^x = -10.000$, $B^y = -17.321$, and $\theta_B = 240.000°$. The results of program 2*b*′ should be $A^x = 5.000$, $A^y = -8.661$, $A = 10.000$, $B^x = -20.000$, $B^y = 0.000$, and $\theta_B = 180.000°$. These should be stored in memories 5, 6, 7, 9, 10, and 12, respectively.

Example 5-7 Develop an algorithm for the solution of case 2*c* of Eq. (5-1), with θ_A and θ_B as unknowns. Assume that data for C^x, C^y, A, and B are already stored in memories 1, 2, 9, and 12, respectively.

SOLUTION The solution for case 2*c* is given by Eqs. (2-44) and (2-45). The algorithm is:

Step 1. Compute $P = (A^2 - B^2 + C^2)/2C^2$ and store in memory 13.
Step 2. Compute $Q = \sqrt{(A/C)^2 - P^2}$ and store in memory 14.
Step 3. Compute $A^x = PC^x \pm QC^y$ and store in memory 5.
Step 4. Compute $A^y = PC^y \mp QC^x$ and store in memory 6.
Step 5. Compute $\theta_A = \tan^{-1}(A^y/A^x)$; use signs of A^x and A^y to give correct quadrant; and store in memory 8.
Step 6. Compute $R = 1 - P$ and store in memory 13.
Step 7. Compute $B^x = RC^x \mp QC^y$ and store in memory 9.
Step 8. Compute $B^y = RC^y \pm QC^x$ and store in memory 10.
Step 9. Compute $\theta_B = \tan^{-1}(B^y/B^x)$; use signs of B^x and B^y to give correct quadrant; and store in memory 12.
Step 10. Stop.

Like case 2*b*, case 2*c* has two solutions and requires two separate programs. One program, using the upper signs of steps 3, 4, 7, and 8, may be called program 2*c*; the other, using the lower signs, may be called program 2*c*′.

The two programs can be checked with the following data: $C^x = 15$, $C^y = -8.661$, $A = 10$, and $B = 20$, stored in memories 1, 2, 7, and 11, respectively. Then program 2*c* should give $A^x = -5.000$, $A^y = 8.661$, $\theta_A = 120.000°$, $B^x = -10.000$, $B^y = -17.321$, and $\theta_B = 240.000°$. The results of program 2*c*′ should be $A^x = 5.000$, $A^y = -8.661$, $\theta_A = -60.000°$, $B^x = -20.000$, $B^y = 0.000$, and $\theta_B = 180.000°$. These should be stored in memories 5, 6, 8, 9, 10, and 12, respectively.

The programs developed in these four examples can be of great benefit in performing the position solution of most planar mechanisms. The procedures for their velocity and acceleration analysis using Chace's method are explained in Secs. 3-9 and 4-8. It is also useful to have available pretested programs for evaluating such vector operations as $\hat{\mathbf{k}} \times \mathbf{A}$, $\mathbf{A} \cdot \mathbf{B}$, and $(\hat{\mathbf{k}} \times \mathbf{A}) \cdot (\hat{\mathbf{k}} \times \mathbf{B})$ and a program for solving two simultaneous linear equations in two unknowns. With these the methods of Secs. 3-9 and 4-8 can be applied directly and evaluated quickly on a calculator.

Although Chace's method has been stressed in this section, it is easy to see how parallel programs could be developed using complex algebra and Raven's method. In fact, once programmed, there is very little difference between the methods and they can be freely intermixed. Their main difference is primarily one of notation and preference of the user. During computations, A^x and A^y, for example, play the role of components of a vector in Chace's approach or of the real and imaginary parts of a complex number in Raven's method.

For three-dimensional mechanism problems, of course, the above algorithms are of little benefit and must be generalized. Exactly parallel procedures can be developed for three dimensions in Chap. 11. However, the computations involved are necessarily more complex and usually go beyond the capabilities of a programmable calculator. Nevertheless, similar algorithms can be used on a digital computer, where memories are larger.

For those who prefer to work with a digital computer rather than a calculator, the above four algorithms can be programmed directly as stated in a language such as FORTRAN or BASIC. The memories mentioned would be replaced by variable names such as A, AX, THETAB, and so on. It is recommended that each algorithm be programmed as a separate procedure to be used within a larger main program written for each problem. For example, using FORTRAN, case 2*a* might be programmed as a SUBROUTINE:

```
SUBROUTINE CASE 2A (CX, CY, THETAA, THETAB, A, B)
```

It could then be called from a FORTRAN main program giving specific values

```
CALL CASE 2A (-15.0, -8.661, -60.0, 180.0, A, B)
```

and the values of A and B would be returned under those two variable names. Each application would then require writing a main program to suit the particular problem. However, the effort would be greatly reduced since these pretested subprograms would be available in a library and could be used as easily as a SIN or COS computation.

5-4 A COMPUTER PROGRAM FOR PLANAR MECHANISMS

As our wishes for more features in a program increase, we soon go beyond the capabilities of a programmable calculator and must use a digital computer. A computer, however, is not always accessible, and often has a cost associated with its use. Therefore, the calculator should be used when possible. Still, with its increased power, there is often sufficient justification to merit use of a computer. In the same way, when programming for a computer rather than a calculator, we should try to make use of as much of its capabilities as possible to make the use of the program more convenient, more flexible, more powerful, more readily understood, and so on. The effort and cost associated with programming a computer may be higher than for a calculator but are justified by the savings in repeated use.

Of course, all the algorithms developed so far in this chapter could be programmed for a digital computer. But, by their nature, they are perhaps better suited to a calculator. They are small programs, intentionally restricted in size to fit a calculator. They also leave a good amount of the burden of analysis to the user. They are algorithms for doing specific small computations, automation of a number of the small steps which often arise in the analysis of a mechanism. They are not algorithms for completely analyzing an entire linkage, as might be done on a computer. In this section we will study an algorithm which is suited to the kinematic analysis of *all* planar linkages in a single program.

When considering programming on a digital computer, it should also be borne in mind that the algorithm, the analysis procedure, should be suited to the capabilities of the computer, not the user, even though the input and output should be suited to the user. Often, the most straightforward procedure for hand solution is not the best for a computer. Such is the case in the algorithm to be explained below, which depends on *numerical iteration* rather than algebraic solution of the position equation.

In order to make the analysis technique more understandable, it will be explained in terms of the example problem shown in Fig. 5-3, but it should be kept in mind throughout that a general procedure and a single program are intended.

For the problem of Fig. 5-3 the loop-closure equation can be written as

$$\boldsymbol{\epsilon} = \mathbf{r}_1 + \mathbf{r}_3 + \mathbf{r}_4 - \mathbf{r}_2 = 0 \tag{a}$$

where the meaning of the symbol $\boldsymbol{\epsilon}$ will become clear as we proceed. In complex polar form this is

$$\boldsymbol{\epsilon} = r_1 e^{j\theta_1} + r_3 e^{j\theta_3} + r_4 e^{j\theta_4} - r_2 e^{j\theta_2} = 0 \tag{b}$$

and, for the general problem with n vectors, is

$$\boldsymbol{\epsilon} = \sum_{i=1}^{n} \pm r_i e^{j\theta_i} = 0 \tag{5-2}$$

It is not difficult to write a subprogram to evaluate the loop-closure equation for any particular mechanism. For our example, using FORTRAN, the subprogram might be written

```
SUBROUTINE LOOPEQ (LOOP)
COMPLEX LOOP (1), R
LOOP (1) = R(1) + R(3) + R(4) - R(2)
RETURN
END
```

This subroutine makes use of another FORTRAN subprogram named R, described below, which will evaluate a vector in complex polar form given its length and angle. When each of the vectors is evaluated and summed, the subroutine LOOPEQ returns the result, which is $\boldsymbol{\epsilon}$ of Eq. (b), in the complex variable named LOOP(1). In the general case, a problem might have several

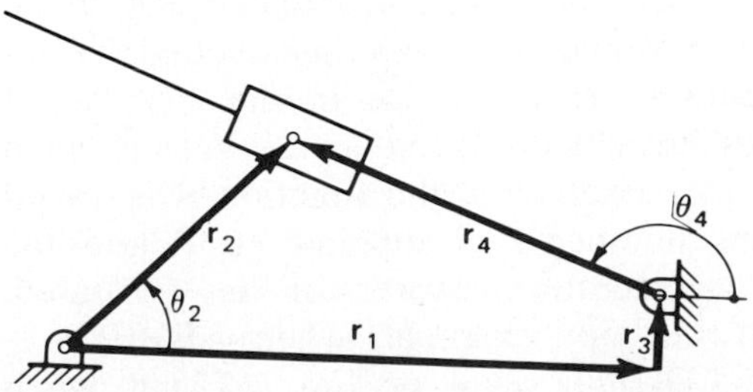

Figure 5-3 Inverted slider-crank mechanism.

loop-closure equations, and they would be programmed as LOOP(1), LOOP(2), and so on, in subroutine LOOPEQ.

In general, the velocities and accelerations will be found from the time derivatives of the loop-closure equation. From Eq. (5-2), these are

$$\dot{\epsilon} = \sum_{i=1}^{n} \pm(\dot{r}_i e^{j\theta_i} + j\dot{\theta}_i r_i e^{j\theta_i}) = 0 \tag{5-3}$$

$$\ddot{\epsilon} = \sum_{i=1}^{n} \pm(\ddot{r}_i e^{j\theta_i} + j2\dot{\theta}_i \dot{r}_i e^{j\theta_i} + j\ddot{\theta}_i r_i e^{j\theta_i} - \dot{\theta}_i^2 r_i e^{j\theta_i}) = 0 \tag{5-4}$$

which, for our example, became

$$\dot{\epsilon} = \dot{r}_4 e^{j\theta_4} + j\dot{\theta}_4 r_4 e^{j\theta_4} - j\dot{\theta}_2 r_2 e^{j\theta_2} = 0 \tag{c}$$

$$\ddot{\epsilon} = \ddot{r}_4 e^{j\theta_4} + j2\dot{\theta}_4 \dot{r}_4 e^{j\theta_4} + j\ddot{\theta}_4 r_4 e^{j\theta_4} - \dot{\theta}_4^2 r_4 e^{j\theta_4} - j\ddot{\theta}_2 r_2 e^{j\theta_2} + \dot{\theta}_2^2 r_2 e^{j\theta_2} = 0 \tag{d}$$

where $\dot{\theta}_2$ and $\ddot{\theta}_2$ are given, and $\dot{r}_4$, $\dot{\theta}_4$, $\ddot{r}_4$, and $\ddot{\theta}_4$ are unknown.

Now, rather than reprogram these expressions for each new problem, let us take advantage of subroutine LOOPEQ. Let us define the subprogram R so that when a certain value called LEVEL is set to 0, 1, or 2, the function R will compute the appropriate position, velocity, or acceleration expression, respectively. In FORTRAN, the function subprogram R might appear as follows:

```
      COMPLEX FUNCTION R(I)
      COMMON LEVEL,RM(20),RA(20),DRM(20),DRA(20),
    & DDRM(20),DDRA(20)
      COMPLEX Z,CMPLX,CEXP
      IF(LEVEL-1)1,2,3
    1 Z = CMPLX(RM(I),0.0)
      GO TO 4
    2 Z = CMPLX(DRM(I), DRA(I)*RM(I))
      GO TO 4
    3 Z = CMPLX(DDRM(I)-RM(I)*DRA(I)**2 +
    &          2.0*DRA(I)*DRM(I) + DDRA(I)*RM(I))
    4 R = Z*CEXP(CMPLX(0.0,RA(I)))
      RETURN
      END
```

When this subprogram is used for R, let us assume that proper data are already calculated for values of the magnitudes RM and angles RA of all vectors, and for their first derivatives DRM and DRA, and their second derivatives DDRM and DDRA with respect to time. Notice in the COMMON statement above that memory spaces have been saved for 20 vectors even though only 4 are used in our example. Each time the function R is called from subroutine LOOPEQ, a vector number I is supplied to R and the appropriate position, velocity, or acceleration expression is evaluated, depending on the value of LEVEL. In our example, when LEVEL = 0, subroutine LOOPEQ will evaluate LOOP(1) as ϵ of Eq. (*b*), but when

LEVEL = 1 or LEVEL = 2, subroutine LOOPEQ will evaluate LOOP(1) as $\dot{\epsilon}$ or $\ddot{\epsilon}$ of Eq. (*c*) or (*d*), respectively. In all cases, if the data are accurate, the result should be LOOP(1) = 0.

We might well question the purpose of calculating LOOP(1) if it is always zero for correct data. But this is exactly the point; if the data do *not* accurately satisfy the loop-closure condition, Eq. (*b*), LOOP(1) or $\boldsymbol{\epsilon}$ will contain a numerical evaluation of the *error* and can be used to numerically adjust the data until it is correct.

Suppose that a main program is written which starts by reading in the lengths and angles of all vectors at some initial position of the linkage. These data would be measured from a drawing, read into the main program, and stored in the arrays RM(20) and RA(20) of COMMON according to their vector numbers. Of course, the data for RM(1), RM(2), RM(3), RA(1), and RA(3) must be measured accurately since they represent fixed dimensions of the linkage. The angle RA(2) represents the input crank angle. RM(4) and RA(4) represent variable quantities which should be calculated by the program, and only approximate values need be given for them. Let us assume that there is some unknown error associated with each of these. Then the accurate values, $\bar{r}_4$ and $\bar{\theta}_4$, are

$$\bar{r}_4 = r_4 + \delta r \tag{e}$$

$$\bar{\theta}_4 = \theta_4 + \delta\theta_4 \tag{f}$$

where δr_4 and $\delta\theta_4$ represent the errors.

Since data are available for all RM and RA variables, the subprogram LOOPEQ could now be called but would return with LOOP(1) having a nonzero value of $\boldsymbol{\epsilon}$. Through a Taylor-series expansion of Eq. (*b*), we can approximate how this error in closure $\boldsymbol{\epsilon}$ is related to δr_4 and $\delta\theta_4$.

$$(r_1 e^{j\theta_1} + r_2 e^{j\theta_2} + r_3 e^{j\theta_3} + r_4 e^{j\theta_4}) + (e^{j\theta_4})\,\delta r_4 + (jr_4 e^{j\theta_4})\,\delta\theta_4 + \cdots = 0$$

Dropping higher-order terms, using Eq. (*b*), and rearranging gives

$$(e^{j\theta_4})\,\delta r_4 + (jr_4 e^{j\theta_4})\,\delta\theta_4 = -\boldsymbol{\epsilon} \tag{g}$$

This is a complex equation, having real and imaginary parts, and can therefore be solved for the two unknown errors δr_4 and $\delta\theta_4$. These can then be added to r_4 and θ_4 and the procedure can be repeated until the errors converge to within acceptable tolerance. At that time accurate values $\bar{r}_1$ and $\bar{\theta}_4$ will be stored in RM(4) and RA(4). This procedure is called the *Newton-Raphson iteration method.*† For the general linkage, the iteration equation is found by a Taylor-series expansion of Eq. (5-2)

$$\sum_{i=1}^{n} \pm (e^{j\theta_i})\,\delta r_i + (jr_i e^{j\theta_i})\,\delta\theta_i = -\boldsymbol{\epsilon} \tag{5-5}$$

† J. J. Uicker, Jr., et al., An Iterative Method for the Displacement Analysis of Spatial Linkages, *J. Appl. Mech.*, vol. 31, *ASME Trans.*, vol. 86, ser. E. pp. 309–314, 1964.

where all δr_i and $\delta\theta_i$ are zero except for those corresponding to the dependent variables. Since there are always twice as many dependent variables as loop-closure equations, there are equal numbers of equations and unknown error terms. The equations are linear in the error terms and can be solved by a matrix-inversion program, available in most digital-computer libraries of standard programs.

You may validly object that a lot of special programming is required to form the coefficients of Eq. (g) and that the form of these coefficients changes for each new problem, thus defeating our goal of writing a general program. However, this can be avoided by a method due to Wengert.† Suppose that we set the velocity data DRM and DRA, representing $\dot{r}_i$ and $\dot{\theta}_i$, to zero for all vectors. Then suppose that we set DRM(4), representing $\dot{r}_4$, to 1. If we use these data for velocities, even though they are not correct, a check of Eq. (c) will show that we obtain

$$\dot{\boldsymbol{\epsilon}} = e^{j\theta_4}$$

Similarly, if all velocities are set to zero except DRA(4) $= \dot{\theta}_4 = 1$, then

$$\dot{\boldsymbol{\epsilon}} = jr_4e^{j\theta_4}$$

Generalizing on this, we see that by setting all velocity data to zero, then setting the velocity of any one of the dependent variables to unity, and then calling our subroutine LOOPEQ with LEVEL = 1, we will obtain precisely the right values returned in the variable LOOP which we need as the coefficients of the iteration equation (5-5) for the column of the matrix corresponding to that dependent variable's error term. Therefore by calling subprogram LOOPEQ once for each dependent variable, we can develop the matrix of coefficients of Eq. (5-5) with no specialized problem-dependent coding. Setting LEVEL back to zero, one more call to LOOPEQ will produce the negative of the column of constants on the right side of Eq. (5-5). The equations can then be solved by matrix inversion using a standard subprogram from the computer library.

Once the iteration procedure explained above converges, this completes the position analysis for the current-input crank position. We must now consider the velocity and acceleration analyses. Rearranging Eqs. (c) and (d), we can put the unknown terms on the left, known terms on the right giving

$$(e^{j\theta_4})\dot{r}_4 + (jr_4e^{j\theta_4})\dot{\theta}_4 = -(-j\dot{\theta}_2r_2e^{j\theta_2}) \tag{h}$$

$$(e^{j\theta_4})\ddot{r}_4 + (jr_4e^{j\theta_4})\ddot{\theta}_4 = -(-j\ddot{\theta}_2r_2e^{j\theta_2} + \dot{\theta}_2^2r_2e^{j\theta_2} + j2\dot{\theta}_4\dot{r}_4e^{j\theta_4} - \dot{\theta}_4^2r_4e^{j\theta_4}) \tag{i}$$

Note that the coefficients on the left side of these equations are identical with those of Eq. (g); indeed, this will always be the case. Therefore the velocity and acceleration analyses can make use of the same inverse matrix

† R. E. Wengert, A Simple Automatic Derivative Evaluation Program, *Commun, ACM*, vol. 7, no. 8, pp. 463–464, 1964.

found in solving Eq. (g) for the position errors. All that is needed for position and velocity analysis are the appropriate columns of constants for Eqs. (h) and (i). These will be found, as before, by judicious use of our subroutine LOOPEQ.

After setting all velocities to zero, resetting the input velocity $\dot{\theta}_2$ to its proper value, and setting LEVEL = 1, a call to LOOPEQ will produce the negative of the column of constants for the velocity equations (h). With the signs reversed these constants can be multiplied by the stored inverse matrix to give the values of the unknown dependent velocities, $\dot{r}_4$ and $\dot{\theta}_4$. Once this is completed, we can do the same for acceleration analysis. After setting all accelerations to zero, resetting the input acceleration $\ddot{\theta}_2$ to its proper value, and setting LEVEL = 2, a call to LOOPEQ will produce the negative of the column of constants of the acceleration equations (i). Reversing the signs, these constants can also be multiplied by the same stored inverse matrix to give the values of the unknown dependent accelerations, $\ddot{r}_4$ and $\ddot{\theta}_4$. The analysis is now complete for this position of the mechanism and results can be printed.

Proceeding to analyze the next position, we can increment the input crank angle θ_2 by $\Delta\theta_2$ and use the data of the last position as initial estimates for the next when we return to repeat the iteration process.

Let us review the process once more by setting down the steps of the algorithm in proper order for programming. Assuming that the program would be in FORTRAN, it would start by defining the data-storage arrays. These would include a COMMON statement similar to that of the function R, as well as arrays for the matrix of coefficients, the column of constants, and their product. If we design the program for maximums of, say, 20 vectors with 10 dependent variables (five loops) and one input variable, the initial data storage might be defined by the statements

```
COMMON LEVEL,RM(20),RA(20),DRM(20),DRA(20),DDRM(20),DDRA(20)
DIMENSION COEFF(10,10),CONST(10),PROD(10)
```

The main program would be written according to the following algorithm:

Step 1. Set all arrays to zero.

Step 2. Accept data for the number of vectors, the number of loops, and the vector numbers and types (length or angle) of the dependent variables and the input variable.

Step 3. Accept data for magnitudes and angles RM(I) and RA(I) of all vectors.

Step 4. Accept data for the input variable increment, the final position, input velocity, and input acceleration.

Step 5. Print out all input data. If program is interactive, allow users to modify any data they wish to.

Step 6. Convert all angles into radians.

Step 7. Set an iteration counter, ITER = 0.

Step 8. Set LEVEL = 0 and CALL LOOPEQ(CONST).
Step 9. Set LEVEL = 1 and J = 1.
Step 10. Set all velocities, DRM(I) and DRA(I), to zero.
Step 11. Set the proper velocity, DRM or DRA, to one for the Jth dependent variable.
Step 12. CALL LOOPEQ(COEFF(J,1)) to compute the Jth column of the coefficient matrix of Eq. (5-5).
Step 13. Increment J and repeat steps 10 to 12 for each dependent variable in turn.
Step 14. Use a library subprogram to invert the matrix COEFF.
Step 15. Test for difficulty (zero determinant) during matrix inversion of step 14. If determinant is zero, print appropriate message and stop.
Step 16. Set LEVEL = 0 and CALL LOOPEQ(CONST).
Step 17. Premultiply the column CONST by the inverse matrix COEFF to form the column PROD of negative errors $-\delta r_i$ and $-\delta\theta_i$.
Step 18. Form the corrected position values $r_i + \delta r_i$ and $\theta_i + \delta\theta_i$ for all dependent variables.
Step 19. If the iteration counter ITER is greater than 10, print an appropriate message and stop.
Step 20. If any of the errors δr_i or $\delta\theta_i$ is larger than an acceptably small tolerance, increment ITER by 1 and return to step 10.
Step 21. Zero the velocity arrays, DRM(I) and DRA(I). Then set the input velocity value into the proper variable.
Step 22. Set LEVEL = 1 and CALL LOOPEQ(CONST).
Step 23. Premultiply the column CONST by the inverse matrix COEFF to form the column PROD of negative velocities $-\dot{r}_i$ and $-\dot{\theta}_i$.
Step 24. Reverse the signs of velocities from step 23 and store in proper DRM or DRA for each dependent variable.
Step 25. Zero the acceleration arrays, DDRM(I) and DDRA(I). Then set the input acceleration into the proper variable.
Step 26. Set LEVEL = 2 and CALL LOOPEQ(CONST).
Step 27. Premultiply the column CONST by the inverse matrix COEFF to form the column PROD of negative accelerations $-\ddot{r}_i$ and $-\ddot{\theta}_i$.
Step 28. Reverse the signs of accelerations from step 27 and store in proper DDRM or DDRA for each dependent variable.
Step 29. Print out the positions (with angles in degrees), velocities, and accelerations of all dependent variables.
Step 30. If the input variable has not yet reached the final position, add the input-variable increment and return to step 7.
Step 31. If interactive, ask if the user wishes to continue. If so, return to step 5.
Step 32. Stop.

To those who have used iterative methods in other areas, it may seem that such a program would be terribly inefficient, requiring a large number of iterations for convergence. However, in kinematic analysis this is not the case. Experience with a wide variety of problems has proved that usually three or four iterations are sufficient to solve the loop-closure equations of even very complicated linkages to accuracies better than the machining tolerances of the link dimensions. Although convergence is slower in positions with poor mechanical advantage, more than five iterations are never required. Therefore, the test of step 19 should never be met unless invalid data have been given for link dimensions or a dead-center position has been reached (see below) or extremely large steps are taken between positions.

Another worry might be that the iteration process might not converge to a solution if either the initial dependent-variable estimates are grossly in error or the increments between position are so large that the values of the last position are not reasonable starting estimates for the next. Again, experience does not show these worries to be valid. Initial dependent-variable values can be estimated without measurement, and input crank-angle changes of 45 to 60° do not cause problems with convergence.

In kinematic analysis the above iteration scheme is extremely efficient and has only one potential source of difficulty. This is when the matrix of coefficients has a zero or near zero determinant, causing trouble in calculating its inverse. As shown in step 15, this will cause the program to halt. However, in analyzing mechanisms designed for real machines, this is an indication of a mechanical difficulty with the device itself; it is at or near to a dead-center position. This can be shown by recalling Eq. (5-3); if the matrix has a zero determinant, there is no finite solution for the dependent velocities, the definition of a dead-center position (see Sec. 3-16).

A program called KAPCA† using the algorithm described above was written by students at the University of Wisconsin and has proved highly efficient as well as easy to use. The program has been extended to draw a picture of the mechanism on a computer graphics display, and several photographs taken from this display are shown in Fig. 5-4. In spite of the age of the computer on which it operates, the speed of the calculations of the above algorithm, even though iterative, is quick enough to display the mechanism in motion. As shown in Fig. 5-4, the program is also equipped to plot the locus of moving points, thus making it easy to display coupler curves. By sitting at the console, watching the display, altering the dimensions of the links, and analyzing again, the user can quickly design a mechanism with the desired kinematic properties.

The only inconvenience of the above algorithm is that the user must write a new subprogram LOOPEQ for each new *type* of mechanism to be analyzed. Although this requires some limited knowledge of FORTRAN, it is easily done, as shown by the previous example, Also, once a few basic subprograms

† The Kinematic Analysis Program using Complex Algebra (KAPCA) was written by R. A. Lund and O. Hanson, with improvements and extensions by L. T. Duong, C. R. Kishline, and R. Lozano Dominguez.

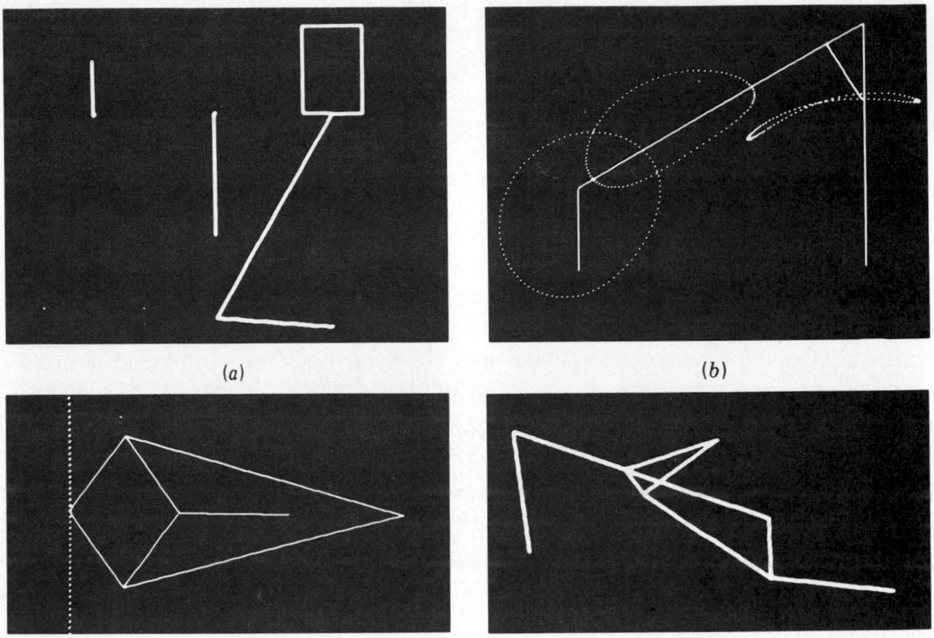

(a) (b) (c) (d)

Figure 5-4 Examples of linkages analyzed by KAPCA: (*a*) slider-crank mechanism with the velocity and acceleration vectors displayed for the piston (displayed left to right); (*b*) four-bar linkage with three of its coupler curves traced; (*c*) Peaucellier mechanism, showing the straight-line coupler curve; (*d*) fork-lift truck mechanism in its *down* position.

are written (one for a four-bar linkage, one for a slider-crank mechanism, and so on), many planar mechanisms are found to be variations of one of these, differing only in the dimensions or choice of input and output links.

5-5 GENERALIZED MECHANISM-ANALYSIS PROGRAMS

As suggested by the program described in the previous section, it is desirable to develop general computer programs with broad ranges of application so that development costs can be justified through repeated use. Also, every computer program requires some initial study and trial-and-error experimentation by the user before its capabilities can be fully utilized; general programs require less training time and cost than using a new program for each new problem.

Although the KAPCA program of the previous section may seem to have a broad range of application, it still has severe limitations which quickly restrict its usefulness in a true industrial-design situation. Probably the most severe limitation of KAPCA is its inability to perform a force analysis of the mechanism being analyzed.

The first widely available general program for mechanism analysis was named KAM (Kinematic Analysis Method) and was written and distributed by IBM. It included the capabilities for position, velocity, acceleration, and force analysis of both planar and spatial mechanisms and was developed around the vector-tetrahedron solutions of Chace (see Chap. 11). This program, first released in 1964, was a landmark achievement, being the first to recognize the need for a general program for mechanical systems exhibiting large geometric changes. Being first, however, it had limitations and has now been superseded by the more powerful programs described below.

Powerful generalized programs have also been developed using finite-element methods; NASTRAN and ANSYS are two examples. These programs have been developed primarily for stress analysis and therefore have excellent capabilities for both static- and dynamic-force analysis of mechanical systems. They also allow the links in a simulated mechanism to deflect under load and are capable of solving statically indeterminate force problems. They are very powerful programs with wide application in industry. Although they are sometimes used for mechanism analysis, they are limited by the inability to simulate the large geometric changes typical of kinematic systems.

There are four large generalized computer programs available for general public use which address the type of problems discussed in this text.† The four programs are named KINSYN, DRAM, ADAMS, and IMP.

KINSYN is the only generalized program available today which is intended primarily for kinematic synthesis. It addresses the synthesis of planar linkages using methods analogous to those described in Chap. 10. This program was developed by Kaufman at the Massachusetts Institute of Technology.

The primary mode of communication between KINSYN and the user is graphical. Users input data describing their motion requirements with an electronic pen on an electronic data tablet; the computer accepts their sketch and provides the required design information on a graphic display screen. Users can gain a good intuitive feeling for the quality of their design by watching as it is animated on the display screen. From this animation they can make judgments regarding clearances, velocities, or forces. An example of the use of KINSYN on a graphic display screen is shown in Fig. 5-5.

The DRAM program, standing for Dynamic Response of Articulated Machinery, is a generalized program for the kinematic and dynamic analysis of planar mechanisms. It was developed by Chace at the University of Michigan. DRAM can be used to simulate even very complex planar mechanisms and provide position, velocity, acceleration, and static- or dynamic-force analyses. The program is interactive, and the user com-

† R. E. Kaufman, Mechanism Design by Computer, *Mach. Des.*, vol. 50, no. 24, pp. 94–100, 1978.

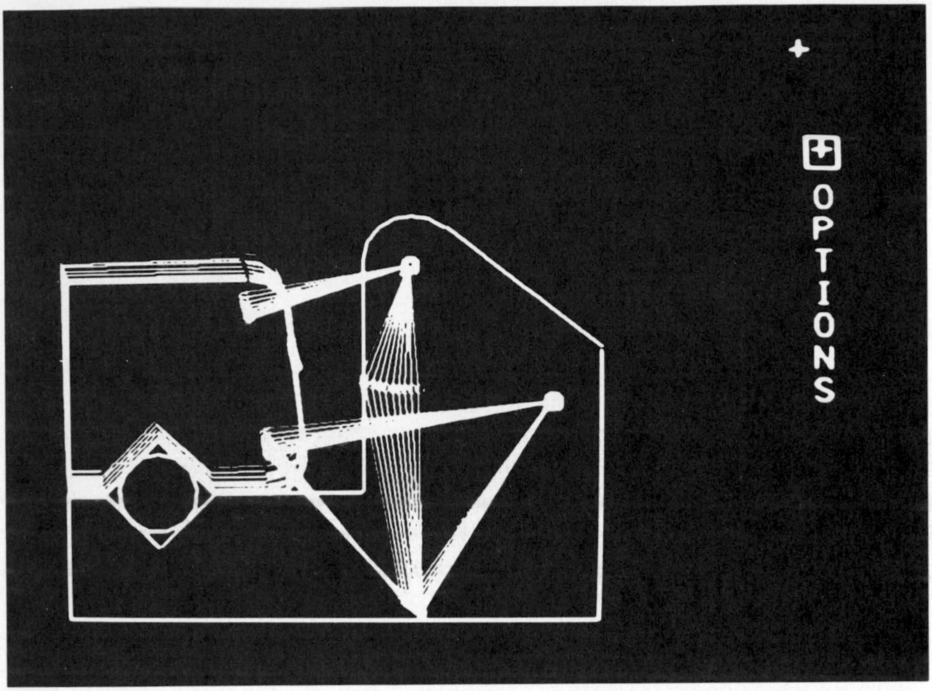

Figure 5-5 Example of a linkage design. This pipe-clamp mechanism was designed in about 15 min using KINSYN III. KINSYN III was developed at the Joint Computer Facility of the Massachusetts Institute of Technology under the direction of Roger E. Kaufman, now Professor of Engineering at The George Washington University. (*Courtesy of Professor Roger E. Kaufman.*)

municates with it by using the special problem-oriented DRAM language, either by a teletype or a graphic display terminal. Special provisions in the program allow it to deal with impact between parts as well as a wide variety of friction effects.

The ADAMS program, standing for Automatic Dynamic Analysis of Mechanical Systems, was also developed by Chace at the University of Michigan. Like DRAM, it is intended for kinematic, static, or dynamic analysis of mechanical systems. However, it allows the simulation of either two- or three-dimensional systems.

IMP, the Integrated Mechanisms Program, was developed by Uicker at the University of Wisconsin. It too can be used to simulate either planar or spatial systems and provide kinematic, static, or dynamic analyses.

Although quite different internally, IMP and ADAMS are comparable from the user's viewpoint. Both are capable of simulating even complex three-dimensional rigid-body systems and give a broad range of analyses, including positions, velocities, accelerations, and static and dynamic forces. Each uses its own problem-oriented language for input data, and both can be used in either a batch or an interactive environment. Either can simulate the

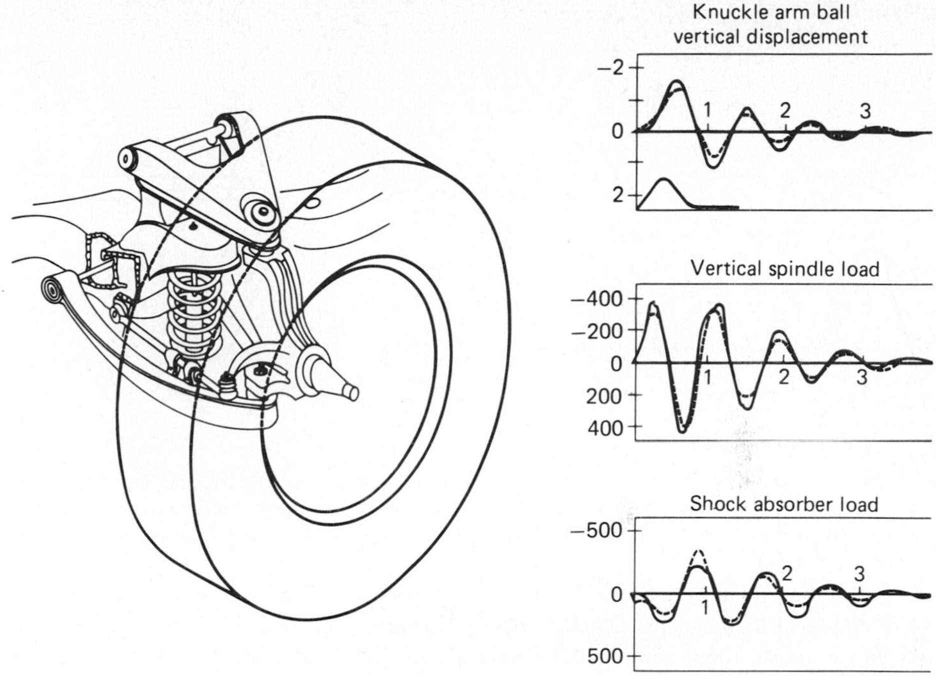

Figure 5-6 Example of a half-front automotive suspension system simulated with both ADAMS and IMP programs. The graphs show the comparison of experimental test data (solid curves) and the numerical simulation results (dashed curves) as the suspension encounters a 1-in hole. The units of the graphs are inches and pounds on the vertical axes vs. time in seconds on the horizontal axes. *(University of Wisconsin, Madison, Wisconsin, and Mechanical Dynamics, Inc., Ann Arbor, Michigan.)*

time history of a mechanical system, starting from some initial configuration and subjected to known force or motion disturbances. Both can also display output on a graphic display terminal. A good application for either program would be the simulation of the automotive front suspension shown in Fig. 5-6. Simulation of this same problem has been performed with both programs, and both compare well with experimental test data.†

PROBLEMS

5-1 Write a calculator or computer program for the analysis of the elliptic trammel mechanism shown in the figure on page 192. The starting position, position increment, and velocity (constant) of link 4 should be accepted as data, and the position, velocity, and acceleration of links 2 and 3 should be displayed.

† These simulations were done for the Strain History Prediction Committee of the Society of Automotive Engineers. Vehicle data and experimental test results were provided by Chevrolet Engineering Division, General Motors Corporation.

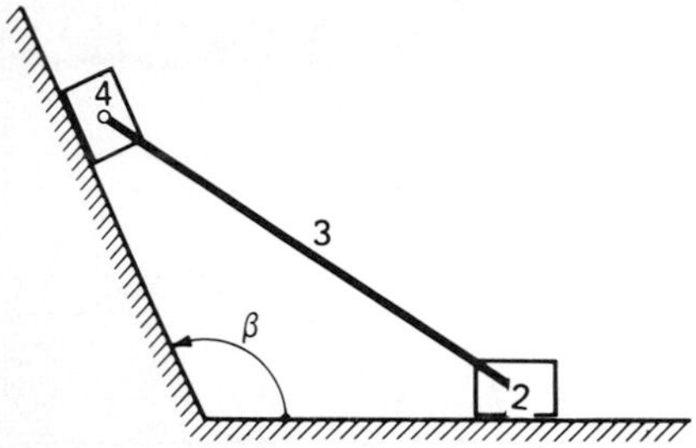

Problem 5-1

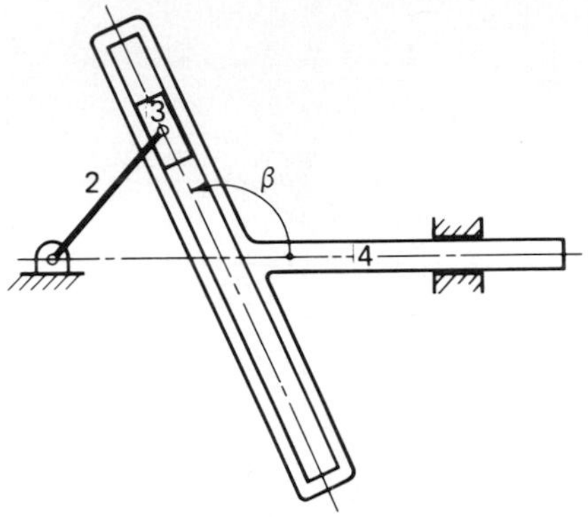

Problem 5-2

5-2 Write a calculator or computer program to analyze the position, velocity, and acceleration of link 4 of the Scotch-yoke mechanism shown in the figure. The position, increment, and velocity (constant) of the crank should be accepted as data, and the analysis should continue over the full cycle of operation.

5-3 Write and verify programs for each of the algorithms of Sec. 5-2.

5-4 Write and verify programs for each of the algorithms of Sec. 5-3.

5-5 Write a computer program using the algorithm described in Sec. 5-4.

5-6 Perform a library search and write a report on computer programs for mechanism design and analysis. This report can expand on the description of the programs mentioned in Sec. 5-5 or can include similar descriptions of other programs.

CHAPTER

SIX

CAM DESIGN

A *cam* is a mechanical element used to drive another element, called the *follower*, through a specified motion by direct contact. Cam-and-follower mechanisms are simple and inexpensive, have few moving parts, and occupy a very small space. Furthermore, follower motions having almost any desired characteristics are not difficult to design. For these reasons cam mechanisms are used extensively in modern machinery.

Much of the material of this chapter is an application of the theory developed in the preceding chapters. In addition one of the more interesting problems treated is how to determine a cam contour that will ultimately deliver a specified motion.

6-1 CLASSIFICATION OF CAMS AND FOLLOWERS

The versatility and flexibility in the design of cam systems are among their more attractive features. Yet this also leads to a wide variety of shapes and forms and the need for terminology to distinguish them.

Cams are classified according to their basic shapes. Figure 6-1 illustrates four different types of cams:

(*a*) A *plate cam*, also called a *disk cam* or a *radial cam*
(*b*) A *wedge cam*
(*c*) A *cylindric cam* or *barrel cam*
(*d*) An *end* or *face cam*

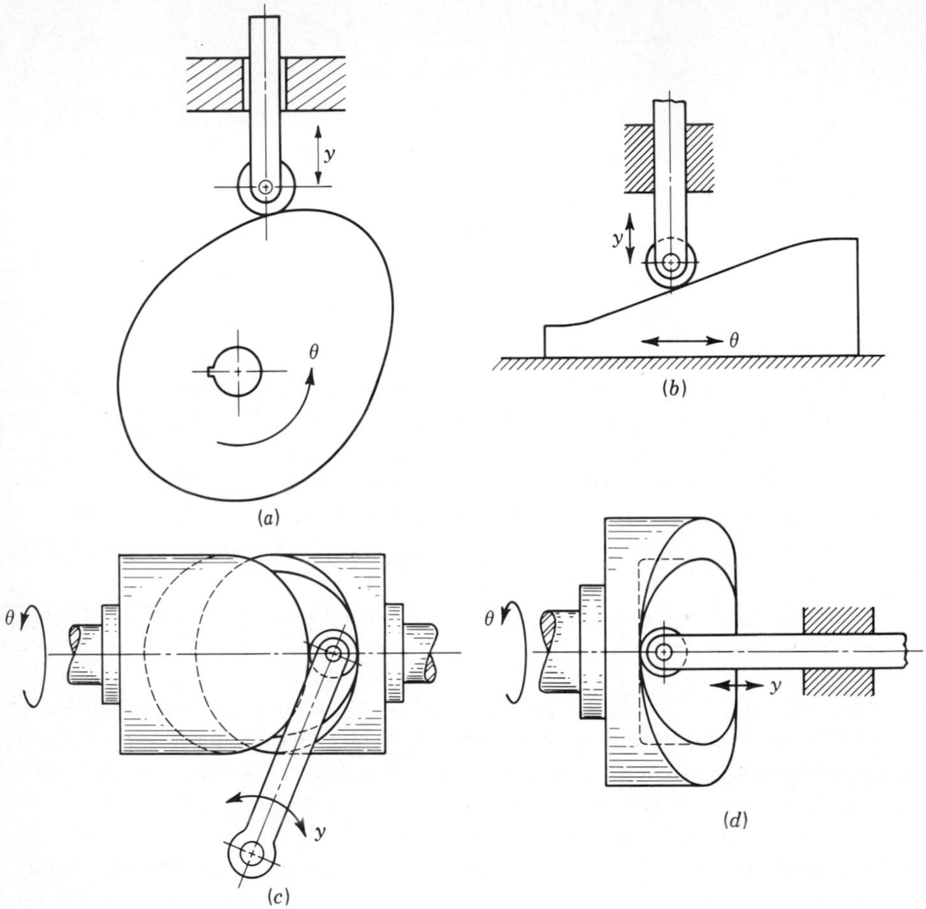

Figure 6-1 Types of cams: (*a*) plate cam, (*b*) wedge cam, (*c*) barrel cam, and (*d*) face cam.

The least common of these in practical applications is the wedge cam because of its need for a reciprocating rather than continuous input motion. By far the most common is the plate cam. For this reason, most of the remainder of this chapter will be specifically addressed to plate cams, although the concepts presented pertain universally.

Cam systems can also be classified according to the basic shape of the follower. Figure 6-2 shows plate cams acting with four different types of followers:

(*a*) A *knife-edge* follower
(*b*) A *flat-face* follower
(*c*) A *roller* follower
(*d*) A *spherical-face* or *curved-shoe* follower

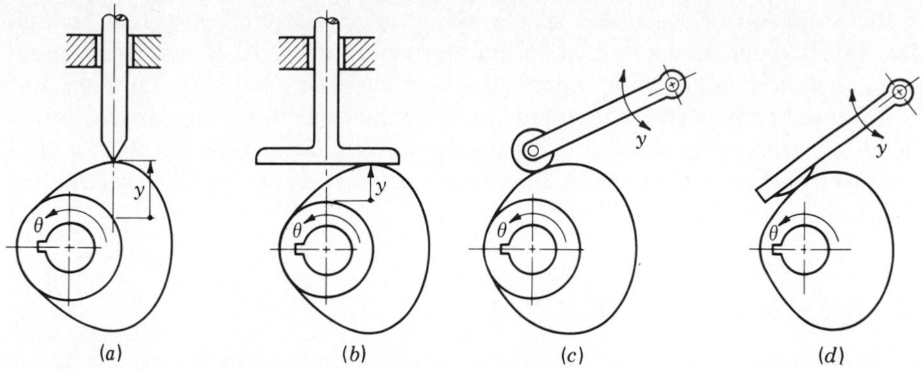

Figure 6-2 Plate cams with (*a*) an offset reciprocating knife-edge follower; (*b*) reciprocating flat-face follower; (*c*) an oscillating roller follower; and (*d*) an oscillating curved-shoe follower.

Notice that the follower face is usually chosen to have a simple geometric shape and the motion is achieved by proper design of the cam shape to mate with it. This is not always the case, and examples of *inverse cams*, where the output element is machined to a complex shape, can be found.

Another method of classifying cams is according to the characteristic output motion allowed between the follower and the frame. Thus, some cams have *reciprocating* (translating) followers, as in Figs. 6-1*a*, *b*, *d* and 6-2*a*,*b*, while others have *oscillating* (rotating) followers, as in Figs. 6-1*c* and 6-2*c*,*d*. Further classification of reciprocating followers distinguishes whether the centerline of the follower stem relative to the center of the cam is *offset*, as in Fig. 6-2*a*, or *radial*, as in Fig. 6-2*b*.

In all cam systems the designer must ensure that the follower maintains contact with the cam at all times. This can be done by depending on gravity,

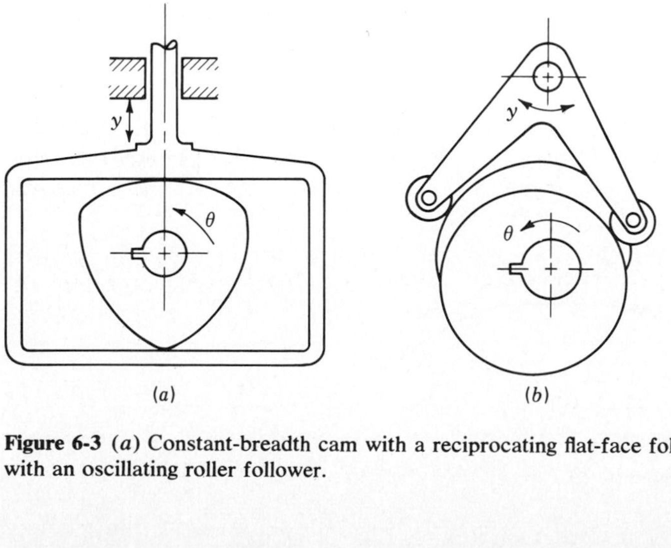

Figure 6-3 (*a*) Constant-breadth cam with a reciprocating flat-face follower. (*b*) Conjugate cams with an oscillating roller follower.

by the inclusion of a suitable spring, or by a mechanical constraint. In Fig. 6-1*c* the follower is constrained by the groove. Figure 6-3*a* shows an example of a *constant-breadth* cam, where two contact points between the cam and the follower provide the constraint. Mechanical constraint can also be introduced by employing *dual* or *conjugate* cams in an arrangement like that illustrated in Fig. 6-3*b*. Here each cam has its own roller, but the rollers are mounted on a common follower.

6-2 DISPLACEMENT DIAGRAMS

In spite of the wide variety of cam types used and their difference in form, they also have certain features in common which allow a systematic approach to their design. Usually a cam system is a single-degree-of-freedom device. It is driven by a known input motion, usually a shaft which rotates at constant speed, and it is intended to produce a certain desired output motion for the follower.

In order to investigate the design of cams in general, we will denote the known input motion by $\theta(t)$ and the output motion by y. Reviewing Figs. 6-1 to 6-3 will demonstrate the definitions of y and θ for various types of cams. These figures also show that y is a translational distance for a reciprocating follower but is an angle for an oscillating follower.

During the rotation of the cam through one cycle of input motion, the follower executes a series of events as shown in graphical form in the *displacement diagram* of Fig. 6-4. In such a diagram the abscissa represents one cycle of the input motion θ (one revolution of the cam) and is drawn to any convenient scale. The ordinate represents the follower travel y and for a reciprocating follower is usually drawn at full scale to help in layout of the cam. On a displacement diagram it is possible to identify a portion of the graph called the *rise*, where the motion of the follower is away from the cam center. The maximum rise is called the *lift*. Periods during which the follower is at rest are referred to as *dwells*, and the *return* is the period in which the motion of the follower is toward the cam center.

Many of the essential features of a displacement diagram, e.g., the total lift or the placement and duration of dwells, are usually dictated by the

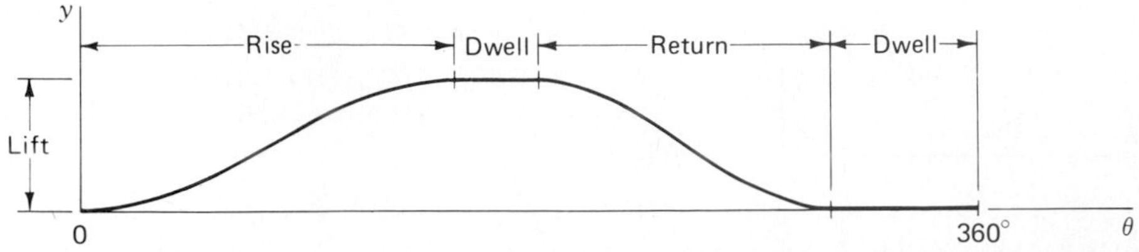

Figure 6-4 Displacement diagram.

requirements of the application. There are, however, many possible follower motions which can be used for the rise and return, and some are preferable to others depending on the situation. One of the key steps in the design of a cam is the choice of suitable forms for these motions. Once the motions have been chosen, i.e., once the exact relationship is set between the input θ and the output y, the displacement diagram can be constructed precisely and is a graphical representation of the functional relationship

$$y = y(\theta)$$

This equation has stored in it the exact nature of the shape of the final cam, the necessary information for its layout and manufacture, and also the important characteristics which determine the quality of its dynamic performance. Before looking further at these topics, however, we will display graphical methods of constructing the displacement diagrams for various rise and return motions.

The displacement diagram for *uniform motion* is a straight line with a constant slope. Thus, for constant input speed, the velocity of the follower is also constant. This motion is not useful for the full lift because of the corners produced at the boundaries with other sections of the displacement diagram. It is often used, however, between other curve sections, thus eliminating the corners.

The displacement diagram for a modified uniform motion is illustrated in Fig. 6-5*a*. The central portion of the diagram, subtended by the cam angle β_2 and the lift L_2, is uniform motion. The ends, angles β_1 and β_3 and corresponding lifts L_1 and L_3, are shaped to deliver *parabolic motion* to the follower. Soon we shall learn that this produces constant acceleration. The diagram shows how to match the slopes of the parabolic motion with that of the uniform motion. With β_1, β_2, β_3, and the total lift L known, the individual lifts L_1, L_2, and L_3 are found by locating the midpoints of the β_1 and β_3 sections and constructing a straight line as shown. Figure 6-5*b* illustrates a graphical

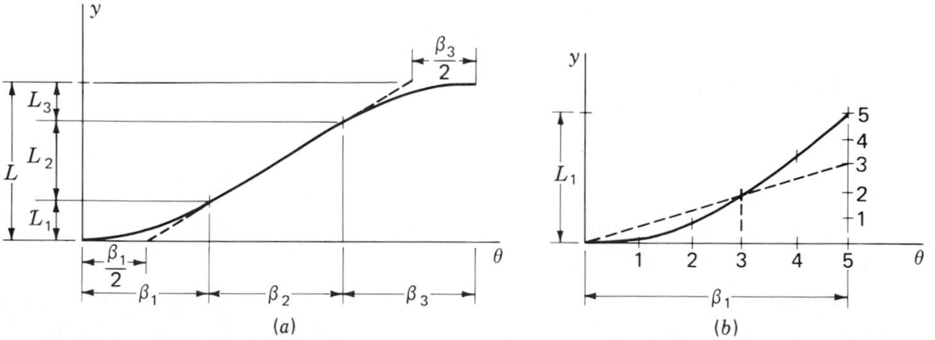

Figure 6-5 Parabolic motion: (*a*) interfaces with uniform motion and (*b*) graphical construction of the displacement diagram.

construction for a parabola to be fit within a given rectangular boundary defined by L_1 and β_1. The abscissa and ordinate are first divided into a convenient but equal number of divisions and numbered as shown. The construction of each point of the parabola then follows that shown by dashed lines for point 3.

In the graphical layout of an actual cam, a great many divisions must be employed to obtain good accuracy. At the same time, the drawing is made to a large scale, perhaps 10 times size. However, for clarity in reading, the figures in this chapter are shown with a minimum number of points to define the curves and illustrate the graphical techniques.

The displacement diagram for *simple harmonic motion* is shown in Fig. 6-6. The graphical construction makes use of a semicircle having a diameter equal to the rise L. The semicircle and abscissa are divided into an equal number of parts and the construction then follows that shown by dashed lines for point 2.

Cycloidal motion obtains its name from the geometric curve called a cycloid. As shown in the left of Fig. 6-7, a circle of radius $L/2\pi$, where L is the total rise, will make exactly one revolution when rolling along the ordinate from the origin to $y = L$. A point P of the circle, originally located at the origin, traces a cycloid as shown. If the circle rolls without slip at a constant rate, the graph of the point's vertical position y vs. time gives the displacement diagram shown at the right of the figure. We find it much more convenient for graphical purposes to draw the circle only once using point B as a center. After dividing the circle and the abscissa into an equal number of parts and numbering them as shown, we project each point of the circle horizontally until it intersects the ordinate; next, from the ordinate, we project parallel to the diagonal OB to obtain the corresponding point on the displacement diagram.

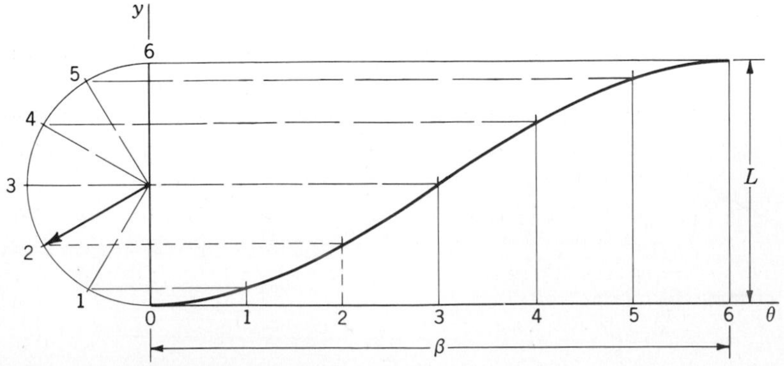

Figure 6-6 Simple harmonic motion.

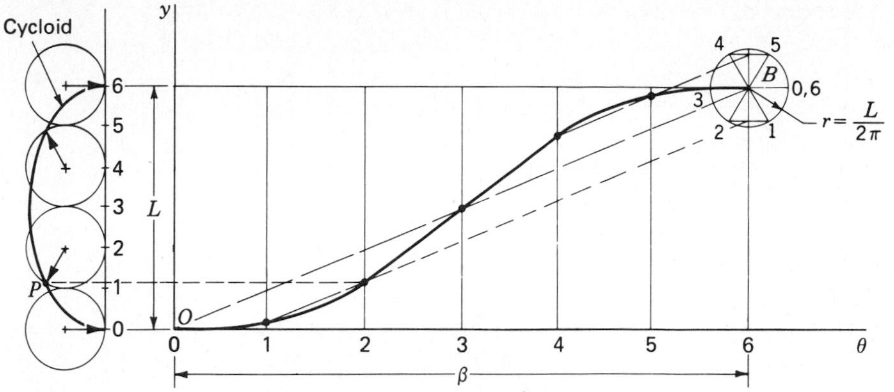

Figure 6-7 Cycloidal motion.

6-3 GRAPHICAL LAYOUT OF CAM PROFILES

Let us now examine the problem of determining the exact shape of a cam surface required to deliver a specified follower motion. We assume here that the required motion has been completely determined, graphically, analytically, or numerically, as discussed in later sections. Thus a complete displacement diagram can be drawn to scale for the entire cam rotation. The problem is now to lay out the proper cam shape to achieve the follower motion represented by this displacement diagram.

We will illustrate for the case of a plate cam as shown in Fig. 6-8. Let us first note some additional nomenclature shown on this figure:

The *trace point* is a theoretical point of the follower; it corresponds to the point of a fictitious knife-edge follower. It is chosen at the center of a roller follower or on the surface of a flat-face follower.

The *pitch curve* is the locus generated by the trace point as the follower moves relative to the cam. For a knife-edge follower the pitch curve and cam surface are identical. For a roller follower they are separated by the radius of the roller.

The *prime circle* is the smallest circle which can be drawn with center at the cam rotation axis and tangent to the pitch curve. The radius of this circle is R_0.

The *base circle* is the smallest circle centered on the cam rotation axis and tangent to the cam surface. For a roller follower it is smaller than the prime circle by the radius of the roller, and for a flat-face follower it is identical with the prime circle.

In constructing the cam profile, we employ the principle of kinematic inversion, imagining the cam to be stationary and allowing the follower to

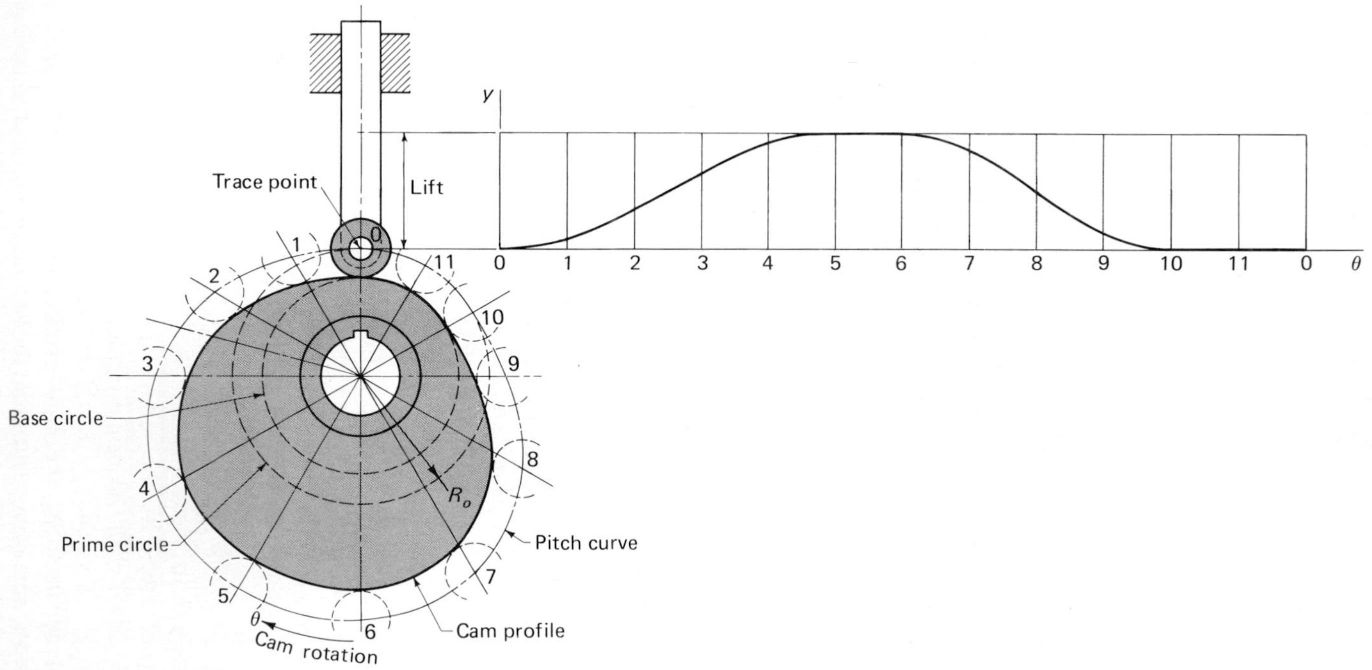

Figure 6-8 Cam nomenclature. Cam surface developed by holding cam stationary and rotating follower from station 0 through stations 1, 2, 3, etc.

rotate *opposite to the direction of cam rotation.* As shown in Fig. 6-8, we divide the prime circle into a number of segments and assign station numbers to the boundaries of these segments. Dividing the displacement-diagram abscissa into corresponding segments, we can then transfer distances, by means of dividers, from the displacement diagram directly onto the cam layout to locate the corresponding positions of the trace point. A smooth curve through these points is the pitch curve. For the case of a roller follower, as in this example, we simply draw the roller in its proper position at each station and then construct the cam profile as a smooth curve tangent to all these roller positions.

Figure 6-9 shows how the method of construction must be modified for an offset roller follower. We begin by constructing an *offset circle,* using a radius equal to the amount of offset. After identifying station numbers around the prime circle, the centerline of the follower is constructed for each station, making it tangent to the offset circle. The roller centers for each station are now established by transferring distances from the displacement diagram directly to these follower centerlines, always measuring outward from the prime circle. An alternative procedure is to identify the points 0′,1′,2′, etc. on a single follower centerline and then to rotate them about the cam center to the corresponding follower centerline positions. In either case the roller

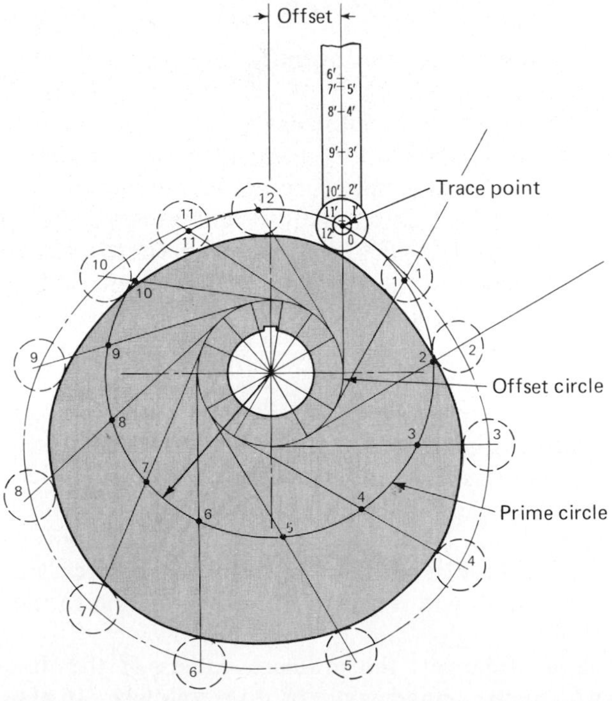

Figure 6-9 Layout of a cam profile for an offset reciprocating roller follower.

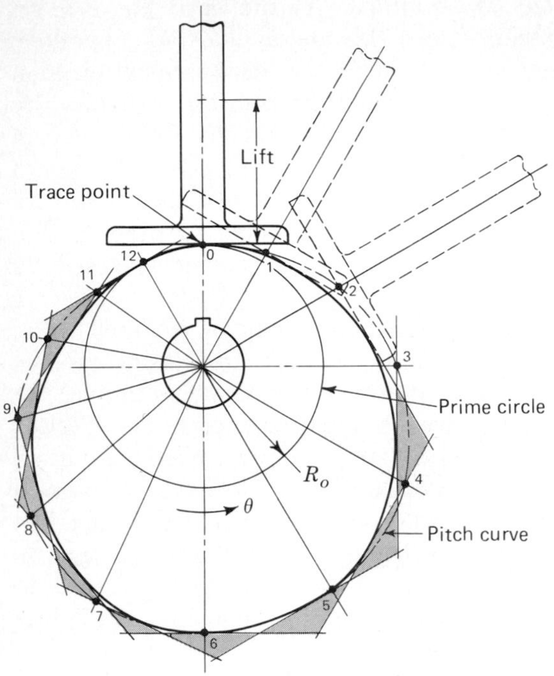

Figure 6-10 Layout of a cam profile for a reciprocating flat-face follower.

circles can be drawn next and a smooth curve tangent to all roller circles is the required cam profile.

Figure 6-10 shows the construction for a plate cam with a reciprocating flat-face follower. The pitch curve is constructed by using a method similar to that used for the roller follower in Fig. 6-8. A line representing the flat face of the follower is then constructed in each position. The cam profile is a smooth curve drawn tangent to all the follower positions. It may be helpful to extend each straight line representing a position of the follower face to form a series of triangles. If these triangles are lightly shaded, as suggested in the illustration, it will be easier to draw the cam profile inside all the shaded triangles and tangent to the inner sides of the triangles.

Figure 6-11 shows the layout of the profile of a plate cam with an oscillating roller follower. In this case we must rotate the fixed pivot center of the follower opposite the direction of cam rotation to develop the cam profile. To perform this inversion, first a circle is drawn about the camshaft center through the fixed pivot of the follower. This circle is then divided and given station numbers to correspond to the displacement diagram. Next, arcs are drawn about each of these centers, all with equal radii corresponding to the length of the follower.

In the case of an oscillating follower, the ordinate values of the displacement diagram represent angular movements of the follower. If the vertical scale of the displacement diagram is properly chosen initially,

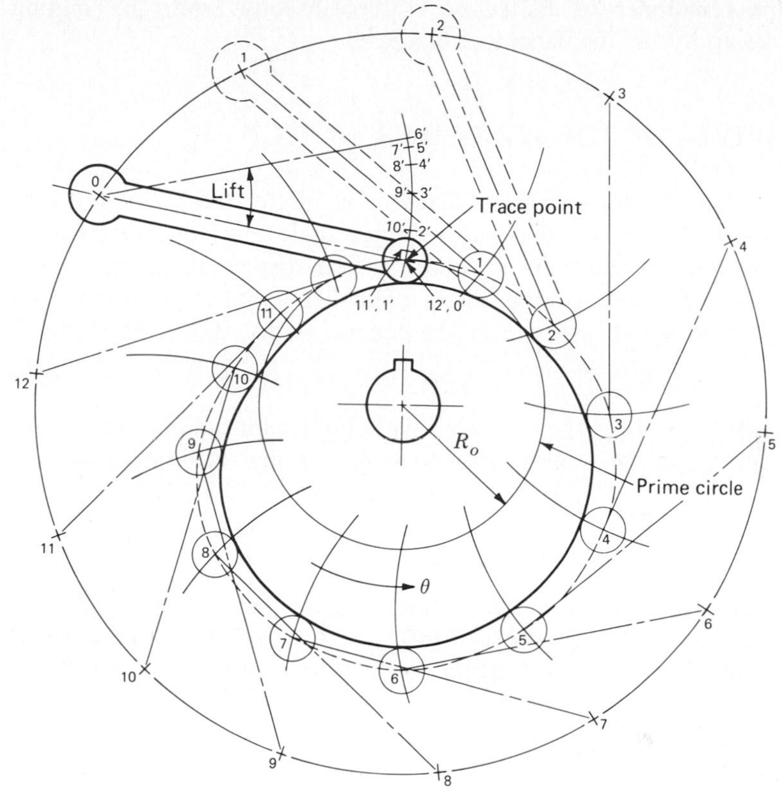

Figure 6-11 Layout of a cam profile for an oscillating roller follower.

however, and if the total lift of the follower is a reasonably small angle, ordinate distances of the displacement diagram at each station can be transferred directly to the corresponding arc traveled by the roller by using dividers and measuring outward along the arc from the prime circle to locate the center of the roller for that station. Finally, circles representing the roller positions are drawn at each station, and the cam profile is constructed as a smooth curve tangent to each of these roller positions.

From the several examples presented in this section it should be clear that each different type of cam and follower system requires its own method of construction to determine the cam profile graphically from the displacement diagram. The examples presented are not intended to be exhaustive of those possible, but they illustrate the general approach. They should also serve to illustrate and reinforce the discussion of the previous section; it should now be clear that much of the detailed shape of the cam itself results directly from the shape of the displacement diagram. Although different types of cams and followers will have different shapes for the same displacement diagram, once a few parameters (such as prime-circle radius) are given to determine the size of a cam, the remainder of its shape results directly from the motion requirements given by the displacement diagram.

6-4 DERIVATIVES OF THE FOLLOWER MOTION

We have seen that the displacement diagram is plotted with the follower motion y as the ordinate and the cam rotation angle θ as the abscissa no matter what the type of the cam or follower. The displacement diagram is therefore a graph representing some mathematical function relating the input and output motions of the cam system. In general terms, this relationship is

$$y = y(\theta) \tag{6-1}$$

If we wished to take the trouble, we could plot additional graphs representing derivatives of y with respect to θ. The first derivative we will denote as y'

$$y'(\theta) = \frac{dy}{d\theta} \tag{6-2}$$

It represents the slope of the displacement diagram at each angle θ. This derivative, although it may now seem of little practical value, is a measure of "steepness" of the displacement diagram. We will find in later sections that it is closely related to the mechanical advantage of the cam system and manifests itself in such things as pressure angle (see Sec. 6-10). If we consider a wedge cam (Fig. 6-1*b*) with a knife-edge follower, the displacement diagram itself is of the same shape as the corresponding cam. Here we can begin to visualize that difficulties will occur if the cam is too steep, i.e., if y' has too high a value.

The second derivative of y with respect to θ is also significant. It is denoted here as y''

$$y''(\theta) = \frac{d^2y}{d\theta^2} \tag{6-3}$$

Although not quite as easy to visualize, this derivative is very closely related to the radius of curvature of the cam at various points along its profile. Since there is an inverse relationship, as y'' becomes very large, the radius of curvature becomes very small; if y'' becomes infinite, the cam profile at that position becomes pointed, a highly unsatisfactory condition from the point of view of contact stresses between the cam and follower surfaces.

The next derivative can also be plotted if desired

$$y'''(\theta) = \frac{d^3y}{d\theta^3} \tag{6-4}$$

Although not easy to describe geometrically it is the rate of the change of y'', and we will see below that this derivative should also be controlled when choosing the detailed shape of the displacement diagram.

Example 6-1 Derive equations to describe the displacement diagram of a cam which rises with parabolic motion from a dwell to another dwell such that the total lift is L and the total cam rotation angle is β. Plot the displacement diagram and its first three derivatives with respect to cam rotation.

SOLUTION As illustrated in Fig. 6-5a, two parabolas will be required, meeting at an inflection point taken here at midrange. For the first half of the motion we choose the general equation of a parabola

$$y = A\theta^2 + B\theta + C \tag{a}$$

which has derivatives

$$y' = 2A\theta + B \tag{b}$$

$$y'' = 2A \tag{c}$$

$$y''' = 0 \tag{d}$$

To match the position and slope with those of the preceding dwell properly, at $\theta = 0$ we have $y(0) = y'(0) = 0$. Thus, Eqs. (a) and (b) show that $B = C = 0$. Looking next at the inflection point, at $\theta = \beta/2$ we want $y = L/2$; Eq. (a) yields

$$A = \frac{2L}{\beta^2}$$

Thus, for the first half of the parabolic motion, the equations are

$$y = 2L\left(\frac{\theta}{\beta}\right)^2 \tag{6-5}$$

$$y' = \frac{4L}{\beta}\frac{\theta}{\beta} \tag{6-6}$$

$$y'' = \frac{4L}{\beta^2} \tag{6-7}$$

$$y''' = 0 \tag{6-8}$$

The maximum slope occurs at the inflection point, where $\theta = \beta/2$. Its value is

$$y'_{max} = \frac{2L}{\beta} \tag{6-9}$$

For the second half of the motion we return to the general equations (a) to (d) for a parabola. Substituting the conditions that at $\theta = \beta$, $y = L$ and $y' = 0$, we have

$$L = A\beta^2 + B\beta + C \tag{e}$$

$$0 = 2A\beta + B \tag{f}$$

Since the slope must match that of the first parabola at $\theta = \beta/2$, we have, from Eqs. (6-9) and (b),

$$\frac{2L}{\beta} = 2A\frac{\beta}{2} + B \tag{g}$$

Solving Eqs. (e) to (g) simultaneously gives

$$A = -\frac{2L}{\beta^2} \qquad B = \frac{4L}{\beta} \qquad C = -L$$

When these constants are substituted into the general forms, we obtain the equations for the second half of the parabolic motion

$$y = L\left[1 - 2\left(1 - \frac{\theta}{\beta}\right)^2\right] \tag{6-10}$$

$$y' = \frac{4L}{\beta}\left(1 - \frac{\theta}{\beta}\right) \tag{6-11}$$

$$y'' = -\frac{4L}{\beta^2} \tag{6-12}$$

$$y''' = 0 \tag{6-13}$$

The displacement diagram for this example is shown in Fig. 6-12 with its three derivatives.

The above discussion relates to the *kinematic derivatives* of the follower motion. These are derivatives with respect to θ and relate to the geometry of the cam system. Let us now consider the derivatives of the follower motions with respect to time. First we will assume that the time history of the input motion $\theta(t)$ is known. Its velocity $\omega = d\theta/dt$, its acceleration $\alpha = d^2\theta/dt^2$, and its next derivative, often called *jerk* or *second acceleration*, $\dot{\alpha} = d^3\theta/dt^3$ are also assumed known. Usually, a plate cam is driven by a constant-speed shaft. In this case, ω is a known constant, $\theta = \omega t$, and $\alpha = \dot{\alpha} = 0$. During start-up of the cam system, however, this is not the case, and we will consider the more general situation first.

From the general equation of the displacement diagram

$$y = y(\theta) \qquad \theta = \theta(t)$$

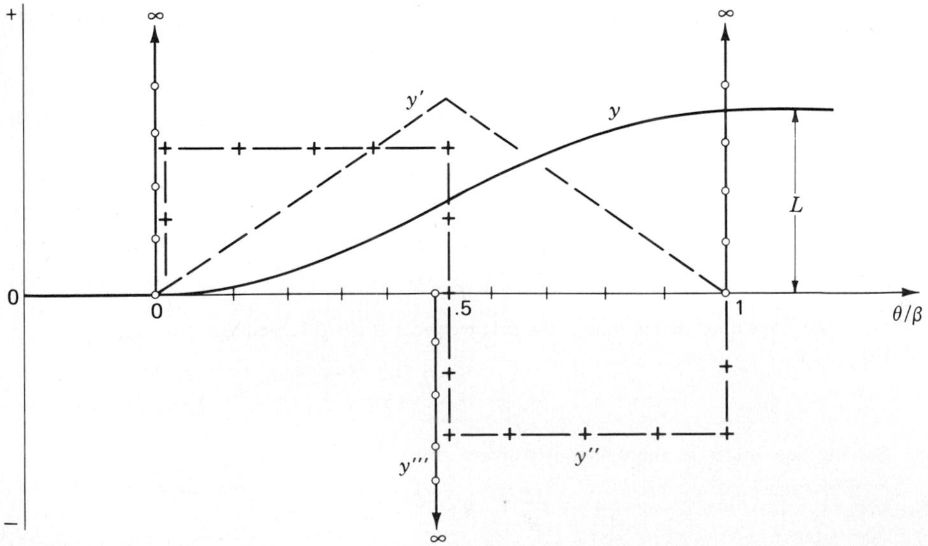

Figure 6-12 Displacement diagram and derivatives for parabolic motion.

We can therefore differentiate to find the time derivatives of the follower motion. The velocity of the follower, for example, is given by

$$\dot{y} = \frac{dy}{dt} = \frac{dy}{d\theta}\frac{d\theta}{dt}$$

$$\dot{y} = y'\omega \tag{6-14}$$

Similarly, the acceleration and jerk of the follower are given by

$$\ddot{y} = \frac{d^2y}{dt^2} = y''\omega^2 + y'\alpha \tag{6-15}$$

and

$$\dddot{y} = \frac{d^3y}{dt^3} = y'''\omega^3 + 3y''\omega\alpha + y'\dot{\alpha} \tag{6-16}$$

When the camshaft speed is constant, these reduce to

$$\dot{y} = y'\omega \qquad \ddot{y} = y''\omega^2 \qquad \dddot{y} = y'''\omega^3 \tag{6-17}$$

For this reason, it has become somewhat common to refer to graphs of the kinematic derivatives y', y'', and y''', such as those shown in Fig. 6-12, as the "velocity," "acceleration," and "jerk" curves for a given motion. They would be appropriate names for a constant-speed cam only, and then only when scaled by ω, ω^2, and ω^3, respectively.† However, it is helpful to use these names for the derivatives when considering the physical implications of a certain choice of displacement diagram. For the parabolic motion of Fig. 6-12, for example, the "velocity" of the follower rises linearly to a maximum and then decreases to zero. The "acceleration" of the follower is zero during the initial dwell and changes abruptly to a constant positive value upon beginning the rise. There are two more abrupt changes in "acceleration" of the follower, at the midpoint and end of the rise. At each of the abrupt changes of "acceleration," the "jerk" of the follower becomes infinite.

6-5 HIGH-SPEED CAMS

Continuing with our discussion of parabolic motion, let us consider briefly the implications of the "acceleration" curve of Fig. 6-12 on the dynamic performance of the cam system. Any real follower must, of course, have some mass and, when multiplied by acceleration, will exert an inertia force (see Chap. 13). Therefore, the "acceleration" curve of Fig. 6-12 can also be thought of as indicating the inertia force of the follower, which, in turn, must be felt at the follower bearings and at the contact point with the cam surface.

† Accepting the word "velocity" literally, for example, leads to consternation when it is discovered that for a plate cam with a reciprocating follower, the units of y' are length per radian. Multiplying these units by radians per second, the units of ω, will give units of length per second, however.

An "acceleration" curve with abrupt changes, such as parabolic motion, will exert abruptly changing contact stresses at the bearings and on the cam surface and lead to noise, surface wear, and eventual failure. Thus it is very important in choosing a displacement diagram to ensure that the first and second derivatives, the "velocity" and "acceleration" curves, are continuous, that is, that they contain no step changes.

Sometimes in low-speed applications compromises are made with the velocity and acceleration relationships. It is sometimes simpler to employ a reverse procedure and design the cam shape first, obtaining the displacement diagram as a second step. Such cams are often composed of some combination of curves such as straight lines and circular arcs which are readily produced by machine tools. Two examples are the *circle-arc cam* and *tangent cam* of Fig. 6-13. The design approach is by iteration. A trial cam is designed and its kinematic characteristics computed. The process is then repeated until a cam with the desired characteristics is obtained. Points A, B, C, and D of the circle-arc and tangent cams are points of tangency or blending points. It is worth noting, as with the parabolic-motion example above, that the acceleration changes abruptly at each of the blending points because of the instantaneous change in radius of curvature.

Although cams with discontinuous acceleration characteristics are sometimes found in low-speed applications, such cams are certain to exhibit major problems as the speed is raised. For any high-speed cam application, it is extremely important that not only the displacement and "velocity" curves but also the "acceleration" curve be made continuous for the entire motion cycle. No discontinuities should be allowed at the boundaries of different sections of the cam.

As shown by Eq. (6-17) the importance of continuous derivatives becomes more serious as the camshaft speed is raised. The higher the speed,

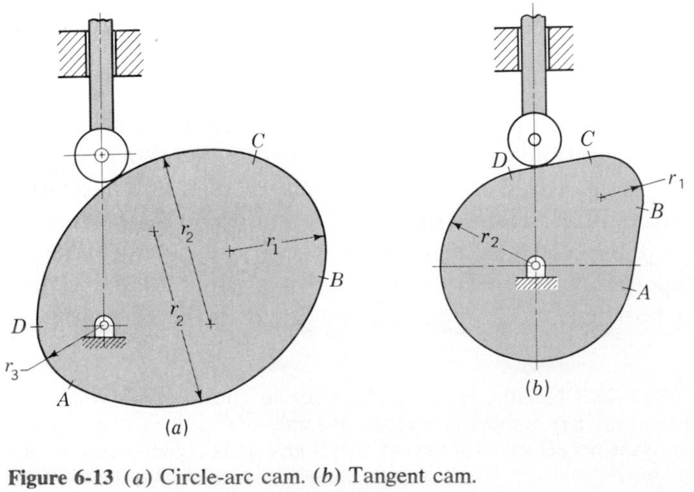

Figure 6-13 (*a*) Circle-arc cam. (*b*) Tangent cam.

the greater the need for smooth curves. At very high speeds it might also be desirable to require that jerk, which is related to rate of change of force, and perhaps even higher derivatives, be made continuous as well. In most applications this is not necessary, however.

How high a speed one must have before considering the application to require high-speed design techniques cannot be given a simple answer. This depends not only on the mass of the follower but also on the stiffness of the return spring, the materials used, the flexibility of the follower, and many other factors.† Further analysis techniques on cam dynamics are presented in Chap. 16. Still, with the methods presented below, it is not difficult to achieve continuous derivative displacement diagrams. Therefore it is recommended that this be done as standard practice. Parabolic-motion cams are no easier to manufacture than cycloidal-motion cams, for example, and there is no good reason for their use. The circle-arc and tangent cams are easier to produce but with modern machining methods cutting of more complex cam shapes is not expensive.

6-6 STANDARD CAM MOTIONS

Example 6-1 gave a detailed derivation of the equations for parabolic motion and its derivatives. Then in Sec. 6-5 reasons were given for avoiding the use of parabolic motion in high-speed cam systems. The purpose of this section is to present equations for a number of standard types of displacement curves which can be used to address most high-speed cam-motion requirements. The derivations parallel those of Example 6-1 and are not given.

The displacement diagram and its derivatives for simple harmonic rise motion are shown in Fig. 6-14. The equations are

$$y = \frac{L}{2}\left(1 - \cos\frac{\pi\theta}{\beta}\right) \qquad \text{(6-18}a\text{)}$$

$$y' = \frac{\pi L}{2\beta}\sin\frac{\pi\theta}{\beta} \qquad \text{(6-18}b\text{)}$$

$$y'' = \frac{\pi^2 L}{2\beta^2}\cos\frac{\pi\theta}{\beta} \qquad \text{(6-18}c\text{)}$$

$$y''' = -\frac{\pi^3 L}{2\beta^3}\sin\frac{\pi\theta}{\beta} \qquad \text{(6-18}d\text{)}$$

Unlike parabolic motion, simple harmonic motion shows no discontinuity at the inflection point.

† A good analysis of this subject is presented in D. Tesar and G. K. Matthew, "The Dynamic Synthesis, Analysis, and Design of Modeled Cam Systems," Heath, Lexington, Mass., 1976.

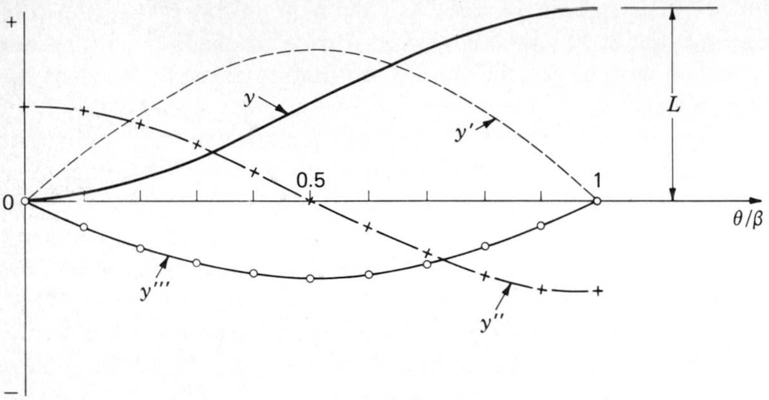

Figure 6-14 Displacement diagram and derivatives for full-rise simple harmonic motion, Eq. (6-18).

The equations for cycloidal rise motion and its derivatives are

$$y = L\left(\frac{\theta}{\beta} - \frac{1}{2\pi}\sin\frac{2\pi\theta}{\beta}\right) \tag{6-19a}$$

$$y' = \frac{L}{\beta}\left(1 - \cos\frac{2\pi\theta}{\beta}\right) \tag{6-19b}$$

$$y'' = \frac{2\pi L}{\beta^2}\sin\frac{2\pi\theta}{\beta} \tag{6-19c}$$

$$y''' = \frac{4\pi^2 L}{\beta^3}\cos\frac{2\pi\theta}{\beta} \tag{6-19d}$$

They are shown in Fig. 6-15.

Figure 6-16 shows the displacement diagram and derivatives for a rise

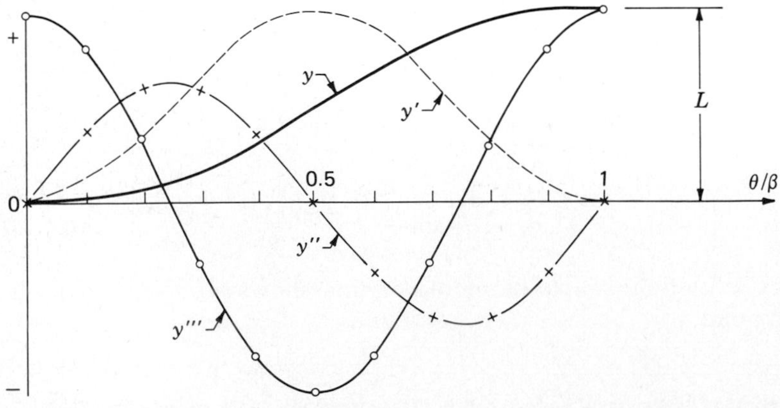

Figure 6-15 Displacement diagram and derivatives for full-rise cycloidal motion, Eq. (6-19).

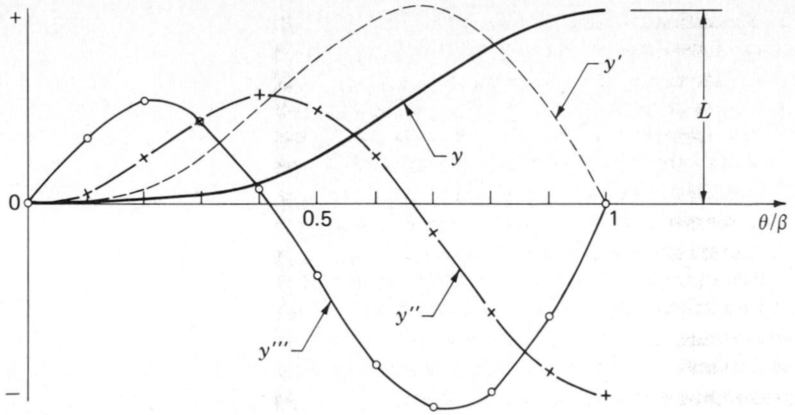

Figure 6-16 Displacement diagram and derivatives for full-rise modified harmonic motion, Eq. (6-20).

motion called *modified harmonic motion.* The equations are

$$y = \frac{L}{2}\left[\left(1 - \cos\frac{\pi\theta}{\beta}\right) - \frac{1}{4}\left(1 - \cos\frac{2\pi\theta}{\beta}\right)\right] \tag{6-20a}$$

$$y' = \frac{\pi L}{2\beta}\left(\sin\frac{\pi\theta}{\beta} - \frac{1}{2}\sin\frac{2\pi\theta}{\beta}\right) \tag{6-20b}$$

$$y'' = \frac{\pi^2 L}{2\beta^2}\left(\cos\frac{\pi\theta}{\beta} - \cos\frac{2\pi\theta}{\beta}\right) \tag{6-20c}$$

$$y''' = -\frac{\pi^3 L}{2\beta^3}\left(\sin\frac{\pi\theta}{\beta} - 2\sin\frac{2\pi\theta}{\beta}\right) \tag{6-20d}$$

The displacement diagrams of simple harmonic, cycloidal, and modified harmonic motions look quite similar at first glance. Each rises through a lift of L in a total cam angle of β. Each begins and ends with a horizontal slope, and for this reason they are all referred to as *full-rise* motions. However, their "acceleration" curves are quite different. Simple harmonic motion has nonzero "acceleration" at both ends of the range; cycloidal motion has zero "acceleration" at both boundaries; and modified harmonic motion has one zero and one nonzero "acceleration" at its ends. This provides the selection necessary when matching these curves with neighboring curves of different types.

Full-return motions of the same three types are shown in Figs. 6-17 to 6-19. The equations for simple harmonic motion are

$$y = \frac{L}{2}\left(1 + \cos\frac{\pi\theta}{\beta}\right) \tag{6-21a}$$

$$y' = -\frac{\pi L}{2\beta}\sin\frac{\pi\theta}{\beta} \tag{6-21b}$$

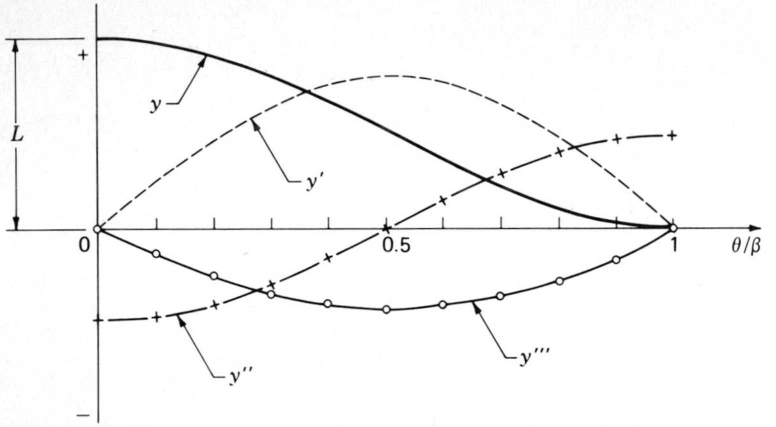

Figure 6-17 Displacement diagram and derivatives for full-return simple harmonic motion, Eq. (6-21).

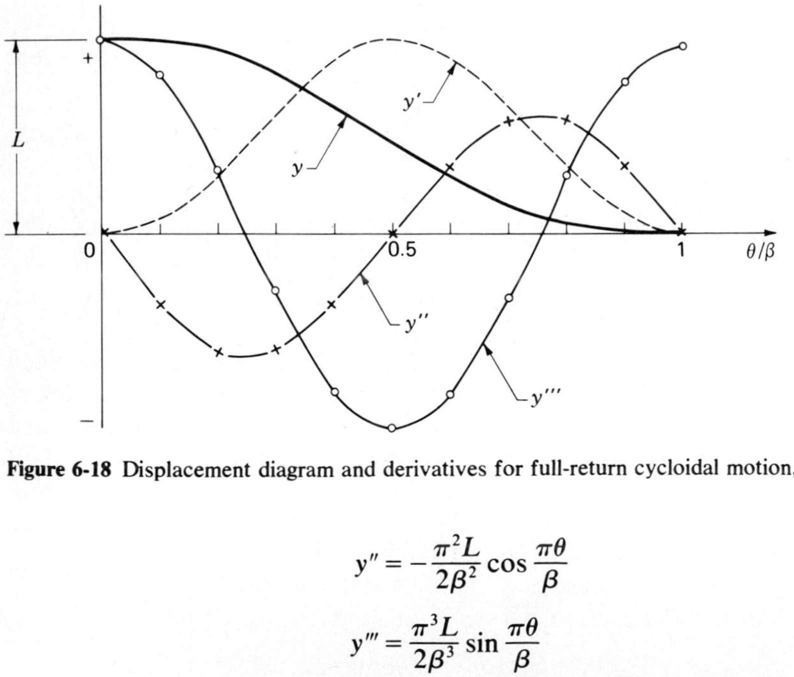

Figure 6-18 Displacement diagram and derivatives for full-return cycloidal motion, Eq. (6-22).

$$y'' = -\frac{\pi^2 L}{2\beta^2}\cos\frac{\pi\theta}{\beta} \tag{6-21c}$$

$$y''' = \frac{\pi^3 L}{2\beta^3}\sin\frac{\pi\theta}{\beta} \tag{6-21d}$$

For full-return cycloidal motion the equations are

$$y = L\left(1 - \frac{\theta}{\beta} + \frac{1}{2\pi}\sin\frac{2\pi\theta}{\beta}\right) \tag{6-22a}$$

$$y' = -\frac{L}{\beta}\left(1 - \cos\frac{2\pi\theta}{\beta}\right) \tag{6-22b}$$

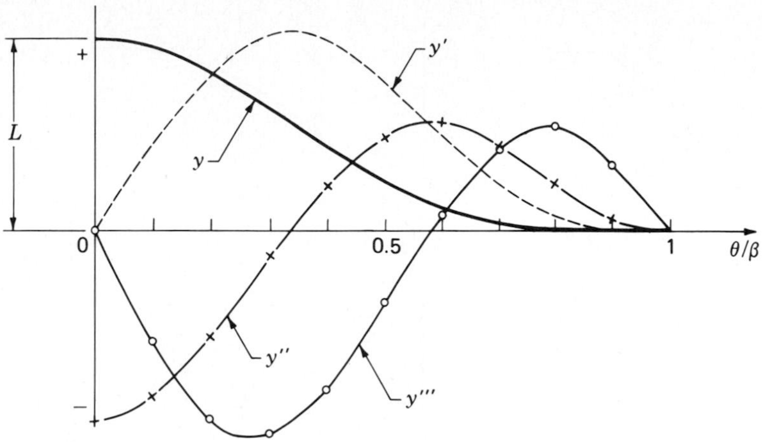

Figure 6-19 Displacement diagram and derivatives for full-return modified harmonic motion, Eq. (6-23).

$$y'' = -\frac{2\pi L}{\beta^2}\sin\frac{2\pi\theta}{\beta} \tag{6-22c}$$

$$y''' = -\frac{4\pi^2 L}{\beta^3}\cos\frac{2\pi\theta}{\beta} \tag{6-22d}$$

The equations for modified harmonic full-return motion are

$$y = \frac{L}{2}\left[\left(1+\cos\frac{\pi\theta}{\beta}\right) - \frac{1}{4}\left(1-\cos\frac{2\pi\theta}{\beta}\right)\right] \tag{6-23a}$$

$$y' = -\frac{\pi L}{2\beta}\left(\sin\frac{\pi\theta}{\beta} + \frac{1}{2}\sin\frac{2\pi\theta}{\beta}\right) \tag{6-23b}$$

$$y'' = -\frac{\pi^2 L}{2\beta^2}\left(\cos\frac{\pi\theta}{\beta} + \cos\frac{2\pi\theta}{\beta}\right) \tag{6-23c}$$

$$y''' = \frac{\pi^3 L}{2\beta^3}\left(\sin\frac{\pi\theta}{\beta} + 2\sin\frac{2\pi\theta}{\beta}\right) \tag{6-23d}$$

In addition to the full-rise and full-return motions presented above, it is often useful to have a selection of standard *half-rise* or *half-return* motions available. These are curves for which one boundary has a nonzero slope and can be used to blend with uniform motion. The displacement diagrams and derivatives for simple harmonic half-rise motions, sometimes called *half-harmonics*, are shown in Fig. 6-20. The equations corresponding to Fig. 6-20*a* are

$$y = L\left(1-\cos\frac{\pi\theta}{2\beta}\right) \tag{6-24a}$$

$$y' = \frac{\pi L}{2\beta}\sin\frac{\pi\theta}{2\beta} \tag{6-24b}$$

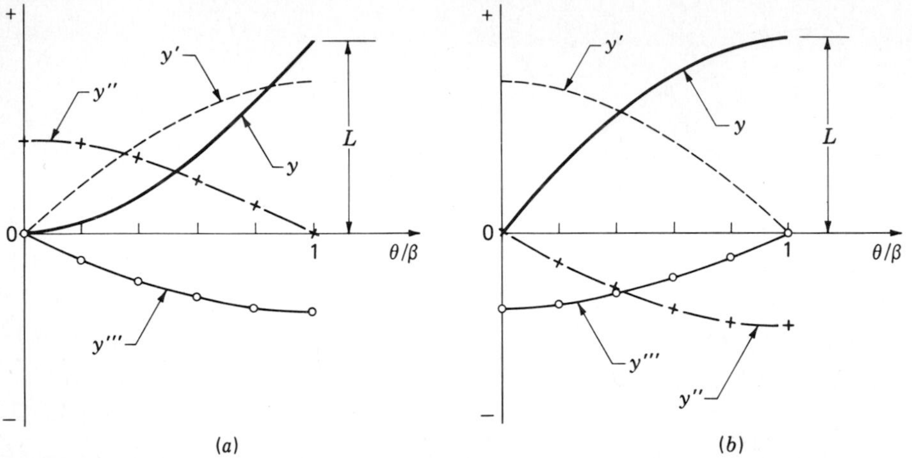

Figure 6-20 Displacement diagrams and derivatives for half-harmonic rise motions: (*a*) Eq. (6-24); (*b*) Eq. (6-25).

$$y'' = \frac{\pi^2 L}{4\beta^2}\cos\frac{\pi\theta}{2\beta} \tag{6-24c}$$

$$y''' = -\frac{\pi^3 L}{8\beta^3}\sin\frac{\pi\theta}{2\beta} \tag{6-24d}$$

For Fig. 6-20*b* the equations are

$$y = L\sin\frac{\pi\theta}{2\beta} \tag{6-25a}$$

$$y' = \frac{\pi L}{2\beta}\cos\frac{\pi\theta}{2\beta} \tag{6-25b}$$

$$y'' = -\frac{\pi^2 L}{4\beta^2}\sin\frac{\pi\theta}{2\beta} \tag{6-25c}$$

$$y''' = -\frac{\pi^3 L}{8\beta^3}\cos\frac{\pi\theta}{2\beta} \tag{6-25d}$$

The half-harmonic curves for half-return motions are shown in Fig. 6-21. The equations corresponding to Fig. 6-21*a* are

$$y = L\cos\frac{\pi\theta}{2\beta} \tag{6-26a}$$

$$y' = -\frac{\pi L}{2\beta}\sin\frac{\pi\theta}{2\beta} \tag{6-26b}$$

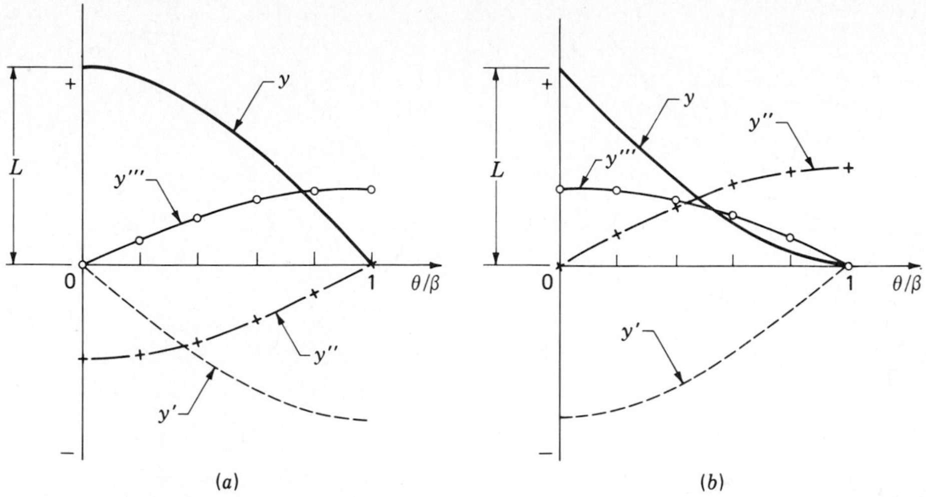

Figure 6-21 Displacement diagrams and derivatives for half-harmonic return motions: (*a*) Eq. (6-26); (*b*) Eq. (6-27).

$$y'' = -\frac{\pi^2 L}{4\beta^2}\cos\frac{\pi\theta}{2\beta} \tag{6-26c}$$

$$y''' = \frac{\pi^3 L}{8\beta^3}\sin\frac{\pi\theta}{2\beta} \tag{6-26d}$$

For Fig. 6-21*b* the equations are

$$y = L\left(1 - \sin\frac{\pi\theta}{2\beta}\right) \tag{6-27a}$$

$$y' = -\frac{\pi L}{2\beta}\cos\frac{\pi\theta}{2\beta} \tag{6-27b}$$

$$y'' = \frac{\pi^2 L}{4\beta^2}\sin\frac{\pi\theta}{2\beta} \tag{6-27c}$$

$$y''' = \frac{\pi^3 L}{8\beta^3}\cos\frac{\pi\theta}{2\beta} \tag{6-27d}$$

Besides the half-harmonics, *half-cycloidal* motions are also useful since their "accelerations" are zero at both boundaries. The displacement diagrams and derivatives for half-cycloidal half-rise motions are shown in Fig. 6-22. The equations corresponding to Fig. 6-22*a* are

$$y = L\left(\frac{\theta}{\beta} - \frac{1}{\pi}\sin\frac{\pi\theta}{\beta}\right) \tag{6-28a}$$

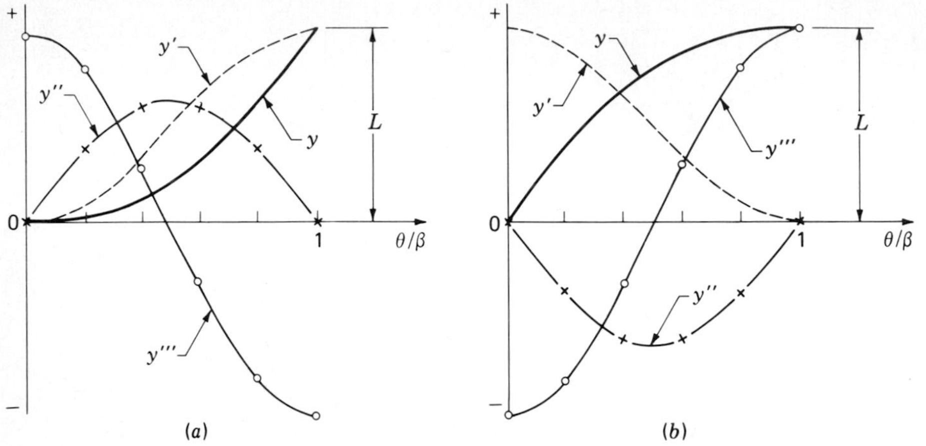

Figure 6-22 Displacement diagrams and derivatives for half-cycloidal rise motions: (*a*) Eq. (6-28); (*b*) Eq. (6-29).

$$y' = \frac{L}{\beta}\left(1 - \cos\frac{\pi\theta}{\beta}\right) \tag{6-28b}$$

$$y'' = \frac{\pi L}{\beta^2}\sin\frac{\pi\theta}{\beta} \tag{6-28c}$$

$$y''' = \frac{\pi^2 L}{\beta^3}\cos\frac{\pi\theta}{\beta} \tag{6-28d}$$

For Fig. 6-22*b* the equations are

$$y = L\left(\frac{\theta}{\beta} + \frac{1}{\pi}\sin\frac{\pi\theta}{\beta}\right) \tag{6-29a}$$

$$y' = \frac{L}{\beta}\left(1 + \cos\frac{\pi\theta}{\beta}\right) \tag{6-29b}$$

$$y'' = -\frac{\pi L}{\beta^2}\sin\frac{\pi\theta}{\beta} \tag{6-29c}$$

$$y''' = -\frac{\pi^2 L}{\beta^2}\cos\frac{\pi\theta}{\beta} \tag{6-29d}$$

The half-cycloidal curves for half-return motions are shown in Fig. 6-23. The equations corresponding to Fig. 6-23*a* are

$$y = L\left(1 - \frac{\theta}{\beta} + \frac{1}{\pi}\sin\frac{\pi\theta}{\beta}\right) \tag{6-30a}$$

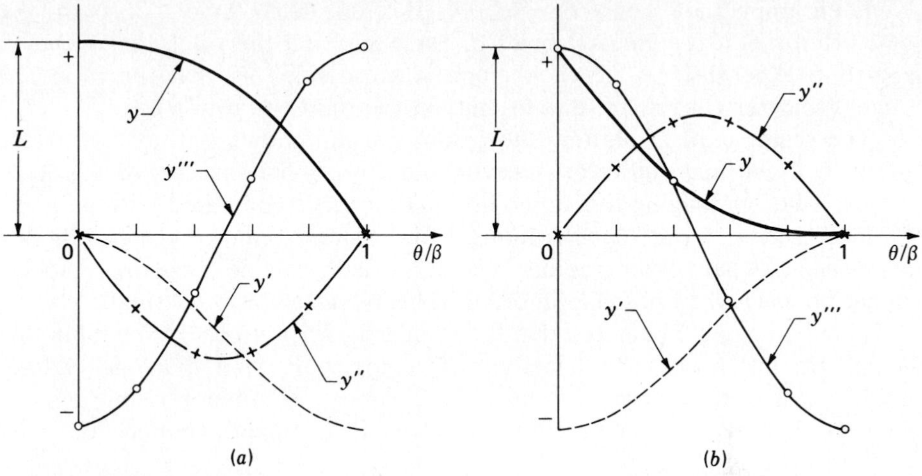

Figure 6-23 Displacement diagrams and derivatives for half-cycloidal return motions: (*a*) Eq. (6-30); (*b*) Eq. (6-31).

$$y' = -\frac{L}{\beta}\left(1 - \cos\frac{\pi\theta}{\beta}\right) \tag{6-30b}$$

$$y'' = -\frac{\pi L}{\beta^2}\sin\frac{\pi\theta}{\beta} \tag{6-30c}$$

$$y''' = -\frac{\pi^2 L}{\beta^3}\cos\frac{\pi\theta}{\beta} \tag{6-30d}$$

For Fig. 6-23*b* the equations are

$$y = L\left(1 - \frac{\theta}{\beta} - \frac{1}{\pi}\sin\frac{\pi\theta}{\beta}\right) \tag{6-31a}$$

$$y' = -\frac{L}{\beta}\left(1 + \cos\frac{\pi\theta}{\beta}\right) \tag{6-31b}$$

$$y'' = \frac{\pi L}{\beta^2}\sin\frac{\pi\theta}{\beta} \tag{6-31c}$$

$$y''' = \frac{\pi^2 L}{\beta^3}\cos\frac{\pi\theta}{\beta} \tag{6-31d}$$

We shall soon see how the graphs and equations presented in this section can greatly reduce the analytical effort involved in designing the full displacement diagram for a high-speed cam. But first we should note a few features of the graphs of Figs. 6-14 to 6-23.

Each graph shows only one section of a full displacement diagram; the total lift *for that section* is labeled L for each, and the total cam travel is labeled β. The abscissa of each graph is normalized so that the ratio θ/β ranges from zero at the left end to unity at the right end ($\theta = \beta$).

The scales used in plotting the graphs are not shown but are consistent for all full-rise and full-return curves and for all half-rise and half-return curves. Thus, in judging the suitability of one curve compared with another, the magnitudes of the "accelerations," for example, can be compared. For this reason, when other factors are equivalent, simple harmonic motion should be used when possible in order to keep "accelerations" small.

Finally, it should be noted that the standard cam motions presented in this section do not form an exhaustive set; cams with good dynamic characteristics can also be formed from a wide variety of other possible motion curves.† The set presented here, however, is complete enough for most applications.

6-7 MATCHING DERIVATIVES OF DISPLACEMENT DIAGRAMS

In the previous section, a great many equations were presented which might be used to represent the different segments of a cam's displacement diagram. In this section we will study how they can be joined together to form the motion specification for a complete cam. The procedure is one of solving for proper values of L and β for each segment so that

1. The motion requirements of the particular application are met.
2. The displacement, "velocity," and "acceleration" diagrams are continuous across the boundaries of the segments. The "jerk" diagram may be allowed discontinuities if necessary, but it must not become infinite; i.e., the "acceleration" curve may contain corners but not discontinuities.
3. The maximum magnitudes of the "velocity" and "acceleration" peaks are kept as low as possible consistent with the above two conditions.

The procedure is best understood through an example.

Example 6-2 A plate cam with a reciprocating follower is to be driven by a constant-speed motor at 150 rpm. The follower is to start from a dwell, accelerate to a uniform velocity of 25 in/s, maintain this velocity for 1.25 in of rise, decelerate to the top of the lift, return, and then dwell for 0.1 s. The total lift is to be 3.0 in. Determine the complete specifications of the displacement diagram.

† H. A. Rothbart, "Cams," Wiley, New York, 1956, a real classic on cams, contains a comparison of 11 different motions on p. 184.

SOLUTION The input shaft speed is

$$\omega = 150 \text{ rpm} = 15.708 \text{ rad/s} \tag{1}$$

Using Eq. (6-14), we can find the slope of the uniform-velocity segment

$$y' = \frac{\dot{y}}{\omega} = \frac{25 \text{ in/s}}{15.708 \text{ rad/s}} = 1.592 \text{ in/rad} \tag{2}$$

and since this is held constant for 1.25 in of rise, the cam rotation in this segment is

$$\beta_2 = \frac{L_2}{y'} = \frac{1.25 \text{ in}}{1.592 \text{ in/rad}} = 0.785 \text{ rad} = 45.000° \tag{3}$$

Similarly, from Eq. (1), we find the cam rotation during the final dwell

$$\beta_5 = 0.067\text{s}\ (15.708 \text{ rad/s}) = 1.047 \text{ rad} = 60.000° \tag{4}$$

From this and the given information, we can sketch the beginnings of the displacement diagram, not necessarily working to scale, but in order to visualize the motion requirements. This gives the general shapes shown by the heavy curves of Fig. 6-24*a*. The lighter sections of the displacement diagram are not yet accurately known, but can also be sketched by producing a smooth curve for visualization. Working from this curve, we can also sketch the general nature of the derivative curves. From the slope of the displacement diagram, we sketch the "velocity" curve, Fig. 6-24*b*, and from *its* slope we find the "acceleration" curve, Fig. 6-24*c*. At this time no attempt is made to produce accurate curves drawn to scale, only to provide some idea of the curve shapes.

Now, using the sketches of Fig. 6-24, we compare the desired motion curves with the various standard curves of Figs. 6-14 to 6-23, in order to choose an appropriate set of equations for each segment of the cam. In the segment *AB*, for example, we find that Fig. 6-22*a* is the only motion curve available with half-rise characteristics, an appropriate slope curve, and the necessary zero "acceleration" at both ends of the segment. Thus we choose the half-cycloidal motion of Eq. (6-28) for that portion of the cam. There are two sets of choices possible for the segments *CD* and *DE*. One set might be the choice of Fig. 6-22*b* matched with Fig. 6-18. However, in order to keep the "jerk" curve as smooth as possible, we will choose Fig. 6-20*b* matched with Fig. 6-19. Thus for the segment *CD* we use the half-harmonic rise curves of Eq. (6-25), and for segment *DE* we choose the modified harmonic return curves of Eq. (6-23).

Choosing the motion curve types, however, is not sufficient to specify the motion characteristics fully. We must also find values for the unknown parameters of the motion equations; these are L_1, L_3, β_1, β_3, and β_4. We do this by equating values at each nonzero boundary of the derivative curves. For example, to match the "velocities" at *B* we must equate the value of y' from Eq. (6-28*b*) at $\theta/\beta = 1$ (its right end) with the value of y' in the *BC* segment

$$y'_B = \frac{2L_1}{\beta_1} = \frac{L_2}{\beta_2} = \frac{1.25 \text{ in}}{0.785 \text{ rad}} = 1.592 \text{ in/rad}$$

or

$$L_1 = 0.796\beta_1 \tag{5}$$

Similarly, to match "velocities" at point *C*, we equate the value of y' of the *BC* segment to that of Eq. (6-25*b*) at $\theta/\beta = 0$ (its left end)

$$y'_C = \frac{L_2}{\beta_2} = \frac{\pi L_3}{2\beta_3} = 1.592 \text{ in/rad}$$

or

$$L_3 = 1.013\beta_3 \tag{6}$$

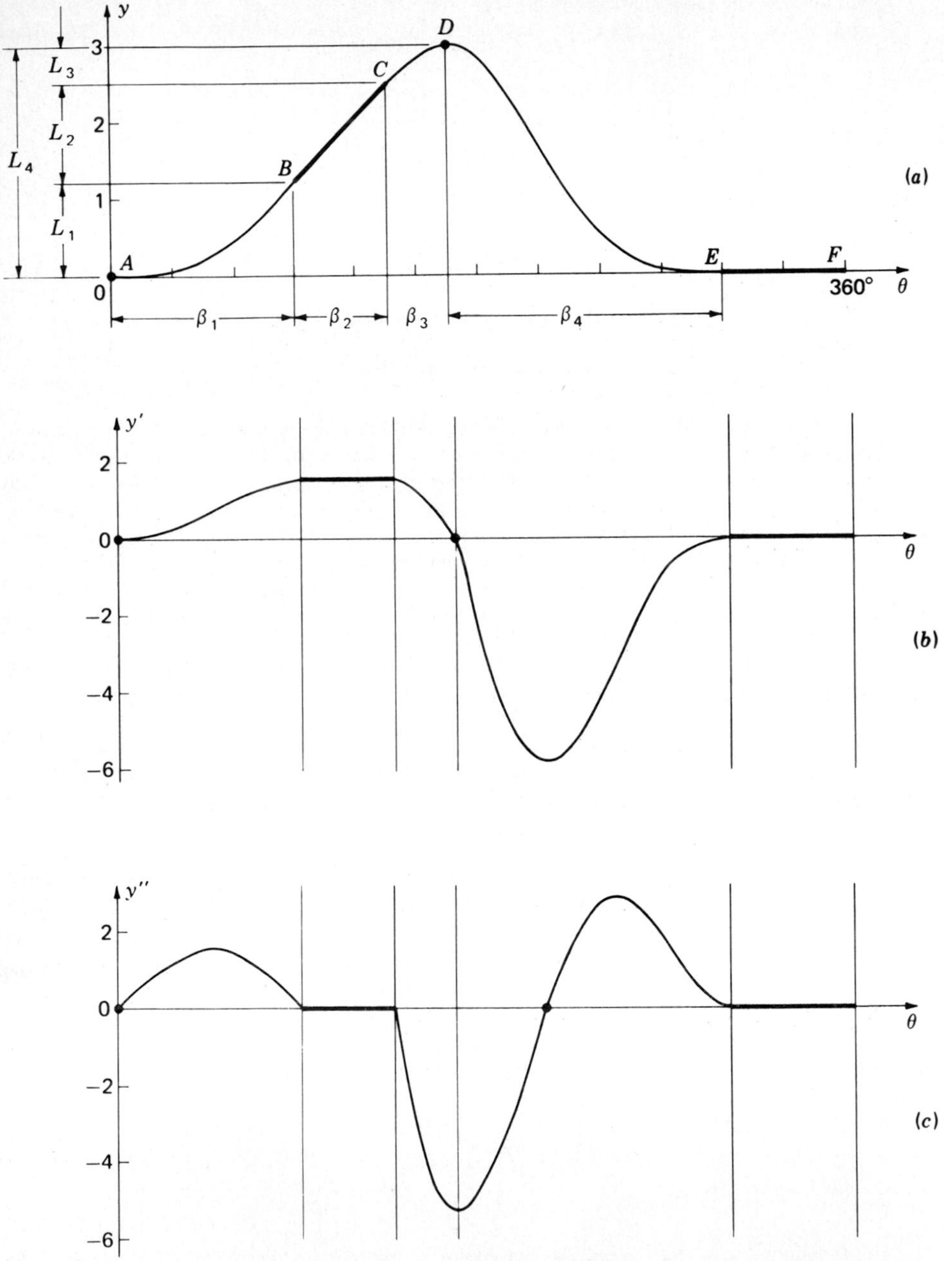

Figure 6-24 Example 6-2: (*a*) displacement diagram, inches; (*b*) "velocity" diagram, inches per radian; and (*c*) "acceleration" diagram, inches per radian squared.

In order to match "accelerations" (curvatures) at point D, we equate the value of y'' of Eq. (6-25c) at $\theta/\beta = 1$ (its right end) with y'' of Eq. (6-23c) at $\theta/\beta = 0$ (its left end). This gives

$$y''_D = -\frac{\pi^2 L_3}{4\beta_3^2} = -\frac{\pi^2 L_4}{\beta_4^2}$$

and, after making use of Eq. (6), yields

$$\beta_3 = 0.0844\beta_4^2 \tag{7}$$

Finally, for geometric compatibility, we have

$$L_1 + L_3 = L_4 - L_2 = 1.75 \text{ in} \tag{8}$$

and

$$\beta_1 + \beta_3 + \beta_4 = 2\pi - \beta_2 - \beta_5 = 4.451 \text{ rad} \tag{9}$$

Solving the five equations (5 to 9) simultaneously for the unknowns L_1, L_3, β_1, β_3, and β_4, we determine the proper values of the remaining parameters. Therefore, in summary, we have

$$\begin{aligned}
L_1 &= 1.264 \text{ in} & \beta_1 &= 1.589 \text{ rad} = 91.04^\circ \\
L_2 &= 1.250 \text{ in} & \beta_2 &= 0.785 \text{ rad} = 45.00^\circ \\
L_3 &= 0.486 \text{ in} & \beta_3 &= 0.479 \text{ rad} = 27.46^\circ \\
L_4 &= 3.000 \text{ in} & \beta_4 &= 2.382 \text{ rad} = 136.50^\circ \\
L_5 &= 0 & \beta_5 &= 1.047 \text{ rad} = 60.00^\circ
\end{aligned} \tag{10}$$

At this time an accurate layout of the displacement diagram and, if desired, its derivatives can be made to replace the original sketches. The curves of Fig. 6-24 have been drawn to scale using these values.

6-8 POLYNOMIAL CAM DESIGN

Although the variety of basic curves studied in previous sections are usually adequate, they certainly do not represent an exhaustive list of motions which might be used in cam design. Another common approach in designing cams is to synthesize appropriate motion curves with polynomial equations. We start with the basic equation

$$y = C_0 + C_1\frac{\theta}{\beta} + C_2\left(\frac{\theta}{\beta}\right)^2 + C_3\left(\frac{\theta}{\beta}\right)^3 + \cdots \tag{6-32}$$

where y and θ are the rise and cam input motion as before. The value of β represents the total travel of θ such that for the cam section under development the ratio θ/β ranges from 0 to 1. The constants C_i depend on the boundary conditions imposed. By proper choice of the boundary conditions and the order of the polynomial a suitable motion can usually be developed.

As an example of the polynomial approach, let us synthesize a full-rise curve with the boundary conditions

$$\begin{aligned}
\theta &= 0 & y &= 0 & y' &= 0 & y'' &= 0 \\
\theta &= \beta & y &= L & y' &= 0 & y'' &= 0
\end{aligned}$$

Since there are six conditions, Eq. (6-32) is written with six unknown constants

$$y = C_0 + C_1\frac{\theta}{\beta} + C_2\left(\frac{\theta}{\beta}\right)^2 + C_3\left(\frac{\theta}{\beta}\right)^3 + C_4\left(\frac{\theta}{\beta}\right)^4 + C_5\left(\frac{\theta}{\beta}\right)^5 \tag{a}$$

The first and second derivatives with respect to θ are

$$y' = \frac{1}{\beta}\left[C_1 + 2C_2\frac{\theta}{\beta} + 3C_3\left(\frac{\theta}{\beta}\right)^2 + 4C_4\left(\frac{\theta}{\beta}\right)^3 + 5C_5\left(\frac{\theta}{\beta}\right)^4\right] \tag{b}$$

$$y'' = \frac{1}{\beta^2}\left[2C_2 + 6C_3\frac{\theta}{\beta} + 12C_4\left(\frac{\theta}{\beta}\right)^2 + 20C_5\left(\frac{\theta}{\beta}\right)^3\right] \tag{c}$$

When the boundary conditions are substituted, the following six equations are obtained

$$0 = C_0 \tag{d}$$

$$L = C_0 + C_1 + C_2 + C_3 + C_4 + C_5 \tag{e}$$

$$0 = C_1 \tag{f}$$

$$0 = C_1 + 2C_2 + 3C_3 + 4C_4 + 5C_5 \tag{g}$$

$$0 = 2C_2 \tag{h}$$

$$0 = 2C_2 + 6C_3 + 12C_4 + 20C_5 \tag{i}$$

When these equations are solved simultaneously, there results

$$C_0 = 0 \qquad C_1 = 0 \qquad C_2 = 0 \qquad C_3 = 10L \qquad C_4 = -15L \qquad C_5 = 6L$$

The displacement equation is obtained by substituting these constants into Eq. (*a*)

$$y = L\left[10\left(\frac{\theta}{\beta}\right)^3 - 15\left(\frac{\theta}{\beta}\right)^4 + 6\left(\frac{\theta}{\beta}\right)^5\right] \tag{6-33a}$$

This is called the 3-4-5 *polynomial* full-rise motion because of the powers of the terms remaining. Its derivatives are

$$y' = \frac{L}{\beta}\left[30\left(\frac{\theta}{\beta}\right)^2 - 60\left(\frac{\theta}{\beta}\right)^3 + 30\left(\frac{\theta}{\beta}\right)^4\right] \tag{6-33b}$$

$$y'' = \frac{L}{\beta^2}\left[60\frac{\theta}{\beta} - 180\left(\frac{\theta}{\beta}\right)^2 + 120\left(\frac{\theta}{\beta}\right)^3\right] \tag{6-33c}$$

$$y''' = \frac{L}{\beta^3}\left[60 - 360\frac{\theta}{\beta} + 360\left(\frac{\theta}{\beta}\right)^2\right] \tag{6-33d}$$

The displacement diagram and its derivatives are plotted in Fig. 6-25. The properties are similar to those of cycloidal motion yet distinctly different.

Equations for the full-return 3-4-5 polynomial motion are derived by a

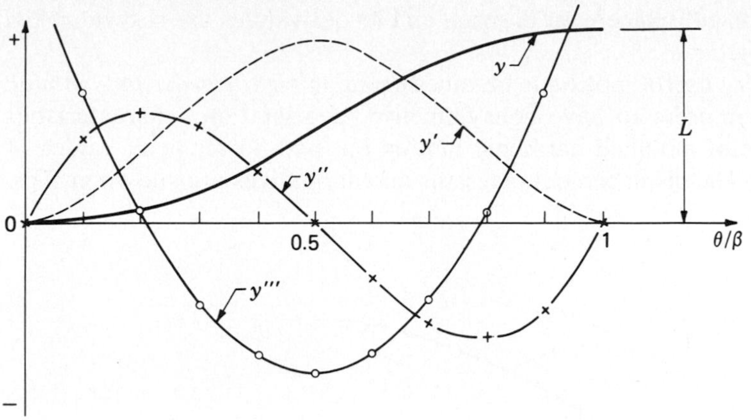

Figure 6-25 Displacement diagram and derivatives for full-rise 3-4-5 polynomial motion, Eq. (6-33).

parallel procedure and are

$$y = L\left[1 - 10\left(\frac{\theta}{\beta}\right)^3 + 15\left(\frac{\theta}{\beta}\right)^4 - 6\left(\frac{\theta}{\beta}\right)^5\right] \tag{6-34a}$$

$$y' = -\frac{L}{\beta}\left[30\left(\frac{\theta}{\beta}\right)^2 - 60\left(\frac{\theta}{\beta}\right)^3 + 30\left(\frac{\theta}{\beta}\right)^4\right] \tag{6-34b}$$

$$y'' = -\frac{L}{\beta^2}\left[60\frac{\theta}{\beta} - 180\left(\frac{\theta}{\beta}\right)^2 + 120\left(\frac{\theta}{\beta}\right)^3\right] \tag{6-34c}$$

$$y''' = -\frac{L}{\beta^3}\left[60 - 360\frac{\theta}{\beta} + 360\left(\frac{\theta}{\beta}\right)^2\right] \tag{6-34d}$$

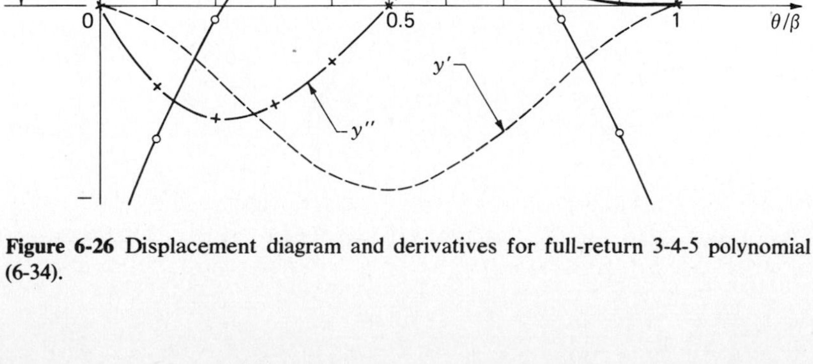

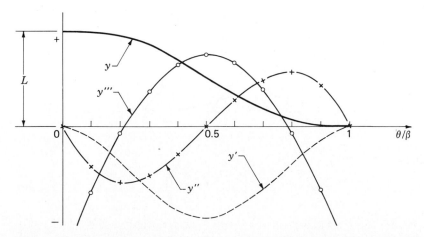

Figure 6-26 Displacement diagram and derivatives for full-return 3-4-5 polynomial motion, Eq. (6-34).

The corresponding displacement diagram and its derivatives are shown in Fig. 6-26.

Another very useful motion is obtained from an *eighth-order polynomial.* It was derived in order to have nonsymmetric "acceleration" characteristics similar to those of modified harmonic motion but with lower peak values of "acceleration." The displacement diagrams and derivatives are shown in Figs.

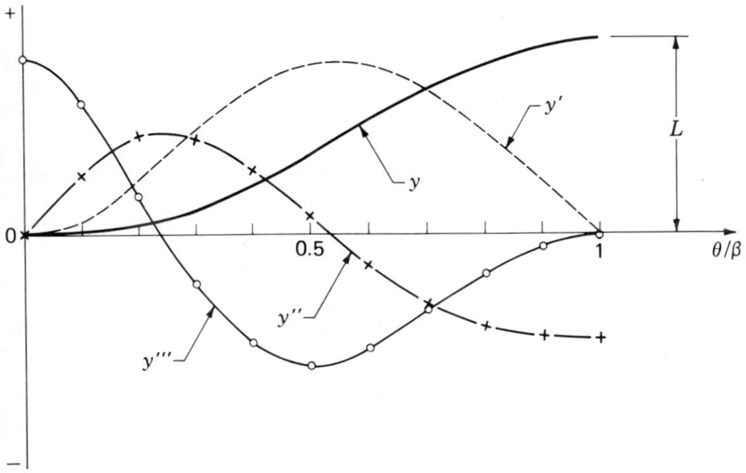

Figure 6-27 Displacement diagram and derivatives for full-rise eighth-order polynomial motion, Eq. (6-35).

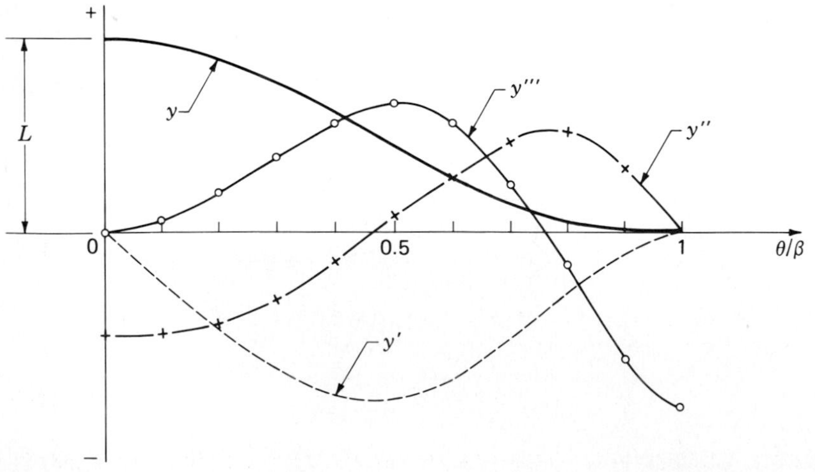

Figure 6-28 Displacement diagram and derivatives for full-return eighth-order polynomial motion, Eq. (6-36).

6-27 and 6-28. For the full-rise motion of Fig. 6-27 the equations are

$$y = L\left[6.097\,55\left(\frac{\theta}{\beta}\right)^3 - 20.780\,40\left(\frac{\theta}{\beta}\right)^5 + 26.731\,55\left(\frac{\theta}{\beta}\right)^6 - 13.609\,65\left(\frac{\theta}{\beta}\right)^7 + 2.560\,95\left(\frac{\theta}{\beta}\right)^8\right] \tag{6-35a}$$

$$y' = \frac{L}{\beta}\left[18.292\,65\left(\frac{\theta}{\beta}\right)^2 - 103.902\,00\left(\frac{\theta}{\beta}\right)^4 + 160.389\,30\left(\frac{\theta}{\beta}\right)^5 - 95.267\,55\left(\frac{\theta}{\beta}\right)^6 + 20.487\,60\left(\frac{\theta}{\beta}\right)^7\right] \tag{6-35b}$$

$$y'' = \frac{L}{\beta^2}\left[36.585\,30\left(\frac{\theta}{\beta}\right) - 415.608\,00\left(\frac{\theta}{\beta}\right)^3 + 801.946\,50\left(\frac{\theta}{\beta}\right)^4 - 571.605\,30\left(\frac{\theta}{\beta}\right)^5 + 143.413\,20\left(\frac{\theta}{\beta}\right)^6\right] \tag{6-35c}$$

$$y''' = \frac{L}{\beta^3}\left[36.585\,30 - 1246.824\,00\left(\frac{\theta}{\beta}\right)^2 + 3207.786\,00\left(\frac{\theta}{\beta}\right)^3 - 2858.026\,50\left(\frac{\theta}{\beta}\right)^4 + 860.479\,20\left(\frac{\theta}{\beta}\right)^5\right] \tag{6-35d}$$

For the full-return eighth-order polynomial motions of Fig. 6.28 the equations are

$$y = L\left[1.000\,00 - 2.634\,15\left(\frac{\theta}{\beta}\right)^2 + 2.780\,55\left(\frac{\theta}{\beta}\right)^5 + 3.170\,60\left(\frac{\theta}{\beta}\right)^6 - 6.877\,95\left(\frac{\theta}{\beta}\right)^7 + 2.560\,95\left(\frac{\theta}{\beta}\right)^8\right] \tag{6-36a}$$

$$y' = -\frac{L}{\beta}\left[5.268\,30\frac{\theta}{\beta} - 13.902\,75\left(\frac{\theta}{\beta}\right)^4 - 19.023\,60\left(\frac{\theta}{\beta}\right)^5 + 48.145\,65\left(\frac{\theta}{\beta}\right)^6 - 20.487\,60\left(\frac{\theta}{\beta}\right)^7\right] \tag{6-36b}$$

$$y'' = -\frac{L}{\beta^2}\left[5.268\,30 - 55.611\,00\left(\frac{\theta}{\beta}\right)^3 - 95.118\,00\left(\frac{\theta}{\beta}\right)^4 + 288.873\,90\left(\frac{\theta}{\beta}\right)^5 - 143.413\,20\left(\frac{\theta}{\beta}\right)^6\right] \tag{6-36c}$$

$$y''' = \frac{L}{\beta^3}\left[166.833\,00\left(\frac{\theta}{\beta}\right)^2 + 380.472\,00\left(\frac{\theta}{\beta}\right)^3 - 1444.369\,50\left(\frac{\theta}{\beta}\right)^4 + 860.479\,20\left(\frac{\theta}{\beta}\right)^5\right] \tag{6-36d}$$

Polynomial displacement equations of much higher order and meeting many more conditions than those presented here are also in common use. Automated procedures for determining the coefficients have been developed by Stoddart,† who also shows how the choice of the coefficients can be made to compensate for the elastic deformation of the follower system under dynamic conditions. Such cams are referred to as *polydyne cams.*

6-9 PLATE CAM WITH RECIPROCATING FLAT-FACE FOLLOWER

Once the displacement diagram of a cam system has been completely determined, as described in Sec. 6-7, the layout of the actual cam shape can be made, as shown in Sec. 6-3. In laying out the cam, however, we recall the need for a few more parameters, depending on the type of cam and follower, e.g., the prime-circle radius, any offset distance, roller radius, and so on. Also, as we will see, each different type of cam can be subject to certain further problems unless these remaining parameters are properly chosen.

In this section we study the problems that may be encountered in the design of a plate cam with a reciprocating flat-face follower. The geometric parameters of such a system which may yet be chosen are the prime-circle radius R_0, the offset ϵ of the follower stem, and the minimum width of the follower face.

† D. A. Stoddart, Polydyne Cam Design, *Mach. Des.*, vol. 25, no. 1, pp. 121–135; vol. 25, no. 2, pp. 146–154; vol. 25, no. 3, pp. 149–164, 1953.

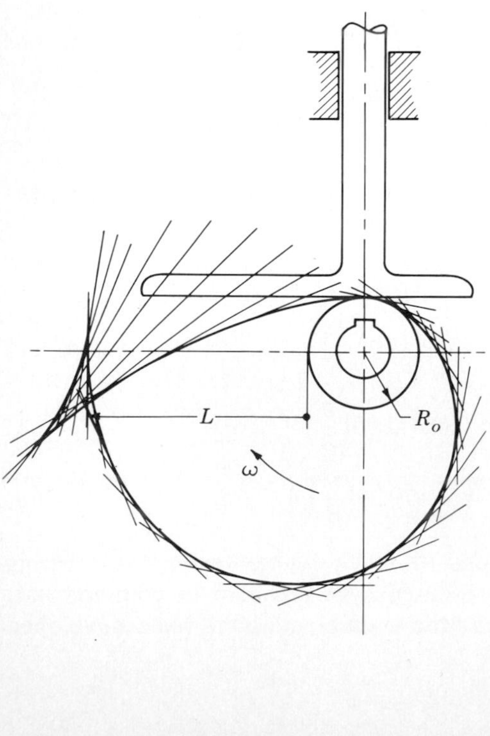

Figure 6-29 Undercut plate-cam layout.

Figure 6-29 shows the layout of a plate cam with a radial reciprocating flat-face follower. In this case the displacement chosen was a cycloidal rise of $L = 100$ mm in $\beta_1 = 90°$ of cam rotation, followed by a cycloidal return in the remaining $\beta_2 = 270°$ of cam rotation. The layout procedure of Fig. 6-10 was followed to develop the cam shape, and a prime circle radius of $R_0 = 25$ mm was used. Obviously, there is a problem since the cam profile crosses over itself. In machining, part of the cam shape would be lost and thereafter the intended cycloidal motion would not be achieved. Such a cam is said to be *undercut*.

Why did undercutting occur in this example, and how can it be avoided? It resulted from attempting to achieve too great a lift in too little cam rotation with too small a cam. One possibility is to decrease the desired lift L or to increase the cam rotation β_1 to avoid the problem. However, this may not be possible while still achieving the design objectives. Another solution is to use the same displacement characteristics but increase the prime circle radius R_0. This will produce a larger cam but with sufficient increase will overcome the undercutting problem.

It will save the effort of a trial-and-error layout procedure, however, if we can predict the minimum radius of the prime circle R_0 to avoid undercutting. This can be done by developing an equation for the radius of curvature of the cam profile. We start by writing the loop-closure equation using the vectors shown in Fig. 6-30. Using complex polar notation, this is

$$re^{j(\theta+\alpha)} + j\rho = j(R_0 + y) + s \tag{a}$$

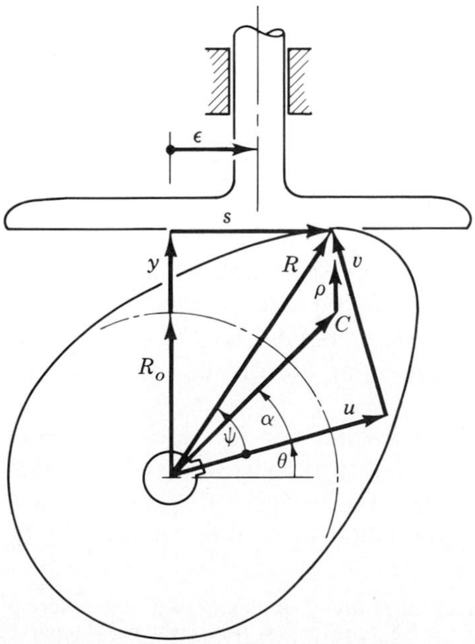

Figure 6-30

Here we have carefully chosen the vectors so that point C is the instantaneous center of curvature and ρ is the radius of curvature corresponding to the current contact point. The line along vector u which separates angles θ and α is fixed on the cam and is horizontal for the cam position $\theta=0$.

Separating the real and imaginary parts of Eq. (a), we have

$$r\cos(\theta+\alpha)=s \tag{b}$$

$$r\sin(\theta+\alpha)+\rho=R_0+y \tag{c}$$

Since the center of curvature C is stationary on the surface of the cam, the magnitudes of r, α, and ρ do not change for small changes in cam rotation;† that is, $dr/d\theta=d\alpha/d\theta=d\rho/d\theta=0$. Therefore, on differentiating Eq. (a) with respect to θ, we get

$$jre^{j(\theta+\alpha)}=jy'+\frac{ds}{d\theta} \tag{d}$$

and this can also be separated into real and imaginary parts

$$-r\sin(\theta+\alpha)=\frac{ds}{d\theta} \tag{e}$$

$$r\cos(\theta+\alpha)=y' \tag{f}$$

Between Eqs. (b) and (f) we can eliminate $\theta+\alpha$ and find

$$s=y' \tag{6-37}$$

Also, after differentiating this with respect to θ

$$\frac{ds}{d\theta}=y'' \tag{g}$$

we can eliminate $\theta+\alpha$ between Eqs. (c) and (e) and then substitute Eq. (g) to give a solution for ρ

$$\rho=R_0+y+y'' \tag{6-38}$$

We should note carefully the usefulness of Eq. (6-38); it states that the radius of curvature of the cam can be found for each value of cam rotation θ directly from the displacement equations *without laying out the cam profile.* All that is needed is a value for R_0 and values of the displacement and its second derivative.

We can use this equation to help choose a value for R_0 which will avoid undercutting. When undercutting occurs, the radius of curvature switches sign from positive to negative. If we are on the verge of undercutting, the cam will come to a point and ρ will be zero for some value of θ. We can say that R_0

† The values of r, α, and ρ are not constant but are currently at stationary values; their higher derivatives are nonzero.

must be chosen large enough that this never is the case. In fact, to avoid high contact stresses, we may wish to ensure that ρ is everywhere larger than some specified value ρ_{min}. Then, from Eq. (6-38), we must require that

$$\rho = R_0 + y + y'' > \rho_{min}$$

Since R_0 and y are always positive, the critical situation occurs where y'' has its largest negative value. Denoting this minimum of y'' as y''_{min} and remembering that y corresponds to the same cam angle θ, we have the condition

$$R_0 > \rho_{min} - y''_{min} - y \tag{6-39}$$

which must be met. This can easily be checked once the displacement equations have been established, and an appropriate value of R_0 can be chosen before the cam layout is attempted.

Returning now to Eq. (6-37), we see from Fig. 6-30 that this can also be of use. It states that the distance of travel of the point of contact to either side of the cam rotation center corresponds precisely with the plot of y'. Thus the minimum face width for the flat-face follower must extend at least y'_{max} to the right and $-y'_{min}$ to the left of the cam center to maintain contact. That is,

$$\text{Face width} > y'_{max} - y'_{min} \tag{6-40}$$

Example 6-3 Assuming that the displacement characteristics found in Example 6-2 are to be achieved by a plate cam with a reciprocating flat-face follower, determine the minimum face width and the minimum prime-circle radius to ensure that the radius of curvature of the cam is everywhere greater than 0.25 in.

SOLUTION From Fig. 6-24*b* we see that the maximum "velocity" occurs in the section *BC* and is

$$y'_{max} = \frac{L_2}{\beta_2} = 1.59 \text{ in/rad}$$

The minimum "velocity" occurs in section *DE* at $\theta/\beta_4 = \frac{1}{3}$. From Eq. (6-23*b*) its value is

$$y'_{min} = -\frac{\pi(1.250)}{2(0.785)}\left(\sin\frac{\pi}{3} + \frac{1}{2}\sin\frac{2\pi}{3}\right) = -3.25 \text{ in/rad}$$

Therefore, from Eq. (6-40), the minimum face width is

$$\text{Face width} > 1.59 + 3.25 = 4.84 \text{ in} \qquad \textit{Ans.}$$

This would be positioned with 1.59 in to the right and 3.25 in to the left of the cam rotation axis, and some appropriate clearance would be added on each side.

The maximum negative "acceleration" occurs at point *D*. Its value can be found from Eq. (6-25*c*) at $\theta/\beta = 1$

$$y''_{min} = -\frac{\pi^2(0.486)}{4(0.479)^2} = -5.23 \text{ in/rad}^2$$

Using this in Eq. (6-39), we find the minimum prime-circle radius

$$R_0 > 0.25 + 5.23 - 3.0 = 2.48 \text{ in} \qquad \textit{Ans.}$$

From this calculation we would then choose the actual prime-circle radius as, say, $R_0 = 2.50$ in.

We see that the eccentricity of the flat-face follower stem does not affect the geometry of the cam. This eccentricity is usually chosen to relieve high bending stresses in the follower.

Looking again at Fig. 6-30, we can write another loop-closure equation

$$ue^{j\theta} + ve^{j(\theta+\pi/2)} = j(R_0 + y) + s$$

where u and v denote the coordinates of the contact point in a coordinate system attached to the cam. Dividing this equation by $e^{j\theta}$, we get

$$u + jv = j(R_0 + y)e^{-j\theta} + se^{-j\theta}$$

which has for its real and imaginary parts,

$$u = (R_0 + y)\sin\theta + y'\cos\theta \qquad (6\text{-}41a)$$

$$v = (R_0 + y)\cos\theta - y'\sin\theta \qquad (6\text{-}41b)$$

These two equations give the coordinates of the cam profile and provide an alternative to the graphical layout procedure of Fig. 6-10. They can be used to generate a table of numeric rectangular coordinate data from which the cam can be machined. Polar coordinate equations for this same data are

$$R = \sqrt{(R_0 + y)^2 + (y')^2} \qquad (6\text{-}42a)$$

and

$$\psi = \frac{\pi}{2} - \theta - \tan^{-1}\frac{y'}{R_0 + y} \qquad (6\text{-}42b)$$

6-10 PLATE CAM WITH RECIPROCATING ROLLER FOLLOWER

Figure 6-31 shows a plate cam with a reciprocating roller follower. We see that three geometric parameters remain to be chosen after the displacement diagram is completed before the cam layout can be accomplished. These are the radius of the prime circle R_0, the eccentricity ϵ, and the radius of the roller R_r. There are also two potential problems to be considered when choosing these parameters; one is undercutting and the other is improper pressure angle.

The *pressure angle* is the angle between the axis of the follower stem and the line of the force exerted by the cam onto the roller follower, the normal to the pitch curve through the trace point. The pressure angle is labeled ϕ in the figure. Only the component of force along the line of motion of the follower is useful in overcoming the output load; the perpendicular component should be kept low to reduce sliding friction between the follower and its guideway. Too high a pressure angle will increase the effect of friction and may cause the translating follower to chatter or perhaps even to jam. Cam pressure angles of up to about 30 to 35° are about the largest that can be used without causing difficulties.

In Fig. 6-31 we see that the normal to the pitch curve intersects the

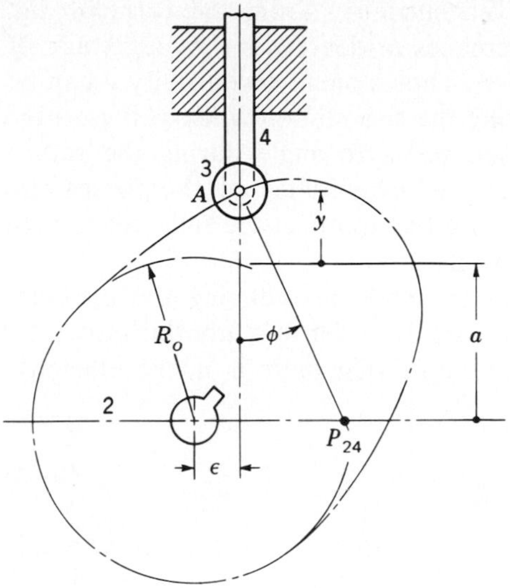

Figure 6-31

horizontal axis at point P_{24}, the instantaneous center of velocity between cam 2 and follower 4. Since the follower is translating, all points of the follower have velocities equal to that of P_{24}. But this must also be equal to the velocity of the coincident point on link 2

$$V_{P_{24}} = \dot{y} = \omega R_{P_{24}O_2}$$

Dividing by ω and using Eq. (6-14) we can reduce this to a strictly geometric relationship,

$$\frac{\dot{y}}{\omega} = y' = R_{P_{24}O_2}$$

This can now be written in terms of the eccentricity and pressure angle

$$y' = \epsilon + (a + y) \tan \phi \qquad (a)$$

where, as shown in Fig. 6-31, a is the vertical distance from the cam axis to the prime circle

$$a = \sqrt{R_0^2 - \epsilon^2} \qquad (b)$$

Substituting this into Eq. (a) and solving for ϕ, we obtain an expression for pressure angle

$$\phi = \tan^{-1} \frac{y' - \epsilon}{\sqrt{R_0^2 - \epsilon^2} + y} \qquad (6\text{-}43)$$

From this equation we see that once the displacement equations y have been determined, two parameters R_0 and ϵ can be adjusted to obtain a suitable pressure angle. We also notice that ϕ is continuously changing as the cam rotates, and therefore we are interested in studying the extreme values of ϕ.

Let us first consider the effect of eccentricity. From the form of Eq. (6-43) we see that increasing ϵ either increases or decreases the magnitude of the numerator depending on the sign of y'. Thus a small eccentricity ϵ can be used to reduce the pressure angle ϕ during the rise motion when y' is positive but only at the expense of an increased pressure angle during the return motion when y' is negative. Still, since the magnitudes of the forces are usually greater during rise, it is common practice to offset the follower to take advantage of this reduction in pressure angle.

A much more significant effect can be made in reducing the pressure angle by increasing the prime-circle radius R_0. In order to study this effect, let us take the conservative approach and assume that there is no eccentricity, $\epsilon = 0$. Equation (6-43) then reduces to

$$\phi = \tan^{-1} \frac{y'}{R_0 + y} \tag{6-44}$$

To find the extremum values of y' it is possible to differentiate this equation with respect to cam rotation and equate it to zero, thus finding the values of θ which yield maximum and minimum pressure angle. This is a tedious mathematical process, however, and can be avoided by using the nomogram of Fig. 6-32. This nomogram was produced by searching out on a digital computer the maximum value of ϕ from Eq. (6-44) for each of the standard full-rise motion curves of Sec. 6-6. With the nomogram it is possible to use the known values of L and β for each segment of the displacement diagram and to read directly the maximum pressure angle occurring in that segment for a particular choice of R_0. Alternatively, a desired maximum pressure angle can be chosen and an appropriate value of R_0 can be found. The process is best illustrated by an example.

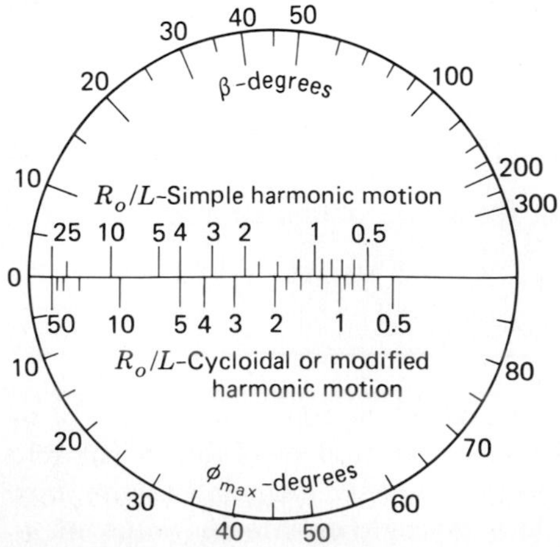

Figure 6-32 Nomogram relating maximum pressure angle ϕ_{max} to prime-circle radius R_0, lift L, and active cam angle β for radial roller follower cams with simple harmonic, cycloidal, or modified harmonic full-rise or full-return motion.

Example 6-4 Assuming that the displacement characteristics of Example 6-2 are to be achieved by a plate cam with a reciprocating radial roller follower, determine the minimum prime-circle radius such that the pressure angle will not exceed 30°.

SOLUTION Each section of the displacement diagram is checked in succession using the nomogram of Fig. 6-32.

For section *AB* of Fig. 6-24, we have half-cycloidal motion with $\beta_1 = 91°$ and $L_1 = 1.264$ in. Since this is a half-rise curve, whereas Fig. 6-32 is for full-rise curves, it is necessary to double both β_1 and L_1, thus pretending that the curve is full-rise; this gives $\beta = 182°$ and $L = 2.53$ in. Next, connecting a straight line from $\beta = 182°$ to $\phi_{max} = 30°$, we read from the lower scale on the center axis of the nomogram a value of $R_0/L = 0.65$, from which

$$R_0 = 0.65(2.53) = 1.64 \text{ in}$$

The segment *BC* need not be checked since its maximum pressure angle will occur at the boundary *B* and cannot be greater than that for the segment *AB*.

The segment *CD* has half-harmonic motion with $\beta_3 = 27.5°$ and $L_3 = 0.486$ in. Again, since this is a half-rise curve, these values are doubled and $\beta = 55°$, $L = 0.972$ in are used instead. Then from the nomogram we find $R_0/L = 2.4$, from which

$$R_0 = 2.4(0.972) = 2.33 \text{ in}$$

Here we must be careful, however. This value is the radius of a prime circle for which the horizontal axis of our doubled curve, the full-harmonic, has $y = 0$; it is not the R_0 we seek since the horizontal axis of our full-harmonic has a nonzero y value of

$$y = 3.00 - 0.972 = 2.03 \text{ in}$$

The appropriate value of R_0 for this situation is

$$R_0 = 2.33 - 2.03 = 0.30 \text{ in}$$

Next we check the segment *DE*, which has modified harmonic motion with $\beta_4 = 136.5°$ and $L_4 = 3.00$ in. Since this is a full-return motion curve, no adjustments are necessary. From the nomogram we find $R_0/L = 1.00$ and

$$R_0 = 1.00(3.00) = 3.00 \text{ in}$$

In order to ensure that the pressure angle does not exceed 30° throughout all segments of the cam, we must choose the prime-circle radius at least as large as the maximum of these predicted values. Remembering the inability to read the nomogram with great precision, we might choose a larger value, such as

$$R_0 = 3.25 \text{ in} \qquad \textit{Ans.}$$

Now that a final value has been chosen, we can use Fig. 6-32 again to find the actual maximum pressure angle in each segment.

AB: $$\frac{R_0}{L} = \frac{3.25}{2.53} = 1.28 \qquad \phi_{max} = 21°$$

CD: $$\frac{R_0}{L} = \frac{5.28}{0.97} = 5.45 \qquad \phi_{max} = 16°$$

DE: $$\frac{R_0}{L} = \frac{3.25}{3.00} = 1.08 \qquad \phi_{max} = 29°$$

Even though the prime circle has been proportioned to give a satisfactory pressure angle, the follower may still not complete the desired motion; if the

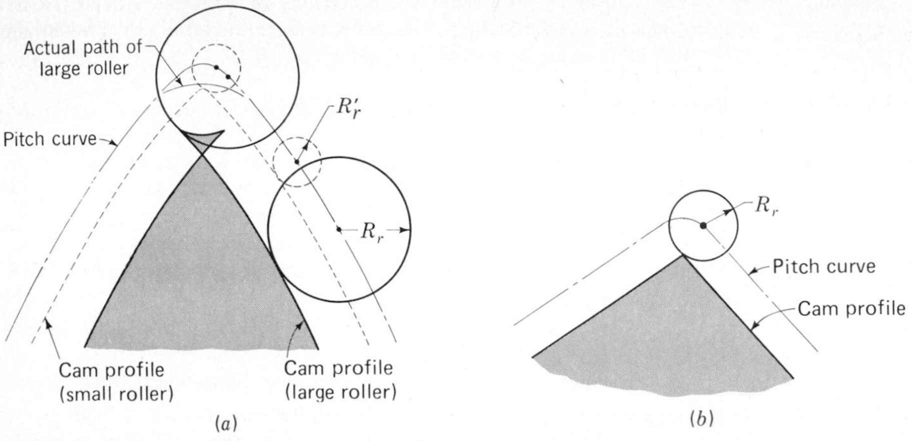

Figure 6-33

curvature of the pitch curve is too sharp, the cam profile may be undercut. Figure 6-33*a* shows a portion of a cam pitch curve and two cam profiles generated by two different size rollers. The cam profile generated by the larger roller is undercut and doubles over itself. The result, after machining, is a pointed cam which does not produce the desired motion. It is also clear from the same figure that a smaller roller moving on the same pitch curve generates a satisfactory cam profile. Similarly, if the prime circle and thus the cam size is increased enough, the larger roller will operate satisfactorily.

In Fig. 6-33*b* we see that the cam profile will be pointed when the radius of the roller R_r is equal to the radius of curvature of the pitch curve. Therefore, to achieve some chosen minimum value ρ_{min} for the minimum radius of curvature of the cam profile, the radius of curvature of the pitch curve must always be greater than this value by the radius of the roller

$$\rho_{pitch} = \rho + R_r \tag{c}$$

Now, for the case of a radial roller follower, the polar coordinates of the pitch curve are θ and

$$R = R_0 + y \tag{d}$$

From any standard text on differential calculus we can write the general expression for radius of curvature of a curve in polar coordinates. This is

$$\rho_{pitch} = \rho + R_r = \frac{[(R_0 + y)^2 + (y')^2]^{3/2}}{(R_0 + y)^2 + 2(y')^2 - (R_0 + y)y''} \tag{6-45}$$

As before, it is possible to differentiate this expression with respect to cam rotation θ and thus to seek out the minimum value of ρ for a particular choice of displacement equation y and a particular prime-circle radius R_0. However, since this would be an extremely tedious calculation to repeat for each new cam design, the minimum radius of curvature has been found by a digital computer program for each of the standard cam motions of Sec. 6-6; the results are presented graphically in Figs. 6-34 to 6-38. Each of these figures shows graphs of $(\rho_{min} + R_r)/R_0$ vs. β for one type of standard-motion curve with various ratios of R_0/L. Since we have solved for the displacement diagram and have chosen a value of R_0, each segment of the cam can be checked to find its minimum radius of curvature.

Saving even more effort, it is not necessary to check those segments of the cam where y'' remains positive throughout the segment, such as the half-rise motions of Eqs. (6-24) and (6-28) or the half-return motions of Eqs. (6-27) and (6-31). Assuming that the "acceleration" curve has been made continuous, the minimum radius of curvature of the cam cannot occur in these segments; Eq. (6-45) yields $\rho_{min} = R_0 - R_r$ for each.

Example 6-5 Assuming that the displacement characteristics of Example 6-2 are to be achieved by a plate cam with a reciprocating roller follower, determine the minimum radius of curvature of the cam profile if a prime-circle radius of $R_0 = 3.25$ in and a roller radius of $R_r = 0.5$ in are used.

SOLUTION: For the segment *AB* of Fig. 6-24, we have no need to check since y'' is positive throughout the segment.

For the segment *CD* we have $\beta_3 = 27.46°$ and $L_3 = 0.486$ in, from which we find

$$\frac{R_0}{L} = \frac{3.25 + (L_1 + L_2)}{L_3} = \frac{5.76}{0.486} = 11.8$$

where R_0 was adjusted by $L_1 + L_2$ since the graphs of Fig. 6-37 were plotted for $y = 0$ at the base of the segment. Now, using Fig. 6-37*b*, we find $(\rho_{min} + R_r)/R_0 = 0.57$ and therefore

$$\rho_{min} = 0.57R_0 - R_r = 0.57(5.76) - 0.50 = 2.78 \text{ in}$$

where, again, the adjusted value of R_0 was used.

For the segment *DE* we have $\beta_4 = 136.5°$ and $L_4 = 3.00$ in, from which $R_0/L = 1.08$. Using Fig. 6-36*a*, we find $(\rho_{min} + R_r)/R_0 = 1.00$ and

$$\rho_{min} = 1.00R_0 - R_r = 3.25 - 0.50 = 2.75 \text{ in}$$

Picking the smaller value, we find the minimum radius of curvature of the entire cam profile to be

$$\rho_{min} = 2.75 \text{ in} \qquad \textit{Ans.}$$

The rectangular coordinates of the cam profile of a plate cam with a reciprocating roller follower are given by

$$u = (\sqrt{R_0^2 - \epsilon^2} + y) \sin\theta + \epsilon \cos\theta + R_r \sin(\phi - \theta) \tag{6-46a}$$

$$v = (\sqrt{R_0^2 - \epsilon^2} + y) \cos\theta - \epsilon \sin\theta - R_r \cos(\phi - \theta) \tag{6-46b}$$

where ϕ is the pressure angle given by Eq. (6-43). The polar coordinates are

$$R = \sqrt{(\sqrt{R_0^2 - \epsilon^2} + y - R_r \cos\phi)^2 + (\epsilon + R_r \sin\phi)^2} \tag{6-47a}$$

$$\psi = -\theta + \tan^{-1} \frac{\sqrt{R_0^2 - \epsilon^2} - R_r \cos\phi}{\epsilon + R_r \sin\phi} \tag{6-47b}$$

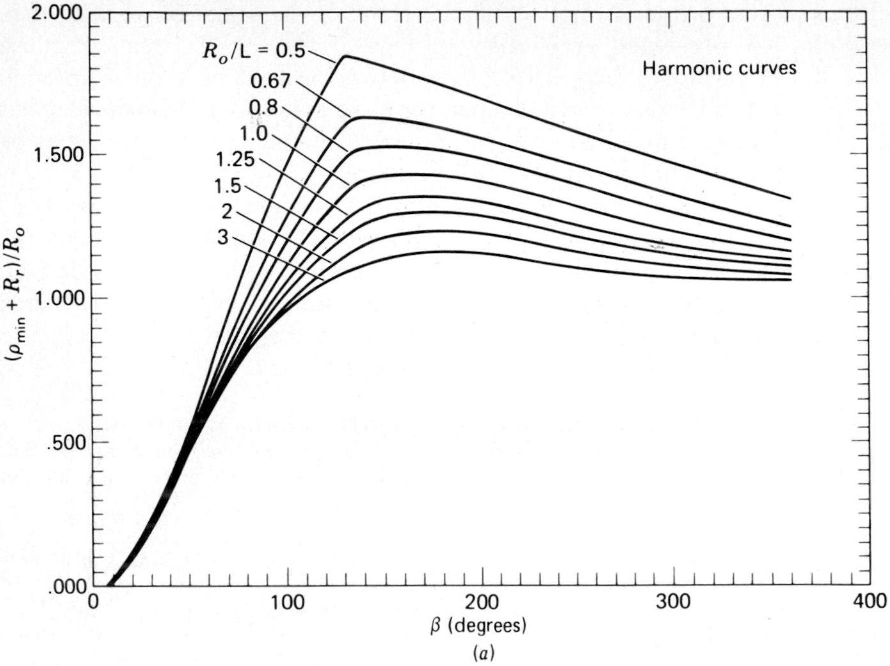

(a)

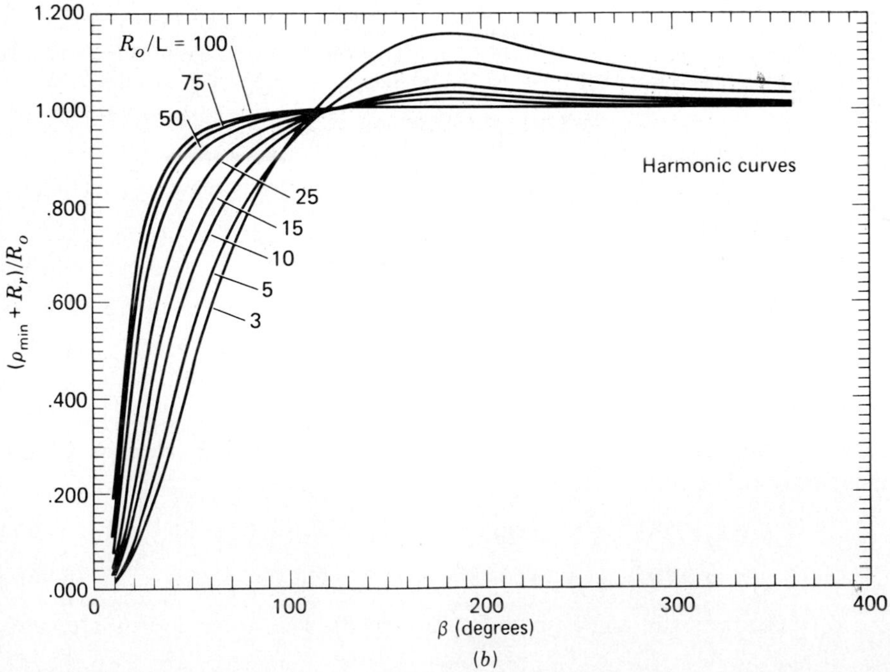

(b)

Figure 6-34 Minimum radius of curvature for radial roller follower cams with full-rise or full-return simple harmonic motion, Eqs. (6-18) and (6-21). (*From M. A. Ganter and J. J. Uicker, Jr., J. Mech. Des., ASME Trans., ser B, vol. 101, no. 3, p. 465–470, 1979, by permission.*)

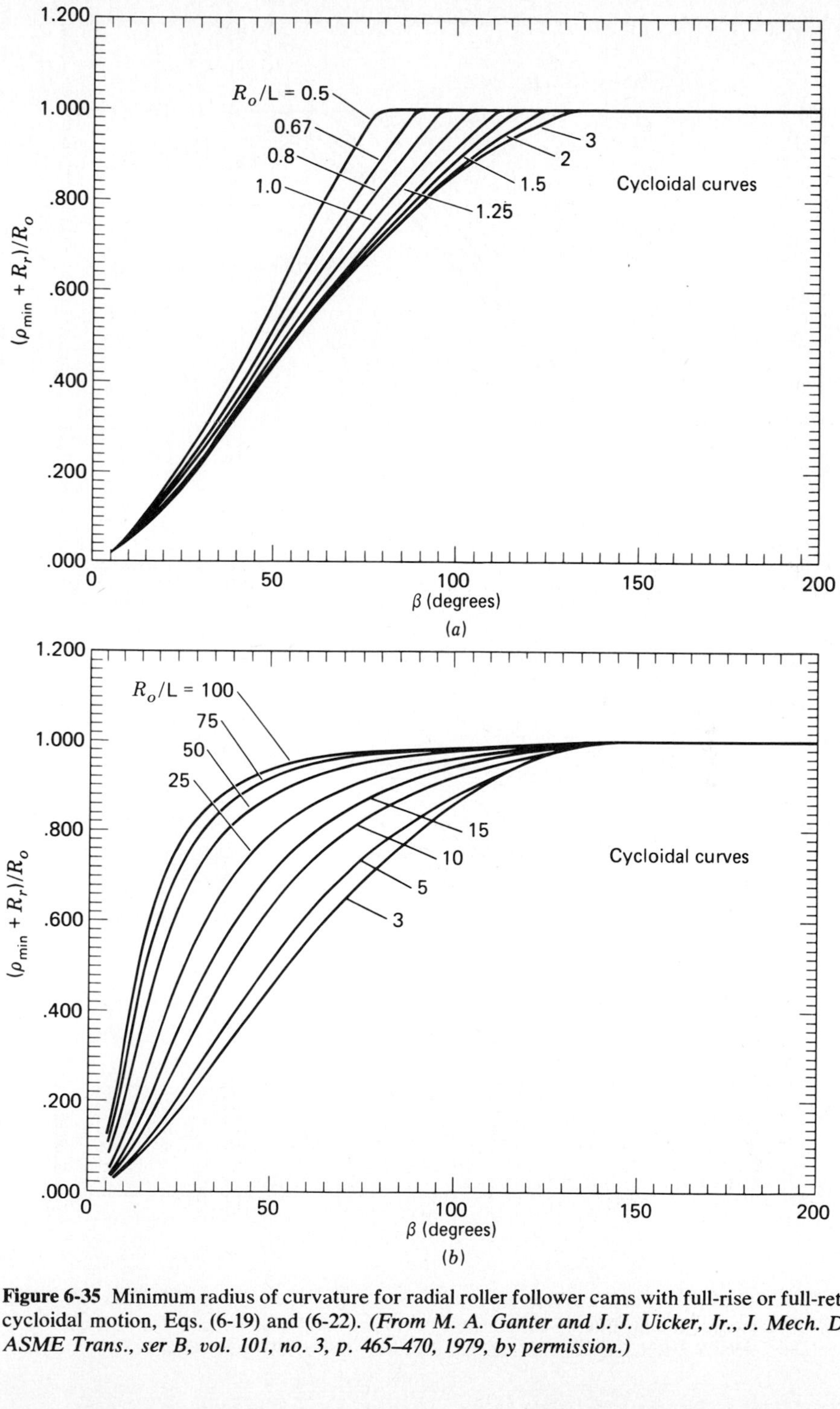

Figure 6-35 Minimum radius of curvature for radial roller follower cams with full-rise or full-return cycloidal motion, Eqs. (6-19) and (6-22). *(From M. A. Ganter and J. J. Uicker, Jr., J. Mech. Des., ASME Trans., ser B, vol. 101, no. 3, p. 465–470, 1979, by permission.)*

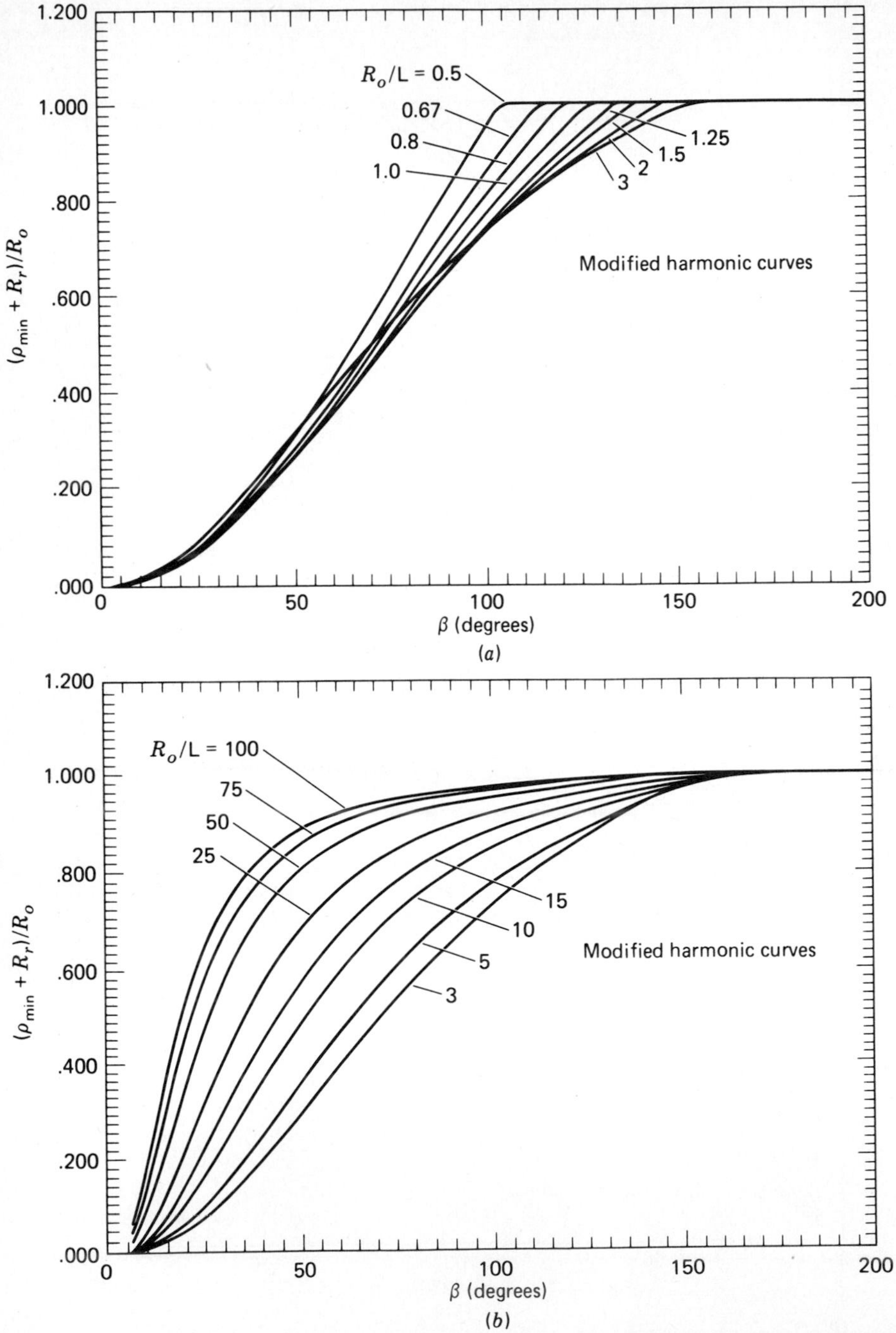

Figure 6-36 Minimum radius of curvature for radial roller follower cams with full-rise or full-return modified harmonic motion, Eqs. (6-20) and (6-23). *(From M. A. Ganter and J. J. Uicker, Jr., J. Mech. Des., ASME Trans., ser. B, vol. 101, no. 3, p. 465–470, 1979, by permission.)*

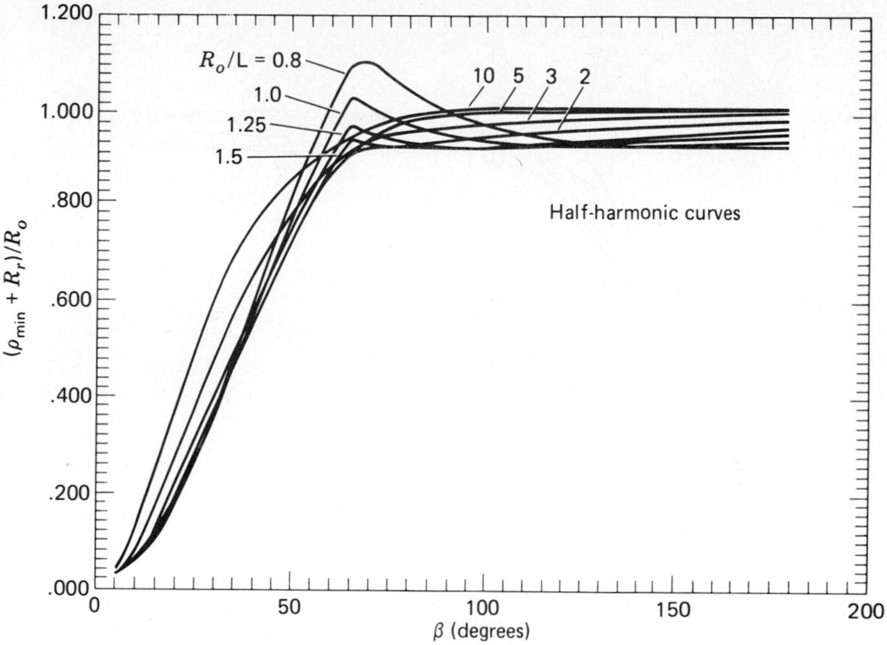

(a)

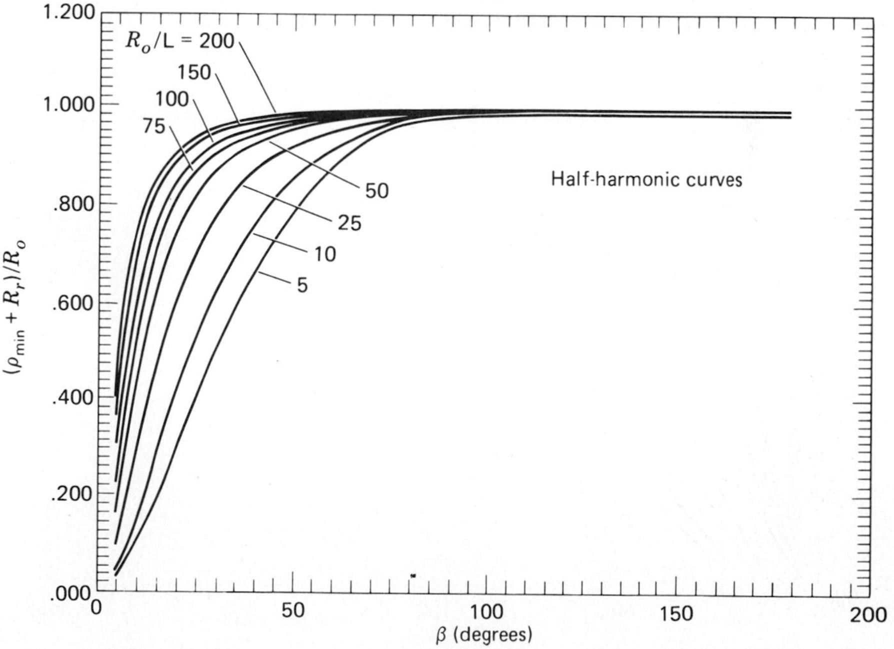

(b)

Figure 6-37 Minimum radius of curvature for radial roller follower cams with half-harmonic motion, Eqs. (6-25) and (6-26). (*From M. A. Ganter and J. J. Uicker, Jr., J. Mech. Des., ASME Trans., ser. B, vol. 101, no. 3, p. 465–470, 1979, by permission.*)

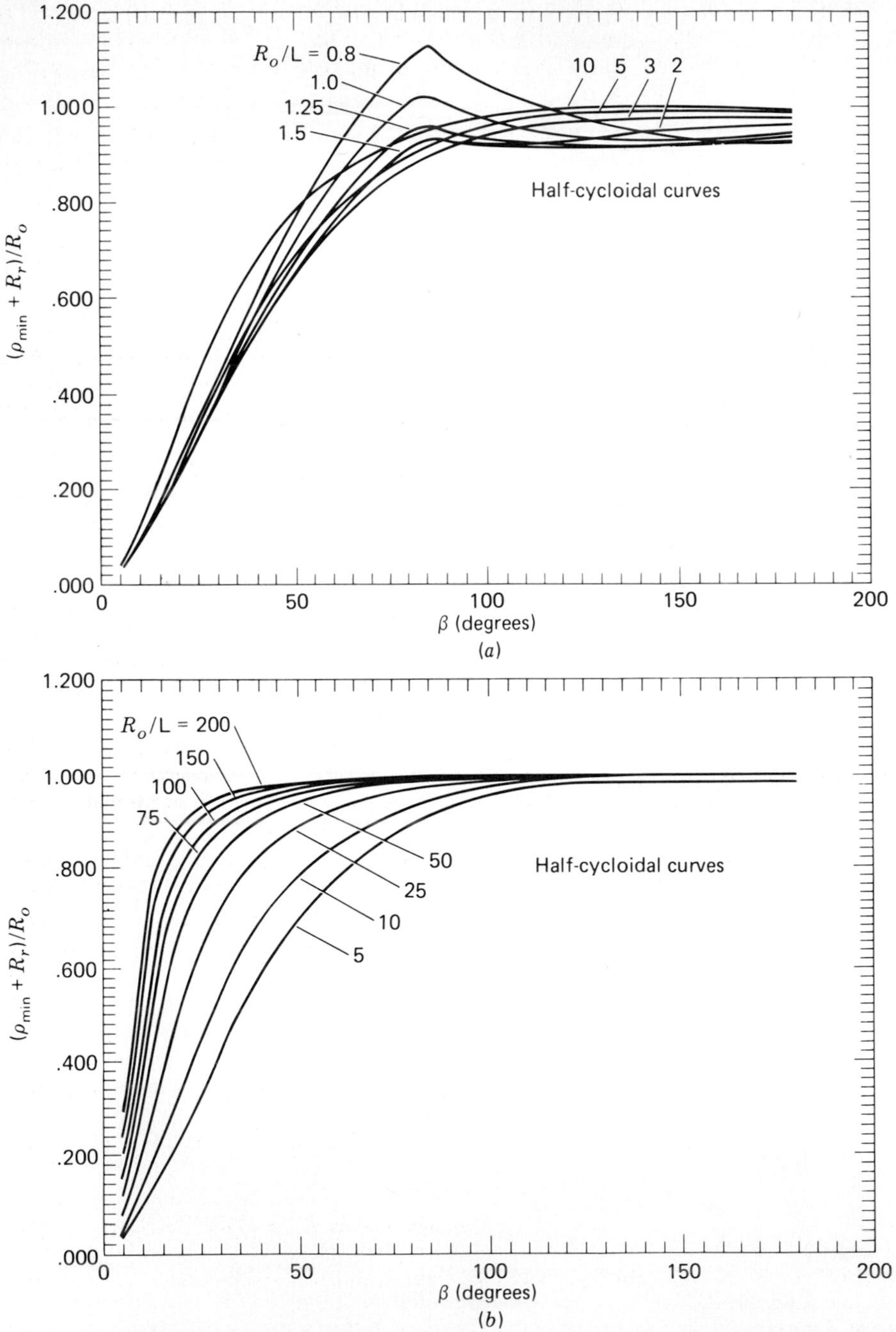

Figure 6-38 Minimum radius of curvature for radial roller follower cams with half-cycloidal motion, Eqs. (6-29) and (6-30). (*From M. A. Ganter and J. J. Uicker, Jr., J. Mech. Des. ASME Trans., ser. B, vol. 101, no. 3, p. 465–470, 1979, by permission.*)

In this and the previous section we have looked at the problems which result from improper choice of prime-circle radius for a plate cam with a reciprocating follower. Although the equations are different for oscillating followers or other types of cams, a similar approach can be used to guard against undercutting† and severe pressure angles.‡ Similar equations can also be developed for cam profile data.§ A good survey of the current literature on cam design has been compiled by Chen.¶

PROBLEMS

6-1 A plate cam's reciprocating radial roller follower is to rise 2 in with simple harmonic motion in 180° of cam rotation and return with simple harmonic motion in the remaining 180°. If the roller radius is 0.375 in and the prime-circle radius is 2 in, construct the displacement diagram, the pitch curve, and the cam profile for clockwise cam rotation.

6-2 A plate cam with a reciprocating flat-face follower is to have the same motion as in Prob. 6-1. The prime-circle radius is to be 1.5 in, and the cam is to rotate counterclockwise. Construct the displacement diagram and the cam profile, offsetting the follower stem by 0.75 in in the direction which reduces the bending stress in the follower during rise.

6-3 Construct the displacement diagram and the cam profile for a plate cam with an oscillating radial flat-face follower which rises through 30° with cycloidal motion in 150° of clockwise cam rotation, then dwells for 30°, returns with cycloidal motion in 120°, and dwells for 60°. Graphically determine the necessary length of the follower face, allowing 5 mm clearance at each end. The prime-circle radius is 30 mm. The follower pivot is 120 mm to the right. Cam rotation is counterclockwise.

6-4 A plate cam with an oscillating roller follower is to produce the same motion as in Prob. 6-3. The prime-circle radius is 60 mm; the length of the follower is 100 mm, and it is pivoted at 125 mm from the cam rotation axis; the roller radius is 10 mm. Construct the pitch curve and the cam profile. Determine the maximum pressure angle. Cam rotation is clockwise.

6-5 For full-rise simple harmonic motion, write the equations for velocity and jerk at the midpoint of the motion. Also determine the acceleration at the beginning and end of the motion.

6-6 For full-rise cycloidal motion, determine the values of θ for which the acceleration is maximum and minimum. What is the formula for acceleration at these points? Find the equations for velocity and jerk at the midpoint of the motion.

6-7 A plate cam with a reciprocating follower is to rotate clockwise at 400 rpm. The follower is to dwell for 60° of cam rotation, after which it rises to a lift of 2.5 in. During 1 in of its return stroke it must have a constant velocity of 40 in/s. Recommend standard cam motions from Sec. 6-6 to be used for high-speed operation and determine the corresponding lifts and cam rotation angles for each segment of the cam.

6-8 Repeat Prob. 6-7 except that the dwell is to be for 20° of cam rotation.

† M. Kloomok and R. V. Muffley, Plate Cam Design: Radius of Curvature, *Prod. Eng.*, vol. 26, no. 9, pp. 186–201, 1955.

‡ M. Kloomok and R. V. Muffley, Plate Cam Design: Pressure Angle Analysis, *Prod. Eng.*, vol. 26, no. 5, pp. 155–171, 1955.

§ See, for example, the excellent book by S. Molian, "The Design of Cam Mechanisms and Linkages," Constable, London, 1968.

¶ F. Y. Chen, A Survey of the State of the Art of Cam System Dynamics, *Mech. Mach. Theory*, vol. 12, no. 3, pp. 201–224, 1977.

6-9 If the cam of Prob. 6-7 is driven at constant speed, determine the time of the dwell and the maximum and minimum velocity and acceleration of the follower for the cam cycle.

6-10 A plate cam with an oscillating follower is to rise through 20° in 60° of cam rotation, dwell for 45°, then rise through an additional 20°, return, and dwell for 60° of cam rotation. Assuming high-speed operation, recommend standard cam motions from Sec. 6-6 to be used and determine the lifts and cam rotation angles for each segment of the cam.

6-11 Determine the maximum velocity and acceleration of the follower for Prob. 6-10 assuming that the cam is driven at a constant speed of 600 rpm.

6-12 The boundary conditions for a polynomial cam motion are as follows: for $\theta = 0$, $y = 0$, and $y' = 0$; for $\theta = \beta$, $y = L$ and $y' = 0$. Determine the appropriate displacement equation and its first three derivatives with respect to θ. Sketch the corresponding diagrams.

6-13 Determine the minimum face width using 0.1 in clearances at each end, and the minimum radius of curvature for the cam of Prob. 6-2.

6-14 Determine the maximum pressure angle and the minimum radius of curvature for the cam of Prob. 6-1.

6-15 A radial reciprocating flat-face follower is to have the motion described in Prob. 6-7. Determine the minimum prime-circle radius if the radius of curvature of the cam is not to be less than 0.5 in. Using this prime-circle radius, what is the minimum length of the follower face allowing a clearance of 0.25 in on each side?

6-16 Graphically construct the cam profile of Prob. 6-15 for clockwise cam rotation.

6-17 A radial reciprocating roller follower is to have the motion described in Prob. 6-7. Using a prime-circle radius of 20 in, determine the maximum pressure angle and the maximum roller radius which can be used without producing undercutting.

6-18 Graphically construct the cam profile of Prob. 6-17 using a roller radius of 0.75 in. Cam rotation is clockwise.

6-19 A plate cam rotates at 300 rpm and drives a reciprocating radial roller follower through a full rise of 75 mm in 180° of cam rotation. Find the minimum radius of the prime circle if simple harmonic motion is used and the pressure angle is not to exceed 25°. Find the maximum acceleration of the follower.

6-20 Repeat Prob. 6-19, except that the motion is cycloidal.

6-21 Repeat Prob. 6-19, except that the motion is modified harmonic.

6-22 Using a roller diameter of 20 mm, determine whether the cam of Prob. 6-19 will be undercut.

6-23 Equations (6-41) and (6-42) describe the profile of a plate cam with a reciprocating flat-face follower. If such a cam is to be cut on a milling machine with cutter radius R_C, determine similar equations for the center of the cutter.

6-24 Write calculator programs for each of the displacement equations of Sec. 6-6.

6-25 Write a computer program to plot the cam profile for Prob. 6-2.

CHAPTER

SEVEN

SPUR GEARS

We study gears because the transmission of rotary motion from one shaft to another occurs in nearly every machine one can imagine. Gears constitute one of the best of the various means available for transmitting this motion.

In the United States the task of converting from U.S. customary units to SI units for the design and manufacture of gears is so huge, so complex, and so costly that complete conversion may never be made. It is for this reason that most of the material in this and in the chapter that follows is in U.S. customary units. Readers of this book in all-SI countries should supplement the material with copies of their own standards.

7-1 TERMINOLOGY AND DEFINITIONS

Spur gears are used to transmit rotary motion between parallel shafts; they are usually cylindrical, and the teeth are straight and parallel to the axis of rotation.

The terminology of gear teeth is illustrated in Fig. 7-1, where most of the following definitions are shown:

The *pitch circle* is a theoretical circle upon which all calculations are usually based. The pitch circles of a pair of mating gears are tangent to each other.

A *pinion* is the smaller of two mating gears. The larger is often called the *gear.*

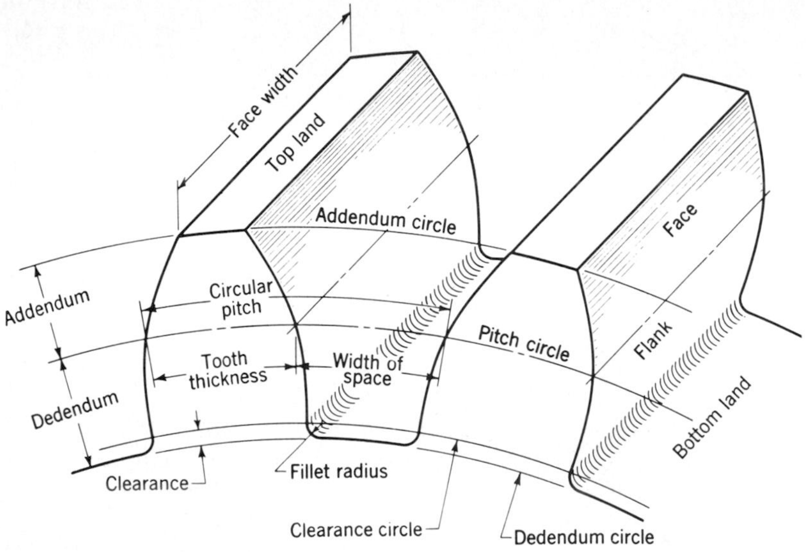

Figure 7-1 Terminology.

The *circular pitch* p_c is the distance, in inches, measured on the pitch circle, from a point on one tooth to a corresponding point on an adjacent tooth.

The *diametral pitch* P is the number of teeth on the gear per inch of pitch diameter. The units of diametral pitch are the reciprocal of inches. Note that the diametral pitch cannot actually be measured on the gear itself.

The *module* m is the ratio of the pitch diameter to the number of teeth. The customary unit of length is the millimetre. The module is the index of tooth size in SI, while the diametral pitch is only used with U.S. customary units.

The *addendum* a is the radial distance between the top land and the pitch circle.

The *dedendum* b is the radial distance from the bottom land to the pitch circle.

The *whole depth* h_t is the sum of the addendum and dedendum.

The *clearance circle* is a circle that is tangent to the addendum circle of the mating gear.

The *clearance* c is the amount by which the dedendum in a given gear exceeds the addendum of its mating gear.

The *backlash* is the amount by which the width of a tooth space exceeds the thickness of the engaging tooth on the pitch circles.

You should satisfy yourself of the validity of the following useful relations:

$$P = \frac{N}{d} \qquad m = \frac{d}{N} \tag{7-1}$$

where P = diametral pitch, teeth per inch
N = number of teeth
d = pitch diameter, in or mm
m = module, mm

$$p_c = \frac{\pi d}{N} = \pi m \tag{7-2}$$

where p_c is the circular pitch in inches or in millimetres

$$p_c P = \pi \tag{7-3}$$

7-2 FUNDAMENTAL LAW OF TOOTHED GEARING

Mating gear teeth acting against each other to produce rotary motion may be likened to a cam and follower. When the tooth profiles (or cam and follower profiles) are shaped so as to produce a constant angular-velocity ratio during meshing, the surfaces are said to be *conjugate.* It is possible to specify any profile for one tooth and then to find a profile for the mating tooth so that the surfaces are conjugate. One of these solutions is the *involute profile,* which, with few exceptions, is in universal use for gear teeth.

The action of a single pair of mating teeth as they pass through an entire phase of action must be such that the ratio of the angular velocity of the driving gear to that of the driven gear remains constant. This is the fundamental criterion which governs the choice of the tooth profiles. If it were not true of gearing, very serious vibration and impact problems would exist, even at low speeds.

In Sec. 3-14 we learned that the angular-velocity-ratio theorem states that the angular-velocity ratio of any mechanism is inversely proportional to the segments into which the common instant center cuts the line of centers. In Fig. 7-2 two profiles are in contact at A; let profile 2 be the driver and 3 the driven. A normal to the profiles at the point of contact A intersects the line of centers O_2O_3 at the instant center P.

In gearing, P is called the *pitch point* and BC the *line of action.* Designating the pitch-point radii of the two profiles as r_2 and r_3, from Eq. (3-25),

$$\frac{\omega_2}{\omega_3} = \frac{r_3}{r_2} \tag{7-4}$$

This equation is frequently used to define the *law of gearing,* which states that *the pitch point must remain fixed on the line of centers.* This means that all the lines of action for every instantaneous point of contact must pass through the pitch point. Our problem now is to determine the shape of the mating surfaces to satisfy the law of gearing.

It should not be assumed that just any shape or profile for which a

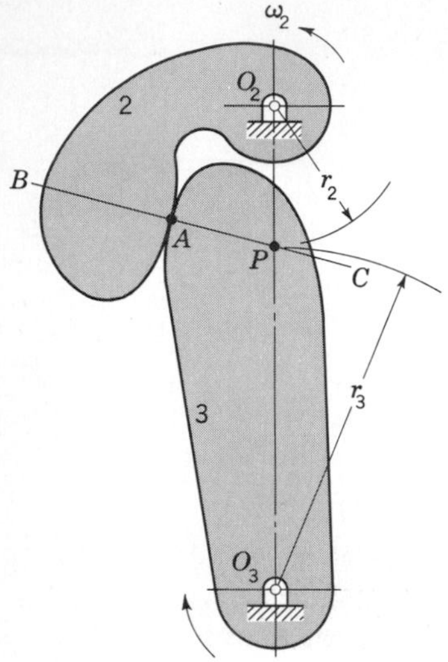

Figure 7-2

conjugate can be found will be satisfactory. Even though conjugate curves are found, the practical problems of reproducing these curves in great quantities on steel gear blanks, as well as other materials, and using existing machinery still exist. In addition, the changes in the shaft centers due to misalignment and large forces must be considered. And finally, the tooth profile selected must be one which can be reproduced economically. A major portion of this chapter is devoted to illustrating how the involute profile fulfills these requirements.

7-3 INVOLUTE PROPERTIES

If mating tooth profiles have the shape of involute curves, the condition that the common normal at all points of contact is to pass through the pitch point is satisfied. An involute curve is the path generated by a tracing point on a cord as the cord is unwrapped from a cylinder called the *base cylinder.* This is shown in Fig. 7-3, where T is the tracing point. Note that the cord AT is normal to the involute at T and that the distance AT is the instantaneous value of the radius of curvature. As the involute is generated from the origin T_0 to T_1, the radius of curvature varies continuously; it is zero at T_0 and greatest at T_1. Thus the cord is the generating line, and it is always normal to the involute.

Let us now examine the involute profile to see how it satisfies the requirement for the transmission of uniform motion. In Fig. 7-4 two gear blanks with fixed centers O_2 and O_3 are shown having base cylinders whose respective radii are O_2A and O_3B. We now imagine that a cord is wound clockwise around the base cylinder of gear 2, pulled tightly between points A

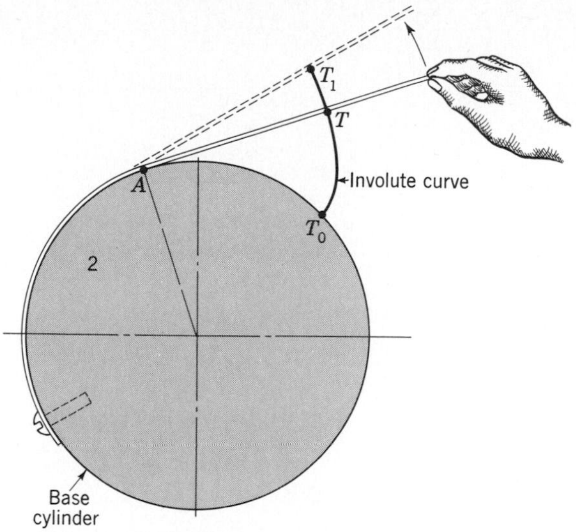

Figure 7-3

and B, and wound counterclockwise around the base cylinder of gear 3. If now the base cylinders are rotated in different directions so as to keep the cord tight, a point T will trace out the involutes CD on gear 2 and EF on gear 3. The involutes thus generated simultaneously by the single tracing point are conjugate profiles.

Next imagine that the involutes of Fig. 7-4 are scribed on plates and the plates cut along the scribed curves and then bolted to the respective cylinders in the same positions. The result is shown in Fig. 7-5. The cord can now be

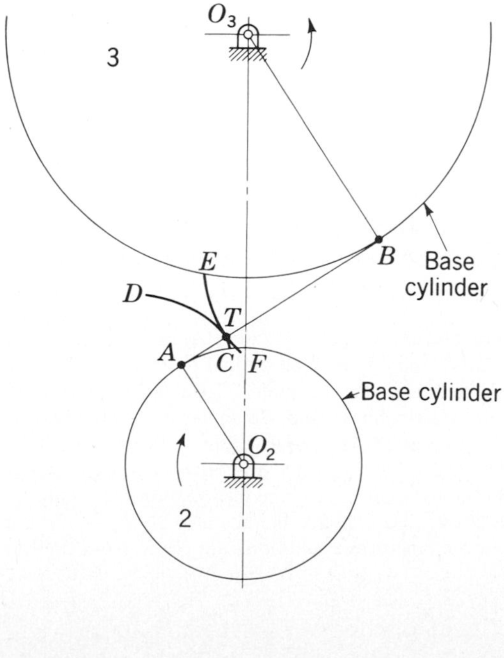

Figure 7-4 Involute action.

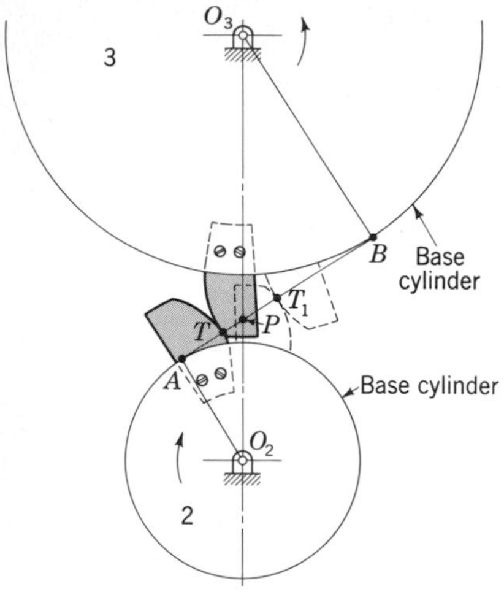

Figure 7-5

removed, and if gear 2 is moved clockwise, gear 3 is caused to move counterclockwise by the camlike action of the two curved plates. The path of contact will be the line *AB* formerly occupied by the cord. Since the line *AB* is the generating line for each involute, it is normal to both profiles at all points of contact. Also, it always occupies the same position because it is tangent to both base cylinders. Therefore point *P* is the pitch point; it does not move; and so the involute curve satisfies the law of gearing.

Before concluding this section you should observe that a change in center distance, which might be caused by incorrect mounting, will have no effect upon the shape of the involute. In addition, the pitch point is still fixed, and so the law of gearing is satisfied.

7-4 INTERCHANGEABLE GEARS; AGMA STANDARDS

A *tooth system* is a *standard*† which specifies the relationships between addendum, dedendum, working depth, tooth thickness, and pressure angle to attain interchangeability of gears of all tooth numbers but of the same pressure angle and pitch. You should be aware of the advantages and

† Standardized by the American Gear Manufacturers Association (AGMA) and the American National Standards Institute (ANSI). The AGMA Standards may be quoted or extracted in their entirety provided an appropriate credit line is included, for example, "Extracted from AGMA Information Sheet—Strength of Spur, Helical, Herringbone, and Bevel Gear Teeth (AGMA 225.01), with permission of the publisher, the American Gear Manufacturers Association, 1330 Massachusetts Avenue, N.W., Washington, DC 20005." These standards have been used extensively in this chapter and in the chapter which follows. In each case the information sheet number is given. Table 7-1 is from AGMA publication 201.02 and 201.02A, but see also 207.04. Write the AGMA for a complete list of standards because changes and additions are made from time to time.

Table 7-1 Standard AGMA and ANSI tooth systems for spur gears†

Quantity	Coarse pitch‡ (up to 20P) full depth		Fine pitch (20P and up) full depth
Pressure angle ϕ	20°	25°	20°
Addendum a	$\dfrac{1.000}{P}$	$\dfrac{1.000}{P}$	$\dfrac{1.000}{P}$
Dedendum b	$\dfrac{1.250}{P}$	$\dfrac{1.250}{P}$	$\dfrac{1.200}{P} + 0.002$ in
Working depth h_k	$\dfrac{2.000}{P}$	$\dfrac{2.000}{P}$	$\dfrac{2.000}{P}$
Whole depth h_t (minimum)	$\dfrac{2.25}{P}$	$\dfrac{2.25}{P}$	$\dfrac{2.200}{P} + 0.002$ in
Circular tooth thickness t	$\dfrac{\pi}{2P}$	$\dfrac{\pi}{2P}$	$\dfrac{1.5708}{P}$
Fillet radius of basic rack r_f	$\dfrac{0.300}{P}$	$\dfrac{0.300}{P}$	Not standardized
Basic clearance c (minimum)	$\dfrac{0.250}{P}$	$\dfrac{0.250}{P}$	$\dfrac{0.200}{P} + 0.002$ in
Clearance c (shaved or ground teeth)	$\dfrac{0.350}{P}$	$\dfrac{0.350}{P}$	$\dfrac{0.3500}{P} + 0.002$ in
Minimum number of pinion teeth	18	12	18
Minimum number of teeth per pair	36	24	
Minimum width of top land t_0	$\dfrac{0.25}{P}$	$\dfrac{0.25}{P}$	Not standardized

† See AGMA Standards 201.02, 201.02A, and 207.04.
‡ But not including 20P.

disadvantages of the various systems so that you can choose the optimum tooth for a given design and have a basis of comparison when departing from a standard tooth profile.

Table 7-1 lists the tooth proportions for completely interchangeable gears in the U.S. customary system and for operation on standard center distances. No standard has been established in the U.S. for tooth systems based wholly upon the use of SI units. In fact it is probable that a number of years will elapse before agreement can be reached; the problems to be solved are complex as well as expensive. Even in England, where they have been ahead of the U.S. in the metric changeover, the inch system is still predominantly used for gearing. Merritt† states that among the reasons is that new standards had been approved and adopted shortly before metrication began.

The addenda listed in Table 7-1 are for gears having tooth numbers equal to or greater than the minimum numbers listed, and for these numbers there

† H. E. Merritt, "Gear Engineering," Wiley, New York, 1971.

Table 7-2 Diametral pitches in general use

Coarse pitch	2, $2\frac{1}{4}$, $2\frac{1}{2}$, 3, 4, 6, 8, 10, 12, 16
Fine pitch	20, 24, 32, 40, 48, 64, 80, 96, 120, 150, 200

will be no undercutting. For fewer numbers of teeth a modification called the *long-and-short-addendum system* should be used. In this system the addendum of the gear is decreased just sufficiently to ensure that contact does not begin before the interference point (see Sec. 7-7). The addendum of the pinion is then increased a corresponding amount. In this modification, there is no change in the pressure angle or in the pitch circles, so the center distance remains the same. The intent is to increase the recess action and decrease the approach action.

The 0.002-in additional dedendum shown in Table 7-1 for fine-pitch gears provides space for the accumulation of dirt at the roots of the teeth.

The addenda and working depths shown in Table 7-1 are for, and define, full-depth teeth; for stub teeth, use $1.60/P$ for the working depths and $0.8/P$ for the addenda.

It should be particularly noted that the standards shown in Table 7-1 are *not* intended to restrict the freedom of the designer. Standard tooth proportions lead to interchangeability and standard cutters which are economical to purchase, but the need for high-performance gears may well dictate considerable deviation from these systems.

Some of the tooth systems which are now obsolete are the two AGMA $14\frac{1}{2}°$ systems, the Fellows 20° stub-tooth system, and the Brown & Sharpe system.

The obsolete systems should not be used for new designs, but you may need to refer to them when redesigning or remodeling existing machinery which uses these older systems.

The diametral pitches listed in Table 7-2 should be used whenever possible in order to keep to a minimum the inventory of gear-cutting tools.

7-5 FUNDAMENTALS OF GEAR-TOOTH ACTION

In order to illustrate the fundamentals of spur gears, we shall proceed, step by step, through the actual layout of a pair of spur gears. The dimensions used will be obtained from Sec. 7-4, which lists standard tooth forms. New terms will be introduced and explained as we progress through the layout.

The purpose of a layout of gear teeth is *not* to use it in the shop but for analysis. In the production of large quantities of gears the shop requires only the drawings of the gear blanks together with a specification (not a drawing) of the tooth form and size. On the other hand, if tools are to be manufactured for cutting gear teeth, drawings of the tooth form and shape must be made.

Sometimes these drawings are made at a scale many times larger than the tooth itself so that accurate dimensions can be obtained.

For given information we shall select a 2-in-diameter 10-diametral-pitch pinion to drive a 50-tooth gear. The tooth form selected is the 20° full depth. The various steps in correct order are illustrated in Figs. 7-6 and 7-7 and are as follows.

Step 1 Calculate the pitch diameters and draw the pitch circles tangent to each other (Fig. 7-6). The numbers 2 and 3 will be used as subscripts to designate the pinion and gear, respectively. From Eq. (7-1) the pitch diameter

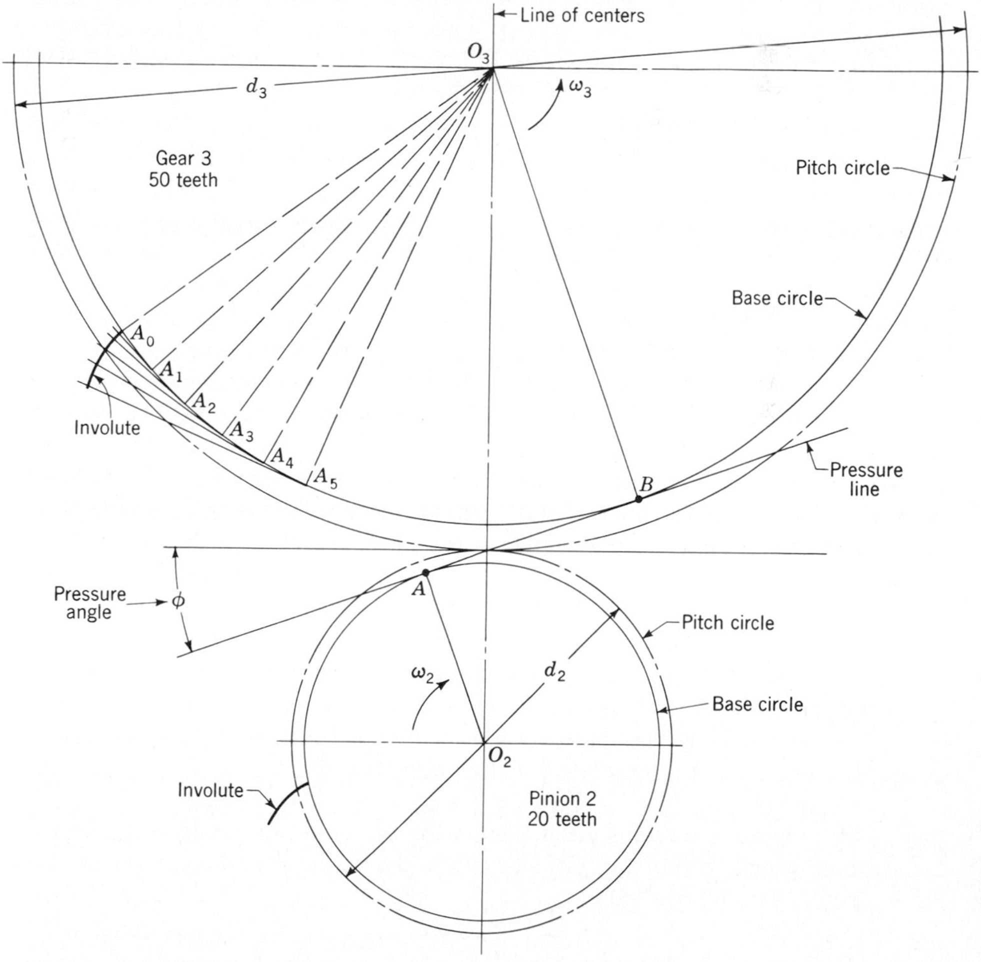

Figure 7-6 Layout of a pair of spur gears.

of the gear is

$$d_3 = \frac{N_3}{P} = \frac{50}{10} = 5 \text{ in}$$

Step 2 Draw a line perpendicular to the line of centers through the pitch point (Fig. 7-6). The pitch point is the point of tangency of the pitch circles. Draw the pressure line at an angle equal to the pressure angle from the perpendicular. The *pressure line* corresponds to the generating line, or the line of action, defined in the previous sections. As shown, it is always normal to the involutes at the point of contact and passes through the pitch point. It is called the pressure line because the resultant tooth force during action is along this line. The *pressure angle* is the angle the pressure line makes with a perpendicular to the line of centers through the pitch point. In this example the pressure angle is 20°.

Step 3 Through the centers of each gear construct perpendiculars O_2A and O_3B to the pressure line (Fig. 7-6). These radial distances, from the centers to the pressure line, are the radii of the two *base circles.* The base circles correspond to the base cylinders of Sec. 7-3. The involute curve originates on these base circles. Draw each base circle.

Step 4 Generate an involute curve on each base circle (Fig. 7-6). This is illustrated on gear 3. First, divide the base circle into equal parts A_0, A_1, A_2, etc. Now construct the radial lines O_3A_0, O_3A_1, O_3A_2, etc. Next, construct perpendiculars to these radial lines. The involute begins at A_0. The second point is obtained by laying the distance A_0A_1 on the perpendicular through A_1. The next point is found by laying off twice A_0A_1 on the perpendicular through A_2, and so on. The curve constructed through these points is the involute. The involute for the pinion is constructed in the same manner on the pinion base circle.

Step 5 Using cardboard or, preferably, a sheet of clear plastic, cut a template for each involute and mark on it the corresponding center of each gear. These templates are then used to draw the involute portion of each tooth. They can be turned over to draw the opposite side of the tooth. In some cases it may be desirable to construct a template for the entire tooth.

Step 6 Calculate the circular pitch. The width of tooth and width of space are constructed equal to half the circular pitch. Mark these distances off on the pitch circles. From Eq. (7-3)

$$p_c = \frac{\pi}{P} = \frac{\pi}{10} = 0.314\,16 \text{ in}$$

so the width of tooth and space is (0.314 16)/2, = 0.157 08 in. These points are marked off on the pitch circles of Fig. 7-7.

Step 7 Draw the addendum and dedendum circles for the pinion and gear (Fig. 7-7). From Table 7-1, the addendum is

$$a = \frac{1}{P} = \frac{1}{10} = 0.10 \text{ in}$$

The dedendum is

$$b = \frac{1.25}{P} = \frac{1.25}{10} = 0.125 \text{ in}$$

Step 8 Now draw the involute portion of the profiles of the teeth on the pinion and gear (Fig. 7-7). The portion of the tooth between the clearance and dedendum circles may be used for a fillet. Notice that the base circle of the gear is smaller than the dedendum circle, and so, except for the fillet, the profile of the tooth is all involute. On the other hand, the pinion base-circle radius is larger than the radius of the dedendum circle. This means that the portion of the tooth below the base circle is not involute. For the present we

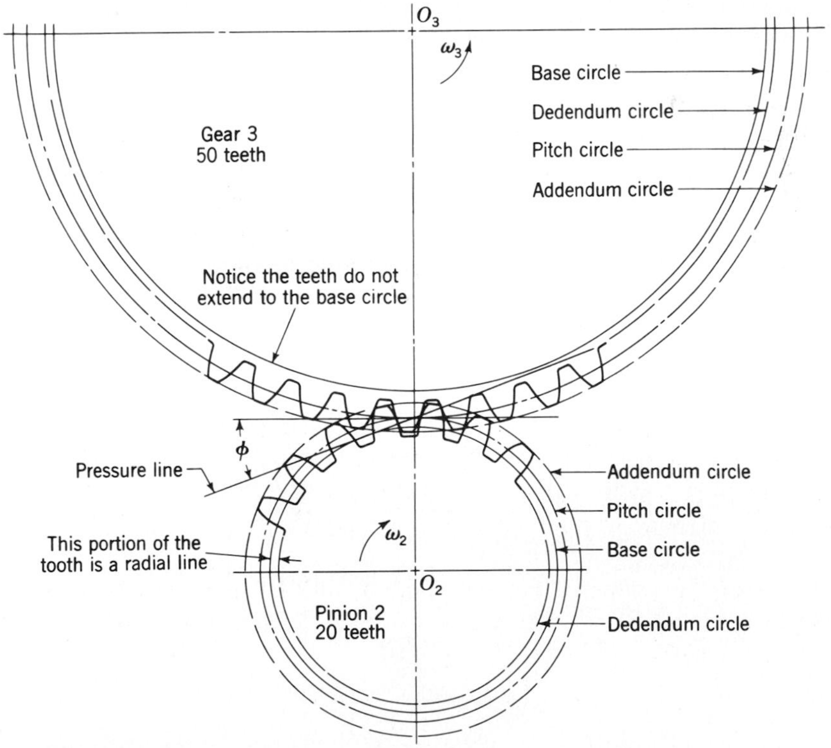

Figure 7-7 Layout of a pair of spur gears (*continued*).

shall construct this portion as a radial line except for the fillet. This completes the construction.

Involute rack We may imagine a *rack* as a spur gear having an infinitely large pitch diameter. Therefore the rack has an infinite number of teeth, and the base circle, too, is located an infinite distance from the pitch point. For involute teeth the sides become straight lines making an angle to the line of centers equal to the pressure angle. Figure 7-8 shows an involute rack in mesh with the pinion of the previous example.

Base pitch Corresponding sides of involute teeth are parallel curves; the *base pitch* is the constant and fundamental distance between them along a common normal (Fig. 7-8). The base pitch and the circular pitch are related as follows.

$$\frac{p_b}{p_c} = \cos \phi \tag{7-5}$$

where p_b is the base pitch and p_c is the circular pitch. We have already learned that the circular pitch is the distance between teeth measured along the pitch circle. The base pitch is a much more basic measurement since it is the distance between teeth measured along the common normal as well as the distance between teeth measured along the base circle.

Internal gear Figure 7-9 depicts the pinion of the preceding example mating with an *internal*, or *annular*, gear. With internal contact both centers are on the same side of the pitch point. Thus the positions of the addendum and dedendum circles with respect to the pitch circle are reversed. As shown in Fig. 7-9, the addendum circle of the internal gear lies *inside* the pitch circle; similarly, the dedendum circle lies outside the pitch circle.

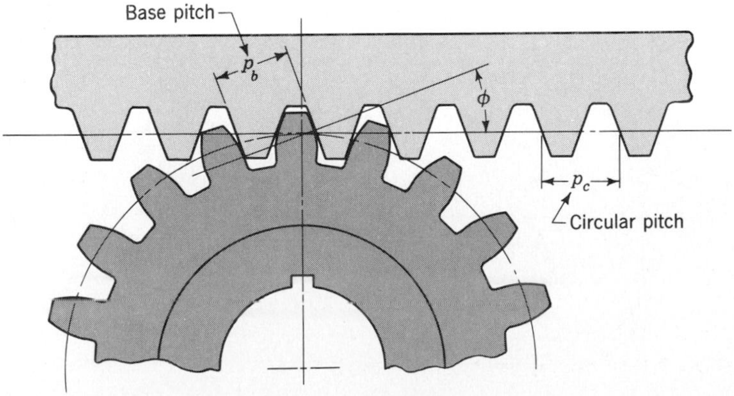

Figure 7-8 Involute pinion and rack.

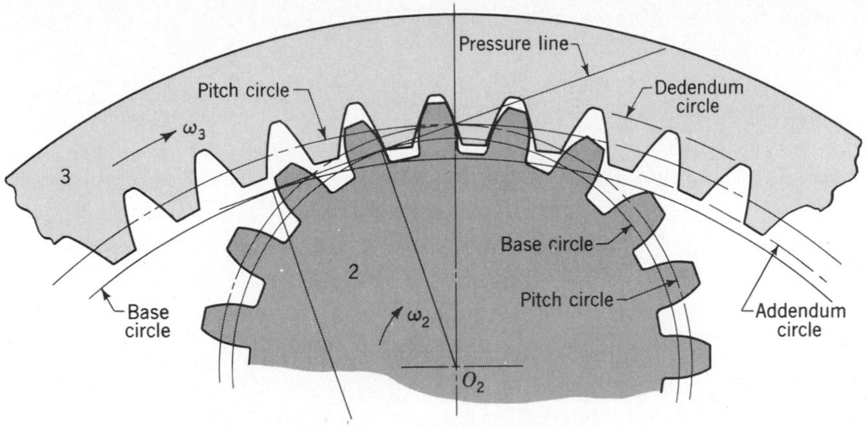

Figure 7-9 Internal gear and pinion.

It is also seen from Fig. 7-9 that the base circle lies inside the pitch circle near the addendum circle.

7-6 THE FORMING OF GEAR TEETH

There are many ways of forming the teeth of gears, e.g., *sand casting, shell molding, investment casting, permanent-mold casting, die casting,* or *centrifugal casting.* They can be formed by using the *powder-metallurgy process*; or, by using *extrusion*, a single bar of aluminum can be formed and then sliced into gears. Gears which carry large loads in comparison with their size are usually made of steel and are cut with either *form cutters* or *generating cutters.* In form cutting, the tooth space takes the exact shape of the cutter. In generating, a tool having a shape different from the tooth profile is moved relative to the gear blank to obtain the proper tooth shape.

Probably the oldest method of cutting gear teeth is *milling.* A form milling cutter corresponding to the shape of the tooth space is used to cut one tooth space at a time, after which the gear is indexed through one circular pitch to the next position. Theoretically, with this method, a different cutter is required for each gear to be cut because, for example, the shape of the space in a 25-tooth gear is different from the space in, say, a 24-tooth gear. Actually, the change in space is not too great, and eight cutters can be used to cut any gear in the range of 12 teeth to a rack with reasonable accuracy. Of course, a separate set of cutters is required for each pitch.

Shaping is a highly favored method of generating gear teeth. The cutting tool may be either a rack cutter or a pinion cutter. The operation is best explained by reference to Fig. 7-10. Here the reciprocating rack cutter is first fed into the blank until the pitch circles are tangent. Then, after each cutting stroke, the gear blank and cutter roll slightly on their pitch circles. When the

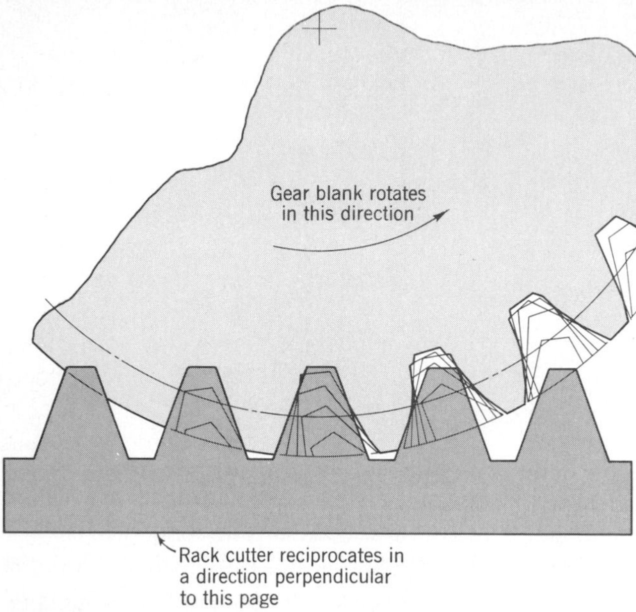

Figure 7-10 Shaping teeth with a rack cutter.

Figure 7-11 Lapping spiral-bevel gears. (*The Falk Corporation, Subsidiary of Sundstrand Corporation, Milwaukee, Wis.*)

blank and cutter have rolled a distance equal to the circular pitch, the cutter is returned to the starting point and the process is continued until all the teeth have been cut.

Hobbing is a method of generating gear teeth which is quite similar to generating them with a rack cutter. The hob is a cylindrical cutter with one or more helical threads quite like a screw-thread tap, and has straight sides like a rack. The hob and the blank are rotated continuously at the proper angular-velocity ratio, and the hob is then fed slowly across the face of the blank from one end of the teeth to the other.

Following the cutting process, *grinding, lapping, shaving,* and *burnishing* are often used as final finishing processes when tooth profiles of very good accuracy and surface finish are desired. Figure 7-11 illustrates the lapping process.

7-7 INTERFERENCE AND UNDERCUTTING

It will be profitable, at this stage, to trace the action of a pair of teeth from the time they begin contact until they leave contact. In Fig. 7-12 we have reproduced the pitch circles of the gears of Sec. 7-5. Let the pinion be the driver, and let it be rotating clockwise. Our problem is to locate the initial and final points of contact as a pair of mating teeth go through the meshing cycle.

To solve the problem we construct the pressure line and the addendum and dedendum circles of both gears. We have seen, for involute teeth, that contact must take place along the pressure line. This explains why the pressure line is also called the line of action. As shown in the figure, contact begins where the addendum circle of the driven gear crosses the line of action. Thus initial contact is on the tip of the gear tooth and on the flank of the pinion tooth.

As the pinion tooth drives the gear tooth, both approach the pitch point. Near the pitch point, contact *slides up* the flank of the pinion tooth and *down* the face of the gear tooth. At the pitch point, contact exists at the pitch circles. Note that the motion is pure rolling only at the pitch point.

As the teeth recede from the pitch point, the point of contact is traveling in the same direction as before. Contact *slides up* the face of the pinion tooth and *down* the flank of the gear tooth. The last point of contact occurs at the tip of the pinion tooth and the flank of the gear tooth. This is located at the intersection of the line of action and the addendum circle of the pinion.

The *approach* phase of the action is the period between initial contact and the pitch point. During the approach phase, contact is sliding down the face of the gear tooth toward the pitch circle. This kind of action may be likened to *pushing* a stick over a surface.

At the pitch point there is no sliding. The action is pure rolling.

The *recess* phase of the action is the period between contact at the pitch point and final contact. During the recess phase, contact is sliding down the

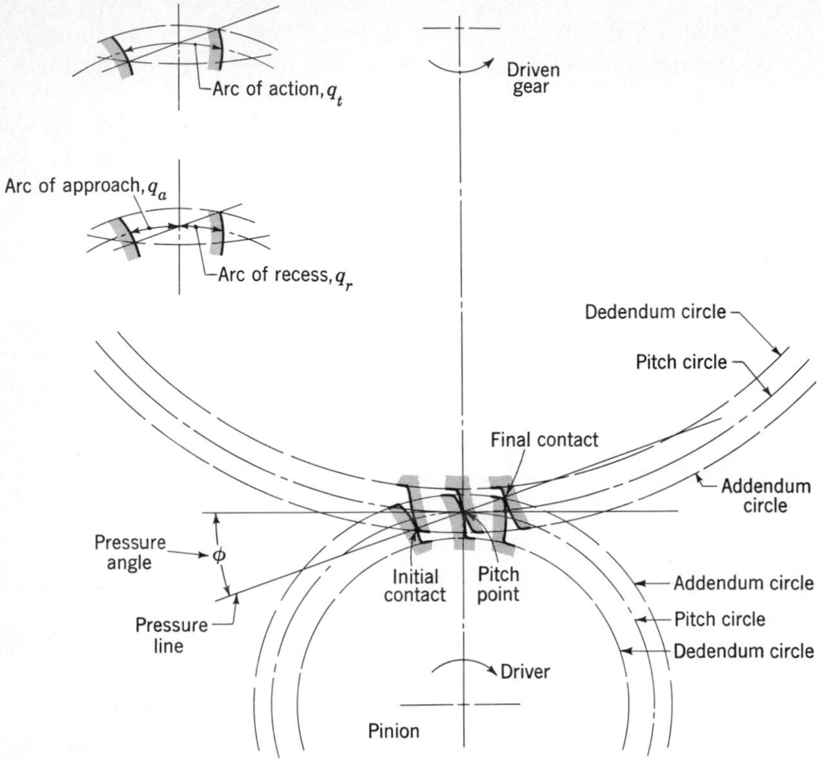

Figure 7-12 Approach and recess phases of gear-tooth action.

flank of the gear tooth, away from the pitch circle. This kind of action may be likened to *pulling* a stick over a surface.

Pinion and gear-tooth profiles are now constructed through the initial and final points of contact in Fig. 7-12. The intersection of these profiles with the pitch circles defines the arcs of action, approach, and recess.

The *arc of action* q_t is the arc of the pitch circle through which a tooth profile moves from the beginning to the end of contact with a mating profile.

The *arc of approach* q_a is the arc of the pitch circle through which a tooth profile moves from its beginning of contact until the point of contact arrives at the pitch point.

The *arc of recess* q_r is the arc of the pitch circle through which a tooth profile moves from contact at the pitch point until contact ends.

The contact of portions of tooth profiles which are not conjugate is called *interference.* Consider Fig. 7-13. Illustrated are two 16-tooth $14\frac{1}{2}°$-pressure-angle gears with full-depth teeth. The driver, gear 2, turns clockwise. The initial and final points of contact are designated A and B, respectively, and are located on the pressure line. Now notice that the points of tangency of the

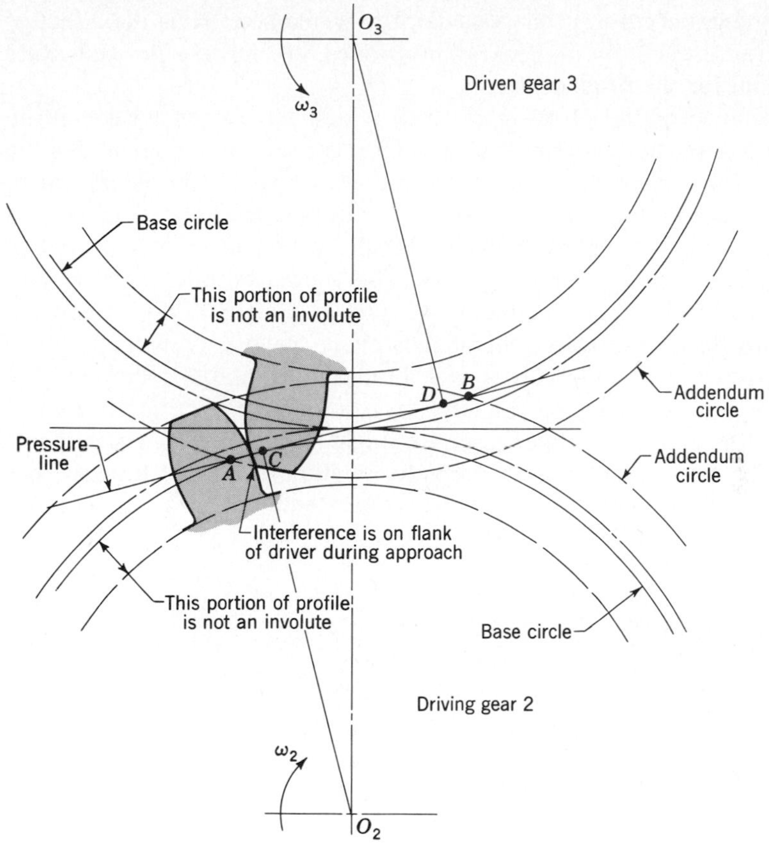

Figure 7-13 Interference in the action of gear teeth.

pressure line with the base circles C and D are located *inside* of points A and B. Interference is present.

The interference is explained as follows. Contact begins when the tip of the driven tooth contacts the flank of the driving tooth. In this case the flank of the driving tooth first makes contact with the driven tooth at point A, and this occurs *before* the involute portion of the driving tooth comes within range. In other words, contact is occurring below the base circle of gear 2 on the *noninvolute* portion of the flank. The actual effect is that the involute tip or face of the driven gear tends to dig out the noninvolute flank of the driver.

In this example the same effect occurs as the teeth leave contact. Contact should end at point D or before. Since it does not end until point B, the effect is for the tip of the driving tooth to dig out, or interfere with, the flank of the driven tooth.

When gear teeth are produced by a generation process, interference is automatically eliminated because the cutting tool removes the interfering portion of the flank. This effect is called *undercutting*; if undercutting is at all pronounced, the undercut tooth is considerably weakened. Thus the effect of eliminating interference by a generation process is merely to substitute another problem for the original one.

The importance of the problem of teeth which have been weakened by undercutting cannot be overemphasized. Of course, interference can be eliminated by using more teeth on the gears. However, if the gears are to transmit a given amount of power, more teeth can be used only by increasing the pitch diameter. This makes the gears larger, which is seldom desirable, and it also increases the pitch-line velocity. This increased pitch-line velocity makes the gears noisier and reduces the power transmission somewhat, although not in direct ratio. In general, however, the use of more teeth to eliminate interference or undercutting is seldom an acceptable solution.

Another method of reducing the interference and the resulting degree of undercutting is to employ a larger pressure angle. The larger pressure angle creates a smaller base circle, so that a greater portion of the tooth profile has an involute shape. In effect this means that fewer teeth can be used, and as a result, gears with a large pressure angle are smaller.

7-8 CONTACT RATIO

The zone of action of meshing gear teeth is shown in Fig. 7-14, where tooth contact begins and ends at the intersections of the two addendum circles with the pressure line. In Fig. 7-14 initial contact occurs at a and final contact at b. Tooth profiles drawn through these points intersect the pitch circle at A and B, respectively. As shown, the distance AP is the arc of approach q_a and the distance PB the arc of recess q_r. The sum of these is the arc of action q_t.

Consider a situation in which the arc of action is exactly equal to the circular pitch; that is, $q_t = p_c$. This means that one tooth and its space will occupy the entire arc AB. In other words, when a tooth is just beginning contact at a, the previous tooth is simultaneously ending its contact at b. Therefore, during the tooth action from a to b there will be exactly one pair of teeth in contact.

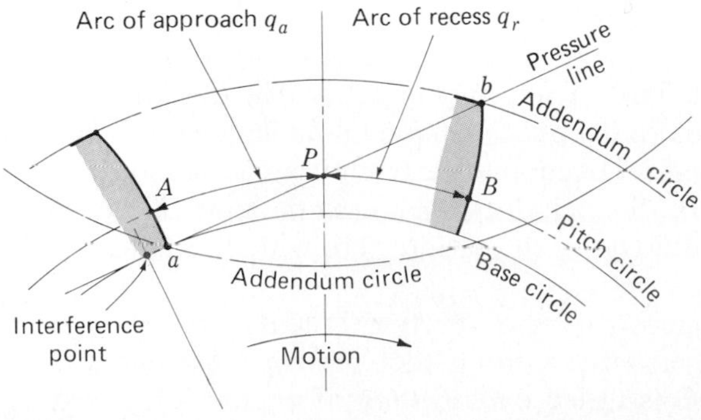

Figure 7-14

Next, consider a situation in which the arc of action is greater than the circular pitch, but not very much greater, say, $q_t \approx 1.2p_c$. This means that when one pair of teeth is just entering contact at *a*, the previous pair, already in contact, will not yet have reached *b*. Thus, for a short time there will be two pairs of teeth in contact, one in the vicinity of *A* and the other near *B*. As the meshing proceeds, the pair near *B* must cease contact, leaving only a single pair of teeth in contact, until the procedure repeats itself.

Because of the nature of this tooth action (one, two, or even more pairs of teeth in contact) it is convenient to define the term *contact ratio* m_c as

$$m_c = \frac{q_t}{p_c} \tag{7-6}$$

a number which indicates the average number of pairs of teeth in contact.

Equation (7-6) is rather inconvenient to use unless a drawing like Fig. 7-14 is produced so that the distances q_a and q_r can be measured. These distances depend upon the diameters of the pitch circles, which may vary since they depend upon the mounting distance between the two gear centers. We can also define contact ratio using the base circle, and this will really be a better definition because the base circle has a fixed diameter.

In Fig. 7-15, with gear 2 as the driver, contact begins at point *B*, where the

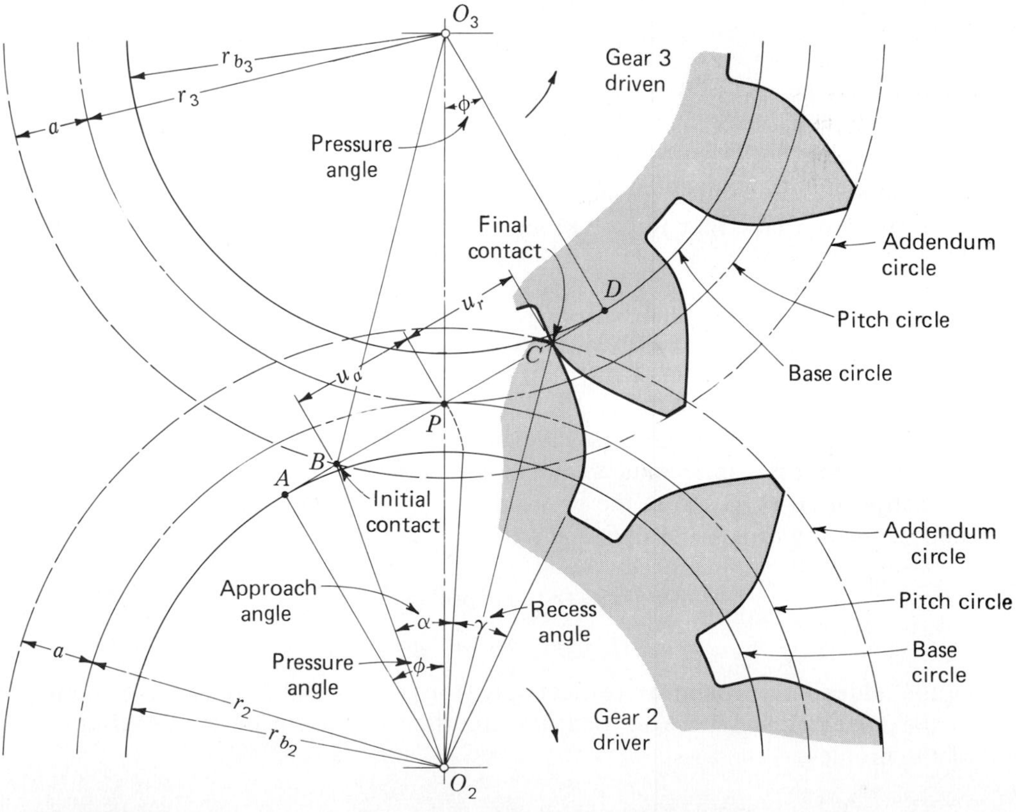

Figure 7-15

addendum circle of the driven gear crosses the line of action, and ends at C, where the addendum circle of the driver crosses the line of action. The length of the path of contact is

$$u = u_a + u_r \tag{a}$$

where the subscripts a and r designate the approach and recess phases, respectively. During approach contact occurs along the line BP and the gear rotates through the angle α, called the *approach angle. This angle subtends an arc of the base circle obtained by constructing tooth profiles through B and P to intersect the base circle.*

During recess, contact occurs along PC while the gear rotates through the angle γ, called the *recess angle.* Note that the recess angle also subtends an arc of the base circle obtained by finding the intersection of tooth profiles through P and C with the base circle.

The base pitch is the distance between corresponding tooth profiles measured on the line of action. Therefore the contact ratio is

$$m_c = \frac{u_a + u_r}{p_b} \tag{7-7}$$

The values of u_a and u_b can be obtained analytically by observing the two right triangles O_2AC and O_3DB in Fig. 7-15. From triangle O_2AC we can write

$$u_a = [(r_3 + a)^2 - r_{b_3}^2]^{1/2} - r_3 \sin \phi \tag{7-8}$$

Similarly, from triangle O_3DB we have

$$u_r = [(r_2 + a)^2 - r_{b_2}^2]^{1/2} - r_2 \sin \phi \tag{7-9}$$

The contact ratio is then obtained by substituting Eqs. (7-8) and (7-9) into (7-7). We might note that Eqs. (7-8) and (7-9), however, are valid only for the conditions

$$u_a \leq r_2 \sin \phi \qquad u_r \leq r_3 \sin \phi \tag{7-10}$$

because contact cannot begin before point A (Fig. 7-15) or end after point D. Thus if the value of u_a or u_r as calculated by Eq. (7-8) or Eq. (7-9) fails to satisfy the inequalities of Eq. (7-10), Eq. (7-10) should be used to calculate u_a or u_r, whichever the case may be, using the equality sign.

The largest possible contact ratio is obtained by adjusting the addendums of each gear so as to utilize the entire distance AD (Fig. 7-15). The action is then defined by triangles O_2AD and O_3AD. Therefore

$$a_2 = [r_{b_2}^2 + (r_2 + r_3)^2 \sin^2 \phi]^{1/2} - r_2 \tag{7-11}$$

$$a_3 = [r_{b_3}^2 + (r_2 + r_3)^2 \sin^2 \phi]^{1/2} - r_3 \tag{7-12}$$

as the addendums a_2 and a_3, respectively, for gears 2 and 3. If either or both of these addendums are exceeded, undercutting will occur during generation of the profiles.

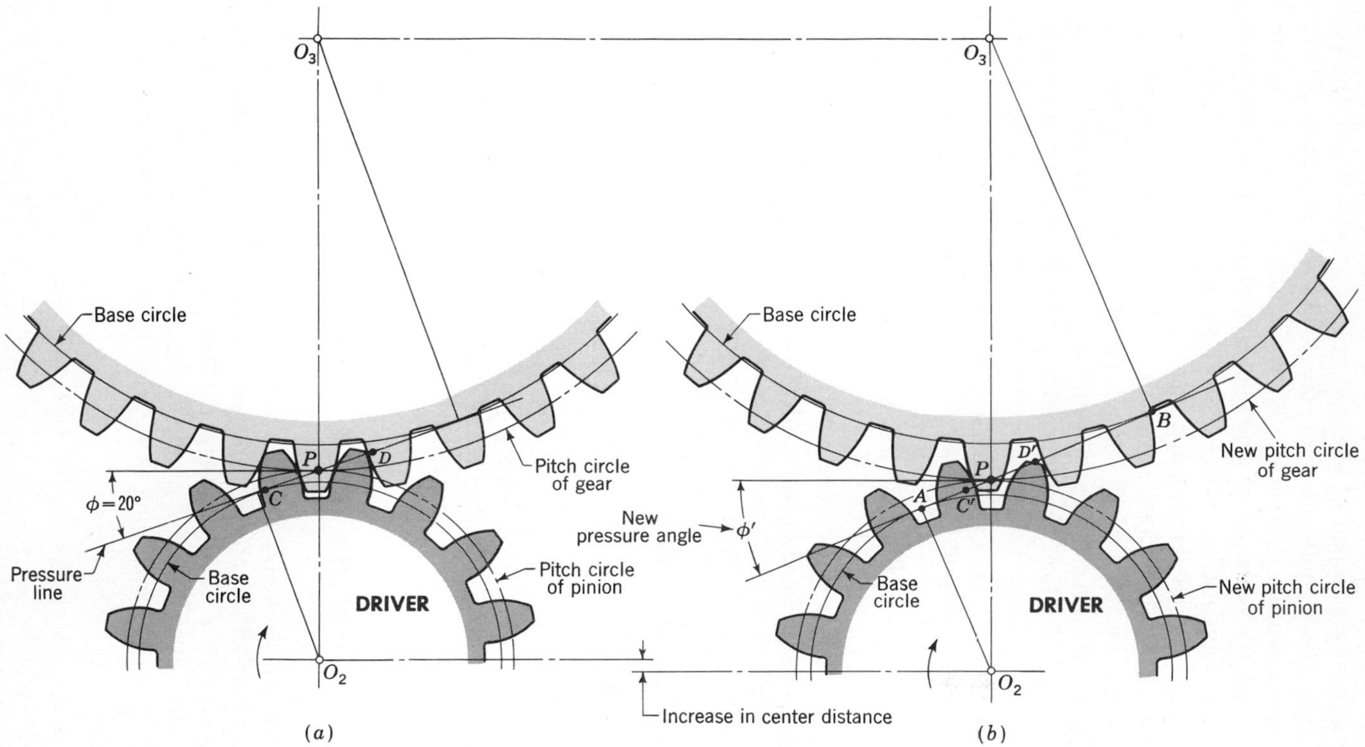

Figure 7-16 Effect of increased center distance upon the action of involute gearing: mounting at (*a*) normal center distance and (*b*) increased center distance.

7-9 VARYING THE CENTER DISTANCE

Figure 7-16*a* illustrates a pair of meshing gears having involute teeth at a 20° pressure angle. Since both sides of the teeth are in contact, the center distance O_2O_3 cannot be shortened without jamming or deforming the teeth.

In Fig. 7-16*b* the same pair of gears have been separated by increasing the center distance slightly. Clearance, or *backlash*, now exists between the teeth, as shown. When the center distance is increased, new pitch circles having larger radii are created because the pitch circles are always tangent to each other. However, the base circles are a constant and fundamental characteristic of the gears. This means that an increase in center distance changes the inclination of the line of action and results in a larger pressure angle. Notice, too, that a tracing point on the new pressure line will still generate the same involutes as in Fig. 7-16*a*, the normal to the tooth profiles still passes through the same pitch point, and hence the law of gearing is satisfied for any center distance.

To see that the velocity ratio has not changed in magnitude, we observe that the triangles O_2AP and O_3BP are similar. Also, since O_2A and O_3B are fixed distances and do not change with varying center distances, the ratio of the pitch radii, O_2P and O_3P, will remain fixed too.

A second effect of increasing the center distance observable in Fig. 7-16 is the shortening of the path of contact. The original path of contact CD has been shortened to $C'D'$. The contact ratio [Eq. (7-7)] can be defined as the ratio of the length of the path of contact to the base pitch. The limiting value of this ratio is unity; otherwise, periods would occur in which there would be no contact at all. Thus the center distance cannot be larger than that corresponding to a contact ratio of unity.

It is interesting to conclude from the preceding discussion that two gears of slightly different tooth numbers can be mounted upon the same axis (though not fixed to each other or to the shaft) and mated with the same pinion or rack provided the limitations discussed above are not exceeded.

7-10 INVOLUTOMETRY

The study of the geometry of the involute is called *involutometry*.† In Fig. 7-17 a base circle with center at O is used to generate the involute BC. AT is the generating line, ρ the instantaneous radius of curvature of the involute, and r the radius to any point T on the curve. If we designate the radius of the base circle as r_b, the generating line AT has the same length as the arc AB and so

$$\rho = r_b(\alpha + \varphi) \tag{a}$$

† Pronounced ĭn′vōl ū tŏ′m e trē.

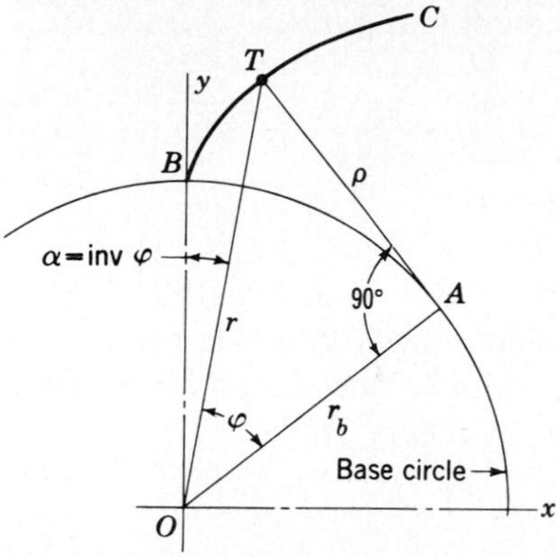

Figure 7-17

where α is the angle between radius vectors defining the origin of the involute and any point, such as T, on the involute and φ is the angle between radius vectors defining any point T on the involute and the origin A at the base circle of the corresponding generating line. Since OTA is a right triangle,

$$\rho = r_b \tan \varphi \tag{7-13}$$

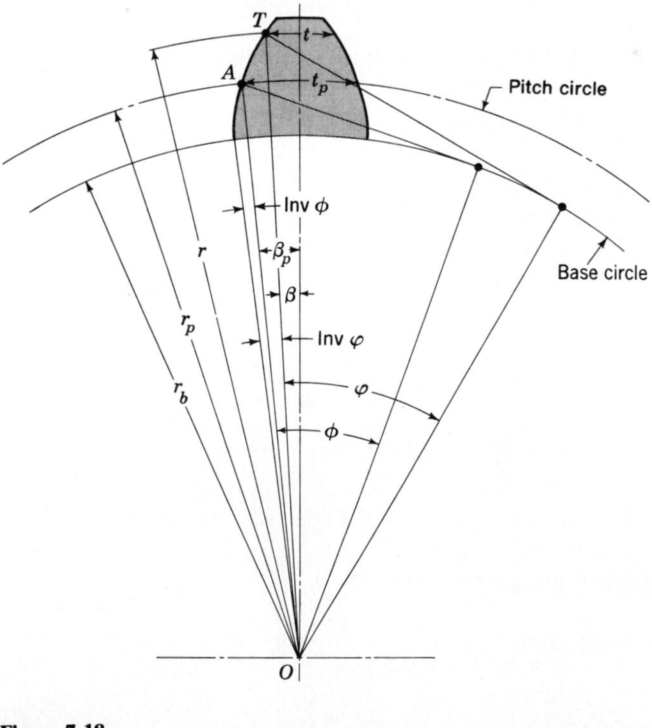

Figure 7-18

Solving Eqs. (a) and (7-13) simultaneously to eliminate ρ gives

$$\alpha = \tan \varphi - \varphi$$

which can be written

$$\text{inv } \varphi = \tan \varphi - \varphi \tag{7-14}$$

and defines the involute function. The angle φ in this equation is the variable involute pressure angle, and it must be specified in radians. If φ is known, inv φ can readily be determined; but tables must be used to find the pressure angle when inv φ is given and φ is to be found (see Appendix Table 6).

Referring again to Fig. 7-17, we see that

$$r = \frac{r_b}{\cos \varphi} \tag{7-15}$$

To illustrate the use of the relations obtained above, the tooth dimensions of Fig. 7-18 will be determined. Here the portion of the tooth profile extending above the base circle has been drawn, and the arc thickness of the tooth t_p at the pitch circle (point A) is given. The problem is to determine the tooth thickness at any other point, say T. The various quantities shown in Fig. 7-18 are identified as follows:

r_b = radius of base circle
r_p = radius of pitch circle
r = radius at which tooth thickness is to be determined
t_p = arc tooth thickness at pitch circle
t = arc thickness to be determined
ϕ = pressure angle corresponding to pitch radius r_p
φ = pressure angle corresponding to any point T
β_p = angular half-tooth thickness at pitch circle
β = angular half-tooth thickness at any point T

The half-tooth thicknesses at points A and T are

$$\frac{t_p}{2} = \beta_p r_p \qquad \frac{t}{2} = \beta r \tag{b}$$

so that

$$\beta_p = \frac{t_p}{2r_p} \qquad \beta = \frac{t}{2r} \tag{c}$$

Now we can write

$$\text{inv } \varphi - \text{inv } \phi = \beta_p - \beta = \frac{t_p}{2r_p} - \frac{t}{2r} \tag{d}$$

The tooth thickness corresponding to any point T is obtained by solving Eq. (d) for t:

$$t = 2r\left(\frac{t_p}{2r_p} + \text{inv } \phi - \text{inv } \varphi\right) \tag{7-16}$$

Example 7-1 A gear has 20° teeth cut full depth, a diametral pitch of 2 teeth per inch, and 22 teeth. (*a*) Calculate the radius of the base circle. (*b*) Find the thickness of the tooth at the base circle, and also at the addendum circle.

SOLUTION By employing Sec. 7-4 and the equations of Sec. 7-1, the following quantities are determined: addendum $a = 0.500$ in, dedendum $b = 0.5785$ in, pitch radius $r_p = 5.500$ in, circular pitch $p_c = 1.571$ in. The radius of the base circle is obtained from Eq. (7-15)

$$r_b = r_p \cos \phi = 5.500 \cos 20° = 5.168 \text{ in}$$

The thickness of the tooth at the pitch circle is

$$t_p = \frac{p_c}{2} = \frac{1.571}{2} = 0.785\,4 \text{ in}$$

Converting the tooth pressure angle 20° to radians gives $\phi = 0.349$ rad. Then

$$\text{inv } \phi = \tan 0.349 - 0.349 = 0.015 \text{ rad}$$

At the base circle $\varphi_b = 0$, so inv $\varphi_b = 0$. By Eq. (7-16) the tooth thickness at the base circle is

$$t_b = 2r_b\left(\frac{t_p}{2r_p} + \text{inv } \phi - \text{inv } \varphi_b\right) = (2)(5.168)\left[\frac{0.785\,4}{(2)(5.500)} + 0.015 - 0\right] = 0.886 \text{ in}$$

The radius of the addendum circle is $r_a = 6.000$ in. The involute pressure angle corresponding to this radius is, from Eq. (7-15),

$$\varphi_a = \cos^{-1}\frac{r_b}{r_a} = \cos^{-1}\frac{5.168}{6.000} = 0.532 \text{ rad}$$

Thus

$$\text{inv } \varphi_a = \tan 0.532 - 0.532 = 0.058 \text{ rad}$$

and Eq. (7-16) gives the tooth thickness at the addendum circle as

$$t_a = 2r_a\left(\frac{t_p}{2r_p} + \text{inv } \phi - \text{inv } \varphi_a\right) = (2)(6.000)\left[\frac{0.785\,4}{(2)(5.500)} + 0.015 - 0.058\right] = 0.341 \text{ in}$$

7-11 NONSTANDARD GEAR TEETH

In this section, we investigate the effects obtained by modifying such things as pressure angle, tooth depth, addendum, or center distance. Some of these modifications do not eliminate interchangeability; all of them are made with the intent of obtaining improved performance or more economical production.

There are three principal reasons for the use of nonstandard gears. The designer is often under great pressure to produce gear designs which are small and yet which will transmit large amounts of power. Consider, for example, a gearset which must have a 4:1 velocity ratio. If the smallest pinion that will carry the load has a pitch diameter of 2 in, the gear will have a pitch diameter of 8 in, making the overall space required for the two gears slightly more than 10 in. On the other hand, if the pitch diameter of the pinion can be reduced by only $\frac{1}{4}$ in, the pitch diameter of the gear is reduced a full 1 in and the overall size of the gearset reduced by $1\frac{1}{4}$ in. This reduction assumes considerable importance when it is realized that the sizes of associated machine elements, such as shafts, bearings, and enclosures, are also reduced. If a tooth of a certain pitch is required to carry the load, the only method of decreasing the pinion diameter is to use fewer teeth. We have already seen that problems involving interference, undercutting, and contact ratio are encountered when the tooth numbers are made less than prescribed minima. Thus the principal reasons for employing nonstandard gears are to eliminate undercutting, to prevent interference, and to maintain a reasonable contact ratio. It should be noted, too, that if a pair of gears is manufactured of the same material, the pinion is the weaker and is subject to greater wear because its teeth are in contact a greater portion of the time. Thus undercutting weakens the tooth, which is already the weaker of the two. So another advantage of nonstandard gears is the tendency toward a better balance of strength between the pinion and gear.

As an involute curve is generated from the base circle, its radius of curvature becomes larger and larger. Near the base circle the radius of curvature is quite small, being exactly zero at the base circle. Contact near this region of sharp curvature should be avoided if possible because of the difficulty of obtaining good cutting accuracy in areas of small curvature and, too, because the contact stresses are likely to be very high. Nonstandard gears present the opportunity of designing to avoid these sensitive areas.

Clearance modifications A larger fillet at the root of the tooth increases the fatigue strength of the tooth and provides extra depth for shaving the tooth profile. Since interchangeability is not lost, the clearance is sometimes increased to $0.400/P$ to obtain this larger fillet.

In some applications a pressure angle of $17\frac{1}{2}°$ has been used with a clearance of $0.300/P$ to produce a contact ratio of 2.

Center-distance modifications When gears of low tooth numbers are to be paired with each other, or when they are to be mated with larger gears, reduction in interference and improvement in the contact ratio can be obtained by increasing the center distance. Although this system changes the tooth proportions and the pressure angle of the gears, the resulting teeth can be generated with rack cutters (or hobs) of standard pressure angles or with standard pinion shapers. Before introducing this system it will be of value to develop additional relations in the geometry of gears.

The first relation to be obtained is that of finding the thickness of a tooth which is cut by a rack cutter (or hob) when the pitch line of the rack has been displaced or offset a distance e from the pitch circle of the gear. What we are doing here is moving the rack cutter farther away from the center of the gear which is being cut. This will produce teeth which are thicker than before, and this thickness must be found. Figure 7-19a shows the problem and Fig. 7-19b the solution. The increase over the standard amount is $2e \tan \phi$, so that

$$t = 2e \tan \phi + \frac{p_c}{2} \tag{7-17}$$

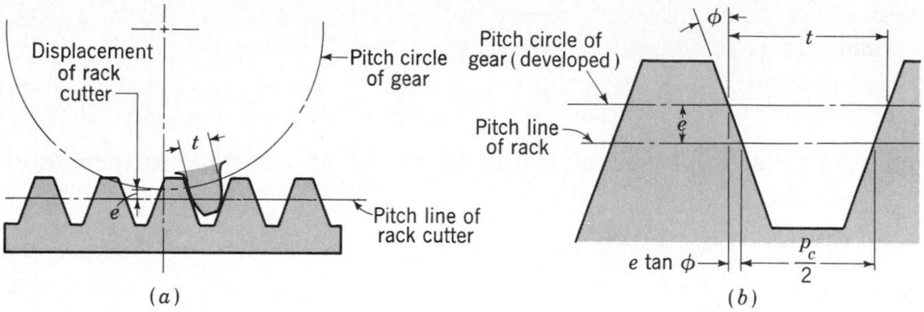

Figure 7-19

where ϕ is the pressure angle of the rack cutter and t is the thickness of the gear tooth on its own pitch circle.

Now suppose two gears of different tooth numbers have been cut with the cutter offset from the pitch circles, as in the previous paragraph. Since the teeth have been cut with an offset cutter, they will mate at a new pressure angle and with new pitch circles and consequently new center distances. The word *new* is used in the sense, here, of not being standard. Our problem is to determine the radius of these new pitch circles and the value of this new pressure angle.

In the following notation the word *standard* refers to values which would have been obtained had the usual, or standard, systems been employed to obtain the dimensions:

ϕ = pressure angle of rack generating cutter
ϕ' = new pressure angle at which gears will mate
r_2 = standard pitch radius of pinion
r_2' = new pitch radius of pinion when meshing with given gear
r_3 = standard pitch radius of gear
r_3' = new pitch radius of gear when meshing with given pinion
t_2 = actual thickness of pinion tooth at standard pitch radius
t_3 = actual thickness of gear tooth at standard pitch radius
t_2' = thickness of pinion tooth at new pitch radius r_2'
t_3' = thickness of gear tooth at new pitch radius r_3'
N_2 = number of teeth on pinion
N_3 = number of teeth on gear

From Eq. (7-16)

$$t_2' = 2r_2'\left(\frac{t_2}{2r_2} + \text{inv }\phi - \text{inv }\phi'\right) \tag{a}$$

$$t_3' = 2r_3'\left(\frac{t_3}{2r_3} + \text{inv }\phi - \text{inv }\phi'\right) \tag{b}$$

The sum of these two thicknesses must be the same as the circular pitch, or, from Eq. (7-2),

$$t_2' + t_3' = p_c = \frac{2\pi r_2'}{N_2} \tag{c}$$

The pitch diameters of a pair of mating gears are proportional to their tooth numbers, and so

$$r_3 = \frac{N_3}{N_2} r_2 \qquad \text{and} \qquad r_3' = \frac{N_3}{N_2} r_2' \tag{d}$$

Substituting Eqs. (*a*), (*b*), and (*d*) in (*c*) and rearranging gives

$$\text{inv}\ \phi' = \frac{N_2(t_2 + t_3) - 2\pi r_2}{2r_2(N_2 + N_3)} + \text{inv}\ \phi \tag{7-18}$$

Equation (7-18) gives the pressure angle ϕ' at which a pair of gears will operate when the tooth thicknesses on their standard pitch circles have been modified to t_2 and t_3.

It has been demonstrated that gears have no pitch circles until a pair of them are brought into contact. Bringing a pair of gears into contact creates a pair of pitch circles which are tangent to each other at the pitch point. Throughout this discussion the idea of a pair of so-called standard-pitch circles has been used in order to define a certain point on the involute curves. These standard-pitch circles, we have seen, are the ones which would have come into existence when the gears were paired *if the gears had not been modified from the standard dimensions.* On the other hand, the base circles are fixed circles which are not changed by tooth modifications. The base circle remains the same whether the tooth dimensions are changed or not; so we can determine the base-circle radius, using either the standard-pitch circle or the new-pitch circle. Equation (7-15) can therefore be written as either

$$r_b = r_2 \cos\phi \qquad \text{or} \qquad r_b = r_2' \cos\phi'$$

Thus

$$r_2' \cos\phi' = r_2 \cos\phi$$

or

$$r_2' = \frac{r_2 \cos\phi}{\cos\phi'} \tag{7-19}$$

Similarly, for the gear,

$$r_3' = \frac{r_3 \cos\phi}{\cos\phi'} \tag{7-20}$$

These equations give the values of the actual pitch radii when the two gears with modified teeth are *brought* into mesh without backlash. The new center distance is, of course, the sum of these radii.

All the necessary relations have now been developed to create nonstandard gears with changes in the center distance. The usefulness of these relations is best illustrated by an example.

Figure 7-20 is a drawing of a 20° 1-pitch 12-tooth pinion generated with a rack cutter with a standard clearance of $0.250/P$. In the 20° full-depth system, interference is severe whenever the number of teeth is less than 14. The resulting undercutting is evident from the drawing. If this pinion were mated with a standard 40-tooth gear, the contact ratio would be 1.41, which can easily be verified by Eq. (7-7).

In an attempt to eliminate the undercutting, improve the tooth action, and increase the contact ratio, let the 12-tooth pinion be cut from a larger blank. Then the resulting pinion will be paired again with the 40-tooth standard gear to determine the degree of improvement. If we designate the pinion as subscript 2 and the gear as 3, the following values will be found:

$$\phi = 20° \qquad r_2 = 6 \text{ in} \qquad r_3 = 20 \text{ in} \qquad P = 1$$

$$p_c = 3.1416 \text{ in} \qquad t_3 = 1.5708 \text{ in} \qquad N_2 = 12 \text{ teeth} \qquad N_3 = 40 \text{ teeth}$$

We will offset the rack cutter so that its addendum line passes through the interference point of the pinion, i.e., the point of tangency of the 20° pressure line and the base circle, as shown in Fig. 7-21. From Eq. (7-15) we have

$$r_b = r_2 \cos \phi \tag{e}$$

Then, from Fig. 7-21

$$e = a + r_b \cos \phi - r_2 \tag{f}$$

Substituting Eq. (e) in (f) gives

$$e = a + r_2 \cos^2 \phi - r_2 = a - r_2 \sin^2 \phi \tag{7-21}$$

For a standard rack the addendum is $a = 1/P$; so $a = 1$ in for this problem.

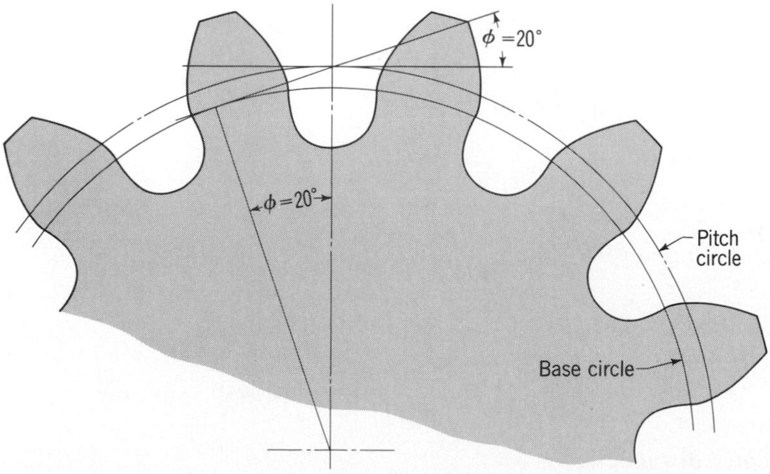

Figure 7-20 Standard 12-tooth 20° full-depth gear showing undercut.

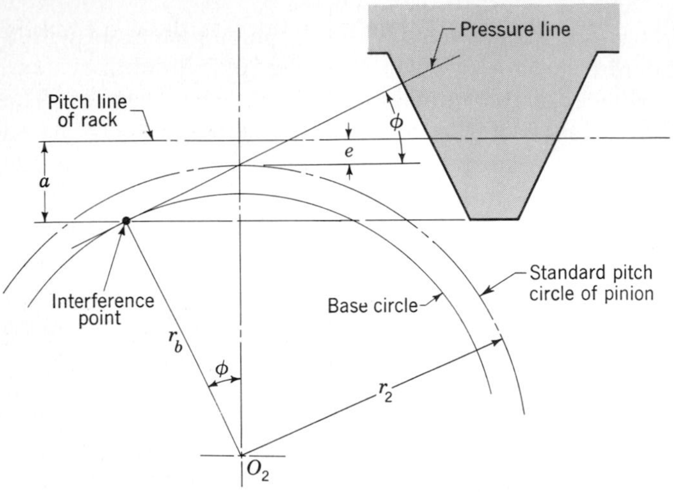

Figure 7-21 Offsetting a rack to cause its addendum line to pass through the interference point.

The offset to be used is

$$e = 1 - 6 \sin^2 20° = 0.2981 \text{ in}$$

Then, solving Eq. (7-17) for the thickness of the pinion tooth at its 6-in-pitch circle, we get

$$t_2 = 2e \tan \phi + \frac{p_c}{2} = (2)(0.2981) \tan 20° + \frac{3.1416}{2} = 1.7878 \text{ in}$$

The pressure angle at which these gears (and only these gears) will operate is found from Eq. (7-18).

$$\text{inv } \phi' = \frac{N_2(t_2 + t_3) - 2\pi r_2}{2r_2(N_2 + N_3)} + \text{inv } \phi$$

$$= \frac{12(1.7878 + 1.5708) - 2\pi 6}{(2)(6)(12 + 40)} + \text{inv } 20° = 0.019\,077 \text{ rad}$$

From Appendix Table 6

$$\phi' = 21.6511°$$

By using Eqs. (7-19) and (7-20) the new pitch radii are found to be

$$r_2' = \frac{r_2 \cos \phi}{\cos \phi'} = \frac{6 \cos 20°}{\cos 21.6511°} = 6.0662 \text{ in}$$

$$r_3' = \frac{r_3 \cos \phi}{\cos \phi'} = \frac{20 \cos 20°}{\cos 21.6511°} = 20.220 \text{ in}$$

So the new center distance is

$$r_2' + r_3' = 6.0662 + 20.220 = 26.286 \text{ in}$$

Notice that the center distance has not increased so much as the offset of the rack cutter.

In the beginning, a clearance of $0.25/P$ was specified, making the standard dedendums equal to $1.25/P$. So the root radii of the two gears are

Root radius of pinion	$6.2981 - 1.25 =$	5.0481 in
Root radius of gear	$20.0000 - 1.25 =$	18.7500 in
Sum of root radii	$=$	23.7981 in

The difference between this sum and the center distance is the working depth plus twice the clearance. Since the clearance is 0.25 in for each gear, the working depth is

$$26.286 - 23.7981 - (2)(0.25) = 1.9879 \text{ in}$$

The outside radius of each gear is the sum of the root radius, the clearance, and the working depth.

$$\text{Outside radius of pinion} = 5.0481 + 0.25 + 1.9879 = 7.2860 \text{ in}$$

$$\text{Outside radius of gear} = 18.75 + 0.25 + 1.9879 = 20.9879 \text{ in}$$

The result is shown in Fig. 7-22, and the pinion is seen to have a stronger-looking form than the one of Fig. 7-20. Undercutting has been completely eliminated. The contact ratio can be obtained from Eqs. (7-7) to

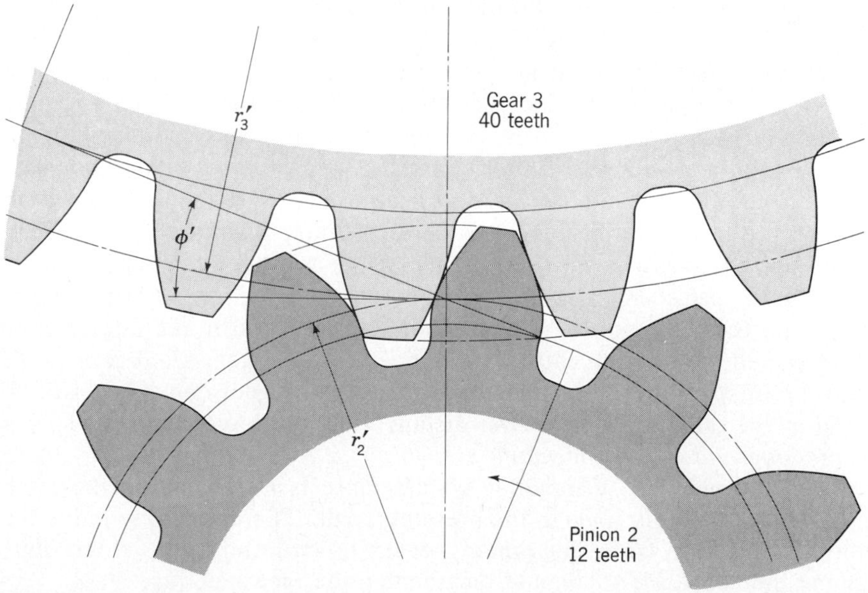

Figure 7-22

(7-9). The following quantities are needed:

$$\text{Outside radius of pinion} = r_2' + a = 7.2860 \text{ in}$$

$$\text{Outside radius of gear} = r_3' + a = 20.9879 \text{ in}$$

$$r_{b_2} = r_2 \cos\phi = 6\cos 20^\circ = 5.6381 \text{ in}$$

$$r_{b_3} = r_3 \cos\phi = 20\cos 20^\circ = 18.7938 \text{ in}$$

$$p_b = p_c \cos\phi = 3.1416\cos 20^\circ = 2.9521 \text{ in}$$

We then have

$$u_a = [(r_3' + a)^2 - r_{b_3}^2]^{1/2} - r_3' \sin\phi'$$

$$= [(20.9879)^2 - (18.7938)^2]^{1/2} - 20.220 \sin 21.6511^\circ$$

$$= 1.8826 \text{ in}$$

$$u_r = [(r_2' + a)^2 - r_{b_2}^2]^{1/2} - r_2' \sin\phi'$$

$$= [(7.2860)^2 - (5.6381)^2]^{1/2} - 6.0662 \sin 21.6511^\circ$$

$$= 2.3247 \text{ in}$$

Finally, from Eq. (7-7), the contact ratio is

$$m_c = \frac{u_a + u_r}{p_b} = \frac{1.8826 + 2.3247}{2.9521} = 1.425$$

Thus, the contact ratio has increased only slightly. The modification, however, is justified because of the elimination of undercutting and results in a very substantial improvement in the strength of the tooth.

Long-and-short-addendum systems It often happens in the design of machinery that the center distance between a pair of gears is fixed by some other characteristic or feature of the machine. In such a case, modifications to obtain improved performance cannot be made by varying the center distance.

In the previous section we have seen that improved action and tooth shape can be obtained by backing out the rack cutter from the pinion blank. The effect of this withdrawal is to create active tooth profile farther away from the base circle. An examination of Fig. 7-22 will reveal that more dedendum on the gear (not the pinion) could be used before the interference point is reached. If the rack cutter is advanced into the gear blank a distance equal to the offset from the pinion blank, more of the gear dedendum will be used and at the same time the center distance will not have changed. This is called *the long-and-short-addendum system.*

In the long-and-short-addendum system there is no change in the pitch circles and consequently none in the pressure angle. The effect is to move the contact region away from the pinion center toward the gear center, thus shortening the approach action and lengthening the recess action.

The characteristics of the long-and-short-addendum system can be

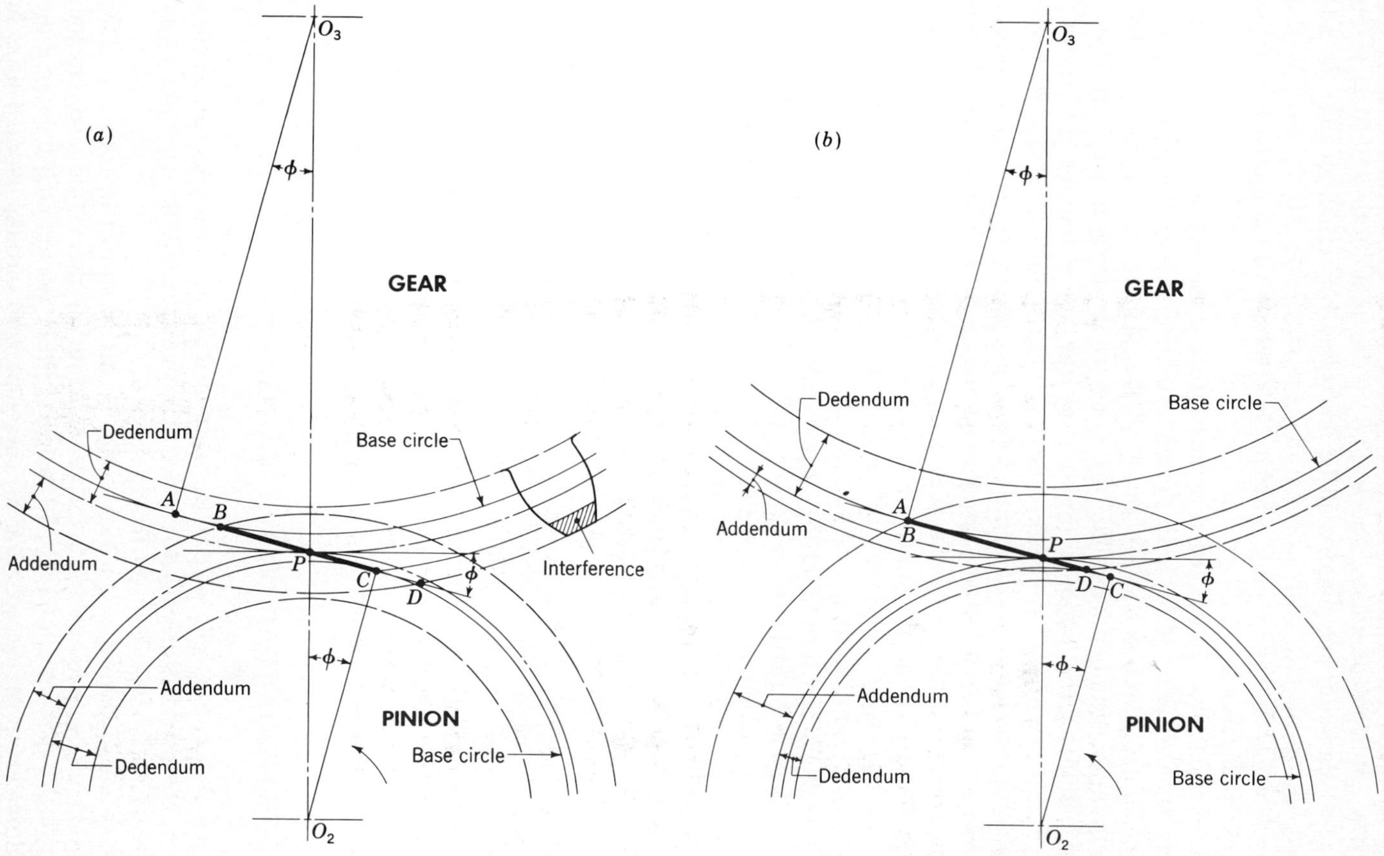

Figure 7-23 Comparison of standard gears and gears cut by the long-and-short addendum system: (*a*) gear and pinion with standard addendum and dedendum; (*b*) gear and pinion with long-and-short addendum.

explained by reference to Fig. 7-23. Figure 7-23*a* illustrates a conventional (standard) set of gears having a dedendum equal to the addendum plus the clearance. Interference exists, and the tip of the gear tooth will have to be relieved as shown or the pinion will be undercut. This is so because the addendum circle of the gear crosses the pressure line at *D*, outside the tangency or interference point *C*; hence the distance *CD* is a measure of the degree of interference.

To eliminate the undercutting or interference, the pinion addendum has been enlarged in Fig. 7-23*b* until the addendum circle of the pinion passes through the interference point (point *A*) of the gear. In this manner we shall be using all the gear-tooth profile. The same whole depth is retained; hence the dedendum of the pinion is reduced by the same amount that the addendum is increased. This means that we must now lengthen the gear dedendum and shorten the addendum. With these changes the path of contact is the line *BD* in Fig. 7-23*b*. It is longer than the path *BC* in Fig. 7-23*a*, and so the contact ratio is higher. Notice, too, that the base circles, the pitch circles, the pressure angle, and the center distance have not changed. Both gears can be cut with standard cutters by advancing the cutter into the gear blank a distance equal to the amount of withdrawal, for this modification, from the pinion blank. Finally, note that the blanks, from which the gears are cut, are now of different diameters than standard blanks.

The tooth dimensions for the long-and-short-addendum system can be determined by using the equations developed in the previous sections.

A less obvious advantage of the long-and-short-addendum system is that more recess action than approach action is obtained. The approach action of gear teeth is analogous to pushing a piece of chalk across the blackboard; the chalk screeches. But when the chalk is pulled across the blackboard, it glides smoothly; this is analogous to recess action. Thus recess action is always preferred because of the smoothness and the lower frictional forces.

The long-and-short-addendum system has no advantage if the mating gears are of the same size. In this case, increasing the addendum of one gear would simply produce more undercutting of the mate. It is also apparent that the smaller gear of the pair should be the driver if the advantages of recess action are to be obtained.

7-12 THE CYCLOIDAL TOOTH PROFILE

The cycloidal tooth profile was extensively used for gear manufacture about a century ago because it is easy to form by casting. It is seldom used today, for reasons which we will discover in this section.

The construction of a cycloidal tooth profile is shown in Fig. 7-24. Two generating circles, shown in dashed lines, roll on the inside and outside, respectively, of the pitch circle and generate the hypocycloidal flank and the epicycloidal face of the gear tooth. The same two circles are also used to

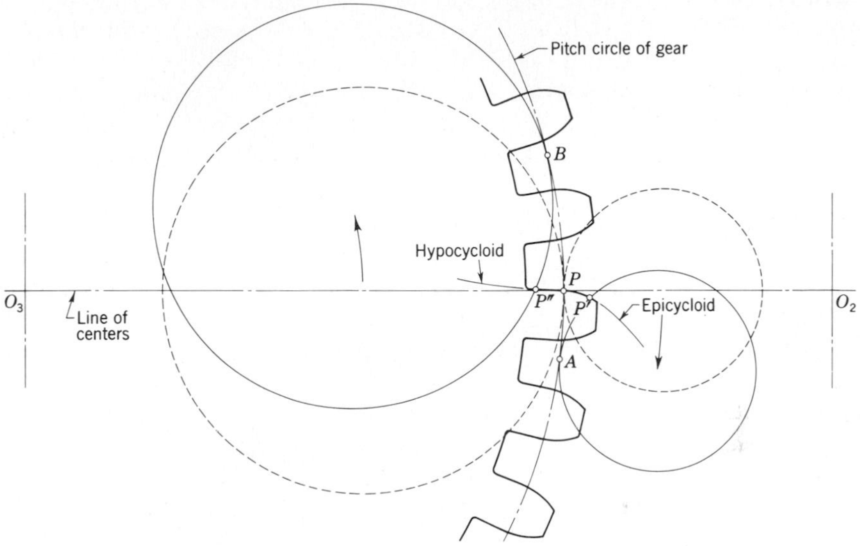

Figure 7-24 Generating cycloidal teeth on a gear.

generate the profile of the mating pinion teeth, but now the role of the generating circles is reversed. The circle which generated the flank of the gear tooth now generates the epicycloidal face of the pinion tooth. And, similarly, the circle which generated the face of the gear tooth now generates the flank of the pinion tooth.

Notice that in generating one side of a tooth, the two generating circles roll in opposite directions.

In Fig. 7-25 the pinion and gear produced by this method have been placed in mesh. Consider the pinion as the driver and let it rotate counterclockwise. The two pitch circles are tangent at the pitch point *P*, and they roll upon each other without slipping. The two generating circles have stationary centers at *A* and *B* and also roll with the moving pitch circles. A point of contact *C* exists at the intersection of the generating circle with center at *A* and the two contacting profiles. Let C_2 be a point on the flank of the pinion tooth and C_3 a point on the face of the gear tooth. As the two pitch circles and the generating circle roll upon each other, a point on the generating circle simultaneously traces out the face of the tooth upon the moving gear and the flank of the tooth on the moving pinion. So point *C* is an instantaneous position of this moving point, and the arc *CP*, of the generating circle, is its path. Initial contact will occur at *D*, where the addendum circle of the driven gear cuts the generating circle. Thus the complete approach path is the arc *DP*. During approach, only the portions of the tooth profiles generated by the circle with center at *A* have been used.

Returning to Fig. 7-25, notice that the pitch point *P* is the instantaneous center of rotation of the generating circle no matter which of the two pitch

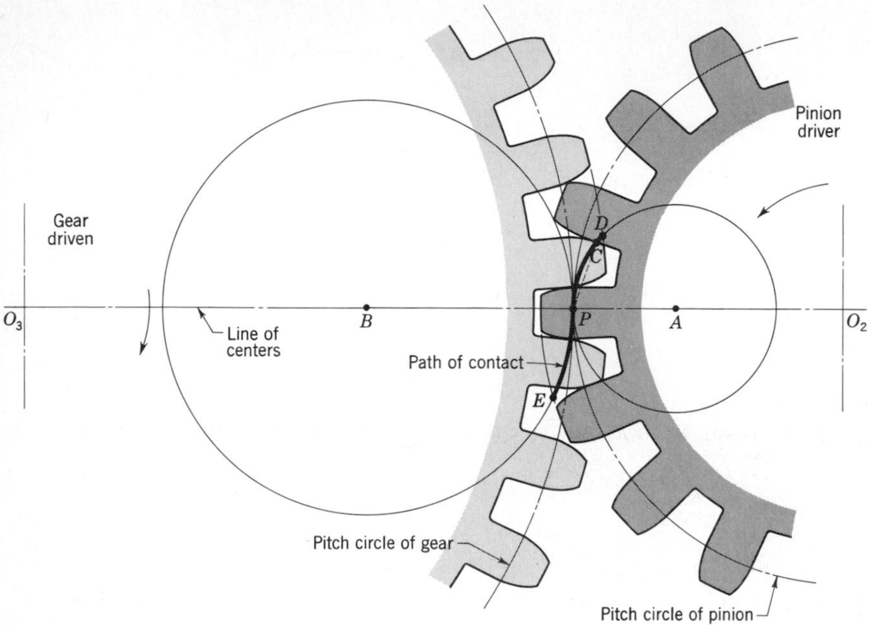

Figure 7-25

circles it is considered to be rolling upon. Thus P is the instantaneous center of rotation of point C, on the generating circle, and consequently the line PC is normal to both tooth profiles; as both gears rotate this will always be so. Thus cycloidal gearing satisfies the law of gearing in that the normal to the tooth profile always passes through the pitch point.

Notice, however, that line PC, which is the pressure line, will not have a constant inclination. As the point of contact nears the pitch point, the pressure line approaches a perpendicularity with the line of centers.

During recess action the generating circle with center at B takes over. Contact occurs on the face of the pinion tooth and the flank of the gear tooth. Notice that each of these profiles is generated by the circle with center at B. During recess the pressure line swings back toward an inclination similar to that during approach. The final point of contact is at E, where the addendum circle of the pinion intersects the generating circle. Thus the path of contact during recess is the arc distance PE.

The varying pressure angle of cycloidal teeth results in additional noise and wear, and also in changing bearing reactions at the shaft supports. Also, the double curvature which frequently occurs introduces problems in the cutting of teeth which are not present with the involute form. In order to run properly, cycloidal gears must be operated at exactly the correct center distance, otherwise the contacting portions of the profiles will not be conjugate. Since deflections due to the transmission of load are bound to occur, it would be virtually impossible to maintain correct center distance under all loading conditions. Thus for most existing applications it appears that the cycloidal tooth form has little to offer over the involute profile.

PROBLEMS

7-1 Find the diametral pitch of a pair of gears whose center distance is 0.362 5 in. The gears have 32 and 84 teeth, respectively.

7-2 Find the number of teeth and the circular pitch of a 6-in-pitch-diameter gear whose diametral pitch is 9.

7-3 Determine the module of a pair of gears whose center distance is 58 mm. The gears have 18 and 40 teeth, respectively.

7-4 Find the number of teeth and the circular pitch of a gear whose pitch diameter is 200 mm if the module is 8 mm per tooth.

7-5 What are the diametral pitch and the pitch diameter of a 40-tooth gear whose circular pitch is 3.50 in?

7-6 The pitch diameters of a pair of mating gears are $3\frac{1}{2}$ and $8\frac{1}{4}$ in, respectively. If the diametral pitch is 16, how many teeth are there on each gear?

7-7 Find the module and the pitch diameter of a gear whose circular pitch is 40 mm if the gear has 36 teeth.

7-8 The pitch diameters of a pair of gears are 60 and 100 mm, respectively. If the module is 2.5 mm per tooth, how many teeth are there on each gear?

7-9 What is the diameter of a 33-tooth gear if the circular pitch is 0.875 in?

7-10 A shaft carries a 30-tooth 3-diametral-pitch gear which drives another gear at a speed of 480 rpm. How fast is the 30-tooth gear rotating if the shaft center distance is 9 in?

7-11 Two gears having an angular velocity ratio of 3 : 1 are mounted on shafts whose centers are 136 mm apart. If the module of the gears is 4 mm, how many teeth are there on each gear?

7-12 A gear having a module of 4 mm per tooth and 21 teeth drives another gear whose speed is 240 rpm. How fast is the 21-tooth gear rotating if the shaft center distance is 156 mm?

7-13 A 4-diametral-pitch, 24-tooth pinion is to drive a 36-tooth gear. The gears are cut on the 20°-full-depth-involute system. Make a drawing of the gears showing one tooth on each gear. Find and tabulate the addendum, dedendum, clearance, circular pitch, tooth thickness, and base-circle diameters; the paths of approach, recess, and action; and the contact ratio and base pitch.

7-14 A 5-diametral-pitch 15-tooth pinion is to mate with a 30-tooth internal gear. The gears are 20° full-depth involute. Make a drawing of the gears showing several teeth on each gear. Can these gears be assembled in a radial direction? If not, what remedy should be used?

7-15 A $2\frac{1}{2}$-diametral-pitch 17-tooth pinion and a 50-tooth gear are paired. The gears are cut on the 20°-full-depth-involute system. Make a drawing of the gears showing one tooth on each gear. Find the arcs of approach, recess, and action and the contact ratio, obtaining the data directly from the drawing.

7-16† A gearset has a module of 5 mm per tooth, full-depth teeth, and a $22\frac{1}{2}°$ pressure angle, and 19 and 31 teeth, respectively. Make a drawing of the gears showing one tooth on each gear. Use 1.0 m for the addendum and 1.35 m for the dedendum. Tabulate the addendum, dedendum, clearance, circular pitch, tooth thickness, base-circle diameter, base pitch, and contact ratio.

7-17† A gear has a module of 8 mm per tooth and 22 teeth and is in mesh with a rack. The addendum and dedendum are 1.0 m and 1.25 m, respectively; pressure angle is 25°. Make a

† In SI, tooth dimensions are given in modules, m. Thus $a = 1.0$ m means 1 module, not 1 metre.

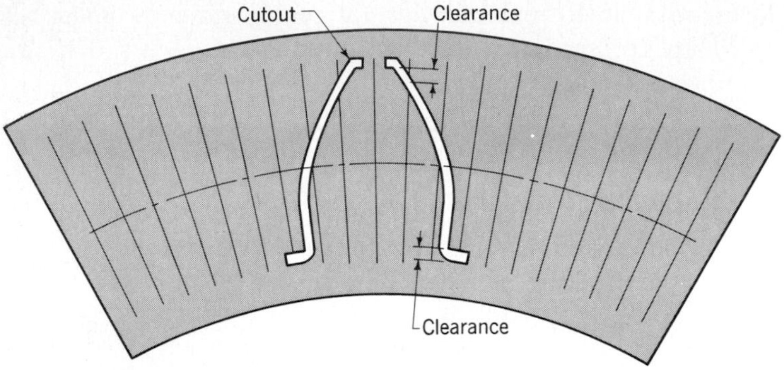

Problems 7-20 to 7-24

drawing showing the teeth in mesh and measure the lengths of the approach path, the recess path, and the total path of contact on the line of action. What is the contact ratio?

7-18 Repeat Prob. 7-15, using the 25°-full-depth system.

7-19 Draw a 2-diametral-pitch 26-tooth gear in mesh with a rack. The gears are 20°-full-depth involute.

(*a*) Find the arcs of approach, recess, and action and the contact ratio.

(*b*) Draw a second rack in mesh with the same gear but offset $\frac{1}{8}$ in away from the gear center. Determine the new contact ratio. Has the pressure angle changed?

7-20 to 7-24 Shaper gear cutters have the advantage that they can be used for either external or internal gears and also that only a small amount of runout is necessary at the end of the stroke. The generating action of a pinion shaper cutter can easily be simulated by employing a sheet of clear plastic. The figure illustrates one tooth of a 16-tooth-pinion cutter with 20° pressure angle as it can be cut from a plastic sheet. To construct the cutter, lay out the tooth on a sheet of drawing paper. Be sure to include the clearance at the top of the tooth. Draw radial lines through the pitch circle spaced at distances equal to one-fourth of the tooth thickness, as shown in the figure. Now fasten the sheet plastic to the drawing and scribe the cutout, the pitch circle, and the radial lines onto the sheet. The sheet is then removed and the tooth outline trimmed with a razor blade. A small piece of fine sandpaper should then be used to remove any burrs.

To generate a gear with the cutter, only the pitch circle and the addendum circle need be drawn. Divide the pitch circle into spaces equal to those used on the template and construct radial lines through them. The tooth outlines are then obtained by rolling the template pitch circle upon that of the gear and drawing the cutter tooth lightly for each position. The resulting generated tooth upon the gear will be evident. The following problems all employ a standard 1-diametral-pitch full-depth template constructed as described above. In each case generate a few teeth and estimate the amount of undercutting.

Problem number	Tooth number
7-20	10
7-21	12
7-22	14
7-23	20
7-24	36

7-25† A 10-mm module gear has 17 teeth, a 20° pressure angle, an addendum of 1.0 m, and a dedendum of 1.25 m. Find the thickness of the teeth at the base circle and at the addendum circle. What is the pressure angle corresponding to the addendum circle?

7-26 A 15-tooth pinion has 1.5 diametral pitch 20°-full-depth teeth. Calculate the thickness of the teeth at the base circle. What are the thickness and pressure angle at the addendum circle?

7-27 A tooth is 0.785 in thick at a radius of 8 in and a pressure angle of 25°. What is the thickness at the base circle?

7-28 A tooth is 1.57 in thick at the pitch radius of 16 in and a pressure angle of 20°. At what radius does the tooth become pointed?

7-29 A 25°-involute 12-diametral-pitch pinion has 18 teeth. Calculate the tooth thickness at the base circle. What are the thickness and pressure angle at the addendum circle?

7-30 A special 10-tooth 8-diametral-pitch pinion is to be cut with a $22\frac{1}{2}°$ pressure angle. What maximum addendum can be used before the teeth become pointed?

7-31 The accuracy of cutting gear teeth can be measured by fitting hardened and ground pins in diametrically opposite tooth spaces and measuring the distance over these pins. A gear has 96 teeth, is 10 diametral pitch, and is cut with the 20°-full-depth-involute system.

(*a*) Calculate the pin diameter which will contact the teeth at the pitch lines if there is to be no backlash.

(*b*) If the gear is accurately cut, what should be the distance measured over the pins?

7-32 A set of interchangeable gears is cut on the 20°-full-depth-involute system, having a diametral pitch of 4. The gears have tooth numbers of 24, 32, 48, and 96. For each gear, calculate the radius of curvature of the tooth profile at the pitch circle and at the addendum circle.

7-33 Calculate the contact ratio of a 17-tooth pinion which drives a 73-tooth gear. The gears are 96 diametral pitch and cut on the 20° fine-pitch system.

7-34 A special 25°-pressure-angle 11-tooth pinion is to drive a 23-tooth gear. The gears have a diametral pitch of 8 and are stub teeth. What is the contact ratio?

7-35 A 22-tooth pinion mates with a 42-tooth gear. The gears are full-depth, have a diametral pitch of 16, and are cut with a $17\frac{1}{2}°$ pressure angle. Find the contact ratio.

7-36 A pair of mating gears are 24 diametral pitch and are generated on the 20° system. If the tooth numbers are 15 and 50, what maximum addendums may they have if interference is not to occur?

7-37 A set of gears is cast with a $17\frac{1}{2}°$ pressure angle and a circular pitch of $4\frac{1}{2}$ in. The pinion has 20 full-depth teeth. If the gear has 240 teeth, what must its addendum be in order to avoid interference?

7-38 Using the method described in Prob. 7-20, cut a 1-diametral-pitch 20°-pressure-angle full-depth rack tooth from a sheet of clear plastic. Use a modified clearance of $0.35/P$ in order to obtain a stronger fillet. This template can be used to simulate the generating action of a hob. Now, using the variable-center-distance system, generate an 11-tooth pinion to mesh with a 25-tooth gear without interference. Record the values found for center distance, pitch radii, pressure angle, gear-blank diameters, cutter offset, and contact ratio. Note that more than one satisfactory solution exists.

7-39 Using the template constructed in Prob. 7-38, generate an 11-tooth pinion to mesh with a 44-tooth gear with the long-and-short-addendum system. Determine and record suitable values for gear and pinion addendum and dedendum and for the cutter offset and contact ratio. Compare the contact ratio with that which would have been obtained with standard gears.

7-40 A standard 20-tooth 20°-pressure-angle full-depth 1-diametral-pitch pinion drives a 48-tooth gear. The speed of the pinion is 500 rpm. Using the length of the path of contact as the abscissa, plot a curve showing the sliding velocity at all points of contact. Notice that the sliding velocity changes sign when the point of contact passes through the pitch point.

† See footnote on p. 279.

CHAPTER

EIGHT

HELICAL, WORM, AND BEVEL GEARS

Most engineers prefer to use spur gears when power is to be transferred between parallel shafts because they are easier to design and often more economical to manufacture, but sometimes the design requirements are such that helical gears are a better choice. This is especially true when loads are heavy, speeds are high, or the noise level must be kept low.

When motion is to be transmitted between shafts which are not parallel, the spur gear cannot be used; the designer must then choose between crossed-helical, worm, bevel, or hypoid gears. Bevel gears have straight teeth, line contact, and high efficiencies. Crossed-helical and worm gears have a much lower efficiency because of the increased sliding action; however, if good engineering is used, crossed-helical and worm gears can be designed with quite acceptable values of efficiency. Hypoid and bevel gears are used for similar applications, and although hypoid gears have inherently stronger teeth, the efficiency is often much less. Worm gears are used when very high velocity ratios are required.

8-1 PARALLEL-AXIS HELICAL GEARS

Helical gears are used for the transmission of motion between nonparallel and parallel shafts. When they are used with nonparallel shafts, they are called *crossed helical gears*; these are considered in Sec. 8-6.

The shape of the tooth of a helical gear is an involute helicoid, as illustrated in Fig. 8-1. If a piece of paper is cut into the shape of a parallelogram and wrapped around a cylinder, the angular edge of the paper becomes a helix. If the paper is then unwound, each point on the angular edge

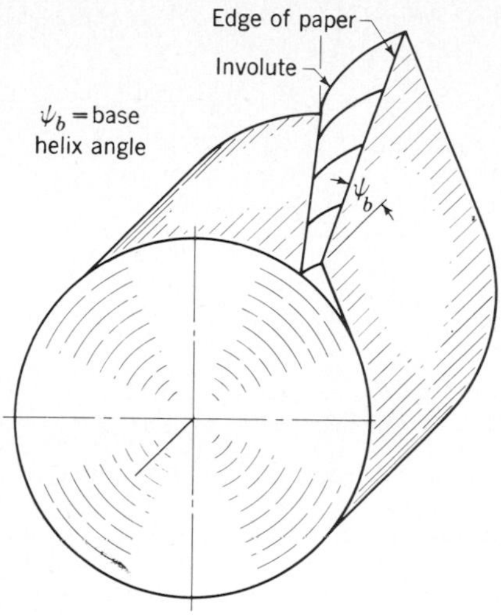

Figure 8-1 An involute helicoid.

generates an involute curve. The surface obtained when every point on the edge generates an involute is called an *involute helicoid.*

The initial contact of spur-gear teeth is a line extending all the way across the face of the tooth. The initial contact of helical-gear teeth is a point which changes into a line as the teeth come into more engagement; in helical gears the line is diagonal across the face of the tooth. It is this gradual engagement of the teeth and the smooth transfer of load from one tooth to another which give helical gears the ability to transmit heavy loads at high speeds.

Double-helical (herringbone) gears are obtained when each gear has both right- and left-hand teeth cut on the same blank, and they operate on parallel axes. The thrust forces of the right- and left-hand halves are equal and opposite and cancel each other.

8-2 HELICAL-GEAR-TOOTH RELATIONS

Figure 8-2 represents a portion of the top view of a helical rack. Lines *AB* and *CD* are the centerlines of two adjacent helical teeth taken on the pitch plane. The angle ψ is the *helix angle* and is to be measured at the pitch diameter unless otherwise specified. The distance *AC* is the ***transverse circular pitch*** p_t in the plane of rotation. The distance *AE* is the ***normal circular pitch*** p_n and is related to the transverse circular pitch as follows:

$$p_n = p_t \cos \psi \tag{8-1}$$

The distance *AD* is called the *axial pitch* p_x and is

$$p_x = \frac{p_t}{\tan \psi} \tag{8-2}$$

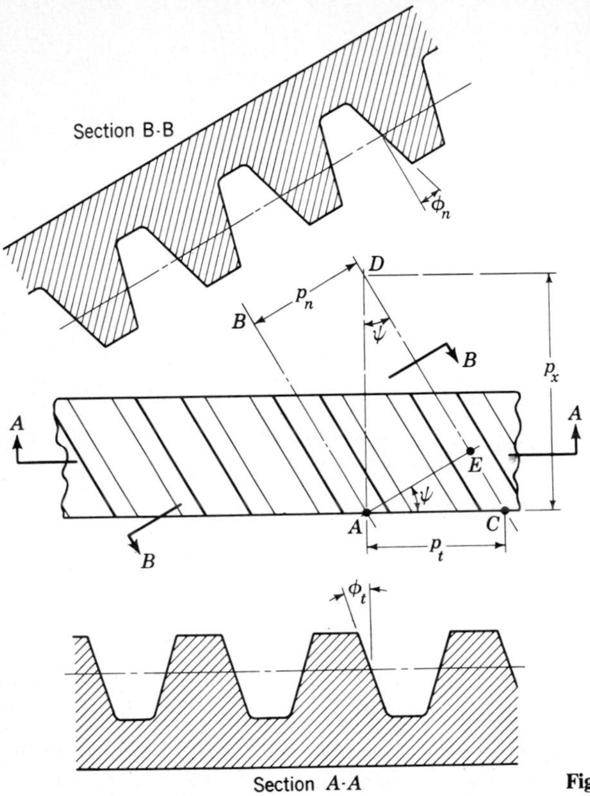

Figure 8-2 Helical-gear-tooth relations.

Since $p_n P_n = \pi$, the *normal diametral pitch* is

$$P_n = \frac{P_t}{\cos \psi} \tag{8-3}$$

where P_t is the *transverse diametral pitch.*

Because of the angularity of the teeth we must define two pressure angles. These are the *transverse pressure angle* ϕ_t and the *normal pressure angle* ϕ_n, as shown in Fig. 8-2. They are related by

$$\cos \psi = \frac{\tan \phi_n}{\tan \phi_t} \tag{8-4}$$

In applying these equations it is convenient to remember that all the equations and relations which are valid for spur gears apply equally for the transverse plane of a helical gear.

A better picture of the tooth relations can be obtained by an examination of Fig. 8-3. In order to obtain the geometric relations a helical gear has been cut by the oblique plane AA at an angle ψ to a right section. For convenience, only the pitch cylinder of radius r is shown. The figure shows that the intersection of the plane and the pitch cylinder produces an ellipse whose

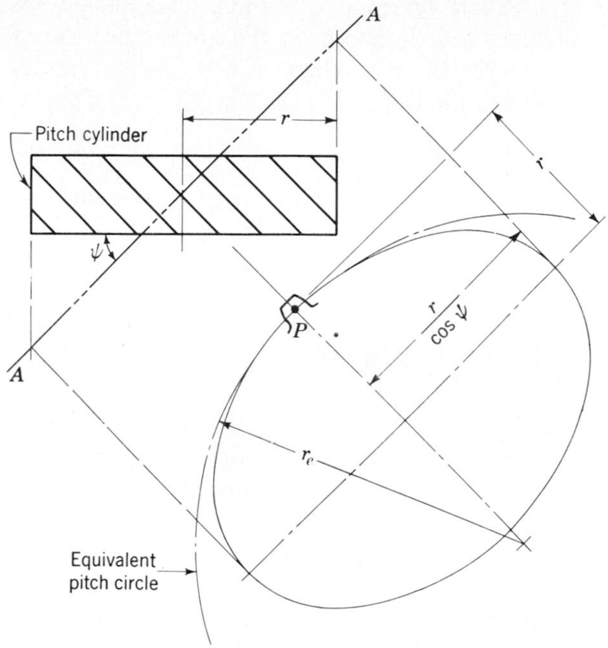

Figure 8-3

radius at the pitch point P is r_e. This is called the *equivalent pitch radius*, and it is the radius of curvature of the pitch surface in the normal cross section. For the condition that $\psi = 0$, this radius of curvature is r. If we imagine the angle ψ to be slowly increased from 0 to 90°, we see that r_e begins at a value of r and increases until, when $\psi = 90°$, $r_e = \infty$.

It can be shown† that

$$r_e = \frac{r}{\cos^2 \psi} \tag{8-5}$$

† The equation of an ellipse with its center at the origin of an xy system with a and b as the semimajor and semiminor axes, respectively, is

$$\frac{x^2}{a^2} + \frac{y^2}{b^2} = 1 \tag{a}$$

Also, the formula for radius of curvature is

$$\rho = \frac{[1 + (dy/dx)^2]^{3/2}}{d^2y/dx^2} \tag{b}$$

By using these two equations, it is not difficult to find the radius of curvature corresponding to $x = 0$ and $y = b$. The result is

$$\rho = \frac{a^2}{b} \tag{c}$$

Now referring to Fig. 8-3, we substitute $a = r/(\cos \psi)$ and $b = r$ into Eq. (c) and obtain Eq. (8-5).

where r is the pitch radius of the helical gear and r_e is the pitch radius of an equivalent spur gear. This equivalent gear is taken on the normal section of the helical gear. Let us define the number of teeth on the helical gear as N and on the equivalent spur gear as N_e. Then

$$N_e = d_e P_n \tag{d}$$

where $d_e = 2r_e$ is the pitch diameter of the equivalent spur gear. We can also write Eq. (d) as

$$N_e = \frac{d}{\cos^2 \psi} \frac{P_t}{\cos \psi} = \frac{N}{\cos^3 \psi} \tag{8-6}$$

8-3 HELICAL-GEAR-TOOTH PROPORTIONS

Except for fine-pitch gears (20-diametral-pitch and finer), there is no standard for the proportions of helical-gear teeth. One reason for this is that it is cheaper to change the design slightly than to purchase special tooling. Since helical gears are rarely used interchangeably anyway, and since many different designs will work well together, there is really little advantage in having them interchangeable.

As a general guide, tooth proportions should be based on a normal pressure angle of 20°. Most of the proportions tabulated in Table 7-1 can then be used. The tooth dimensions should be calculated by using the normal diametral pitch. These proportions are suitable for helix angles from 0 to 30°, and all helix angles can be cut with the same hob. Of course the normal diametral pitch of the hob and the gear must be the same.

An optional set of proportions can be based on a transverse pressure angle of 20° and the use of the transverse diametral pitch. For these the helix angles are generally restricted to 15, 23, 30, or 45°. Angles greater than 45° are not recommended. The normal diametral pitch must still be used to compute the tooth dimensions. The proportions shown in Table 7-1 will usually be satisfactory.

Many authorities recommend that the face width of helical gears be at least twice the axial pitch to obtain true helical-gear action. Exceptions to this rule are automotive gears, which have a face width considerably less, and marine reduction gears, which often have a face width much greater.

We further note that in a parallel helical gearset, the two gears must have the same helix angle and pitch and must be of opposite hand. The velocity ratio is determined the same as for spur gears.

8-4 CONTACT OF HELICAL-GEAR TEETH

Mating spur-gear teeth contact each other in a line which is parallel to their axes of rotation. As shown in Fig. 8-4, the contact between helical-gear teeth is a diagonal line.

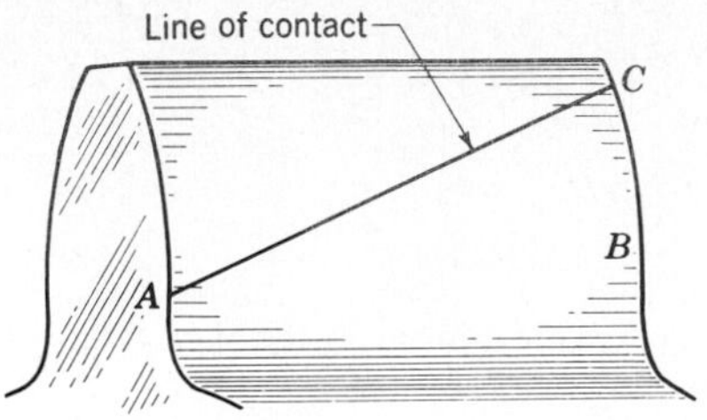

Figure 8-4 While contact at A is just beginning, contact at the other end of the tooth has already progressed from B to C.

There are several kinds of contact ratios used in evaluating the performance of helical gears. The *transverse contact ratio* is designated by m and is the contact ratio in the transverse plane. It is obtained exactly the same as for spur gears.

The *normal contact ratio* m_n is the contact ratio in the normal section. It is also found exactly the same as for spur gears, but the *equivalent spur gears must be used in the determination.* The *base helix angle* ψ_b and the *pitch helix angle* ψ, for helical gears, are related by

$$\tan \psi_b = \tan \psi \cos \phi \tag{8-7}$$

Then the transverse and normal contact ratios are related by

$$m_n = \frac{m}{\cos^2 \psi_b} \tag{8-8}$$

The *axial contact ratio*, also called the *face contact ratio*, is the ratio of the face width of the gear to the axial pitch. It is given by

$$m_x = \frac{F}{p_x} = \frac{F \tan \psi}{p_t} \tag{8-9}$$

where F is the face width. Note that the face contact ratio depends only on the geometry of a single gear, while the transverse and normal contact ratios depend upon the geometry of a pair of mating gears.

The *total contact ratio* m_t is the sum of the face and transverse contact ratios. In a sense it gives the average total number of teeth in contact.

8-5 HERRINGBONE GEARS

Double-helical or *herringbone gears* consist of teeth having a right- and a left-hand helix cut on the same blank, as illustrated schematically in Fig. 8-5. One of the disadvantages of the single-helical gear is the existence of axial thrust loads (see Fig. 12-11). They are eliminated by the herringbone configuration because the thrust force of the right-hand half is balanced by that of the left-hand half. However, one of the members of a herringbone gearset should always be mounted with some axial play or float to accommodate slight tooth errors and mounting allowances.

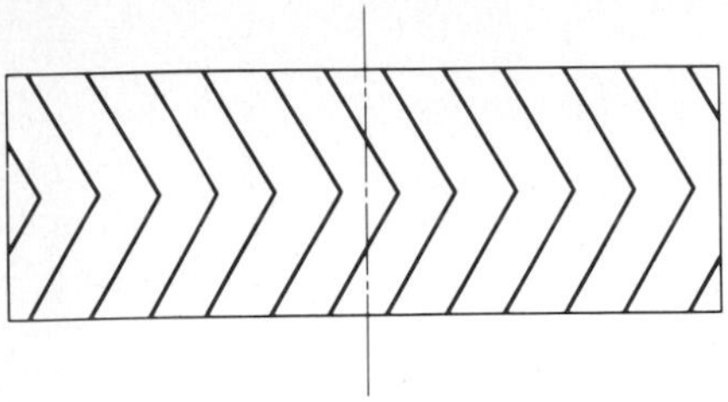

Figure 8-5 Schematic drawing of the pitch cylinder of a double-helical gear.

Helix angles are usually larger for herringbone gears than for single-helical gears because of the absence of the thrust reactions.

8-6 CROSSED-AXIS HELICAL GEARS

Crossed-helical, or spiral, gears are used sometimes when the shaft centerlines are neither parallel nor intersecting. They are essentially nonenveloping worm gears because the gear blanks have a cylindrical form.

The teeth of crossed-helical gears have *point contact* with each other, which changes to *line contact* as the gears wear in. For this reason they will carry only very small loads. Because of the point contact, however, they need not be mounted accurately; either the center distance or the shaft angle may be varied slightly without affecting the amount of contact.

There is no difference between a crossed-helical gear and a helical gear until they are mounted in mesh with each other. They are manufactured in the same way. A pair of meshed crossed-helical gears usually have the same hand; i.e., a right-hand driver goes with a right-hand driven. The relation between thrust, hand, and rotation for crossed-helical gears is shown in Fig. 8-6.

When tooth sizes are specified, the normal pitch should always be used. The reason for this is that when different helix angles are used for the driver and driven, the transverse pitches are not the same. The relation between the shaft and helix angles is

$$\Sigma = \psi_2 \pm \psi_3 \tag{8-10}$$

where Σ is the shaft angle. The plus sign is used when both helix angles are of the same hand and the minus sign when they are of opposite hand. Opposite-hand crossed-helical gears are used when the shaft angle is small.

The pitch diameter is obtained from

$$d = \frac{N}{P_n \cos \psi} \tag{8-11}$$

where N = number of teeth
P_n = normal diametral pitch
ψ = helix angle

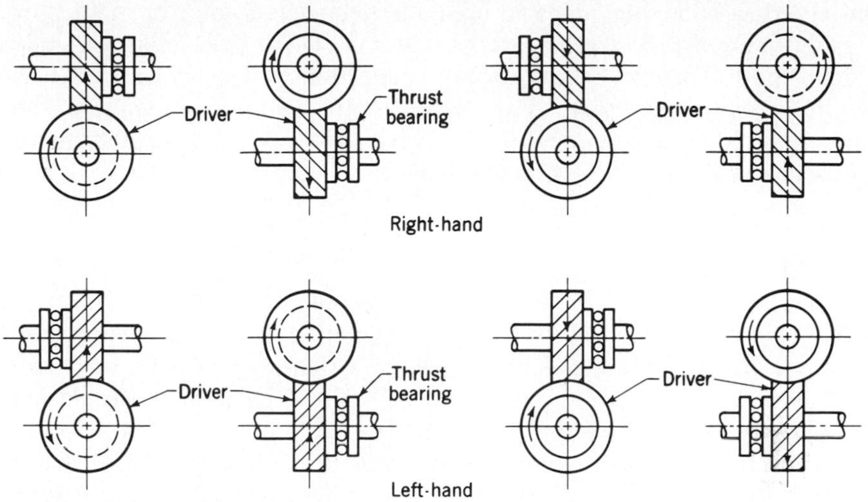

Figure 8-6 Thrust, rotation, and hand relations for crossed-helical gearing. (*Boston Gear Works, Inc., North Quincy, Mass.*)

Since the pitch diameters are not directly related to the tooth numbers, they cannot be used to obtain the angular-velocity ratio. This ratio must be obtained from the ratio of the tooth numbers.

Crossed-helical gears will have the lowest sliding velocity at contact when the helix angles of the two gears are equal. If the helix angles are not equal, the gear with the largest helix angle should be used as the driver if both gears have the same hand.

There is no standard for crossed-helical gear-tooth proportions. Many different proportions give good tooth action. Since the teeth are in point

Table 8-1 Tooth proportions for crossed-axis helical gears

Normal diametral pitch $P_n = 1$; working depth = 2.400 in; whole depth = 2.650 in; addendum = 1.200 in

Driver		Driven	Normal
Helix angle ψ_2, deg	Minimum tooth number N_2	helix angle ψ_3, deg	pressure angle ϕ_n, deg
45	20	45	14.50
60	9	30	17.50
75	4	15	19.50
86	1	4	20

contact, an effort should be made to obtain a contact ratio of 2 or more. For this reason, crossed-helical gears are usually cut with a low pressure angle and a deep tooth. Dudley† lists the tooth proportions shown in Table 8-1 as representative of good design. The driver tooth numbers shown are the minimum required to avoid undercut. The driven gear should have 20 or more teeth if a contact ratio of 2 is to be obtained.

8-7 WORM GEARING

Figure 8-7 shows a worm and worm-gear application. These gears are used with nonintersecting shafts which are usually at a shaft angle of 90°, but there is no reason why shaft angles other than 90° cannot be used if the design demands it.

The worm is the member having the screwlike thread, and worm teeth are frequently spoken of as threads. Worms in common use have from 1 to 8 teeth, and, as we shall see, there is no definite relation between the number of

† Darle W. Dudley, "Practical Gear Design," p. 111, McGraw-Hill, New York, 1954.

Figure 8-7 A single-enveloping worm and worm gear. (*The Falk Corporation, Subsidiary of Sundstrand Corporation, Milwaukee, Wis.*)

teeth and the pitch diameter of a worm. Worms may be designed with a cylindrical pitch surface, as shown in Fig. 8-8, or they may have an hourglass shape, such that the worm wraps around or partially encloses the worm gear.

The worm gear is normally the driven member of the pair and is made to envelop, or wrap around, the worm. If the worm gear is mated with a cylindrical worm, the gearset is said to be *single-enveloping*. When the worm is hourglass-shaped, the worm gearset is said to be *double-enveloping* because each member then wraps around the other.

A worm and worm-gear combination is similar to a pair of mating crossed-helical gears except that the worm gear partially encloses the worm. For this reason they have line contact, instead of the point contact found in crossed-helical gears, and are thus able to transmit more power. When a double-enveloping gearset is used, even more power can be transmitted, theoretically at least, because contact occurs over an area of the tooth surfaces.

In the single-enveloping worm gearset it does not make any difference whether the worm rotates on its own axis and drives the gear by a screw action or whether the worm is translated along its axis and drives the gear through rack action. The resulting motion and contact are the same. For this

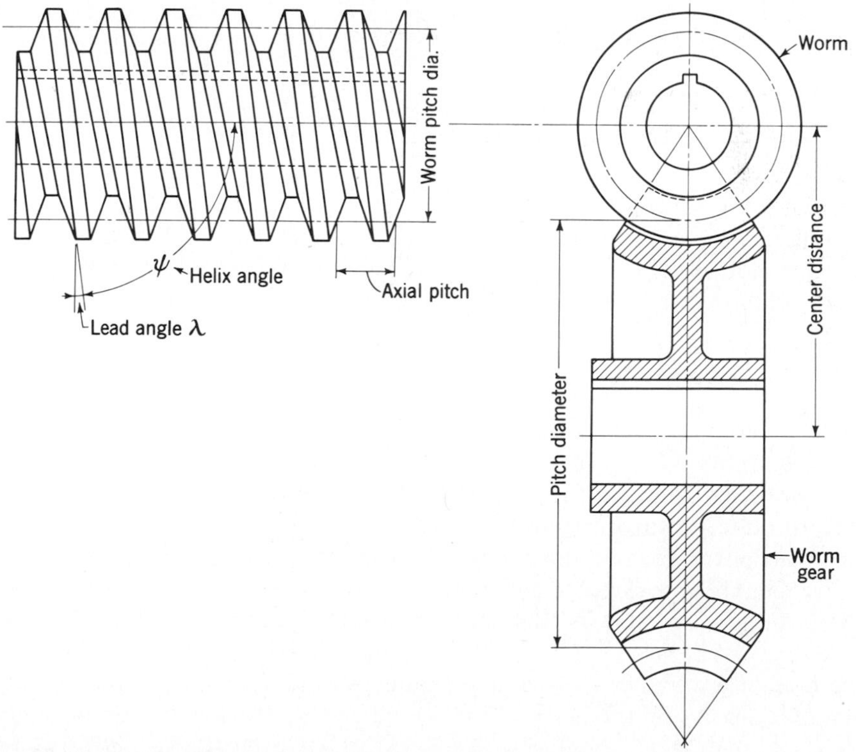

Figure 8-8 Nomenclature of a single-enveloping worm gearset.

reason the worm need not be accurately mounted upon its shaft. However, the worm gear should be correctly mounted along its axis of rotation; otherwise the two pitch surfaces will not be concentric about the worm axis.

In a double-enveloping worm gearset both members are throated and must therefore be accurately mounted in every direction in order to obtain correct action.

The nomenclature of a single-enveloping gearset is shown in Fig. 8-8.

A mating worm and worm gear with a 90° shaft angle have the same hand of helix, but the helix angles are usually quite different. On the worm, the helix angle is quite large (at least, for one or two teeth) and very small on the gear. Because of this, it is customary to specify the *lead angle* for the worm and the helix angle for the gear. This is convenient because the two angles are equal for a 90° shaft angle. The worm lead angle is the complement of the worm helix angle, as shown in Fig. 8-8.

In specifying the pitch of worm gearsets, specify the axial pitch of the worm and the circular pitch of the gear. These are equal if the shaft angle is 90°. It is quite common to employ even fractions for the circular pitch, like $\frac{1}{4}$, $\frac{3}{8}$, $\frac{1}{2}$, $\frac{3}{4}$, 1, $1\frac{1}{4}$ in, etc. There is no reason, however, why standard diametral pitches, like those used for spur gears, cannot be used. The pitch diameter of the gear is the same as that for spur gears:

$$d_3 = \frac{N_3 p}{\pi} \tag{8-12}$$

where d_3 = pitch diameter
N_3 = number of teeth
p = circular pitch

all taken with reference to the worm gear.

The pitch diameter of the worm may be any value, but it should be the same as the hob used to cut the worm-gear teeth. AGMA recommends the following relation between the pitch diameter of the worm and the center distance:

$$d_2 = \frac{(r_2 + r_3)^{0.875}}{2.2} \tag{8-13}$$

where the quantity $r_2 + r_3$ is the center distance. This equation gives a set of proportions which will result in a good power capacity. Equation (8-13) does not have to be used; other proportions will work well too, and in fact power capacity may not always be the primary consideration. However, there are a lot of variables in worm-gear design, and the equation is helpful in obtaining trial dimensions. The AGMA Standard† also states that the denominator of Eq. (8-13) may vary from 1.7 to 3 without appreciably affecting the capacity.

The *lead* of a worm has the same meaning as for a screw thread and is the distance that a point on the helix will move when the worm is turned through

† AGMA Standard 213.02, 1952.

one revolution. Thus, for a one-tooth worm, the lead is equal to the axial pitch. In equation form,

$$l = p_x N_2 \tag{8-14}$$

where l is the lead in inches and N_2 is the number of teeth on the worm. The lead and the lead angle are related as follows:

$$\lambda = \tan^{-1} \frac{l}{\pi d_2} \tag{8-15}$$

where λ is the lead angle, as shown in Fig. 8-8.

The teeth on worms are usually cut on a milling machine or on a lathe. Worm-gear teeth are most often produced by hobbing. Except for clearance at the tip of the hob teeth, the worm should be an exact duplicate of the hob in order to obtain conjugate action. This means, too, that where possible the worm should be designed using the dimensions of existing hobs.

The pressure angles used on worm gearsets vary widely and should depend approximately on the value of the lead angle. Good tooth action will be obtained if the pressure angle is made large enough to eliminate undercutting of the worm-gear tooth on the side at which contact ends. Buckingham recommends the values shown in Table 8-2.

A satisfactory tooth depth which remains in about the right proportion to the lead angle can be obtained by making the depth a proportion of the normal circular pitch. With an addendum of $1/P$ for full-depth spur gears, we obtain the following proportions for worm and worm gears:

$$\text{Addendum} = 0.3183p_n$$

$$\text{Whole depth} = 0.6366p_n$$

$$\text{Clearance} = 0.050p_n$$

The face width of the worm gear should be obtained as shown in Fig. 8-9.

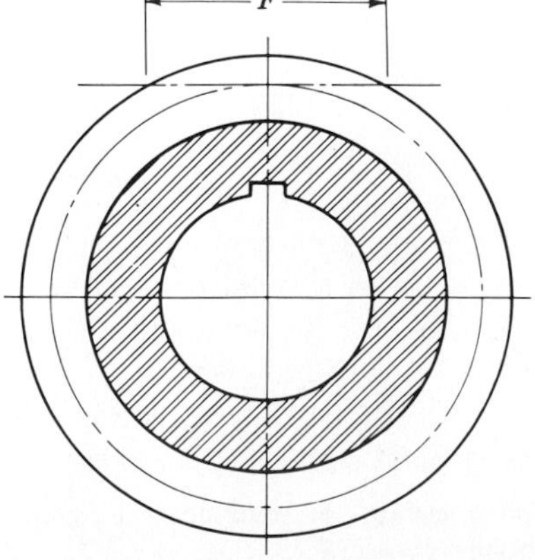

Figure 8-9

Table 8-2 Pressure angles recommended for worm gearsets

Lead angle λ, deg	Pressure angle ϕ, deg
0–16	$14\frac{1}{2}$
16–25	20
25–35	25
35–45	30

This makes the face of the worm gear equal to the length of a tangent to the worm pitch circle between its points of intersection with the addendum circle.

8-8 STRAIGHT-TOOTH BEVEL GEARS

When motion is to be transmitted between shafts whose axes intersect, some form of bevel gear is required. Although bevel gears are often made for a shaft angle of 90°, they can be produced for almost any angle. The most accurate teeth are obtained by generation.

Bevel gears have pitch surfaces which are cones; these cones roll together without slipping, as shown in Fig. 8-10. The gears must be mounted so that the apexes of both pitch cones are coincident because the pitch of the teeth depends upon the radial distance from the apex.

The true shape of a bevel-gear tooth is obtained by taking a spherical section through the tooth, where the center of the sphere is at the common apex, as shown in Fig. 8-11. Thus, as the radius of the sphere increases, the same number of teeth must exist on a larger surface; so the size of the teeth increases as larger and larger spherical sections are taken. We have seen that the action and contact conditions of spur-gear teeth may be viewed on a plane surface taken at right angles to the axes of the spur gears. For bevel-gear teeth, the action and contact conditions should be viewed on a spherical

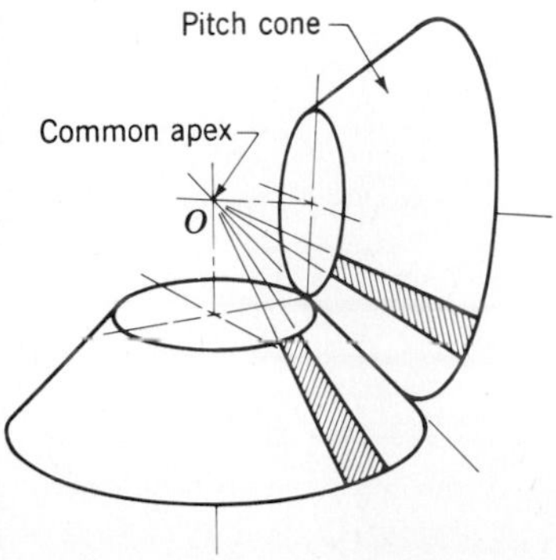

Figure 8-10 The pitch surfaces of bevel gears are cones which have pure rolling contact.

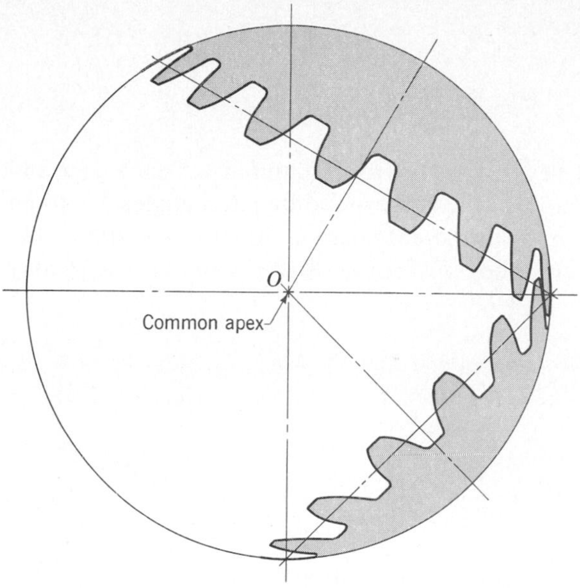

Figure 8-11 A spherical section of bevel-gear teeth.

surface (instead of a plane surface). We can even think of spur gears as a special case of bevel gears in which the spherical radius is infinite, thus producing a plane surface on which the tooth action is viewed.

It is standard practice to specify the pitch diameter of bevel gears at the large end of the teeth. In Fig. 8-12 the pitch cones of a pair of bevel gears are drawn and the pitch radii given as r_2 and r_3, respectively, for the pinion and gear. The angles γ_2 and γ_3 are defined as the *pitch angles*, and their sum is equal to the shaft angle Σ. The velocity ratio is obtained in the same manner

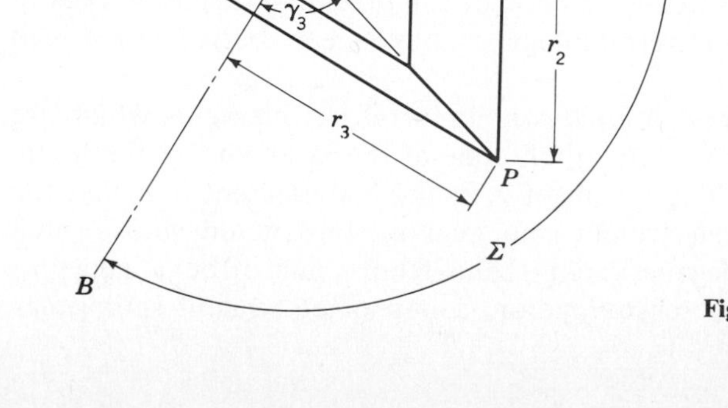

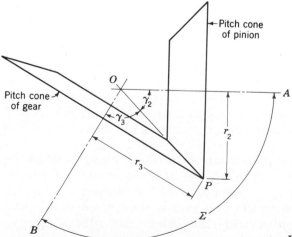

Figure 8-12

as for spur gears and is

$$\frac{\omega_2}{\omega_3} = \frac{r_3}{r_2} = \frac{N_3}{N_2} \tag{8-16}$$

In the kinematic design of bevel gears the tooth numbers of each gear and the shaft angle are usually given, and the corresponding pitch angles are to be determined. Although they can easily be found by a graphical method, the analytical approach gives exact values. From Fig. 8-12 the distance OP may be written

$$OP = \frac{r_2}{\sin \gamma_2} \qquad \text{or} \qquad OP = \frac{r_3}{\sin \gamma_3}$$

so that

$$\sin \gamma_2 = \frac{r_2}{r_3} \sin \gamma_3 = \frac{r_2}{r_3} \sin (\Sigma - \gamma_2) \tag{a}$$

or

$$\sin \gamma_2 = \frac{r_2}{r_3} (\sin \Sigma \cos \gamma_2 - \sin \gamma_2 \cos \Sigma) \tag{b}$$

Dividing both sides of Eq. (*b*) by $\cos \gamma_2$ and rearranging gives

$$\tan \gamma_2 = \frac{\sin \Sigma}{(r_3/r_2) + \cos \Sigma} = \frac{\sin \Sigma}{(N_3/N_2) + \cos \Sigma} \tag{8-17}$$

Similarly,

$$\tan \gamma_3 = \frac{\sin \Sigma}{(N_2/N_3) + \cos \Sigma} \tag{8-18}$$

For a shaft angle of 90° the above expressions reduce to

$$\tan \gamma_2 = \frac{N_2}{N_3} \tag{8-19}$$

and

$$\tan \gamma_3 = \frac{N_3}{N_2} \tag{8-20}$$

The projection of bevel-gear teeth on the surface of a sphere would indeed be a difficult and time-consuming problem. Fortunately, an approximation is available which reduces the problem to that of ordinary spur gears. This method is called *Tredgold's approximation*, and as long as the gear has eight or more teeth, it is accurate enough for practical purposes. It is in almost universal use, and the terminology of bevel-gear teeth has evolved around it.

In using Tredgold's method, a *back cone* is formed of elements which are perpendicular to the elements of the pitch cone at the large end of the teeth. This is shown in Fig. 8-13. The length of a back-cone element is called the *back-cone radius*. Now an equivalent spur gear is constructed whose pitch radius r_e is equal to the back-cone radius. Thus from a pair of bevel gears we can obtain, using Tredgold's approximation, a pair of equivalent spur gears,

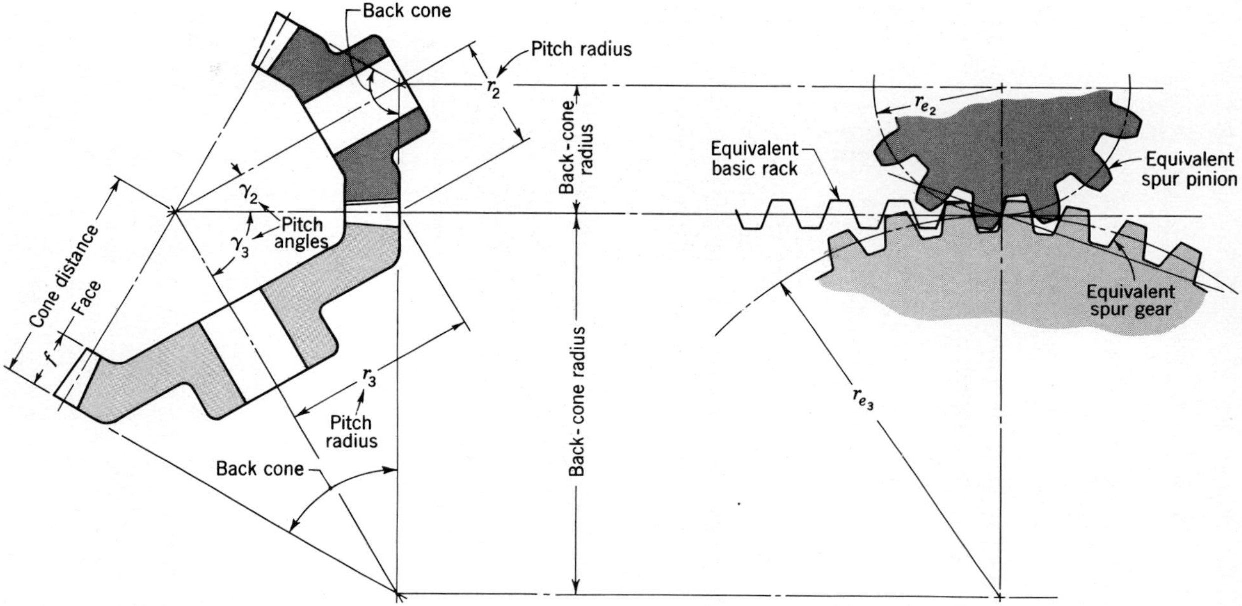

Figure 8-13 Tredgold's approximation.

which are then used to define the tooth profiles; they can also be used to determine the tooth action and contact conditions exactly the same as for ordinary spur gears, and the results will correspond closely to those for the bevel gears. From the geometry of Fig. 8-13, the equivalent pitch radii are

$$r_{e_2} = \frac{r_2}{\cos \gamma_2} \qquad r_{e_3} = \frac{r_3}{\cos \gamma_3} \tag{8-21}$$

The number of teeth on the equivalent spur gear is

$$N_e = \frac{2\pi r_e}{p} \tag{8-22}$$

where p is the circular pitch of the bevel gear measured at the large end of the teeth. In the usual case the equivalent spur gears will *not* have an integral number of teeth.

8-9 TOOTH PROPORTIONS FOR BEVEL GEARS

Practically all straight-tooth bevel gears manufactured today use the 20° pressure angle. It is not necessary to use the interchangeable tooth form because bevel gears cannot be interchanged anyway. For this reason the long-and-short-addendum system, described in Sec. 7-11, is used. These proportions are tabulated in Table 8-3.

Table 8-3 Tooth proportions for 20° straight bevel-gear teeth

Item	Formula
Working depth	$h_k = \frac{2.0}{P}$
Clearance	$c = \frac{0.188}{P} + 0.002$ in
Addendum of gear	$a_G = \frac{0.54}{P} + \frac{0.460}{P(m_{90})^2}$
Gear ratio	$m_G = \frac{N_G}{N_P}$
Equivalent 90° ratio	$m_{90} = \begin{cases} m_G & \text{when } \Sigma = 90° \\ \sqrt{m_G \frac{\cos \gamma_P}{\cos \gamma_G}} & \text{when } \Sigma \neq 90° \end{cases}$
Face width	$F = \frac{l}{3}$ or $F = \frac{10}{P}$ whichever is smaller
Minimum number of teeth	Pinion: 16, 15, 14, 13; Gear: 16, 17, 20, 30

Bevel gears are usually mounted on the outboard side of the bearings because the shaft axes intersect, and this means that the effect of shaft deflection is to pull the small end of the teeth away from mesh, causing the large end to take most of the load. Thus the load across the tooth is variable, and for this reason it is desirable to design a fairly short tooth. As shown in Table 8-3, the face width is usually limited to about one-third of the cone distance. We note also that a short face width simplifies the tooling problems in cutting bevel-gear teeth.

Figure 8-14 defines additional terms characteristic of bevel gears. Note that a constant clearance is maintained by making the elements of the face cone parallel to the elements of the root cone of the mating gear. This explains why the face-cone apex is not coincident with the pitch-cone apex in Fig. 8-14. This permits a larger fillet at the small end of the teeth than would otherwise be possible.

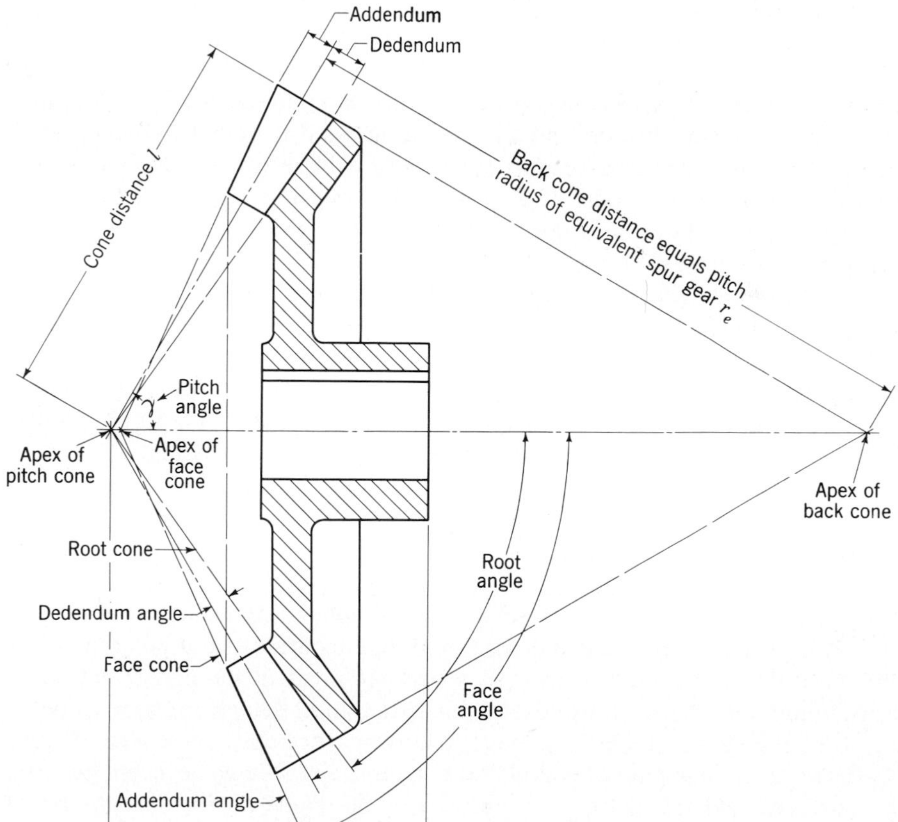

Figure 8-14

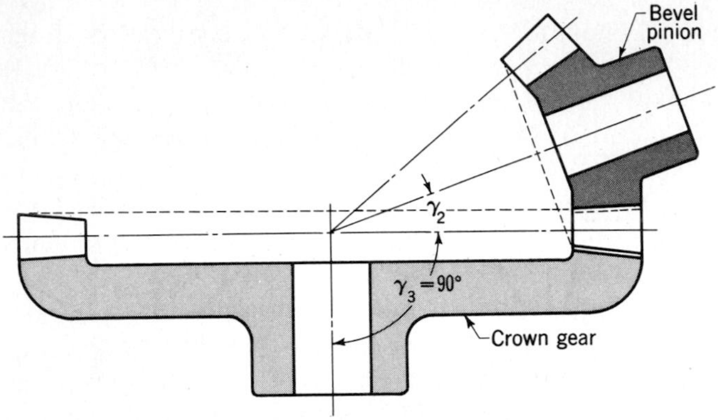

Figure 8-15 A crown gear and bevel pinion.

8-10 CROWN AND FACE GEARS

If the pitch angle of one of a pair of bevel gears is made equal to 90°, the pitch cone becomes a flat surface and the resulting gear is called a *crown gear.* Figure 8-15 shows a crown gear in mesh with a bevel pinion. Notice that a crown gear is the counterpart of a rack in spur gearing. The back cone of a crown gear is a cylinder, and the resulting involute teeth have straight sides, as indicated in Fig. 8-13.

A pseudo-bevel gearset can be obtained by using a face gear in mesh with a spur gear. The shaft angle is at 90°. In order to secure the best tooth action, the spur pinion should be a duplicate of the shaper cutter used to cut the face gear, except, of course, for the additional clearance at the tips of the cutter teeth. The face width of the teeth on the face gear must be held quite short; otherwise the top land will become pointed at the larger diameter.

8-11 SPIRAL BEVEL GEARS

Straight bevel gears are easy to design and simple to manufacture and give very good results in service if they are mounted accurately and positively. As in the case of spur gears, however, they become noisy at the higher values of the pitch-line velocity. In these cases it is often good design practice to go to the spiral bevel gear, which is the bevel counterpart of the helical gear. Figure 8-16 shows a mating pair of spiral bevel gears, and it can be seen that the pitch surfaces and the nature of contact are the same as for straight bevel gears except for the differences brought about by the spiral-shaped teeth.

Spiral-bevel-gear teeth are conjugate to a basic crown rack, which is generated as shown in Fig. 8-17 by using a circular cutter. The spiral angle ψ is measured at the mean radius of the gear. As with helical gears, spiral bevel

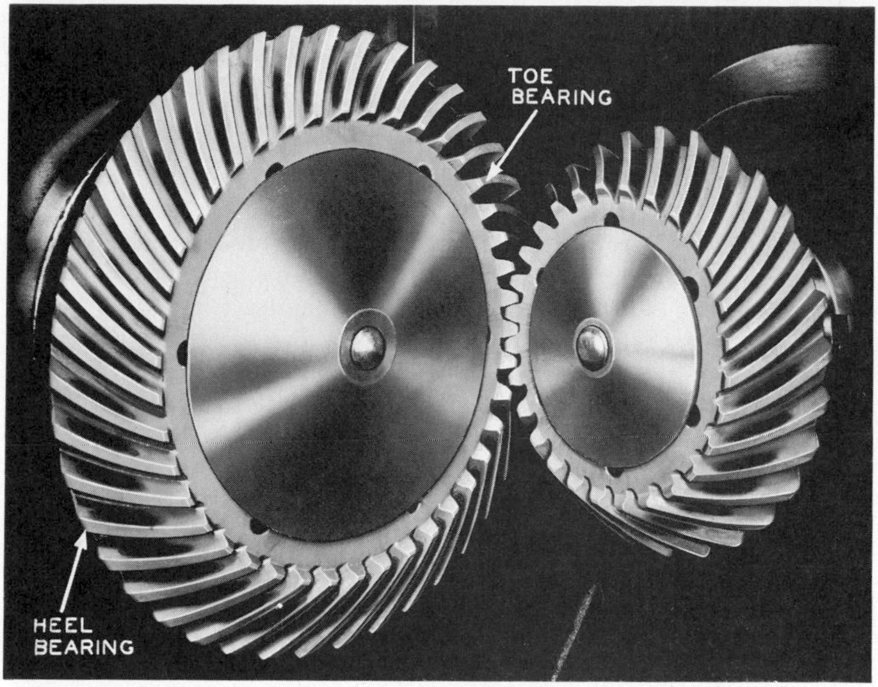

Figure 8-16 Spiral bevel gears. (*Gleason Works, Rochester, N.Y.*)

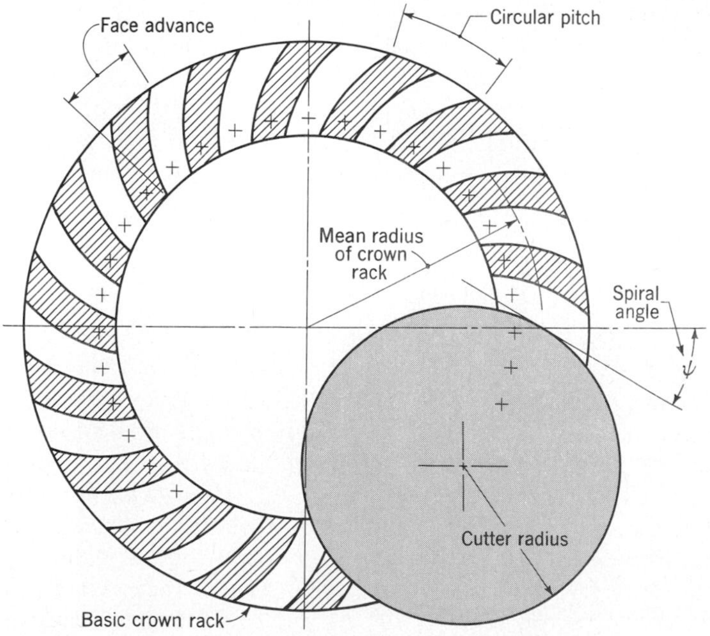

Figure 8-17 Cutting spiral-gear teeth on the basic crown rack.

gears give a much smoother tooth action than straight bevel gears and hence are useful where high speeds are encountered. In order to obtain true spiral-tooth action the face contact ratio should be at least 1.25.

Pressure angles used with spiral bevel gears are generally $14\frac{1}{2}$ to 20°, while the spiral angle is about 30 or 35°. As far as the tooth action is concerned, the hand of the spiral may be either right or left; it makes no difference. However, looseness in the bearings might result in the teeth's jamming or separating, depending upon the direction of rotation and the hand of the spiral. Since jamming of the teeth would do the most damage, the hand of the spiral should be such that the teeth will tend to separate.

Zerol bevel gears The Zerol bevel gear is a patented gear which has curved teeth but a zero-degree spiral angle. It has no advantage in tooth action over the straight bevel gear and is designed simply to take advantage of the cutting machinery used for cutting spiral bevel gears.

8-12 HYPOID GEARS

It is frequently desirable, as in the case of automotive-differential applications, to have a gear similar to a bevel gear but with the shafts offset. Such gears are called hypoid gears because their pitch surfaces are hyper-

Figure 8-18 Hypoid gears. (*Gleason Works, Rochester, N.Y.*)

boloids of revolution. The tooth action between such gears is a combination of rolling and sliding along a straight line and has much in common with that of worm gears. Figure 8-18 shows a pair of hypoid gears.

PROBLEMS

8-1 A pair of parallel helical gears has $14\frac{1}{2}°$ normal pressure angle, 6 diametral pitch, and 45° helix angle. The pinion has 15 teeth, and the gear 24 teeth. Calculate the transverse and normal circular pitch, the normal diametral pitch, the pitch diameters, and the equivalent tooth numbers.

8-2 A pair of parallel helical gears are cut with a 20° normal pressure angle and a 30° helix angle. They are 16 diametral pitch and have 16 and 40 teeth, respectively. Find the transverse pressure angle, the normal circular pitch, the axial pitch, and the pitch radii of the equivalent spur gears.

8-3 A parallel helical gearset is made with a 20° transverse pressure angle and a 35° helix angle. The gears are 10 diametral pitch and have 15 and 25 teeth, respectively. If the face width is $\frac{3}{4}$ in, calculate the base helix angle and the axial contact ratio.

8-4 A pair of helical gears is to be cut for parallel shafts whose center distance is to be about $3\frac{1}{2}$ in to give a velocity ratio of approximately 1.80. The gears are to be cut with a standard 20°-pressure-angle hob whose diametral pitch is 8. Using a helix angle of 30°, determine the transverse values of the diametral and circular pitch and the tooth numbers, pitch diameters, and center distance.

8-5 A 16-tooth helical pinion is to run at 1800 rpm and drive a helical gear on a parallel shaft at 400 rpm. The centers of the shafts are to be spaced 11 in apart. Using a helix angle of 23° and a pressure angle of 20°, determine values for the tooth numbers, pitch diameters, normal circular and diametral pitch, and face width.

8-6 The catalog description of a pair of helical gears is as follows: $14\frac{1}{2}°$ normal pressure angle, 45° helix angle, 8 diametral pitch, 1-in face width, 11.31 normal diametral pitch. The pinion has 12 teeth and a 1.500-in pitch diameter, and the gear 32 teeth and 4.000-in pitch diameter. Both gears have full-depth teeth, and they may be purchased either right- or left-handed. If a right-hand pinion and a left-hand gear are placed in mesh, find the transverse contact ratio, the normal contact ratio, the axial contact ratio, and the total contact ratio.

8-7 In a medium-sized truck transmission a 22-tooth clutch-stem gear meshes continuously with a 41-tooth countershaft gear. The data are 7.6 normal diametral pitch, $18\frac{1}{2}°$ normal pressure angle, $23\frac{1}{2}°$ helix angle, and 1.12-in face width. The clutch-stem gear is cut with a left-hand helix and the countershaft gear with a right-hand helix. Determine the normal and total contact ratio if the teeth are cut full-depth with respect to the normal diametral pitch.

8-8 A helical pinion is right-hand, has 12 teeth, a 60° helix angle, and is to drive another gear at a velocity ratio of 3. The shafts are at a 90° angle, and the normal diametral pitch of the gears is 8. Find the helix angle and the number of teeth on the mating gear. What is the center distance?

8-9 A right-hand helical pinion is to drive a gear at a shaft angle of 90°. The pinion has 6 teeth and a 75° helix angle and is to drive the gear at a velocity ratio of 6.5. The normal diametral pitch of the gears is 12. Calculate the helix angle and the number of teeth on the mating gear. Also, determine the pitch diameter of each gear.

8-10 Gear 2 in the figure (p. 304) is to rotate clockwise and drive gear 3 counterclockwise at a velocity ratio of 2. Use a normal diametral pitch of 5, a center distance of about 10 in, and the same helix angle for both gears. Find the tooth numbers, the helix angles, and the exact center distance.

8-11 A worm having 4 teeth and a lead of 1 in drives a worm gear at a velocity ratio of $7\frac{1}{2}$. Determine the pitch diameters of the worm and worm gear for a center distance of $1\frac{3}{4}$ in.

8-12 Specify a suitable worm and worm-gear combination for a velocity ratio of 60 and a center distance of $6\frac{1}{2}$ in. Use an axial pitch of 0.500 in.

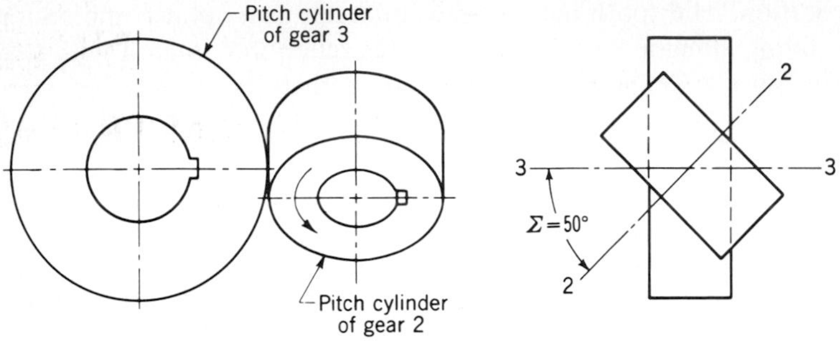

Problem 8-10

8-13 A 3-tooth worm drives a worm gear having 40 teeth. The axial pitch is $1\frac{1}{4}$ in, and the pitch diameter of the worm is $1\frac{3}{4}$ in. Calculate the lead and lead angle of the worm. Find the helix angle and pitch diameter of the worm gear.

8-14 A pair of straight-tooth bevel gears is to be manufactured for a shaft angle of 90°. If the driver is to have 18 teeth and the velocity ratio is 3, what are the pitch angles?

8-15 A pair of straight-tooth bevel gears has a velocity ratio of 1.5 and a shaft angle of 75°. What are the pitch angles?

8-16 A pair of straight bevel gears is to be mounted at a shaft angle of 120°. The pinion and gear are to have 15 and 33 teeth, respectively. What are the pitch angles?

8-17 A pair of 2-diametral-pitch straight bevel gears have 19 teeth and 28 teeth, respectively. The shaft angle is 90°. Determine the pitch diameters, pitch angles, addendum, dedendum, face width, and the pitch diameters of the equivalent spur gears.

8-18 A pair of 8-diametral-pitch straight bevel gears has 17 teeth and 28 teeth, respectively, and a shaft angle of 105°. For each gear calculate the pitch diameter, pitch angle, addendum, dedendum, face width, and the equivalent tooth numbers. Make a sketch of the two gears in mesh. Use the standard tooth proportions as for a 90° shaft angle.

CHAPTER
NINE

MECHANISM TRAINS

Mechanisms arranged in various series and parallel combinations so that the driven member of one mechanism is the driver for another mechanism are called *mechanism trains.* With certain exceptions, to be explored, the analysis of such trains can proceed in chain fashion by using the methods of analysis developed in the previous chapters.

9-1 PARALLEL-AXIS GEAR TRAINS AND DEFINITIONS

In Chap. 3 we learned that *angular-velocity ratio* is a term used to describe the quantity that results when the angular velocity of a driven member is divided by the angular velocity of the driving member. Thus, in a four-bar linkage, with link 2 as the driving, or input, member and link 4 considered as the driven, or output, member, the angular-velocity ratio is

$$e = \frac{\omega_{41}}{\omega_{21}} \tag{a}$$

In this chapter we drop the second subscripts in Eq. (*a*) to simplify the notation. Also, in the case of gearing, it is more convenient to deal with speed, and so we shall employ the symbol n to describe the speed in revolutions per minute (rpm) or in some cases in revolutions per second (r/s or 1/s). Thus we prefer to write Eq. (*a*) as

$$e = \frac{n_L}{n_F} \tag{9-1}$$

where n_L is the speed of the *last* gear in a train and n_F is the speed of the first gear in the same train. Usually the last gear is the output and is the driven gear, and the first is the driving, or input, gear.

The term e defined by Eq. (9-1) is called *speed ratio* by some and *train value* by others. Both terms are completely appropriate. The equation is often written in the more convenient form

$$n_L = en_F \tag{9-2}$$

Now consider a pinion 2 driving a gear 3. The speed of the driven gear is

$$n_3 = \frac{N_2}{N_3} n_2 = \frac{d_2}{d_3} n_2 \tag{b}$$

where N is the number of teeth, d is the pitch diameter, and n may be either the revolutions per minute or the total number of turns. For parallel-shaft gearing, the directions can be kept track of by specifying that the speed is positive or negative, depending upon whether the direction is counterclockwise or clockwise. This approach fails when the gear shafts are not parallel to each other, as in bevel, crossed-helical, or worm gearing, for example. For these reasons it is often simpler to keep track of the directions by using a sketch of the train.

The gear train shown in Fig. 9-1 is made up of five gears. Using Eq. (b) in chain fashion, we find the speed of gear 6 to be

$$n_6 = \frac{N_2}{N_3}\frac{N_4}{N_5}\frac{N_5}{N_6} n_2 \tag{c}$$

Here we notice that gear 5 is an idler that its tooth numbers cancel in Eq. (c), and hence the only purpose served by gear 5 is to change the direction of rotation of gear 6. We furthermore notice that gears 2, 4, and 5 are *drivers*, while 3, 5, and 6 are *driven members*. Thus Eq. (9-1) can also be written

$$e = \frac{\text{product of driving tooth numbers}}{\text{product of driven tooth numbers}} \tag{9-3}$$

Note from Eq. (b) that pitch diameters can be used in Eq. (9-3) as well. For parallel-shaft gearing we shall use the following *sign convention*. If the last gear rotates in the same sense as the first gear, e is made positive; if the last gear rotates in the opposite sense to the first gear, e is made negative.

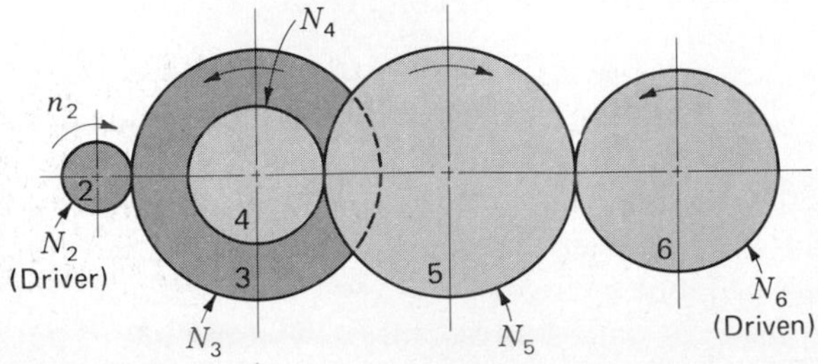

Figure 9-1

9-2 EXAMPLES OF GEAR TRAINS

In speaking of gear trains it is often convenient to describe a *simple gear train* as one having only one gear on each axis. Then a *compound gear train* is one which, like that in Fig. 9-1, has two or more gears on one or more axes.

Figure 9-2 shows a transmission for small- and medium-sized trucks; it has four speeds forward and one in reverse.

The gear train shown in Fig. 9-3 is composed of bevel, helical, and spur gears. The helical gears are crossed, and so the direction of rotation depends upon the hand of the helical gears.

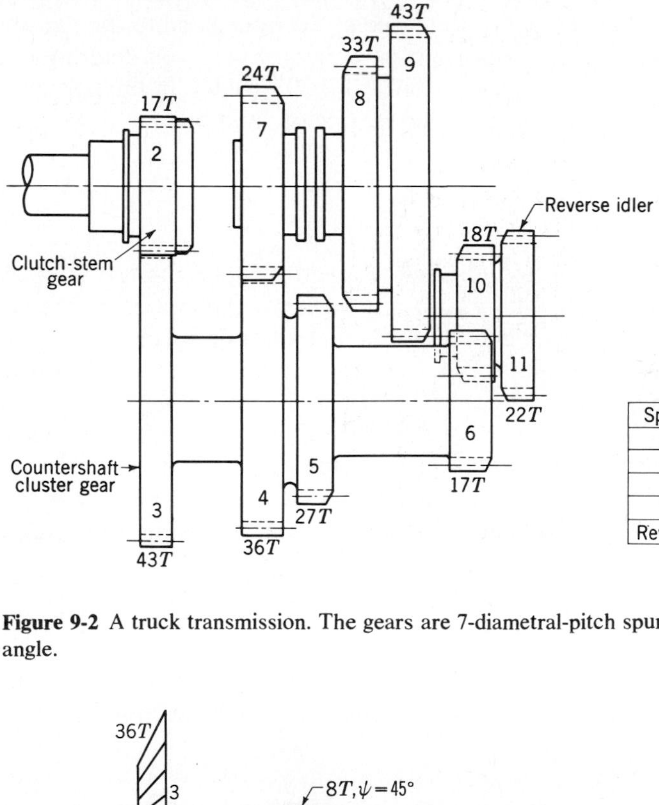

Speed	Drive
1	2-3-6-9
2	2-3-5-8
3	2-3-4-7
4	Straight through
Reverse	2-3-6-10-11-9

Figure 9-2 A truck transmission. The gears are 7-diametral-pitch spur gears with a 22.5° pressure angle.

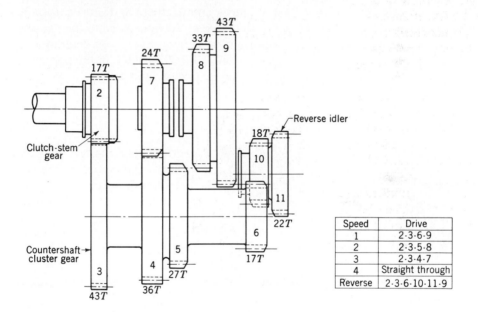

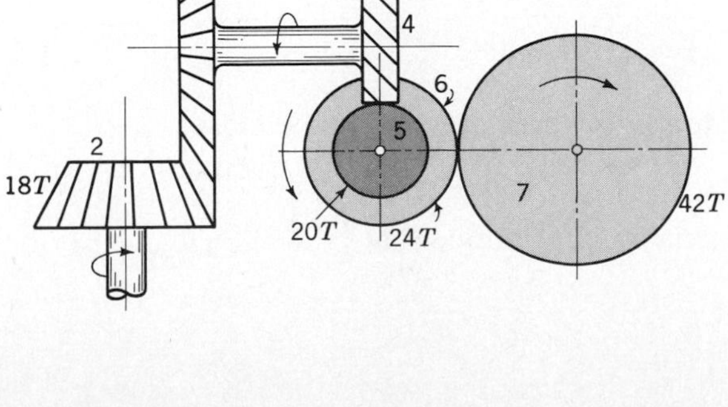

Figure 9-3 A gear train composed of bevel, crossed-helical, and spur gears.

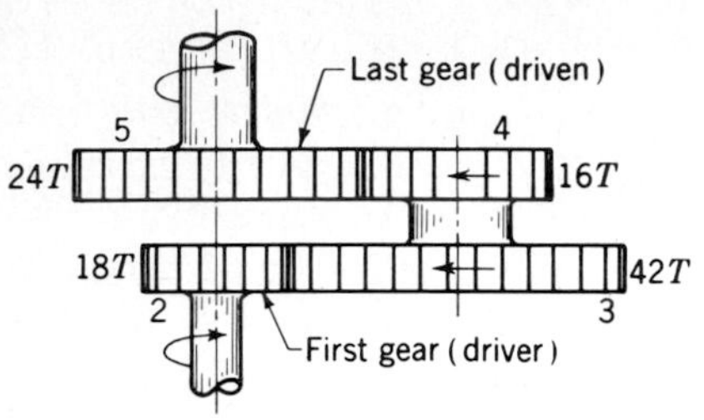

Figure 9-4 A reverted gear train.

A *reverted gear train* (Fig. 9-4) is one in which the first and last gears are on the same axis. This arrangement produces compactness and is used in such applications as speed reducers, clocks (to connect the hour hand to the minute hand), and machine tools. As an exercise it is suggested that you determine a suitable set of diametral pitches for each pair of gears shown in the figure so that the first and last gears will have the same axis of rotation.

9-3 DETERMINING TOOTH NUMBERS

If much power is being transmitted through a speed-reduction unit, the pitch of the last pair of mating gears will be larger than that of the first pair because the torque is greater at the output end. In a given amount of space more teeth can be used on smaller-pitch gears; hence a greater speed reduction can be obtained at the high-speed end.

Without going into the problem of tooth strength, suppose we wish to use two pairs of gears in a train to obtain a train value of $\frac{1}{12}$. Let us also impose the restriction that the tooth numbers must not be less than 15 and that the reduction obtained in the first pair of gears should be about twice that obtained in the second pair. This means that

$$e = \frac{N_2}{N_3}\frac{N_4}{N_5} = \frac{1}{12} \tag{a}$$

where N_2/N_3 is the train value of the first pair, and N_4/N_5 that of the second. Since the train value of the first pair must be half that of the second,

$$\frac{N_4}{2N_5}\frac{N_4}{N_5} = \frac{1}{12} \tag{b}$$

or

$$\frac{N_4}{N_5} = \sqrt{\frac{1}{6}} = 0.4082 \tag{c}$$

to four places. The following tooth numbers are seen to be close:

$$\frac{15}{37} \qquad \frac{16}{39} \qquad \frac{18}{44} \qquad \frac{20}{49} \qquad \frac{22}{54}$$

Of these, $\frac{20}{49}$ is the best approximation, but notice that

$$e = \frac{N_2}{N_3}\frac{N_4}{N_5} = \frac{20}{98}\frac{20}{49} = \frac{200}{2401}$$

which is not quite $\frac{1}{12}$. On the other hand, the combination of $\frac{22}{108}$ for the first reduction and $\frac{18}{44}$ for the second gives a train value of exactly $\frac{1}{12}$. Thus

$$e = \left(\frac{22}{108}\right)\left(\frac{18}{44}\right) = \frac{1}{12}$$

In this case, the reduction in the first pair is not exactly twice that in the second. However, this consideration is usually of only minor importance.

The problem of specifying the tooth numbers and the number of pairs of gears to give a train value within any specified degree of accuracy has interested many persons. Consider, for instance, the problem of specifying a set of gears to have a train value of $\pi/10$ accurate to eight decimal places.

9-4 EPICYCLIC GEAR TRAINS

Figure 9-5 shows the elementary epicyclic gear train together with the simplified designation for these used by Lévai.† The train consists of a *central gear* 2 and an *epicyclic gear* 4, which produces epicyclic motion by rolling around the periphery of the central gear. A *crank arm* 3 contains the bearings for the epicyclic gear to maintain the two gear wheels in mesh.

These trains are also called *planetary* or *sun-and-planet* gear trains. In this nomenclature gear 2 of Fig. 9-5 is the *sun gear*, gear 4 is the *planet gear*, and crank 3 is called the *planet carrier*. Figure 9-6 shows the train of Fig. 9-5 with two redundant planet gears added. This produces a better force balance because adding more planet gears increases the number of forces; but the additional planets contribute nothing to the kinematic performance. For this reason we shall generally show only single planet gears in the illustrations and problems in this book even though an actual machine would probably be constructed with the planets in trios.

The simple epicyclic gear train together with the corresponding simplified

† Literature devoted to the subject of epicyclic gear trains is indeed scarce. For a comprehensive study in English see Z. L. Lévai, "Theory of Epicyclic Gears and Epicyclic Change-Speed Gears," Technical University of Building, Civil and Transport Engineering, Budapest, 1966. This book lists 104 references.

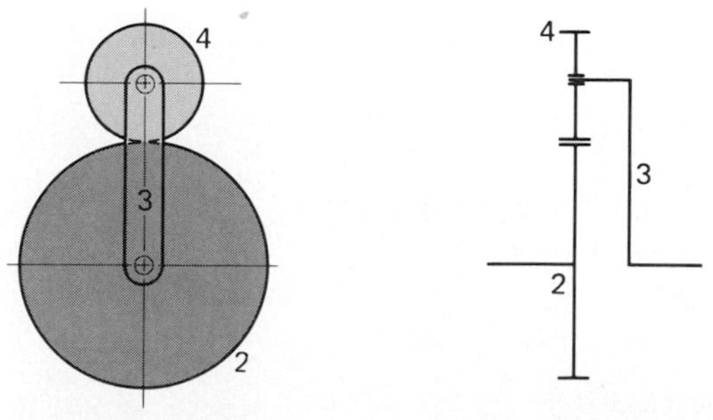

(a) (b)

Figure 9-5 (*a*) The elementary epicyclic gear; (*b*) simplified designation.

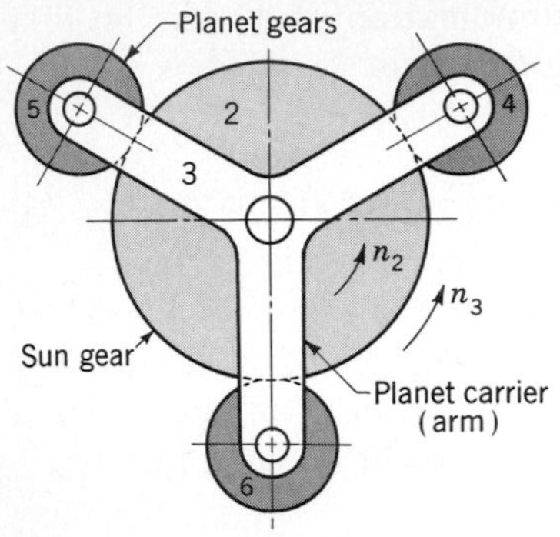

Figure 9-6 A planetary gearset.

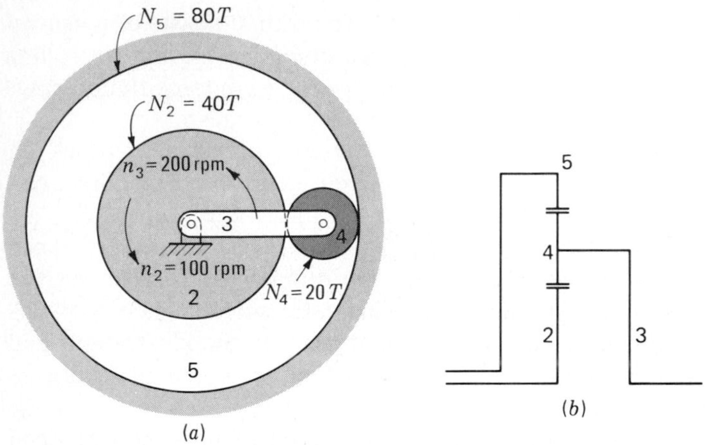

Figure 9-7 (*a*) The simple epicyclic gear train; (*b*) simplified designation.

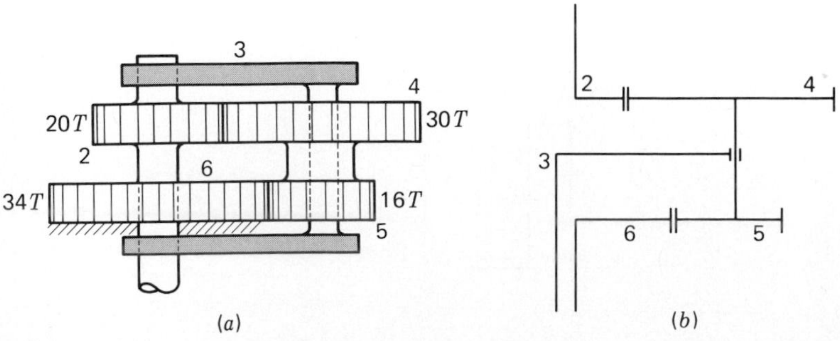

Figure 9-8 Simple epicyclic gear train with double planets.

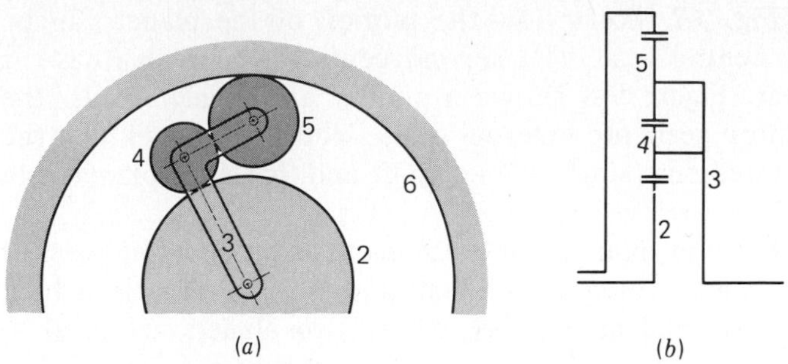

Figure 9-9 Epicyclic train with two planets.

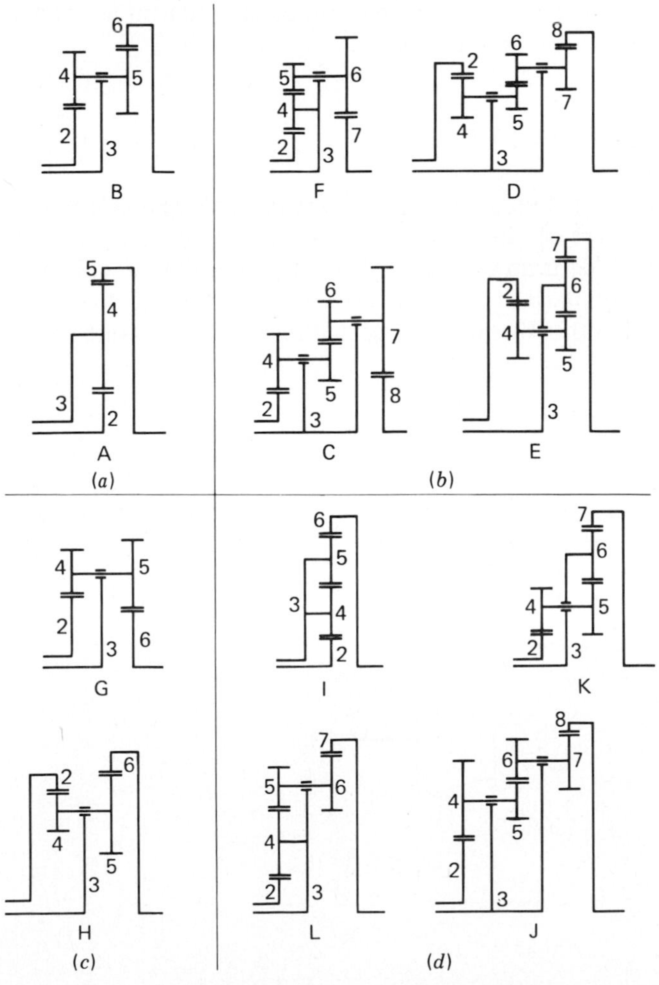

Figure 9-10 All 12 possible epicyclic gear types according to Lévai.

designation shown in Fig. 9-7 shows how the motion of the planet can be transmitted to another central gear. The second central gear in this case is gear 5, an internal gear. Figure 9-8 shows a similar arrangement with the difference that both central gears are external gears. Note, in Fig. 9-8, that the double planets are mounted on a single planet shaft and that each planet is in mesh with a sun gear.

In any case, no matter how many planets are used, only one planet carrier or arm may be used. This principle is illustrated in Fig. 9-6, in which redundant planets are used, and in Fig. 9-9, where two planets are used to alter the kinematic performance.

According to Lévai, 12 variations are possible; they are all shown in simplified form in Fig. 9-10 as Lévai arranged them. Those in Fig. 9-10*a* and *c* are the simple trains in which the planets mesh with both sun gears. The trains shown in Fig. 9-10*b* and *d* have planet pairs that are partly in mesh with each other and partly in mesh with the sun gears.

9-5 BEVEL-GEAR EPICYCLIC TRAINS

The bevel-gear train shown in Fig. 9-11 is called *Humpage's reduction gear.* Bevel-gear epicyclic trains are used quite frequently but they are the same as spur-gear epicyclic trains. The train of Fig. 9-11 is, in fact, a double epicyclic train, and the spur-gear counterpart of each can be found in Fig. 9-10. We shall find in the next section that the analysis of such trains is the same as for spur-gear trains.

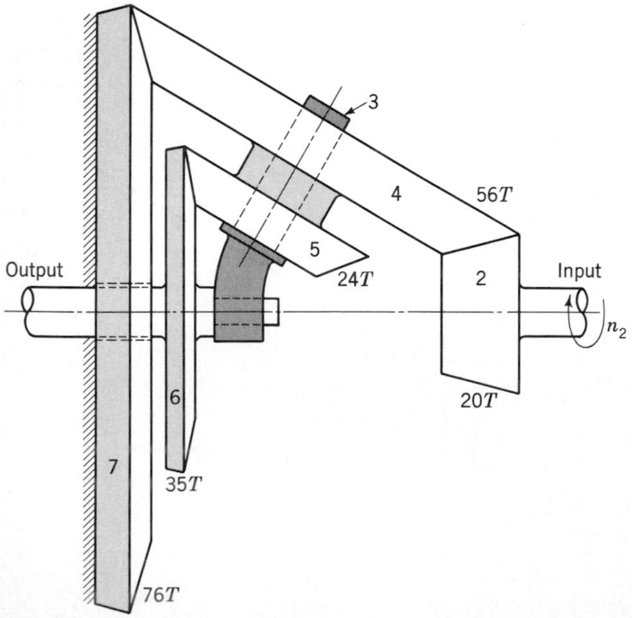

Figure 9-11

9-6 SOLUTION OF PLANETARY TRAINS BY FORMULA

Figure 9-12 shows a planetary gear train which is composed of a sun gear 2, and arm 3, and planet gears 4 and 5. Using (3-10), we can write that the velocity of gear 2 relative to the arm is

$$n_{23} = n_2 - n_3 \tag{a}$$

Also, the speed of gear 5 relative to the arm is

$$n_{53} = n_5 - n_3 \tag{b}$$

Dividing Eq. (*b*) by (*a*) gives

$$\frac{n_{53}}{n_{23}} = \frac{n_5 - n_3}{n_2 - n_3} \tag{c}$$

Equation (*c*) expresses the ratio of the relative velocity of gear 5 to that of gear 2, and both velocities are taken relative to the arm. This ratio is the same and is proportional to the tooth numbers, whether the arm is rotating or not. It is the train value. Therefore we can write

$$e = \frac{n_5 - n_3}{n_2 - n_3} \tag{d}$$

Equation (*d*) is all that we need in order to solve any planetary train. It is convenient to express it in the form

$$e = \frac{n_L - n_A}{n_F - n_A} \tag{9-4}$$

where n_F = speed of first gear in train, rpm
n_L = speed of last gear in train, rpm
n_A = speed of arm, rpm

The following examples will illustrate the use of Eq. (9-4).

Example 9-1 Figure 9-8 shows a reverted planetary train. Gear 2 is fastened to its shaft and is driven at 250 rpm in a clockwise direction. Gears 4 and 5 are planet gears which are joined but are free to turn on the shaft carried by the arm. Gear 6 is stationary. Find the speed and direction of rotation of the arm.

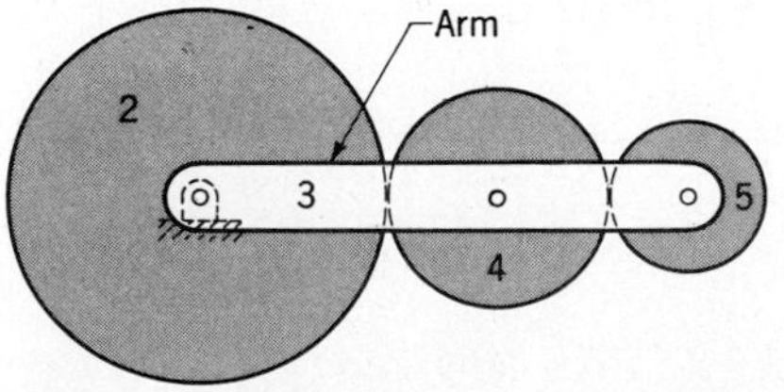

Figure 9-12

SOLUTION We must first decide which gears to designate as the first and last members of the train. Since the speeds of gears 2 and 6 are given, either may be used as the first. The choice makes no difference in the results, but once the decision is made, it cannot be changed. We shall choose gear 2 as the first; then gear 6 will be the last. Thus

$$n_F = n_2 = -250 \text{ rpm} \qquad n_L = n_6 = 0 \text{ rpm} \qquad e = \left(\tfrac{20}{30}\right)\left(\tfrac{16}{34}\right) = \tfrac{16}{51}$$

Substituting these values in Eq. (9-4) gives

$$\frac{16}{51} = \frac{0 - n_A}{-250 - n_A} \qquad n_A = 114 \text{ rpm ccw}$$

Example 9-2 In the bevel-gear train shown in Fig. 9-11 the input is to gear 2, and the output from gear 6, which is connected to the output shaft. The arm 3 turns freely on the output shaft and carries the planets 4 and 5. Gear 7 is fixed to the frame. What is the output speed if gear 2 rotates at 2000 rpm?

SOLUTION The problem is solved in two steps. In the first step we consider the train to be made up of gears 2, 4, and 7 and calculate the velocity of the arm. Thus

$$n_F = n_2 = 2000 \text{ rpm} \qquad n_L = n_7 = 0 \text{ rpm} \qquad e = \left(-\tfrac{20}{56}\right)\left(\tfrac{56}{76}\right) = -\tfrac{5}{19}$$

Substituting in Eq. (9-4) and solving for the velocity of the arm gives

$$-\frac{5}{19} = \frac{0 - n_A}{2000 - n_A} \qquad n_A = 416.7 \text{ rpm}$$

Now consider the train as composed of gears 2, 4, 5, and 6. Then $n_F = n_2 = 2000$ rpm, as before, and $n_L = n_6$, which is to be found. The train value is

$$e = \left(-\tfrac{20}{56}\right)\left(\tfrac{24}{35}\right) = -\tfrac{12}{49}$$

Substituting in Eq. (9-4) again and solving for n_L, since n_A is now known, gives

$$-\frac{12}{49} = \frac{n_L - 416.7}{2000 - 416.7}$$

$$n_L = n_6 = 28.91 \text{ rpm}$$

The output shaft rotates in the same direction as gear 2 with a reduction of 2000:28.91, or 69.2:1.

9-7 TABULAR ANALYSIS OF PLANETARY TRAINS

Figure 9-7 illustrates a planetary gear train composed of a sun gear 2, a planet carrier (arm) 3, a planet gear 4, and an internal gear 5 which is in mesh with the planet. We might reasonably give certain values for the revolutions per minute of the sun gear and of the arm and wish to determine the revolutions per minute of the internal gear.

The analysis is carried out in the following three steps:

1. Lock all the gears to the arm and rotate the arm one turn. Tabulate the resulting turns of the arm and of each gear.
2. Fix the arm and rotate one or more of the sun gears. Tabulate the resulting turns of each gear.
3. Add the turns of each gear in steps 1 and 2 so that the given conditions are satisfied.

Table 9-1 Solution by tabulation, rpm

Step number	Arm 3	Gear 2	Gear 4	Gear 5
1. Gears locked	+200	+200	+200	+200
2. Arm fixed	0	−100	+200	+ 50
3. Results	+200	+100	+400	+250

As an example of such a solution, assign tooth numbers as shown in Fig. 9-7, and also let the speed of the sun gear and arm be 100 and 200 rpm, respectively, both in the positive direction. The solution is shown in Table 9-1. In step 1 the gears are locked to the arm and the arm is given 200 counterclockwise turns. This results in 200 counterclockwise turns for gears 2, 4, and 5, too. In step 2 the arm is fixed. Now determine the turns that gear 2 must make such that when they are added to those in step 1, the result, in this case, will be +100 rpm. This is −100 turns, as shown. To complete step 2, use gear 2 as the driver and find the number of turns of gears 4 and 5. Thus

$$n_4 = (-100)(-\tfrac{40}{20}) = +200 \text{ rpm}$$

and

$$n_5 = (-100)(-\tfrac{40}{20})(\tfrac{20}{80}) = +50 \text{ rpm}$$

These values are entered in the appropriate columns, and steps 1 and 2 are added together to obtain the result.

The following worked-out examples will help give a better understanding of this method.

Example 9-3 Find the speed of the external gear of Fig. 9-7 if gear 2 rotates 100 rpm in the clockwise direction instead and arm 3 rotates 200 rpm counterclockwise.

SOLUTION The results are tabulated below. In step 1 the gears are locked to the arm and the arm is given 200 turns in a counterclockwise direction. This causes gears 2, 4, and 5 to move through 200 counterclockwise turns also.

In step 2 the arm is fixed; so we record zero for the turns of the arm in the first column. In the second column, gear 2 must run in such a way that when its turns are added to those in step 1, the result will be 100 clockwise turns. For this reason we specify −300 turns for gear 2. Now, treating gear 2 as a driver, the turns of gears 4 and 5 are:

Step number	Arm 3	Gear 2	Gear 4	Gear 5
1. Gears locked	+200	+200	+200	+200
2. Arm fixed	0	−300	+600	+150
3. Results	+200	−100	+800	+350

$$n_4 = (-300)(-\tfrac{40}{20}) = +600 \text{ rpm}$$

$$n_5 = (-300)(-\tfrac{40}{20})(\tfrac{20}{80}) = +150 \text{ rpm}$$

When the columns are summed, the result is seen to be

$$n_5 = 350 \text{ rpm ccw}$$

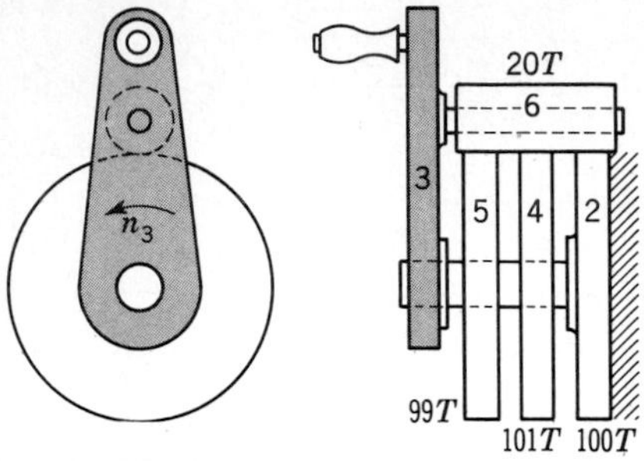

Figure 9-13 Ferguson's paradox.

Example 9-4 The planetary gear train shown in Fig. 9-13 is called Ferguson's paradox. Gear 2 is stationary by virtue of being fixed to a frame. The arm 3 and gears 4 and 5 are free to turn upon the shaft. Gears 2, 4, and 5 have tooth numbers of 100, 101, and 99, respectively, all cut from gear blanks of the same diameter so that the planet 6 meshes with all of them. Find the turns of gears 4 and 5 if the arm is given one counterclockwise turn.

SOLUTION The results are shown in the table below.

Step number	Arm 3	Gear 2	Gear 4	Gear 5
1. Gears locked	+1	+1	+1	+1
2. Arm fixed	0	−1	−100/101	−100/99
3. Results	+1	0	+1/101	−1/99

In order for gear 2 to be fixed, it must be given one clockwise turn in step 2. The results show that as the arm is turned, gear 4 rotates very slowly in the same direction, while gear 5 rotates very slowly in the opposing direction.

Example 9-5 The overdrive unit shown in Fig. 9-14 is used behind a standard transmission to reduce engine speed. Determine the percentage reduction that will be obtained when the overdrive is "in."

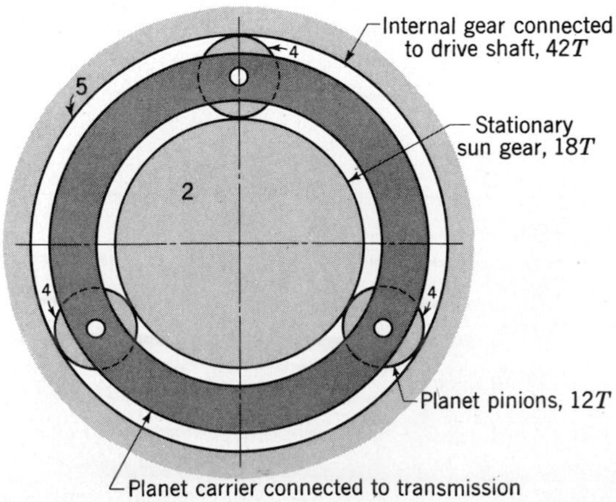

Figure 9-14 Overdrive unit.

SOLUTION It is convenient to use one turn for the arm. This gives the results shown in the table below. The engine speed corresponds to the speed of the arm, and the drive-shaft speed to that of gear 5. Thus

$$\text{Reduction in engine speed} = \frac{1.429 - 1}{1.429}(100) = 30\%$$

Step number	Arm 3	Gear 2	Gear 4	Gear 5
1. Gears locked	+1	+1	+1	+1
2. Arm fixed	0	−1	+1.5	+0.429
3. Results	+1	0	+2.5	+1.429

9-8 DIFFERENTIALS

The class of planetary gear trains known as differentials is used so widely that it deserves special attention. The operation of a differential is illustrated by the schematic drawing of the automotive differential shown in Fig. 9-15. The drive-shaft pinion and the ring gear are normally hypoid gears. The ring gear acts as the planet carrier, and its speed can be calculated as for a simple gear train when the speed of the drive shaft is given. Gears 5 and 6 are connected, respectively, to each rear wheel, and when the car is traveling in a straight line, these two gears rotate in the same direction with exactly the same speed.

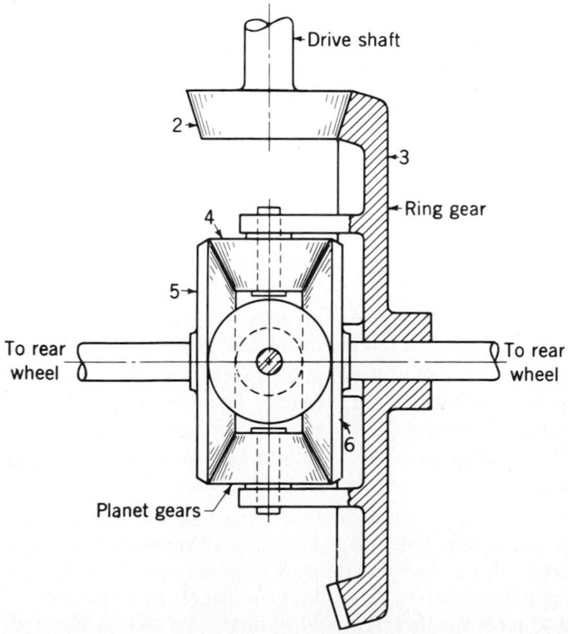

Figure 9-15 Schematic drawing of a bevel-gear automotive differential.

Thus for straight-line motion of the car, there is no relative motion between the planet gears and gears 5 and 6. The planet gears, in effect, serve only as keys to transmit motion from the planet carrier to both wheels.

When the vehicle is making a turn, the wheel on the inside of the turn makes fewer revolutions than the wheel with a longer turning radius. Unless this difference in speed is accommodated in some manner, one or both of the tires would have to slide in order to make the turn. The differential permits each wheel to rotate at different velocities while at the same time delivering power to both. During a turn, the planet gears rotate about their own axes, thus permitting gears 5 and 6 to revolve at different velocities.

The purpose of a differential is to differentiate between the speeds of the two wheels. In the usual passenger-car differential, the torque is divided equally whether the car is traveling in a straight line or on a curve. Sometimes the road conditions are such that the tractive effort developed by the two wheels is unequal. In this case the total tractive effort available will be only twice that at the wheel having the least traction, because the differential divides the torque equally. If one wheel should happen to be resting on snow or ice, the total effort available is very small and only a small torque will be required to cause the wheel to spin.

PROBLEMS

9-1 Find the speed and direction of rotation of gear 8 in the figure. What is the speed ratio of the train?

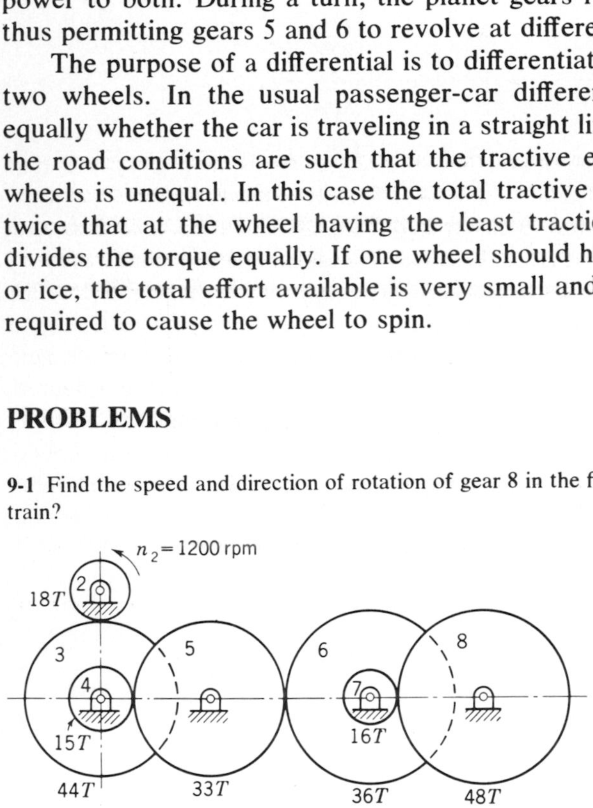

Problem 9-1

9-2 Part (a) of the figure gives the pitch diameters of a set of spur gears forming a train. Compute the speed ratio of the train. Determine the speed and direction of rotation of gears 5 and 7.

9-3 Part (b) of the figure shows a gear train consisting of bevel gears, spur gears, and a worm and worm gear. The bevel pinion is mounted on a shaft which is driven by a V belt on pulleys. If pulley 2 rotates 1200 rpm in the direction shown, find the speed and direction of rotation of gear 9.

9-4 Use the truck transmission of Fig. 9-2 and an input speed of 3000 rpm. Find the drive-shaft speed for each forward gear and for reverse gear.

9-5 The figure illustrates the gears in a speed-change gearbox used in machine-tool applications. By sliding the cluster gears on shafts B and C nine speed changes can be obtained. The problem of the machine-tool designer is to select tooth numbers for the various gears so as to produce a reasonable distribution of speeds for the output shaft. The smallest and largest gears are gears 2 and 9, respectively. Using 20 teeth and 45 teeth for these gears, determine a set of suitable tooth

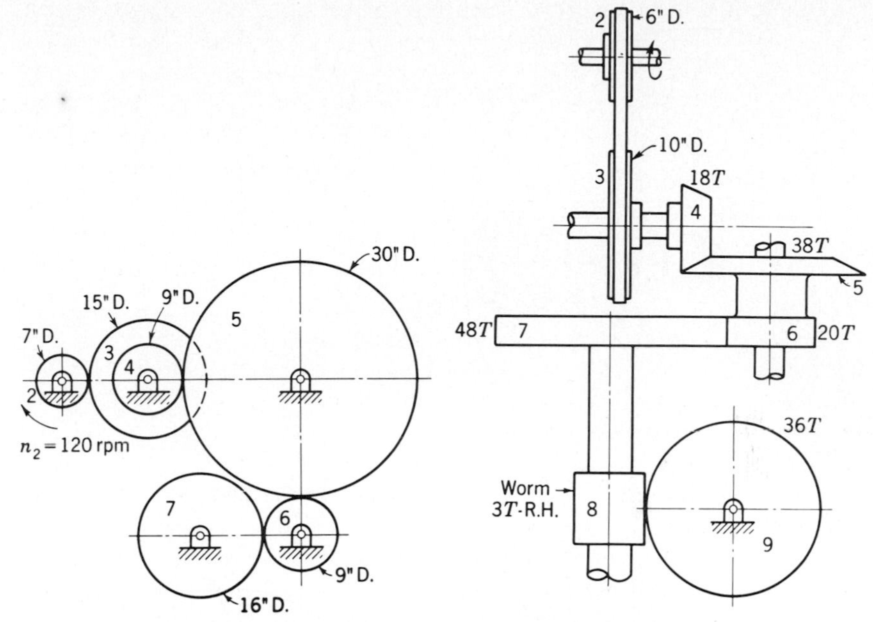

(a) (b)

Problems 9-2 and 9-3

numbers for the remaining gears. What are the corresponding speeds of the output shaft? Notice that the problem has many solutions.

9-6 The internal gear (gear 7) in the figure turns at 60 rpm ccw. What are the speed and direction of rotation of arm 3?

9-7 If the arm in the figure rotates ccw at 300 rpm, find the speed and direction of rotation of internal gear 7.

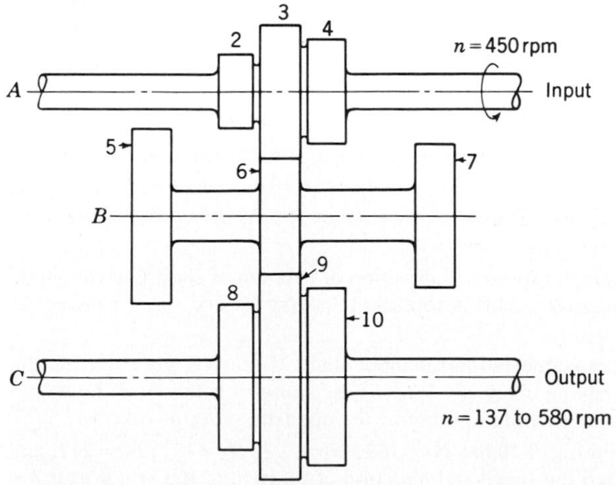

Problem 9-5

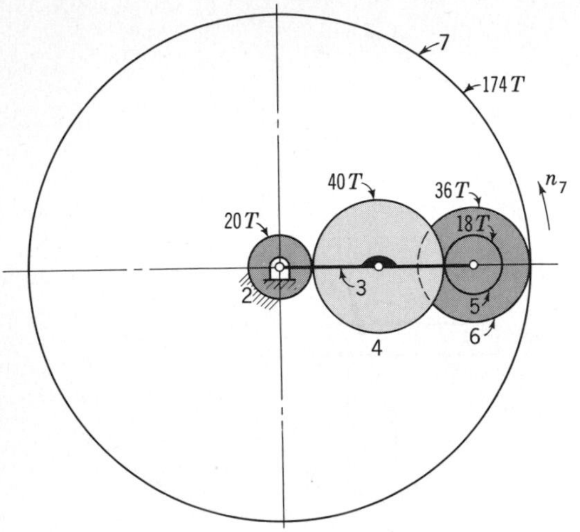

Problems 9-6 and **9-7**

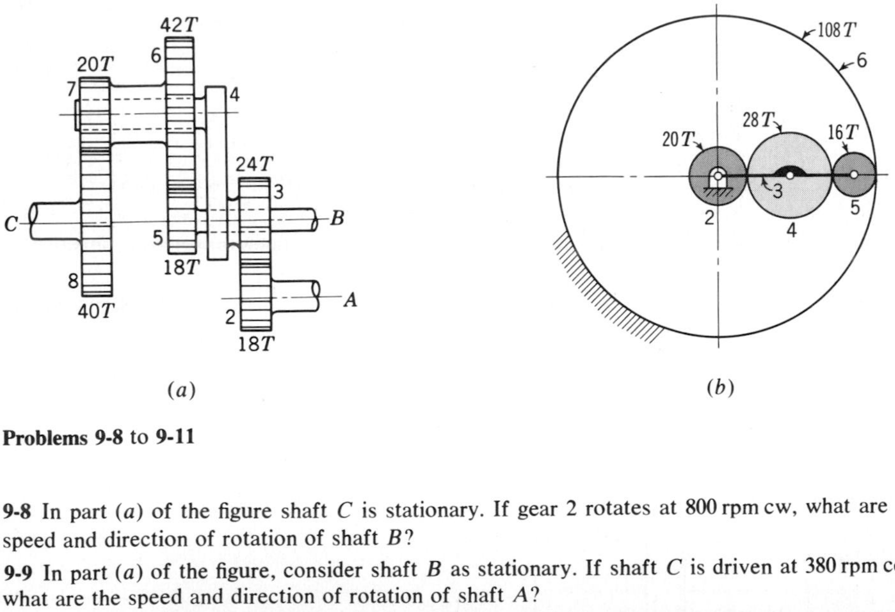

(*a*) (*b*)

Problems 9-8 to **9-11**

9-8 In part (*a*) of the figure shaft *C* is stationary. If gear 2 rotates at 800 rpm cw, what are the speed and direction of rotation of shaft *B*?

9-9 In part (*a*) of the figure, consider shaft *B* as stationary. If shaft *C* is driven at 380 rpm ccw, what are the speed and direction of rotation of shaft *A*?

9-10 In part (*a*) of the figure determine the speed and direction of rotation of shaft *C* if (*a*) shafts *A* and *B* both rotate at 360 rpm ccw and (*b*) shaft *A* rotates at 360 rpm cw and shaft *B* rotates at 360 rpm ccw.

9-11 In part (*b*) of the figure gear 2 is connected to the input shaft. If arm 3 is connected to the output shaft, what speed reduction can be obtained? What is the sense of rotation of the output shaft? What changes could be made in the train to produce the opposite sense of rotation?

9-12 The Lévai type L train shown in Fig. 9-10 has $N_2 = 16T$, $N_4 = 19T$, $N_5 = 17T$, $N_6 = 24T$, and $N_7 = 95T$. Internal gear 7 is fixed. Find the speed and direction of rotation of the arm if gear 2 is driven at 100 rpm cw.

9-13 The Lévai type A train of Fig. 9-10 has $N_2 = 20T$ and $N_4 = 32T$.

(*a*) Find the number of teeth on gear 5 and the crank arm radius if the module is 6 mm.

(*b*) What is the speed and direction of rotation of the arm if gear 2 is fixed and internal gear 5 rotates at 10 rpm ccw?

9-14 The tooth numbers for the automotive differential shown in Fig. 9-15 are $N_2 = 17$, $N_3 = 54$, $N_4 = 11$, and $N_5 = N_6 = 16$. The drive shaft turns at 1200 rpm. What is the speed of the right wheel if it is jacked up and the left wheel is resting on the road surface?

9-15 A vehicle using the differential shown in Fig. 9-15 turns to the right at a speed of 30 mi/h on a curve of 80 ft radius. Use the same tooth numbers as in Prob. 9-14. The tire diameter is 15 in. Use 60 in as the center-to-center distance between treads.

(*a*) Calculate the speed of each rear wheel.

(*b*) What is the speed of the ring gear?

CHAPTER

TEN

SYNTHESIS OF LINKAGES

By *kinematic synthesis* we mean the design or the creation of a mechanism to yield a desired set of motion characteristics. Because of the very large number of techniques available, some of which may be quite frustrating, we here present some of the more useful of the approaches to illustrate the application of the theory.†‡

10-1 TYPE, NUMBER, AND DIMENSIONAL SYNTHESIS

Type synthesis refers to the kind of mechanism selected; it might be a linkage, a geared system, belts and pulleys, or a cam system. This beginning phase of the total design problem usually involves design factors such as manufacturing processes, materials, safety, reliability, space, and economics. The study of kinematics is usually only slightly involved in type synthesis.

† Extensive bibliographies may be found in K. Hain (transl. by T. P. Goodman et al., "Applied Kinematics," 2d ed., pp. 639–727, McGraw-Hill, 1967, and Ferdinand Freudenstein and George N. Sandor, Kinematics of Mechanisms, in Harold A. Rothbart (ed.), "Mechanical Design and Systems Handbook," pp. 4-56 to 4-68, McGraw-Hill, New York, 1964.

‡ In the English language, the following are the most useful references on kinematic synthesis: Rudolf A. Beyer (transl. by Herbert Kuenzel), "Kinematic Synthesis of Mechanisms," McGraw-Hill, New York, 1963; Alexander Cowie, "Kinematics and Design of Mechanisms," International Textbook, Scranton, Pa., 1961; Hain, op. cit.; Allen S. Hall, Jr., "Kinematics and Linkage Design," Prentice-Hall, Englewood Cliffs, N.J., 1961; R. S. Hartenberg and Jacques Denavit, "Kinematic Synthesis of Linkages," McGraw-Hill, New York, 1964; Jeremy Hirschhorn, "Kinematics and Dynamics of Plane Mechanisms," McGraw-Hill, New York, 1962; D. C. Tao, "Fundamentals of Applied Kinematics," Addison-Wesley, Reading, Mass., 1967; A. H. Soni, "Mechanism Synthesis and Analysis," McGraw-Hill, 1974.

Number synthesis deals with the number of links and the number of joints or pairs that are required to obtain a certain mobility (see Sec. 1-6). Number synthesis is the second step in design following type synthesis.

The third step in design, determining the dimensions of the individual links, is called *dimensional synthesis.* This is the subject of the balance of this chapter.

10-2 FUNCTION GENERATION, PATH GENERATION, AND BODY GUIDANCE

A major classification of synthesis problems that arises in the design of linkages is called function generation. A frequent requirement in design is that of causing an output member to rotate, oscillate, or reciprocate according to a specified function of time or function of the input motion. This is called *function generation.* A simple example is that of synthesizing a four-bar linkage to generate the function $y = f(x)$. In this case x would represent the motion of the input crank and the linkage would be designed so that the motion of the output rocker would approximate the function y. Other examples of function generation follow:

1. In a conveyor line the output member of a mechanism must move at the constant velocity of the conveyor while performing some operation, e.g., bottle capping, return, pick up the next cap, and repeat the operation.
2. The output member must pause or stop during its motion cycle to provide time for another event. The second event might be a sealing, stapling, or fastening operation of some kind.
3. The output member must rotate at a specified non-uniform velocity function because it is geared to another mechanism that requires such a rotating motion.

A second type of synthesis problem is that in which a coupler point is to generate a path having a prescribed shape. Common requirements are that a portion of the path be a circle arc, elliptical, or a straight line. Sometimes it is required that the path cross over itself, as in a figure-of-eight.

The third general class of synthesis problems is called body guidance. Here we are interested in moving an object from one position to another. The problem may be a simple translation or a combination of translation and rotation. In the construction industry, for example, heavy parts such as scoops and bulldozer blades must be moved through a series of prescribed positions.

Two kinds of defects, called *branch defect* and *order defect,* may arise in synthesis to confound the designer. The branch defect refers to a developed linkage that meets all the position requirements but has coupler points on both branches of the coupler curve. The order defect refers to a developed linkage that meets all the position requirements but not in the correct order.†

10-3 PRECISION POSITIONS; CHEBYCHEV SPACING

If θ_2 is the angular position of link 2 in a four-bar linkage and θ_4 the angular position of link 4, then one of the problems in kinematic synthesis is to find the dimensions of the linkage such that

$$\theta_4 = f(\theta_2) \tag{a}$$

where f is any desired functional relation.

Though this problem has not been solved, it is possible to specify up to five values for θ_2, called *precision points*, and sometimes find a linkage which will satisfy the desired relationship for these points. The process usually consists in plotting a graph of the function and then selecting two to five precision points from the graph for use in the synthesis. If the process is successful, the functional relation is satisfied for these points, but deviations will occur elsewhere. For many functions the greatest error can be held to less than 4 percent.

Between the points deviations, called *structural errors*, will occur. One of the problems of linkage design is to select a set of precision points for use in the synthesis that will minimize the structural error.

For the first trial the best spacing of these points is called *Chebychev spacing*. For n points in the range $x_0 \leq x \leq x_{n+1}$ the Chebychev spacing, according to Freudenstein and Sandor,‡ is

$$x_j = \frac{1}{2}(x_0 + x_{n+1}) - \frac{1}{2}(x_{n+1} - x_0)\cos\frac{\pi(2j-1)}{2n} \qquad j = 1, 2, \ldots, n \tag{10-1}$$

where x_j are the precision points.

As an example, suppose we wish to devise a linkage to generate the function

$$y = x^{0.8} \tag{b}$$

for the range $1 \leq x \leq 3$ using three precision points. Then, from Eq. (10-1), the three values of x are

$$x_1 = \frac{1}{2}(1+3) - \frac{1}{2}(3-1)\cos\frac{\pi(2-1)}{(2)(3)} = 2 - \cos\frac{\pi}{6} = 1.134$$

$$x_2 = 2 - \cos\frac{3\pi}{6} = 2.000$$

$$x_3 = 2 - \cos\frac{5\pi}{6} = 2.866$$

† See K. J. Waldron and E. N. Stevensen, Jr., Elimination of Branch, Grashof, and Order Defects in Path-Angle Generation and Function Generation Synthesis, ASME Paper No. 78-DET-16.

‡ Op. cit., p. 4-27.

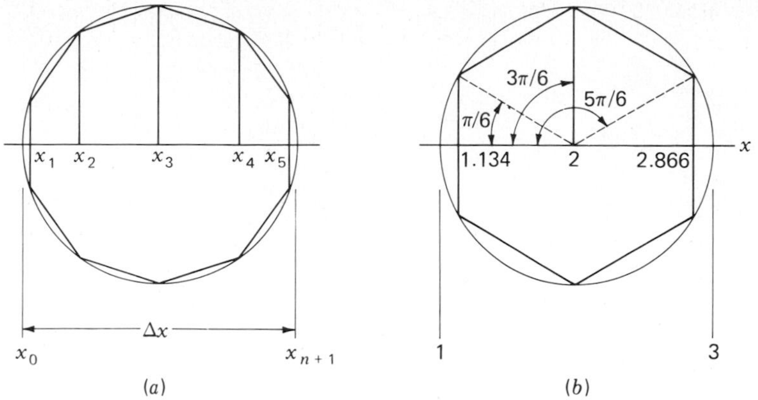

Figure 10-1 Graphical determination of Chebychev spacing.

We find the corresponding values of y, from Eq. (b), to be

$$y_1 = 1.106 \qquad y_2 = 1.741 \qquad y_3 = 2.322$$

These accuracy points are easily obtained by using the graphical approach in Fig. 10-1. The method is shown in Fig. 10-1a, where a circle is first constructed whose diameter is the range Δx, given by the equation

$$\Delta x = x_{n+1} - x_0 \tag{c}$$

In this circle inscribe a regular polygon having $2n$ sides. Perpendiculars dropped from each corner will intersect Δx in the precision points. Figure 10-1b illustrates the construction for the numerical example.

In closing this section it should be noted that the Chebychev spacing is the best first approximation; depending upon the accuracy requirements of the problem, it may be satisfactory. If additional accuracy is required, then by plotting a curve of the structural error vs. x one can usually determine visually the adjustments to be made in the precision points for the next trial.

10-4 POSITION SYNTHESIS OF THE GENERAL SLIDER-CRANK MECHANISM

The centered slider-crank mechanism of Fig. 10-2a has a stroke B_1B_2 equal to twice the crank radius r_2. As shown, the extreme positions B_1 and B_2, also called limit positions, of the slider are found by constructing circle arcs through O_2 of length $r_3 - r_2$ and $r_3 + r_2$, respectively.

In general, the centered slider-crank mechanism must have r_3 larger than r_2. However, the special case of $r_3 = r_2$ results in the *isosceles slider-crank mechanism* in which the slider reciprocates through O_2 and the stroke is 4 times the crank radius. All points on the coupler of the isosceles slider crank generate elliptic paths. The paths generated by points on the coupler of the

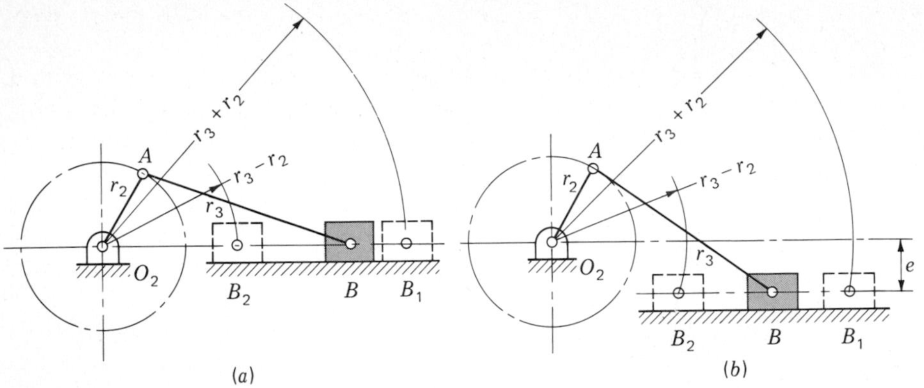

Figure 10-2 (a) Centered slider-crank mechanism; (b) general or offset slider-crank mechanism.

slider crank of Fig. 10-2a are not elliptical, but they are always symmetrical about the sliding axis O_2B.

The linkage of Fig. 10-2b is called the *general* or *offset slider-crank mechanism.* Certain special effects can be obtained by changing the offset distance e. For example, the stroke B_1B_2 is always greater than twice the crank radius. Also, the crank angle required to execute the forward stroke is different from that for the backward stroke. This feature can be used to synthesize quick-return mechanisms where a slower-working stroke is desired. Note from Fig. 10-2b that the limiting positions B_1 and B_2 of the slider are found in the same manner as for the centered slider-crank.

10-5 SYNTHESIS OF CRANK-AND-ROCKER MECHANISMS

The limiting positions of the rocker in a crank-and-rocker mechanism are shown as points B_1 and B_2 in Fig. 10-3. Note that these positions are found

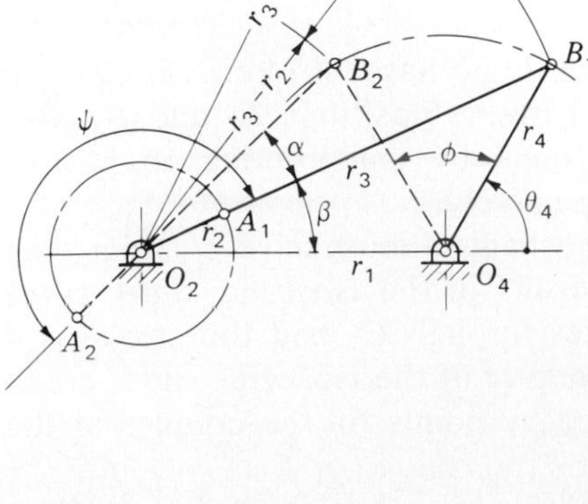

Figure 10-3 The extreme positions of the crank-and-rocker mechanism.

the same as for the slider-crank linkage. Note too that the crank and coupler form a single straight line at each extreme position.

In this particular case the crank executes the angle ψ while the rocker moves from B_1 to B_2 through the angle ϕ. Note, on the back stroke the rocker swings from B_2 back to B_1 through the same angle ϕ but the crank moves through the angle $360° - \psi$.

There are many cases in which a crank-and-rocker mechanism is superior to a cam-and-follower system. Among the advantages over cam systems are smaller forces involved, the elimination of the retaining spring, and the closer clearances because of the use of revolute pairs.

If $\psi > 180°$ in Fig. 10-3, then $\alpha = \psi - 180°$, where α can be obtained from the equation for the *time ratio* (see Sec. 1-12)

$$Q = \frac{180° + \alpha}{180° - \alpha} \tag{10-2}$$

of the forward and backward motions of the rocker. The first problem that arises in the synthesis of crank-and-rocker linkages is how to obtain the dimensions or geometry that will cause the mechanism to generate a specified output angle ϕ when the time ratio is specified.†

To synthesize a crank-and-rocker mechanism for specified values of ϕ and α, locate point O_4 in Fig. 10-4*a* and choose any desired rocker length r_4. Then draw the two positions O_4B_1 and O_4B_2 of link 4 separated by the angle ϕ as given. Through B_1 construct any line X. Then, through B_2 construct the line Y at the given angle α to X. The intersection of these two lines defines the location of the crank pivot O_2. Since any line X was originally chosen, there are an infinite number of solutions to this problem.

Next, as shown in Figs. 10-3 and 10-4*a*, the distance B_2C is $2r_2$, twice the

† The method to be described appears in Hall, op. cit., p. 33, and Soni, op. cit., p. 257. Both Tao, op. cit., p. 241, and Hain, op. cit., p. 317, describe another method that gives different results.

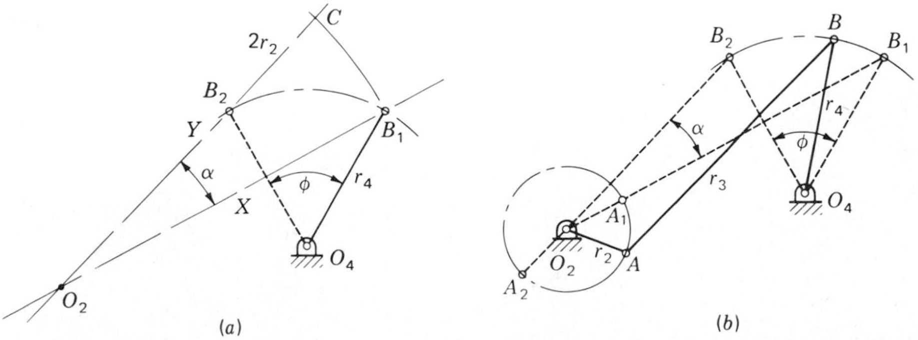

Figure 10-4 Synthesis of a four-bar linkage to generate the rocker angle ϕ.

crank length. So, bisect this distance to find r_2. Then the coupler length is $r_3 = O_2B_1 - r_2$. The completed linkage is shown in Fig. 10-4*b*.

10-6 CRANK-ROCKER MECHANISMS WITH OPTIMUM TRANSMISSION ANGLE

Brodell and Soni† have developed an analytical method of synthesizing the crank-rocker linkage in which the time ratio is $Q = 1$. The design also satisfies the condition

$$\gamma_{min} = 180° - \gamma_{max} \tag{a}$$

where γ is the transmission angle (see Sec. 1-10).

To develop the method, use Fig. 10-3 and the law of cosines to write the two equations

$$\cos(\theta_4 + \phi) = \frac{r_1^2 + r_4^2 - (r_2 + r_3)^2}{2r_1r_4} \tag{b}$$

$$\cos\theta_4 = \frac{r_1^2 + r_4^2 - (r_3 - r_2)^2}{2r_1r_4} \tag{c}$$

Then, from Fig. 10-5,

$$\cos\gamma_{min} = \frac{r_3^2 + r_4^2 - (r_1 - r_2)^2}{2r_3r_4} \tag{d}$$

$$\cos\gamma_{max} = \frac{r_3^2 + r_4^2 - (r_1 + r_2)^2}{2r_3r_4} \tag{e}$$

† R. Joe Brodell and A. H. Soni, Design of the Crank-Rocker Mechanism with Unit Time Ratio, *J. Mech.*, vol. 5, no. 1, p. 1, 1970.

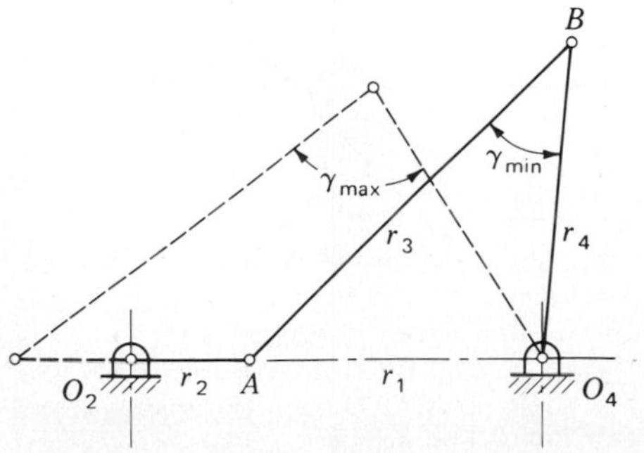

Figure 10-5

Equations (a) to (e) are now solved simultaneously; the results are the link ratios

$$\frac{r_3}{r_1} = \sqrt{\frac{1-\cos\phi}{2\cos^2\gamma_{min}}} \tag{10-3}$$

$$\frac{r_4}{r_1} = \sqrt{\frac{1-(r_3/r_1)^2}{1-(r_3/r_1)^2\cos^2\gamma_{min}}} \tag{10-4}$$

$$\frac{r_2}{r_1} = \sqrt{\left(\frac{r_3}{r_1}\right)^2 + \left(\frac{r_4}{r_1}\right)^2 - 1} \tag{10-5}$$

Brodell and Soni plot these results as a design chart, as shown in Fig. 10-6. They state that the transmission angle should be larger than 30° for a good "quality" motion and even larger if high speeds are involved.

The synthesis of a crank-rocker mechanism for optimum transmission angle when the time-ratio is not unity is more difficult. An orderly method of accomplishing this is explained by Hall† and also by Soni.‡ The first step in the approach is illustrated in Fig. 10-7. Here the two points O_2 and O_4 are selected and the points C and C', symmetrical about O_2O_4 and defined by the angles $(\phi/2)-\alpha$ and $\phi/2$, are found. Then, using C as a center and the

† Op. cit., pp. 36-42.
‡ Op. cit., p. 258.

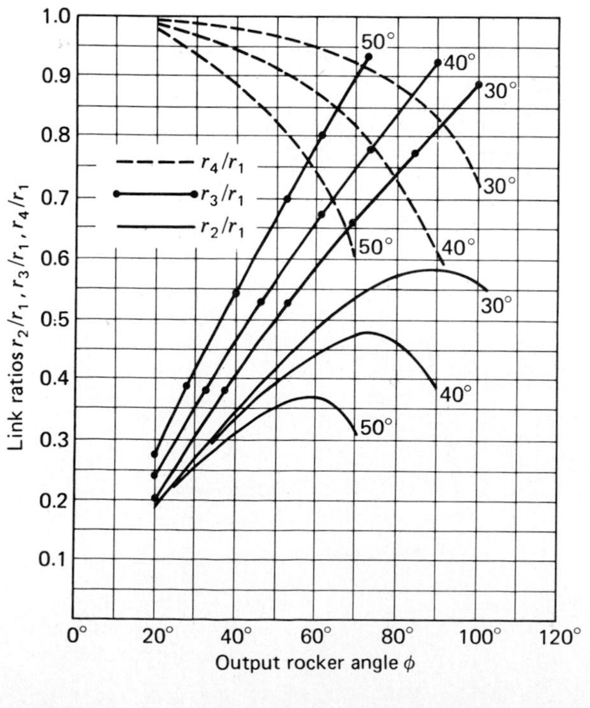

Figure 10-6 The Brodell-Soni chart for the design of the crank-rocker linkage with optimum transmission angle and unity time ratio. The angles shown on the graphs are γ_{min}.

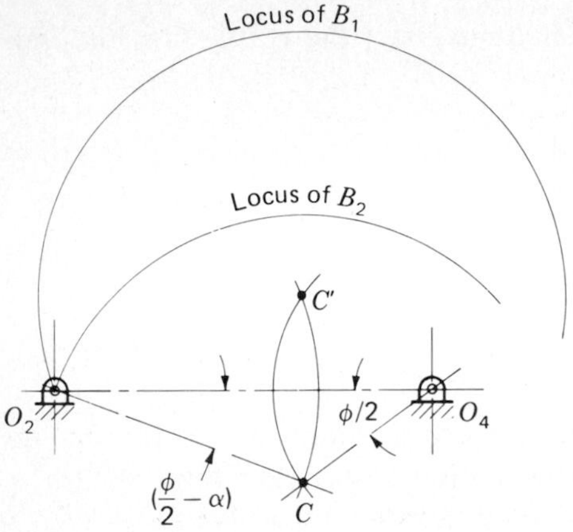

FIGURE 10-7 Layout showing all possible locations of B_1 and B_2.

distance from C to O_2 as the radius, draw the circle arc which is the locus of B_2. Next, using C' as the center and the same radius, draw another circle arc which is the locus of B_1.

One of the many possible crank-rocker linkages has been synthesized in Fig. 10-8. To get the dimensions, choose any point B_1 on the locus of B_1 and swing an arc about O_4 to locate B_2 on the locus of B_2. With these two points defined, the methods of the preceding section are used to locate points A_1 and A_2 together with the link lengths r_2 and r_3.

The resulting linkages should always be checked to ensure that crank 2 is capable of rotating in a complete circle.

To obtain a linkage with an optimum transmission angle, choose a variety of points B_1 on the locus of B_1, synthesizing a linkage for each point.

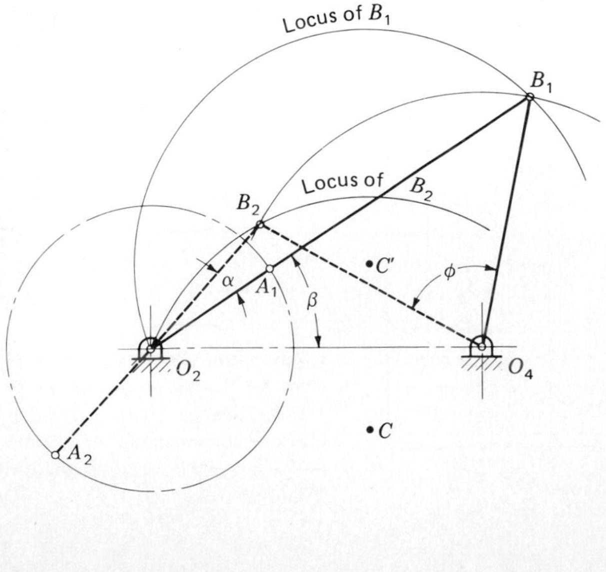

Figure 10-8 Determination of the link lengths for one of the possible crank-rocker mechanisms.

Determine the angles $|90° - \gamma_{min}|$ and $|90° - \gamma_{max}|$ for each of these linkages. Then plot these data on a chart using the angle β (Fig. 10-8) as the abscissa to obtain two curves. The mechanism having the best transmission angle is then defined by the low point on one of the curves.

10-7 THREE-POSITION SYNTHESIS

In Fig. 10-9*a* motion of the input rocker O_2A through the angle ψ_{12} causes a motion of the output rocker O_4B through the angle ϕ_{12}. To define inversion as a technique of synthesis let us hold O_4B stationary and permit the remaining links, including the frame, to occupy the same relative positions as in Fig. 10-9*a*. The result (Fig. 10-9*b*) is called *inverting on the output rocker.* Note that A_1B_1 is positioned the same in Fig. 10-9*a* and *b*. Therefore the inversion is made on the O_4B_1 position. Since O_4B_1 is to be fixed, the frame will have to move in order to get the linkage to the A_2B_2 position. In fact the frame will have to move *backward* through the angle ϕ_{12}. The second position is therefore $O_2'A_2'B_2'O_4$.

Figure 10-10 illustrates a problem and the synthesized linkage in which it

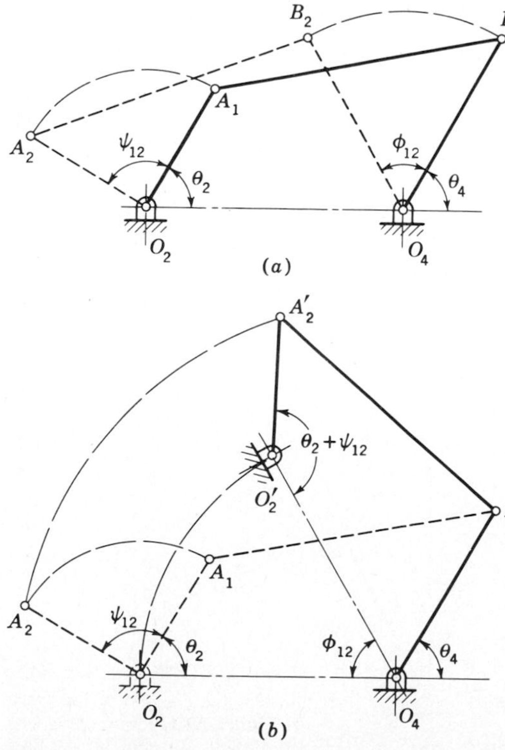

Figure 10-9 (*a*) Rotation of input rocker O_2A through the angle ψ_{12} causes the output rocker O_4B to rock through the angle ϕ_{12}. (*b*) Linkage inverted on the O_4B position.

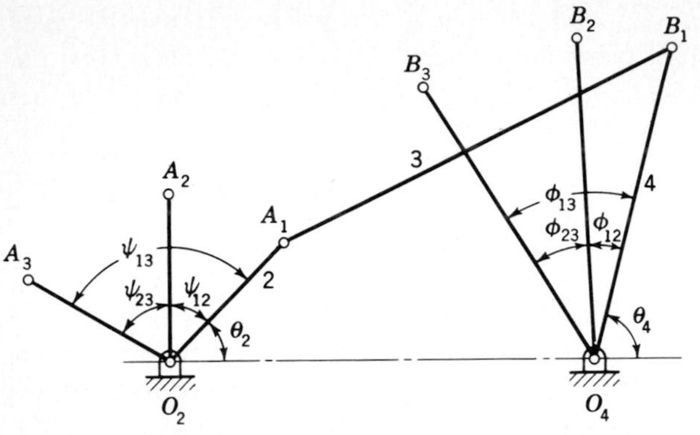

Figure 10-10

is desired to determine the dimensions of a linkage in which the output lever is to occupy three specified positions corresponding to three given positions of the input lever. In Fig. 10-10 the starting angle of the input lever is θ_2; and ψ_{12}, ψ_{23}, and ψ_{13} are the swing angles, respectively, between the design positions 1 and 2, 2 and 3, and 1 and 3. Corresponding angles of swing ϕ_{12}, ϕ_{23}, and ϕ_{13} are desired for the output lever. The length of link 4 and its starting position θ_4 are to be determined.

The solution to the problem is illustrated in Fig. 10-11 and is based on inverting the linkage on link 4. Draw the input rocker O_2A in its three specified positions and locate a desired position for O_4. Since we will invert on link 4 in the first design position, draw a ray from O_4 to A_2 and rotate it backward through the angle ϕ_{12} to locate A_2'. Similarly, draw another ray O_4A_3 and rotate it backward through the angle ϕ_{13} to locate A_3'. Since we are

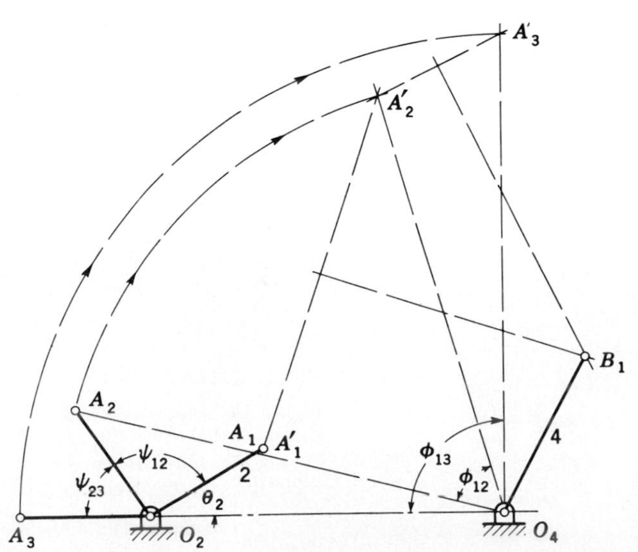

Figure 10-11

inverting on the first design position, A_1 and A_1' are coincident. Now draw midnormals to the lines $A_1'A_2'$ and $A_2'A_3'$. These intersect at B_1 and define the length of the coupler 3 and the length and starting position of link 4.

10-8 POINT-POSITION REDUCTION; FOUR PRECISION POINTS

In point-position reduction the linkage is made symmetrical about the frame centerline O_2O_4 so as to cause two of the A' points to be coincident. The effect of this is to produce three equivalent A' points through which a circle can be drawn as in three-position synthesis. It is best illustrated by an example.

Let us synthesize a linkage to generate the function $y = \log x$ for $10 \leq x \leq 60$ using an input crank range of 120° and an output range of 90°.

To simplify the presentation we shall not employ Chebychev spacing. The angle ψ is evaluated for the four design positions from the equation $\psi =$

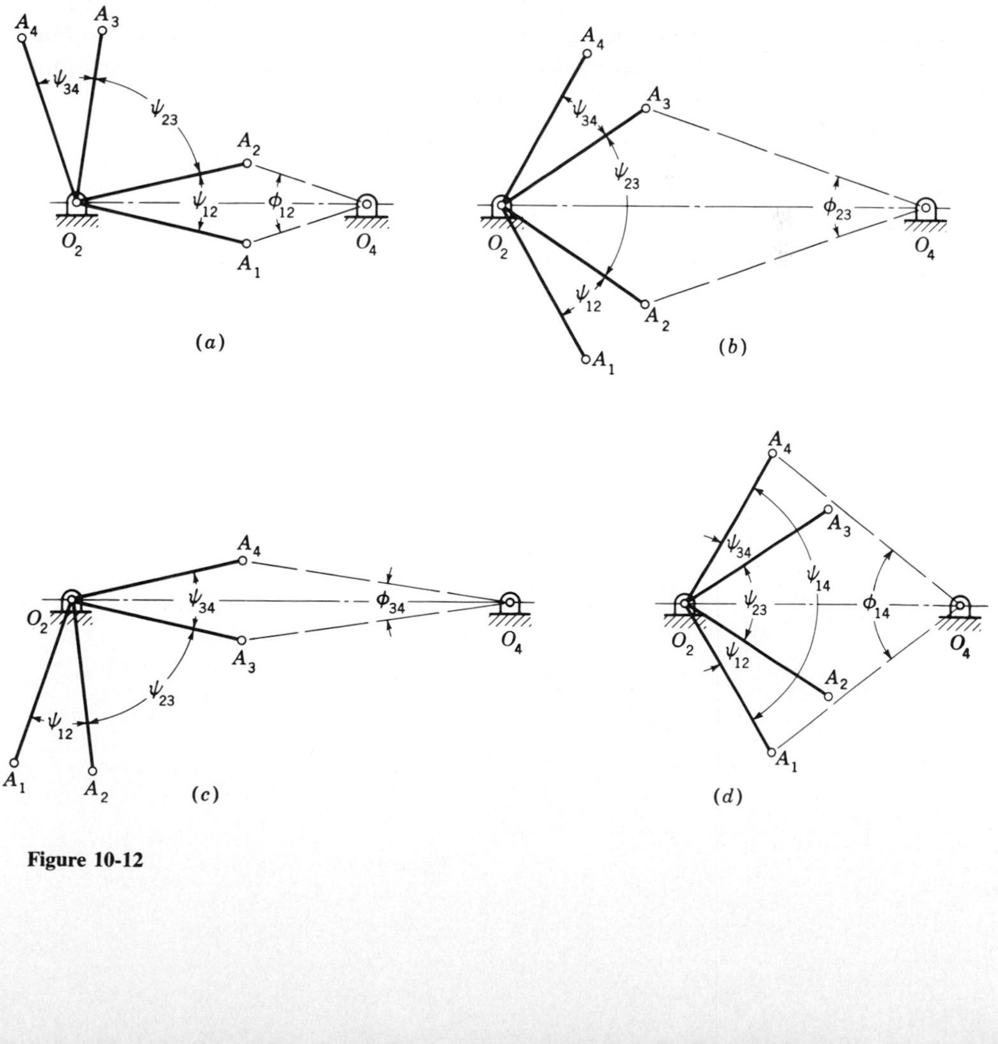

Figure 10-12

Table 10-1†

Position	x	ψ, deg	y	ϕ, deg
1	10	0	2.30	0
2	20	24	3.00	35
3	45	94	3.80	75
4	60	120	4.10	90

† $\psi_{12} = 24°$ $\phi_{12} = 35°$
$\psi_{23} = 70°$ $\phi_{23} = 40°$
$\psi_{34} = 26°$ $\phi_{34} = 15°$

$ax + b$ and from the boundary conditions $\psi = 0$ when $x = 10$ and $\psi = 120°$ when $x = 60$. This gives $\psi = 2.40x - 24$. The angle ϕ is evaluated in exactly the same manner; thus, we get $\phi = 50y - 115$. The results of this preliminary work are shown in Table 10-1.

For the starting position a choice of four arrangements is shown in Fig. 10-12. In a the line O_2O_4 bisects both ψ_{12} and ϕ_{12}; and so, if the output member is turned counterclockwise from the O_4B_2 position, A_1' and A_2' will be coincident and at A_1. The inversion would then be based on the O_4B_1 position.

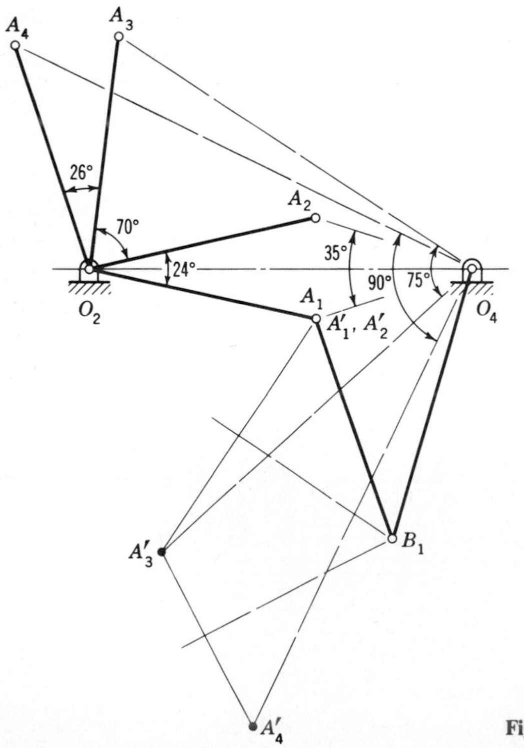

Figure 10-13

Then A_3 would be rotated through the angle ϕ_{13} about O_4 counterclockwise to A'_3, and A_4 through the angle ϕ_{14} to A'_4.

In Fig. 10-12*b* the line O_2O_4 bisects ψ_{23} and ϕ_{23}, while in *d* the angles ψ_{14} and ϕ_{14} are bisected. In obtaining the inversions for each case great care must be taken to ensure that rotation is made in the correct direction and with the correct angles.

When point-position reduction is used, only the length of the input rocker O_2A can be specified in advance. The distance O_2O_4 is dependent upon the values of ψ and ϕ, as indicated in Fig. 10-12. Note that each synthesis position gives a different value for this distance. This is really quite convenient because it is not at all unusual to synthesize a linkage which is not workable. When this happens, one of the other arrangements can be tried.

The synthesized linkage is shown in Fig. 10-13. The procedure is exactly the same as that for three precision points, except as previously noted. Point B_1 is obtained at the intersection of the midnormals to $A'_1A'_3$ and $A'_3A'_4$. In this example the greatest error is less than 3 percent.

10-9 THE OVERLAY METHOD

Synthesis of a function generator, say, using the overlay method, is the easiest and quickest of all methods to use. It is not always possible to obtain a solution, and sometimes the accuracy is rather poor. Theoretically, however, one can employ as many points as are desired in the process.

Let us design a function generator to solve the equation

$$y = x^{0.8} \qquad 1 \le x \le 3 \tag{a}$$

Suppose we choose six positions of the linkage for this example and use uniform spacing of the output rocker. Table 10-2 shows the values of x and y, rounded, and the corresponding angles selected for the input and output rockers.

The first step in the synthesis is shown in Fig. 10-14*a*. Use a sheet of tracing paper and construct the input rocker O_2A in all its positions. This requires a choice for the length of O_2A. Also, on this sheet, choose a length

Table 10-2

Position	x	ψ, deg	y	ϕ, deg
1	1	0	1	0
2	1.366	22.0	1.284	14.2
3	1.756	45.4	1.568	28.4
4	2.16	69.5	1.852	42.6
5	2.58	94.8	2.136	56.8
6	3.02	121.0	2.420	71.0

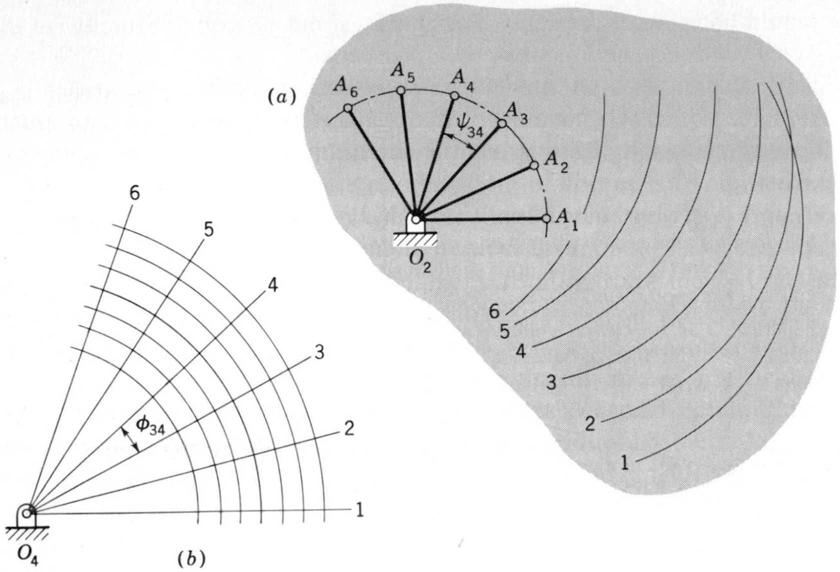

Figure 10-14

for the coupler AB and draw arcs numbered 1 to 6 using A_1 to A_6, respectively, as centers.

Now, on another sheet of paper, construct the output rocker, whose length is unknown, in all its positions, as shown in Fig. 10-14*b*. Through O_4 draw a number of equally spaced arcs intersecting the lines O_41, O_42, etc.; these represent possible lengths of the output rocker.

The final step is to lay the tracing over the drawing and manipulate it in an effort to find a fit. In this case a fit was found, and the result is shown in Fig. 10-15.

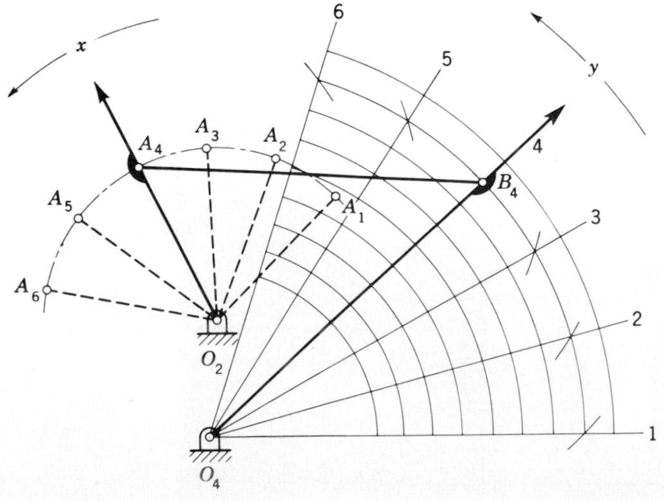

Figure 10-15

10-10 COUPLER-CURVE SYNTHESIS†

In this section we use the method of point-position reduction to synthesize a four-bar linkage so that a tracing point on the coupler will trace any previously specified path when the linkage is moved. Then, in sections to follow we will discover that paths having certain characteristics are particularly useful in synthesizing linkages having dwells of the output member for certain periods of rotation of the input member.

In synthesizing a linkage to generate a path we can choose up to six precision points on the path. If the synthesis is successful, the tracing point will pass through each precision point. The final result may or may not approximate the desired path.

Two positions of a four-bar linkage are shown in Fig. 10-16. Link 2 is the input member; it is connected at A to coupler 3, containing the tracing point C, and connected to output link 4 at B. Two phases of the linkage are illustrated by the subscripts 1 and 3. Points C_1 and C_3 are two positions of the tracer on the path to be generated. In this example C_1 and C_3 have been especially selected so that the midnormal c_{13} passes through O_4. Note, for the selection of points, that the angle $C_1O_4C_3$ is the same as the angle $A_1O_4A_3$, as indicated on the figure.

The advantage of making these two angles equal is that when the linkage is finally synthesized, the triangles $C_3A_3O_4$ and $C_1A_1O_4$ are congruent. Thus, if the tracing point is made to pass through C_1 on the path, then it will also pass through C_3.

†The methods presented here were devised by Hain and are presented in Hain, op. cit., chap. 17.

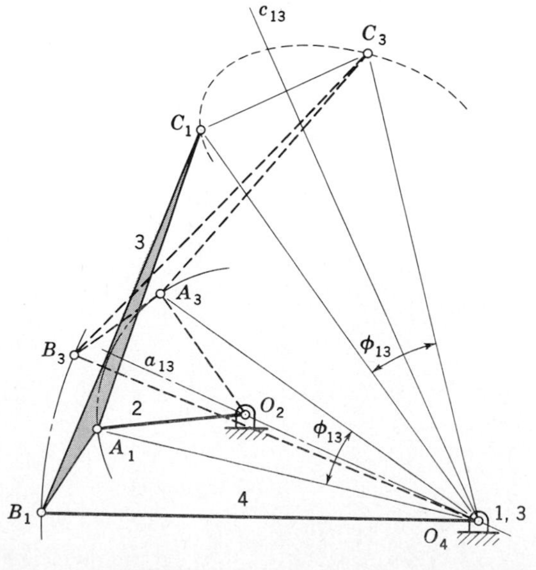

Figure 10-16

To synthesize a linkage so that the coupler will pass through four precision points, we locate any four points C_1, C_2, C_3, C_4 on the desired path (Fig. 10-17). Choosing C_1 and C_3, say, we first locate O_4 anywhere on the midnormal c_{13}. Then, with O_4 as a center and any radius R, construct a circle arc. Next, with centers at C_1 and C_3 and any other radius r, strike arcs to intersect the arc of radius R. These two intersections define points A_1 and A_3 on the input link. Construct the midnormal a_{13} to A_1A_3 and note that it passes through O_4. Locate O_2 anywhere on a_{13}. This provides an opportunity to choose a convenient length for the input rocker. Now use O_2 as a center and draw the crank circle through A_1 and A_3. Points A_2 and A_4 on this circle are obtained by striking arcs of radius r again about C_2 and C_4. This completes the first phase of the synthesis; we have located O_2 and O_4 relative to the desired path and hence defined the distance O_2O_4. We have also defined the length of the input member and located its positions relative to the four precision points on the path.

Our next task is to locate point B, the point of attachment of the coupler and output member. Any one of the four locations of B can be used; in this example we use the B_1 position.

Before beginning the final step we note that the linkage is now defined. Four decisions were made: the location of O_4, the radii R and r, and the location of O_2. Thus an infinite number of solutions are possible.

Referring to Fig. 10-18, locate point 2 by making triangles $C_2A_2O_4$ and C_1A_12 congruent. Locate point 4 by making $C_4A_4O_4$ and C_1A_14 congruent. Points 4, 2, and O_4 lie on a circle whose center is B_1. So B_1 is found at the intersection of the midnormals of O_42 and O_44. Note that the procedure used causes points 1 and 3 to coincide with O_4. With B_1 located, the links can be

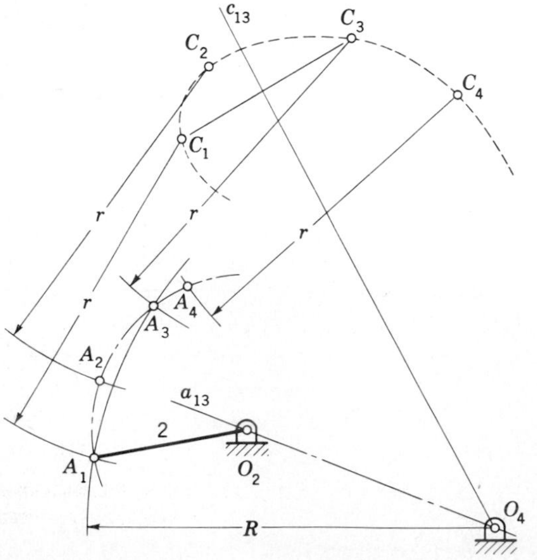

Figure 10-17

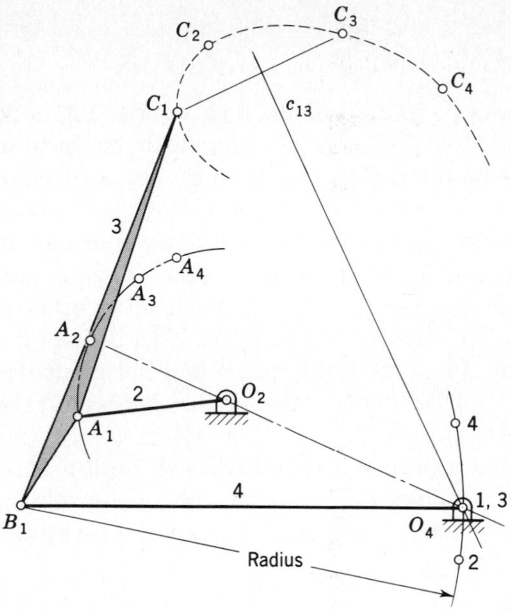

Figure 10-18

drawn in place and the mechanism tested to see how well it traces the prescribed path.

To synthesize a linkage to generate a path through five precision points, it is necessary to make two point reductions. Begin by choosing five points C_1 to C_5 on the path to be traced. Choose two pairs of these for reduction purposes. In Fig. 10-19 we have chosen the pairs C_1C_5 and C_2C_3. Other pairs which

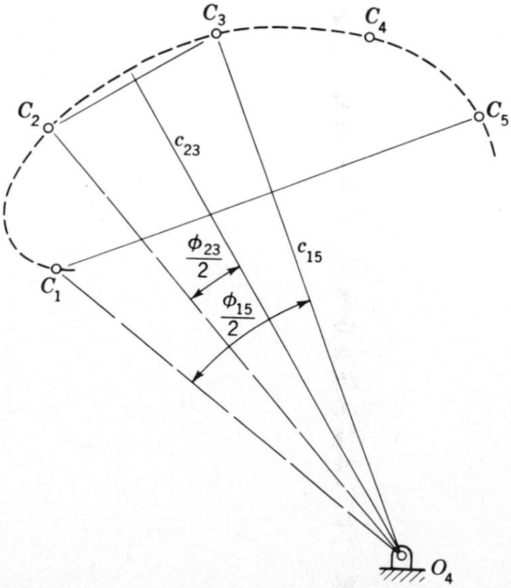

Figure 10-19

could have been used are

$$C_1C_5,\ C_2C_4 \qquad C_1C_5,\ C_3C_4 \qquad C_1C_4,\ C_2C_3 \qquad C_2C_5,\ C_3C_4$$

Construct the perpendicular bisectors c_{23} and c_{15} of the lines connecting each pair. These intersect at point O_4. Note that O_4 can therefore be located conveniently by a judicious choice of the pairs to be used as well as by the choice of the positions of the points C_i on the path.

The next step is best performed by using a scrap of tracing paper as an overlay. Secure the tracing paper to the drawing and mark upon it the center O_4, the midnormal c_{23}, and another line from O_4 to C_2. Such an overlay is shown in Fig. 10-20*a* with the line O_4C_2 designated as O_4C_2'. This defines the angle $\phi_{23}/2$. Now rotate the overlay about O_4 until the midnormal coincides with c_{15} and repeat for point C_1. This defines the angle $\phi_{15}/2$ and the corresponding line O_4C_1'.

Now pin the overlay at O_4, using a thumbtack, and rotate it until a good position is found. It is helpful to set the compass for some convenient radius r and draw circles about each point C_i. The intersection of these circles with the

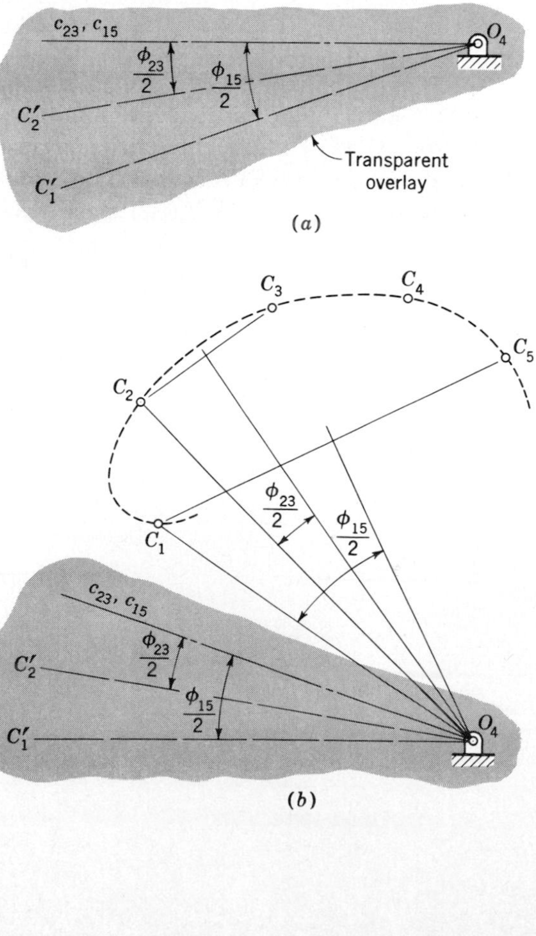

Figure 10-20

lines O_4C_1' and O_4C_2' on the overlay, and with each other, will reveal which areas will be profitable to investigate. See Fig. 10-20*b*.

The final steps in the solution are shown in Fig. 10-21. Having located a good position for the overlay, transfer the three lines to the drawing and remove the overlay. Now draw a circle arc of radius r to intersect O_4C_1' and locate A_1. Another arc of the same radius r from C_2 intersects O_4C_2' at A_2. With A_1 and A_2 located, draw the midnormal a_{12}; it intersects a_{23} at O_2, giving the length of the input rocker. A circle through A_1 about O_2 will contain all the design positions of A; use the same radius r and locate A_3, A_4, and A_5 on arcs about C_3, C_4, and C_5.

We have now located everything except point B_1, and this is found as before. A double point 2,3 exists because of the choice of O_4 on the midnormal c_{23}. To locate this point, strike an arc from C_1 of radius C_2O_4. Then strike another arc from A_1 of radius A_2O_4. These intersect at point 2,3. To locate point 4, strike an arc from C_1 of radius C_4O_4, and another from A_1 of radius A_4O_4. Note that points O_4 and the double points 1,5 are coincident because the synthesis is based on inversion on the O_4B_1 position. Points O_4, 4 and double points 2,3 lie on a circle whose center is B_1, as shown in Fig. 10-21. The linkage is completed by drawing in the coupler and the follower in the first design position.

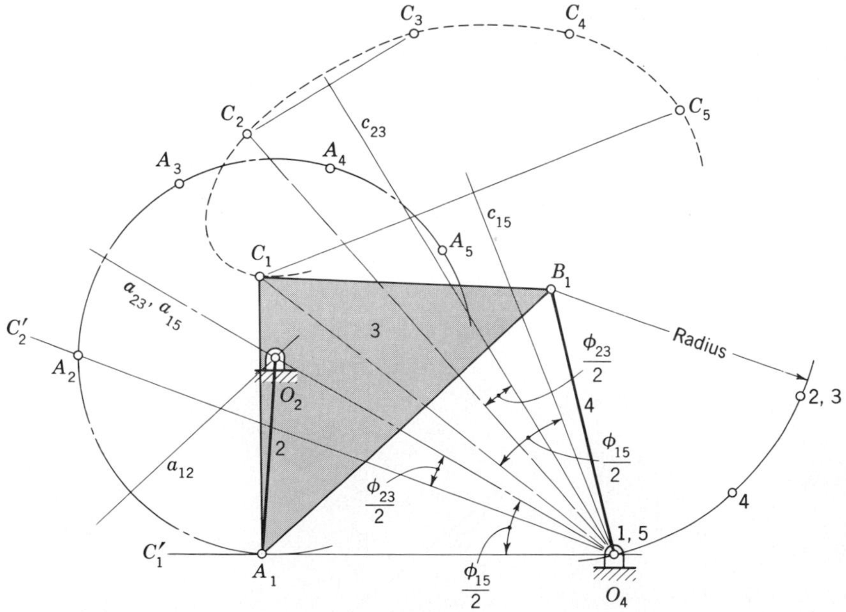

Figure 10-21

10-11 COGNATE LINKAGES; THE ROBERTS-CHEBYCHEV THEOREM

One of the unusual properties of the planar four-bar linkage is that there are not one but three four-bar linkages which will generate the same coupler curve. It was discovered by Roberts† in 1875 and by Chebychev in 1878 and hence is known as the Roberts-Chebychev theorem. Though mentioned in an English publication in 1954,‡ it did not appear in the American literature until it was presented independently and almost simultaneously by Richard S. Hartenberg and Jacques Denavit of Northwestern University and by Rolland T. Hinkle of Michigan State University.§

In Fig. 10-22 let O_1ABO_2 be the original four-bar linkage with a coupler point P attached to AB. The remaining two linkages defined by the Roberts-Chebychev theorem were termed *cognate linkages* by Hartenberg and Denavit. Each of the cognate linkages is shown in Fig. 10-22, one using short dashes for the links and the other using long dashes. The construction is evident by observing that there are four similar triangles, each containing the angles α, β, and γ, and three different parallelograms.

† By S. Roberts, a mathematician; he was not the same Roberts of the approximate-straight-line generator (Fig. 1-12*b*).

‡ P. Grodzinski and E. M'Ewan, Link Mechanisms in Modern Kinematics, *Proc. Inst. Mech. Eng.*, vol. 168, no. 37, pp. 877–896, 1954.

§ R. S. Hartenberg and Jacques Denavit, The Fecund Four-Bar, *Trans. 5th Conf. Mech., Purdue University, Lafayette, Ind., 1958*, p. 194. R. T. Hinkle, Alternate Four-Bar Linkages, *Prod. Eng.*, vol. 29, p. 54, October 1958.

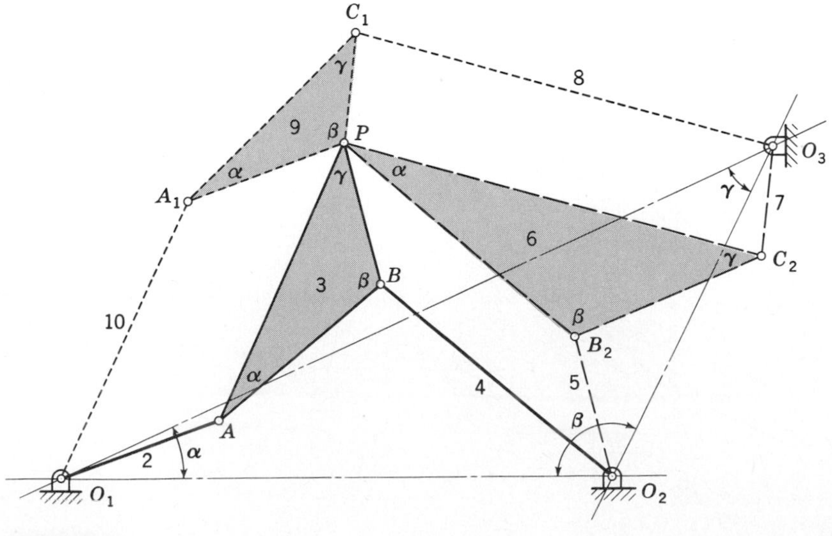

Figure 10-22

A good way to obtain the dimensions of the two cognate linkages is to imagine that the frame connections O_1, O_2, and O_3 can be unfastened. Then "pull" O_1, O_2, and O_3 away from each other until a straight line is formed by the crank, coupler, and follower of each linkage. We get Fig. 10-23 if we do this to Fig. 10-22. Note that the frame distances are incorrect, but all the movable links are of the correct length and all the angles are correct. Given any four-bar linkage and its coupler point, one can create a drawing like Fig. 10-23 and obtain the dimensions of the other two cognate linkages. This approach was discovered by A. Cayley and is called the *Cayley diagram.*†

The advantage of the Roberts-Chebychev theorem is that one of the other two cognates may have better motion characteristics or a better transmission angle, or it may fit into a smaller space.

If the tracing point P is on the straight line AB or its extension, a figure like Fig. 10-23 is of little help because all three linkages are compressed into a single straight line. An example is shown in Fig. 10-24, where O_1ABO_2 is the original linkage having a coupler point P on an extension of AB. To find the cognate linkages, locate O_3 on an extension of O_1O_2 in the same ratio as AB is to BP. Then construct, in order, the parallelograms O_1A_1PA, O_2B_2PB, and $O_3C_1PC_2$.

Hartenberg and Denavit show that the angular-velocity relations between the links in Fig. 10-22 are

$$\omega_9 = \omega_2 = \omega_7 \qquad \omega_{10} = \omega_3 = \omega_5 \qquad \omega_8 = \omega_4 = \omega_6 \tag{10-6}$$

They also observe that if crank 2 is driven at a constant angular velocity and

† A. Cayley, On Three-Bar Motion, *Proc. Lond. Math. Soc.*, vol. 7, pp. 136–166, 1876. In Cayley's time a four-bar linkage was described as a three-bar mechanism because the idea of a kinematic chain had not yet been conceived.

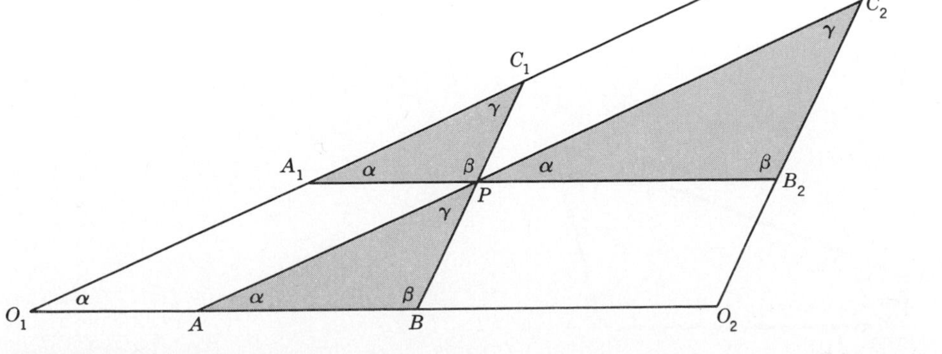

Figure 10-23 The Cayley diagram.

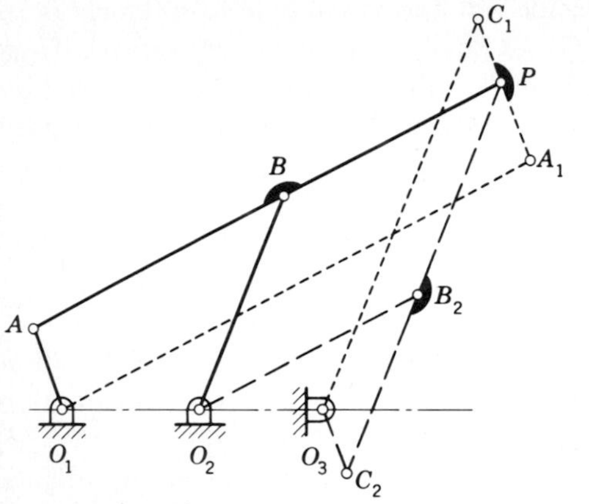

Figure 10-24

if the velocity relationships are to be preserved during generation of the coupler curve, the cognate mechanisms will have to be driven at variable angular velocities.

10-12 ANALYTICAL SYNTHESIS USING COMPLEX ALGEBRA

Sometimes a research paper is published that is a classic in its simplicity and cleverness. Such a paper, written by the Russian kinematician Bloch,† has sparked an entire generation of research. We present the method here more for the additional ideas the method may generate than for its intrinsic value, and also for its historical interest.

In Fig. 10-25 replace the links of a four-bar linkage by position vectors and write the vector equation

$$\mathbf{r}_1 + \mathbf{r}_2 + \mathbf{r}_3 + \mathbf{r}_4 = 0 \tag{a}$$

† S. Sch. Bloch, On the Synthesis of Four-Bar Linkages (in Russian), *Bull. Acad. Sci. USSR*, pp. 47–54, 1940.

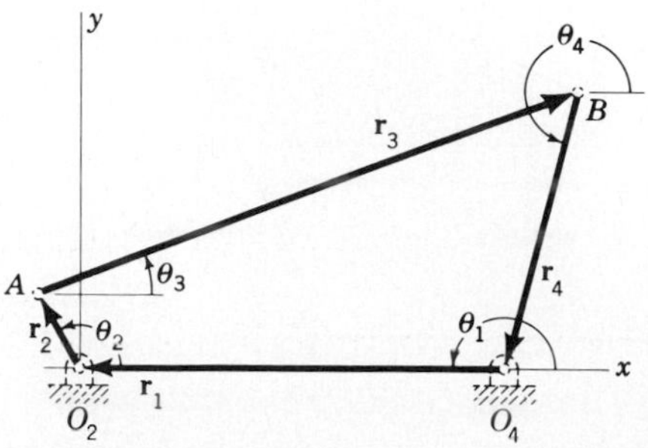

Figure 10-25

In complex polar notation Eq. (*a*) is written

$$r_1 e^{j\theta_1} + r_2 e^{j\theta_2} + r_3 e^{j\theta_3} + r_4 e^{j\theta_4} = 0 \tag{b}$$

The first and second derivatives of this equation are

$$r_2\omega_2 e^{j\theta_2} + r_3\omega_3 e^{j\theta_3} + r_4\omega_4 e^{j\theta_4} = 0 \tag{c}$$

$$r_2(\alpha_2 + j\omega_2^2)e^{j\theta_2} + r_3(\alpha_3 + j\omega_3^2)e^{j\theta_3} + r_4(\alpha_4 + j\omega_4^2)e^{j\theta_4} = 0 \tag{d}$$

If we now transform Eqs. (*b*), (*c*), and (*d*) back into vector notation, we obtain the simultaneous equations

$$\begin{aligned} \mathbf{r}_1 \quad + \mathbf{r}_2 \quad + \mathbf{r}_3 \quad + \mathbf{r}_4 &= 0 \\ \omega_2\mathbf{r}_2 + \omega_3\mathbf{r}_3 + \omega_4\mathbf{r}_4 &= 0 \\ (\alpha_2 + j\omega_2^2)\mathbf{r}_2 + (\alpha_3 + j\omega_3^2)\mathbf{r}_3 + (\alpha_4 + j\omega_4^2)\mathbf{r}_4 &= 0 \end{aligned} \tag{e}$$

This is a set of homogeneous vector equations having complex numbers as coefficients. Bloch specified desired values of all the angular velocities and angular accelerations and then solved the equations for the relative linkage dimensions.

Solving Eqs. (*e*) for $\mathbf{r}_2$ gives

$$\mathbf{r}_2 = \frac{\begin{vmatrix} -1 & 1 & 1 \\ 0 & \omega_3 & \omega_4 \\ 0 & \alpha_3 + j\omega_3^2 & \alpha_4 + j\omega_4^2 \end{vmatrix}}{\begin{vmatrix} 1 & 1 & 1 \\ \omega_2 & \omega_3 & \omega_4 \\ \alpha_2 + j\omega_2^2 & \alpha_3 + j\omega_3^2 & \alpha_4 + j\omega_4^2 \end{vmatrix}} \tag{f}$$

Similar expressions will be obtained for $\mathbf{r}_3$ and $\mathbf{r}_4$. It turns out that the denominators for all three expressions, i.e., for $\mathbf{r}_2$, $\mathbf{r}_3$, and $\mathbf{r}_4$, are complex numbers and are equal. In division, we divide the magnitudes and subtract the angles. Since these denominators are all alike, the effect of the division would be to change the magnitudes of $\mathbf{r}_2$, $\mathbf{r}_3$, and $\mathbf{r}_4$ by the same factor and to shift all the directions by the same angle. For this reason, we make all the denominators unity; the solutions then give dimensionless vectors for the links. When the determinants are evaluated, we find

$$\begin{aligned} \mathbf{r}_2 &= \omega_4(\alpha_3 + j\omega_3^2) - \omega_3(\alpha_4 + j\omega_4^2) \\ \mathbf{r}_3 &= \omega_2(\alpha_4 + j\omega_4^2) - \omega_4(\alpha_2 + j\omega_2^2) \\ \mathbf{r}_4 &= \omega_3(\alpha_2 + j\omega_2^2) - \omega_2(\alpha_3 + j\omega_3^2) \\ \mathbf{r}_1 &= -\mathbf{r}_2 - \mathbf{r}_3 - \mathbf{r}_4 \end{aligned} \tag{10-7}$$

Example 10-1 Synthesize a four-bar linkage to give the following values for the angular velocities and accelerations:

$$\omega_2 = 200 \text{ rad/s} \qquad \omega_3 = 85 \text{ rad/s} \qquad \omega_4 = 130 \text{ rad/s}$$

$$\alpha_2 = 0 \text{ rad/s}^2 \qquad \alpha_3 = -1000 \text{ rad/s}^2 \qquad \alpha_4 = -16\,000 \text{ rad/s}^2$$

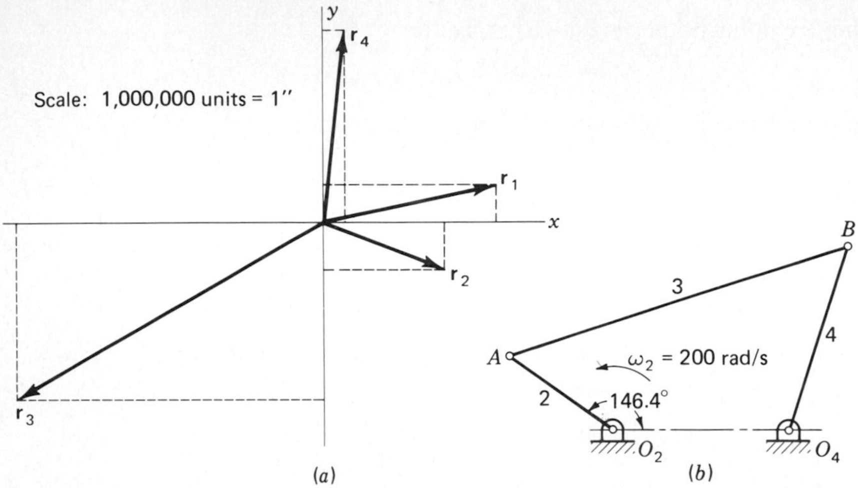

Figure 10-26 $O_2A = 1.33$ in; $AB = 3.69$ in; $O_4B = 1.965$ in; $O_2O_4 = 1.81$ in.

SOLUTION Substituting the given values into Eqs. (10-7) gives

$$\mathbf{r}_2 = 130[-1000 + j(85^2)] - 85[-16\,000 + j(130)^2]$$
$$= 1\,230\,000 - j497\,000 = 1\,330\,000\underline{/-22°} \text{ units}$$
$$\mathbf{r}_3 = 200[-16\,000 + j(130)^2] - 130[0 + j(200)^2]$$
$$= -3\,200\,000 - j1\,820\,000 = 3\,690\,000\underline{/-150.4°} \text{ units}$$
$$\mathbf{r}_4 = 85[0 + j(200)^2] - 200[-1000 + j(85)^2]$$
$$= 200\,000 + j1\,955\,000 = 1\,965\,000\underline{/84.15°} \text{ units}$$
$$\mathbf{r}_1 = -(1\,230\,000 - j497\,000) - (-3\,200\,000 - j1\,820\,000)$$
$$-(200\,000 + j1\,955\,000)$$
$$= 1\,770\,000 + j362\,000 = 1\,810\,000\underline{/11.6°} \text{ units}$$

In Fig. 10-26*a* these four vectors are plotted to a scale of 10^6 units per inch. In order to make $\mathbf{r}_1$ horizontal and in the $-x$ direction, the entire vector system must be rotated counterclockwise $180 - 11.6 = 168.4°$. The resulting linkage can then be constructed by using $\mathbf{r}_1$ for link 1, $\mathbf{r}_2$ for link 2, etc., as shown in Fig. 10-26*b*. This mechanism has been dimensioned in inches and, if analyzed, will show that the conditions of the example have been fulfilled.

10-13 FREUDENSTEIN'S EQUATION†

If Eq. (*b*) of the preceding section is transformed into complex rectangular form, and if the real and the imaginary components are separated, we obtain

† Ferdinand Freudenstein, Approximate Synthesis of Four-Bar Linkages, *Trans. ASME*, vol. 77, no. 6, pp. 853–861, 1955.

the two algebraic equations

$$r_1 \cos \theta_1 + r_2 \cos \theta_2 + r_3 \cos \theta_3 + r_4 \cos \theta_4 = 0 \tag{a}$$

$$r_1 \sin \theta_1 + r_2 \sin \theta_2 + r_3 \sin \theta_3 + r_4 \sin \theta_4 = 0 \tag{b}$$

From Fig. 10-25, $\sin \theta_1 = 0$ and $\cos \theta_1 = -1$; therefore

$$-r_1 + r_2 \cos \theta_2 + r_3 \cos \theta_3 + r_4 \cos \theta_4 = 0 \tag{c}$$

$$r_2 \sin \theta_2 + r_3 \sin \theta_3 + r_4 \sin \theta_4 = 0 \tag{d}$$

In order to eliminate the coupler angle θ_3 from the equations, move all terms except those involving r_3 to the right-hand side and square both sides. This gives

$$r_3^2 \cos^2 \theta_3 = (r_1 - r_2 \cos \theta_2 - r_4 \cos \theta_4)^2 \tag{e}$$

$$r_3^2 \sin^2 \theta_3 = (-r_2 \sin \theta_2 - r_4 \sin \theta_4)^2 \tag{f}$$

Now expand the right-hand sides of both equations and add them together. The result is

$$r_3^2 = r_1^2 + r_2^2 + r_4^2 - 2r_1r_2 \cos \theta_2 - 2r_1r_4 \cos \theta_4 + 2r_2r_4(\cos \theta_2 \cos \theta_4 + \sin \theta_2 \sin \theta_4) \tag{g}$$

Now, note that $\cos \theta_2 \cos \theta_4 + \sin \theta_2 \sin \theta_4 = \cos (\theta_2 - \theta_4)$. If we make this replacement, divide by the term $2r_2r_4$, and rearrange again, we obtain

$$\frac{r_3^2 - r_1^2 - r_2^2 - r_4^2}{2r_2r_4} + \frac{r_1}{r_4} \cos \theta_2 + \frac{r_1}{r_2} \cos \theta_4 = \cos (\theta_2 - \theta_4) \tag{h}$$

Freudenstein writes Eq. (h) in the form

$$K_1 \cos \theta_2 + K_2 \cos \theta_4 + K_3 = \cos (\theta_2 - \theta_4) \tag{10-8}$$

with

$$K_1 = \frac{r_1}{r_4} \tag{10-9}$$

$$K_2 = \frac{r_1}{r_2} \tag{10-10}$$

$$K_3 = \frac{r_3^2 - r_1^2 - r_2^2 - r_4^2}{2r_2r_4} \tag{10-11}$$

We have already learned graphical methods of synthesizing a linkage so that the motion of the output member is coordinated with that of the input member. Freudenstein's equation enables us to perform this same task by analytical means. Thus, suppose we wish the output lever of a four-bar linkage to occupy the positions ϕ_1, ϕ_2, and ϕ_3 corresponding to the angular positions ψ_1, ψ_2, and ψ_3 of the input lever. In Eq. (10-8) we simply replace θ_2 with ψ, θ_4 with ϕ, and write the equation three times, once for each position.

This gives

$$K_1 \cos \psi_1 + K_2 \cos \phi_1 + K_3 = \cos (\psi_1 - \phi_1)$$
$$K_1 \cos \psi_2 + K_2 \cos \phi_2 + K_3 = \cos (\psi_2 - \phi_2) \qquad (i)$$
$$K_1 \cos \psi_3 + K_2 \cos \phi_3 + K_3 = \cos (\psi_3 - \phi_3)$$

Equations (i) are solved simultaneously for the three unknowns K_1, K_2, and K_3. Then a length, say r_1, is selected for one of the links and Eqs. (10-9) to (10-11) solved for the dimensions of the other three. The method is best illustrated by an example.

Example 10-2 Synthesize a function generator to solve the equation

$$y = \frac{1}{x} \qquad 1 \le x \le 2$$

using three precision points.

SOLUTION Choosing Chebychev spacing, we find, from Eq. (10-1), the values of x and corresponding values of y to be

$$x_1 = 1.067 \qquad y_1 = 0.937$$
$$x_2 = 1.500 \qquad y_2 = 0.667$$
$$x_3 = 1.933 \qquad y_3 = 0.517$$

We must now choose starting angles for the input and output levers as well as total swing angles for each. These are arbitrary decisions and may not result in a good linkage in the sense that the structural errors between the precision points may be large or the transmission angles may be poor. Sometimes, in such a synthesis, it is even found that one of the pivots must be removed in order to get from one precision point to another. Generally, some trial-and-error work is necessary to discover the best starting positions and swing angles.

Table 10-3

x	ψ, deg	y	ϕ, deg
1.000	30.00	1.000	240.00
1.067	36.03	0.937	251.34
1.500	75.00	0.667	300.00
1.933	113.97	0.517	326.94
2.000	120.00	0.500	330.00

For the input lever we choose a 30° starting position and a 90° total swing angle. For the output lever, choose the starting position at 240° and a range of 90° total travel too. With these decisions made, the first and last rows of Table 10-3 can be completed.

Next, to obtain the values of ψ and ϕ corresponding to the precision points, write

$$\psi = ax + b \qquad \phi = cy + d \qquad (1)$$

and use the data in the first and last rows of Table 10-3 to evaluate the constants a, b, c, and d. When this is done, we find Eqs. (1) are

$$\psi = 90x - 60 \qquad \phi = -180y + 420 \qquad (2)$$

These equations can now be used to compute the data for the remaining rows in Table 10-3 and to determine the scales of the input and output levers of the synthesized linkage.

Now take the values of ψ and ϕ from the second line of Table 10-3 and substitute them for θ_2 and θ_4 in Eq. (10-8). Repeat this for the third and fourth lines. We then have the three equations

$$\begin{aligned} K_1 \cos 36.03 + K_2 \cos 251.34 + K_3 &= \cos(36.03 - 251.34) \\ K_1 \cos 75.00 + K_2 \cos 300.00 + K_3 &= \cos(75.00 - 300.00) \\ K_1 \cos 113.97 + K_2 \cos 326.94 + K_3 &= \cos(113.97 - 326.94) \end{aligned} \tag{3}$$

When the indicated operations are carried out, we have

$$\begin{aligned} 0.8087K_1 - 0.3200K_2 + K_3 &= -0.8160 \\ 0.2588K_1 + 0.5000K_2 + K_3 &= -0.7071 \\ -0.4062K_1 + 0.8381K_2 + K_3 &= -0.8389 \end{aligned} \tag{4}$$

Upon solving Eqs. (4) we obtain

$$K_1 = 0.4032 \qquad K_2 = 0.4032 \qquad K_3 = -1.0130$$

Using $r_1 = 1$ unit, we get from Eq. (10-9),

$$r_4 = \frac{r_1}{K_1} = \frac{100}{0.4032} = 2.48 \text{ units}$$

Similarly, from Eqs. (10-10) and (10-11), we learn

$$r_2 = 2.48 \text{ units} \qquad r_3 = 0.968 \text{ unit}$$

The result is the crossed linkage shown in Fig. 10-27.

Freudenstein offers the following suggestions which will be helpful in synthesizing such generators:

1. The total swing angles of the input and output members should be less than 120°.
2. Avoid the generation of symmetric functions such as $y = x^2$ in the range $-1 \leq x \leq 1$.
3. Avoid the generation of functions having abrupt changes in slope.

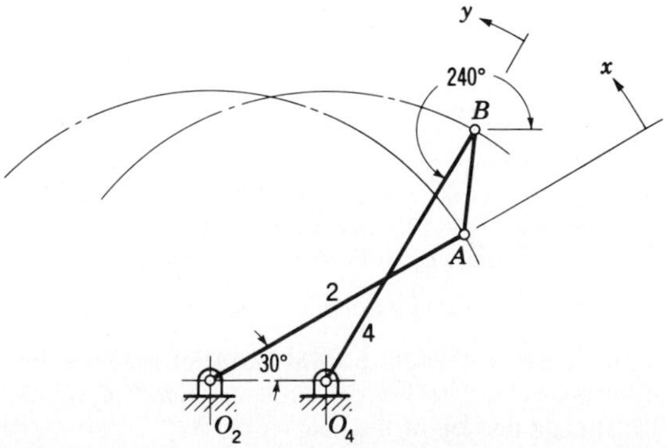

Figure 10-27

10-14 SYNTHESIS OF DWELL MECHANISMS

One of the most interesting uses of coupler curves having straight-line or circle-arc segments is in the synthesis of mechanisms having a substantial dwell during a portion of their operating period. By using segments of coupler curves it is not difficult to synthesize linkages having a dwell at either or both of the extremes of its motion or at an intermediate point.

In Fig. 10-28*a* a coupler curve having approximately an elliptical shape is selected from the Hrones and Nelson atlas so that a substantial portion of the curve approximates a circle arc. Connecting link 5 is then given a length equal to the radius of this arc. Thus, in the figure, points D_1, D_2, and D_3 are stationary while coupler point C moves through positions C_1, C_2, and C_3. The length of output link 6 and the location of the frame point O_6 depend upon the desired angle of oscillation of this link. The frame point should also be positioned for optimum transmission angle.

When segments of circle arcs are desired for the coupler curve, an organized method of searching the Hrones and Nelson atlas should be employed. The overlay, shown in Fig. 10-29, is made on a sheet of tracing paper and can be fitted over the paths in the atlas very quickly. It reveals immediately the radius of curvature of the segment, the location of pivot point D, and the swing angle of the connecting link.

Figure 10-28*b* shows a dwell mechanism employing a slider. A coupler curve having a straight-line segment is used, and the pivot point O_6 placed on an extension of this line.

The arrangement shown in Fig. 10-30*a* has a dwell at both extremes of the motion. A practical arrangement of this mechanism is rather difficult to achieve, however, because link 6 has such a high velocity when the slider is near the pivot O_6.

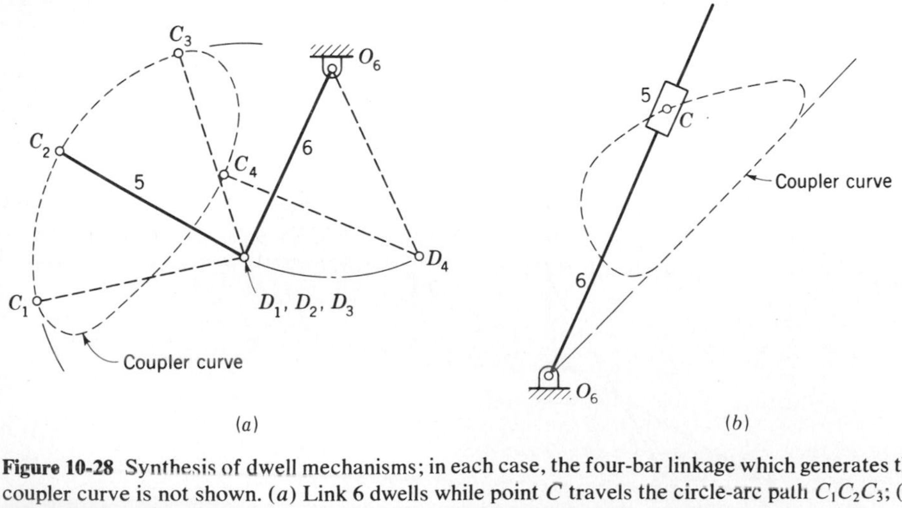

Figure 10-28 Synthesis of dwell mechanisms; in each case, the four-bar linkage which generates the coupler curve is not shown. (*a*) Link 6 dwells while point C travels the circle-arc path $C_1C_2C_3$; (*b*) link 6 dwells while point C travels along the straight portion of the coupler curve.

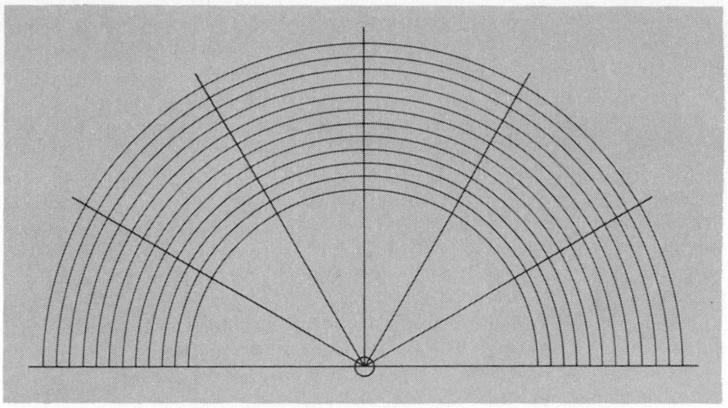

Figure 10-29 Overlay for use with the Hrones and Nelson atlas.

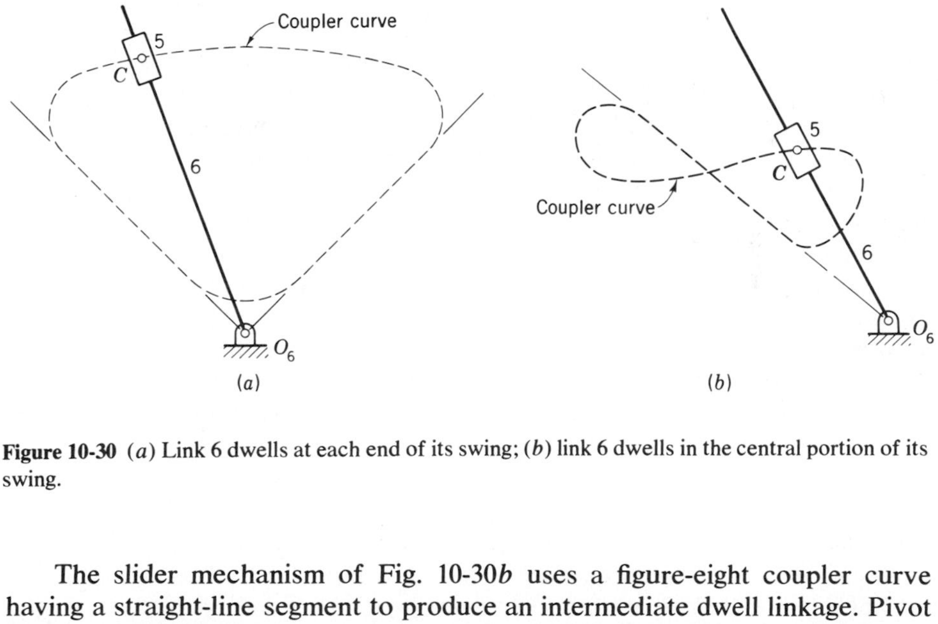

Figure 10-30 (*a*) Link 6 dwells at each end of its swing; (*b*) link 6 dwells in the central portion of its swing.

The slider mechanism of Fig. 10-30*b* uses a figure-eight coupler curve having a straight-line segment to produce an intermediate dwell linkage. Pivot O_6 must be located on an extension of the straight-line segment, as shown.

10-15 INTERMITTENT ROTARY MOTION

The *Geneva wheel*, or *Maltese cross*, is a camlike mechanism which provides intermittent rotary motion and is widely used in both low- and high-speed machinery. Although originally developed as a stop to prevent overwinding of watches, it is now extensively used in automatic machinery, e.g., where a

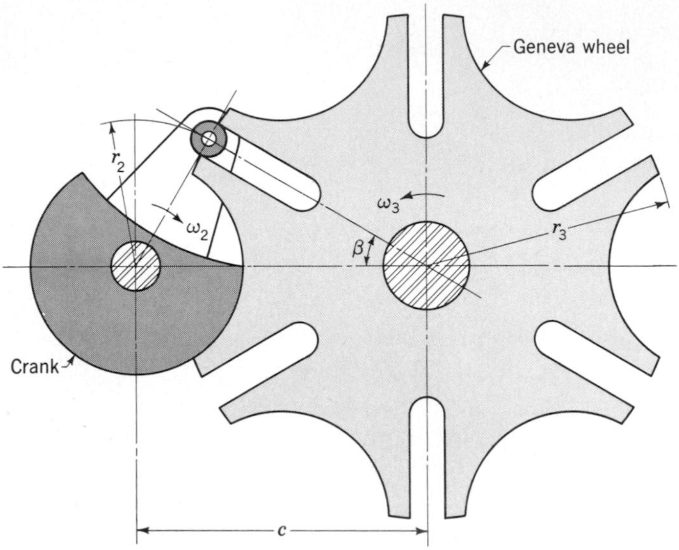

Figure 10-31 The Geneva mechanism.

spindle, turret, or worktable must be indexed. It is also used in motion-picture projectors to provide the intermittent advance of the film.

A drawing of a six-slot Geneva mechanism is shown in Fig. 10-31. Notice that the centerlines of the slot and crank are mutually perpendicular at engagement and at disengagement. The crank, which usually rotates at a uniform angular velocity, carries a roller to engage with the slots. During one revolution of the crank the Geneva wheel rotates a fractional part of a revolution, the amount of which is dependent upon the number of slots. The circular segment attached to the crank effectively locks the wheel against rotation when the roller is not in engagement and also positions the wheel for correct engagement of the roller with the next slot.

The design of a Geneva mechanism is initiated by specifying the crank radius, the roller diameter, and the number of slots. At least three slots are necessary, but most problems can be solved with wheels having from four to twelve slots. The design procedure is shown in Fig. 10-32. The angle β is half the angle subtended by adjacent slots; i.e.,

$$\beta = \frac{360}{2n} \tag{a}$$

where n is the number of slots in the wheel. Then, defining r_2 as the crank radius, we have

$$c = \frac{r_2}{\sin \beta} \tag{b}$$

where c is the center distance. Note, too, from Fig. 10-32 that the actual

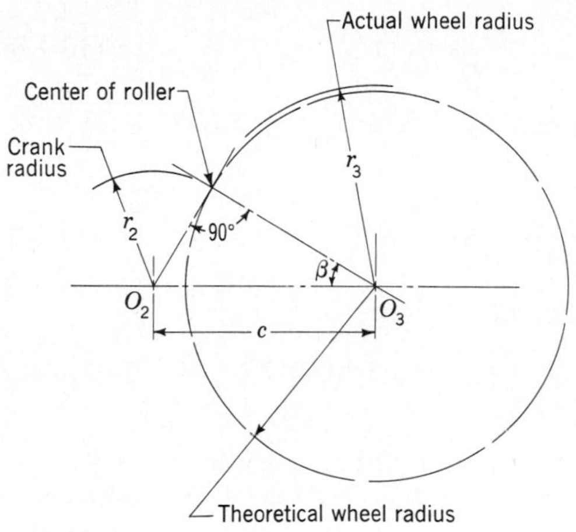

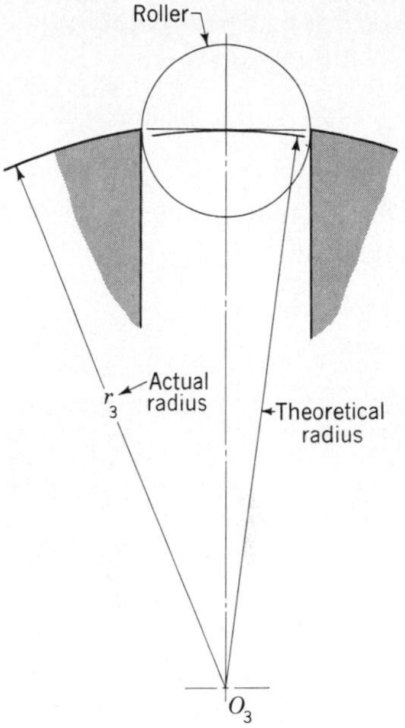

Figure 10-32 Design of a Geneva wheel.

Geneva-wheel radius is more than that which would be obtained by a zero-diameter roller. This is due to the difference between the sine and the tangent of the angle subtended by the roller, measured from the wheel center.

After the roller has entered the slot and is driving the wheel, the geometry is that of Fig. 10-33. Here θ_2 is the crank angle and θ_3 the wheel angle. They are related trigonometrically by

$$\tan \theta_3 = \frac{\sin \theta_2}{(c/r_2) - \cos \theta_2} \qquad (c)$$

We can determine the angular velocity of the wheel for any value of θ_2 by differentiating Eq. (c) with respect to time. This produces

$$\omega_3 = \omega_2 \frac{(c/r_2) \cos \theta_2 - 1}{1 + (c^2/r_2^2) - 2(c/r_2) \cos \theta_2} \qquad (10\text{-}12)$$

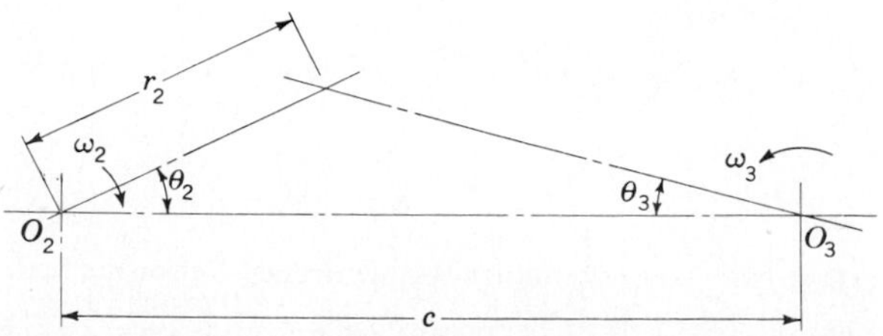

Figure 10-33

The maximum wheel velocity occurs when the crank angle is zero. Substituting $\theta_2 = 0$ therefore gives

$$\omega_3 = \omega_2 \frac{r_2}{c - r_2} \tag{10-13}$$

The angular acceleration is obtained by differentiating Eq. (10-12) with respect to time. It is

$$\alpha_3 = \omega_2^2 \frac{(c/r_2)\sin\theta_2(1 - c^2/r_2^2)}{[1 + (c/r_2)^2 - 2(c/r_2)\cos\theta_2]^2} \tag{10-14}$$

The angular acceleration reaches a maximum when

$$\theta_2 = \cos^{-1}\left\{\pm\sqrt{\left[\frac{1+(c^2/r_2^2)}{4(c/r_2)}\right]^2 + 2} - \frac{1+(c/r_2)^2}{4(c/r_2)}\right\} \tag{10-15}$$

This occurs when the roller has advanced about 30 percent into the slot.

Several methods have been employed to reduce the wheel acceleration in order to reduce inertia forces and the consequent wear on the sides of the slot. Among these is the idea of using a curved slot. This does reduce the acceleration, but it increases the deceleration and consequently the wear on the other side of the slot.

Another method uses the Hrones-Nelson synthesis. The idea is to place the roller on the connecting link of a four-bar linkage. During the period in which it drives the wheel, the path of the roller should be curved and should have a low value of acceleration. Figure 10-34 shows one solution and

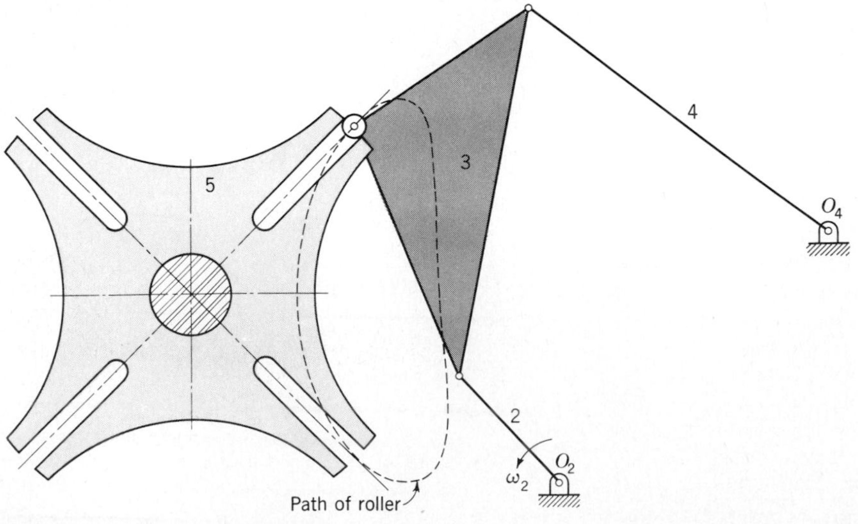

Figure 10-34 Geneva wheel driven by a four-bar linkage synthesized by the Hrones-Nelson method. Link 2 is the driving crank.

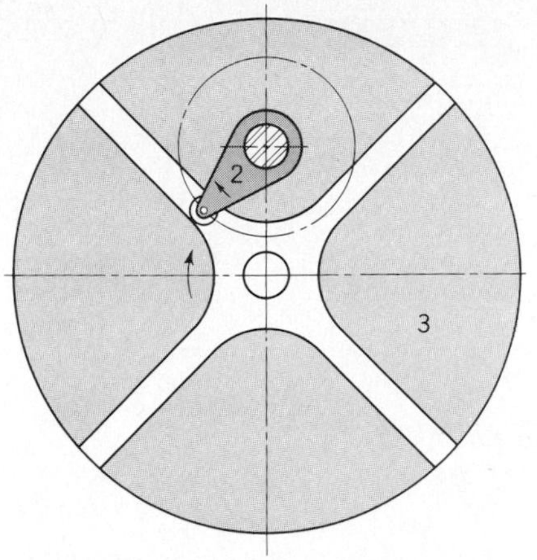

Figure 10-35 The inverse Geneva mechanism.

includes the path taken by the roller. This is the path which is sought for in leafing through the book.

The inverse Geneva mechanism of Fig. 10-35 enables the wheel to rotate in the same direction as the crank and requires less radial space. The locking device is not shown, but this can be a circular segment attached to the crank, as before, which locks by wiping against a built-up rim on the periphery of the wheel.

PROBLEMS

10-1 A function varies from 0 to 10. Find the Chebychev spacing for two, three, four, five, and six precision points.

10-2 Determine the link lengths of a slider-crank linkage to have a stroke of 600 mm and a time ratio of 1.20.

10-3 Determine a set of link lengths for a slider-crank linkage such that the stroke is 16 in and the time ratio is 1.25.

10-4 The rocker of a crank-rocker linkage is to have a length of 500 mm and swing through a total angle of 45° with a time ratio of 1.25. Determine a suitable set of dimensions for r_1, r_2, and r_3.

10-5 A crank-and-rocker mechanism is to have a rocker 6 ft in length and a rocking angle of 75°. If the time ratio is to be 1.32, what are a suitable set of link lengths for the remaining three links?

10-6 Design a crank and coupler to drive rocker 4 in the figure such that slider 6 will reciprocate through a distance of 16 in with a time ratio of 1.20. Use $a = r_4 = 16$ in and $r_5 = 24$ in with r_4 vertical at midstroke. Record the location of O_2 and the dimensions r_2 and r_3.

10-7 Design a crank and rocker for a six-link mechanism such that the slider in the figure for Prob. 10-6 reciprocates through a distance of 800 mm with a time ratio of 1.12. Use $a = r_4 = 1200$ mm and $r_5 = 1800$ mm. Locate O_4 such that rocker 4 is vertical when the slider is at midstroke. Find suitable coordinates for O_2 and lengths for r_2 and r_3.

10-8 Design a crank-rocker mechanism with optimum transmission angle, a unit time ratio, and a rocker angle of 45° using a rocker 250 mm in length. Use the chart of Fig. 10-6 and $\gamma_{min} = 50°$. Make a drawing of the linkage to find and verify γ_{min}, γ_{max}, and ϕ.

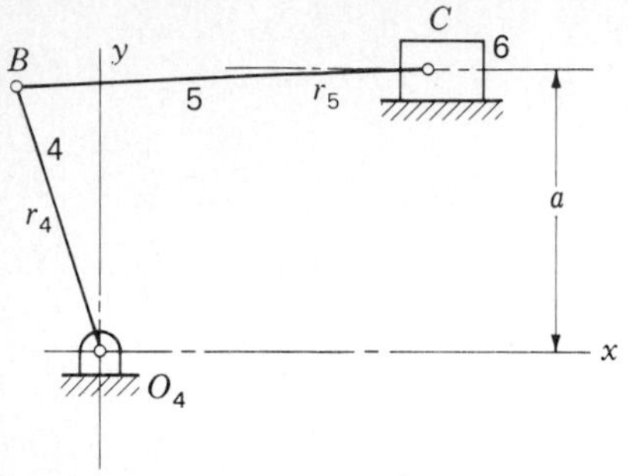

Problem 10-6

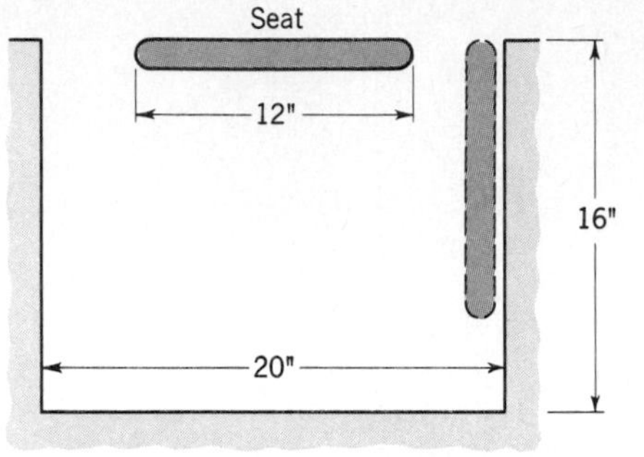

Problem 10-9

10-9 The figure shows two positions of a folding seat used in the aisles of buses to accommodate extra seated passengers. Design a four-bar linkage to support the seat so that it will lock in the open position and fold to a stable closing position along the side of the aisle.

10-10 Design a spring-operated four-bar linkage to support a heavy lid like the trunk lid of an automobile. The lid is to swing through an angle of 80° from the closed to the open position. The springs are to be mounted so that the lid will be held closed against a stop, and they should also hold the lid in a stable open position without the use of a stop.

10-11 In part (a) of the figure synthesize a linkage to move AB from position 1 to position 2 and return.

10-12 In part (b) of the figure synthesize a mechanism to move AB successively through positions 1, 2, and 3.

10-13 to 10-22† The figure shows a function-generator linkage in which the motion of rocker 2 corresponds to x and the motion of rocker 4 to the function $y = f(x)$. Use four precision points and Chebychev spacing and synthesize a linkage to generate the functions shown in the accompanying table. Plot a curve of the desired function and a curve of the actual function which the linkage generates. Compute the maximum error between them in percent.

† Digital-computer solutions for these problems were obtained by F. Freudenstein of Columbia University; see ibid.

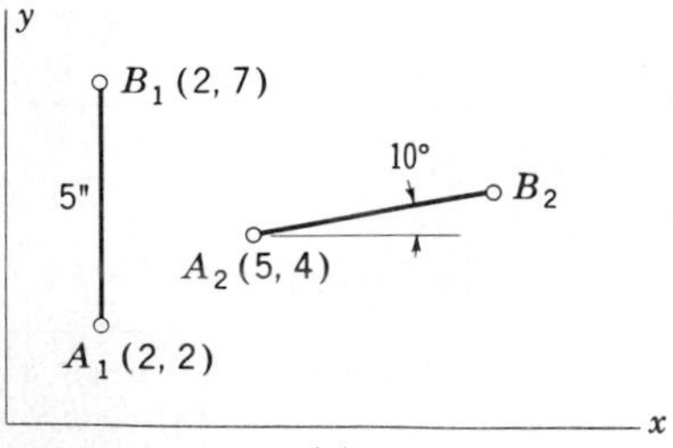

(a)

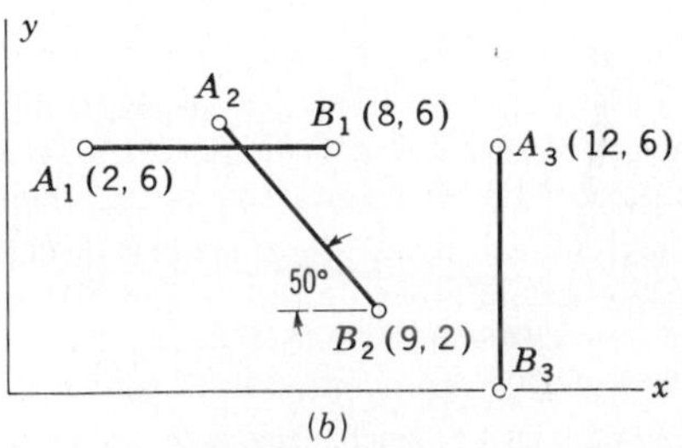

(b)

Problems 10-11 and 10-12

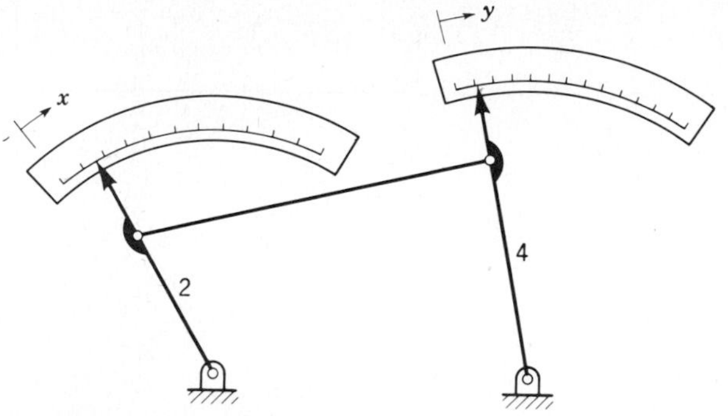

Problems 10-13 to **10-22**

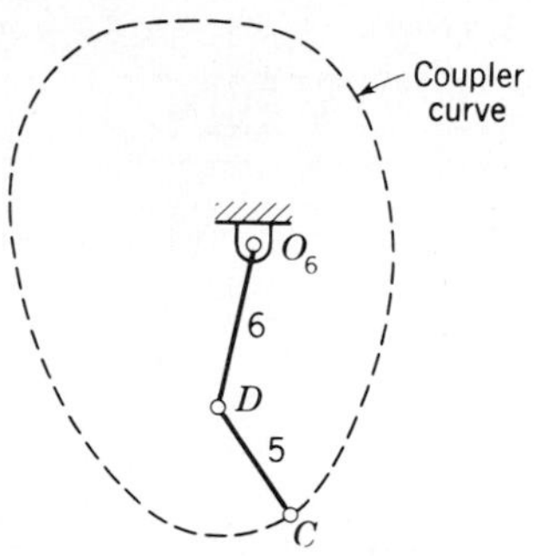

Problem 10-33

Problem number	Function y	Range of x
10-13, 10-23	$\log_{10} x$	$1 \leq x \leq 2$
10-14, 10-24	$\sin x$	$0 \leq x \leq \pi/2$
10-15, 10-25	$\tan x$	$0 \leq x \leq \pi/4$
10-16, 10-26	e^x	$0 \leq x \leq 1$
10-17, 10-27	$1/x$	$1 \leq x \leq 2$
10-18, 10-28	$x^{1.5}$	$0 \leq x \leq 1$
10-19, 10-29	x^2	$0 \leq x \leq 1$
10-20, 10-30	$x^{2.5}$	$0 \leq x \leq 1$
10-21, 10-31	x^3	$0 \leq x \leq 1$
10-22, 10-32	x^2	$-1 \leq x \leq 1$

10-23 to 10-32 Repeat Probs. 10-13 to 10-22 using the overlay method.

10-33 The figure illustrates a coupler curve which can be generated by a four-bar linkage (not shown). Link 5 is to be attached to the coupler point, and link 6 is to be a rotating member with O_6 as the frame connection. In this problem we wish to find a coupler curve from the Hrones and Nelson atlas or, by point-position reduction, such that, for an appreciable distance, point C moves through an arc of a circle. Link 5 is then proportioned so that D lies at the center of curvature of this arc. The result is then called a *hesitation motion* because link 6 will hesitate in its rotation for the period during which point C traverses the approximate circle arc. Make a drawing of the complete linkage and plot the velocity-displacement diagram for 360° of displacement of the input link.

10-34 Synthesize a four-bar linkage to obtain a coupler curve having an approximate straight-line segment. Then, using the suggestion included in Fig. 10-28*b* or 10-30*b*, synthesize a dwell motion. Using an input crank angular velocity of unity, plot the velocity of rocker 6 vs. the input crank displacement.

10-35 Synthesize a dwell mechanism using the idea suggested in Fig. 10-28*a* and the Hrones and Nelson atlas. Rocker 6 is to have a total angular displacement of 60°. Using this displacement as the abscissa, plot a velocity diagram of the motion of the rocker to illustrate the dwell motion.

CHAPTER

ELEVEN

SPATIAL MECHANISMS

11-1 INTRODUCTION TO SPATIAL LINKAGES

As we learned in Sec. 1-5, the large majority of mechanisms in use today are *planar mechanisms*; i.e., the motions of all points produce paths which lie in parallel planes. Although this is the usual case, it is not a necessity, and mechanisms having more general, three-dimensional point paths are called *spatial mechanisms*. Another special category consists of *spherical mechanisms*, in which all point paths lie on concentric spherical surfaces.

Although these definitions have been raised in Chap. 1, almost all the examples of the previous chapters have dealt with planar mechanisms. This is justified because of their very wide use in practical situations. Although a few nonplanar mechanisms, such as universal joints and bevel gears, have been known for centuries, it is only relatively recently that kinematicians have been interested in developing design procedures for other spatial mechanisms.

Although we have concentrated on examples of planar motion so far, a brief review will show that most of the previous theory has been derived in sufficient generality for either planar or spatial motion. Examples have been planar since they can be better visualized and require less tedious computation than the three-dimensional case. Still, most of the previous theory extends directly to spatial mechanisms.

In Sec. 1-6 we learned that the *mobility* of a kinematic chain can be obtained from the Kutzbach criterion. The three-dimensional form of the criterions was given in Eq. (1-3).

$$m = 6(n-1) - 5j_1 - 4j_2 - 3j_3 - 2j_4 - j_5 \tag{11-1}$$

where m = mobility of mechanism
n = number of links
j_i = number of joints having i degrees of freedom

One of the solutions to Eq. (11-1) is $n = 7$, $j_1 = 7$, $j_2 = j_3 = j_4 = j_5 = 0$. Harrisberger calls this a *mechanism type*,† in particular, the $7j_1$ type. Other combinations of j_i's produce other types of mechanisms. For example, the $3j_1 + 2j_2$ type has five links, while the $1j_1 + 2j_3$ type has only three links.

Each mechanism type contains a finite number of *kinds* of mechanisms; there are as many kinds of mechanisms in each type as there are ways of combining the different kinds of joints. In Table 1-1 we saw that three of the six lower pairs have one degree of freedom. These are the revolute *R*, the prismatic *P*, and the screw *S*. Thus, by using any 7 of these lower pairs we get 36 kinds of type $7j_1$ mechanisms. All together, Harrisberger lists 435 kinds which satisfy the Kutzbach criterion. Not all these types or kinds are likely to have practical value, however. Consider for example, the $7j_1$ type with all revolute pairs; this defines a linkage with seven links and seven revolute joints.

For mechanisms defined by the mobility criterion as having mobility of 1, Harrisberger has selected nine kinds from two types which appear to be useful; these are illustrated in Fig. 11-1. They are all spatial four-bar linkages having four joints with either rotating or sliding input and output members. The designations in the legend, such as *RGCS* for Fig. 11-1*f* for example, identify the kinematic pair types (see Table 1-1) beginning with the input link and proceeding through the coupler and output member back to the frame. Thus, for the *RGCS*, the input crank is pivoted to the frame by the revolute *R* and to the coupler by the globular pair *G*. The coupler is paired to the output member by the cylinder *C*. The motion of the output member is determined by the screw pair *S*. The freedoms of these pairs, from Table 1-1, are $R = 1$, $G = 3$, $C = 2$, and $S = 1$.

The linkages of Fig. 11-1*a* to *c* are described by Harrisberger as type 1 mechanisms. Each is composed of one single-freedom pair and three double-freedom pairs and hence is a $1j_1 + 3j_2$ type of mechanism. The remaining linkages of Fig. 11-1 are type 2 linkages having two single-freedom pairs, one double-freedom pair, and one triple-freedom pair. Thus, they are of the $2j_1 + 1j_2 + 1j_3$ type.

11-2 SPECIAL MECHANISMS

Curiously enough, the most useful space mechanisms which have been found are those which *are not* defined as mechanisms by the mobility criterion.

† L. Harrisberger, A Number Synthesis Survey of Three-Dimensional Mechanisms, *J. Eng. Ind., ASME Trans.*, ser. B, vol. 87, no. 2, 1965.

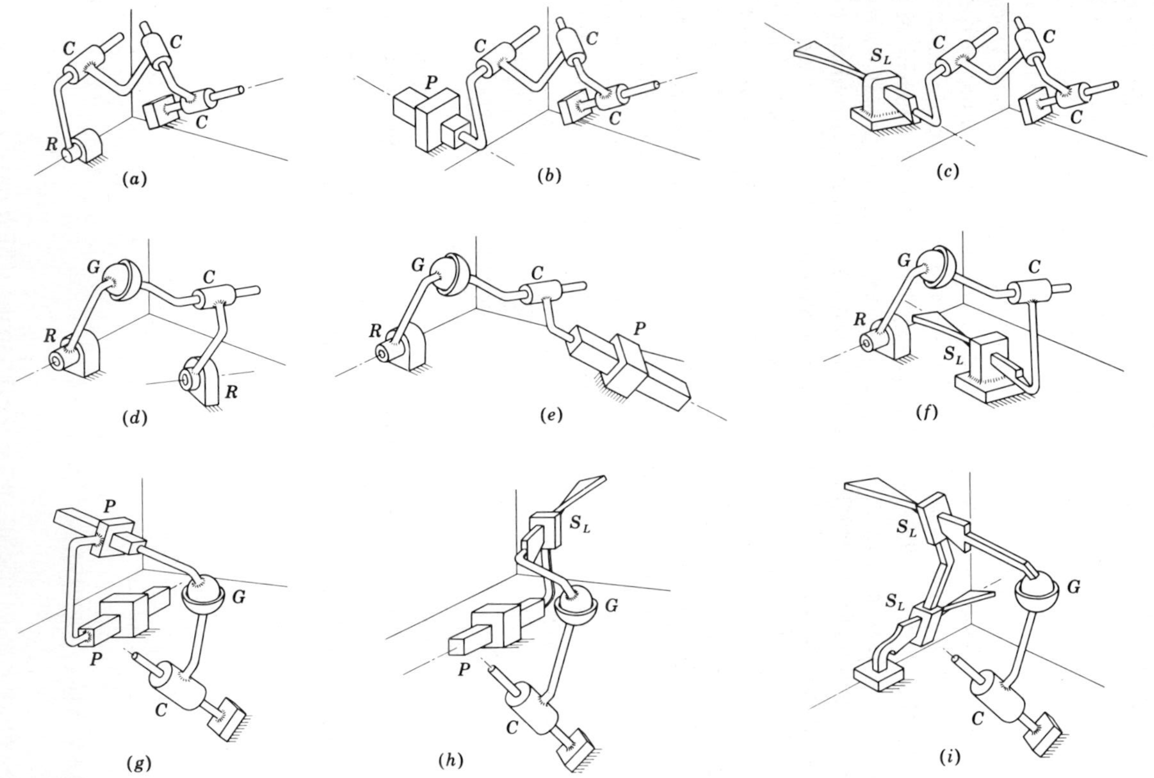

Figure 11-1 Four-bar spatial linkages having mobility of 1: (*a*) *RCCC*; (*b*) *PCCC*; (*c*) S_LCCC; (*d*) *RGCR*; (*e*) $RGC\dot{P}$; (*f*) $RGCS_L$; (*g*) *PPGC*; (*h*) PS_LGC; (*i*) S_LS_LGC. (*From L. Harrisberger, A Number Synthesis Survey of Three-Dimensional Mechanisms, J. Eng. Ind., ser. B, vol. 87, no. 2, May, 1965, published with the permission of the ASME and the author of the paper.*) In this book a screw pair is designated by the symbol *S* but Harrisberger uses S_L; presumably the subscript refers to the lead of a screw.

Because of certain geometric conditions, such as a particular ratio of link lengths or the orientation of single-freedom-pair axes, idle freedoms or idle constraints may be introduced.

At least two of the known space linkages which violate the Kutzbach criterion are four-link *RRRR* mechanisms. Thus $n = 4$, $j_1 = 4$, and Eq. (11-1) gives $m = -2$; so we conclude that there are three idle constraints. One of these mechanisms is the spherical four-bar space linkage shown in Fig. 11-2. The axes of all four revolutes intersect at the center of a sphere, and the links may be regarded as great-circle arcs existing on the surface of the sphere. Their lengths are then designated as spherical angles. By properly proportioning these angles it is possible to design all the spherical counterparts of the plane four-bar mechanism such as the spherical crank-and-rocker linkage and the spherical drag linkage. The spherical four-bar linkage is easy to design and manufacture and hence one of the most useful of all space mechanisms. The well-known Hooke, or Cardan, joint, which is the basis of the universal joint, is a special case of the spherical mechanism having input and output cranks which subtend the same angle at the center of the sphere. The *wobble-plate mechanism*, shown in Fig. 11-3, is also a special case.

Bennett's *RRRR* mechanism, shown in Fig. 11-4, is probably one of the most useless of the known space linkages. In this mechanism the opposite links are twisted the same amount and also have equal lengths. The twist angles α_1 and α_2 must also be proportioned to the lengths of the links a_1 and a_2 according to the equation

$$\frac{\sin \alpha_1}{a_1} = \frac{\sin \alpha_2}{a_2} \tag{11-2}$$

The spatial four-link *RGGR* mechanism of Fig. 11-5 is another important and useful linkage. Since $n = 4$, $j_1 = 2$, and $j_3 = 2$, the mobility criterion of Eq. (11-1) predicts $m = 2$. Although this might appear to be another exception at first glance, on closer examination we find that the extra degree of freedom

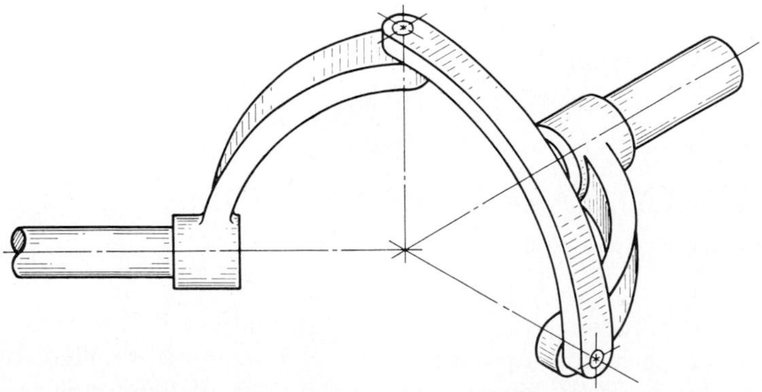

Figure 11-2 The spherical four-bar linkage.

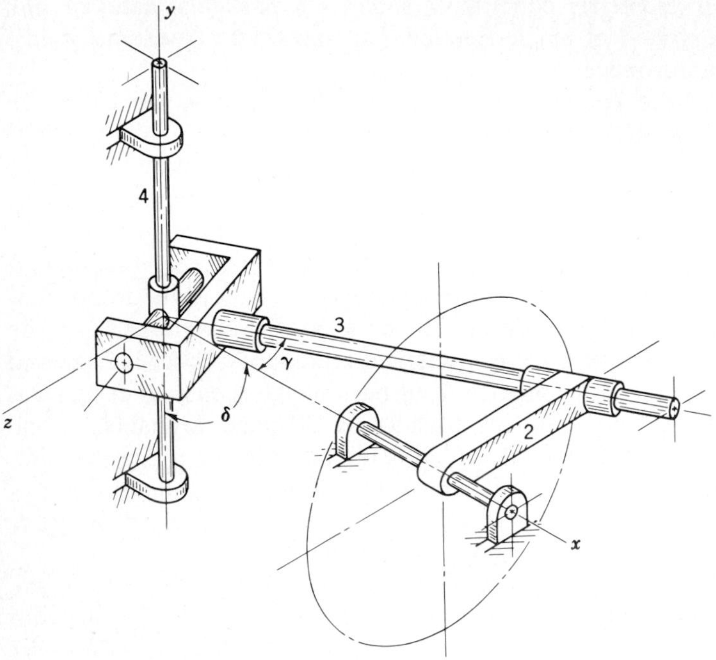

Figure 11-3 The wobble-plate mechanism; input crank 2 rotates and output shaft 4 oscillates. When $\delta \neq 90°$ the mechanism is called a spherical slide oscillator. If $\gamma > \delta$, the output shaft rotates.

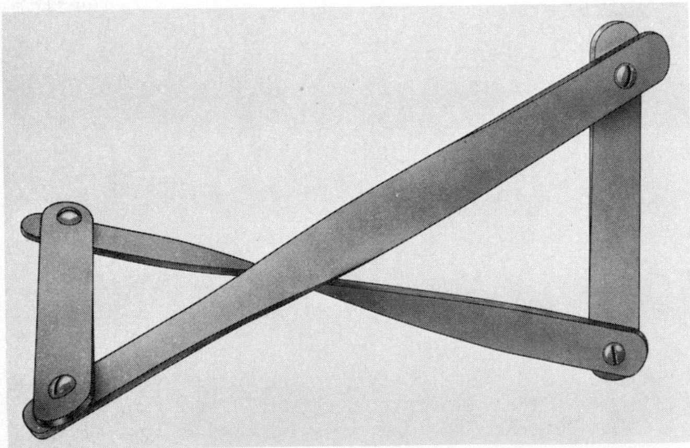

Figure 11-4 The Bennett four-link mechanism.

actually exists; it is the freedom of the coupler to spin about its own axis. Since it does not affect the input-output kinematic relationship it is called an *idle* freedom. This extra freedom does no harm if the mass of the coupler is distributed along its axis; in fact, it may be an advantage because it is easy to manufacture and the rotation of the coupler about its axis should equalize the wear on the two ball-and-socket joints. If the mass center of the coupler lies off-axis, however, this extra freedom is not idle dynamically and can cause quite erratic performance at high speed.

Still other exceptions to the mobility criterion are the Goldberg (not Rube) five-bar five-revolute mechanism and the Bricard six-bar six-revolute linkage.† Again, it is doubtful that these mechanisms have any practical value.

Harrisberger and Soni have sought to identify all spatial linkages having one general constraint.‡ They have identified 8 types and 212 kinds, and found 7 new mechanisms which may have usefulness.

11-3 THE POSITION PROBLEM

Like planar mechanisms, a spatial mechanism is usually connected to form a closed loop. Thus, following methods similar to those of Sec. 2-6, a loop-closure equation can be written which will define the kinematic relationships of the mechanism. A number of different mathematical forms can be used including vectors, dual numbers, and quaternians,§ as well as matrices.¶ In keeping with the remainder of the book, we will use vector notation. The loop-closure condition for a spatial linkage such as the mechanism of Fig. 11-5 can be defined by a vector equation of the form

$$\mathbf{r}+\mathbf{s}+\mathbf{t}+\mathbf{C}=0 \tag{11-3}$$

This equation is called the *vector tetrahedron equation* because the individual vectors can be thought of as defining four of the six edges of a tetrahedron.

The vector tetrahedron equation is three-dimensional and hence can be solved for three scalar unknowns. These can exist in any combination in vectors **r**, **s**, and **t**. The vector **C** is the sum of all known vectors in the loop. By using spherical coordinates, each of the vectors **r**, **s**, and **t** can be expressed as a magnitude and two angles. Vector **r**, for example, is defined once its magnitude r and two angles θ_r and ϕ_r are known. Thus, in Eq. (11-3), any three of the nine quantities r, θ_r, ϕ_r, s, θ_s, ϕ_s, t, θ_t, and ϕ_t can be unknown. When these are solved, there are just nine combinations of unknowns which

† For pictures of these see R. S. Hartenberg and J. Denavit, "Kinematic Synthesis of Linkages," McGraw-Hill, New York, 1964, pp. 85–86.

‡ L. Harrisberger and A. H. Soni, A Survey of Three-Dimensional Mechanisms with One General Constraint, *ASME pap.* 66-MECH-44, October 1966. This paper contains 45 references on spatial mechanisms.

§ A. T. Yang and F. Freudenstein, Application of Dual-Number and Quaternian Algebra to the Analysis of Spatial Mechanisms, *J. Appl. Mech., ASME Trans.*, ser. E, vol. 86, pp. 300–308, 1964.

¶ J. J. Uicker, Jr., J. Denavit, and R. S. Hartenberg, An Iterative Method for the Displacement Analysis of Spatial Mechanisms, *J. Appl. Mech., ASME Trans.*, ser. E, vol. 87, pp. 309–314, 1965.

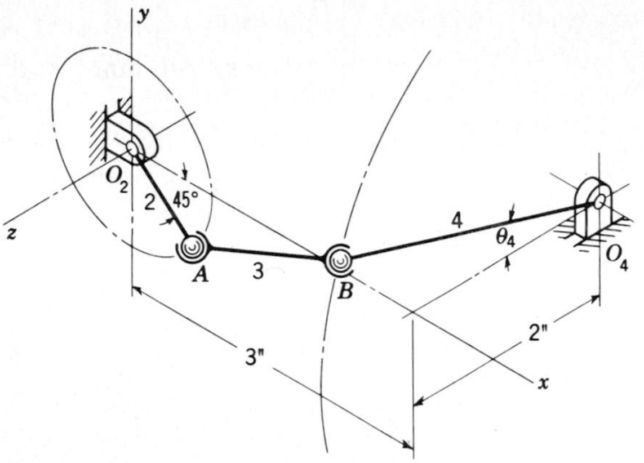

Figure 11-5 The *RGGR* linkage. $R_{AO_2} = 1$ in, $R_{BA} = 3.5$ in, $R_{BO_4} = 4$ in.

result in different solutions. Chace† has solved these nine cases by first reducing each to a polynomial. He classifies the solutions depending upon whether the unknowns occur in one, two, or three vectors and tabulates the forms of the solutions as shown in Table 11-1. In this table, the unit vectors $\hat{\boldsymbol{\omega}}_r$, $\hat{\boldsymbol{\omega}}_s$, and $\hat{\boldsymbol{\omega}}_t$ are *known* directions of axes from which the *known* angles ϕ_r, ϕ_s, and ϕ_t are measured. In case 1, vectors s and **t** are completely known and are summed into the vector **C**. In cases 2*a*, 2*b*, 2*c*, and 2*d*, vector **t** is known and summed into **C**. Cases 3*a*, 3*b*, 3*c*, and 3*d* have unknowns in each of the vectors **r**, s, and **t**.

One major advantage of the Chace solutions to the vector tetrahedron equations is that since they provide known forms for the solutions of the nine cases, it is easy to write a family of nine subprograms for computer or

† M. A. Chace, Vector Analysis of Linkages, *J. Eng. Ind.*, *ASME Trans.*, ser. B, vol. 85, no. 3, pp. 289–297, 1963.

Table 11-1 Classification of the solutions to the vector tetrahedron equation

Case number	Unknowns	Known quantities: Vectors	Known quantities: Scalars	Degree of polynomial
1	r, θ_r, ϕ_r	$\mathbf{C}$		1
2*a*	r, θ_r, s	$\mathbf{C}, \hat{\mathbf{s}}, \hat{\boldsymbol{\omega}}_r$	ϕ_r	2
2*b*	r, θ_r, θ_s	$\mathbf{C}, \hat{\boldsymbol{\omega}}_r, \hat{\boldsymbol{\omega}}_s$	ϕ_r, s, ϕ_s	4
2*c*	θ_r, ϕ_r, s	$\mathbf{C}, \hat{\mathbf{s}}$	r	2
2*d*	$\theta_r, \phi_r, \theta_s$	$\mathbf{C}, \hat{\boldsymbol{\omega}}_s$	r, s, ϕ	2
3*a*	r, s, t	$\mathbf{C}, \hat{\mathbf{r}}, \hat{\mathbf{s}}, \hat{\mathbf{t}}$		1
3*b*	r, s, θ_t	$\mathbf{C}, \hat{\mathbf{r}}, \hat{\mathbf{s}}, \hat{\boldsymbol{\omega}}_t$	t, ϕ_t	2
3*c*	r, θ_s, θ_t	$\mathbf{C}, \hat{\mathbf{r}}, \hat{\boldsymbol{\omega}}_s, \hat{\boldsymbol{\omega}}_t$	s, ϕ_s, t, ϕ_t	4
3*d*	$\theta_r, \theta_s, \theta_t$	$\mathbf{C}, \hat{\boldsymbol{\omega}}_r, \hat{\boldsymbol{\omega}}_s, \hat{\boldsymbol{\omega}}_t$	$r, \phi_r, s, \phi_s, t, \phi_t$	8

calculator evaluation. These would follow the same general approach as described in Sec. 5-3 for the equivalent planar equations. All nine cases except case $3d$ have been reduced to explicit closed-form solutions for the unknowns and can therefore be quickly evaluated. Only case $3d$, involving the solution of an eighth-order polynomial, must be solved by iterative techniques.

Although the vector tetrahedron equation and its nine case solutions can be used to solve most practical spatial mechanisms, we recall from Sec. 11-1 that the Kutzbach criterion predicts the existence of up to seven j_1 joints in a single-loop mechanism with one degree of freedom. A case such as the $7R$ mechanism, for example, would have six unknowns to be solved for from the loop-closure equation. This is not possible from the vector form of the loop-closure equation, and dual quaternians or matrices must be used instead. Such problems also lead to very high-order polynomials and require iterative solution for their final evaluation. Anyone attemping the solution of such equations by hand algebraic techniques quickly appreciates that *position* analysis, not velocity or acceleration, is the most difficult problem in kinematics.

11-4 POSITION ANALYSIS OF THE *RGGR* MECHANISM

Solving the polynomials of Chace's vector tetrahedron equation turns out to be equivalent to finding the intersections of straight lines or circles with various surfaces of revolution. Such problems can usually be easily and quickly solved by the graphical methods of descriptive geometry. The graphic approach has the additional advantage that the geometric nature of the problem is not concealed in a multiplicity of mathematical operations.

Let us use a four-link *RGGR* crank-and-rocker mechanism in which the knowns are the position and plane of rotation of the input link, the plane of rotation of the output link, and the dimensions of all four links. Such a mechanism is shown in Fig 11-5. The position problem consists in finding the position of the coupler and rocker, links 3 and 4. If we treat link 4 as a vector, then the only unknown is one angle, because the magnitude and the plane of oscillation are given. Similarly, if link 3 is a vector, its magnitude is known but there exist two unknowns which are the two angular directions in spherical coordinates. We identify this as case $2d$ in Table 11-1, requiring the solution of a second-degree polynomial and hence yielding two solutions.

This problem is solved by using only two orthographic views, the front and profile. If we imagine, in Fig. 11-5, that the coupler is disconnected from B and permitted to occupy all positions relative to A, then B must lie on the surface of a sphere whose center is at A. With the coupler still disconnected, the motion of B on link 4 is a circle about O_4 in a plane parallel to the yz plane. Therefore, to solve this problem we need only find the two points of intersection of a circle with a sphere.

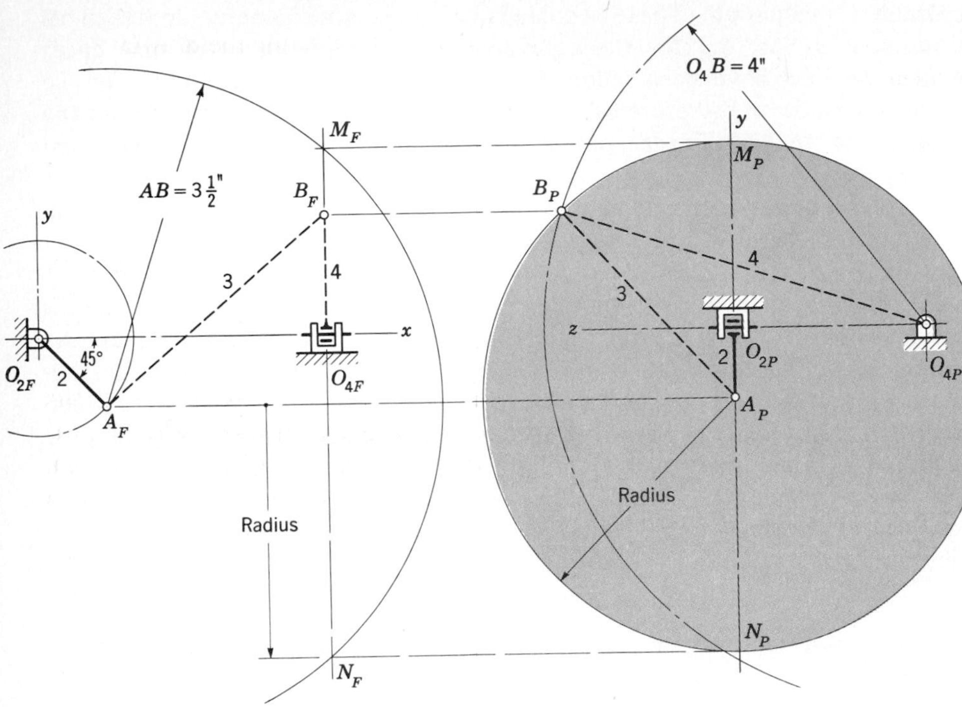

Figure 11-6 Graphic position analysis of the *RGGR* mechanism.

The solution is shown in Fig. 11-6. The subscripts F and P denote projections on the frontal and profile planes, respectively. First locate O_2, A, and O_4 on both views. On the profile view draw a circle of radius $O_4B = 4$ in about O_{4P}; this is the path of point B. This circle appears as the vertical line $M_FO_{4F}N_F$ on the front view. Next, on the front view construct the outline of a sphere with A_F as a center and the coupler length $AB = 3\frac{1}{2}$ in as the radius. If $M_FO_{4F}N_F$ is regarded as the trace of a plane normal to the frontal plane, the intersection of this plane with the sphere appears as the shaded circle on the profile view of diameter M_PN_P. The arc of radius O_4B intersects the circle in two points, yielding two solutions. One of these points is chosen for B_P and projected back to the front view to locate B_F. Links 3 and 4 are now drawn in, as dashed lines in this case, in the front and profile views.

By simply measuring the x, y, and z projections from the graphic solution, one can write the vector expressions for each link:

$$\begin{aligned}
\mathbf{r}_1 &= 3\hat{\mathbf{i}} - 2\hat{\mathbf{k}} \\
\mathbf{r}_2 &= 0.707\hat{\mathbf{i}} - 0.707\hat{\mathbf{j}} \\
\mathbf{r}_3 &= 2.30\hat{\mathbf{i}} + 1.95\hat{\mathbf{j}} + 1.77\hat{\mathbf{k}} \\
\mathbf{r}_4 &= 1.22\hat{\mathbf{j}} + 3.81\hat{\mathbf{k}}
\end{aligned} \tag{11-4}$$

where $\mathbf{r}_1$, $\mathbf{r}_2$, $\mathbf{r}_3$, and $\mathbf{r}_4$ are directed from O_2 to O_4, O_2 to A, A to B, and O_4 to

B, respectively. The components shown above were obtained from a full-size solution; better accuracy would result, of course, by making the drawings 2 or 4 times actual size.

The four-revolute spherical four-link mechanism, shown in Fig. 11-2, is case 2*d* of the vector tetrahedron equation and can be solved in the same manner when the position of the input link is given.

11-5 VELOCITY AND ACCELERATION ANALYSIS OF THE *RGGR* LINKAGE

Once the positions of all the members of a spatial mechanism have been found, the velocities and accelerations can be determined by using the methods of Chaps. 3 and 4. In analyzing planar mechanisms the angular velocities and accelerations were always perpendicular to the plane of motion and hence had only one nonzero vector component. In the analysis of spatial linkages these terms may have three components since their axes may be skew in space. Otherwise the methods of analysis are the same. The following example will serve to illustrate these differences.

Example 11-1 The angular velocity of link 2 of the four-bar *RGGR* linkage of Fig. 11-7 is $\boldsymbol{\omega}_2 = 40\hat{\mathbf{k}}$ rad/s. Find the angular velocity and acceleration of links 3 and 4 and the velocity and acceleration of point *B*.

SOLUTION Using descriptive geometry to solve the position problem as explained in Sec. 11-4 results in the three-view drawing of the linkage shown in Fig. 11-8. We now replace O_2A, AB, and O_4B with vectors $\mathbf{r}_2$, $\mathbf{r}_3$, and $\mathbf{r}_4$, respectively. The components can be read directly from Fig. 11-8:

$$\mathbf{r}_2 = 2\hat{\mathbf{i}} + 3.46\hat{\mathbf{j}} \qquad \mathbf{r}_3 = 10\hat{\mathbf{i}} + 2.71\hat{\mathbf{j}} + 10.89\hat{\mathbf{k}} \qquad \mathbf{r}_4 = 6.17\hat{\mathbf{j}} + 7.89\hat{\mathbf{k}}$$

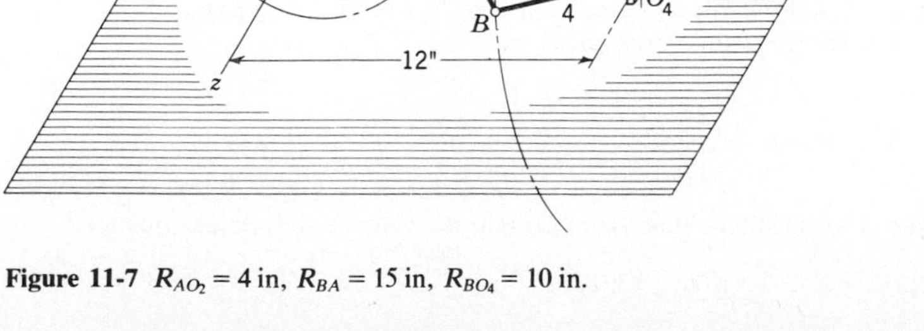

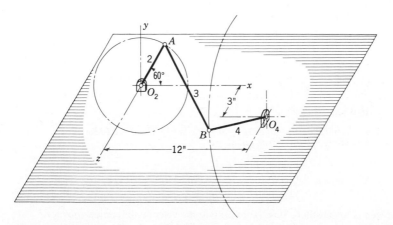

Figure 11-7 $R_{AO_2} = 4$ in, $R_{BA} = 15$ in, $R_{BO_4} = 10$ in.

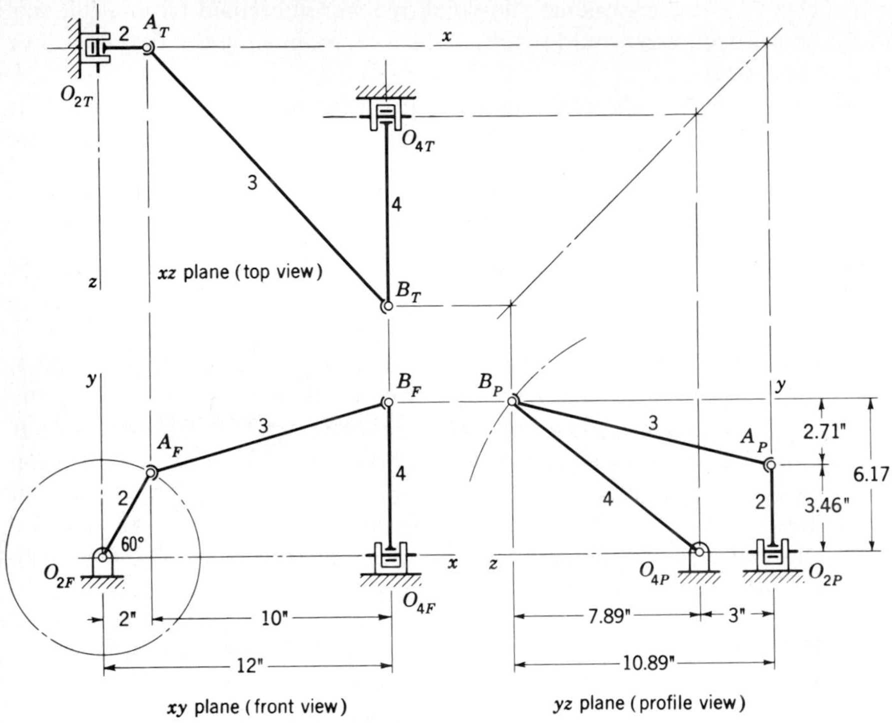

Figure 11-8 Example 11-1: position analysis.

From the constraints we see that the angular velocities and accelerations can be written as

$$\boldsymbol{\omega}_2 = 40\hat{\mathbf{k}} \qquad \boldsymbol{\omega}_3 = \omega_3^x\hat{\mathbf{i}} + \omega_3^y\hat{\mathbf{j}} + \omega_3^z\hat{\mathbf{k}} \qquad \boldsymbol{\omega}_4 = \omega_4\hat{\mathbf{i}}$$

$$\boldsymbol{\alpha}_2 = 0 \qquad \boldsymbol{\alpha}_3 = \alpha_3^x\hat{\mathbf{i}} + \alpha_3^y\hat{\mathbf{j}} + \alpha_3^z\hat{\mathbf{k}} \qquad \boldsymbol{\alpha}_4 = \alpha_4\hat{\mathbf{i}}$$

First, we find $\mathbf{V}_A$ as the velocity difference from point O_2. Thus

$$\mathbf{V}_A = \boldsymbol{\omega}_2 \times \mathbf{r}_2 = \tfrac{1}{12}\begin{vmatrix} \hat{\mathbf{i}} & \hat{\mathbf{j}} & \hat{\mathbf{k}} \\ 0 & 0 & 40 \\ 2 & 3.46 & 0 \end{vmatrix} = -11.53\hat{\mathbf{i}} + 6.67\hat{\mathbf{j}} \tag{1}$$

Similarly,

$$\mathbf{V}_{BA} = \boldsymbol{\omega}_3 \times \mathbf{r}_3 = \tfrac{1}{12}\begin{vmatrix} \hat{\mathbf{i}} & \hat{\mathbf{j}} & \hat{\mathbf{k}} \\ \omega_3^x & \omega_3^y & \omega_3^z \\ 10 & 2.71 & 10.89 \end{vmatrix}$$

$$= (0.908\omega_3^y - 0.226\omega_3^z)\hat{\mathbf{i}} + (0.833\omega_3^z - 0.908\omega_3^x)\hat{\mathbf{j}} + (0.226\omega_3^x - 0.833\omega_3^y)\hat{\mathbf{k}} \tag{2}$$

And, finally,

$$\mathbf{V}_B = \boldsymbol{\omega}_4 \times \mathbf{r}_4 = \tfrac{1}{12}\begin{vmatrix} \hat{\mathbf{i}} & \hat{\mathbf{j}} & \hat{\mathbf{k}} \\ \omega_4 & 0 & 0 \\ 0 & 6.17 & 7.89 \end{vmatrix} = -0.658\omega_4\hat{\mathbf{j}} + 0.514\omega_4\hat{\mathbf{k}} \tag{3}$$

The next step is to substitute Eqs. (1) to (3) into the velocity-difference equation

$$\mathbf{V}_B = \mathbf{V}_A + \mathbf{V}_{BA} \tag{4}$$

When this is done, the $\hat{\mathbf{i}}$, $\hat{\mathbf{j}}$, and $\hat{\mathbf{k}}$ components can be separated to obtain three algebraic equations

$$0.908\omega_3^y - 0.226\omega_3^z = 11.53 \tag{5}$$

$$-0.908\omega_3^x + 0.833\omega_3^z + 0.658\omega_4 = -6.67 \tag{6}$$

$$0.226\omega_3^x - 0.833\omega_3^y - 0.514\omega_4 = 0 \tag{7}$$

We note, however, that there are four unknowns, ω_3^x, ω_3^y, ω_3^z, and ω_4. This would not normally occur in most problems but does here because of the idle freedom of the coupler to spin about its own axis. Since this spin will not affect the input-output relationship, we would get the same result for ω_4 no matter what the spin. We can therefore set one component of ω_3 to zero and proceed. Another approach is to set the spin velocity to zero by requiring that

$$\boldsymbol{\omega}_3 \cdot \mathbf{r}_3 = 0$$

$$0.833\omega_3^x + 0.226\omega_3^y + 0.908\omega_3^z = 0 \tag{8}$$

Equations (5) through (8) can now be solved simultaneously for the four unknowns. The result is

$$\boldsymbol{\omega}_3 = -1.72\hat{\mathbf{i}} + 13.6\hat{\mathbf{j}} + 3.70\hat{\mathbf{k}} \text{ rad/s} \qquad \textit{Ans.}$$

$$\boldsymbol{\omega}_4 = -25.5\hat{\mathbf{i}} \text{ rad/s} \qquad \textit{Ans.}$$

Substituting into Eq. (3), we get

$$\mathbf{V}_B = 16.8\hat{\mathbf{j}} - 13.1\hat{\mathbf{k}} \text{ ft/s} \qquad \textit{Ans.}$$

Turning next to the acceleration analysis, we compute the following components:

$$\mathbf{A}^n_{AO_2} = \boldsymbol{\omega}_2 \times (\boldsymbol{\omega}_2 \times \mathbf{r}_2) = \boldsymbol{\omega}_2 \times \mathbf{V}_A$$

$$= \begin{vmatrix} \hat{\mathbf{i}} & \hat{\mathbf{j}} & \hat{\mathbf{k}} \\ 0 & 0 & 40 \\ -11.53 & 6.67 & 0 \end{vmatrix} = -267\hat{\mathbf{i}} - 461\hat{\mathbf{j}} \tag{9}$$

$$\mathbf{A}^t_{AO_2} = \boldsymbol{\alpha}_2 \times \mathbf{r} = 0$$

$$\mathbf{A}^n_{BA} = \boldsymbol{\omega}_3 \times (\boldsymbol{\omega}_3 \times \mathbf{r}_3) = \boldsymbol{\omega}_3 \times \mathbf{V}_{BA} \tag{10}$$

$$= \begin{vmatrix} \hat{\mathbf{i}} & \hat{\mathbf{j}} & \hat{\mathbf{k}} \\ -7.72 & 13.6 & 3.70 \\ 11.51 & 10.09 & -13.07 \end{vmatrix} = -215\hat{\mathbf{i}} - 58\hat{\mathbf{j}} - 234\hat{\mathbf{k}} \tag{11}$$

$$\mathbf{A}^t_{BA} = \boldsymbol{\alpha}_3 \times \mathbf{r}_3$$

$$= \frac{1}{12}\begin{vmatrix} \hat{\mathbf{i}} & \hat{\mathbf{j}} & \hat{\mathbf{k}} \\ \alpha_3^x & \alpha_3^y & \alpha_3^z \\ 10 & 2.71 & 10.89 \end{vmatrix}$$

$$= (0.908\alpha_3^y - 0.226\alpha_3^z)\hat{\mathbf{i}} + (0.833\alpha_3^z - 0.908\alpha_3^x)\hat{\mathbf{j}}$$

$$+ (0.226\alpha_3^x - 0.833\alpha_3^y)\hat{\mathbf{k}} \tag{12}$$

$$\mathbf{A}^n_{BO_4} = \boldsymbol{\omega}_4 \times (\boldsymbol{\omega}_4 \times \mathbf{r}_4) = \boldsymbol{\omega}_4 \times \mathbf{V}_B$$

$$= \begin{vmatrix} \hat{\mathbf{i}} & \hat{\mathbf{j}} & \hat{\mathbf{k}} \\ -23.8 & 0 & 0 \\ 0 & 16.8 & 13.1 \end{vmatrix} = -312\hat{\mathbf{j}} - 400\hat{\mathbf{k}} \tag{13}$$

$$\mathbf{A}^t_{BO_4} = \boldsymbol{\alpha}_4 \times \mathbf{r}_4$$

$$= \frac{1}{12}\begin{vmatrix} \hat{\mathbf{i}} & \hat{\mathbf{j}} & \hat{\mathbf{k}} \\ \alpha_4 & 0 & 0 \\ 0 & 6.17 & 7.89 \end{vmatrix} = -0.658\alpha_4\hat{\mathbf{j}} + 0.514\alpha_4\hat{\mathbf{k}} \tag{14}$$

These quantities are substituted into the acceleration-difference equation

$$\mathbf{A}^n_{BO_4} + \mathbf{A}^t_{BO_4} = \mathbf{A}^n_{AO_2} + \mathbf{A}^t_{AO_2} + \mathbf{A}^n_{BA} + \mathbf{A}^t_{BA} \tag{15}$$

and, along with the condition $\boldsymbol{\alpha}_3 \cdot \mathbf{r}_3 = 0$ on the spin of the idle freedom, the results can be obtained exactly as above:

$$\boldsymbol{\alpha}_3 = -569\hat{\mathbf{i}} + 623\hat{\mathbf{j}} + 368\hat{\mathbf{k}} \text{ rad/s}^2 \qquad \textit{Ans.}$$

$$\boldsymbol{\alpha}_4 = -937\hat{\mathbf{i}} \text{ rad/s}^2 \qquad \textit{Ans.}$$

$$\mathbf{A}_B = \mathbf{A}^n_{BO_4} + \mathbf{A}^t_{BO_4} = 304\hat{\mathbf{j}} - 881\hat{\mathbf{k}} \text{ ft/s}^2 \qquad \textit{Ans.}$$

The determination of the velocities and accelerations of a space mechanism by graphical means is conducted in the same manner as for a plane-motion mechanism. However, the velocity and acceleration vectors which appear on the standard front, top, and profile views are not usually seen in their true length; i.e., they are foreshortened. This means that the last step in constructing the vector polygon must be completed in an auxiliary view in which the unknown vector does appear in its true length.

The directions of the vectors depend upon the directions of the elements of the mechanism; for this reason, it is necessary to project one of the links of the mechanism into the auxiliary view or views too. Also, for this reason, we choose to connect the poles of the vector polygons to a point on one of the links so that the relationship between the vectors and one of the links is evident in all the views.

Once the unknown vector or vectors are obtained in the auxiliary views, they can be projected back into the standard three-view orthogonal system and the lengths of the x, y, and z projections measured directly. The procedure is best illustrated by an example.

Example 11-2 Construct the velocity and acceleration polygons for the graphical solution of Example 11-1.

SOLUTION The velocity solution is shown in Fig. 11-9, and the notation corresponds with that used in many works on descriptive geometry. The letters F, T, and P designate the front, top, and profile planes, and the numbers 1 and 2 the first and second auxiliary planes of projection. Points projected upon these planes bear the subscripts F, T, P, etc. The steps in obtaining the velocity solution are as follows:

1. Construct the front, profile, and top views of the linkage, and designate each point.
2. Calculate $\mathbf{V}_A$, and place this vector in position with the origin at A on the three views. The velocity of A shows in its true length on the front view. Designate the terminus of $\mathbf{V}_A$ as a_F, and project to the top and profile views.
3. The velocity of B is unknown, but not its direction. The direction is perpendicular to link 4 and in the direction that 4 rotates. When the problem is solved, $\mathbf{V}_B$ will show in its true length in the profile view. Construct a line in the profile view to correspond with the known direction of $\mathbf{V}_B$. Locate any point d_P on this line, and project to the front and top views.
4. The equation to be solved is

$$\mathbf{V}_B = \mathbf{V}_A + \mathbf{V}_{BA} \tag{16}$$

where $\mathbf{V}_A$ and the directions of $\mathbf{V}_B$ and $\mathbf{V}_{BA}$ are known. Note that $\mathbf{V}_{BA}$ is perpendicular to

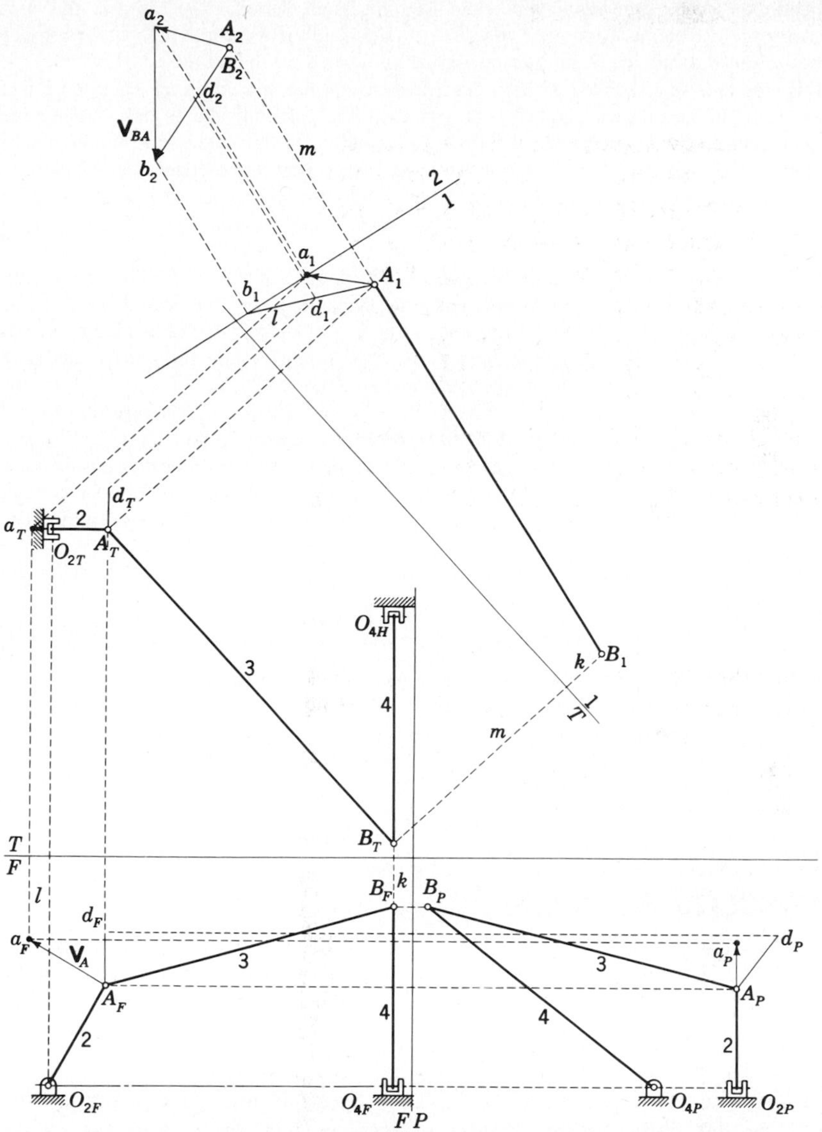

Figure 11-9 Example 11-2: velocity analysis.

link 3, but its magnitude is unknown. In space the lines perpendicular to link 3 are like the spokes of a wheel, link 3 being the axis of rotation of that wheel. There are therefore an infinite number of lines perpendicular to link 3, but we are interested in only one of them. The line we require must originate at the terminus of $\mathbf{V}_A$ and terminate by intersecting the line Ad or its extension. In order to choose this line from the infinite number of available ones we need to examine AB in the direction in which it appears as a point. Therefore in this step we must project AB upon a plane which shows its true length; so we construct the edge view of plane 1 parallel to A_TB_T, and project AB to this plane. In making this projection, note that the distances k and l in the front view are the same in this first auxiliary view. The auxiliary view of AB is A_1B_1, which is its true length. Also project points a and d to this view, but the remaining links need not be projected.

5. In this step choose a second auxiliary plane 2 such that the projection of AB upon it is a point. Then all lines drawn parallel to the plane will be perpendicular to link 3. The edge view of such a plane is perpendicular to A_1B_1 extended. In this example it is convenient to choose this plane so that it contains point a; so construct the edge view of plane 2 through point a_1 perpendicular to A_1B_1 extended. Now project points A, B, a, and d upon this plane. Note that the distances, m, for example, of points from plane 1 must be the same with respect to plane 2.
6. Extend line A_1d_1 until it intersects the edge view of plane 2 at b_1, and find the projection b_2 of this point in plane 2. Now both a and b lie in plane 2; any line drawn in plane 2 is perpendicular to link 3. Therefore the line ab is $\mathbf{V}_{BA}$, and the view of it in the second auxiliary is its true length. The line A_2b is the projection of $\mathbf{V}_B$ on the second auxiliary plane, but not in its true length because A is not in plane 2.
7. (In order to simplify reading of the drawing, step 7 is not shown; if you carefully follow the first six steps you will have no difficulty with the seventh.) Project the three vectors back to the front, top, and profile views. $\mathbf{V}_B$ can then be measured from its profile view because it appears in its true length in this view. When all the vectors have been projected back to these three views, the x, y, and z projections can be measured directly.

The solution to the acceleration problem is obtained in an identical manner by using the same two auxiliary planes. The equation to be solved is

$$\mathbf{A}_B^n + \mathbf{A}_B^t = \mathbf{A}_A^n + \mathbf{A}_A^t + \mathbf{A}_{BA}^n + \mathbf{A}_{BA}^t \tag{17}$$

where the vectors $\mathbf{A}_B^n$, $\mathbf{A}_A^n$, $\mathbf{A}_A^t$, and $\mathbf{A}_{BA}^n$ are known or can be found once the velocity polygon is completed. Also, upon comparing Eq. (17) with Eq. (16), we see that $\mathbf{A}_B^t$ and $\mathbf{A}_{BA}^t$ have the same directions, respectively, as $\mathbf{V}_B$ and $\mathbf{V}_{BA}$. The solution can therefore proceed exactly as for the velocity polygon. The only difference in the approach is that there are more known vectors.

11-6 THE EULERIAN ANGLES

We saw in Sec. 3-2 that angular velocity is a vector quantity; hence, like all vectors, it can be resolved into rectangular components

$$\boldsymbol{\omega} = \omega^x\hat{\mathbf{i}} + \omega^y\hat{\mathbf{j}} + \omega^z\hat{\mathbf{k}} \tag{a}$$

Unfortunately, we also saw in Fig. 3-2 that three-dimensional angular displacements do not behave like vectors. Therefore, we cannot find a set of three angles which specify the orientation of a rigid body and which also have ω^x, ω^y, and ω^z as their time derivatives.

To clarify the problem further, we visualize a rigid body rotating in space about a fixed point O at the origin of a ground or *absolute reference system* xyz. A moving reference system $x'y'z'$ is then defined so that it is fixed to the rotating body. The axes of the $x'y'z'$ system are called *body-fixed axes*. We might define the orientation of $x'y'z'$ by using direction cosines, but nine of them would be required, and they would be related by six orthogonality relations.

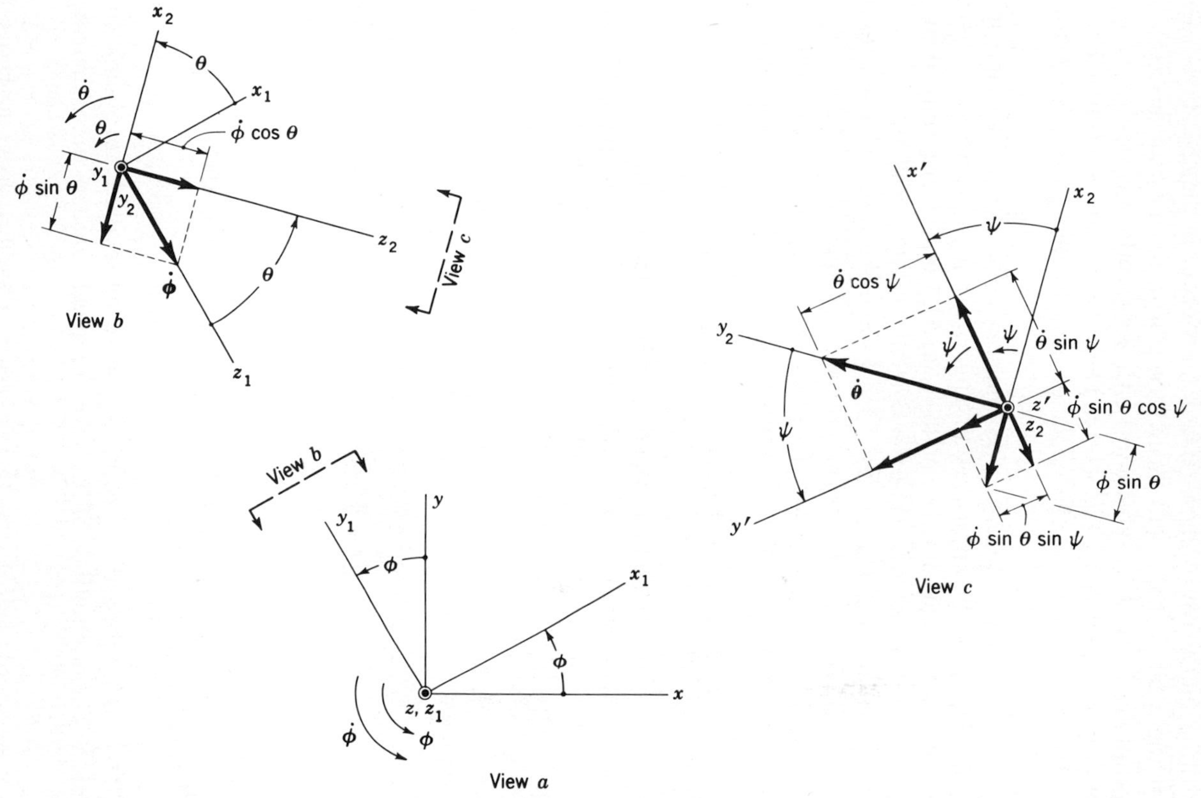

Figure 11-10 Orthographic views of the three successive rotations which define the eulerian angles.

Three angles, called the *eulerian angles*, can be used to specify the orientation of the body-fixed axes. To illustrate the eulerian angles, we begin with the body-fixed axes coincident with the absolute reference axes. We then specify three successive rotations, which must occur in the specified order, to arrive at the $x'y'z'$ orientation.† A pictorial three-dimensional description of these rotations is most unsatisfactory. Consequently we shall employ the three orthographic views of Fig. 11-10. These views are all arranged so that the axes are in the plane of the paper or are directed positively out of the paper.

The first rotation is through the angle ϕ about the z axis and in the positive direction, as shown in view *a*. This rotation yields the $x_1y_1z_1$ system. Thus x rotates through ϕ to x_1, y to y_1 and z and z_1 are coincident. We visualize an angular-velocity vector $\dot{\boldsymbol{\phi}}$ coincident with z and z_1.

The next step is to construct view *b* by projecting orthographically along the positive y_1 axis. The second rotation is through the angle θ about the y_1 axis and in the positive direction, as shown. This rotation yields the $x_2y_2z_2$ system with z_1 rotating through θ to z_2 and x_1 to x_2. Note that y_1 and y_2 are coincident and that we can visualize another angular-velocity vector $\dot{\boldsymbol{\theta}}$ directed along the positive y_2 axis. Note also that the vector $\dot{\boldsymbol{\phi}}$ has been resolved into components along the x_2 and z_2 axes.

The last step is started by projecting orthographically from the positive z_2 axis in view *b* to obtain view *c*. This makes the vector $\dot{\boldsymbol{\theta}}$ appear on the positive y_2 axis, and the z_2 axis directed positively out of the figure. The third rotation is through the angle ψ about the z_2 axis. This yields the desired orientation and the $x'y'z'$ axes. The angular velocities are then resolved again into components along the $x'y'z'$ axes. By using views *b* and *c*, the components can be summed to yield

$$\omega^{x'} = \dot{\theta} \sin \psi - \dot{\phi} \sin \theta \cos \psi \tag{11-5}$$

$$\omega^{y'} = \dot{\theta} \cos \psi + \dot{\phi} \sin \theta \sin \psi \tag{11-6}$$

$$\omega^{z'} = \dot{\psi} + \dot{\phi} \cos \theta \tag{11-7}$$

11-7 A THEOREM ON ANGULAR VELOCITIES AND ACCELERATIONS

Figure 11-11 is a schematic drawing of the seven-revolute seven-link spatial mechanism. The orientations of the seven revolute pair axes are schematically

† There appears to be little agreement among authors on how these angles should be defined. Here we employ the definition given by H. Yeh and J. I. Abrams, "Principles of Mechanics of Solids and Fluids," vol. 1, McGraw-Hill, New York, 1960, pp. 131–133, and by J. L. Synge and B. A. Griffith, "Principles of Mechanics," 3d ed., McGraw-Hill, New York, 1959, pp. 259–261. A wide variety of other definitions, differing in the axes about which the successive rotations are measured, are to be found in other references.

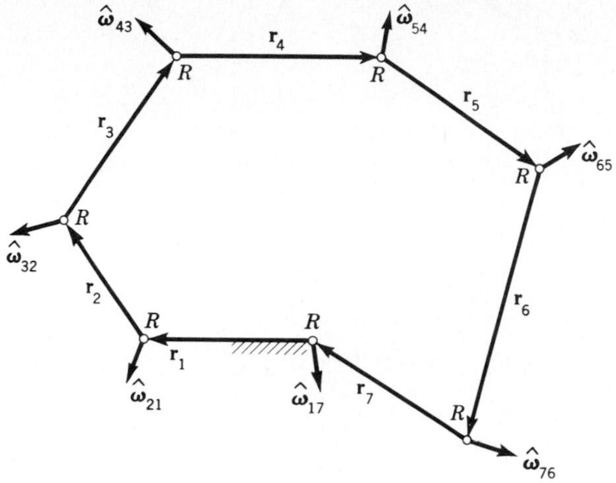

Figure 11-11

represented by the unit apparent-velocity vectors $\hat{\boldsymbol{\omega}}_{ij}$, which are directed along the axes of the pairs. We assume that there are no special geometric relationships and that the linkage therefore has a mobility of 1.

To develop the theorem on angular velocities, we note that

$$\boldsymbol{\omega}_{31} = \boldsymbol{\omega}_{21} + \boldsymbol{\omega}_{32} \qquad (a)$$

which is the apparent-angular-velocity equation (3-11).† It is convenient to rewrite Eq. (a) as

$$\boldsymbol{\omega}_{21} - \boldsymbol{\omega}_{31} + \boldsymbol{\omega}_{32} = 0 \qquad (b)$$

and then, proceeding in a similar manner around the loop, we have

$$\boldsymbol{\omega}_{31} - \boldsymbol{\omega}_{41} + \boldsymbol{\omega}_{43} = 0 \qquad (c)$$

$$\boldsymbol{\omega}_{41} - \boldsymbol{\omega}_{51} + \boldsymbol{\omega}_{54} = 0 \qquad (d)$$

$$\boldsymbol{\omega}_{51} - \boldsymbol{\omega}_{61} + \boldsymbol{\omega}_{65} = 0 \qquad (e)$$

$$\boldsymbol{\omega}_{61} - \boldsymbol{\omega}_{71} + \boldsymbol{\omega}_{76} = 0 \qquad (f)$$

$$\boldsymbol{\omega}_{71} - \boldsymbol{\omega}_{11} + \boldsymbol{\omega}_{17} = 0 \qquad (g)$$

Noting that $\boldsymbol{\omega}_{11} = 0$ by definition, and adding together Eqs. (b) to (g), we obtain

$$\boldsymbol{\omega}_{21} + \boldsymbol{\omega}_{32} + \boldsymbol{\omega}_{43} + \boldsymbol{\omega}_{54} + \boldsymbol{\omega}_{65} + \boldsymbol{\omega}_{76} + \boldsymbol{\omega}_{17} = 0 \qquad (h)$$

which states that *the sum of the relative angular velocities around a closed loop in a single-freedom system is zero.* Expressed mathematically, this

† For a rigorous proof see L. A. Pars, "A Treatise on Analytical Dynamics," Heinemann, London, 1965, p. 102.

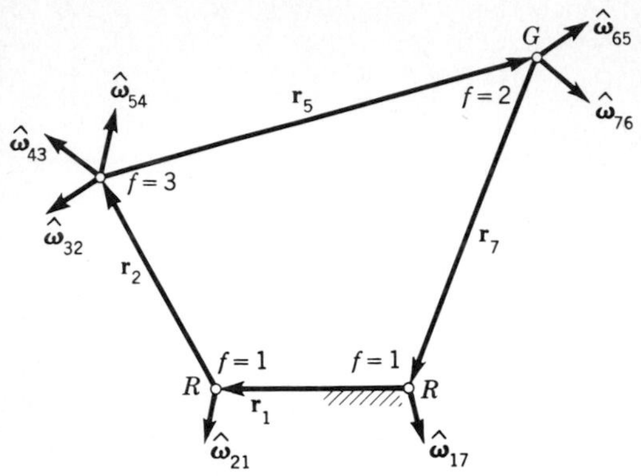

Figure 11-12

theorem reads

$$\sum_{i=1}^{n} \boldsymbol{\omega}_{i+1,i} = 0 \qquad n+1=1 \tag{11-8}$$

This relative-angular-velocity theorem is particularly useful for spatial linkages having two- and three-freedom pairs; see Prob. 11-15, for example. Special care must be taken, however, to eliminate any idle freedoms before Eq. (11-8) is applied.

The approach for the *RGGR* linkage is illustrated in Fig. 11-12. Notice that the diagram shows the multiple rotation axes of the globular joints as separate freedoms and that the idle freedom has been eliminated. The directions of $\hat{\boldsymbol{\omega}}_{32}$, $\hat{\boldsymbol{\omega}}_{43}$, and $\hat{\boldsymbol{\omega}}_{54}$, corresponding to the rotation axes of the first globular pair, need not be orthogonal; in fact, any convenient directions can be assigned as long as they are independent.

The *relative-angular-acceleration theorem* can be developed in the same manner. It is written

$$\sum_{i=1}^{n} \boldsymbol{\alpha}_{i+1,i} = 0 \qquad n+1=1 \tag{11-9}$$

Since

$$\frac{d}{dt}(\omega\hat{\boldsymbol{\omega}}) = \alpha\hat{\boldsymbol{\omega}} + \omega\dot{\hat{\boldsymbol{\omega}}}$$

the direction of $\boldsymbol{\alpha}$ is *not* necessarily the same as that of $\hat{\boldsymbol{\omega}}$. Therefore, care must be used in the application of Eq. (11-9).

11-8 THE HOOKE UNIVERSAL JOINT

Figure 11-13 shows the well-known Hooke, or Cardan, joint. It consists of two yokes, which are the driving and driven members, and a cross, which is

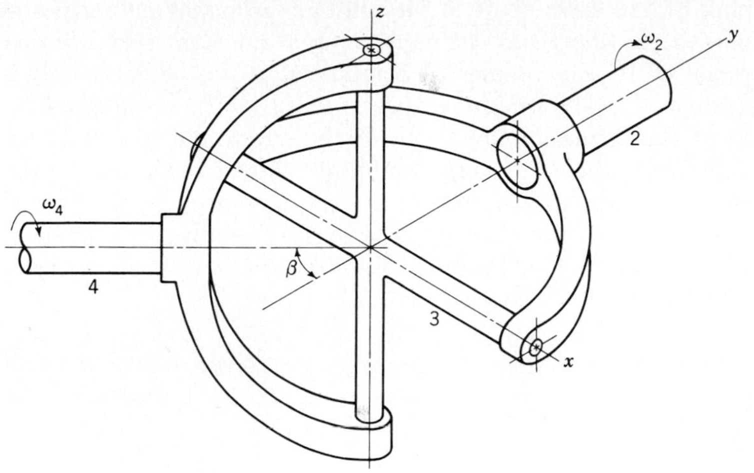

Figure 11-13 The Hooke, or Cardan, universal joint.

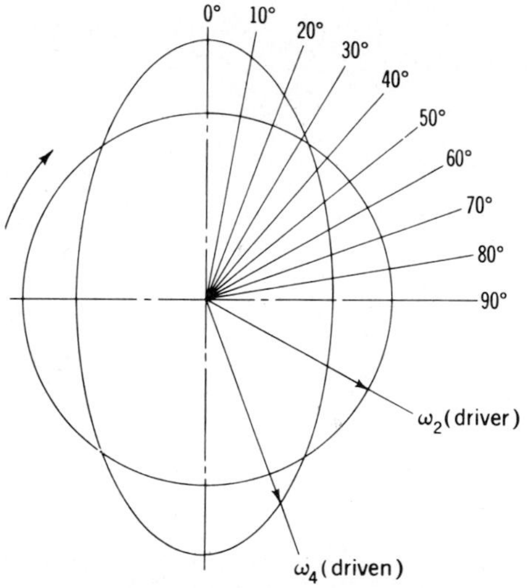

Figure 11-14

the connecting link. One of the disadvantages of this joint is that the velocity ratio is not constant during rotation. Figure 11-14 is a polar angular-velocity diagram which shows the angular velocity of both the driver and the driven links for one complete revolution of the joint. Since the driving member is assumed to have a constant angular velocity, its polar diagram is a circle. But the diagram for the driven member is an ellipse, which crosses the circle at four places. This means that there are four instants during a single rotation when the angular velocities of the two shafts are equal. During the remaining time, the driven shaft rotates faster for part of the time and slower for part of the time.

We may think of the drive shaft of an automobile as having an inertia load at each end—the flywheel and engine rotating at constant speed at one end, and the weight of the car running at high speed at the other. If only a single universal joint, working at a finite angle, were used in an automobile, either the speed of the engine or the speed of the car would have to vary during each revolution of the drive shaft. Both inertias resist this, and so the effect would be for the tires to slip and for the parts composing the line of power transmission to be highly stressed. Figure 11-15 shows two arrangements of universal joints which will provide a uniform velocity ratio between the input and output ends.

Analysis In Fig. 11-16 the driving shaft 2 connects the driven shaft 4 through the connecting cross 3. The shaft centerlines intersect at O, producing the

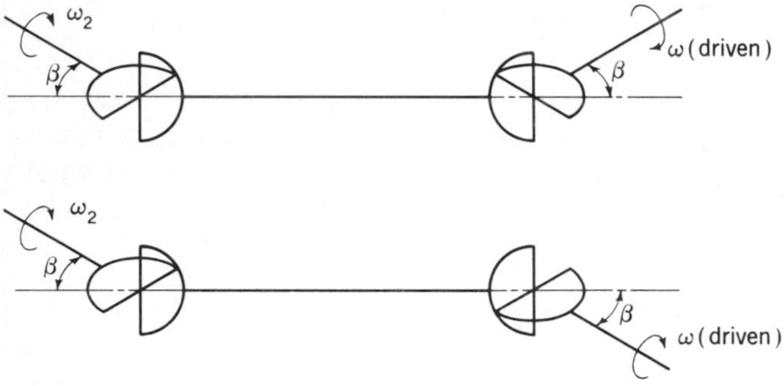

Figure 11-15

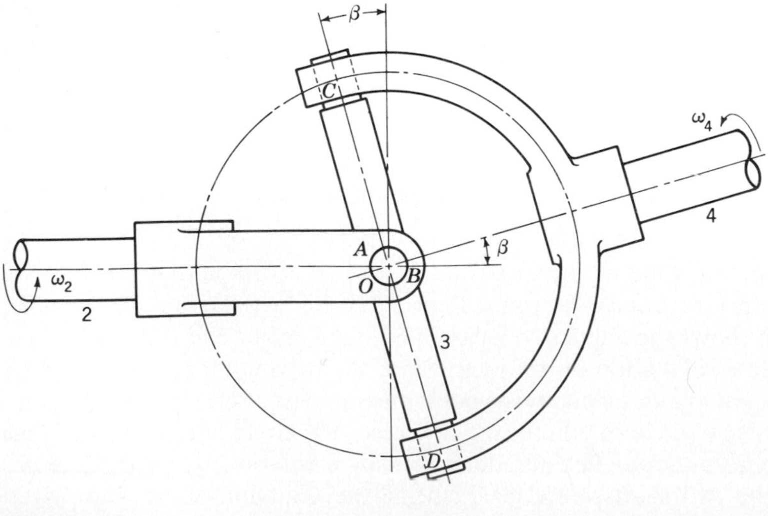

Figure 11-16

shaft angle β. The ends of the crosspiece connect to the driving yoke at points A and B and to the driven yoke at C and D. During motion the line AB describes a circle in a vertical plane perpendicular to the drawing and the line CD another circle in a plane at an angle β to the vertical plane. These two circles are great circles of the same sphere, the center being at O. Points A and C always remain the same distance apart, that is, 90° of great-circle arc. The maximum deviation in the angular-velocity ratio occurs when either point A or point C lies at the intersection of the great circles.

The two great circles on which A and C travel are illustrated again in Fig. 11-17. The circles intersect at D and are shown separated by the shaft angle β. Let point A travel a distance θ from the point of intersection. Then point C will be located on the great-circle arc AC 90° behind A. Now locate C' 90° ahead of C on the great circle which C travels. The triangles $AC'D$ and $AC'C$ are spherical triangles. Arcs AC and $C'C$ are both 90°, and therefore angles $C'AC$ and $AC'C$ are both right spherical angles.† We then have the right spherical triangle $AC'D$ in which angle $AC'D$ is a right angle, $C'DA$ is the

† The sides and angles of a spherical triangle may have any values from 0 to 360°. If one or more of the parts is greater than 180°, the triangle is called a *general spherical triangle.* A triangle having each part less than 180° is called a *spherical triangle.* A spherical right triangle is a spherical triangle one of whose angles is a right angle. The other parts may have any values from 0 to 180°.

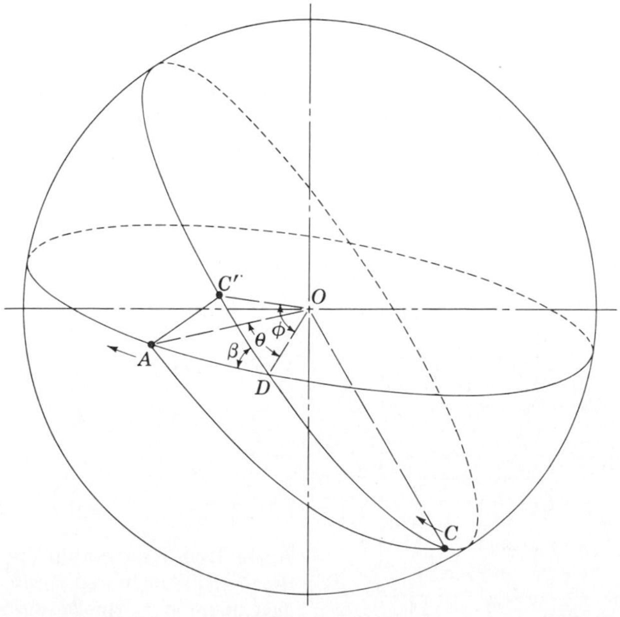

Figure 11-17

shaft angle β, arc AD is the angle through which the driving shaft turns, and arc $C'D$, designated ϕ, is the arc through which the driven shaft turns. According to a right-triangle formula from spherical trigonometry,

$$\cos\beta = \tan\phi \cot\theta \tag{11-10}$$

In order to obtain the relationship between the angular velocities, the equation is rearranged to

$$\tan\phi = \cos\beta \tan\theta \tag{a}$$

Differentiating, we get

$$\dot{\phi}\sec^2\phi = \dot{\theta}\cos\beta\sec^2\theta \tag{b}$$

Since $\dot{\phi} = \omega_4$, the angular velocity of the driven, and $\dot{\theta} = \omega_2$, the angular velocity of the driver, the ratio of these velocities is

$$\frac{\omega_4}{\omega_2} = \frac{\cos\beta\sec^2\theta}{\sec^2\phi} = \frac{\cos\beta\sec^2\theta}{1+\tan^2\phi} \tag{c}$$

It is convenient to eliminate ϕ; substituting Eq. (a) in (c) gives

$$\frac{\omega_4}{\omega_2} = \frac{\cos\beta}{1-\sin^2\theta\sin^2\beta} \tag{11-11}$$

If we assume the shaft angle β a constant, the maximum value of Eq. (11-11) occurs when $\sin\theta = 1$, that is, when $\theta = 90°$, $270°$, etc. The denominator is greatest when $\sin\theta = 0$, and this condition gives the minimum ratio of the velocities.

If the difference between the maximum and minimum ratios of Eq. (11-11) is expressed in percent and plotted against the shaft angle, a curve useful in the evaluation of universal joints is obtained. Figure 11-18 was obtained in this manner for shaft angles up to 28°.

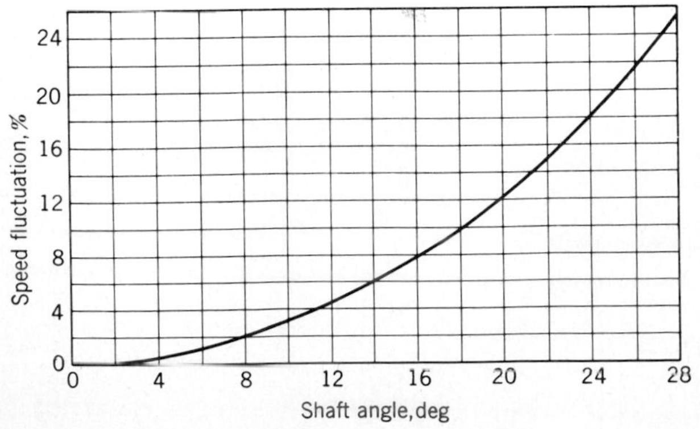

Figure 11-18 Relationship between shaft angle and speed fluctuation in a Hooke universal joint.

PROBLEMS

11-1 Determine the mobility of the *GGC* chain shown in the figure. Identify any idle freedoms and state how they can be removed. What is the nature of the path described by point *B*?

11-2 Using the linkage of Prob. 11-1, with $R_{BA} = R_{O_3O} = 75$ mm, $R_{BO_3} = 150$ mm, and $\theta_2 = 30°$, express the position of each link in vector form.

11-3 Using $\mathbf{V}_A = -50\hat{\mathbf{j}}$ mm/s, find the angular velocities of links 2 and 3 and the velocity of point *B* of the mechanism of Prob. 11-2. Use vector analysis.

11-4 Solve Prob. 11-2 and 11-3 by graphical techniques.

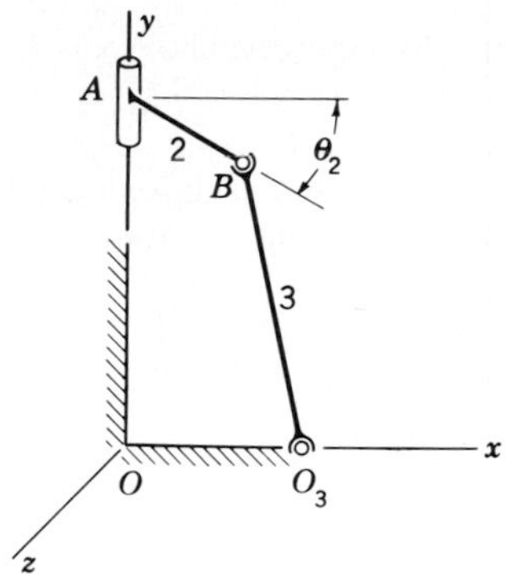

Problems 11-1 and 11-5

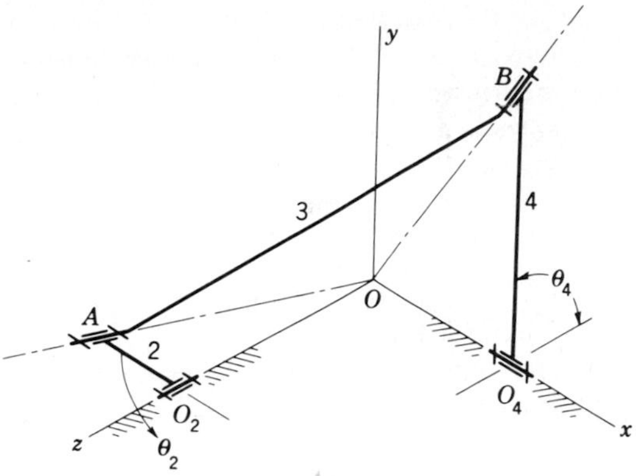

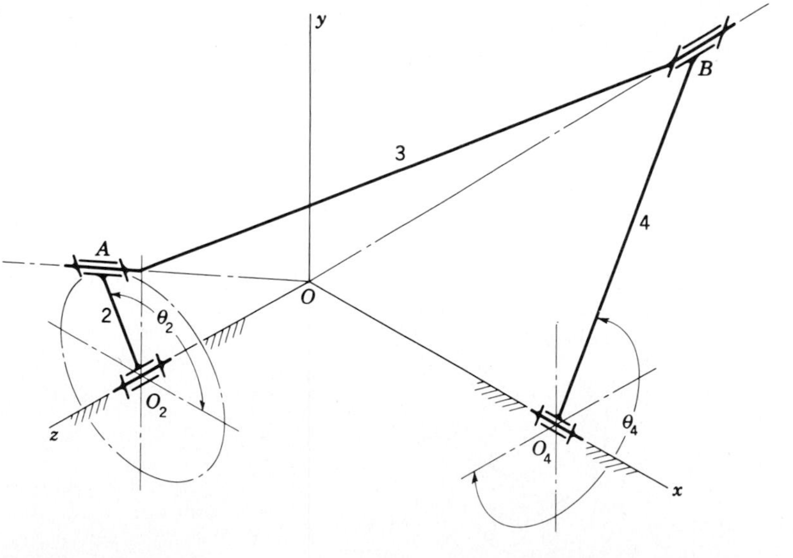

Problem 11-9

11-5 The spherical 4R linkage shown in the figure has the following dimensions: $R_{AO_2} = 3$ in, $R_{O_2O} = 7$ in, $R_{O_4O} = 2$ in, and $R_{BO_4} = 9$ in. Link 2 is shown in the xz plane and link 4 in the xy plane. For better presentation, the figure is not drawn to scale. Express the position of link 3 in vector notation. With $\omega_2 = -60\hat{k}$ rad/s use vector algebra to make a complete velocity and acceleration analysis of the linkage at the position shown.

11-6 Solve Prob. 11-5 by graphical means.

11-7 Determine the advance- to return-time ratio for Prob. 11-5. What is the total angle of oscillation of link 4?

11-8 Repeat Prob. 11-5 except with $\theta_2 = 90°$.

11-9 The spherical four-bar linkage shown has $R_{AO_2} = 37.5$ mm, $R_{O_2O} = 150$ mm, $R_{O_4O} = 225$ mm, $R_{BA} = 412$ mm, and $R_{BO_4} = 262$ mm. The position shown is for $\theta_2 = 120°$. Determine whether the crank 2 is free to rotate through a complete turn. If so, find the angle of oscillation of link 4 and the advance-to return-time ratio.

11-10 With $\omega_2 = 36\hat{k}$ rad/s, use vector analysis to make a complete velocity and acceleration analysis of the mechanism of Prob. 11-9.

11-11 Solve Prob. 11-10 by graphical methods.

11-12 The figure shows the top, front, and auxiliary views of a spatial slider-crank linkage with two spheric joints. The dimensions are $R_{AO} = 2$ in and $R_{BA} = 6$ in. In the construction of many such mechanisms provision is made to vary the angle β. Thus the stroke of slider 4 is adjustable from zero, when $\beta = 0$, to twice the crank length, when $\beta = 90°$. In this example $\beta = 30°$, $\theta_2 = 240°$, and $\omega_2 = 24$ rad/s. Express the links in vector form and use vector algebra to make a complete velocity analysis of the linkage.

11-13 Solve Prob. 11-12 by graphical means.

11-14 Solve Prob. 11-12 with $\beta = 60°$.

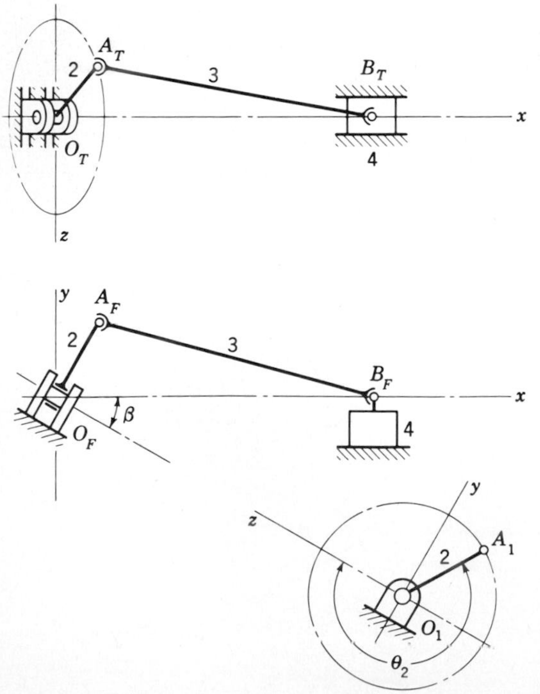

Problem 11-12

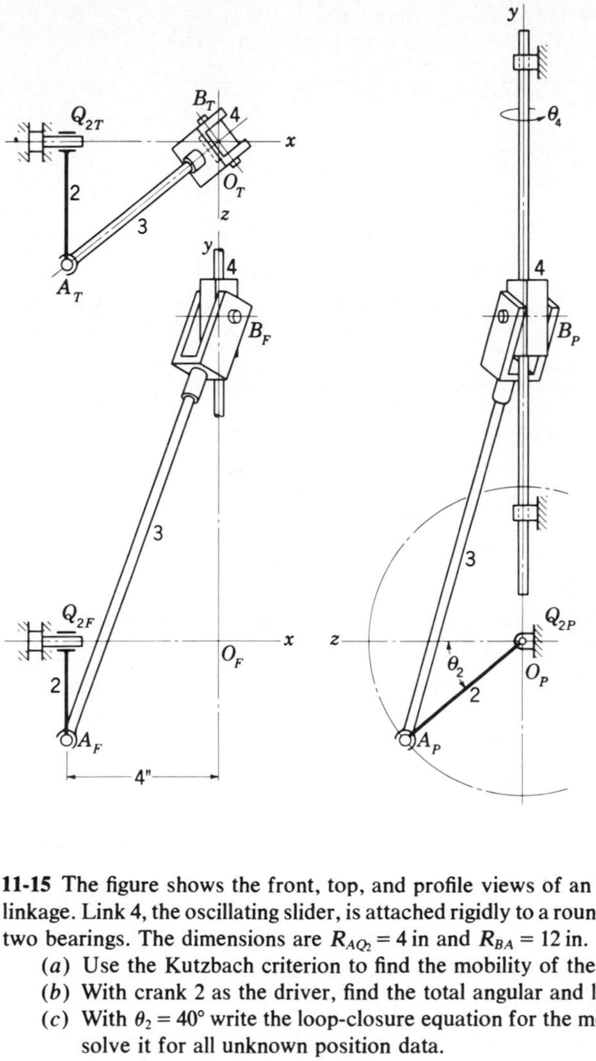

Problem 11-15

11-15 The figure shows the front, top, and profile views of an *RGRC* crank and oscillating-slider linkage. Link 4, the oscillating slider, is attached rigidly to a round rod which rotates and slides in the two bearings. The dimensions are $R_{AQ_2} = 4$ in and $R_{BA} = 12$ in.

(*a*) Use the Kutzbach criterion to find the mobility of the linkage.

(*b*) With crank 2 as the driver, find the total angular and linear travel of link 4.

(*c*) With $\theta_2 = 40°$ write the loop-closure equation for the mechanism and use vector algebra to solve it for all unknown position data.

11-16 With $\boldsymbol{\omega}_2 = -48\hat{\mathbf{i}}$ rad/s for Prob. 11-15, find $\mathbf{V}_B$, $\boldsymbol{\omega}_3$, and $\boldsymbol{\omega}_4$.

CHAPTER

TWELVE

STATIC FORCES

We are now ready for a study of the dynamics of machines and systems. Such a study is simplified by starting with the statics of such systems. In our studies of kinematic analysis we were concerned only with the geometry of the motions and with the relationships between displacement and time. The forces that produced the motion or the motions that would result from the application of a given force system were completely neglected.

Consideration of a problem in the design of a machine in which only the units of *length* and *time* are involved is a tremendous simplification. It frees the mind of the complicating influence of many other factors that ultimately enter into the problem and permits our attention to be focused on the primary problem, that of designing a mechanism to obtain a desired motion.

The fundamental units in kinematic analysis are *length* and *time*; in dynamic analysis they are *length*, *time*, and *force*.

Forces are transmitted into machine members through mating surfaces, e.g., from a gear to a shaft or from one gear through meshing teeth to another gear, from a connecting rod through a bearing to a lever, from a V belt to a pulley, from a cam to a follower, or from a brake drum to a brake shoe. It is necessary to know the magnitudes of these forces for a variety of reasons. The distribution of the forces at the boundaries or mating surfaces must be reasonable, and their intensities must be within the working limits of the materials composing the surfaces. For example, if the force operating on a sleeve bearing becomes too high, it will squeeze out the oil film and cause metal-to-metal contact, overheating, and rapid failure of the bearing. If the forces between gear teeth are too large, the oil film may be squeezed out from between them. This could result in flaking and spalling of the metal, noise, rough motion, and eventual failure. In our study of dynamics we are principally interested in determining the magnitude, direction, and location of the forces but not in sizing the members on which they act.†

12-1 INTRODUCTION

Some of the new terms used in this phase of our studies are defined below.

Force Our earliest ideas concerning forces arose because of our desire to push, lift, or pull various objects. So force is the action of one body acting on another. Our intuitive concept of force includes such ideas as *place of application*, *direction*, and *magnitude*, and these are called the *characteristics of a force.*

Matter Matter is any material or substance; if it is completely enclosed, it is called a body.

Mass Newton defined mass as *the quantity of matter of a body as measured by its volume and density.* This is not a very satisfactory definition because density *is* the mass of a unit volume. We can excuse Newton by surmising that he perhaps did not mean it to be a definition. Nevertheless, he recognized the fact that all bodies possess some inherent property that is different from weight. Thus, a moon rock has a certain constant amount of substance, even though its moon weight is different from its earth weight. This constant amount of substance, or quantity of matter, is called the *mass* of the rock.

Inertia Inertia is the property of mass that causes it to resist any effort to change its motion.

Weight Weight is the force of gravity acting upon a mass. The following quotation is pertinent:

> The great advantage of SI units is that there is one, and only one unit for each physical quantity—the metre for length, the kilogram for mass, the newton for force, the second for time, etc. To be consistent with this unique feature, it follows that a given unit or word should not be used as an accepted technical name for two physical quantities. However, for generations the term "weight" has been used in both technical and nontechnical fields to mean either the force of gravity acting on a body or the mass of a body itself. The reason for this double use of the term "weight" for two different physical quantities—force and mass—is attributed to the dual use of the pound units in our present customary gravitational system in which we often use weight to mean both force and mass.‡

† The determination of the sizes of machine members is the subject of books usually titled machine design or mechanical design. See Joseph E. Shigley, "Mechanical Engineering Design," 3d ed., McGraw-Hill, New York, 1977.

‡ From S.I., The Weight/Mass Controversy, *Mech. Eng.*, vol. 99, no. 9, p. 40, September 1977, and vol. 101, no. 3, p. 42, March 1979.

In this book we shall always use *weight* to mean gravitational *force.*

Particle A particle is a body whose dimensions are so small that they may be neglected.

Rigid body All bodies are either elastic or plastic and will be deformed if acted upon by forces. When the deformation of such bodies is small, they are frequently assumed to be rigid, i.e., incapable of deformation, in order to simplify the analysis.

Deformable body The rigid-body assumption cannot be used when internal stresses and strains due to the applied forces are to be analyzed. Thus we consider the body to be capable of deforming. Such analysis is frequently called *elastic-body analysis*, using the additional assumption that the body remains elastic within the range of the applied forces.

Newton's laws As stated in the "Principia," Newton's three laws are

> [Law 1] Every body perseveres in its state of rest or of uniform motion in a straight line, except in so far as it is compelled to change that state by impressed forces.
>
> [Law 2] Change of motion is proportional to the moving force impressed, and takes place in the direction of the straight line in which such force is impressed.
>
> [Law 3] Reaction is always equal and opposite to action; that is to say, the actions of two bodies upon each other are always equal and directly opposite.

For our purposes, it is convenient to restate these laws:

Law 1 If all the forces acting on a particle are balanced, the particle will either remain at rest or will continue to move in a straight line at a uniform velocity.

Law 2 If the forces acting on a particle are not balanced, the particle will experience an acceleration proportional to the resultant force and in the direction of the resultant force.

Law 3 When two particles react, a pair of interacting forces come into existence; these forces have the same magnitudes and opposite senses, and they act along the straight line common to the two particles.

12-2 SYSTEMS OF UNITS

Newton's first two laws can be summarized by the equation

$$\mathbf{F} = m\mathbf{A} \tag{12-1}$$

which is called the *equation of particle motion.* In this equation $\mathbf{A}$ is the acceleration experienced by a particle of mass m when it is acted upon by the force **F**. Both **F** and **A** are vector quantities.

An important use of Eq. (12-1) occurs in the standardization of systems of units. Let us employ the following symbols to designate units:

Force F
Mass M
Length L
Time T

These symbols are to stand for any unit we may choose. Thus, possible choices for L are inches, kilometres, or miles. The symbols F, M, L, and T are not numbers, but they can be substituted into Eq. (12-1) as if they were. The equality sign then implies that the symbols on one side are equivalent to those on the other. Making the indicated substitution then gives

$$F = MLT^{-2} \tag{12-2}$$

because the acceleration A has units of length divided by time squared. Equation (12-2) expresses an equivalency between the four units force, mass, length, and time. We are free to choose units for three of these and then the units used for the fourth depend upon the first three. It is for this reason that the first three units chosen are called the *basic units* while the fourth is called the *derived unit*.

When force, length, and time are chosen as the basic units, the mass is the derived unit and the system that results is called a *gravitational system of units*.

When mass, length, and time are chosen as the basic units, force is the derived unit and the system that results is called an *absolute system of units*.

In the English-speaking countries the *U.S. customary foot-pound-second system* (fps) and the *inch-pound-second system* (ips) are the two standard gravitational systems most used by engineers.† In the fps system the unit of mass is

$$M = \frac{FT^2}{L} = \frac{\text{(pound force)(second)}^2}{\text{foot}} = \text{lbf} \cdot \text{s}^2/\text{ft} = \text{slug} \tag{12-3}$$

Thus, length, time, and force are the three basic units in the fps gravitational system.

The unit of time in the fps system is the second, abbreviated s.

The unit of force in the fps system is the pound, more properly the *pound force*. We shall seldom abbreviate this unit as lbf; the abbreviation lb is permissible, since we shall be dealing only with the U.S. customary gravita-

† Most engineers prefer to use gravitational systems; this helps to explain some of the resistance to the use of SI units since the International System (SI) is an absolute system.

tional systems.† In some branches of engineering it is useful to represent 1000 lb as a kilopound and to abbreviate it as kip. Many writers add the letter *s* to kip to obtain the plural, but to be consistent with the practice of using only singular units we shall not do so here. Thus 1 kip and 3 kip are used to designate 1000 and 3000 lb, respectively.

Finally, we note in Eq. (12-3) that the derived unit of mass in the fps gravitational system is the lb · s²/ft, called a *slug*; there is no abbreviation for slug.

The unit of mass in the ips gravitational system is

$$M = \frac{FT^2}{L} = \frac{(\text{pound force})(\text{second})^2}{\text{inch}} = \text{lb} \cdot \text{s}^2/\text{in} \tag{12-4}$$

Note that this unit of mass has *not* been given a special name.

The International System of Units (SI) is an absolute system. The basic units are the metre, the kilogram mass, and the second. The unit of force is derived and is called a *newton* to distinguish it from the kilogram, which, as indicated, is the unit of mass. The units of the newton (N) are

$$F = \frac{ML}{T^2} = \frac{(\text{kilogram})(\text{metre})}{(\text{second})^2} = \text{kg} \cdot \text{m/s}^2 = \text{N} \tag{12-5}$$

The weight of an object is the force exerted upon it by gravity. Designating the weight as W and the acceleration due to gravity as g, Eq. (12-1) becomes

$$W = mg \tag{12-6}$$

In the fps system standard gravity is $g = 32.1740\ \text{ft/s}^2$. For most cases this is rounded off to 32.2. Thus the weight of a mass of 1 slug in the fps system is

$$W = mg = (1\ \text{slug})(32.2\ \text{ft/s}^2) = 32.2\ \text{lb}$$

In the ips system, standard gravity is 386.088 or about 386 in/s². Thus, in this system, a unit mass weighs

$$W = (1\ \text{lb} \cdot \text{s}^2/\text{in})(386\ \text{in/s}^2) = 386\ \text{lb}$$

With SI units, standard gravity is 9.806 m/s² or about 9.80 m/s². Thus, the weight of a 1 kg mass is

$$W = (1\ \text{kg})(9.80\ \text{m/s}^2) = 9.80\ \text{N}$$

It is convenient to remember that a large apple weighs about 1 N.

† The abbreviation lb for pound comes from Libra the balance, the seventh sign of the zodiac, which is represented as a pair of scales.

12-3 APPLIED AND CONSTRAINT FORCES

When a number of bodies are connected together to form a group or system, the forces of action and reaction between any two of the connecting bodies are called *constraint forces.* These forces constrain the bodies to behave in a specific manner. Forces external to this system of bodies are called *applied forces.*

Electric, magnetic, and gravitational forces are examples of forces that may be applied without actual physical contact. A great many, if not most, of the forces with which we shall be concerned occur through direct physical or mechanical contact.

As indicated earlier, the *characteristics of a force* are its *magnitude,* its *direction,* and its *point of application.* The direction of a force includes the concept of a line, along which the force is directed, and a *sense.* Thus, a force is directed positively or negatively along a *line of action.*

Sometimes the point of application is not important, e.g., when we are studying the equilibrium of a rigid body. Thus in Fig. 12-1*a* it does not matter whether we diagram the force pair $\mathbf{F}_1\mathbf{F}_2$, as if they compress the link, or if we diagram them as putting the link in tension provided we are interested only in the equilibrium of the link. Of course, if we are interested in the internal link stresses, the forces cannot be interchanged.

The notation to be used for force vectors is shown in Fig. 12-1*b*. Boldface letters are used for force vectors and lightface letters for their magnitudes. Thus the components of a force vector are

$$\mathbf{F} = F^x\hat{\mathbf{i}} + F^y\hat{\mathbf{j}} + F^z\hat{\mathbf{k}} \tag{a}$$

Note that the directions of the components in this book are indicated by superscripts, not subscripts.

Two equal and opposite forces acting along two noncoincident parallel straight lines in a body cannot be combined to obtain a single resultant force.

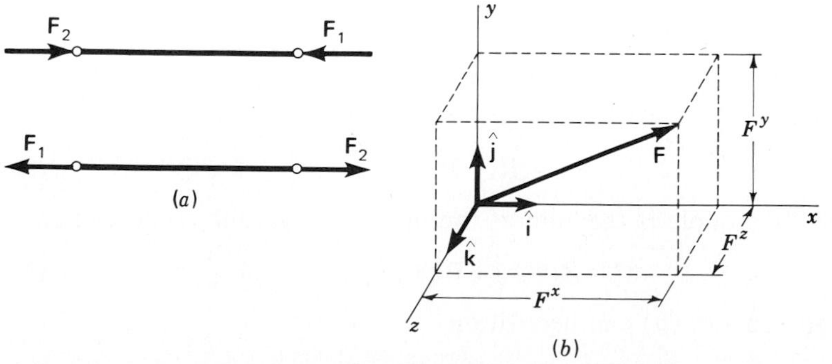

Figure 12-1 (*a*) Points of application of $\mathbf{F}_1$ and $\mathbf{F}_2$ to a rigid body may or may not be important. (*b*) The rectangular components of a force vector.

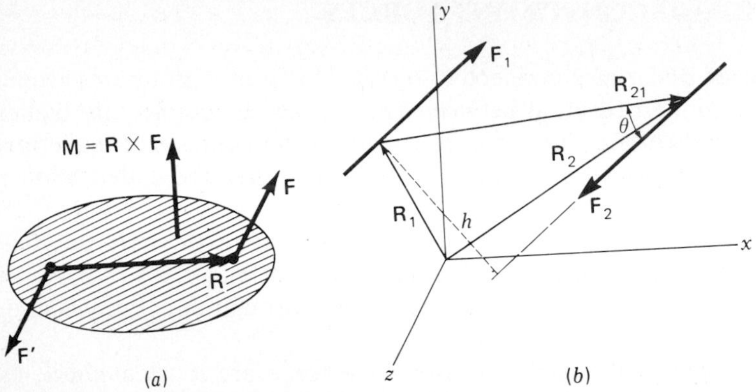

Figure 12-2 (*a*) **R** is a position vector, but **F** and **F′** are force vectors; the free vector **M** is the moment of the couple formed by **F** and **F′**. (*b*) A force couple formed by $\mathbf{F}_1$ and $\mathbf{F}_2$.

Any two such forces acting on a body constitute a *couple.* The *arm of the couple* is the perpendicular distance between their lines of action, and the *plane of the couple* is the plane containing the two lines of action.

The *moment of a couple* is another vector **M** directed normal to the plane of the couple; the sense of **M** is in accordance with the right-hand rule for rotation. The magnitude of the moment is the product of the arm of the couple and the magnitude of one of the forces. Thus

$$M = hF \tag{12-7}$$

where h is the moment arm.

As shown in Fig. 12-2*a* the moment vector is the cross product of the relative-position vector **R** and the force vector **F**, and so it is defined by the equation

$$\mathbf{M} = \mathbf{R} \times \mathbf{F} \tag{12-8}$$

Some of the interesting properties of couples can be determined by an examination of Fig. 12-2*b*. Here $\mathbf{F}_1$ and $\mathbf{F}_2$ are two equal, opposite, and parallel forces. Choose any point on each line of action and define these points by the position vectors $\mathbf{R}_1$ and $\mathbf{R}_2$. Then the relative-position vector, or position-difference vector, is

$$\mathbf{R}_{21} = \mathbf{R}_2 - \mathbf{R}_1 \tag{a}$$

The moment of the couple is the sum of the moments of each force and is

$$\mathbf{M} = \mathbf{R}_1 \times \mathbf{F}_1 + \mathbf{R}_2 \times \mathbf{F}_2 \tag{b}$$

But $\mathbf{F}_1 = -\mathbf{F}_2$, and so Eq. (*b*) can be written

$$\mathbf{M} = (\mathbf{R}_2 - \mathbf{R}_1) \times \mathbf{F}_2 = \mathbf{R}_{21} \times \mathbf{F}_2 \tag{c}$$

Equation (*c*) shows that:

1. The value of the moment of the couple is independent of the choice of the center about which the moments are taken because the vector $\mathbf{R}_{21}$ is the same for all positions of the origin.
2. Since $\mathbf{R}_1$ and $\mathbf{R}_2$ define any set of points on the lines of action, the vector $\mathbf{R}_{21}$ is not restricted to perpendicularity with $\mathbf{F}_1$ and $\mathbf{F}_2$. This is a very important result of the vector product because it means that the value of the moment is independent of how $\mathbf{R}_{21}$ is chosen. The magnitude of the moment can be obtained as follows. Resolve $\mathbf{R}_{21}$ into two components $\mathbf{R}_{21}^t$ and $\mathbf{R}_{21}^n$, parallel and perpendicular, respectively, to $\mathbf{F}_2$. Then

$$\mathbf{M} = \mathbf{R}_{21}^t \times \mathbf{F}_2 + \mathbf{R}_{21}^n \times \mathbf{F}_2 \tag{d}$$

But $\mathbf{R}_{21}^n$ is the perpendicular distance between the lines of action and $\mathbf{R}_{21}^t$ is parallel to $\mathbf{F}_2$. Therefore $\mathbf{R}_{21}^t \times \mathbf{F}_2 = 0$ and

$$\mathbf{M} = \mathbf{R}_{21}^n \times \mathbf{F}_2$$

is the moment of the couple. Since $R_{21}^n = R_{21} \sin \theta$, where θ is the angle between $\mathbf{R}_{21}$ and $\mathbf{F}_2$, the magnitude of the moment is

$$M = (R_{21} \sin \theta) F_2 \tag{e}$$

3. The moment vector $\mathbf{M}$ is independent of any particular origin or line of application and is thus a *free vector.*
4. The forces of a couple can be rotated together within their plane, keeping their magnitudes and the distance between their lines of action constant, or they can be translated to any parallel plane without changing the magnitude or sense of the moment vector. Also, two couples are equal if they have the same moment vectors, regardless of the forces or moment arms. This means that it is the vector product of the two that is significant, not the individual values.

12-4 CONDITIONS FOR EQUILIBRIUM

A rigid body is in static equilibrium if:

1. *The vector sum of all forces acting upon it is zero.*
2. *The sum of the moments of all the forces acting about any single axis is zero.*

Mathematically these two statements are expressed as

$$\sum \mathbf{F} = 0 \qquad \sum \mathbf{M} = 0 \tag{12-9}$$

Note how these statements are a result of Newton's first and third laws, it being understood that a body constitutes a collection of particles.

Many problems have forces acting in a single plane. When this is true, it is convenient to choose the xy plane for use. Then Eqs. (12-9) can be simplified to

$$\sum F^x = 0 \qquad \sum F^y = 0 \qquad \sum M = 0$$

where the z direction for the moment M is implied by the fact that the forces exist only in xy.

12-5 FREE-BODY DIAGRAMS

The term "body" as used here may be an entire machine, several connected parts of a machine, a single part, or a portion of a part. A *free-body diagram* is a sketch or drawing of the body, isolated from the machine, on which the forces and moments are shown in action. It is usually desirable to include on the diagram the known magnitudes and directions as well as other pertinent information.

The diagram so obtained is called "free" because the part or portion of the body has been freed from the remaining machine elements and their effects have been replaced by forces and moments. If the free-body diagram is of an entire machine part, the forces shown on it are the external forces (applied forces) and moments exerted by adjoining or connected parts. If the diagram is a portion of a part, the forces and moments acting on the cut portion are the *internal* forces and moments exerted by the part that has been cut away.

The construction and presentation of clear and neatly drawn free-body diagrams represent the heart of engineering communication. This is true because they represent a part of the thinking process, whether they are actually placed on paper or not, and because the construction of these diagrams is the *only* way the results of thinking can be communicated to others. You should acquire the habit of drawing free-body diagrams no matter how simple the problem may appear to be. They are a means of storing one thought while concentrating on the next step in a problem. Construction of the diagrams speeds up the problem-solving process and greatly decreases the chances of making mistakes.

The advantages of using free-body diagrams can be summarized as follows:

1. They make it easy for one to translate words and thoughts and ideas into physical models.
2. They assist in seeing and understanding all facets of a problem.
3. They help in planning the attack on the problem.
4. They make mathematical relations easier to see or find.
5. Their use makes it easy to keep track of one's progress and helps in making simplifying assumptions.
6. The methods used in the solution may be stored for future reference.

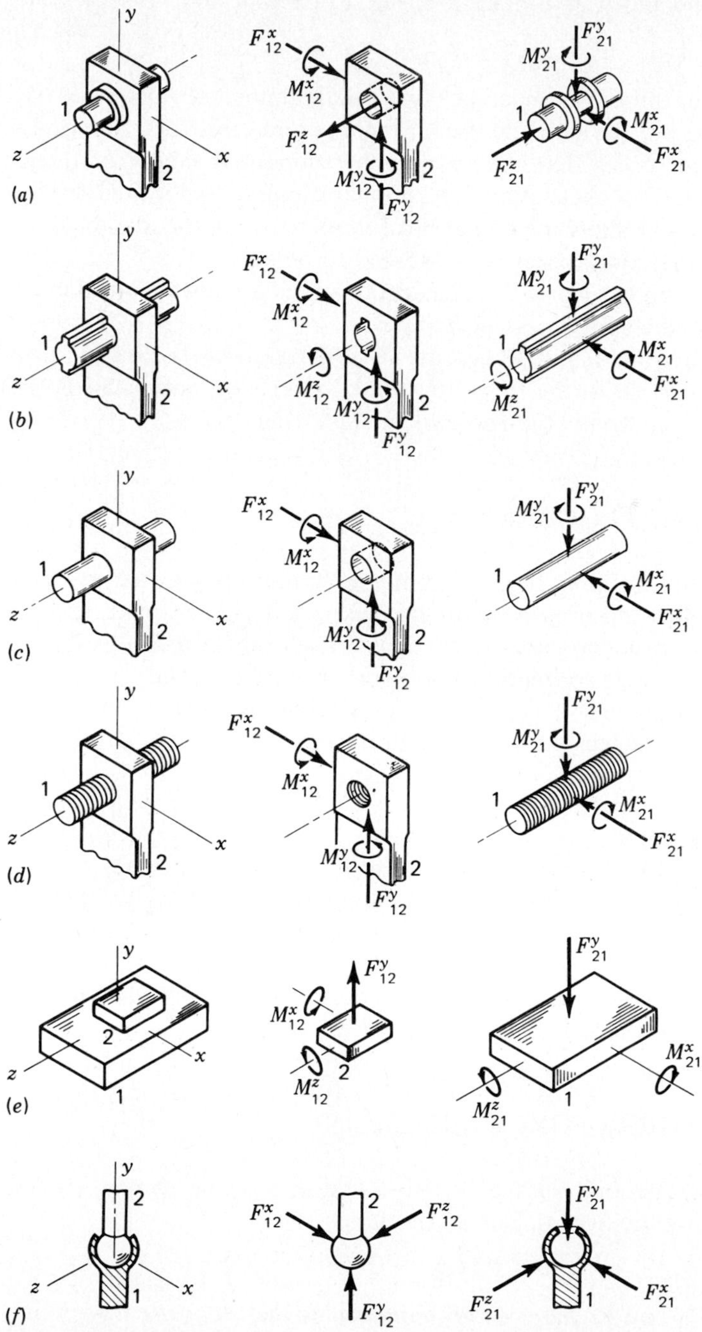

Figure 12-3 All the lower pairs and their constraint forces: (*a*) revolute or turning pair; pair variable θ; (*b*) prismatic pair; pair variable z; (*c*) cylindric pair; pair variables z, θ; (*d*) screw pair; pair variables z or θ; (*e*) planar pair; pair variables x, z, θ; (*f*) globular pair; pair variables θ, ϕ, ψ.

7. They assist your memory and make it easier to explain and present your work to others.

In analyzing the forces in machines we shall almost always need to separate the machine into its individual components and construct free-body diagrams showing the forces that act upon each component. Many of these parts will be connected to each other by kinematic pairs. Accordingly, Fig. 12-3 has been prepared to show the constraint forces between the members of the lower pairs when friction forces are assumed to be zero.

In the case of the higher pairs, the constraint forces are always normal to the contact surfaces when friction is neglected.

The notation shown in Fig. 12-3 will be used throughout the balance of this book. $\mathbf{F}_{21}$, for example, is the force that link 2 exerts on link 1. So $\mathbf{F}_{12}$ is the reaction to this force and is the force of link 1 acting on link 2.

12-6 CALCULATING PROGRAMS

If you have access to any kind of programmable computational facility, you should create the following programs for use, especially in the next several chapters. Since these problems are short, they can all be formed as subroutines and stored on a single magnetic card or other storage facility. Program flags or conditional transfers facilitate entering the various subroutines. The following problems are suggested for inclusion:

1. Given $R\underline{/\theta}$; find $x\hat{\mathbf{i}}+y\hat{\mathbf{j}}$.
2. Given $x\hat{\mathbf{i}}+y\hat{\mathbf{j}}$; find $R\underline{/\theta}$.
3. Given θ; find $\hat{\mathbf{R}}=\bar{x}\hat{\mathbf{i}}+\bar{y}\hat{\mathbf{j}}$, where $\bar{x}$ and $\bar{y}$ are the direction cosines.
4. Given $\mathbf{F}_1$, $\mathbf{F}_2$, $\mathbf{F}_3, \ldots$ in x, y, and z components; find $\Sigma\,\mathbf{F}$.
5. Given $\mathbf{C}$ and $\mathbf{C}'$ in x, y, and z components; find $\mathbf{C}\times\mathbf{C}'$.

These programs should be set up so that zero components are entered automatically without requiring that positive action be taken.

12-7 TWO- AND THREE-FORCE MEMBERS

The equilibrium or nonequilibrium of a two-force member is shown in Fig. 12-4*a* and *b*. Applying the first of Eqs. (12-9) gives

$$\sum \mathbf{F}=\mathbf{F}_A+\mathbf{F}_B=0 \tag{a}$$

This requires that $\mathbf{F}_A$ and $\mathbf{F}_B$ have *equal magnitudes* and *opposite directions.* The second of Eqs. (12-9), $\Sigma\,\mathbf{M}=0$, requires that $\mathbf{F}_A$ and $\mathbf{F}_B$ have the *same line of action*; otherwise the two moments would not sum to zero.

The equilibrium or nonequilibrium of a three-force member is shown in

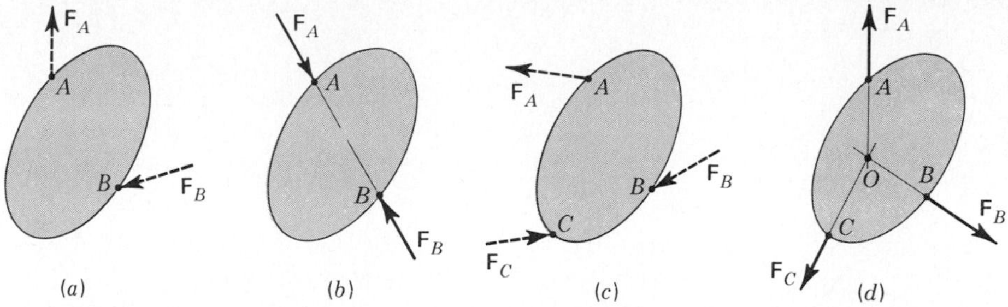

Figure 12-4 (*a*) Two-force member not in equilibrium; (*b*) two-force member is in equilibrium if $\mathbf{F}_A$ and $\mathbf{F}_B$ are equal, opposite, and have the same line of action; (*c*) three-force member not in equilibrium; (*d*) three-force member is in equilibrium if $\mathbf{F}_A$, $\mathbf{F}_B$, and $\mathbf{F}_C$ are coplanar, if their lines of action intersect at a common point *O*, and if their vector sum is zero.

Fig. 12-4*c* and *d*. Suppose two of the forces, say $\mathbf{F}_A$ and $\mathbf{F}_B$, intersect at some point *O*. They can be summed to form the single vector $\mathbf{F}_A + \mathbf{F}_B$. Since the line of action of this sum passes through point *O*, its moment about *O* is zero. Applying $\Sigma\,\mathbf{M} = 0$ to all three forces shows that the moment of $\mathbf{F}_C$ about *O* must also be zero. Thus the lines of action of the three forces intersect at a common point; i.e., the forces are concurrent. This explains why a three-force member can be solved only for two force magnitudes even though there are three equations: the moment equation has already been used in finding the directions of the lines of action.

The case frequently arises, e.g., in beams, in which the three forces are parallel; this is the limiting case, and the common point of intersection of the three lines of action lies at infinity.

The equation $\Sigma\,\mathbf{F} = 0$ for a three-force member requires that the forces be coplanar and that their vector sum be zero.

Example 12-1 The four-bar linkage of Fig. 12-5*a* has crank 2 driven by an input torque $\mathbf{M}_{12}$; an external load $\mathbf{P} = 120\underline{/220°}$ lb acts at *Q* on link 4. For the particular position of the linkage shown, find all the link forces and their reactions.

GRAPHICAL SOLUTION

1. Select a space scale *S*. The space scale for Fig. 12-5 is about $S = 9$ in/in. This means that 1 in of drawing represents 9 in of the linkage.
2. Select a force scale S_F. The force scale for Fig. 12-5 is about 80 lb/in. Thus, a vector 1 in long represents a force of 80 lb.
3. Draw the mechanism and the given force or forces to the proper scales, as shown in Fig. 12-5*a*.
4. Select a link or links from which the analysis can be started and construct the free-body diagram. In this example we are beginning with link 4, as shown in Fig. 12-5*b*, because *P* is given. Since link 3 is a two-force member, it can support only tension or compression. Thus the line of action of $\mathbf{F}_{34}$ lies along link 3. Link 4 is a three-force member. Neither the direction nor the magnitude of the frame reaction $\mathbf{F}_{14}$ is known. One method of determining the unknown vector forces acting on link 4 is to use Eqs. (12-9). Thus, in Fig.

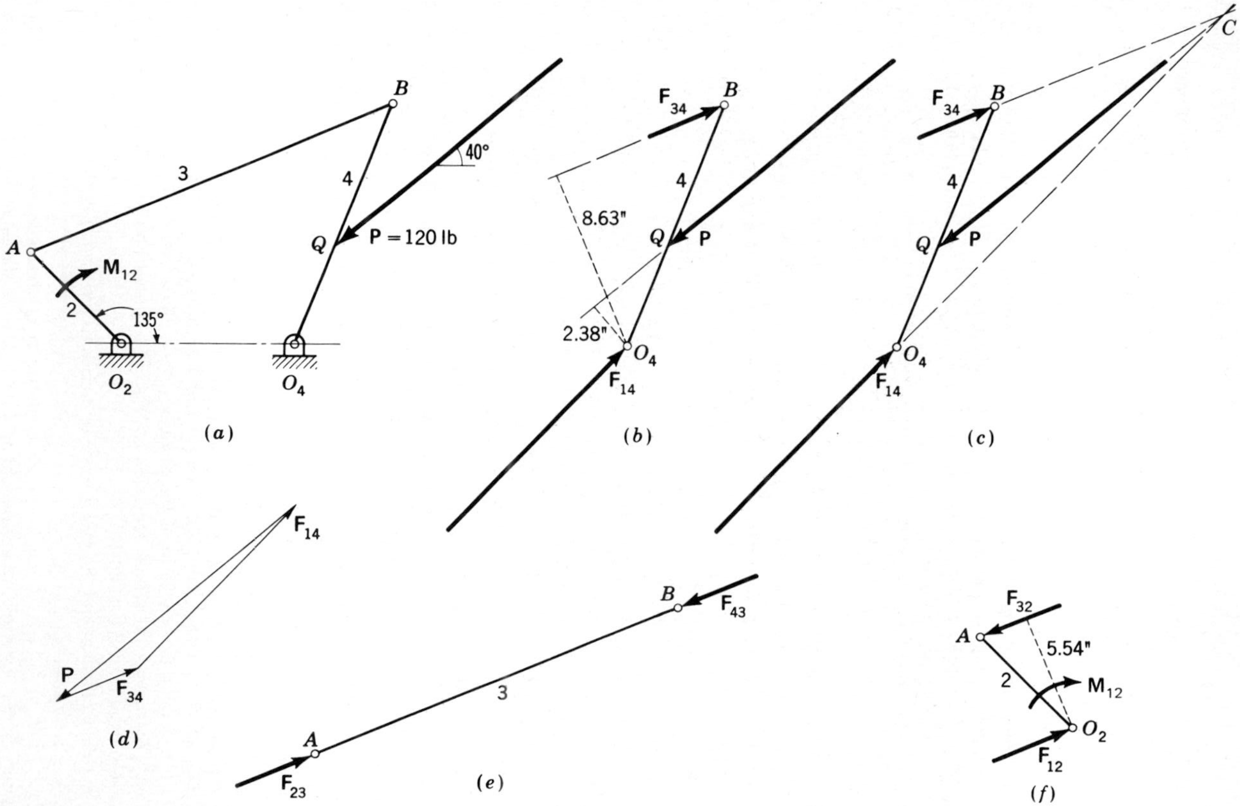

Figure 12-5 $O_2A = 6$ in, $AB = 18$ in, $O_4B = 12$ in, $O_2O_4 = 8$ in, $O_4Q = 5$ in.

12-5b, draw and measure the moment arms of $\mathbf{P}$ and $\mathbf{F}_{34}$ about O_4. These are found to be 2.38 and 8.63 in, respectively. Summing the moments of these two forces about O_4 gives the magnitude of $\mathbf{F}_{34}$.

$$\sum M_{O_4} = 2.38(120) + 8.63F_{34} = 0$$

A solution gives $F_{34} = -33.1$ lb, where the minus sign indicates that the moment of F_{34} about O_4 is clockwise, as shown.

5. Link 4 is a three-force member, and so the direction of $\mathbf{F}_{14}$ can be found using the point of concurrency. When the lines of action of $\mathbf{P}$ and $\mathbf{F}_{34}$ are extended, they intersect at C, the point of concurrency, as shown in Fig. 12-5c.
6. The force polygon, shown in Fig. 12-5d, is the graphical solution to the equation

$$\sum \mathbf{F} = \mathbf{P} + \mathbf{F}_{34} + \mathbf{F}_{14} = 0$$

Note that both steps 4 and 5 are unneccessary. The force polygon can be used to solve for the unknowns using either step 4 first or step 5 first.

7. In Fig. 12-5e we construct the free-body diagram of link 3, noting that $\mathbf{F}_{23} = -\mathbf{F}_{43} = \mathbf{F}_{34}$.
8. The free-body diagram of link 2 is shown in Fig. 12-5f. Here $\mathbf{F}_{32} = -\mathbf{F}_{23}$ and $\mathbf{F}_{12} = -\mathbf{F}_{32}$. When the moment arm of F_{32} about O_2 is measured, it is found to be 5.54 in. Therefore

$$M_{12} = -33.1(5.54) = -183 \text{ lb}\cdot\text{in} \qquad \textit{Ans.}$$

where the negative sign indicates that the moment is clockwise.

9. A free-body diagram of the frame, link 1, is not shown. If drawn, a force $\mathbf{F}_{21} = -\mathbf{F}_{12}$ would be shown at O_2, a force $\mathbf{F}_{41} = -\mathbf{F}_{14}$ at O_4, and a moment $\mathbf{M}_{21} = -\mathbf{M}_{12}$ would be shown.

ANALYTICAL SOLUTION First make a position analysis of the linkage to determine the angular location of each link. The results are shown in Fig. 12-6a. Referring to Fig. 12-6b, we begin by summing moments about an axis through O_4. Thus

$$\sum \mathbf{M}^z_{O_4} = \mathbf{R}_Q \times \mathbf{P} + \mathbf{R}_B \times \mathbf{F}_{34} = 0 \tag{1}$$

The vectors in Eq. (1) are

$$\mathbf{R}_Q = 5\underline{/68.4^\circ} = 1.84\hat{\mathbf{i}} + 4.65\hat{\mathbf{j}}$$
$$\mathbf{P} = 120\underline{/220^\circ} = -91.9\hat{\mathbf{i}} - 77.1\hat{\mathbf{j}}$$
$$\mathbf{R}_B = 12\underline{/68.4^\circ} = 4.42\hat{\mathbf{i}} + 11.16\hat{\mathbf{j}}$$
$$\mathbf{F}_{34} = F_{34}\underline{/22.4^\circ} = (0.924\hat{\mathbf{i}} + 0.381\hat{\mathbf{j}})F_{34}$$

Performing the cross-product operation, we find the first term of Eq. (1) to be $\mathbf{R}_Q \times \boldsymbol{P} = 285.5\hat{\mathbf{k}}$. The second term is $\mathbf{R}_B \times \mathbf{F}_{34} = -8.63F_{34}\hat{\mathbf{k}}$. Substituting these two terms into Eq. (1) and solving yields $F_{34} = 33.1$ lb; and so

$$\mathbf{F}_{34} = 33.1\underline{/22.4^\circ} = 30.6\hat{\mathbf{i}} + 12.6\hat{\mathbf{j}} \text{ lb} \qquad \textit{Ans.}$$

The frame reaction is then found from the equation

$$\sum \mathbf{F} = \mathbf{F}_{34} + \mathbf{P} + \mathbf{F}_{14} = (30.6\hat{\mathbf{i}} + 12.6\hat{\mathbf{j}}) + (-91.9\hat{\mathbf{i}} - 77.1\hat{\mathbf{j}}) + \mathbf{F}_{14} = 0$$

Solving gives

$$\mathbf{F}_{14} = 61.3\hat{\mathbf{i}} + 64.5\hat{\mathbf{j}} = 89.0\underline{/46.5^\circ} \text{ lb} \qquad \textit{Ans.}$$

Next, from Fig. 12-6c, for link 2 we write

$$\sum \mathbf{M}^z_{O_2} = \mathbf{M}_{12} + \mathbf{R}_A \times \mathbf{F}_{32} = 0 \tag{2}$$

Since $\mathbf{R}_A = 6\underline{/135^\circ} = -4.24\hat{\mathbf{i}} + 4.24\hat{\mathbf{j}}$ and $\mathbf{F}_{32} = -\mathbf{F}_{34} = -30.6\hat{\mathbf{i}} - 12.6\hat{\mathbf{j}}$, we find $\mathbf{R}_A \times \mathbf{F}_{32} = 183.2\hat{\mathbf{k}}$ lb · in. Thus

$$\mathbf{M}_{12} = -183.2\hat{\mathbf{k}} \text{ lb}\cdot\text{in} \qquad \textit{Ans.}$$

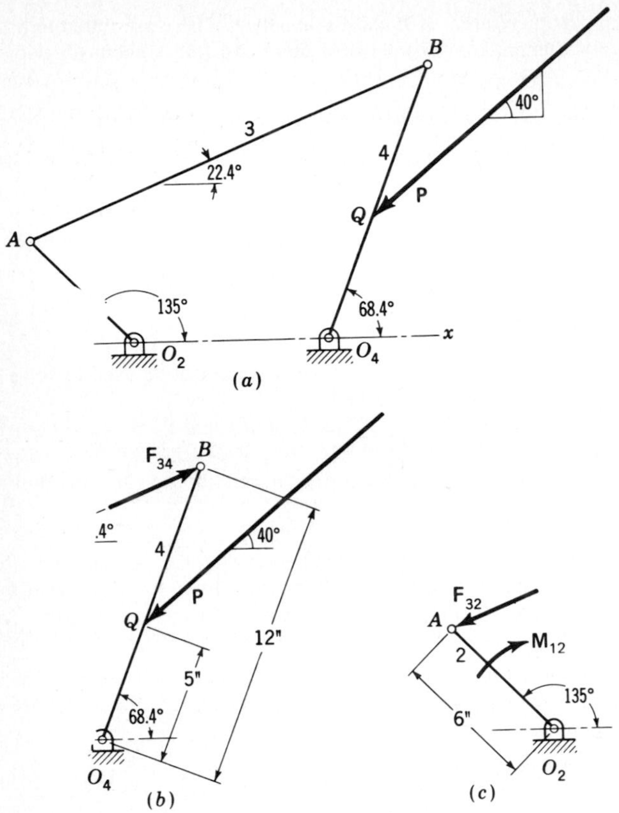

Figure 12-6

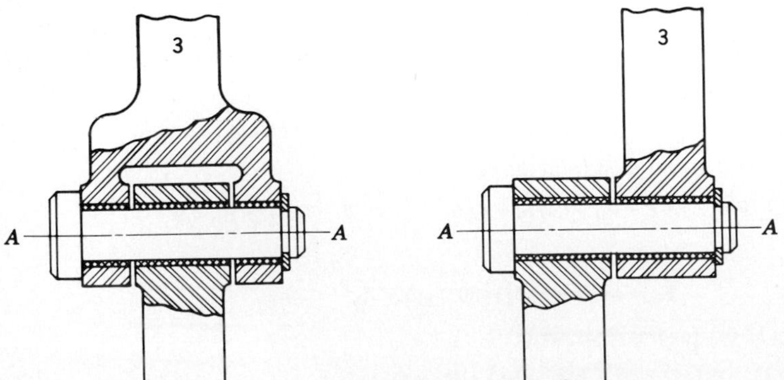

Figure 12-7 (*a*) A balanced connection. (*b*) This connection produces a turning moment on the pin and on each link.

In the preceding example it has been assumed that the forces all act in the same plane. For the connecting link 3 it was also assumed that the line of action of the forces and the centerline of the link were coincident. A careful machine designer will sometimes go to extreme measures in order to approach these conditions as nearly as possible. Note that if the pin connections are arranged as shown in Fig. 12-7*a*, such conditions are theoretically obtained. On the other hand, if the connection is like the one of Fig. 12-7*b*, the pin itself as well as each link will have turning couples acting upon them. If the forces are not in the same plane, couples exist whose moments are proportional to the distance between the force planes.

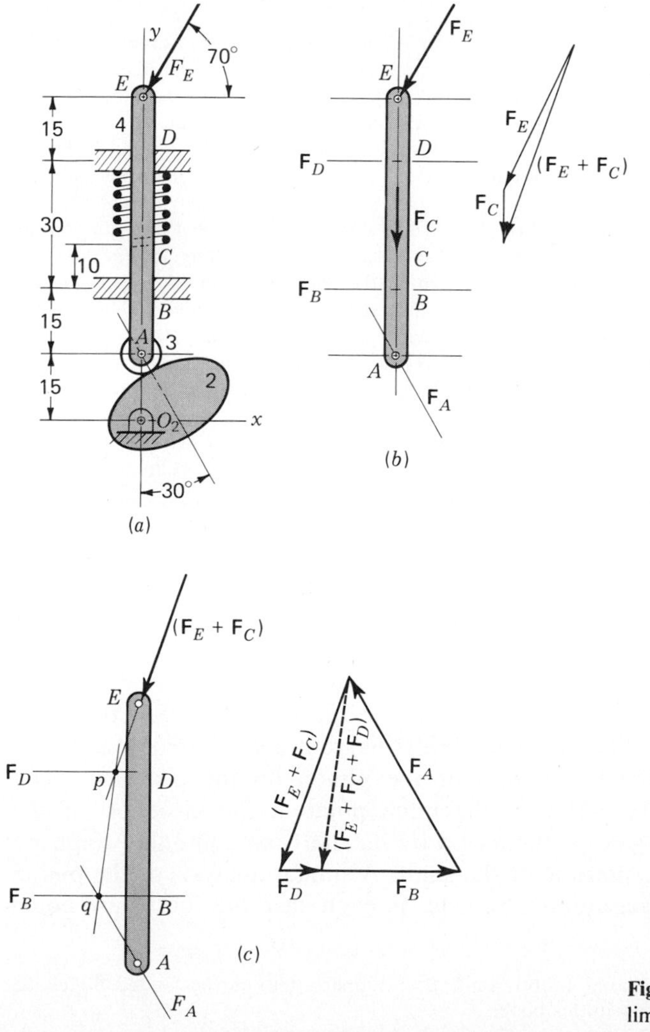

Figure 12-8 Dimensions in millimetres.

12-8 FOUR-FORCE MEMBERS

The most general case of a system of forces is one in which the forces are neither concurrent nor parallel. Such a system can always be replaced by a single resultant force acting at an arbitrary point and a resultant couple. A body acted upon by such a general system of forces is in equilibrium only if both the resultant couple and the resultant force are zero. Equation (12-9) expresses these conditions in mathematical form.

Example 12-2 A cam and reciprocating follower are shown in Fig. 12-8*a*. The follower is held in contact with the cam by a spring pushing downward at C with a spring force $F_C = 12$ N for this particular position. Also, an external load $F_E = 35$ N acts at E on the follower in the direction shown. Determine the follower pin force at A and the bearing reactions at B and D. Assume no friction and a weightless follower.

SOLUTION A free-body diagram of the follower is shown in Fig. 12-8*b*. Forces $\mathbf{F}_E$ and $\mathbf{F}_C$ are known, and their sum $\mathbf{F}_E + \mathbf{F}_C$ is obtained graphically in this figure. The diagram shows the lines of action of the three unknowns $\mathbf{F}_A$, $\mathbf{F}_B$, and $\mathbf{F}_D$. Thus the problem reduces to one known force, the resultant $\mathbf{F}_E + \mathbf{F}_C$, and three forces of unknown magnitudes.

In Fig. 12-8*c* we show the resultant $\mathbf{F}_E + \mathbf{F}_C$ with its point of application at E. This is permissible by sliding $\mathbf{F}_C$ along its line of action. If $\mathbf{F}_D$ were known, it could be added to $\mathbf{F}_E + \mathbf{F}_C$ to produce the resultant $\mathbf{F}_E + \mathbf{F}_C + \mathbf{F}_D$, which would then act through point p.

Now consider the moment equation. If we write $\Sigma \mathbf{M}_q = 0$, we see that the equation can be satisfied only if the resultant $\mathbf{F}_E + \mathbf{F}_C + \mathbf{F}_D$ has pq as its line of action. This is therefore the basis for the graphical solution. As shown in the force polygon of Fig. 12-8*c*, the resultant $\mathbf{F}_E + \mathbf{F}_C + \mathbf{F}_D$ acting along the line pq is first used to find the force $\mathbf{F}_D$. The polygon is completed by finding $\mathbf{F}_A$ and $\mathbf{F}_B$ since their lines of action are known.

Note that this approach defines a general concept, useful in the analytical approach too: when there are three unknowns, choose a point such as q where the lines of action of two of the unknown forces cross and write the moment equation $\Sigma \mathbf{M}_q = 0$. This equation will have only one unknown, $\mathbf{F}_D$ in this example, and can be solved directly. Only then should $\Sigma \mathbf{F} = 0$ be written, since the problem has now been reduced to two unknowns.

An analytical solution for this problem yields $F_A = 51.8$ N, $F_B = 32.8$ N, and $F_D = 5.05$ N, rounded to three places.

12-9 SPUR- AND HELICAL-GEAR FORCE ANALYSIS†

Figure 12-9*a* shows a pinion with center at O_2 rotating clockwise at n_2 rpm and driving a gear with center at O_3 at n_3 rpm. The reactions between the teeth occur along the pressure line AB. Free-body diagrams of the pinion and gear are shown in Fig. 12-9*b*. The action of the pinion on the gear has been replaced by the force $\mathbf{W}$ acting at the pitch point in the direction of the pressure line. Since the gear is supported by its shaft, an equal and opposite force $\mathbf{F}$ must act at the centerline of the shaft. A similar analysis of the pinion shows that the same observations are true. In each case the forces are equal

† Since gear standards are based entirely on U.S. customary units in the United States, SI units are seldom used for gearing in this book.

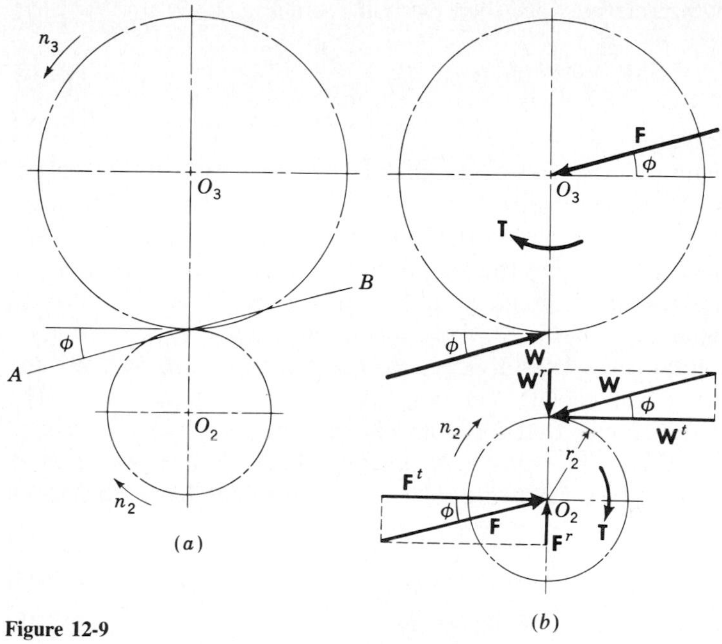

Figure 12-9

in magnitude, opposite in direction, parallel, and in the same plane. They therefore constitute a couple.

Note that the free-body diagram of the pinion has the forces resolved into components. Here we employ the superscripts r and t to indicate the radial and tangential directions with respect to the pitch circle. It is expedient to use the same superscripts for the components of the force **F** which the shaft exerts on the gear. The moment of the couple $\mathbf{W}^t$ and $\mathbf{F}^t$ is the torque which must be applied to drive the gearset. When the pitch radius of the pinion is designated as r_2, the torque is

$$T = r_2 W^t \tag{12-11}$$

where T is the applied torque, positive for the counterclockwise direction and W^t is the magnitude of the force vector $\mathbf{W}^t$.

Note that the radial force W^r serves no useful purpose as far as the transmission of power is concerned. For this reason $\mathbf{W}^t$ is frequently called the *transmitted* force.

If the horsepower and speed of the pinion are given, the tangential force W^t can be obtained from the equation

$$W^t = \frac{(33\,000)(12)\,\text{hp}}{2\pi r_2 n_2} \tag{12-12}$$

where r_2 is the pitch radius in inches and n_2 is the speed in revolutions per minute. The following relations are then evident from Fig. 12-9:

$$W^r = W^t \tan \phi \qquad W = \frac{W^t}{\cos \phi} \tag{12-13}$$

where ϕ is the pressure angle.

In the treatment of the forces on helical gears it is convenient to determine the axial force, work with it independently, and treat the remaining force components the same as for straight spur gears. Figure 12-10 is a drawing of a helical gear with half the face removed to show the forces acting at the pitch point. The gear is imagined to be rotating clockwise. The driving gear has been removed and its effect replaced by the forces shown acting on the teeth. The resultant force **W** is divided into three components $\mathbf{W}^a$, $\mathbf{W}^r$, and $\mathbf{W}^t$, which are the axial, radial, and tangential forces, respectively. The tangential force is the transmitted force and is the one which is effective in transmitting torque. When the transverse pressure angle is designated as ϕ_t and the helix angle as ψ, the following relations are evident from Fig. 12-10:

$$\mathbf{W} = \mathbf{W}^a + \mathbf{W}^r + \mathbf{W}^t \tag{12-14}$$

$$W^a = W^t \tan \psi \tag{12-15}$$

$$W^r = W^t \tan \phi_t \tag{12-16}$$

It is also expedient to make use of the resultant of $\mathbf{W}^r$ and $\mathbf{W}^t$. We shall designate this force as $\mathbf{W}^\phi$. It is defined by the equation

$$\mathbf{W}^\phi = \mathbf{W}^r + \mathbf{W}^t \tag{12-17}$$

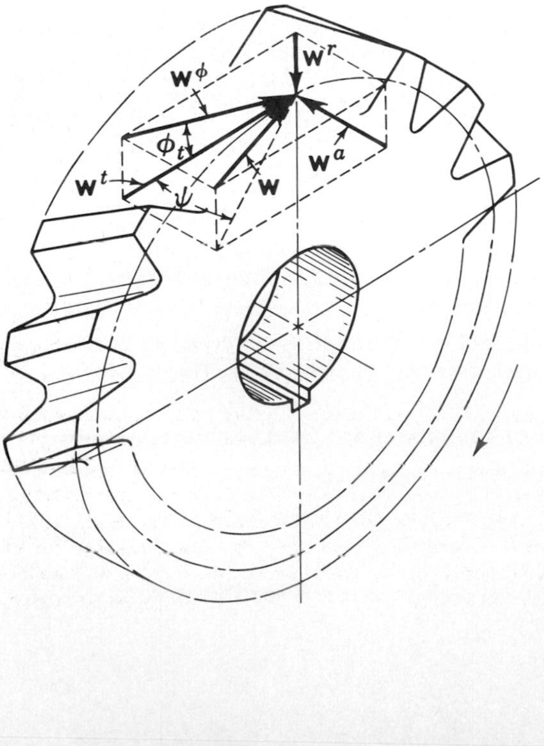

Figure 12-10

Example 12-3 A gear train is composed of three helical gears with the shaft centers in line. The driver is a right-hand helical gear having a pitch radius of 2 in, a transverse pressure angle of 20°, and a helix angle of 30°. An idler gear in the train has the teeth cut left hand and has a pitch radius of 3.25 in. The idler transmits no power to its shaft. The driven gear in the train has the teeth cut right hand and has a pitch radius of 2.50 in. If the transmitted force is 600 lb, find the shaft forces acting on each gear.

SOLUTION First, we consider only the axial components as previously suggested. For each mesh the axial component of the reaction is, from Eq. (12-15),

$$W^a = W^t \tan \psi = 600 \tan 30° = 347 \text{ lb}$$

Figure 12-11*a* is a top view of the three gears looking down on the plane formed by the three axes of rotation. For each gear, rotation is considered to be about the z axis for this problem. In Fig. 12-11*b* free-body diagrams of each of the three gears are drawn in perspective and the three coordinate axes are shown. As indicated, the idler exerts a force $\mathbf{W}_{32}^a$ on the driver. This is resisted by the axial shaft force $\mathbf{F}_{12}^a$. The forces $\mathbf{F}_{12}^a$ and $\mathbf{W}_{32}^a$ form a couple which is resisted by the moment $\mathbf{T}_{12}^a$. Note that this moment is negative about the y axis, and consequently it is a moment which tends to rotate the driver shaft end over end. The magnitude of this moment is

$$T_{12}^a = W_{32}^a r_2 = (347)(2) = 694 \text{ lb} \cdot \text{in}$$

Turning next to the idler, we see from Fig. 12-11*a* and *b* that the axial force of the shaft on the idler is zero. The axial component of the driver on the idler is $\mathbf{W}_{23}^a$, and that of the driven gear on the idler is $\mathbf{W}_{43}^a$. These two forces are equal and form a couple, tending to turn the shaft end over end, which is resisted by the moment $\mathbf{T}_{13}^a$ of magnitude

$$T_{13}^a = W_{23}^a(2r_3) = (347)(2)(3.25) = 2260 \text{ lb} \cdot \text{in}$$

The driven gear has the axial force component $\mathbf{W}_{34}^a$, due to the idler acting at its pitch line, which is resisted by the axial shaft reaction $\mathbf{F}_{14}^a$. As shown, these forces are equal and form a couple tending to turn the shaft end over end, which is resisted by the moment $\mathbf{T}_{14}^a$. Since $W_{34}^a = 347$ lb, the magnitude of this moment, which is negative about the y axis, is

$$T_{14}^a = W_{34}^a r_4 = (347)(2.5) = 867 \text{ lb} \cdot \text{in}$$

It is emphasized again that the three resisting moments $\mathbf{T}_{12}^a$, $\mathbf{T}_{13}^a$, and $\mathbf{T}_{14}^a$ are due solely to the axial components of the reactions between the gear teeth. They produce static bearing reactions and have no effect on the amount of power transmitted.

Now that all the reactions due to the axial components have been found, we turn our attention to the remaining force components and examine their effect as if they were operating independently of the axial forces.

Free-body diagrams showing the forces in the plane of rotation for the driver, idler, and driven gears are shown, respectively, in Fig. 12-11*c*, *d*, and *e*. The forces can be obtained graphically as shown or by applying Eqs. (12-11) and (12-12). It is not necessary to combine the components to find the resultant forces because, in machine design, the component forces are exactly those which are desired.

Example 12-4 The planetary gear train in Fig. 12-12*a* has shaft *a* driven by an input torque of $-100\hat{\mathbf{k}}$ lb · in. Note that shaft *a* is connected directly to gear 2 and that the planetary arm 3 is connected directly to shaft *b* and is separately loosely, but with a minimum clearance, on shaft *a*. Gear 6 is fixed to the stationary frame 1 (not shown). All gears have a diametral pitch of 10 teeth per inch and a pressure angle of 20°. Assuming that the forces act in a single plane and that the centrifugal forces on the planet gears can be neglected, make a complete force analysis of the parts of the train and compute the magnitude and direction of the output torque delivered by shaft *b*.

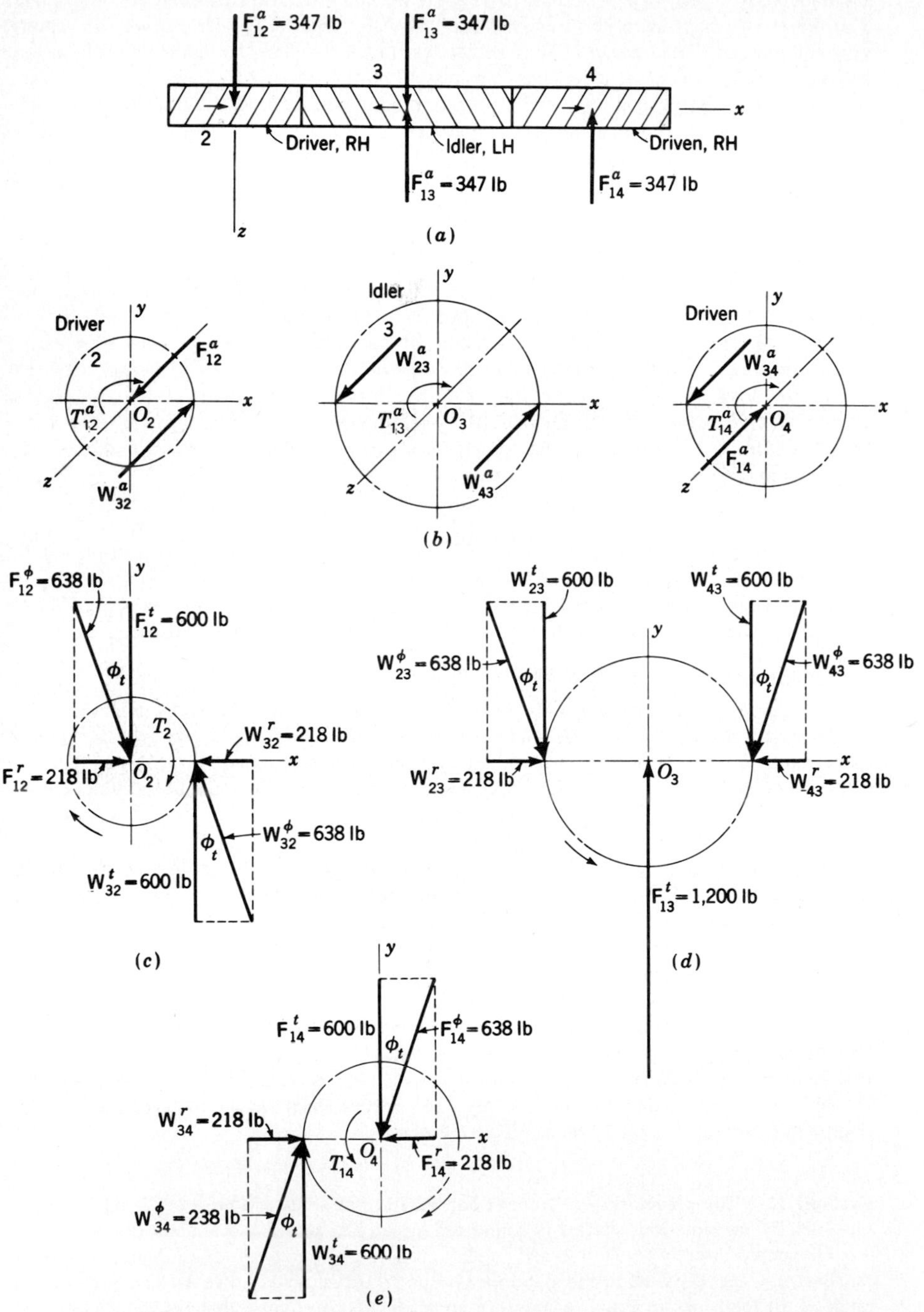

Figure 12-11 (*a*) and (*b*) Axial forces; (*c*) driver; (*d*) idler; (*e*) driven.

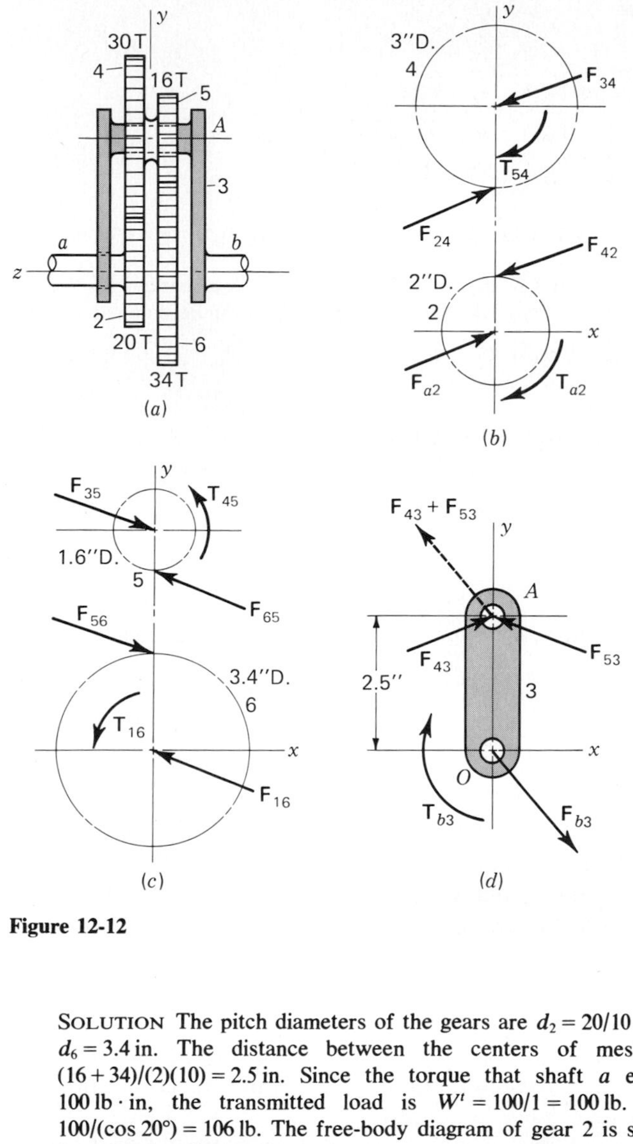

Figure 12-12

SOLUTION The pitch diameters of the gears are $d_2 = 20/10 = 2$ in, $d_4 = 3$ in, $d_5 = 1.6$ in, and $d_6 = 3.4$ in. The distance between the centers of meshing gears is $(N_5 + N_6)/2P = (16 + 34)/(2)(10) = 2.5$ in. Since the torque that shaft a exerts against gear 2 is $T_{a2} = 100\ \text{lb} \cdot \text{in}$, the transmitted load is $W^t = 100/1 = 100$ lb. Therefore $F_{42} = W^t/(\cos \phi) = 100/(\cos 20°) = 106$ lb. The free-body diagram of gear 2 is shown in Fig. 12-12*b*. In vector form, the results are

$$\mathbf{F}_{a2} = -\mathbf{F}_{42} = 106\underline{/20°}\ \text{lb}$$

Figure 12-12*b* also shows the free-body diagram of gear 4. The forces are

$$\mathbf{F}_{24} = -\mathbf{F}_{34} = 106\underline{/20°}\ \text{lb}$$

where $\mathbf{F}_{34}$ is the force of the shaft on planet arm 3 acting against gear 4. Gears 4 and 5 are connected to each other but turn freely on the planet-arm shaft. Thus $\mathbf{T}_{54}$ is the torque exerted by gear 5 on gear 4. This torque is $T_{54} = W^t r_4 = 100(\frac{3}{2}) = 150\ \text{lb} \cdot \text{in}$.

Turning next to the free-body diagram of gear 5 in Fig. 12-12*c*, we first find $F^t_{65} = T_{45}/r_5 = 150/0.8 = 187.5$ lb. Therefore $F_{65} = 187.5/\cos 20° = 200$ lb. In vector form the results for gear 5 are summarized as

$$\mathbf{F}_{65} = -\mathbf{F}_{35} = 200\underline{/160^\circ}\text{ lb} \qquad \mathbf{T}_{45} = 150\hat{\mathbf{k}}\text{ lb}\cdot\text{in}$$

For gear 6 in Fig. 12-12*c*, we have

$$\mathbf{F}_{16} = -\mathbf{F}_{56} = 200\underline{/160^\circ}\text{ lb}$$

$$\mathbf{T}_{16} = \frac{d_6}{2} F_{56} \cos\phi\,\hat{\mathbf{k}} = \frac{3.4(200)(\cos 20^\circ)\hat{\mathbf{k}}}{2} = 319\hat{\mathbf{k}}\text{ lb}\cdot\text{in}$$

Note that $\mathbf{F}_{16}$ and $\mathbf{T}_{16}$ are the force and the torque, respectively, exerted by the frame on gear 6.

The free-body diagram of arm 3 is shown in Fig. 12-12*d*. As noted earlier, the forces are assumed to act in a single plane. Thus, the two forces $\mathbf{F}_{43}$ and $\mathbf{F}_{53}$ can be summed and are

$$\mathbf{F}_{43} = -\mathbf{F}_{34} = 106\underline{/20^\circ}\text{ lb} \qquad \mathbf{F}_{53} = -\mathbf{F}_{35} = 200\underline{/160^\circ}\text{ lb}$$

Then the sum turns out to be

$$\mathbf{F}_{43} + \mathbf{F}_{53} = 137\underline{/130.2^\circ}\text{ lb}$$

We now find the shaft reaction to be

$$\mathbf{F}_{b3} = -\mathbf{F}_{43} - \mathbf{F}_{53} = 137\underline{/-49.8^\circ}\text{ lb}$$

Using $\mathbf{r}_{AO} = 2.5\hat{\mathbf{j}}$ and the equation

$$\sum \mathbf{M}_O = \mathbf{T}_{b3} + \mathbf{r}_{AO} \times (\mathbf{F}_{43} + \mathbf{F}_{53}) = 0$$

we find $\mathbf{T}_{b3} = -221\hat{\mathbf{k}}$ lb · in. Therefore, the output shaft torque is $\mathbf{T}_b = +221\hat{\mathbf{k}}$ lb · in.

12-10 STRAIGHT BEVEL GEARS

In determining the tooth forces on bevel gears it is customary to use the forces that would occur at the midpoint of the tooth on the pitch cone. The resultant tangential force probably occurs somewhere between the midpoint and the large end of the tooth, but there will be only a small error in making this assumption. The tangential or transmitted force is given by

$$W^t = \frac{T}{r} \tag{12-18}$$

where r is the average radius of the pitch cone as shown in Fig. 12-13 and T is the torque.

Figure 12-13 also shows all the components of the resultant force acting at the midpoint of the tooth. The following relationships can be derived by inspection of the figure:

$$\mathbf{W} = \mathbf{W}^a + \mathbf{W}^r + \mathbf{W}^t \tag{12-19}$$

$$W^r = W^t \tan\phi \cos\gamma \tag{12-20}$$

$$W^a = W^t \tan\phi \sin\gamma \tag{12-21}$$

Note, as in the case of helical gears, that the axial force $\mathbf{W}^a$ results in a couple on the shaft which tends to turn it end over end.

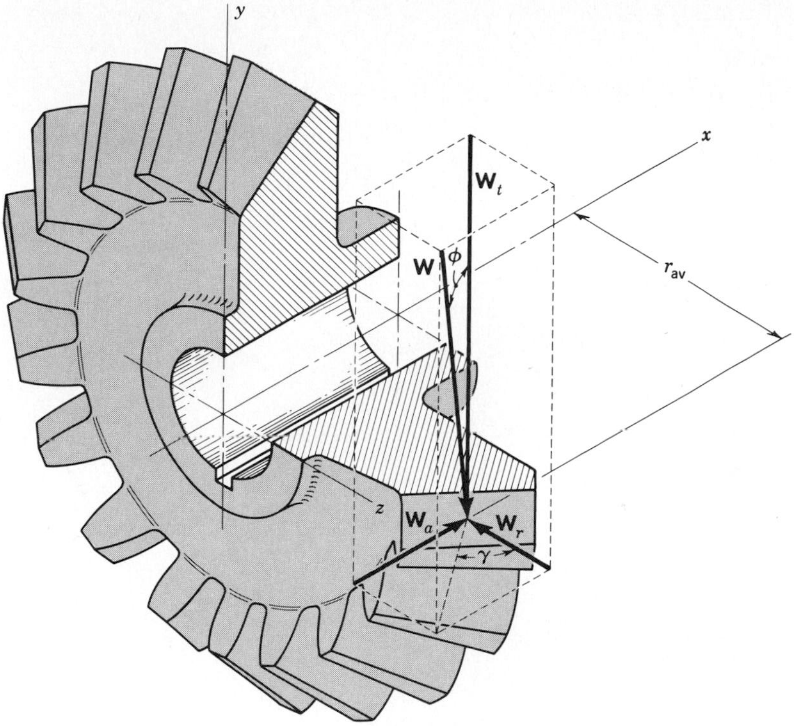

Figure 12-13

Example 12-5 The bevel pinion shown in Fig. 12-14 rotates at 600 rpm in the direction shown and transmits 5 hp to the gear. The mounting distances are shown, together with the location of the bearings on each shaft. Bearings *A* and *C* are capable of taking both radial and axial loads, while bearings *B* and *D* are built to receive only pure radial loads. The teeth of the gears have a 20° pressure angle. Find the components of the forces which the bearings exert on the shafts in the *x*, *y*, and *z* directions.

SOLUTION The pitch angles for the pinion and gear are

$$\gamma = \tan^{-1}\tfrac{3}{9} = 18.4^\circ \qquad \Gamma = \tan^{-1}\tfrac{9}{3} = 71.6^\circ$$

The radii to the midpoint of the teeth are shown on the drawing and are $r_2 = 1.293$ in and $r_3 = 3.88$ in for the pinion and gear, respectively.

Let us determine the forces acting upon the pinion first. The tangential force is

$$W^t = \frac{(33\,000)(12)\text{ hp}}{2\pi r_2 n_2} = \frac{(33\,000)(12)(5)}{2\pi(1.293)(600)} = 406\text{ lb}$$

This force acts in the negative *z* direction. (In Fig. 12-14 the *z* axis is positive out of the paper for a right-handed system.) The radial and axial components are obtained from Eqs. (12-20) and (12-21).

$$W^r = W^t \tan\phi \cos\gamma = 406 \tan 20^\circ \cos 18.4^\circ = 140\text{ lb}$$
$$W^a = W^t \tan\phi \sin\gamma = 406 \tan 20^\circ \sin 18.4^\circ = 46.6\text{ lb}$$

Here W^r acts in the positive *y* direction and W^a in the positive *x* direction.

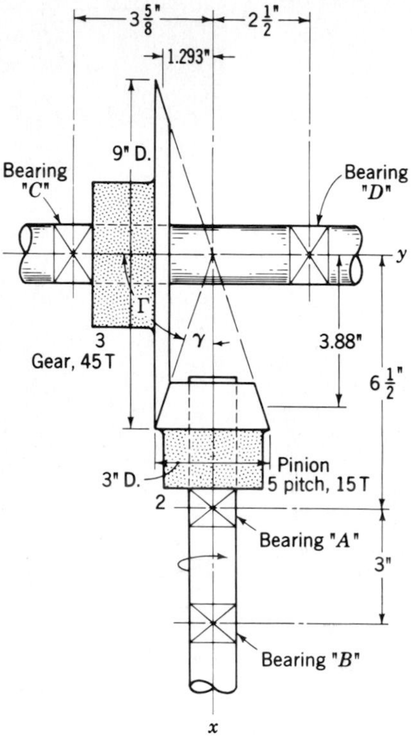

Figure 12-14

These three forces are components of the force **W**. Thus

$$\mathbf{W} = 46.6\hat{\mathbf{i}} + 140\hat{\mathbf{j}} - 406\hat{\mathbf{k}}$$

The torque applied to the pinion shaft must be

$$\mathbf{T}_2 = -406(1.293)\hat{\mathbf{i}} = -525\hat{\mathbf{i}} \text{ lb} \cdot \text{in}$$

A free-body diagram of the pinion and shaft is shown schematically in Fig. 12-15*a*. The bearing reactions $\mathbf{F}_A$ and $\mathbf{F}_B$ are to be determined; the dimensions, the torque $\mathbf{T}_2$, and the force **W** are the given elements of the problem. In order to find $\mathbf{F}_B$ we shall sum moments about *A*. This requires two relative position vectors, defined as

$$\mathbf{R}_{PA} = -2.62\hat{\mathbf{i}} - 1.293\hat{\mathbf{j}} \qquad \mathbf{R}_{BA} = 3\hat{\mathbf{i}}$$

Summing moments about *A* gives

$$\sum \mathbf{M}_A = \mathbf{T}_2 + \mathbf{R}_{BA} \times \mathbf{F}_B + \mathbf{R}_{PA} \times \mathbf{W} = 0 \tag{1}$$

The second and third terms for Eq. (1) are, respectively

$$\mathbf{R}_{BA} \times \mathbf{F}_B = 3\hat{\mathbf{i}} \times (F_B^y\hat{\mathbf{j}} + F_B^z\hat{\mathbf{k}}) = -3F_B^z\hat{\mathbf{j}} + 3F_B^y\hat{\mathbf{k}} \tag{2}$$

$$\mathbf{R}_{PA} \times \mathbf{W} = (-2.62\hat{\mathbf{i}} - 1.293\hat{\mathbf{j}}) \times (46.6\hat{\mathbf{i}} + 140\hat{\mathbf{j}} - 406\hat{\mathbf{k}})$$
$$= 525\hat{\mathbf{i}} - 1064\hat{\mathbf{j}} - 308\hat{\mathbf{k}} \tag{3}$$

Placing the value of $\mathbf{T}_2$ and Eqs. (2) and (3) into (1) and solving gives

$$\mathbf{F}_B = 102\hat{\mathbf{j}} - 355\hat{\mathbf{k}} \text{ lb} \qquad \textit{Ans.}$$

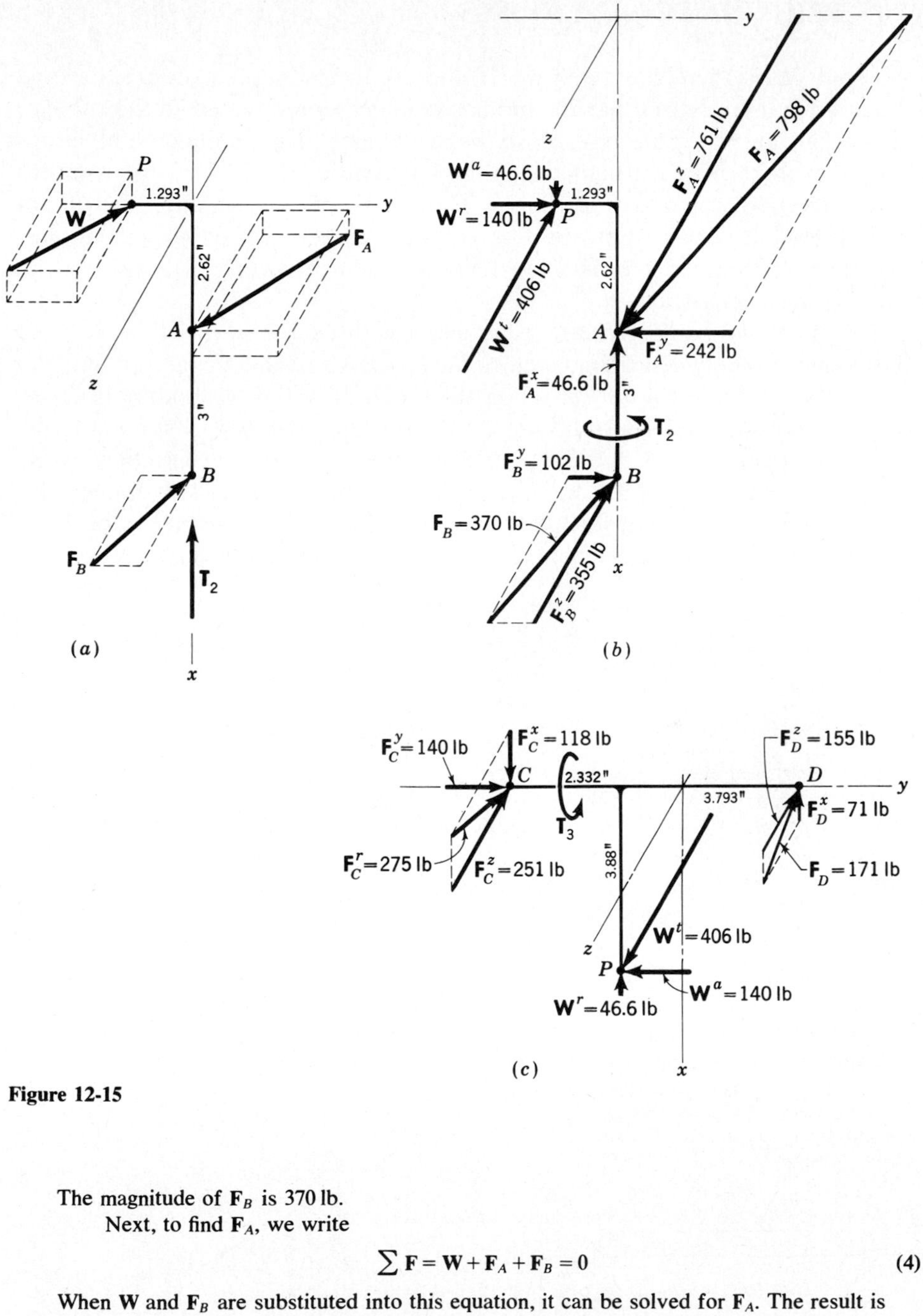

Figure 12-15

The magnitude of $\mathbf{F}_B$ is 370 lb.

Next, to find $\mathbf{F}_A$, we write

$$\sum \mathbf{F} = \mathbf{W} + \mathbf{F}_A + \mathbf{F}_B = 0 \tag{4}$$

When $\mathbf{W}$ and $\mathbf{F}_B$ are substituted into this equation, it can be solved for $\mathbf{F}_A$. The result is

$$\mathbf{F}_A = -46.6\hat{\mathbf{i}} - 242\hat{\mathbf{j}} + 761\hat{\mathbf{k}}\ \text{lb} \qquad \textit{Ans.}$$

The magnitude is $F_A = 798$ lb. The results are shown in Fig. 12-15b. A similar procedure is used for the gear shaft. The results are displayed in Fig. 12-15c.

12-11 FRICTION-FORCE MODELS

In recent years there has been much interest in the subject of friction and wear, and many research papers and books have been devoted to the subject. It is not our purpose here to explore the mechanics of the subject at all but to present well-known mathematical simplifications which can be used to analyze the performance of machines. The results of such an analysis will not be theoretically exact, but they correspond closely to experimental performance, so that reliable decisions can be made regarding a design and its operating characteristics.

Consider two bodies forced into contact with each other, with or without relative motion between them, such as the block 3 and the surface of link 2 in Fig. 12-16*a*. A force $\mathbf{F}_{43}$ is exerted on the block 3 by link 4, tending to cause the block to slide relative to the slot 2. Without the presence of friction in the surface between links 2 and 3, the block would slide in the direction of the horizontal component of $\mathbf{F}_{43}$ and equilibrium would not be possible unless $\mathbf{F}_{43}$ were perpendicular to the slot. With friction, however, a resisting force $\mathbf{F}_{23}^t$ is developed at the contact surface as shown in the free-body diagrams of Fig.

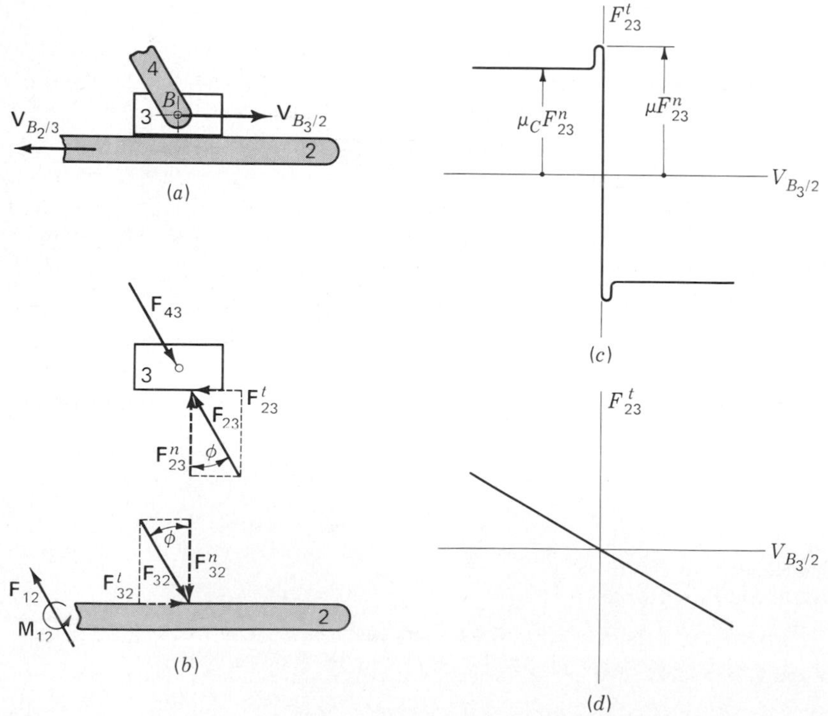

Figure 12-16 Mathematical representation of friction forces: (*a*) physical system; (*b*) free-body diagrams; (*c*) static and Coulomb friction; (*d*) viscous friction.

12-16*b*. This friction force $\mathbf{F}_{23}^t$ acts in addition to the usual constraint force $\mathbf{F}_{23}^n$ across the surface of the sliding joint, and together the forces $\mathbf{F}_{23}^n$ and $\mathbf{F}_{23}^t$ form a total force $\mathbf{F}_{23}$ which balances $\mathbf{F}_{43}$ to keep the block in equilibrium. The reaction forces $\mathbf{F}_{32}^n$ and $\mathbf{F}_{32}^t$, of course, are also acting simultaneously on link 2, as shown in the other free-body diagram of Fig. 12-16*b*. The force $\mathbf{F}_{23}^t$ and its reaction $\mathbf{F}_{32}^t$ are called *friction forces.*

Depending on the materials of links 2 and 3, there is a limit to the size of the force $\mathbf{F}_{23}^t$ which can be developed by friction while still maintaining equilibrium. This limit is given by the relationship

$$F_{23}^t \leq \mu F_{23}^n \tag{12-22}$$

where μ, defined as the *coefficient of static friction,* is a characteristic property of the materials in contact. Values of the coefficient μ have been determined experimentally for many materials and can be found in most engineering handbooks.†

If the force $\mathbf{F}_{43}$ is tipped too far, so that its horizontal component and therefore $\mathbf{F}_{23}^t$ are too large to satisfy Eq. (12-22), equilibrium is not possible and the block will slide relative to link 2 with an apparent velocity $\mathbf{V}_{B_3/2}$. When sliding takes place, the friction force takes on the value

$$F_{23}^t = \mu_c F_{23}^n \tag{12-23}$$

where μ_c is the *coefficient of sliding friction.* Sliding friction is also frequently called *Coulomb friction,* and we shall use this term frequently. The coefficient μ_c can also be found experimentally and is slightly less than μ for most materials.

Figure 12-16*c* shows a graph of the friction force F_{23}^t vs. the apparent velocity $V_{B_3/2}$. Here it can be seen that when the sliding velocity is zero, the friction force $\mathbf{F}_{23}^t$ can have any magnitude between μF_{23}^n and $-\mu F_{23}^n$. When the velocity is not zero, the friction force $\mathbf{F}_{23}^t$ drops slightly in magnitude to the value $\mu_c F_{23}^n$ and has a direction which opposes the sliding motion $\mathbf{V}_{B_3/2}$.

Looking at the total force $\mathbf{F}_{23}$ in Fig. 12-16*b*, we see that it is tipped at an angle ϕ to be equal and opposite to $\mathbf{F}_{43}$ whenever the system is in equilibrium. When $\mathbf{F}_{43}$ is tipped so that the block is just on the verge of sliding, the angle ϕ is given by

$$\tan\phi = \frac{F_{23}^t}{F_{23}^n} = \mu\,\frac{F_{23}^n}{F_{23}^n} = \mu$$

or

$$\phi = \tan^{-1}\mu \tag{12-24}$$

The angle ϕ, called the *friction angle,* defines the maximum angle through which $\mathbf{F}_{23}$ can tip from the normal to the surface before equilibrium will be destroyed and sliding will take place. Notice that ϕ does not depend on the

† See, for example, D. B. Dallas (ed.), "Tool and Manufacturing Engineers Handbook," 3d ed., McGraw-Hill, New York, 1976, p. 41-12.

magnitude of the force $\mathbf{F}_{23}$ but only on the coefficient of friction for the materials.

Even though the resisting forces in a machine may be predominantly Coulomb friction, it is sometimes more convenient to analyze the machine's performance using another kind of resisting force, called *viscous friction* or *viscous damping*. The situation is very much the same as for the free-body diagrams of Fig. 12-16*b*. However, in the case of viscous friction, the friction force $\mathbf{F}_{23}^t$ is assumed to be given by

$$\mathbf{F}_{23}^t = -c\mathbf{V}_{B_3/2} \tag{12-25}$$

where c is the *coefficient of viscous damping*, sometimes called the *damping factor* or *viscous damping constant*. As shown in the graph of Fig. 12-16*d*, this friction force has a linear relationship with velocity. This is especially useful when the analysis of the dynamic response of a machine or system leads to one or more differential equations. The nonlinear relationship of Coulomb friction, shown in Fig. 12-16*c*, leads to a nonlinear differential equation which is more difficult to handle.

Whether the friction effect comes from viscous, Coulomb, or static friction, it is important to recognize the sense of the friction force. As mnemonic, the rule is often stated that "the friction force opposes the motion," as shown by the free-body diagram of link 3, Fig. 12-16*b*, where the sense of $\mathbf{F}_{23}^t$ is opposite that of $\mathbf{V}_{B_3/2}$. This rule of thumb is not wrong if carefully applied, but it can be dangerous. It will be noted in Fig. 12-16*a* that there are *two* motions which might be thought of, $\mathbf{V}_{B_3/2}$ and $\mathbf{V}_{B_2/3}$; there are also two friction forces $\mathbf{F}_{23}$ and $\mathbf{F}_{32}^t$. Careful examination of Fig. 12-16*b* will show that $\mathbf{F}_{23}^t$ opposes the sense of $\mathbf{V}_{B_3/2}$ while $\mathbf{F}_{32}^t$ opposes the sense of $\mathbf{V}_{B_2/3}$. In machine systems, where both sides of a sliding joint are often in motion, it is important to understand which friction force opposes which motion.

12-12 STATIC-FORCE ANALYSIS WITH FRICTION

We will show the effect of including friction on our previous methods of static-force analysis by presenting an example.

Example 12-6 Repeat the static-force analysis of the cam-follower system analyzed in Example 12-2, Fig. 12-8, assuming that there is a coefficient of static friction of 0.15 between links 1 and 4 at both sliding bearings B and D. Friction in all other joints is considered negligible. Determine the minimum force necessary at A to hold the system in equilibrium.

SOLUTION As always when we begin a force analysis with friction, it is necessary first to solve the entire problem without friction. The purpose is to find the direction of each normal force, $\mathbf{F}_B^n$ and $\mathbf{F}_D^n$ in our case. This was done in Example 12-2, where $\mathbf{F}_B$ and $\mathbf{F}_D$ were both found to act to the right in Fig. 12-8*c*.

The next step in the solution is to consider the problem statement carefully and to decide the direction of the impending motion. As given, the problem asks for the minimum force at A necessary for equilibrium; i.e., if $\mathbf{F}_A$ were any smaller, the system would move downward. Thus, the impending motion is with velocities $\mathbf{V}_{D_4/1}$ and $\mathbf{V}_{B_4/1}$ downward, and

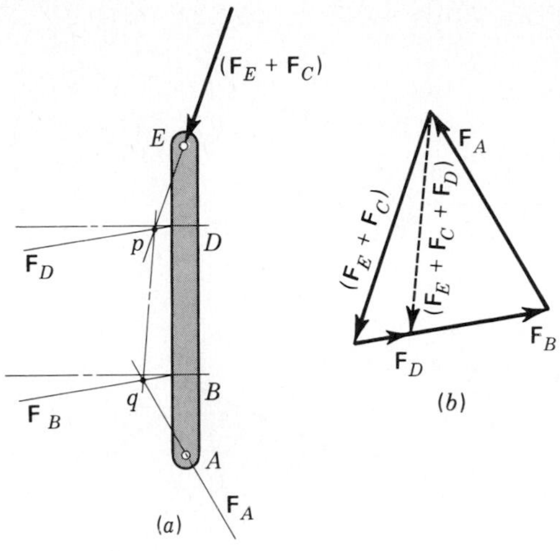

Figure 12-17 Example 12-6.

therefore the two friction forces at B and D must act upward on link 4. Notice that if the problem statement had asked for the maximum force at A, the impending motion of link 4 would be upward and the friction forces downward on link 4.

Next we redraw the free-body diagram of link 4, Fig. 12-8*c*, and include the friction forces as shown in Fig. 12-17*a*. Here, because of static friction, the lines of action of $\mathbf{F}_B$ and $\mathbf{F}_D$ are both shown tipped through the angle ϕ, which can be calculated from Eq. (12-24)

$$\phi = \tan^{-1} 0.15 = 8.5° \tag{1}$$

In deciding the direction of tip of the angles ϕ it was necessary to know both the direction of the friction forces (upward) and the direction of the normal forces (to the right) at B and D. This explains why the solution without friction must be done first.

Now that the new lines of action of the forces $\mathbf{F}_B$ and $\mathbf{F}_D$ are known, the solution can proceed exactly as in Example 12-2. The graphical solution with friction is shown in Fig. 12-17*b*, where it is found that

$$F_A = 45.8 \text{ N} \quad \textit{Ans.} \quad F_B = 28.7 \text{ N} \quad \text{and} \quad F_D = 6.57 \text{ N}$$

Notice that the normal components of the two forces at B and D are now $F_B^n = 28.4$ and $F_B^t = 6.50$ and are different from the values without friction. Therefore, whether one is solving a friction-force problem graphically or analytically, it is incorrect simply to multiply the frictionless normal forces by the coefficient of friction to find the friction forces. All forces may change size when friction is included, and the problem must be completely solved from the beginning with friction included. The effect of friction *cannot* be added in afterward by superposition.

PROBLEMS†

12-1 The figure shows four mechanisms and the external forces or torques exerted on or by the mechanisms. Sketch the free-body diagram of each part of each mechanism, including the frame. Do not attempt to show the magnitudes of the forces, except roughly, but do sketch them in the proper directions.

† Unless otherwise stated, solve all problems without friction.

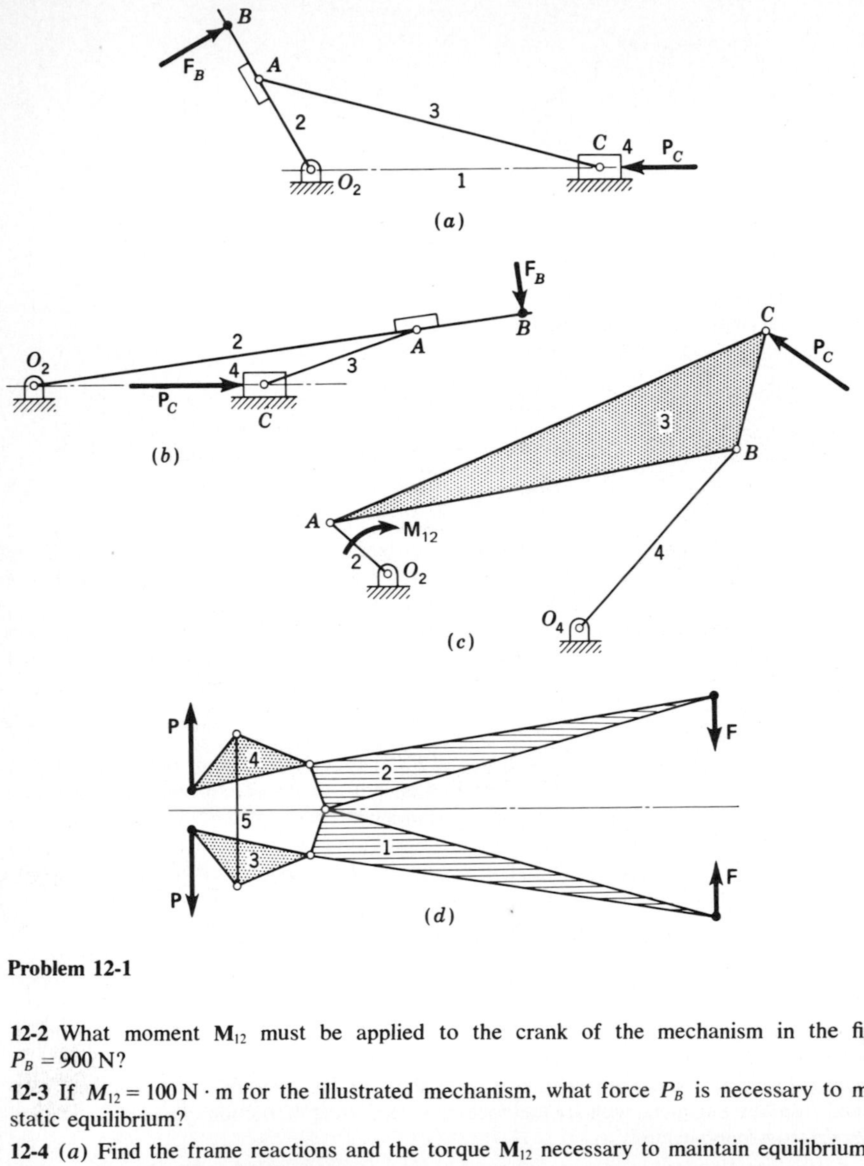

Problem 12-1

12-2 What moment $\mathbf{M}_{12}$ must be applied to the crank of the mechanism in the figure if $P_B = 900$ N?

12-3 If $M_{12} = 100$ N · m for the illustrated mechanism, what force P_B is necessary to maintain static equilibrium?

12-4 (*a*) Find the frame reactions and the torque $\mathbf{M}_{12}$ necessary to maintain equilibrium of the four-bar linkage shown in the figure.

(*b*) What torque must be applied to link 2 of the illustrated mechanism to maintain static equilibrium? Draw complete free-body diagrams of links 1 and 4.

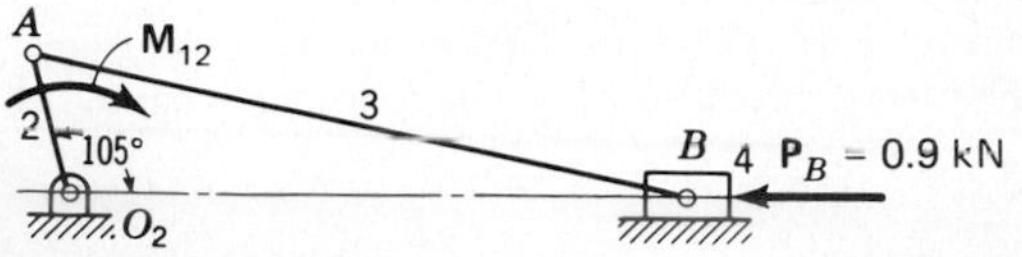

Problems 12-2 and 12-3 $O_2A = 75$ mm; $AB = 350$ mm.

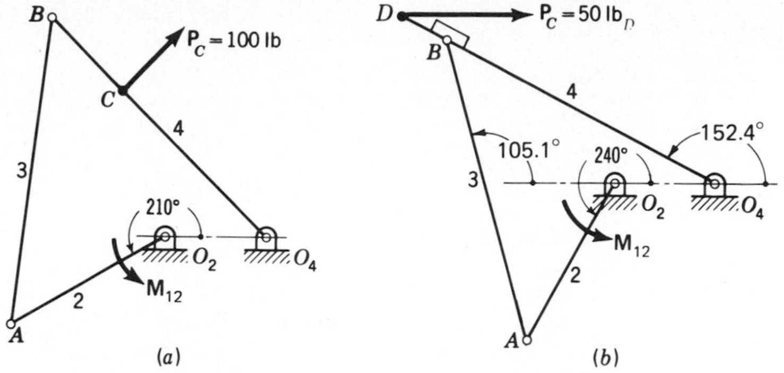

Problem 12-4 (*a*) and (*b*) $O_2A = 3.5$ in; $AB = O_4B = 6$ in; $O_4C = 4$ in; $O_4D = 7$ in; $O_2O_4 = 2$ in.

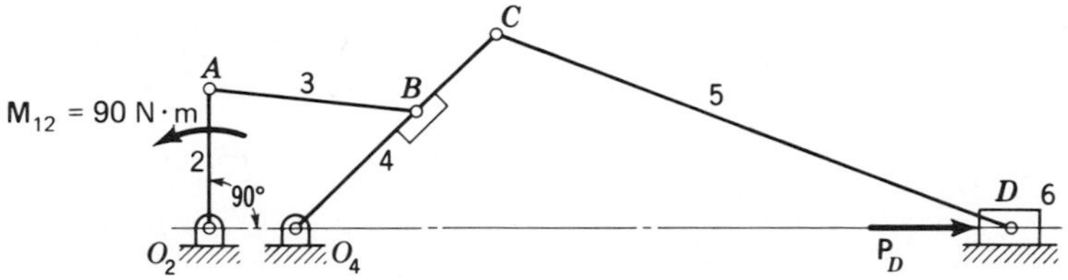

Problem 12-5 $O_2A = 100$ mm; $AB = 150$ mm; $O_4B = 125$ mm; $O_4C = 200$ mm; $CD = 400$ mm; $O_2O_4 = 60$ mm.

12-5 What force P is necessary for equilibrium of the linkage shown? Sketch a complete free-body diagram of each link.

12-6 (*a*) Determine the torque $\mathbf{M}_{12}$ required to drive slider 6 in the figure against a load $P = 100$ lb for a crank angle $\theta = 30°$, or as specified by your instructor.

(*b*) What torque $\mathbf{M}_{12}$ must be applied to link 2 of the illustrated four-bar linkage to retain static equilibrium? Sketch free-body diagrams of links 1 and 3 and find the forces that act.

12-7 Find the magnitude and direction of the moment that must be applied to link 2 to drive the linkage against the forces shown. Sketch a free-body diagram of each link and show all the forces acting.

12-8 The figure shows a four-bar linkage with external forces applied at points B and C. Find the couple that must be applied to link 2 to maintain equilibrium. Draw a free-body diagram of each link including the frame, and show all the forces acting upon each.

12-9 Draw a free-body diagram of each of the members of the mechanism shown in the figure, and find the magnitude and direction of all the forces and moments. Compute the magnitude and direction of the couple that must be applied to link 2 to retain static equilibrium.

12-10 Determine the magnitude and direction of the forces that must be applied to link 2 to maintain static equilibrium.

12-11 In each case shown, pinion 2 is the driver, gear 3 is an idler, the gears are 6 diametral pitch and have a 20° pressure angle. For each case sketch a free-body diagram of gear 3 and show all forces acting.

(*a*) Pinion 2 rotates at 600 rpm and transmits 18 hp to the gearset.

(*b*) and (*c*) Pinion 2 rotates at 900 rpm and transmits 25 hp to the gearset.

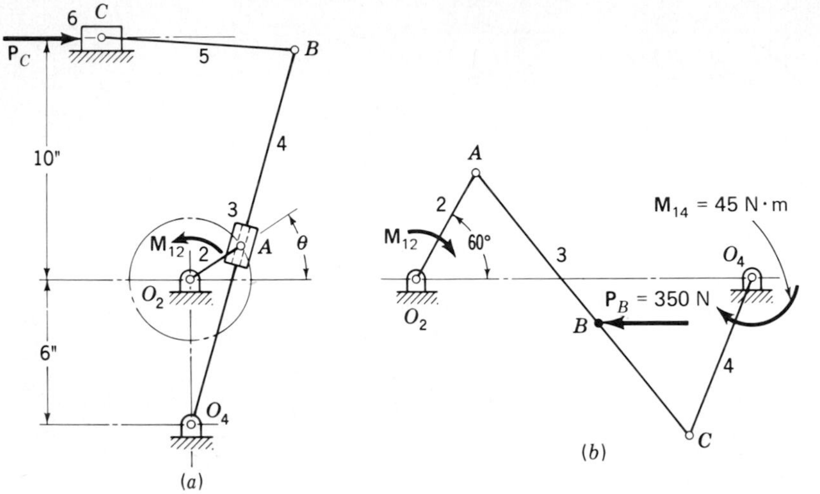

Problem 12-6 (a) $O_2A = 2.5$ in; $O_4B = 16$ in; $BC = 8$ in. (b) $O_2A = 250$ mm; $AB = 400$ mm; $AC = O_2O_4 = 700$ mm; $O_4C = 350$ mm.

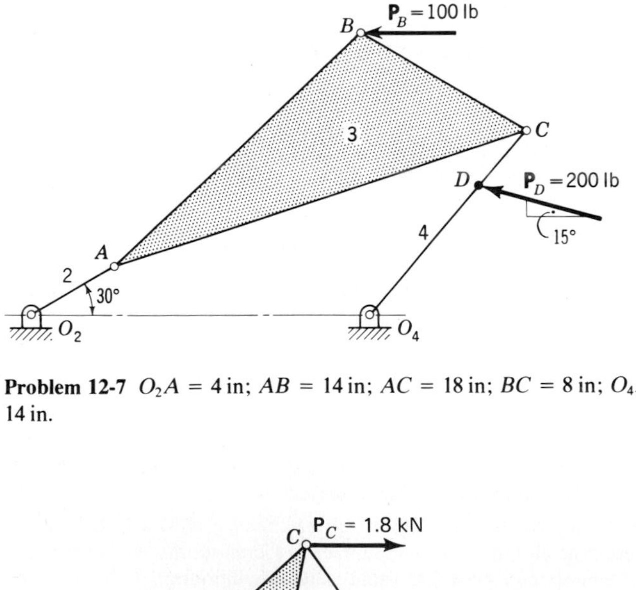

Problem 12-7 $O_2A = 4$ in; $AB = 14$ in; $AC = 18$ in; $BC = 8$ in; $O_4D = 7$ in; $O_4C = 10$ in; $O_2O_4 = 14$ in.

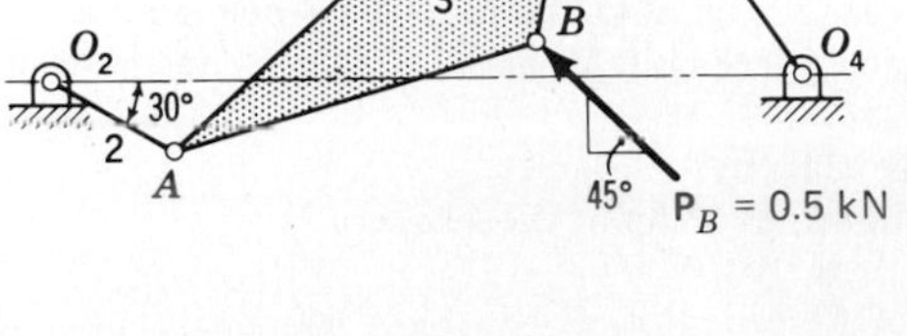

Problem 12-8 $O_2A = 75$ mm; $AB = O_4C = 200$ mm; $AC = 300$ mm; $BC = 150$ mm; $O_2O_4 = 400$ mm.

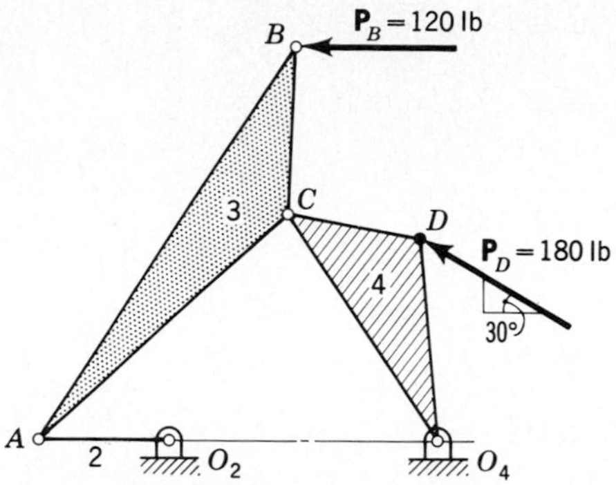

Problem 12-9 $O_2A = 4$ in; $AB = 14$ in; $AC = 10$ in; $BC = 5$ in; $O_4C = O_2O_4 = 8$ in; $CD = 4$ in; $O_4D = 6$ in.

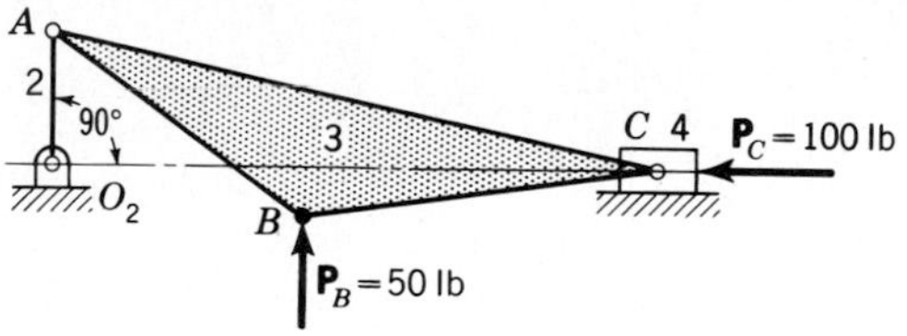

Problem 12-10 $O_2A = 3$ in; $AB = 7$ in; $AC = 14$ in; $BC = 8$ in.

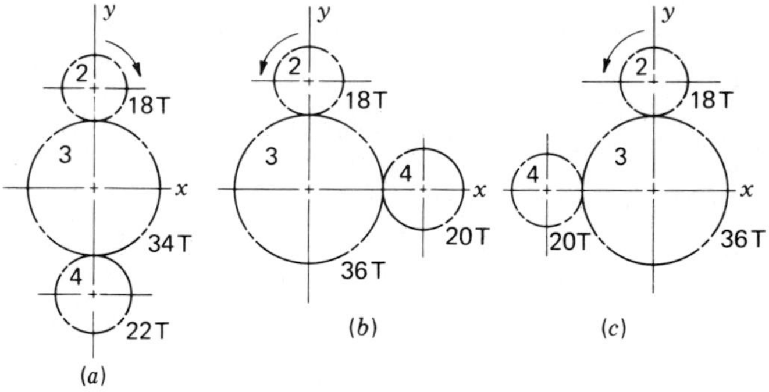

Problem 12-11

12-12 A 15-tooth spur pinion has a diametral pitch of 5 and a 20° pressure angle, rotates at 600 rpm and drives a 60-tooth gear. The horsepower transmitted is 25. Construct a free-body diagram of each gear showing upon it the tangential and radial components of the forces and their proper directions.

12-13 The 16-tooth pinion on shaft 2 rotates at 1720 rpm and transmits 5 hp to the double-reduction gear train. All gears have a 20° pressure angle. The distances between the centers of the bearings and gears for shaft 3 are shown in the figure. Find the magnitude and direction of the radial force that each bearing exerts against this shaft.

12-14 Solve Prob. 12-11 if each pinion has right-hand helical teeth with a 30° helix angle and a normal pressure angle of 20°. All gears in the train are, of course, helical, and the normal diametral pitch is 6 teeth per inch for each case.

12-15 Analyze the gear shaft of Example 12-5 and find the bearing reactions $\mathbf{F}_C$ and $\mathbf{F}_D$.

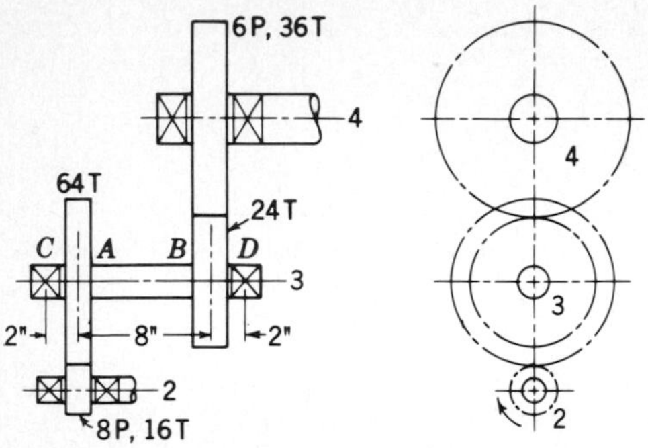

Problem 12-13

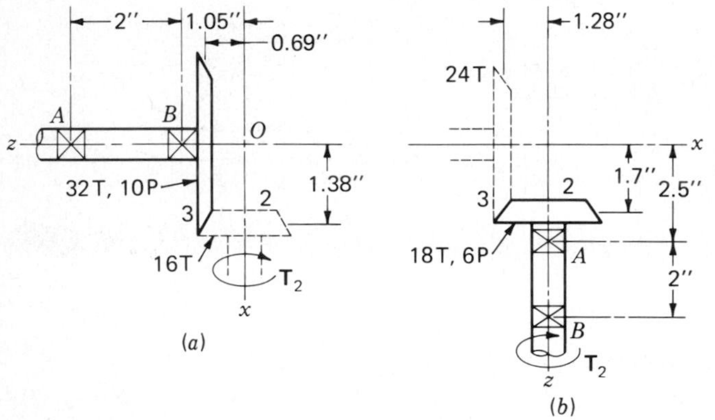

Problem 12-16

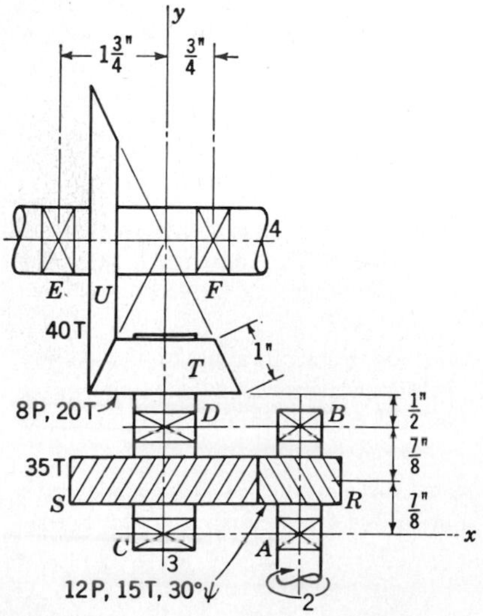

Problem 12-17

12-16 In each of the bevel-gear drives shown in the figure, bearing A takes both thrust load and radial load, while bearing B takes only a pure radial component. Compute these bearing loads. The teeth are cut with a 20° pressure angle.

(*a*) $\mathbf{T}_2 = -180\hat{\mathbf{i}}$ lb · in.

(*b*) $\mathbf{T}_2 = -240\hat{\mathbf{k}}$ lb · in.

12-17 The figure shows a gear train composed of a pair of helical gears and a pair of straight bevel gears. Shaft 4 is the output of the train, and it delivers 6 hp to the load at a speed of 370 rpm. The bevel gears have a pressure angle of 20°. If bearing E is to take both thrust and

(*a*)

Problem 12-19 (*a*) Figee floating crane with lemniscate boom configuration; (*b*) schematic diagram (see page 420). The dimensions in metres are $O_2A = 14.7$, $O_4B = 19.3$, $AB = 6.5$, $AC = 22.3$, $BC = 16$. (*Photograph and dimensional details by permission from B. V. Machinefabriek Figee, Haarlem, Holland.*)

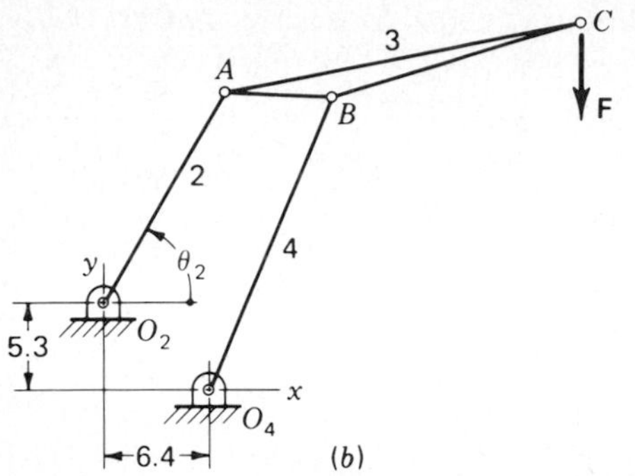

Problem 12-19 (*cont.*)

radial load while bearing F is to take only radial, determine the force which these bearings exert against shaft 4.

12-18 Using the data of Prob. 12-17 find the forces exerted by bearings C and D on shaft 3. Which of these bearings should take the thrust load if the shaft is to be loaded in compression? The helical gears are 20° transverse pressure angle.

12-19 The photograph is of the new Figee floating crane with lemniscate boom configuration. Also shown is a schematic diagram of the crane. The lifting capacity is 16 t (1 t = 1 metric ton = 1000 kg) including the grab; contents of the grab, about 10 t. The maximum outreach is 30 m, corresponding to $\theta_2 = 49°$. Minimum outreach is 10.5 m at $\theta_2 = 132°$. Other dimensions are shown in the figure caption. For the maximum-outreach position and a grab load of 10 t, find the bearing reactions at A, B, O_2, and O_4 and the moment M_{12} required. Note that the photograph shows a counterweight on link 2. Neglect this weight and also the weight of the members.

12-20 Repeat Prob. 12-19 for the minimum-outreach position.

12-21 Repeat Prob. 12-6a assuming coefficients of Coulomb friction $\mu_c = 0.20$ between links 1 and 6 and $\mu_c = 0.10$ between links 3 and 4. Determine the torque $\mathbf{M}_{12}$ necessary to drive the system, including friction against the load P.

12-22 Repeat Prob. 12-10 assuming a coefficient of static friction of $\mu = 0.15$ between links 1 and 4. Determine the torque $\mathbf{M}_{12}$ necessary to overcome friction.

CHAPTER

THIRTEEN

DYNAMIC FORCES

13-1 RIGID- AND ELASTIC-BODY FORCE ANALYSIS

A bell rings when it is struck by a clapper. The characteristics of the ringing, such as frequency, loudness, duration, and tone, depend upon the geometry of the bell and its material. A bell made of an inelastic material like lead or putty will not ring. Bells are therefore made of very elastic materials such as glass or hard steel. The analysis of the ringing of a bell and of other vibrating systems is called *elastic-body analysis.* We use elastic-body analysis when we are interested in such things as deflection, deformation, extension, or the motions of the various particles of the body.

In contrast, we use *rigid-body analysis* when we are interested in the whole or overall motion of a body. The entire body is assumed to be absolutely rigid and incapable of deforming in any manner. Our studies in this chapter are concerned only with rigid-body analysis.

13-2 CENTROIDS AND CENTER OF MASS

In solving engineering problems, we frequently find that forces are distributed in some manner over a line, an area, or a volume. The resultant of these distributed forces is usually not too difficult to find. In order to have the same effect, this resultant must act at the *centroid* of the system. Thus, *the centroid of a system is a point at which a system of distributed forces may be considered concentrated with exactly the same effect.*

Instead of a system of distributed forces, we may have a distributed mass. Then, by *center of mass* we mean *the point at which the mass may be considered concentrated so that the same effect is obtained.*

In Fig. 13-1*a* a series of concentrated masses are located on a line. The center of mass G or centroid is located at

$$\bar{x} = \frac{\sum_{i=1}^{N} m_i x_i}{\sum_{i=1}^{N} m_i} = \frac{m_1 x_1 + m_2 x_2 + m_3 x_3}{m_1 + m_2 + m_3} \tag{13-1}$$

In Fig. 13-1*b* the masses are located on a plane. The x coordinates of the center of mass G can be obtained from Eq. (13-1). The y coordinate is

$$\bar{y} = \frac{\sum_{i=1}^{N} m_i y_i}{\sum_{i=1}^{N} m_i} = \frac{m_1 y_1 + m_2 y_2 + m_3 y_3 + m_4 y_4}{m_1 + m_2 + m_3 + m_4} \tag{13-2}$$

This procedure can be extended to masses concentrated in a volume by simply writing an equation like (13-1) for the z axis.

When mass is distributed in a plane, the center of mass can often be found by symmetry. Figure 13-2 shows the location for a circle, a rectangle, and a triangle. Note that the intersection of the medians locates G for the triangle.

The plane area in Fig. 13-3 is a composite shape made up of a rectangular area plus a triangular area minus a circular area. The location of the centroids of the parts G_1, G_2, and G_3 can be found with the help of Fig. 13-2. The center of mass G of the composite area is then located using the equation

$$\bar{x} = \frac{\sum_{i=1}^{N} A_i x_i}{\sum_{i=1}^{N} A_i} = \frac{A_1 \bar{x}_1 + A_2 \bar{x}_2 - A_3 \bar{x}_3}{A_1 + A_2 - A_3} \tag{13-3}$$

where the expression for $\bar{y}$ is similar.

A more general set of relations for the location of a centroid in a plane

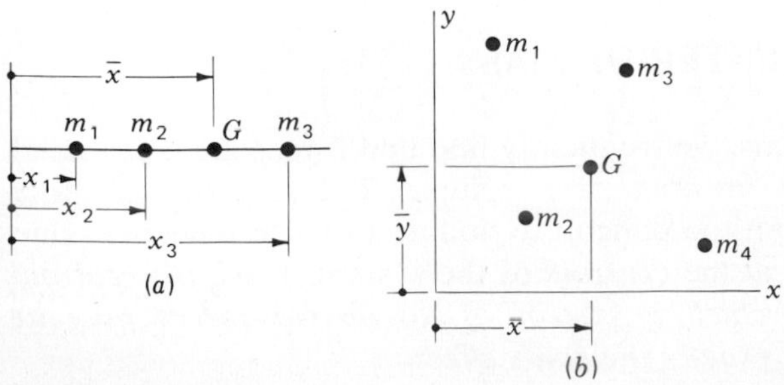

Figure 13-1 (*a*) Concentrated masses on a line; (*b*) concentrated masses in a plane.

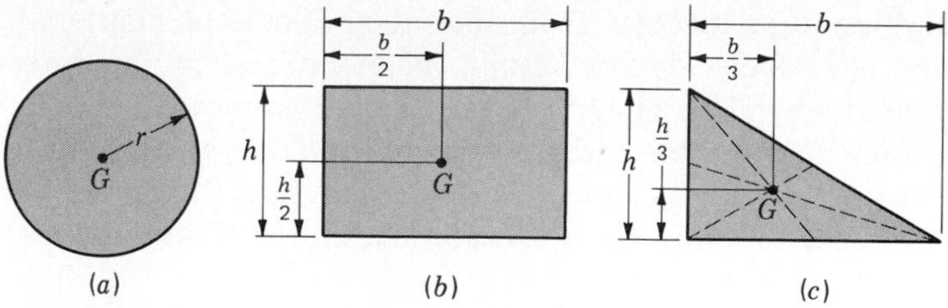

Figure 13-2

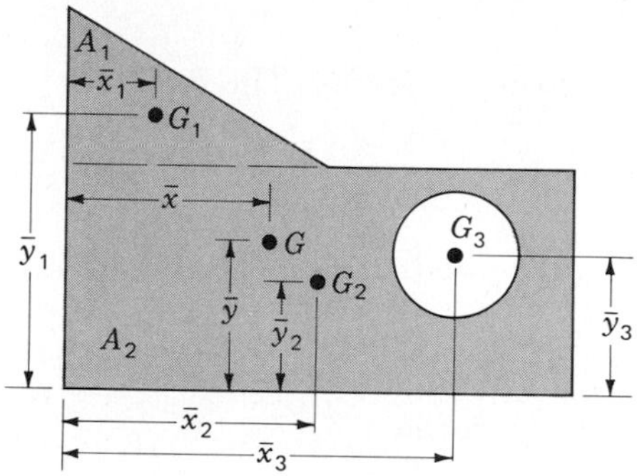

Figure 13-3 A composite shape.

can be obtained by using integration instead of summation. The relations then become

$$\bar{x} = \frac{\int x'\,dA}{\int dA} = \frac{1}{A}\int x'\,dA$$
$$\bar{y} = \frac{\int y'\,dA}{\int dA} = \frac{1}{A}\int y'\,dA \qquad (13\text{-}4)$$

where x' and y' are the distances to the centroid of the area dA measured parallel to the x and y axes, respectively.

For three-dimensional bodies, Eqs. (13-4) can be written in terms of masses instead of areas. The equations then become

$$\bar{x} = \frac{1}{m}\int x'\,dm \qquad \bar{y} = \frac{1}{m}\int y'\,dm \qquad \bar{z} = \frac{1}{m}\int z'\,dm \qquad (13\text{-}5)$$

13-3 MOMENT OF INERTIA

Another problem that often arises when forces are distributed over an area is that of calculating their moment about a specified axis. Sometimes the force intensity varies according to its distance from the moment axis. A mathematical analysis of such a problem always results in an integral of the form $\int (\text{distance})^2 \times$ differential area. This integral is called *area moment of inertia.* Some authorities prefer to call this integral the *second moment of area* because an area cannot possess inertia and hence the term "moment of inertia" for such an integral is a misnomer. Nevertheless, the term is widely used, and we must learn to live with it.

The formulas for the second moments of areas about the x and y axes are, respectively,

$$I_x = \int y^2\,dA \qquad \text{and} \qquad I_y = \int x^2\,dA \tag{13-6}$$

Here I_x and I_y are called the *rectangular moments of inertia.* The integral

$$J_z = \int r^2\,dA \tag{13-7}$$

is called the *polar area moment of inertia.* A relation between Eqs. (13-6) and (13-7) is

$$J_z = I_x + I_y \tag{13-8}$$

Sometimes moment of inertia is expressed in the form

$$I = k^2 A \tag{13-9}$$

where, of course,

$$k = \sqrt{\frac{I}{A}} \tag{13-10}$$

Here k is called the *radius of gyration*; it is a quantitative measure of the distribution of the area from the moment axis.

Equations (13-6) to (13-10) have been solved for the most common shapes, and the results are shown in Appendix Table 4. To obtain a moment of inertia at any specified distance d from the centroidal axis, use the *transfer formulas*:

$$I_x = \bar{I}_x + Ad_x^2 \qquad I_y = \bar{I}_y + Ad_y^2 \qquad J_z = \bar{J}_z + Ad^2 \tag{13-11}$$

The moment of inertia of a volume is a true moment of inertia because a volume has mass. But to distinguish it from area, it is frequently called *mass moment of inertia.* For a volume, the inertia integrals are

$$I_x = \int (y^2 + z^2)\,dm \qquad I_y = \int (x^2 + z^2)\,dm \qquad I_z = \int (x^2 + y^2)\,dm \tag{13-12}$$

Another set of integrals, called *products of inertia*, may also arise in mathematical analysis:

$$I_{xy} = \int xy\,dm \qquad I_{yz} = \int yz\,dm \qquad I_{zx} = \int zx\,dm \tag{13-13}$$

Equations (13-13) are useful because when these integrals become zero, they define the three coordinate axes of a body called the *principal axes*. The corresponding values of Eqs. (13-12) are then called the *principal mass moments of inertia*. Appendix Table 5 has been obtained by solving Eqs. (13-12) for a variety of geometric solids. These moments of inertia are all given about the principal axes, and hence the products of inertia vanish.

The general form of the *transfer*, or *parallel-axis, formula* for mass moment of inertia is written

$$I = I_G + md^2 \tag{13-14}$$

where I_G is the principal moment of inertia and I is the moment of inertia about a parallel axis at distance d from the original axis. Equation (13-14) must only be used when the inertia axes are translated. The rotation of these axes results in the introduction of product terms.

The term *radius of gyration* is also used with mass moment of inertia. The relations are

$$I_G = k^2 m \qquad k = \sqrt{\frac{I_G}{m}} \tag{13-15}$$

Example 13-1 Figure 13-4 shows a steel prism welded to a thin rod to form a pendulum. Assuming that the rod is weightless, compute the moment of inertia of the pendulum about O. Use $\rho = 7.80\ \text{Mg/m}^3$ as the mass density of steel.

SOLUTION The mass of the prism is computed as follows:

$$m = abc\rho = 75(100)(12)(7.8)\frac{1000\ \text{kg/Mg}}{(1000\ \text{mm/m})^3} = 0.702\ \text{kg}$$

Then, from Appendix Table 5 we find the mass moment of inertia of the prism about its own center of mass to be

$$I_G = \frac{m}{12}(a^2 + c^2) = \frac{0.702}{12}[(75)^2 + (100)^2] = 914\ \text{kg}\cdot\text{mm}^2$$

Equation (13-14) is now used to transfer to the axis through O. Thus

$$I_O = I_G + md^2 = 914 + (0.702)(250)^2 = 44\,800\ \text{kg}\cdot\text{mm}^2$$

or, in basic units,

$$I_O = (44\,800)\frac{1}{(1000\ \text{mm/m})^2} = 0.0448\ \text{kg}\cdot\text{m}^2 \qquad \textit{Ans.}$$

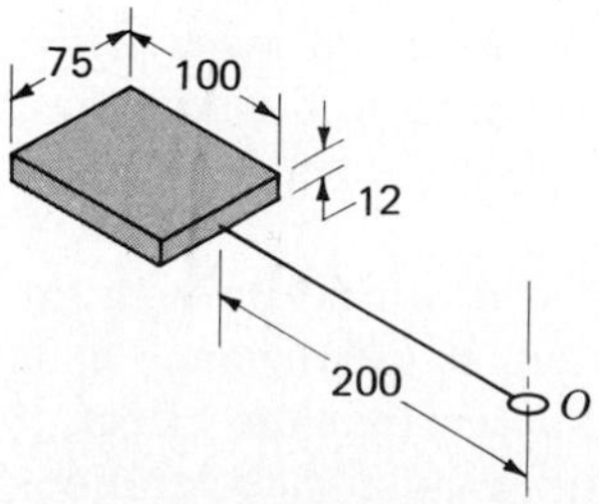

Figure 13-4 Dimensions in millimetres.

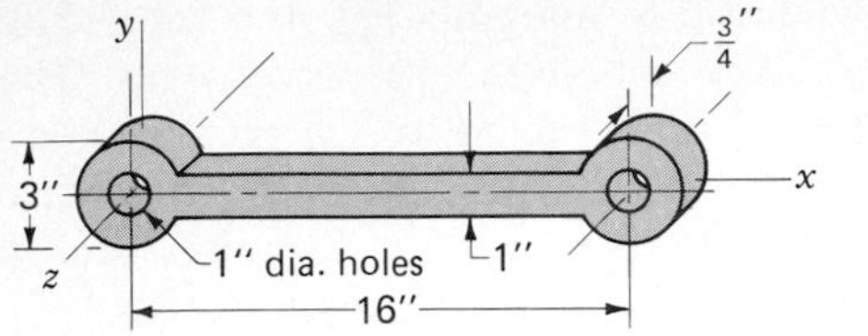

Figure 13-5

Example 13-2 Figure 13-5 shows a cast-iron connecting rod. Find the mass moment of inertia of the rod about the z axis in ips gravitational units. Use $w = 0.260\ \text{lb/in}^3$ as the unit weight of cast iron.

SOLUTION We solve by finding the moment of inertia of each of the cylinders at the ends and of the central prism about their own mass centers. Then we use the transfer formulas to transfer these to the z axis.

The mass of each cylinder is

$$m_{\text{cyl}} = \frac{\pi l w}{4g}(d_o^2 - d_i^2) = \frac{\pi(0.75)(0.260)}{4(386)}[(3)^2 - (1)^2]$$
$$= 0.003\,17\ \text{lb} \cdot \text{s}^2/\text{in}$$

From Appendix Table 5 we find the moment of inertia of each cylinder to be

$$I_{G,\text{cyl}} = \frac{m}{8}(d_o^2 + d_i^2) = \frac{0.003\,17}{8}[(3)^2 + (1)^2] = 0.003\,96\ \text{lb} \cdot \text{s}^2 \cdot \text{in}$$

The mass of the central prism is

$$m_{\text{pr}} = \frac{abcw}{g} = \frac{0.75(1)(13)(0.260)}{386} = 0.006\,57\ \text{lb} \cdot \text{s}^2/\text{in}$$

Then the moment of inertia of the prism about its mass center is

$$I_{G,\text{pr}} = \frac{m}{12}(b^2 + c^2) = \frac{0.006\,57}{12}[(1)^2 + (3)^2] = 0.005\,48\ \text{lb} \cdot \text{s}^2 \cdot \text{in}$$

Using the transfer formula, we finally get

$$I_z = I_{G,\text{cyl}} + (I_{G,\text{cyl}} + m_{\text{cyl}} d_{\text{cyl}}^2) + (I_{G,\text{pr}} + m_{\text{pr}} d_{\text{pr}}^2)$$
$$= 0.003\,96 + [0.003\,96 + (0.003\,17)(16)^2] + [0.005\,48 + (0.006\,57)(8)^2]$$
$$= 1.25\ \text{lb} \cdot \text{s}^2 \cdot \text{in} \qquad \textit{Ans.}$$

13-4 INERTIA FORCES AND D'ALEMBERT'S PRINCIPLE

Consider a moving rigid body of mass m acted upon by any system of forces, say $\mathbf{F}_1$, $\mathbf{F}_2$, and $\mathbf{F}_3$, as shown in Fig. 13-6a. Designate the center of mass of the body as point G, and find the resultant of the force system from the equation

$$\sum \mathbf{F} = \mathbf{F}_1 + \mathbf{F}_2 + \mathbf{F}_3 \qquad (a)$$

In the general case the line of action of this resultant will *not* be through the mass center but will be displaced by some distance, such as the distance h, as shown in the figure. In the study of mechanics it is shown that the effect of this unbalanced force system is to produce linear and angular accelerations

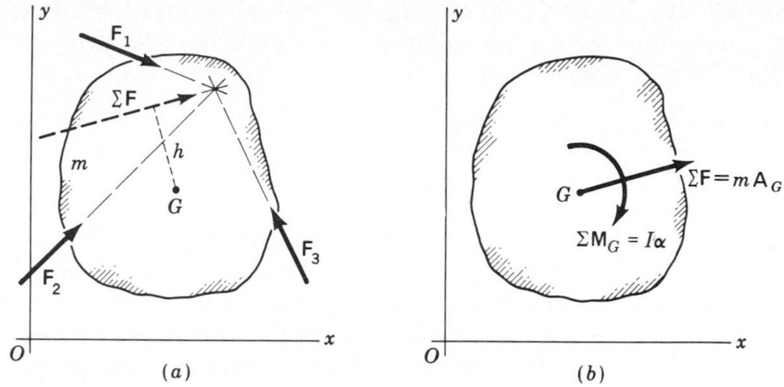

Figure 13-6

whose values are given by

$$\sum \mathbf{F} = m\mathbf{A}_G \tag{13-16}$$

$$\sum \mathbf{M}_G = I\boldsymbol{\alpha} \tag{13-17}$$

in which $\mathbf{A}_G$ is the acceleration of the mass center and $\boldsymbol{\alpha}$ is the angular acceleration of m (Fig. 13-6*b*). The quantity $\Sigma\,\mathbf{F}$ is the resultant of all the external forces acting upon the body, and $\Sigma\,\mathbf{M}_G$ is the sum of the external moments together with the moments of the external forces taken about G in the plane of motion. The mass moment of inertia is designated as I and is taken with reference to the mass center G also.

Equations (13-16) and (13-17) show that when an unbalanced system of forces acts upon a rigid body, the body experiences a linear acceleration $\mathbf{A}_G$ of its mass center in the same direction as the resultant force $\Sigma\,\mathbf{F}$; that the body also experiences an angular acceleration $\boldsymbol{\alpha}$, due to the moments of the forces and torques about the center of mass, in the same direction as the resultant moment $\Sigma\,\mathbf{M}_G$. If the forces and moments are known, Eqs. (13-16) and (13-17) can be used to determine the resulting accelerations.

In engineering design the motion of the machine members is usually specified in advance by other machine requirements. The problem then is: given the motion of the machine elements, what forces are required to produce these motions? The problem therefore requires (1) a kinematic analysis in order to determine the linear and angular accelerations of the various members and (2) a definition of the actual shape, dimensions, and material of the members; otherwise the masses and moments of inertia could not be determined. In the examples to be demonstrated here only the results of the kinematic analysis will be presented. The selection of the materials, shape, and many of the dimensions of machine members is a subject of machine design and will not be discussed here either.

Since, in the dynamic analysis of machines, the acceleration vectors are usually known, an alternative form of Eq. (13-16) and (13-17) is often convenient in determining the forces required to produce these known accelerations. Thus, we can write

$$\sum \mathbf{F} - m\mathbf{A}_G = 0 \tag{13-18}$$

$$\sum \mathbf{M}_G - I\boldsymbol{\alpha} = 0 \tag{13-19}$$

Both these equations are vector equations applying to the plane motion of a rigid body. Equation (13-18) states that the vector sum of all the external forces acting upon the body plus the fictitious force $-m\mathbf{A}_G$ is zero. The fictitious force $-m\mathbf{A}_G$ is called an *inertia force.* It has the same line of action as $\mathbf{A}_G$ but is opposite in sense. Equation (13-19) states that the sum of the moments of all the external forces about an axis through G perpendicular to the plane of motion and the external torques acting upon the body plus a fictitious torque $-I\boldsymbol{\alpha}$ is zero. The fictitious torque $-I\boldsymbol{\alpha}$ is called an *inertia torque.* The inertia torque is opposite in sense to the angular acceleration vector $\boldsymbol{\alpha}$. Equations (13-18) and (13-19) are extremely useful in studying the dynamics of machinery because they enable us to add inertia forces and torques to the external system of forces and to solve the resulting problem using the methods of statics.

The equations above are known as *D'Alembert's principle* because he was the first to call attention to the fact that addition of inertia forces to the real force system enabled a solution to be obtained from equations of equilibrium. We might note that the equations can also be written

$$\sum \mathbf{F} = 0 \qquad \sum \mathbf{M} = 0 \tag{13-20}$$

where it is to be understood that both the external and the inertia forces and torques are to be included as terms in $\Sigma\,\mathbf{F}$ and $\Sigma\,\mathbf{M}$. Equation (13-20) is useful because it permits us to take a summation of the moments about any axis perpendicular to the plane of motion.

D'Alembert's principle is summarized as follows. *The vector sum of all the external forces and the inertia forces acting upon a rigid body is zero. The vector sum of all the external moments and the inertia torques acting upon a rigid body is also separately zero.*

Equations (13-20) can be combined when a graphical solution is desired by a force polygon. In Fig. 13-7*a* a member is acted upon by the two external forces $\mathbf{F}_{43}$ and $\mathbf{F}_{23}$. The resultant $\mathbf{F}_{43} + \mathbf{F}_{23}$ produces an acceleration $\mathbf{A}_G$ of the mass center of the member and an angular acceleration $\boldsymbol{\alpha}_3$ because the line of action of the resultant does not pass through the mass center. Representing the inertia torque $-I\boldsymbol{\alpha}_3$ as a couple, as shown in Fig. 13-7*b*, we intentionally choose the two forces of this couple to be $\pm m\mathbf{A}_G$. In order for the moment of the couple to be of magnitude $-I\alpha_3$, the distance between the forces must be

$$h = \frac{I\alpha_3}{mA_G} \tag{13-21}$$

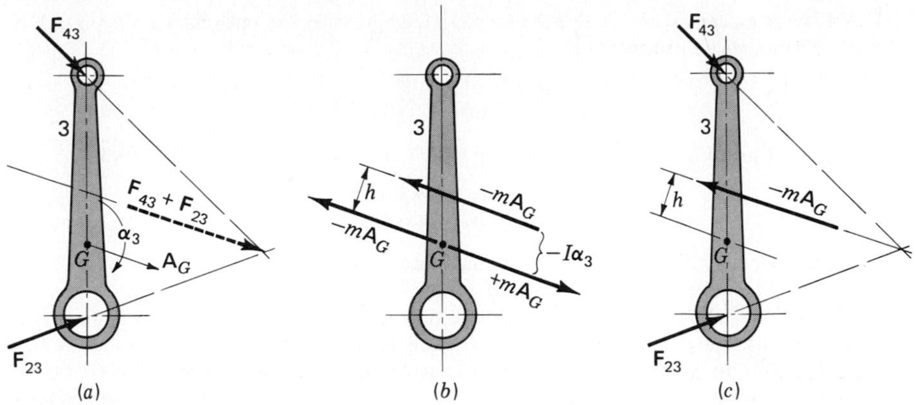

Figure 13-7

Because of this particular choice of the couple, one force of the couple exactly cancels the inertia force itself and leaves only a single force, as shown in Fig. 13-7*c*, which includes the combined effects of the inertia force and the inertia torque.

Example 13-3 Determine the force F_A required to produce a velocity $V_A = 12.6$ ft/s for the mechanism shown in Fig. 13-8*a*. Assume that the linkage is in the horizontal plane so that gravity acts normal to the plane of motion; also assume no friction. Link 3 weighs 2.20 lb, and $I_3 = 0.0479$ lb · s² · in.

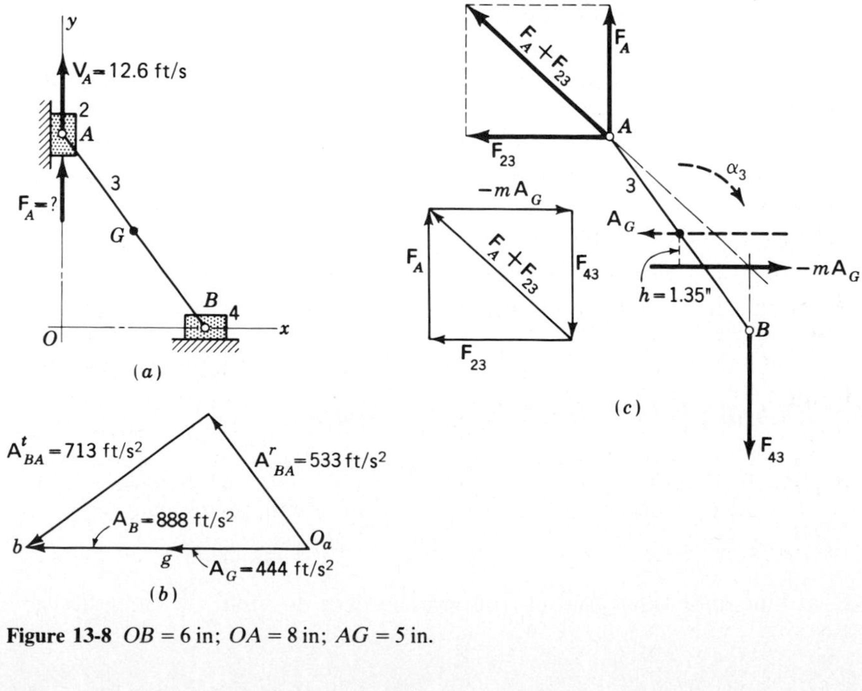

Figure 13-8 $OB = 6$ in; $OA = 8$ in; $AG = 5$ in.

SOLUTION A kinematic analysis of the accelerations provides the information shown in Fig. 13-8*b*. The angular acceleration is

$$\alpha_3 = \frac{A_{BA}^t}{R_{BA}} = \frac{713}{10/12} = 856 \text{ rad/s cw}$$

The mass of link 3 is $m = W/g = 2.20/386 = 0.0057 \text{ lb} \cdot \text{s}^2/\text{in}$. Then Eq. (13-21) gives

$$h = \frac{I_{G_3}\alpha_3}{mA_G} = \frac{(0.0479)(856)}{(0.0057)(444)(12)} = 1.35 \text{ in}$$

The free-body diagram and the resulting force polygon are shown in Fig. 13-8*c*. Notice that the inertia force $-m\mathbf{A}_G$ is displaced from G by the distance h, so as to produce a moment of $-I\boldsymbol{\alpha}_3$ about G, and that $-m\mathbf{A}_G$ is in the opposite sense to $\mathbf{A}_G$. The reaction at B is $\mathbf{F}_{43}$ and is vertically downward because friction is neglected. The forces at A are the actuating force $\mathbf{F}_A$ and the block reaction $\mathbf{F}_{23}$, which is horizontal, also because friction is neglected. The point of concurrency is the intersection of $-m\mathbf{A}_G$ and $\mathbf{F}_{43}$, the directions of both being known. The line of action $\mathbf{F}_A + \mathbf{F}_{23}$ of the total force at A, must pass through the point of concurrency. This fact permits construction of the force polygon. Then the unknown forces $\mathbf{F}_A$ and $\mathbf{F}_{23}$ are known in direction and are found as components of $\mathbf{F}_A + \mathbf{F}_{23}$, as shown in the figure. The actuating force is found by measurement to be

$$\mathbf{F}_A = 27\hat{\mathbf{j}} \text{ lb} \qquad \textit{Ans.}$$

13-5 THE PRINCIPLE OF SUPERPOSITION

Linear systems are those in which effect is proportional to cause. This means that the response or output of a system depends directly upon the drive, or input, to the system. An example of a linear system is a spring; the deflection of a spring (output) is proportional to the force (input) exerted on the spring.

The principle of superposition is used to solve problems by considering separately each of the drives or inputs to a system. If the system is linear, the responses to each of these inputs can be summed or superposed on each other to determine the total response of the system. Thus, the principle of superposition states that for linear systems *the individual responses to several disturbances or driving functions can be superposed on each other to obtain the total response.*

Some examples of nonlinear systems to which the principle of superposition does not apply are springs that get stiffer as you deflect them, Coulomb friction in systems, and systems with clearance or backlash.

13-6 AN EXAMPLE OF GRAPHIC ANALYSIS

We have now demonstrated all the principles necessary for making a complete dynamic-force analysis of a plane-motion mechanism. The steps in making such an analysis can be summarized as follows:

1. Make a kinematic analysis of the mechanism to find the angular ac-

celeration of each link or element. Locate the center of mass of each link and determine the accelerations of these points.

2. Using the given value or values of the force or torque which the follower must deliver, make a complete static-force analysis of the mechanism. The results of this analysis will then include the magnitudes and directions of the forces and torques acting upon each member. Observe particularly that this is a static-force analysis and that it does not include inertia forces or torques.
3. Employing the given values of the masses and moments of inertia and the angular and linear accelerations found in step 1, calculate the inertia forces and inertia torques for each link or element of the mechanism. Taking these as applied forces, make a free-body analysis of each member of the entire mechanism to find the total effect of all the inertia forces and torques.
4. Vectorially add the results of steps 2 and 3 to obtain the resultant forces and torques for every machine element.

Example 13-4 Make a complete dynamic-force analysis of the four-bar linkage illustrated in Fig. 13-9. The given quantities are included in the figure caption.

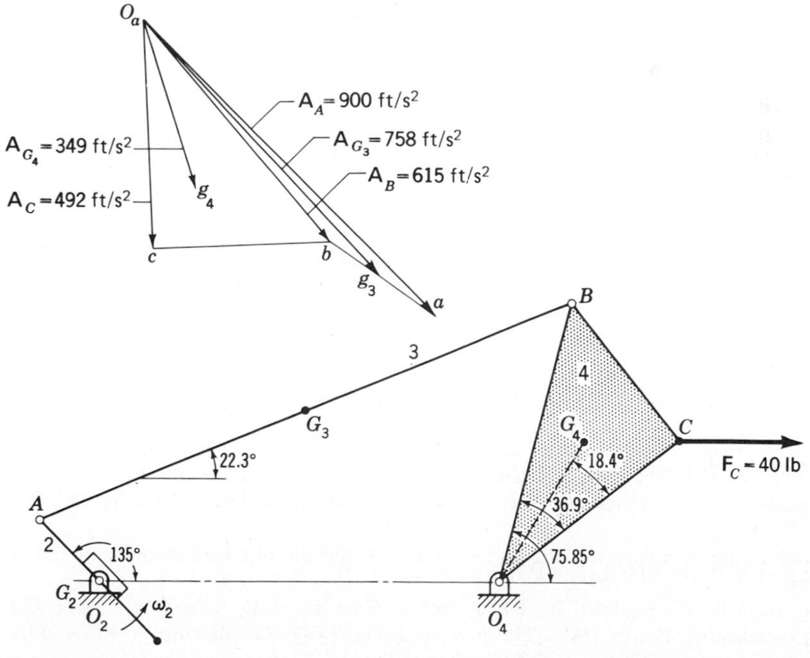

Figure 13-9 $O_2A = 3$ in, $AB = 20$ in, $O_4B = 10$ in, $O_2O_4 = 14$ in, $O_4G_4 = 5.69$ in, $AG_3 = 10$ in, $BC = 6$ in, $O_4C = 8$ in, $\omega_2 = 60$ rad/s, $\alpha_2 = 0$ rad/s², $W_3 = 7.13$ lb, $I_{G_3} = 0.625$ lb·s²·in, $W_4 = 3.42$ lb, $I_{G_4} = 0.037$ lb·s²·in. The angular positions of the various links have been calculated for the given position of link 2 and are given on the figure.

SOLUTION The first step is to make the kinematic analysis of the mechanism. This step is not included here, but the acceleration polygon resulting from the analysis is shown in Fig. 13-9. The numerical results are shown on the polygon in case you wish to verify them. By the methods of Chap. 4 the angular acceleration of links 3 and 4 are found to be

$$\alpha_3 = 148 \text{ rad/s}^2 \text{ ccw} \qquad \alpha_4 = 604 \text{ rad/s}^2 \text{ cw}$$

A major portion of the analysis is concerned with links 3 and 4 because the center of mass of link 2 is located at O_2. Free-body diagrams of links 4 and 3 are shown separately in Figs. 13-10 and 13-11, respectively. Notice also that these diagrams are arranged in equation form for simpler reading. Thus, in each illustration, the forces in (*a*) plus those in (*b*) and (*c*) produce the resultants shown in (*d*). The two sets of illustrations are also correlated; for example, $\mathbf{F}'_{34}$ in Fig. 13-10*a* is equal to $-\mathbf{F}'_{43}$ in Fig. 13-11*a*, etc. The following analysis is not difficult but is complicated; read it slowly and examine the illustrations carefully, detail by detail.

Let us start with link 4 in Fig. 13-10*a*. Proceeding in accordance with our earlier investigations, we make the following calculations:

$$I_{G_4}\alpha_4 = 0.037(604) = 22.3 \text{ lb} \cdot \text{in}$$

$$m_4 A_{G_4} = \frac{3.42}{32.2}(349) = 37.1 \text{ lb} \qquad h_4 = \frac{I_{G_4}\alpha_4}{m_4 A_{G_4}} = \frac{22.3}{37.1} = 0.602 \text{ in}$$

Now the force $-m_4 A_{G_4} = 37.1$ lb is placed on the free-body diagram opposite in direction to $\mathbf{A}_{G_4}$ and offset from G_4 by the distance h_4. The direction of the offset is that required to produce a torque about G_4 opposed to $I_{G_4}\boldsymbol{\alpha}_4$. The direction of $\mathbf{F}'_{34}$ is taken along link 3. The intersection of $\mathbf{F}'_{34}$ and $-m_4\mathbf{A}_{G_4}$ gives the point of concurency and established the direction of $\mathbf{F}'_{14}$. The force polygon can now be construced and the magnitudes of $\mathbf{F}'_{34}$ and $\mathbf{F}'_{14}$ found. These values are given in the figure caption.

Proceed next to Fig. 13-11*a*. The forces $\mathbf{F}'_{43}$ and $\mathbf{F}'_{23}$ are now known from the preceding analysis.

Now go to Fig. 13-11*b* and link 3, and make the calculations

$$I_{G_3}\alpha_3 = 0.625(148) = 92.5 \text{ lb} \cdot \text{in}$$

$$m_3 A_{G_3} = \frac{7.13}{32.2}(758) = 168 \text{ lb} \qquad h_3 = \frac{I_{G_3}\alpha_3}{m_3 A_{G_3}} = \frac{92.5}{168} = 0.550 \text{ in}$$

Locate the inertia force $-m_3\mathbf{A}_{G_3} = 168$ lb on the free-body diagram opposite in direction to $\mathbf{A}_{G_3}$ and offset a distance h_3 from G_3 so as to produce a torque about G_3 opposite in direction to α_3. The direction of $\mathbf{F}''_{43}$ is along the line BO_4. The forces $-m_3\mathbf{A}_{G_3}$ and $\mathbf{F}''_{43}$ intersect to determine the point of concurrency. Thus the direction of $\mathbf{F}''_{23}$ is known and the force polygon can be constructed. The resulting values of $\mathbf{F}''_{43}$ and $\mathbf{F}''_{23}$ are included in the caption.

In Fig. 13-10*b* the forces $\mathbf{F}''_{34}$ and $\mathbf{F}''_{14}$ acting on link 4 are now known from the preceding analysis.

Figures 13-10*c* and 13-11*c* show the results of the static-force analysis with $F_c = 40$ lb as the given quantity. The force polygon in Fig. 13-10*c* determines the values of the forces acting on link 4, and from these the direction and magnitude of the forces operating on link 3 are found.

The next step is a vector addition of these results already obtained as shown in (*d*) of each figure.

The analysis is completed by taking the resultant force $\mathbf{F}_{23}$ from Fig. 13-11*d* and applying its negative, $\mathbf{F}_{32}$, to link 2. This is done in Fig. 13-12. The distance h_2 is found by measurement. The external torque to be applied to link 2 is

$$T_{12} = h_2 F_{32} = 1.56(145) = 226 \text{ lb} \cdot \text{in cw}$$

Note that this torque is opposite in direction to the rotation of link 2.

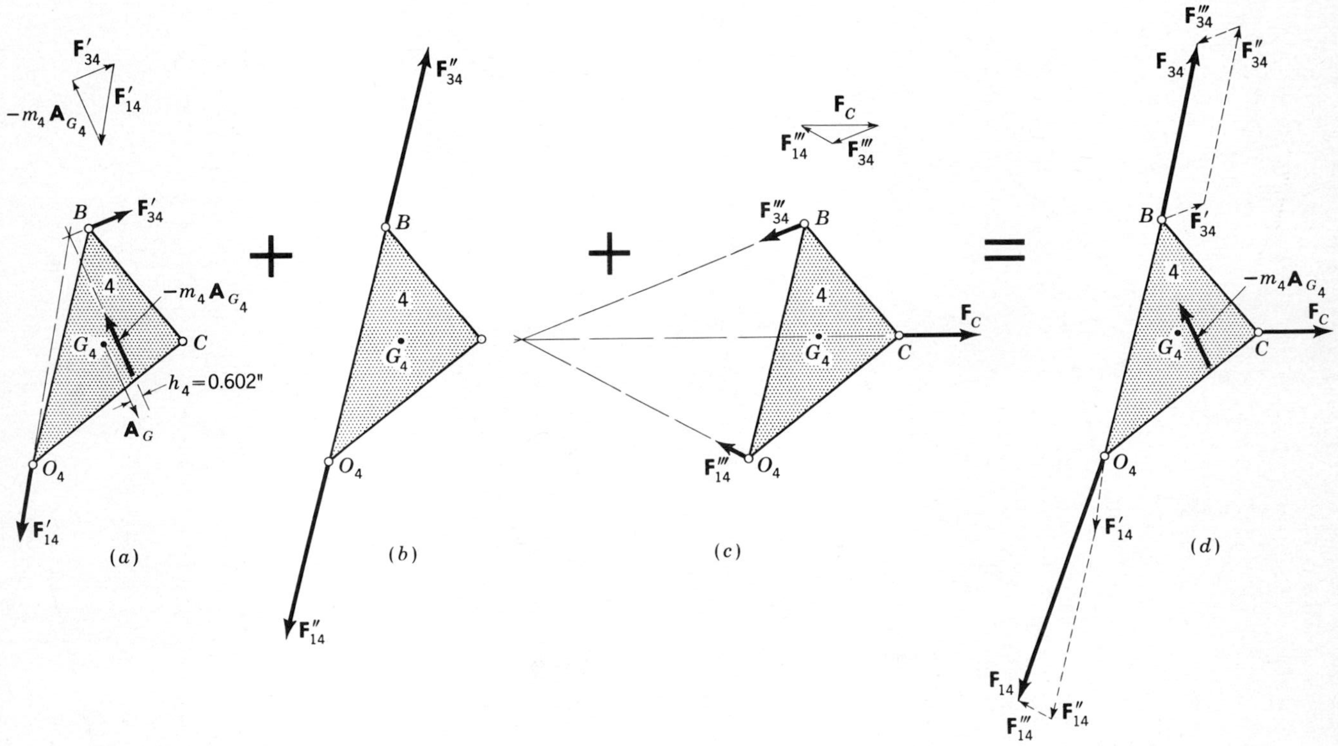

Figure 13-10 Free-body diagrams of link 4; $-m_4\mathbf{A}_{G_4}$ = 37.1 lb, $\mathbf{F}'_{34}$ = 24.3 lb, $\mathbf{F}'_{14}$ = 44.3 lb, $\mathbf{F}''_{34} = -\mathbf{F}''_{14} = -\mathbf{F}''_{43}$ = 94.8 lb, $\mathbf{F}_C$ = 40 lb, $\mathbf{F}'''_{34}$ = 25 lb, $\mathbf{F}'''_{14}$ = 19.3 lb, $\mathbf{F}_{34}$ = 94.3 lb, $\mathbf{F}_{14}$ = 132 lb.

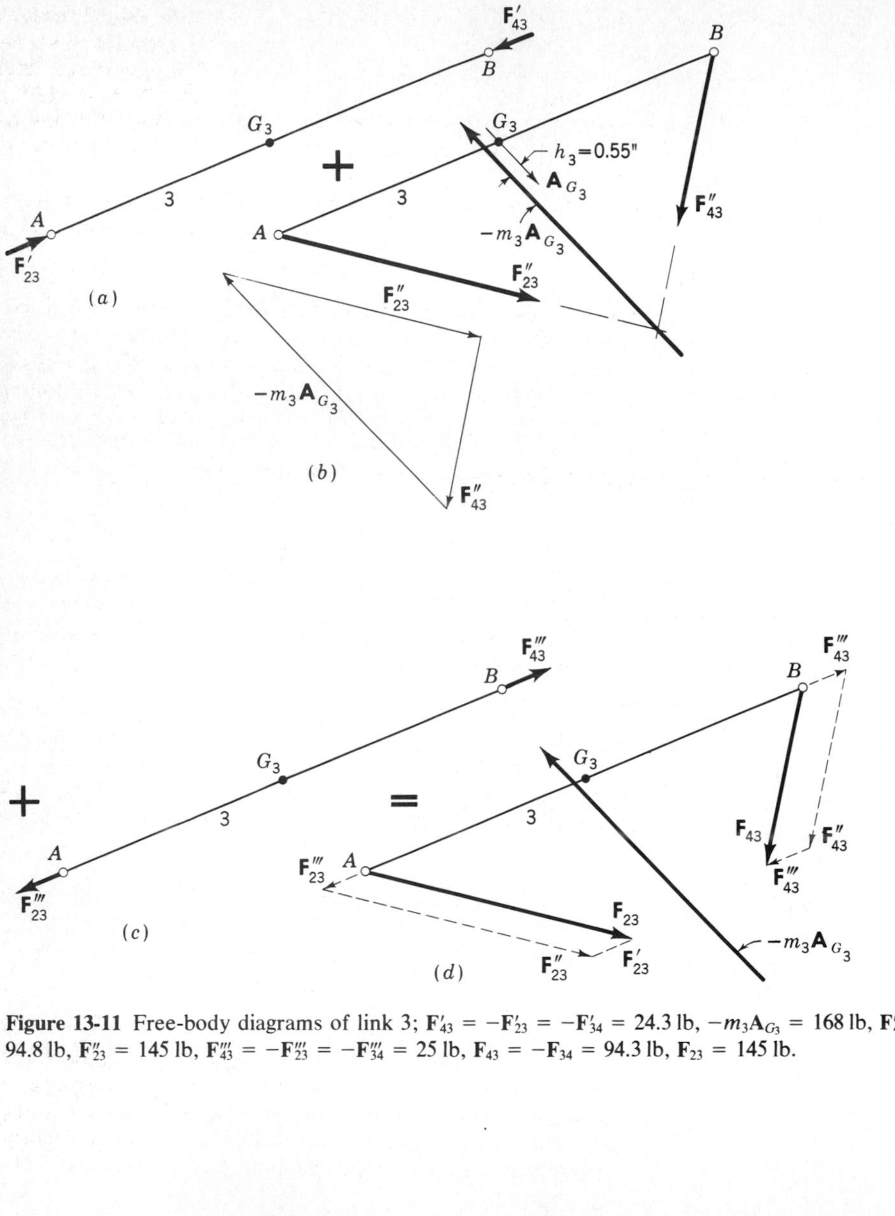

Figure 13-11 Free-body diagrams of link 3; $\mathbf{F}'_{43} = -\mathbf{F}'_{23} = -\mathbf{F}'_{34} = 24.3$ lb, $-m_3\mathbf{A}_{G_3} = 168$ lb, $\mathbf{F}''_{43} = 94.8$ lb, $\mathbf{F}''_{23} = 145$ lb, $\mathbf{F}'''_{43} = -\mathbf{F}'''_{23} = -\mathbf{F}'''_{34} = 25$ lb, $\mathbf{F}_{43} = -\mathbf{F}_{34} = 94.3$ lb, $\mathbf{F}_{23} = 145$ lb.

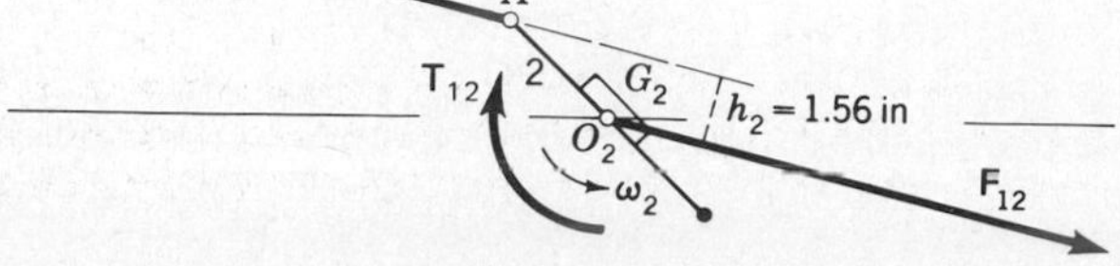

Figure 13-12 A free-body diagram of link 2; $\mathbf{F}_{32} = -\mathbf{F}_{23} = -\mathbf{F}_{12} = 145$ lb, $\mathbf{T}_{12} = 226$ lb·in.

13-7 ROTATION ABOUT A FIXED CENTER

The previous sections have dealt with the general case of dynamic forces for a rigid body having a combined motion of translation and rotation. It is important to emphasize that the equations and the methods of analysis investigated in these sections are general and apply to *all* problems of plane motion. It will be interesting now to study the application of these methods to a rigid body rotating about a *fixed* center.

Let us suppose a rigid body constrained to rotate about some fixed center O not coincident with the center of mass G (Fig. 13-13a). A system of forces (not shown) is to be applied to the body, causing it to have an angular acceleration α. We also include the fact that the body is rotating with an angular velocity ω. This motion of the body means that the mass center will have transverse and radial components of acceleration $\mathbf{A}_G^t$ and $\mathbf{A}_G^r$ whose magnitudes are $r_G\alpha$ and $r_G\omega^2$ respectively. Thus, if we resolve the resultant

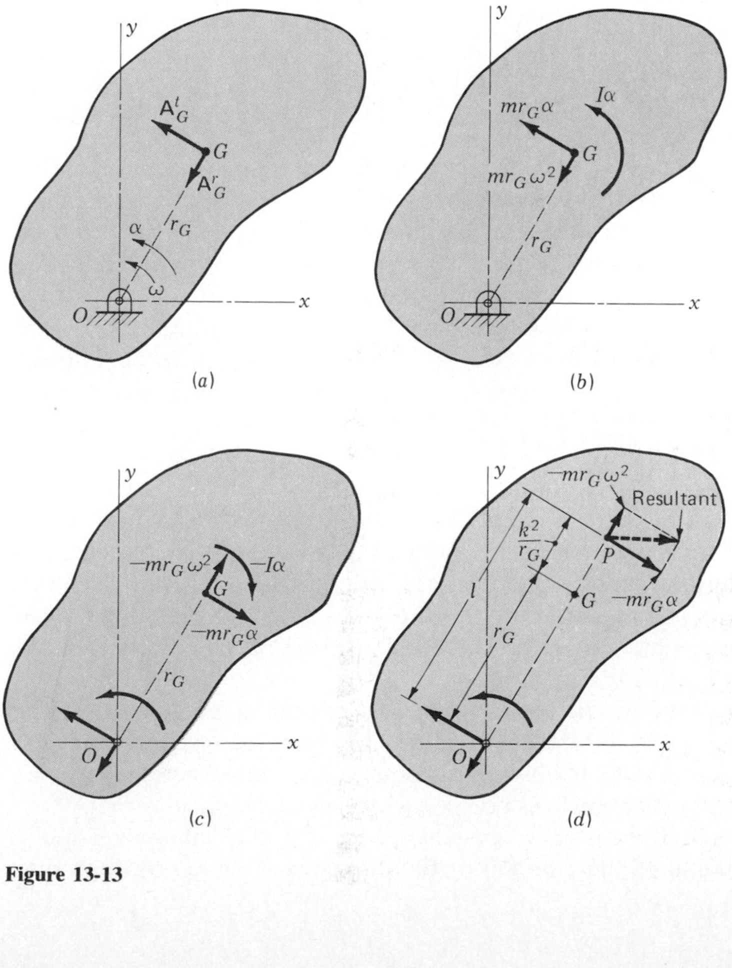

Figure 13-13

external force into transverse and radial components, these components must have magnitudes

$$\sum F^t = mr_G\alpha \qquad \text{and} \qquad \sum F^r = mr_G\omega^2 \tag{a}$$

in accordance with Eq. (13-16). In addition, Eq. (13-17) states that an external torque must exist to create the angular acceleration and that the magnitude of this torque is $T_G = I\alpha$. If we now sum the moments of these forces about O, we have

$$\sum M_O = I\alpha + r_G(mr_G\alpha) = (I + mr_G^2)\alpha \tag{b}$$

But the quantity in parentheses in Eq. (*b*) is identical with Eq. (13-14) and transfers the moment of inertia to another axis not coincident with the mass center. Therefore Eq. (*b*) can be written in vector form as

$$\sum \mathbf{M}_O = I_O\boldsymbol{\alpha}$$

Equations (13-18) and (13-19) then become

$$\sum \mathbf{F} - m\mathbf{A}_G = 0 \tag{13-23}$$

$$\sum \mathbf{M}_O - I_O\boldsymbol{\alpha} = 0 \tag{13-24}$$

by including the inertia force $-m\mathbf{A}_G$ and inertia torque $-I_O\boldsymbol{\alpha}$ (Fig. 13-13*c*). We observe particularly that the system of forces does *not* reduce to a single couple because of the existence of the inertia-force component $-mr_G\omega^2$, which has no moment arm about O. Thus both Eqs. (13-23) and (13-24) are necessary.

A particular case arises when $\alpha = 0$. Then the external moment M_O is zero and the only inertia force is, from Fig. 13-13*c*, the centrifugal force $-mr_G\omega^2$.

A second special case exists under starting conditions when $\omega = 0$, but α is not zero. Under these conditions the only inertia force is $-mr_G\alpha$, and the system reduces to a single couple.

When a rigid body has a motion of pure translation, the resultant inertia force and the resultant external force have the same line of action, which passes through the mass center of the body. When a rigid body has rotation and angular acceleration, the resultant inertia force and the resultant external force have the same line of action but this line *does not* pass through the mass center. Let us now locate a point on the line of action of the resultant of the inertia forces of Fig. 13-13*c*.

The resultant of the inertia forces will pass through some point P on the line OG of Fig. 13-13*c* or an extension of it. This force can be resolved into two components, one of which will be the component $-mr_G\omega^2$ acting along the line OG. The other component will be $-mr_G\alpha$ acting perpendicular to OG but not through point G. The distance, designated as l, to the unknown point P can be found by equating the moment of the component $-mr_G\alpha$ through P to the sum of the inertia torque and the moment of the inertia forces which act through G. Thus, taking moments about O, we have

$$(-mr_G\alpha)l = -I\alpha + (-mr_G\alpha)r_G$$

or

$$l = \frac{I}{mr_G} + r_G \tag{c}$$

Substituting the value of I from Eq. (13-15) gives

$$l = \frac{k^2}{r_G} + r_G \tag{13-25}$$

The point P located by Eq. (13-25) and shown in Fig. 13-13d is called the *center of percussion.* As shown, the resultant inertia force passes through P, and consequently the inertia force has zero moment about the center of percussion. If an external force is applied at P, perpendicular to OG, an angular acceleration α will result but the bearing reaction at O will be zero except for the radial component due to the inertia force $-mr_G\omega^2$. It is the usual practice in shock-testing machines to apply the force at the center of percussion in order to eliminate the transverse bearing reaction due to the external force.

Equation (13-25) shows that the location of the center of percussion is independent of the values of ω and α.

If the axis of rotation is coincident with the mass center, $r_G = 0$ and Eq. (13-25) shows that $l = \infty$. Under these conditions there is no resultant inertia force but a resultant inertia couple $-I\alpha$ instead.

In closing this section we observe that the transverse and radial components of the acceleration of G can be written

$$\mathbf{A}^t_G = \boldsymbol{\alpha} \times \mathbf{r}_G \tag{13-26}$$

$$\mathbf{A}^r_G = \boldsymbol{\omega} \times (\boldsymbol{\omega} \times \mathbf{r}_G) \tag{13-27}$$

where $\mathbf{r}_G$ is the position vector of point G. Equations (a) can now be expressed in vector form:

$$\sum \mathbf{F}^t = m\boldsymbol{\alpha} \times \mathbf{r}_G \tag{13-28}$$

$$\sum \mathbf{F}^r = m\boldsymbol{\omega} \times (\boldsymbol{\omega} \times \mathbf{r}_G) \tag{13-29}$$

The resultant external force defined in terms of the transverse and radial components as given by these equations is often useful in analysis.

13-8 MEASUREMENT OF MOMENT OF INERTIA

Frequently the shape of a body is so complex that it is impossible to compute the moment of inertia. Consider, for example, the problem of finding the moment of inertia of an automobile about a vertical axis through its center of mass. For such problems it is usually possible to determine the moment of inertia by observing the dynamic behavior of the body to a known input.

Many bodies, e.g., connecting rods and cranks, are shaped so that their mass can be assumed to lie in a single plane. If these bodies can be weighed and their mass centers located, they can be suspended like a pendulum and caused to oscillate. The moment of inertia of such bodies can then be computed from an observation of their period or frequency of oscillation. As shown in Fig. 13-14*a*, the part should be suspended fairly near the mass center but not coincident with it.

It is not usually necessary to drill a hole to suspend the body; e.g., a spoked wheel or gear can be suspended on a knife-edge at the rim.

When the body of Fig. 3-14*a* is displaced through an angle θ, a gravity force mg acts at G. Summing moments about O gives

$$\sum M_O = -mg(r_G \sin\theta) - I_O\ddot{\theta} = 0 \tag{a}$$

We intend that the pendulum be swung only through small angles, so that $\sin\theta$ can be replaced by θ. Equation (a) can then be written

$$\ddot{\theta} + \frac{mgr_G}{I_O}\theta = 0 \tag{b}$$

This differential equation has the well-known solution

$$\theta = C_1 \sin\sqrt{\frac{mgr_G}{I_O}}\,t + C_2\cos\sqrt{\frac{mgr_G}{I_O}}\,t \tag{c}$$

where C_1 and C_2 are the constants of integration. We shall start the pendulum motion by displacing it through a small angle θ_0 and releasing it from this position. Thus at $t = 0$, $\theta = \theta_0$, and $\dot{\theta} = 0$. Substituting these conditions into Eq. (c) and its first derivative enables us to evaluate the constants. They are found to be $C_1 = 0$ and $C_2 = \theta_0$. Therefore

$$\theta = \theta_0 \cos\sqrt{\frac{mgr_G}{I_O}}\,t \tag{13-30}$$

Since a cosine function repeats itself every 360°, the period of the motion in

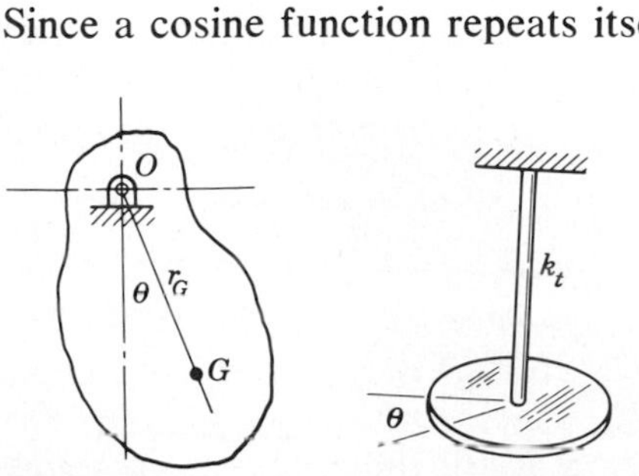

(a) (b) **Figure 13-14**

seconds is

$$\tau = 2\pi \sqrt{\frac{I_O}{mgr_G}} \tag{d}$$

Therefore

$$I_O = mgr_G\left(\frac{\tau}{2\pi}\right)^2 \tag{13-31}$$

This equation shows that the body must be weighed to get mg, the distance r_G must be measured, and then the pendulum must be suspended and oscillated so that the period τ can be observed. Equation (13-31) can then be solved to give the moment of inertia I_O about O. If the moment of inertia about the mass center is desired, it can be obtained by using the transfer formula (13-14).

Figure 13-14b shows how the moment of inertia can be determined without actually weighing the body. The inertia I is connected to a wire or slender rod at the mass center of the inertia. A torsional stiffness k_t of the rod or wire is defined as the torque necessary to twist the rod through a unit angle. If the inertia of Fig. 13-14b is turned through any angle θ and released, the equation of motion becomes

$$\ddot{\theta} + \frac{k_t}{I_G}\theta = 0$$

This is similar to Eq. (b) and with the same starting conditions has the solution

$$\theta = \theta_0 \cos \sqrt{\frac{k_t}{I_G}}\, t \tag{13-32}$$

The period of oscillation is then

$$\tau = 2\pi \sqrt{\frac{I_G}{k_t}}$$

or

$$I_G = k_t\left(\frac{\tau}{2\pi}\right)^2 \tag{13-33}$$

The torsional stiffness is usually known or can be computed from a knowledge of the size of the rod and its material. Then the oscillation of the unknown inertia I_G is observed and Eq. (13-33) used to compute I_G. Alternately, when k_t is unknown, a known inertia can be mounted on the rod and Eq. (13-33) used to determine k_t.

The *trifilar pendulum*, also called the *three-string torsional pendulum*, illustrated in Fig. 13-15, can be a very accurate method of measuring mass moment of inertia. Three strings of equal length support a lightweight platform and are equally spaced about the center of it. A round platform would serve just as well as the triangular one shown. The part whose moment of inertia is to be determined is carefully placed on the platform so that the center of mass of the object coincides with the platform center. The platform

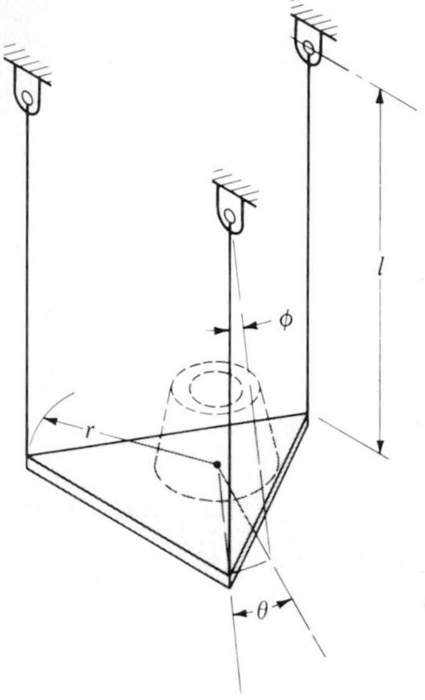

Figure 13-15

is then made to oscillate, and the number of oscillations is counted over a specified period of time.†

The notation for the three-string-pendulum analysis is as follows:

m = mass of part
m_p = mass of platform
I_G = moment of inertia of part
I_p = moment of inertia of platform
r = platform radius
θ = platform angle
l = string length
ϕ = string angle
z = vertical axis through center of platform

We begin by writing Eq. (13-19) for the z axis. This gives

$$\sum M_z = -r(m + m_p)g \sin \phi - (I_G + I_p)\ddot{\theta} = 0 \qquad (e)$$

Since we are dealing with small motions, the sines of angles can be equated to the angles themselves. Therefore

$$\phi = \frac{r}{l}\theta \qquad (f)$$

† Additional details can be found in F. E. Fisher and H. H. Alvord, "Instrumentation for Mechanical Analysis," The University of Michigan Summer Conferences, Ann Arbor, Michigan, 1977, p. 129. The analysis presented here is by permission of the authors.

and Eq. (*e*) becomes

$$\ddot{\theta} + \frac{(m + m_p)r^2 g}{l(I_G + I_p)}\theta = 0 \tag{g}$$

This equation can be solved in the same manner as Eq. (*b*). The result is

$$I_G + I_p = \frac{(m + m_p)r^2 g}{l}\left(\frac{\tau}{2\pi}\right)^2 \tag{13-34}$$

This equation should be used first with an empty platform. With I_p and m_p known, the equation can readily be solved for the unknown inertia I_G.

13-9 ANALYSIS OF A FOUR-BAR MECHANISM

Example 13-5 As an example of dynamic analysis using SI units we use the four-bar linkage of Fig. 13-16. The required data, based on a complete kinematic analysis, are shown on the figure and in the caption.

SOLUTION We start with the following kinematic information:

$$\boldsymbol{\alpha}_3 = -119\hat{\mathbf{k}}\ \text{rad/s}^2 \qquad \boldsymbol{\alpha}_4 = -625\hat{\mathbf{k}}\ \text{rad/s}^2$$

$$\mathbf{A}_{G_3} = 162\underline{/-73.2^\circ}\ \text{m/s}^2 \qquad \mathbf{A}_{G_4} = 104\underline{/233^\circ}\ \text{m/s}^2$$

The two inertia forces are

$$-m_3\mathbf{A}_{G_3} = \frac{1.5(162)}{1000}\underline{/-73.2 + 180^\circ} = -0.070\hat{\mathbf{i}} + 0.233\hat{\mathbf{j}}\ \text{kN}$$

and

$$-m_4\mathbf{A}_{G_4} = \frac{5(104)}{1000}\underline{/233 - 180^\circ} = 0.313\hat{\mathbf{i}} + 0.415\hat{\mathbf{j}}\ \text{kN}$$

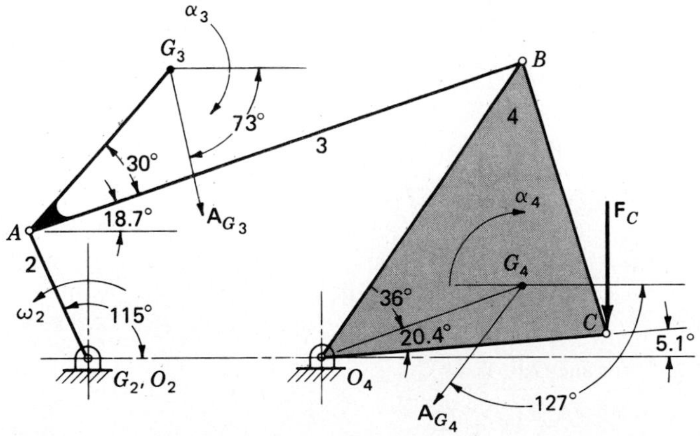

Figure 13-16 Dimensions in millimetres; $O_2A = 60$, $O_2O_4 = 100$, $AB = 220$, $O_4B = 150$, $AG_3 = 90$, $O_4C = BC = 120$, $O_4G_4 = 90$, $\omega_2 = 48$ rad/s, $m_3 = 1.5$ kg, $m_4 = 5$ kg, $I_3 = 0.012$ kg·m², $I_4 = 0.054$ kg·m², $\boldsymbol{\alpha}_3 = -119\hat{\mathbf{k}}$ rad/s², $\boldsymbol{\alpha}_4 = -625\hat{\mathbf{k}}$ rad/s², $A_{G_3} = 162$ m/s², $A_{G_4} = 104$ m/s², $\mathbf{F}_C = -0.8\hat{\mathbf{j}}$ kN.

Of course, $\mathbf{F}_C = -0.8\hat{\mathbf{j}}$ kN. We require the following position vectors (see Fig. 13-16; note that the dimensions are in millimetres):

$$\mathbf{R}_A = 60\underline{/115°} = -25.4\hat{\mathbf{i}} + 54.4\hat{\mathbf{j}}$$
$$\mathbf{R}_{G_3A} = 90\underline{/48.7°} = 59.4\hat{\mathbf{i}} + 67.6\hat{\mathbf{j}}$$
$$\mathbf{R}_{BA} = 220\underline{/18.7°} = 208\hat{\mathbf{i}} + 70.5\hat{\mathbf{j}}$$
$$\mathbf{R}_B = 150\underline{/56.4°} = 83.0\hat{\mathbf{i}} + 125\hat{\mathbf{j}}$$
$$\mathbf{R}_{G_4} = 90\underline{/20.4°} = 84.4\hat{\mathbf{i}} + 31.4\hat{\mathbf{j}}$$
$$\mathbf{R}_C = 120\underline{/5.1°} = 120\hat{\mathbf{i}} + 10.7\hat{\mathbf{j}}$$

We begin the analysis with link 4 and determine the $\mathbf{F}'$ forces. These are due to $\mathbf{F}_C$ and $-m_4\mathbf{A}_{G_4}$ with the effects of $-m_3\mathbf{A}_{G_3}$ and $-I_3\boldsymbol{\alpha}_3$ neglected. Taking moments about O_4 gives the equation

$$\sum \mathbf{M}_{O_4} = \mathbf{R}_{G_4O_4} \times (-m_4\mathbf{A}_{G_4}) + (-I_4\boldsymbol{\alpha}_4) + \mathbf{R}_{CO_4} \times \mathbf{F}_C + \mathbf{R}_B \times \mathbf{F}'_{34} = 0 \quad (1)$$

The first three terms are found to be

$$\mathbf{R}_{G_4O_4} \times (-m_4\mathbf{A}_{G_4}) = 25.2\hat{\mathbf{k}}$$
$$-I_4\boldsymbol{\alpha}_4 = -0.054(-625\hat{\mathbf{k}}) = 33.8\hat{\mathbf{k}}$$
$$\mathbf{R}_{CO_4} \times \mathbf{F}_C = -95.6\hat{\mathbf{k}}$$

The force $\mathbf{F}'_{34}$ has the same direction as link 3; thus

$$\mathbf{F}'_{34} = F'_{34}\underline{/18.7°} = (0.947\hat{\mathbf{i}} + 0.321\hat{\mathbf{j}})F'_{34}$$

Then $$\mathbf{R}_{BO_4} \times \mathbf{F}'_{34} = -91.7F'_{34}\hat{\mathbf{k}}$$

These four terms must now be substituted into Eq. (1). Upon solving, we find $F'_{34} = -0.400$ kN. Thus

$$\mathbf{F}'_{34} = -0.400\underline{/18.7°} = -0.378\hat{\mathbf{i}} - 0.128\hat{\mathbf{j}} = 0.400\underline{/198.7°}\text{ kN}$$

Next, summing forces on link 4 yields the equation

$$\sum \mathbf{F}_4 = \mathbf{F}'_{34} + (-m_4\mathbf{A}_{G_4}) + \mathbf{F}_C + \mathbf{F}'_{14} = 0 \quad (2)$$

All the terms are known except $\mathbf{F}'_{14}$. Upon solving, we find

$$\mathbf{F}'_{14} = 0.0655\hat{\mathbf{i}} + 0.513\hat{\mathbf{j}} = 0.517\underline{/82.7°}\text{ kN}$$

Turning next to link 3, we assume that the $\mathbf{F}''$ forces are due only to $-m_3\mathbf{A}_{G_3}$. Thus the effects of $\mathbf{F}_C$, $-I_4\boldsymbol{\alpha}_4$, and $-m_4\mathbf{A}_{G_4}$ are neglected. Taking moments about A yields

$$\sum \mathbf{M}_A = \mathbf{R}_{G_3A} \times (-m_3\mathbf{A}_{G_3}) + (-I_3\boldsymbol{\alpha}_3) + \mathbf{R}_{BA} \times \mathbf{F}''_{43} = 0 \quad (3)$$

The first two terms are found to be

$$\mathbf{R}_{G_3A} \times (-m_3\mathbf{A}_{G_3}) = 18.6\hat{\mathbf{k}}$$

and $$-I_3\boldsymbol{\alpha}_3 = -0.012(-119\hat{\mathbf{k}}) = 1.43\hat{\mathbf{k}}$$

The force $\mathbf{F}''_{43}$ is taken along the line O_4B. Thus

$$\mathbf{F}''_{43} = F''_{43}\underline{/56.4°} = (0.553\hat{\mathbf{i}} + 0.833\hat{\mathbf{j}})F''_{43}$$

Then $$\mathbf{R}_{BA} \times \mathbf{F}''_{43} = 134F''_{43}\hat{\mathbf{k}}$$

Substituting these three terms into Eq. (3) and solving gives $F''_{43} = -0.149$ kN. Therefore

$$\mathbf{F}''_{43} = -0.149\underline{/56.4°} = -0.0824\hat{\mathbf{i}} - 0.124\hat{\mathbf{j}} = 0.149\underline{/236.4°}\text{ kN}$$

Next, summing forces on link 3 and solving for $\mathbf{F}''_{23}$ yields

$$\mathbf{F}''_{23} = m_3\mathbf{A}_{G_3} - \mathbf{F}''_{43} = 0.153\hat{\mathbf{i}} - 0.109\hat{\mathbf{j}} = 0.187\underline{/-35.5^\circ}\ \text{kN}$$

The third step in the analysis is to find the vector sums of the $\mathbf{F}'$ and $\mathbf{F}''$ forces at A, B and O_4. At A we have

$$\mathbf{F}_{23} = \mathbf{F}'_{23} + \mathbf{F}''_{23} = \mathbf{F}'_{34} + \mathbf{F}''_{23}$$

The result is found to be

$$\mathbf{F}_{23} = -0.225\hat{\mathbf{i}} - 0.237\hat{\mathbf{j}} = 0.327\underline{/226.5^\circ}\ \text{kN}$$

Also

$$\mathbf{F}_{32} = -\mathbf{F}_{23} = 0.225\hat{\mathbf{i}} + 0.237\hat{\mathbf{j}} = 0.327\underline{/46.5^\circ}\ \text{kN}$$

Next

$$\mathbf{F}_{43} = \mathbf{F}'_{43} + \mathbf{F}''_{43} = -\mathbf{F}'_{34} + \mathbf{F}''_{43}$$

Solving this equation gives $\mathbf{F}_{43}$ and $\mathbf{F}_{34}$ as

$$\mathbf{F}_{43} = 0.296\hat{\mathbf{i}} + 0.004\hat{\mathbf{j}} = 0.296\underline{/0.8^\circ}\ \text{kN}$$

$$\mathbf{F}_{34} = -0.296\hat{\mathbf{i}} - 0.004\hat{\mathbf{j}} = 0.296\underline{/180.8^\circ}\ \text{kN}$$

At O_4 we have

$$\mathbf{F}_{14} = \mathbf{F}'_{14} + \mathbf{F}''_{14} = \mathbf{F}'_{14} + \mathbf{F}''_{43}$$

The solution is

$$\mathbf{F}_{14} = -0.0169\hat{\mathbf{i}} + 0.389\hat{\mathbf{j}} = 0.389\underline{/92.5^\circ}\ \text{kN}$$

For link 2, we have

$$\mathbf{F}_{12} = -\mathbf{F}_{32} = -0.225\hat{\mathbf{i}} - 0.237\hat{\mathbf{j}} = 0.327\underline{/226.5^\circ}\ \text{kN}$$

Also,

$$\sum \mathbf{M}_{O_2} = \mathbf{R}_A \times \mathbf{F}_{32} + \mathbf{T}_2 = 0$$

Solving gives

$$\mathbf{T}_2 = -\mathbf{R}_A \times \mathbf{F}_{32} = 18.3\hat{\mathbf{k}}\ \text{N} \cdot \text{m}$$

13-10 SHAKING FORCES AND MOMENTS

Of especial interest to the designer are the forces transmitted to the frame or foundation of the machine owing to the inertia of the moving links and other machine members. When these forces vary in magnitude or direction, they tend to shake or vibrate the machine, and consequently such effects are called *shaking forces and moments.*

If we consider a four-bar linkage, as an example, with links 2, 3, and 4 as the moving members and link 1 as the frame, the inertia forces associated with the moving members are $-m_2\mathbf{A}_{G_2}$, $-m_3\mathbf{A}_{G_3}$, and $-m_4\mathbf{A}_{G_4}$. Taking the moving members as a free body, one can immediately write

$$\sum \mathbf{F} = \mathbf{F}_{12} + \mathbf{F}_{14} + (-m_2\mathbf{A}_{G_2}) + (-m_3\mathbf{A}_{G_3}) + (-m_4\mathbf{A}_{G_4}) = 0 \qquad (a)$$

Using $\mathbf{F}_S$ as the resultant shaking force gives

$$\mathbf{F}_S = \mathbf{F}_{21} + \mathbf{F}_{41} \tag{b}$$

Therefore $$\mathbf{F}_S = -(m_2\mathbf{A}_{G_2} + m_3\mathbf{A}_{G_3} + m_4\mathbf{A}_{G_4}) \tag{13-35}$$

To determine the shaking moment we write

$$\sum \mathbf{M}_{O_2} = \mathbf{R}_{G_2} \times (-m_2\mathbf{A}_{G_2}) + \mathbf{R}_{G_3} \times (-m_3\mathbf{A}_{G_3}) + \mathbf{R}_{G_4O_2} \times (-m_4\mathbf{A}_{G_4}) - I_2\alpha_2 - I_3\alpha_3 - I_4\alpha_4 + \mathbf{M}_{12} = 0 \tag{c}$$

Then

$$\mathbf{M}_S = \mathbf{M}_{21} = -(\mathbf{R}_{G_2} \times m_2\mathbf{A}_{G_2} + \mathbf{R}_{G_3} \times m_3\mathbf{A}_{G_3} + \mathbf{R}_{G_4O_2} \times m_4\mathbf{A}_{G_4} + I_2\alpha_2 + I_3\alpha_3 + I_4\alpha_4) \tag{13-36}$$

13-11 COMPUTER ANALYSIS

In this section we present the steps necessary for a general computer or programmable-calculator solution to the kinematics and dynamics of the four-bar mechanism. The approach presented here is probably not the optimum solution because almost every programmer approaches the problem in a different manner. Nevertheless the program can be used as a guide for more complex problems and to generate ideas. This program has been checked using the Texas Instrument TI-59 programmable calculator.

The equations are presented without development; they are all based on the fundamentals already covered in the book. Since there are so many equations, they are presented in text form, rather than displayed, to save space.

It is recommended that basic units always be used in a computer analysis. Thus, if ips gravitational units are used, forces and weights should be in pounds force and dimensions in inches, with $g = 386$ in/s^2. If SI units are used, forces should be in newtons, masses in kilograms, and distances in metres. Results can always be expressed using prefixes such as kilo- or milli- at the conclusion of the program.

The notation to be used is customary, and most of it is shown on Fig. 13-17. Three subroutines are necessary and should be programmed first. These are $\mathbf{A} \times \mathbf{B} = (x_A y_B - y_A x_B)\hat{\mathbf{k}}$, a two-dimensional cross-product routine, $\mathbf{F}_C = \mathbf{f}_1(\theta_2) + \mathbf{f}_2(\theta_3)$, and $\mathbf{F}_D = \mathbf{f}_3(\theta_2) + \mathbf{f}_4(\theta_4)$. The last two must be formulated from the original problem statement and will probably change from problem to problem.

Three kinds of storage are needed, permanent storage for the initial or given values, temporary storage for certain terms that occur frequently and are used during the computation and then discarded, and permanent storage for all answers of interest.

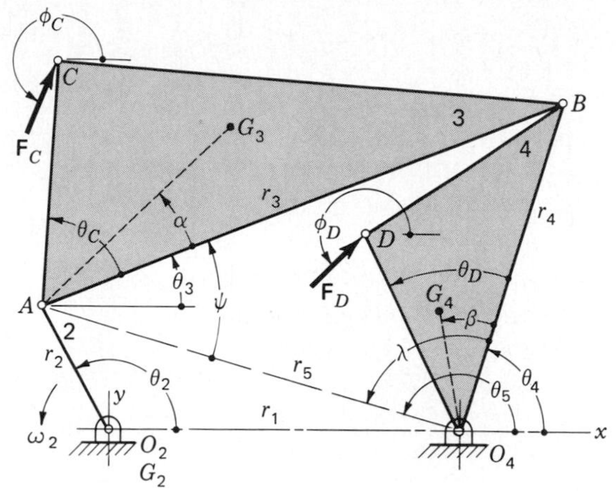

Figure 13-17

Initial permanent storage Store θ_2, $\Delta\theta_2$, r_1, r_2, r_3, r_4, R_{G_3A}, R_{G_4}, R_{CA}, R_D, α, θ_C, θ_D, β, ω_2, m_3, m_4, I_3, and I_4. Note that $\Delta\theta_2$ is the increment through which the crank is advanced after every solution.

Temporary storage θ_5, r_5, ψ, $|F'_{34}|$, $|F''_{34}|$, $\Sigma\,(\mathbf{A}\times\mathbf{B})$, x_A, y_A, x_B, y_B. It may be desirable to store other quantities temporarily too, such as the arguments of trigonometric terms that arise often.

Final permanent storage θ_3, θ_4, ω_3, ω_4, α_3, α_4, $A^x_{G_3}$, $A^y_{G_3}$, $A^x_{G_4}$, $A^y_{G_4}$, F^x_C, F^y_C, F^x_D, F^y_D, F^x_{23}, F^y_{23}, F^x_{34}, F^y_{34}, F^x_{14}, F^y_{14}, T_2.

Step 1. Equation (1), $r_1 + r_5 \cos\theta_5 = r_2 \cos\theta_2$; Eq. (2), $r_5 \sin\theta_5 = r_2 \sin\theta_2$. Solve for θ_5 and r_5. Note that $r_2 \sin\theta_2$ and $r_2\cos\theta_2 - r_1$ form the legs of a right triangle with r_5 as the hypotenuse and θ_5 as one of the angles. If solving on the calculator, use the polar-rectangular conversion key to obtain θ_5 and r_5.

Step 2. Solve Eq. (3), $\psi = \cos^{-1}[(r_3^2 + r_5^2 - r_4^2)/2r_3r_5]$.

Step 3. Solve Eq. (4), $\lambda = \cos^{-1}[(r_5 - r_3\cos\psi)/r_4]$; Eq. (5), $\theta_4 = \theta_5 - \lambda$, and $\theta_3 = \psi + \theta_5 - 180$.

Step 4. Solve Eq. (6), $\omega_3 = [r_2\omega_2 \sin(\theta_2 - \theta_4)]/[r_3 \sin(\theta_4 - \theta_3)]$.

Step 5. Solve Eq. (7), $\omega_4 = [r_2\omega_2 \sin(\theta_2 - \theta_3)]/[r_4 \sin(\theta_4 - \theta_3)]$.

Step 6. Solve Eq. (8), $\alpha_3 = [r_2\omega_2^2\cos(\theta_2 - \theta_4) + r_3\omega_3^2\cos(\theta_3 - \theta_4) - r_4\omega_4^2]/[r_3\sin(\theta_4 - \theta_3)]$.

Step 7. Solve Eq. (9), $\alpha_4 = [r_2\omega_2^2\cos(\theta_2 - \theta_3) - r_4\omega_4^2\cos(\theta_3 - \theta_4) + r_3\omega_3^2]/[r_4\sin(\theta_4 - \theta_3)]$.

Step 8. Solve Eq. (10), $A^x_{G_3} = r_2\omega_2^2\cos(\theta_2 + 180) + R_{G_3A}\alpha_3\cos(\theta_3 + \alpha + 90) + R_{G_3A}\omega_3^2\cos(\theta_3 + \alpha + 180)$.

Step 9. Solve Eq. (11), $A^y_{G_3} = r_2\omega_2^2\sin(\theta_2 + 180) + R_{G_3A}\alpha_3\sin(\theta_3 + \alpha + 90) + R_{G_3A}\omega_3^2\sin(\theta_3 + \alpha + 180)$.

Step 10. Solve Eq. (12), $A^x_{G_4} = R_{G_4}\alpha_4 \cos(\theta_4+\beta+90) + R_{G_4}\omega_4^2 \cos(\theta_4+\beta+180)$.

Step 11. Solve Eq. (13), $A^y_{G_4} = R_{G_4}\alpha_4 \sin(\theta_4+\beta+90) + R_{G_4}\omega_4^2 \sin(\theta_4+\beta+180)$. This step ends the kinematic analysis.

Step 12. Solve Eq. (14), $|\mathbf{F}'_{34}| = |\Sigma(\mathbf{A}\times\mathbf{B})|/|\mathbf{R}_B \times \hat{\mathbf{F}}'_{34}|$, where $\Sigma(\mathbf{A}\times\mathbf{B}) = \mathbf{R}_{G_4}\times m_4\mathbf{A}_{G_4} - \mathbf{R}_D\times\mathbf{F}_D + I_4\boldsymbol{\alpha}_4$, and $\hat{\mathbf{F}}'_{34} = \cos\theta_3\hat{\mathbf{i}} + \sin\theta_3\hat{\mathbf{j}}$.

Step 13. Solve Eq. (15), $\mathbf{F}'_{14} = -\mathbf{F}'_{34} + m_4\mathbf{A}_{G_4} - \mathbf{F}_D$.

Step 14. Solve Eq. (16), $|\mathbf{F}''_{34}| = |\Sigma(\mathbf{A}\times\mathbf{B})|/|\mathbf{R}_{BA}\times\hat{\mathbf{F}}''_{34}|$, where $\Sigma(\mathbf{A}\times\mathbf{B}) = \mathbf{R}_{G_3A}\times(-m_3\mathbf{A}_{G_3}) + \mathbf{R}_{CA}\times\mathbf{F}_C - I_3\boldsymbol{\alpha}_3$ and $\hat{\mathbf{F}}''_{34} = \cos\theta_4\hat{\mathbf{i}} + \sin\theta_4\hat{\mathbf{j}}$.

Step 15. Solve Eq. (17), $F^x_{34} = F'_{34}\cos\theta_3 + F''_{34}\cos\theta_4$. $F^y_{34} = F'_{34}\sin\theta_3 + F''_{34}\sin\theta_4$; Eq. (18), $F^x_{14} = -F'_{34}\cos\theta_3 - F''_{34}\cos\theta_4$, $F^y_{14} = -F'_{34}\sin\theta_3 - F''_{34}\sin\theta_4$.

Step 16. Solve Eq. (19), $\mathbf{F}_{23} = -\mathbf{F}_C + m_3\mathbf{A}_{G_3} - \mathbf{F}_{43}$.

Step 17. Solve Eq. (20), $\mathbf{T}_2 = \mathbf{r}_2\times\mathbf{F}_{23}$.

Step 18. Solve Eq. (21), $\theta_2 = \theta_2 + \Delta\theta_2$, and return to step 1.

PROBLEMS

13-1 The steel bell crank shown in the figure is used as an oscillating cam follower. Find the mass moment of inertia of the lever about an axis through *O*. Use $w = 0.282\ \text{lb/in}^3$ for the unit weight of steel.

13-2 A 5- by 50- by 300-mm steel bar has two round steel disks each 50 mm in diameter and 20 mm long welded to one end as shown. A small hole is drilled 25 mm from the end. Find the mass moment of inertia of this weldment about an axis through the hole. The mass density of steel is $7.80\ \text{Mg/m}^3$.

13-3 Find the external torque that must be applied to link 2 of the mechanism illustrated in the figure to drive it at the given velocity.

13-4 Crank 2 of the four-bar linkage shown in the figure is balanced. For the given angular velocity of link 2, find the forces acting at each pin joint and the external torque that must be applied to link 2.

13-5 For the given angular velocity of crank 2 in the figure, find the reactions at each pin joint and the external torque to be applied to the crank.

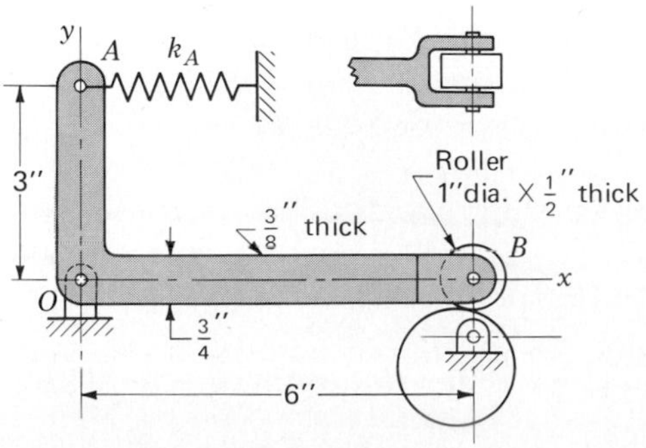

Problem 13-1

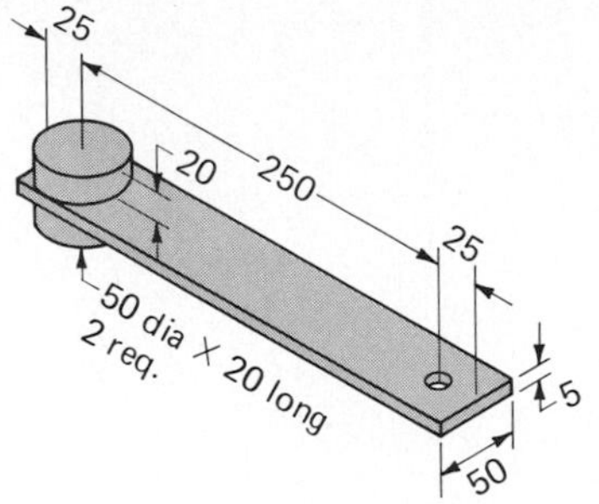

Problem 13-2 Dimensions in millimetres.

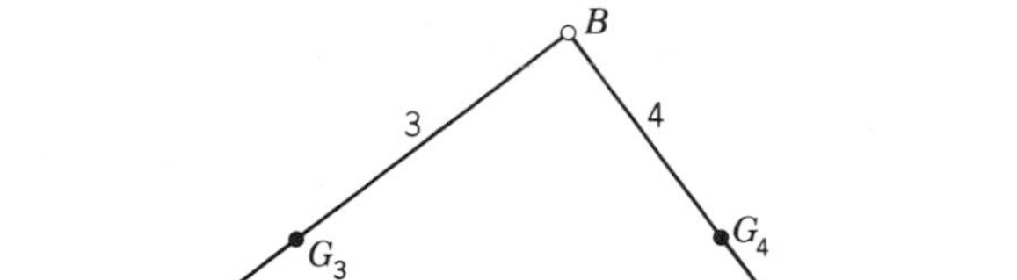

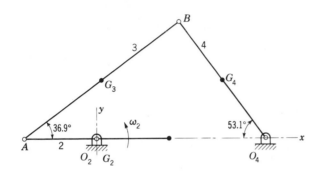

Problem 13-3 O_2A = 3 in, AG_3 = 4 in, AB = 8 in, O_4G_4 = 3 in, O_4B = 6 in, O_2O_4 = 7 in, ω_2 = $180\hat{\mathbf{k}}$ rad/s, $W_3 = 0.708$ lb, W_4 = 0.780 lb, I_3 = 0.0154 lb · s² · in, I_4 = 0.0112 lb · s² · in, $\boldsymbol{\alpha}_2$ = 0 rad/s², $\boldsymbol{\alpha}_3$ = $4950\hat{\mathbf{k}}$ rad/s², $\boldsymbol{\alpha}_4$ = $-8900\hat{\mathbf{k}}$ rad/s², $\mathbf{A}_{G_3}$ = $6320\hat{\mathbf{i}}$ + $750\hat{\mathbf{j}}$ ft/s², $\mathbf{A}_{G_4}$ = $2280\hat{\mathbf{i}}$ + $750\hat{\mathbf{j}}$ ft/s².

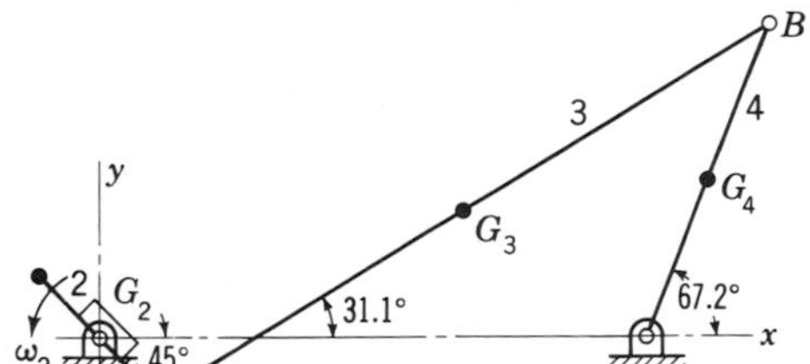

Problem 13-4 O_2A = 2 in, AG_3 = 8.50 in, AB = 17 in, O_4G_4 = 4 in, O_4B = 8 in, O_2O_4 = 13 in, $\boldsymbol{\omega}_2$ = $200\hat{\mathbf{k}}$ rad/s, W_3 = 2.65 lb, W_4 = 6.72 lb, I_3 = 0.0606 lb · s² · in, I_4 = 0.531 lb · s² · in, $\boldsymbol{\alpha}_2$ = 0 rad/s², $\boldsymbol{\alpha}_3$ = $-6530\hat{\mathbf{k}}$ rad/s², $\boldsymbol{\alpha}_4$ = $-240\hat{\mathbf{k}}$ rad/s², $\mathbf{A}_{G_3}$ = $-3160\hat{\mathbf{i}} + 262\hat{\mathbf{j}}$ ft/s², $\mathbf{A}_{G_4} = -800\hat{\mathbf{i}} - 2110\hat{\mathbf{j}}$ ft/s².

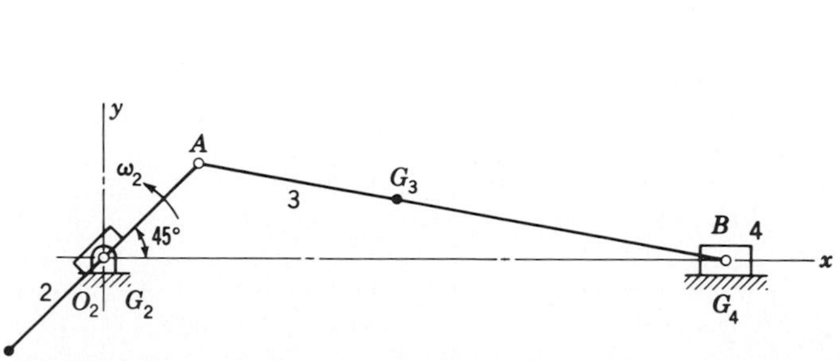

Problem 13-5 O_2A = 3 in, AG_3 = 4.5 in, AB = 12 in, $\boldsymbol{\omega}_2$ = $210\hat{\mathbf{k}}$ rad/s, $\boldsymbol{\omega}_3$ = $-37.7\hat{\mathbf{k}}$ rad/s, $\boldsymbol{\alpha}_2$ = 0 rad/s², $\boldsymbol{\alpha}_3 = 7670\hat{\mathbf{k}}$ rad/s², $\mathbf{A}_{G_3} = -7820\hat{\mathbf{i}} - 4876\hat{\mathbf{j}}$ ft/s², $\mathbf{A}_B = -7850\hat{\mathbf{i}}$ ft/s², $W_3 = 3.40$ lb, W_4 = 2.86 lb, I_3 = 0.1085 lb · s² · in.

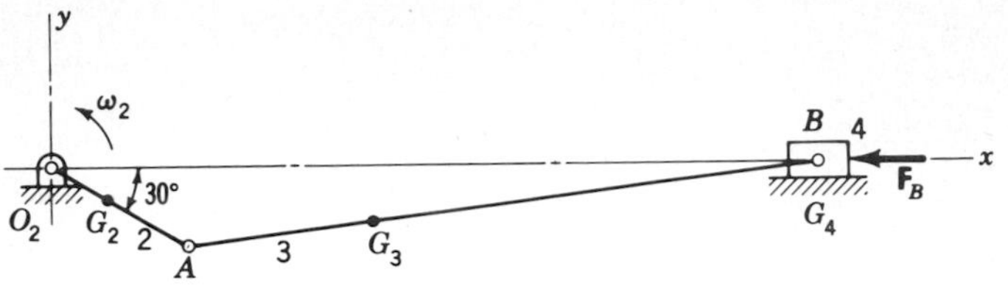

Problem 13-6 $O_2G_2 = 1.25$ in, $O_2A = 3$ in, $AG_3 = 3.5$ in, $AB = 12$ in, $\boldsymbol{\omega}_2 = 160\hat{\mathbf{k}}$ rad/s, $\boldsymbol{\omega}_3 = -35\hat{\mathbf{k}}$ rad/s, $\boldsymbol{\alpha}_2 = 0$ rad/s^2, $\boldsymbol{\alpha}_3 = -3090\hat{\mathbf{k}}$ rad/s^2, $\mathbf{A}_{G_2} = 2640\angle 150^\circ$ ft/s^2, $\mathbf{A}_{G_3} = 6130\angle 158.3^\circ$ ft/s^2, $\mathbf{A}_B = 6280\angle 180^\circ$ ft/s^2, $W_2 = 0.95$ lb, $W_3 = 3.50$ lb, $W_4 = 2.50$ lb, $I_2 = 0.003\,69$ lb · s^2 · in, $I_3 = 0.110$ lb · s^2 · in, $\mathbf{F}_B = 800\angle 180^\circ$ lb.

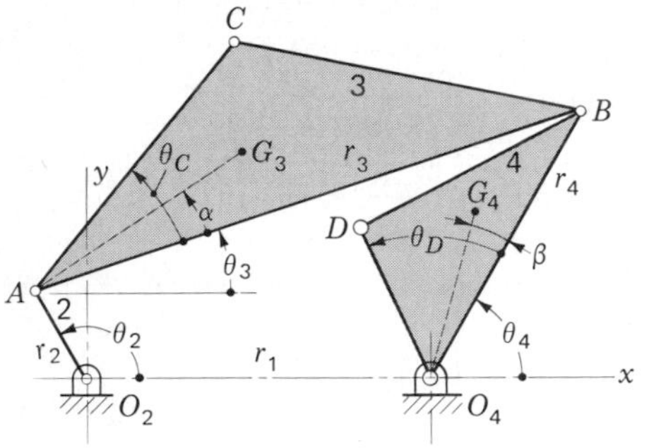

Problem 13-7

13-6 The figure shows an engine mechanism with an external force F_B applied to the piston. For the given crank velocity find all the pin reactions and the crank torque.

13-7 The following data, all in basic SI units, pertain to the four-bar linkage shown in the accompanying figure: $r_1 = 0.9$, $r_2 = 0.3$, $r_3 = 1.5$, $r_4 = 0.8$, $AG_3 = 0.65$, $O_4G_4 = 0.45$, $AC = 0.85$, $O_4D = 0.4$, $\alpha = 16^\circ$, $\theta_C = 33^\circ$, $\beta = 17^\circ$, $\theta_D = 53^\circ$, $m_3 = 65.8$, $m_4 = 21.8$, $I_3 = 4.2$, $I_4 = 0.51$. Link 2 is balanced. A kinematic analysis at $\theta_2 = 60^\circ$ and $\omega_2 = 12$ rad/s gave $\theta_3 = 0.7^\circ$, $\theta_4 = 20.4^\circ$, $\alpha_3 = -85.6$ rad/s^2, $\alpha_4 = -172$ rad/s^2, $\mathbf{A}_{G_3} = 96.4\angle 259^\circ$ m/s^2, and $\mathbf{A}_{G_4} = 97.8\angle 270^\circ$ m/s^2. Make a complete dynamic analysis and find all the pin reactions and the torque to be applied to link 2.

13-8 Repeat Prob. 13-7 if an external force $\mathbf{F}_D = 12\angle 0^\circ$ kN acts at point D.

13-9 Make a complete kinematic and dynamic analysis of the linkage of Prob. 13-7 using the same data but with $\theta_2 = 170^\circ$, $\omega_2 = 12$ rad/s, and an external force $\mathbf{F}_D = 8.94\angle 63.4^\circ$ kN.

13-10 Repeat Prob. 13-9 using $\theta_2 = 200^\circ$, $\omega_2 = 12$ rad/s, and an external force $\mathbf{F}_C = 8.49\angle 45^\circ$kN.

13-11 At $\theta_2 = 270^\circ$ and $\omega_2 = 18$ rad/s, a kinematic analysis of the linkage whose geometry is given in Prob. 13-7 gives $\theta_3 = 46.6^\circ$, $\theta_4 = 80.5^\circ$, $\alpha_3 = -178$ rad/s^2, $\alpha_4 = -256$ rad/s^2, $\mathbf{A}_{G_3} = 112\angle 22.7^\circ$ m/s^2, $\mathbf{A}_{G_4} = 119\angle 352.5^\circ$ m/s^2. An external force $\mathbf{F}_D = 8.60\angle 215.5^\circ$ kN acts at point D. Make a complete dynamic analysis of the linkage.

13-12 The following data applies to the four-bar linkage illustrated for Prob. 13-7: $r_1 = 300$ mm, $r_2 = 120$ mm, $r_3 = 320$ mm, $r_4 = 250$ mm, $AG_3 = 200$ mm, $O_4G_4 = 125$ mm, $AC - 360$ mm, $O_4D = 0$, $\alpha = 8^\circ$, $\theta_C = 15^\circ$, $\beta = \theta_D = 0$. A kinematic analysis at $\theta_2 = 90^\circ$ and $\omega_2 = 32$ rad/s gave results of $\theta_3 = 23.9^\circ$, $\theta_4 = 91.7^\circ$, $\alpha_3 = 221$ rad/s^2, $\alpha_4 = 122$ rad/s^2, $\mathbf{A}_{G_3} = 88.6\angle 255^\circ$ m/s^2, and $\mathbf{A}_{G_4} = 32.6\angle 244^\circ$ m/s^2. Also, $m_3 = 4$ kg, $I_3 = 0.011$ kg · s^2 · m, $m_4 = 1.5$ kg, and $I_4 = 0.0023$ kg · s^2 · m. Using an external force $\mathbf{F}_C = 632\angle 342^\circ$ N, make a complete dynamic analysis of the system.

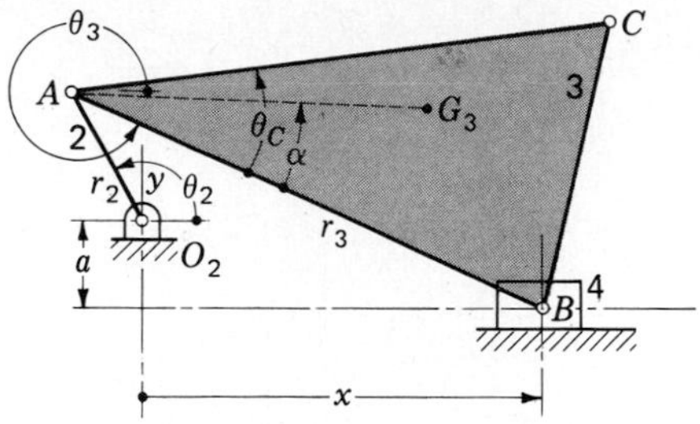

Problem 13-15

13-13 Repeat Prob. 13-12 if $\theta_2 = 260°$. Analyze both the kinematics and dynamics of the system at this position.

13-14 Repeat Prob. 13-13 if $\theta_2 = 300°$.

13-15 Analyze the dynamics of the offset slider-crank linkage shown in the figure using the following data: $\theta_2 = 120°$, $a = 0.06$ m, $r_2 = 0.1$ m, $r_3 = 0.38$ m, $AC = 0.4$ m, $AG_3 = 0.26$ m, $\omega_2 = -18$ rad/s, $\alpha = 22°$, $\theta_C = 32°$, $m_3 = 7.4$ kg, $m_4 = 3.2$ kg, $I_3 = 0.0136$ kg · s^2 · m, $\mathbf{F}_C = -1000\hat{\mathbf{i}}$ N, $\mathbf{F}_{14}^x = -2000\hat{\mathbf{i}}$ N. Assume a balanced crank and no friction forces.

13-16 Analyze the system of Prob. 13-15 for a complete rotation of the crank. Use $F_C = 0$ and $F_{14}^x = -1000$ N when $\dot{x}$ is positive and $F_{14}^x = 0$ when $\dot{x}$ is negative. Assume that the crank is balanced. Plot a graph of T_2 and F_{14}^y vs. θ_2.

13-17 A slider-crank linkage similar to that for Prob. 13-15 has zero offset and $r_2 = 0.10$ m, $r_3 = 0.45$ m, $AC = 0$, $AG_3 = 0.20$ m, $\omega_2 = -24$ rad/s, $\alpha = \theta_C = 0$, $m_3 = 3.5$ kg, $m_4 = 1.2$ kg, $I_3 = 0.060$ kg · s^2 · m, and $T_2 = 60$ N · m. Corresponding to $\theta_2 = 135°$, a kinematic analysis gave $\theta_3 = -9.0°$, $\alpha_3 = 89.3$ rad/s^2, $x = 0.374$ m, $\ddot{x} = 40.6$ m/s^2, and $\mathbf{A}_{G_3} = 40.6\hat{\mathbf{i}} - 22.6\hat{\mathbf{j}}$ m/s^2. Find $\mathbf{F}_{14}$ and $\mathbf{F}_{23}$. Assume link 2 is balanced.

13-18 Repeat Prob. 13-17 if $\theta_2 = 240°$. The results of a kinematic analysis are $\theta_3 = 11.1°$, $\alpha_3 = -112$ rad/s^2, $x = 0.392$ m, $\ddot{x} = 35.2$ m/s^2, $\mathbf{A}_{G_3} = 31.6\hat{\mathbf{i}} + 27.7\hat{\mathbf{j}}$ m/s^2.

13-19 An offset slider-crank linkage, as in Prob. 13-15, has $a = 0.08$ m, $r_2 = 0.25$ m, $r_3 = 1.25$ m, $AC = 1.0$ m, $AG_3 = 0.75$ m, $\omega_2 = 6$ rad/s, $\alpha = -18°$, $\theta_C = -38°$, $m_3 = 140$ kg, $m_4 = 50$ kg, and $I_3 = 8.42$ kg · s^2 · m. Make a complete kinematic and dynamic analysis of this system at $\theta_2 = 25°$ using $\mathbf{F}_C = 80\angle{-60°}$ kN and $F_{14}^x = -50$ kN. Assume a balanced crank.

13-20 Cranks 2 and 4 of the crossed linkage shown in the figure are balanced. The dimensions of the linkage are $O_2A = 6$ in, $AB = 18$ in, $AG = 12$ in, $AC = 24$ in, $O_2O_4 = 18$ in, and $O_4B = 6$ in. Corresponding to the position shown, and with $\omega_2 = 10$ rad/s, a kinematic analysis gave as results $\omega_3 = -1.43$ rad/s, $\omega_4 = -11.43$ rad/s, $\alpha_3 = \alpha_4 = 84.7$ rad/s^2, and $\mathbf{A}_{G_3} = 47.6\hat{\mathbf{i}} + 70.3\hat{\mathbf{j}}$ ft/s^2. Also $W_3 = 4$ lb, $I_3 = 0.497$ lb · s^2 · in, and $I_4 = 0.063$ lb · s^2 · in. If $\mathbf{F}_C = -30\hat{\mathbf{j}}$ lb and link 2 is the driver, find the driving torque and the pin reactions.

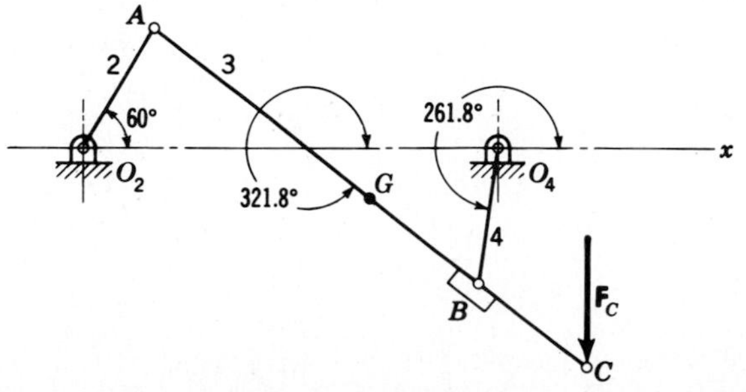

Problem 13-20

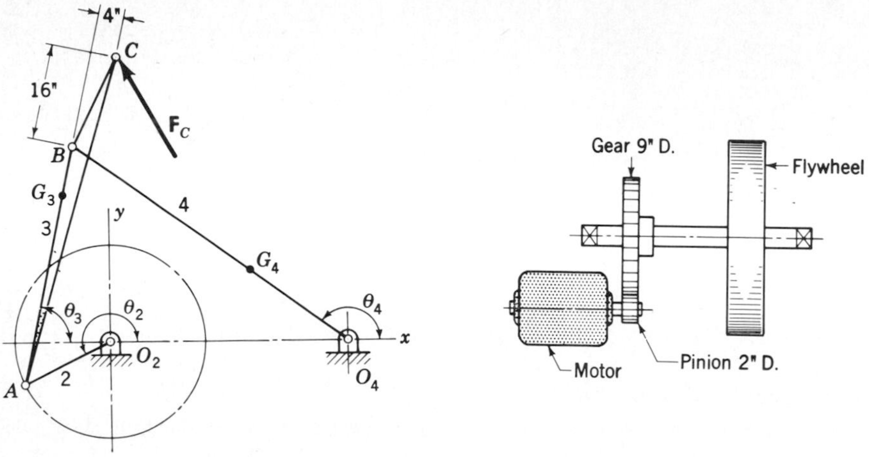

Problems 13-23 and 13-24

13-21 Find the driving torque and pin reactions for the mechanism of Prob. 13-20 if crank 4 is the driver.

13-22 A kinematic analysis of the mechanism of Prob. 13-20 at $\theta_2 = 210°$ gave $\theta_3 = 14.7°$, $\theta_4 = 164.7°$, $\omega_3 = 4.73$ rad/s, $\omega_4 = -5.27$ rad/s, $\alpha_3 = \alpha_4 = -10.39$ rad/s^2, and $\mathbf{A}_{G_3} = 26\underline{/20.85°}$ ft/s^2. Compute T_2 and the pin reactions for this phase of the motion using the same force $\mathbf{F}_C$ as in Prob. 13-20.

13-23 Part (*a*) of the figure shows a linkage with an extended coupler having an external force $\mathbf{F}_C$ acting during a portion of the stroke. The dimensions of the linkage are $O_2A = 16$ in, $AG_3 = 32$ in, $AB = O_2O_4 = 40$ in, $O_4G_4 = 20$ in, and $O_4B = 56$ in. Make a kinematic and dynamic analysis for a complete rotation of the crank using $\omega_2 = 10$ rad/s and $\mathbf{F}_C = -500\hat{\mathbf{i}} + 866\hat{\mathbf{j}}$ lb for $90° \le \theta_2 \le 300°$ with $F_C = 0$ for other angles. Use $W_3 = 222$ lb, $W_4 = 208$ lb, $I_3 = 226$ lb · s^2 · in, $I_4 = 264$ lb · s^2 · in and assume a balanced crank.

13-24 Part (*b*) of the figure shows a motor geared to a shaft on which a flywheel is mounted. The moments of inertia of the parts are as follows: flywheel, $I = 2.73$ lb · s^2 · in; flywheel shaft, $I = 0.0155$ lb · s^2 · in; gear, $I = 0.172$ lb · s^2 · in; pinion, $I = 0.003\,49$ lb · s^2 · in; motor, $I = 0.0864$ lb · s^2 · in. If the motor has a starting torque of 75 lb · in, what is the angular acceleration of the flywheel shaft at the instant the motor switch is turned on?

CHAPTER
FOURTEEN
DYNAMICS OF RECIPROCATING ENGINES

The purpose of this chapter is to apply fundamentals—kinematic and dynamic analysis—in a complete investigation of a particular group of machines. The reciprocating engine has been selected for this purpose because it has reached such a high state of development and is of more general interest than other machines. For our purpose, however, any machine or group of machines involving interesting dynamical situations would serve just as well. The primary objective is to demonstrate methods of applying fundamentals to the analysis of any machine.

14-1 ENGINE TYPES

The description and characteristics of all the engines which have been conceived and constructed would fill many books. Here our purpose is to outline very briefly a few of the engine types which are currently popular and in general use. The exposition is not intended to be complete. Furthermore, since you are expected to be mechanically inclined and generally familiar with internal combustion engines, the primary purpose of this section is merely to record things which you know and to furnish a nomenclature for the balance of the chapter.

We also include in this section, so as to have it all in one place, the descriptions and specifications of some of the more interesting engines. The material will then be easily available for use in problems and examples.

In this chapter we classify engines according to their intended use, the combustion cycle used, and the number and arrangement of the cylinders.

Thus, we refer to aircraft engines, automotive engines, marine engines, and stationary engines, for example, all so named because of the purpose for which they were designed. Similarly, one might have in mind an engine designed on the basis of the *Otto cycle*, in which the fuel and air are mixed before compression and in which combustion takes place with no excess air, or the *diesel engine*, in which the fuel is injected near the end of compression and combustion takes place with much excess air. The Otto-cycle engine uses quite volatile fuels, and ignition is by spark, but the diesel-cycle engine operates on fuels of low volatility and ignition occurs because of compression.

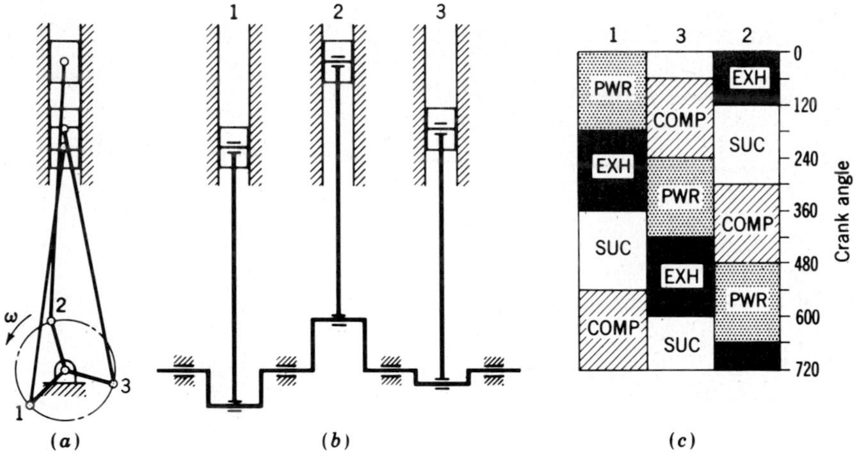

Figure 14-1 A three-cylinder in-line engine: (*a*) front view, (*b*) side view, (*c*) firing order.

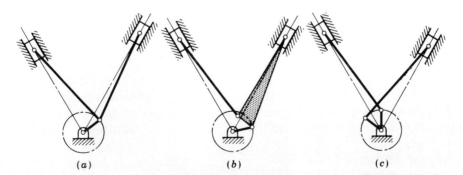

Figure 14-2 Crank arrangements of V engines: (*a*) single crank per pair of cylinders; connecting rods interlock with each other and are of *fork-and-blade* design; (*b*) single crank per pair of cylinders; the *master connecting rod* carries a bearing for the *articulated rod*; (*c*) separate cranks connect to staggered pistons.

The diesel- and Otto-cycle engines may be either *two-stroke-cycle* or *four-stroke-cycle*, depending upon the number of piston strokes required for the complete combustion cycle. Many outboard marine engines use the two-stroke-cycle (or simply the two-cycle) process, in which the piston uncovers exhaust ports in the cylinder wall near the end of the expansion stroke and permits the exhaust gases to flow out. Soon after the exhaust ports are opened, the inlet ports open too and permit entry of a precompressed fuel-air mixture which also assists in expelling the remaining exhaust gases. The ports are then closed by the piston moving upward, and the fuel mixture

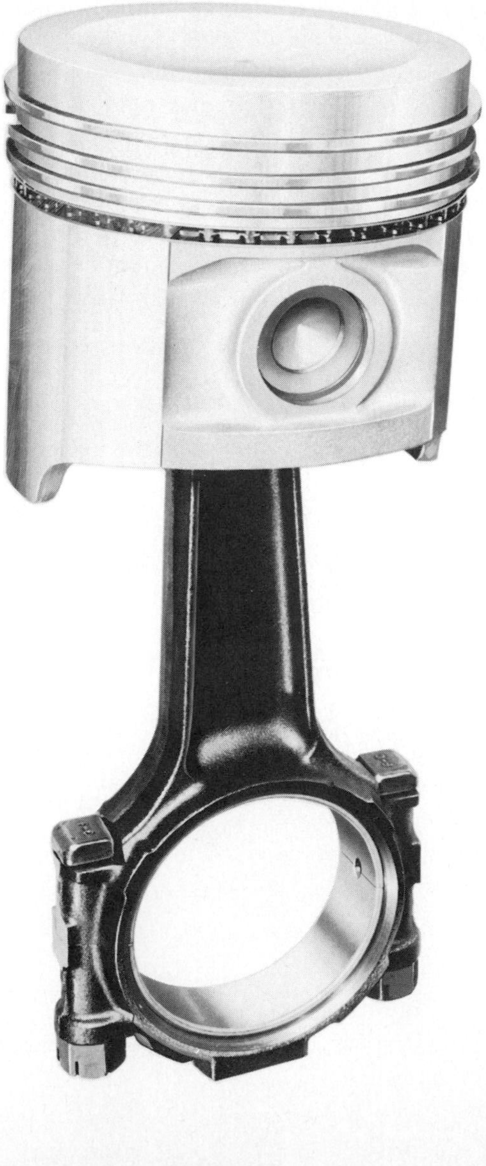

Figure 14-3 A piston-connecting-rod assembly for a 351-in^3 V6 truck engine. (*GMC Truck and Coach Division, General Motors Corporation, Pontiac, Michigan.*)

again compressed. Then the cycle begins again. Note that the two-cycle engine has an expansion and a compression stroke and that they occur during one revolution of the crank.

The four-cycle engine has four piston strokes in a single combustion cycle corresponding to two revolutions of the crank. The events corresponding to the four strokes are (1) expansion, or power, stroke, (2) exhaust, (3) suction, or intake, stroke, (4) compression.

Multicylinder engines are broadly classified according to how the cylinders are arranged with respect to each other and the crankshaft. Thus, an *in-line* engine is one in which the piston axes form a single plane coincident with the crankshaft and in which the pistons are all on the same side of the crankshaft. Figure 14-1 is a schematic drawing of a three-cylinder in-line engine with the cranks spaced 120°; a firing-order diagram for four-cycle operation is included for interest.

A V-type engine uses two banks of one or more in-line cylinders each and a single crankshaft. Figure 14-2 illustrates several common crank arrangements. The pistons in the right and left banks of (*a*) and (*b*) are in the same plane, but those in (*c*) are in different planes.

If the V angle is increased to 180°, the result is called an *opposed-piston* engine. An opposed engine may have the two piston axes coincident or offset, and the rods may connect to the same crank or to separate cranks spaced 180° apart.

Figure 14-4 A cast crankshaft for a 305-in^3 V6 truck engine. (*GMC Truck and Coach Division, General Motors Corporation, Pontiac, Michigan.*)

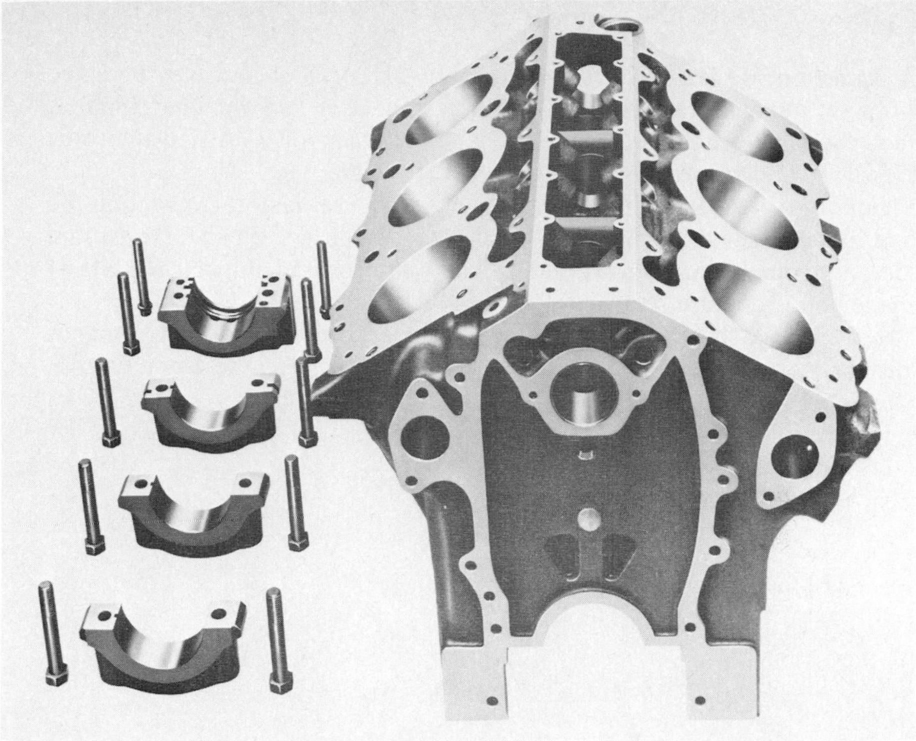

Figure 14-5 Block for a 305-in^3 V6 truck engine. The same casting is used for a 351-in^3 engine by boring for larger pistons. (*GMC Truck and Coach Division, General Motors Corporation, Pontiac, Michigan.*)

Figure 14-6 Model HM 80 single-cylinder engine. (*Tecumseh Products Company, Lauson Engine Division, New Holstein, Wisconsin.*)

A *radial* engine is one having the pistons arranged in a circle about the crank center. Radial engines use a master connecting rod for one cylinder, and the remaining pistons are connected to the master rod by articulated rods somewhat the same as for the V engine of Fig. 14-2*b*.

Figures 14-3 to 14-5 illustrate, respectively, the piston-connecting-rod assembly, the crankshaft, and the block of a V6 truck engine. These are included as typical of modern design to show the form of important parts of an engine and for future reference.

The following specifications will give a general idea of the performance and design of modern engines, together with the sizes of the parts used in them.

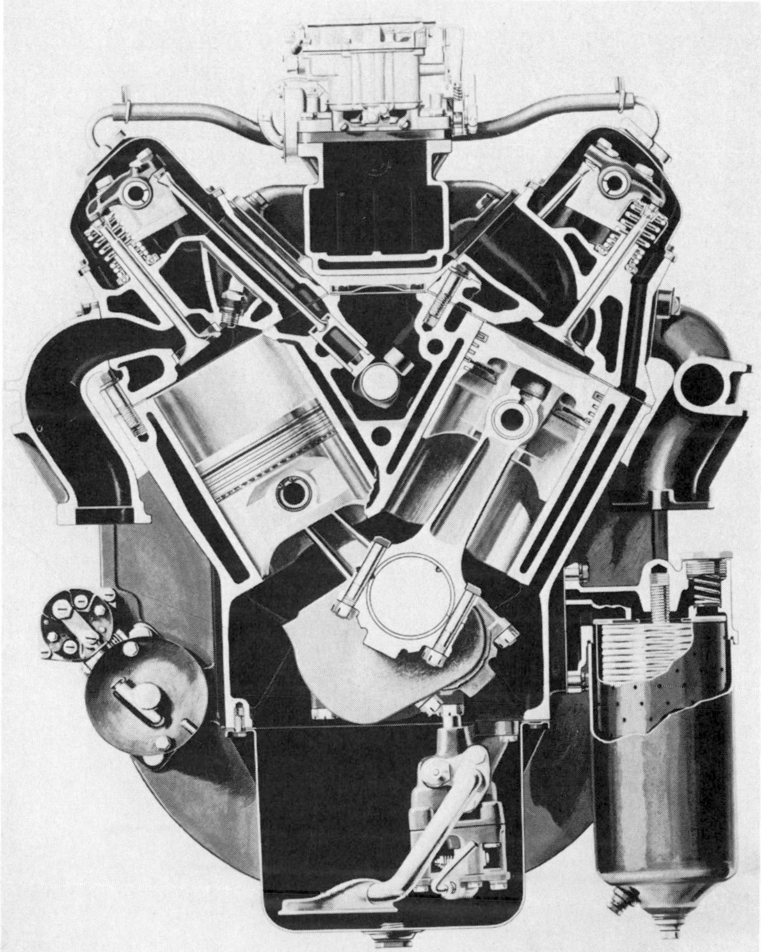

Figure 14-7 Cross-sectional view of the 401-in^3 V6 truck engine. (*GMC Truck and Coach Division, General Motors Corporation, Pontiac, Michigan.*)

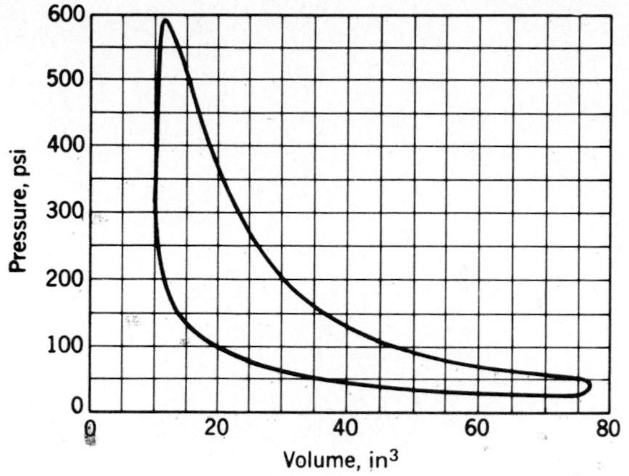

Figure 14-8 Typical *indicator diagram* for a 401-in^3 V6 truck engine; conditions unknown. (*GMC Truck and Coach Division, General Motors Corporation, Pontiac, Michigan.*)

Tecumseh Products Company, Lauson Engine Division, New Holstein, Wisconsin. Model HM 80 single-cylinder engine, shown in Fig. 14-6, has the following specifications: 5.0 hp at 2200 rpm, 6.9 hp at 2900 rpm, 8.0 hp at 3600 rpm, recoil starter, 46 lb net weight, $3\frac{1}{8}$-in bore (79.38 mm), $2\frac{17}{32}$-in stroke (64.31 mm), 19.41 in^3 (318.27 mL) displacement, four-stroke cycle, air-cooled, counterclockwise rotation viewed from the power-takeoff side, weight of piston assembly 0.530 lb (0.2405 kg), weight of connecting-rod assembly 0.365 lb (0.1655 kg), connecting-rod length, 3.956 in, 1.34 in from crank-pin bearing to mass center of connecting rod, flywheel $Wr^2 =$ 69.6 lb · in^2.

GMC Truck and Coach Division, General Motors Corporation, Pontiac, Michigan. One of the V6 truck engines is illustrated in Fig. 14-7. These engines are manufactured in four displacements, and they include one model, a V12 (702 in^3), which is described as a twin six because many of the V6 parts are interchangeable with it. Data to be included here are restricted to the 401-in^3 engine. Typical performance curves are exhibited in Figs. 14-8, 14-9, and 14-10. The specifications are as follows: bore, 4.875 in; stroke, 3.56 in; 60° vee design; 7.50:1 compression ratio; cylinders numbered from front to rear, 1, 3, 5, on the left bank, 2, 4, 6 on the right bank; firing order, 1, 6, 5, 4, 3, 2; crank arrangement, Fig. 14-11; connecting-rod length, 7.19 in.

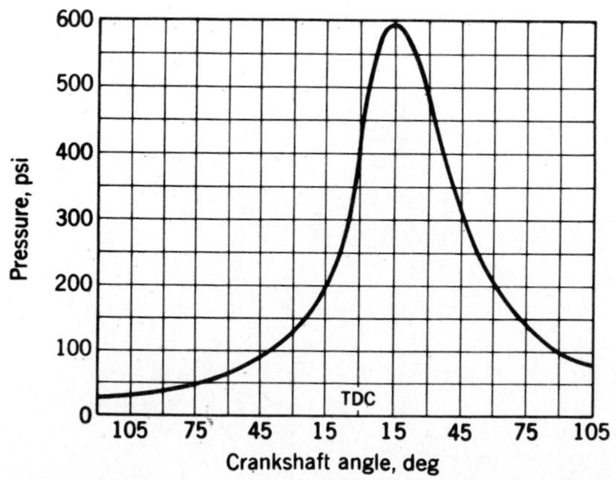

Figure 14-9 A pressure-time curve for the 401-in^3 V6 truck engine. These data were taken from a running engine. (*GMC Truck and Coach Division, General Motors Corporation, Pontiac, Michigan.*)

Reciprocating weights:	Grams
Piston	1 560
Piston pin	317.5
Piston rings	127.0
Retainers	0.34
Rod	360.0
Total	2 364.84
Reciprocating weight balanced	1 182.42
Rotating weights:	
Rod	926.00
Bearings	101.28
Total	2 209.70

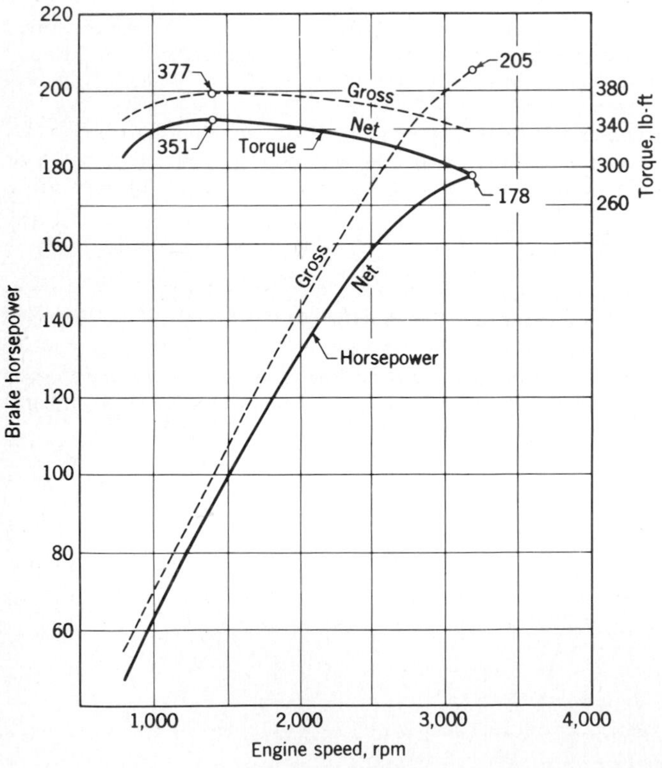

Figure 14-10 Horsepower and torque characteristics of the V6 401-in^3 truck engine. The solid curve is the net output as installed; dashed curve is the maximum output without accessories. Notice that the maximum torque occurs at a very low engine speed. (*GMC Truck and Coach Division, General Motors Corporation, Pontiac, Michigan.*)

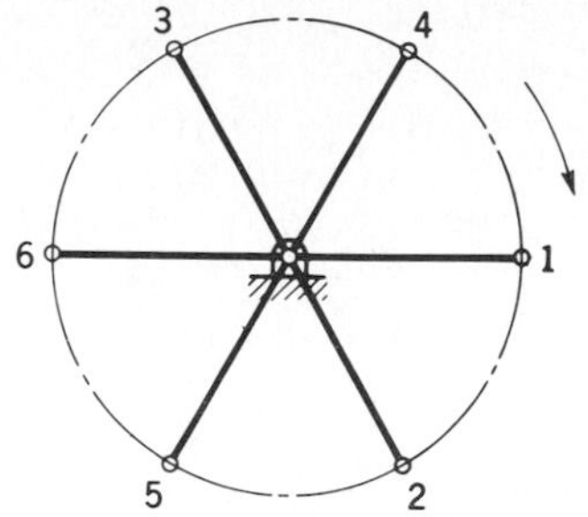

Figure 14-11 Front view of V6 truck engine showing crank arrangement and direction of rotation.

14-2 INDICATOR DIAGRAMS

Experimentally, an instrument called an *engine indicator* is used to measure the variation in pressure within a cylinder. The instrument constructs a graph, during operation of the engine, which is known as an *indicator diagram*. Known constants of the indicator make it possible to study the diagram and determine the relationship between the gas pressure and the crank angle for the particular set of running conditions in existence at the time the diagram was taken.

When an engine is in the design stage, it is necessary to estimate a diagram from theoretical considerations. From such an approximation a pilot model of the proposed engine can be designed and built and the actual indicator diagram taken and compared with the theoretically devised one. This provides much useful information for the design of the production model.

An indicator diagram for the ideal air-standard cycle is shown in Fig. 14-12 for a four-stroke-cycle engine. During compression the cylinder volume changes from v_1 to v_2 and the cylinder pressure from p_1 to p_2. The relationship at any point of the stroke is given by the polytropic gas law as

$$p_x v_x^k = p_1 v_1^k = \text{const} \tag{14-1}$$

In an actual indicator card the corners at points 2 and 3 are rounded and the line joining these points is curved. This is explained by the fact that combustion is not instantaneous and that ignition occurs before the end of the compression stroke. An actual card is also rounded at points 4 and 1 because the valves do not operate instantaneously.

The polytropic exponent k in Eq. (14-1) is often taken to be 1.30 for both compression and expansion, although differences probably do exist.

The relationship between the horsepower developed and the dimensions of the engine is given by

$$\text{bhp} = \frac{p_b l a n}{(33\,000)(12)} \tag{14-2}$$

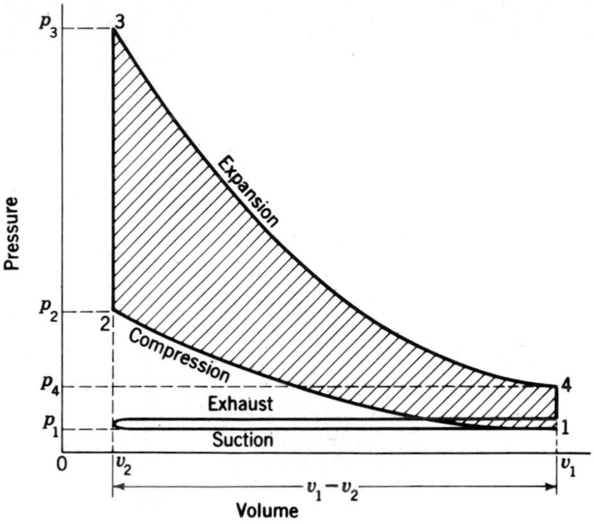

Figure 14-12 An ideal indicator diagram for a four-cycle engine.

where bhp = brake horsepower per cylinder
p_b = brake mean effective pressure, psi
l = length of stroke, in
a = piston area, in^2
n = number of working strokes per minute

The amount of horsepower that can be obtained from 1 in^3 of piston displacement varies considerably, depending upon the engine type. For automotive engines it ranges from about 0.55 up to 1.00 hp/in^3, with an average of perhaps 0.70 at the present time. On the other hand many marine diesel engines have ratios varying from 0.10 to 0.20 hp/in^3. About the best that can be done in designing a new engine is to use standard references to discover what others have done with the same types of engines and then to choose a value which seems to be reasonably attainable.

For many engines the ratio of bore to stroke varies from about 0.75 to 1.00. The tendency in automotive-engine design seems to be toward shorter-stroke engines in order to reduce engine height.

Decisions on the bore-stroke ratio and horsepower per unit displacement volume will be helpful in solving Eq. (14-2) to obtain suitable dimensions when the horsepower, speed, and number of cylinders have been decided upon.

The ratio of the brake mean effective pressure p_b to the indicated mean effective pressure p_i, obtained experimentally from an indicator card, is the mechanical efficiency e_m

$$e_m = \frac{p_b}{p_i} \tag{14-3}$$

Differences between a theoretical and an experimentally determined indicator diagram can be accounted for by applying a correction called a *card factor.* The card factor is defined by

$$f_c = \frac{p_i}{p_i'} \tag{14-4}$$

where p_i' is the theoretical indicated mean effective pressure and f_c is the card factor, usually about 0.90 to 0.95.

If the compression ratio (Fig. 14-12) is defined as

$$r = \frac{v_1}{v_2} \tag{14-5}$$

the work done during compression is

$$U_c = \int_{v_2}^{v_1} p\,dv = p_1 v_1^k \int_{v_2}^{v_1} \frac{dv}{v^k} = \frac{p_1 v_1}{k-1}(r^{k-1} - 1) \tag{a}$$

The displacement volume can be written

$$v_1 - v_2 = v_1 - \frac{v_1}{r} = \frac{v_1(r-1)}{r} \tag{b}$$

Substituting v_1 from Eq. (*b*) into Eq. (*a*) yields

$$U_c = \frac{p_1(v_1 - v_2)}{k-1}\frac{r^k - r}{r-1} \tag{c}$$

The work done during expansion is the area under the curve between points 3 and 4 of Fig. 14-12. This is found in the same manner; the result is

$$U_e = \frac{p_4(v_1 - v_2)}{k-1}\frac{r^k - r}{r-1} \tag{d}$$

The net work accomplished in a cycle is the difference in the amounts given by Eqs. (*c*) and (*d*), and it must be equal to the product of the indicated mean effective pressure and the displacement volume. Thus

$$\begin{aligned} U &= U_e - U_c = p_i'(v_1 - v_2) \\ &= \frac{p_4(v_1 - v_2)}{k-1}\frac{r^k - r}{r-1} - \frac{p_1(v_1 - v_2)}{k-1}\frac{r^k - r}{r-1} \end{aligned} \tag{14-6}$$

If the exponent is the same for expansion as for compression, Eq. (14-6) can be solved to give

$$p_4 = p_i'(k-1)\frac{r-1}{r^k - r} + p_1 \tag{e}$$

Substituting p_i from Eq. (14-4) produces

$$p_4 = (k-1)\frac{r-1}{r^k - r}\frac{p_i}{f_c} + p_1 \tag{14-7}$$

Equations (14-1) and (14-7) can be used to create the theoretical indicator diagram. The corners are then rounded off so that the pressure at point 3 is made about 75 percent of that given by Eq. (14-1). As a check, the area of the diagram can be measured and divided by the displacement volume. The result should equal the indicated mean effective pressure.

14-3 DYNAMIC ANALYSIS—GENERAL

The balance of this chapter is devoted to an analysis of the dynamics of the single-cylinder engine. To simplify this work, the gas forces and the inertia forces are found in separate sections. Then in other sections these forces are combined, using the principle of superposition to obtain the bearing forces and the crankshaft torque.

The subject of engine balancing is treated in Chap. 15, and flywheel dynamics is treated in Chap. 17.

14-4 GAS FORCES

In this section we assume that the moving parts are weightless so that the inertia forces and inertia torques are zero and there is no friction. These assumptions make it possible to trace the effect of the gas pressure from the piston to the crankshaft without the complicating effects of other forces.

In Chap. 12 both graphical and vector methods for analyzing the forces in any mechanism were presented. Either of these approaches can be used to solve the gas-force problem. The advantage of the vector method is that it can be programmed for automatic calculator or computer solution. But the graphical solution must be repeated for each crank position until a complete cycle of operation (720° for a four-cycle engine) is completed. Since we prefer not to duplicate the studies of Chap. 12, we present here an algebraic approach.

In Fig. 14-13 designate the crank angle as ωt, positive in the ccw direction, and the connecting-rod angle as ϕ, positive in the direction shown. A relation

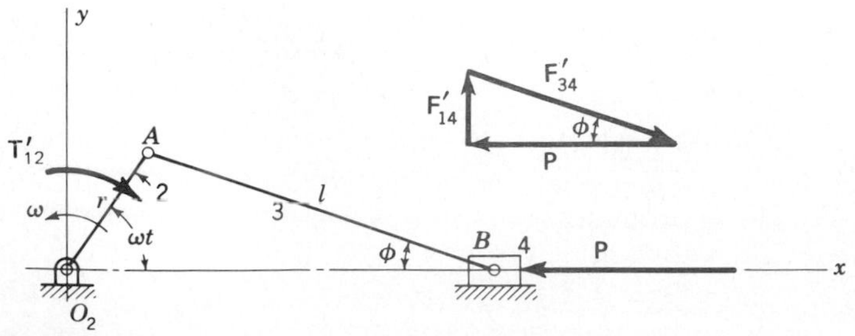

Figure 14-13

between these two angles is

$$r \sin \omega t = l \sin \phi \tag{a}$$

Designating the piston position from O_2 by the coordinate x, we find

$$x = r \cos \omega t + l \cos \phi = r \cos \omega t + l\sqrt{1 - \left(\frac{r}{l} \sin \omega t\right)^2} \tag{14-8}$$

For most engines the ratio r/l is about $\frac{1}{4}$, and so the maximum value of the second term under the radical is about $\frac{1}{16}$, or perhaps less. If we expand the radical using the binomial theorem and neglect all but the first two terms, there results

$$\sqrt{1 - \left(\frac{r}{l} \sin \omega t\right)^2} = 1 - \frac{r^2}{2l^2} \sin^2 \omega t \tag{b}$$

Since

$$\sin^2 \omega t = \frac{1 - \cos 2\omega t}{2} \tag{c}$$

Eq. (14-8) becomes

$$x = l - \frac{r^2}{4l} + r\left(\cos \omega t + \frac{r}{4l} \cos 2\omega t\right) \tag{14-9}$$

Differentiating successively to obtain the velocity and acceleration gives

$$\dot{x} = -r\omega\left(\sin \omega t + \frac{r}{2l} \sin 2\omega t\right) \tag{14-10}$$

$$\ddot{x} = -r\alpha\left(\sin \omega t + \frac{r}{2l} \sin 2\omega t\right) - r\omega^2\left(\cos \omega t + \frac{r}{l} \cos 2\omega t\right) \tag{14-11}$$

Referring again to Fig. 14-13, we designate a gas-force vector **P** as defined or obtained using the methods of Sec. 14-2. Reactions due to this force are designated using a single prime. Thus $\mathbf{F}'_{14}$ is the force of the cylinder wall acting against the piston. $\mathbf{F}'_{34}$ is the force of the connecting rod acting against the piston at the piston pin. The force polygon in Fig. 14-13 shows the relation between **P**, $\mathbf{F}'_{14}$, and $\mathbf{F}'_{34}$. Thus, we have

$$\mathbf{F}'_{14} = P \tan \phi \, \hat{\mathbf{j}} \tag{14-12}$$

The quantity $\tan \phi$ appears frequently in expressions throughout this chapter. It is therefore convenient to develop an expression in terms of the crank angle ωt. Thus

$$\tan \phi = \frac{(r/l) \sin \omega t}{\cos \phi} = \frac{(r/l) \sin \omega t}{\sqrt{1 - [(r/l) \sin \omega t]^2}} \tag{d}$$

Now, using the binomial theorem again, we find that

$$\frac{1}{\sqrt{1 - [(r/l) \sin \omega t]^2}} = 1 + \frac{r^2}{2l^2} \sin^2 \omega t \tag{e}$$

where only the first two terms have been retained. Equation (d) now becomes

$$\tan\phi = \frac{r}{l}\sin\omega t\left(1+\frac{r^2}{2l^2}\sin^2\omega t\right) \tag{14-13}$$

The trigonometry of Fig. 14-13 shows that the wrist-pin (piston-pin) bearing force has a magnitude of

$$F'_{34} = \frac{P}{\cos\phi} = \frac{P}{\sqrt{1-[(r/l)\sin\omega t]^2}} = P\left(1+\frac{r^2}{2l^2}\sin^2\omega t\right) \tag{f}$$

or, in vector notation,

$$\mathbf{F}'_{34} = P\hat{\mathbf{i}} - F'_{14}\hat{\mathbf{j}} = P\hat{\mathbf{i}} - P\tan\phi\,\hat{\mathbf{j}} \tag{14-14}$$

By taking moments about the crank center we find that the torque $\mathbf{T}'_{21}$ delivered by the crank to the shaft is the product of the force $\mathbf{F}'_{14}$ and the piston coordinate x. Using Eqs. (14-9), (14-12), and (14-13) yields

$$\mathbf{T}'_{21} = F'_{14}x\hat{\mathbf{k}} = P\left(\frac{r}{l}\sin\omega t\right)\left(1+\frac{r^2}{2l^2}\sin^2\omega t\right)\left[l-\frac{r^2}{4l}+r\left(\cos\omega t+\frac{r}{4l}\cos 2\omega t\right)\right]\hat{\mathbf{k}} \tag{g}$$

When the terms of Eq. (g) are multiplied, we can neglect those containing second or higher powers of r/l with only a very small error. Equation (g) then becomes

$$\mathbf{T}'_{21} = Pr\sin\omega t\left(1+\frac{r}{l}\cos\omega t\right)\hat{\mathbf{k}} \tag{14-15}$$

This is the torque delivered to the crankshaft by the gas force; the counterclockwise direction is positive.

14-5 EQUIVALENT MASSES

Problems 13-5 and 13-6 are examples of engine mechanisms the dynamics of which are to be analyzed using the methods of that chapter, consisting of either a graphical or a vector analysis. The results are exact, except for rounding errors, no matter which method is employed.

In this chapter we are concerned with the same problem. Certain simplifications, however, are customarily used to reduce the problem to an algebraic form. These simplifications introduce certain errors into the analysis, and it is these errors and simplifications that are the subject of this section.

In analyzing the inertia forces due to the connecting rod of an engine it is often convenient to concentrate a portion of the mass at the crankpin A and the remaining portion at the wrist pin B (Fig. 14-14). The reason for this is that the crankpin moves on a circle and the wrist pin on a straight line. Both of these motions are quite easy to analyze. However, the center of gravity G

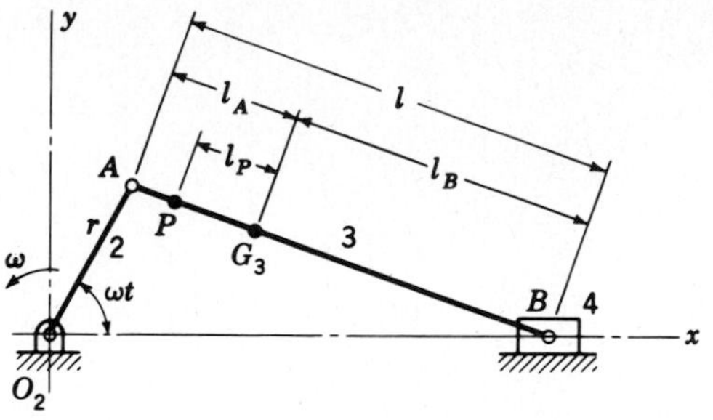

Figure 14-14

is somewhere between the crankpin and wrist pin, and its motion is more complicated and consequently more difficult to determine in algebraic form.

The mass of the connecting rod m_3 is assumed to be concentrated at the center of gravity G_3. We divide this mass into two parts; one, m_{3B}, is then concentrated at the wrist pin B. The other part, m_{3P}, is concentrated at the *center of percussion P* for oscillation of the rod about B. This disposition of the mass of the rod is dynamically equivalent to the original rod if the total mass is the same, if the position of the center of gravity G_3 is unchanged, and if the moment of inertia is the same. Writing these three conditions, respectively, in equation form produces

$$m_3 = m_{3B} + m_{3P} \tag{a}$$

$$m_{3B} l_B = m_{3P} l_P \tag{b}$$

$$I_G = m_{3B} l_B^2 + m_{3P} l_P^2 \tag{c}$$

Solving Eqs. (a) and (b) simultaneously gives the portion of mass to be concentrated at each point

$$m_{3B} = m_3 \frac{l_P}{l_B + l_P} \qquad m_{3P} = m_3 \frac{l_B}{l_B + l_P} \tag{14-16}$$

Substituting Eqs. (14-16) into (c) gives

$$I_G = m_3 \frac{l_P}{l_B + l_P} l_B^2 + m_3 \frac{l_B}{l_B + l_P} l_P^2 = m_3 l_P l_B \tag{d}$$

or

$$l_P l_B = \frac{I_G}{m_3} \tag{14-17}$$

Equation (14-17) shows that the two distances l_P and l_B are dependent on each other. Thus, if l_B is specified in advance, l_P is fixed in length by Eq. (14-17).

In the usual connecting rod the center of percussion is close to the crankpin and it is assumed that they are coincident. Thus, letting $l_A = l_P$, Eqs. (14-16) reduce to

$$m_{3B} = \frac{m_3 l_A}{l} \qquad m_{3A} = \frac{m_3 l_B}{l} \tag{14-18}$$

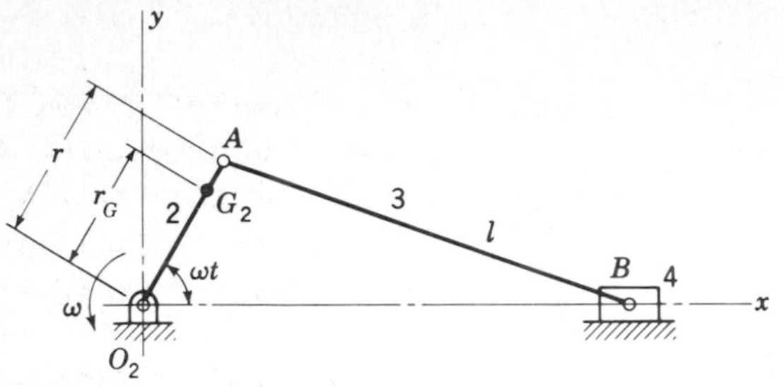

Figure 14-15

We note again that the equivalent masses, obtained by Eqs. (14-18), are not exact because of the assumption made but are close enough for ordinary connecting rods. The approximation, for example, is not valid for the master connecting rod of a radial engine because the crankpin end has bearings for all of the other connecting rods as well as its own bearing.

For estimating and checking purposes, about two-thirds of the mass should be concentrated at A and the remaining third at B.

Figure 14-15 illustrates an engine linkage in which the mass of the crank m_2 is not balanced, as evidenced by the fact that the center of gravity G_2 is displaced outward along the crank a distance r_G from the axis of rotation. In the inertia-force analysis, simplification is obtained by locating an equivalent mass m_{2A} at the crankpin. Thus, for equivalence

$$m_2 r_G = m_{2A} r \qquad \text{or} \qquad m_{2A} = m_2 \frac{r_G}{r} \tag{14-19}$$

14-6 INERTIA FORCES

Using the methods of the preceding section, we begin by locating equivalent masses at the crankpin and at the wrist pin. Thus

$$m_A = m_{2A} + m_{3A} \tag{14-20}$$

$$m_B = m_{3B} + m_4 \tag{14-21}$$

Equation (14-20) states that the mass m_A located at the crankpin is made up of the equivalent masses m_{2A} of the crank and m_{3A} of part of the connecting rod. Of course, if the crank is balanced, all its mass is assumed to be located at the axis of rotation and m_{2A} is then zero. Equation (14-21) indicates that the reciprocating mass m_B located at the wrist pin is composed of the equivalent mass m_{3B} of the other part of the connecting rod and the mass m_4 of the piston assembly.

Figure 14-16 shows the slider-crank mechanism with masses m_A and m_B located at points A and B, respectively. If we designate the angular velocity of the crank as ω and the angular acceleration as α, the position vector of the

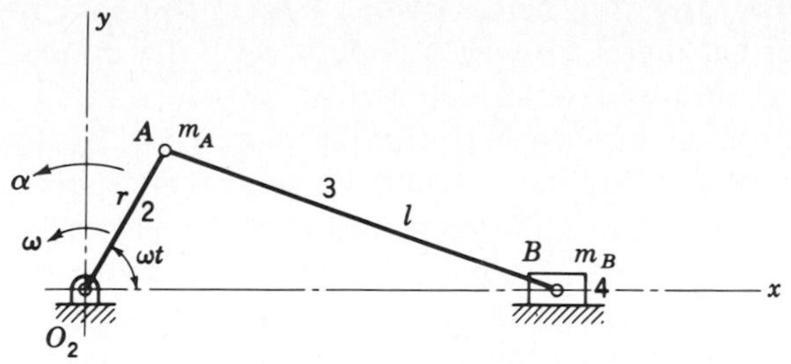

Figure 14-16

crankpin relative to the origin O_2 is

$$\mathbf{R}_A = r \cos \omega t \; \hat{\mathbf{i}} + r \sin \omega t \; \hat{\mathbf{j}} \tag{a}$$

Differentiating twice to obtain the acceleration gives

$$\mathbf{A}_A = (-r\alpha \sin \omega t - r\omega^2 \cos \omega t)\hat{\mathbf{i}} + (r\alpha \cos \omega t - r\omega^2 \sin \omega t)\hat{\mathbf{j}} \tag{14-22}$$

The inertia force of the rotating parts is then

$$-m_A\mathbf{A}_A = m_A r(\alpha \sin \omega t + \omega^2 \cos \omega t)\hat{\mathbf{i}} + m_A r(-\alpha \cos \omega t + \omega^2 \sin \omega t)\hat{\mathbf{j}} \tag{14-23}$$

Since the analysis is usually made at constant angular velocity ($\alpha = 0$), Eq. (14-23) reduces to

$$-m_A\mathbf{A}_A = m_A r\omega^2 \cos \omega t \; \hat{\mathbf{i}} + m_A r\omega^2 \sin \omega t \; \hat{\mathbf{j}} \tag{14-24}$$

The acceleration of the piston has already been determined [Eq. (14-11)] and is repeated here for convenience in a slightly different form:

$$\mathbf{A}_B = \left[-r\alpha\left(\sin \omega t + \frac{r}{2l} \sin 2\omega t\right) - r\omega^2\left(\cos \omega t + \frac{r}{l} \cos 2\omega t\right)\right]\hat{\mathbf{i}} \tag{14-25}$$

The inertia force of the reciprocating parts is therefore

$$-m_B\mathbf{A}_B = \left[m_B r\alpha\left(\sin \omega t + \frac{r}{2l} \sin 2\omega t\right) + m_B r\omega^2\left(\cos \omega t + \frac{r}{l} \cos 2\omega t\right)\right]\hat{\mathbf{i}} \tag{14-26}$$

or, for constant angular velocity,

$$-m_B\mathbf{A}_B = m_B r\omega^2\left(\cos \omega t + \frac{r}{l} \cos 2\omega t\right)\hat{\mathbf{i}} \tag{14-27}$$

Adding Eqs. (14-24) and (14-27) gives the total inertia force for all the moving parts. The components in the x and y directions are

$$F^x = (m_A + m_B)r\omega^2 \cos \omega t + \left(m_B \frac{r}{l}\right) r\omega^2 \cos 2\omega t \tag{14-28}$$

$$F^y = m_A r\omega^2 \sin \omega t \tag{14-29}$$

It is customary to refer to the portion of the force occurring at the circular frequency ω rad/s as the *primary inertia force* and the portion occurring at

2ω rad/s as the *secondary inertia force.* We note that the vertical component has only a primary part and that it therefore varies directly with the crankshaft speed. On the other hand, the horizontal component, which is in the direction of the cylinder axis, has a primary part varying directly with the crankshaft speed and a secondary part moving at twice the crankshaft speed.

We proceed now to a determination of the inertia torque. As shown in Fig. 14-17, the inertia force due to the mass at A has no moment arm about O_2 and therefore produces no torque. Consequently, we need consider only the inertia force given by Eq. (14-27) due to the reciprocating part of the mass. From the force polygon of Fig. 14-17 the inertia torque exerted by the engine on the crankshaft is

$$\mathbf{T}_{21}'' = -(-m_B\ddot{x}\tan\phi)x\hat{\mathbf{k}} \tag{b}$$

Expressions for x, $\ddot{x}$, and $\tan\phi$ appear in Sec. 14-4. Making appropriate substitutions for these yields the following for the torque:

$$\mathbf{T}_{21}'' = -m_B r\omega^2\left(\cos\omega t + \frac{r}{l}\cos 2\omega t\right)$$
$$\left[l - \frac{r^2}{4l} + r\left(\cos\omega t + \frac{r}{4l}\cos 2\omega t\right)\right]\frac{r}{l}\sin\omega t\left(1 + \frac{r^2}{2l^2}\sin^2\omega t\right)\hat{\mathbf{k}} \tag{c}$$

Terms which are proportional to the second or higher powers of r/l can be neglected in performing the indicated multiplication. Equation (c) then can be written

$$\mathbf{T}_{21}'' = -m_B r^2\omega^2\sin\omega t\left(\frac{r}{2l} + \cos\omega t + \frac{3r}{2l}\cos 2\omega t\right)\hat{\mathbf{k}} \tag{d}$$

Then, using the identities

$$2\sin\omega t\cos 2\omega t = \sin 3\omega t - \sin\omega t \tag{e}$$

and

$$2\sin\omega t\cos\omega t = \sin 2\omega t \tag{f}$$

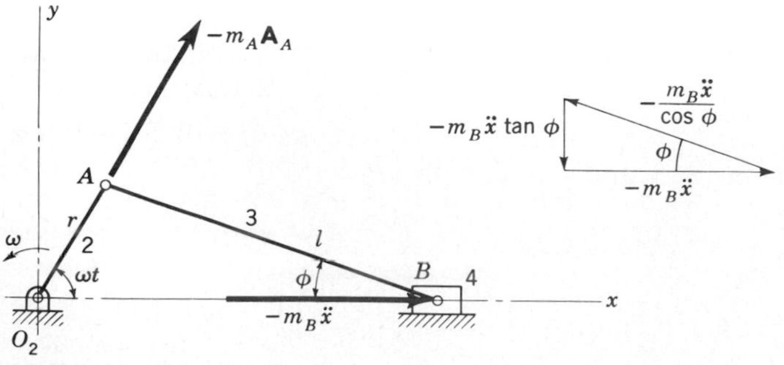

Figure 14-17

results in an equation having only sine terms, and Eq. (*d*) finally becomes

$$\mathbf{T}''_{21} = \frac{m_b}{2} r^2\omega^2\left(\frac{r}{2l}\sin \omega t - \sin 2\omega t - \frac{3r}{2l}\sin 3\omega t\right)\hat{\mathbf{k}} \qquad (14\text{-}30)$$

This is the inertia torque exerted by the engine on the shaft in the positive direction. A clockwise or negative inertia torque of the same magnitude is, of course, exerted on the frame of the engine.

The assumed distribution of the connecting-rod mass results in a moment of inertia which is greater than the true value. Consequently, the torque given by Eq. (14-30) is not the exact value. In addition, terms proportional to the second- or higher-order powers of r/l were dropped in simplifying Eq. (*c*). These two errors are about the same magnitude and are quite small for ordinary connecting rods having r/l ratios near $\frac{1}{4}$.

14-7 BEARING LOADS IN THE SINGLE-CYLINDER ENGINE

The designer of a reciprocating engine must know the values of the forces acting upon the bearings and how these forces vary in a cycle of operation. This is necessary to proportion and select the bearings properly, and it is also needed for the design of other engine parts. This section is an investigation of the force exerted by the piston against the cylinder wall and the forces acting against the piston pin and against the crankpin. Main bearings forces will be investigated in a later section because they depend upon the action of all the cylinders of the engine.

The resultant bearing loads are made up of the following components:

1. The gas-force components, designated by a single prime
2. Inertia force due to the weight of the piston assembly, designated by a double prime
3. Inertia force of that part of the connecting rod assigned to the piston-pin end, triple-primed
4. Connecting-rod inertia force at the crankpin end, quadruple-primed

Equations for the gas-force components have been determined in Sec. 14-4, and reference will be made to them in finding the total bearing loads.

Figure 14-18 is a graphical analysis of the forces in the engine mechanism with a zero gas force and subjected to an inertia force resulting only from the weight of the piston assembly. Figure 14-18*a* shows the position of the mechanism selected for analysis, and the inertia force $-m_4\mathbf{A}_B$ is shown acting upon the piston. In Fig. 14-18*b* the free-body diagram of the piston forces is shown together with the force polygon from which they were obtained. Figure 14-18*c* to *e* illustrates, respectively, the free-body diagrams of forces acting upon the connecting rod, crank, and frame.

In Fig. 14-18*e* notice that the torque $\mathbf{T}''_{21}$ balances the force couple formed

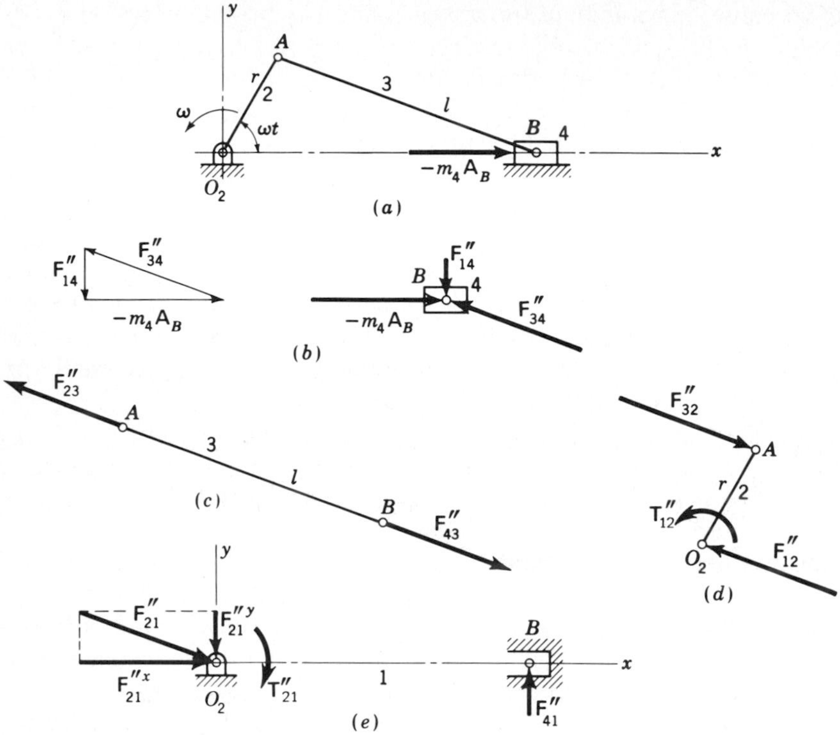

Figure 14-18 Analysis of the forces in the engine mechanism when only the inertia force due to the weight of the piston assembly is considered.

by the forces $\mathbf{F}''_{41}$ and $\mathbf{F}''^{y}_{21}$. But the force $\mathbf{F}''^{x}_{21}$ at the crank center remains unopposed by any other force. This is a very important observation which we shall reserve for discussion in a separate section.

The following forces are of interest to us:

1. The force $\mathbf{F}''_{41}$ of the piston against the cylinder wall
2. The force $\mathbf{F}''_{34}$ of the connecting rod against the piston pin
3. The force $\mathbf{F}''_{32}$ of the connecting rod against the crankpin
4. The force $\mathbf{F}''_{12}$ of the shaft against the crank

By methods similar to those used earlier in this chapter the analytical expressions are found to be

$$\mathbf{F}''_{41} = -m_4\ddot{x} \tan \phi \, \hat{\mathbf{j}} \tag{14-31}$$

$$\mathbf{F}''_{34} = m_4\ddot{x}\underline{\mathbf{i}} - m_4\ddot{x} \tan \phi \, \hat{\mathbf{j}} \tag{14-32}$$

$$\mathbf{F}''_{32} = -\mathbf{F}''_{34} \tag{14-33}$$

$$\mathbf{F}''_{12} = -\mathbf{F}''_{32} = \mathbf{F}''_{34} \tag{14-34}$$

where $\ddot{x}$ is the acceleration of the piston as given by Eq. (14-11) and m_4 is the mass of the piston assembly. The quantity tan ϕ can be evaluated in terms of the crank angle with the use of Eq. (14-13).

In Fig. 14-19 we neglect all forces except those which result because of that part of the mass of the connecting rod which is assumed to be located at the piston-pin center. Thus Fig. 14-19*b* is a free-body diagram of the connecting rod showing the inertia force $-m_{3B}\mathbf{A}_B$ acting at the piston-pin end.

We now note that it is incorrect to add m_{3B} and m_4 together and then to compute a resultant inertia force in finding the bearing loads, although such a procedure would seem to be simpler. The reason for this is that m_4 is the mass of the piston assembly and the corresponding inertia force acts on the piston side of the wrist pin. But m_{3B} is part of the connecting-rod mass, and hence its inertia force acts on the connecting-rod side of the wrist pin. Thus, adding the two will yield correct results for the crankpin load and the force of the piston against the cylinder wall but will give *incorrect* results for the piston-pin load.

The forces on the piston pin, the crank, and the frame are illustrated in Fig. 14-19*c*, *d*, and *e*, respectively. The equations for these forces for a crank having uniform angular velocity are found to be

$$\mathbf{F}_{41}''' = -m_{3B}\ddot{x} \tan \phi \, \hat{\mathbf{j}} \tag{14-35}$$

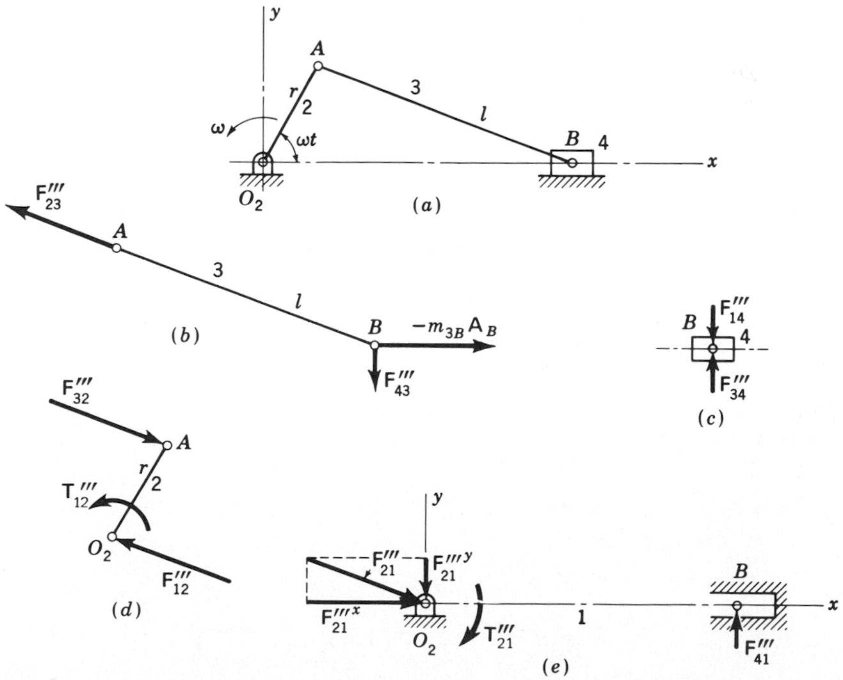

Figure 14-19 Graphical analysis of forces resulting solely from the mass of the connecting rod, assumed to be concentrated at the wrist-pin end.

$$\mathbf{F}'''_{34} = \mathbf{F}'''_{41} \tag{14-36}$$

$$\mathbf{F}'''_{32} = -m_{3B}\ddot{x}\hat{\mathbf{i}} + m_{3B}\ddot{x}\tan\phi\,\hat{\mathbf{j}} \tag{14-37}$$

$$\mathbf{F}'''_{12} = -\mathbf{F}'''_{32} \tag{14-38}$$

Figure 14-20 illustrates the forces which result because of that part of the connecting-rod mass which is concentrated at the crankpin end. While a counterweight attached to the crank balances the reaction at O_2, it cannot make $\mathbf{F}''''_{32}$ zero. Thus the crankpin force exists no matter whether the rotating mass of the connecting rod is balanced or not. This force is

$$\mathbf{F}''''_{32} = m_{3A}r\omega^2(\cos\omega t\,\hat{\mathbf{i}} + \sin\omega t\,\hat{\mathbf{j}}) \tag{14-39}$$

The last step is to sum these expressions to obtain the resultant bearing loads. The total force of the piston against the cylinder wall, for example, is found by summing Eqs. (14-12), (14-31), and (14-35), with due regard for subscripts and signs. When simplified, the answer is

$$\mathbf{F}_{41} = \mathbf{F}'_{41} + \mathbf{F}''_{41} + \mathbf{F}'''_{41} = -[(m_{3B} + m_4)\ddot{x} + P]\tan\phi\,\hat{\mathbf{j}} \tag{14-40}$$

The forces on the piston pin, the crankpin, and the crankshaft are found in a similar manner and are

$$\mathbf{F}_{34} = (m_4\ddot{x} + P)\hat{\mathbf{i}} - [(m_{3B} + m_4)\ddot{x} + P]\tan\phi\,\hat{\mathbf{j}} \tag{14-41}$$

$$\begin{aligned}\mathbf{F}_{32} = {} & [m_{3A}r\omega^2\cos\omega t - (m_{3B} + m_4)\ddot{x} - P]\hat{\mathbf{i}} \\ & + \{m_{3A}r\omega^2\sin\omega t + [(m_{3B} + m_4)\ddot{x} + P]\tan\phi\}\hat{\mathbf{j}}\end{aligned} \tag{14-42}$$

$$\mathbf{F}_{21} = \mathbf{F}_{32} \tag{14-43}$$

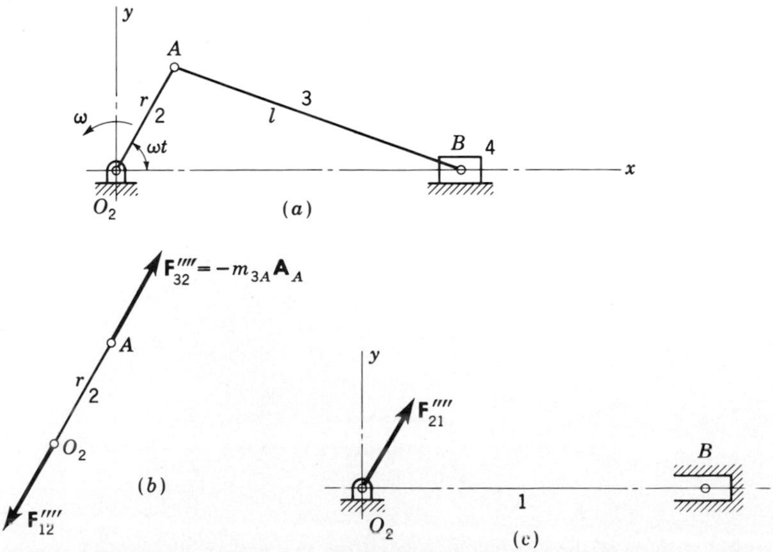

Figure 14-20 Graphical analysis of forces resulting solely from the mass of the connecting rod, assumed to be concentrated at the crankpin end.

14-8 CRANKSHAFT TORQUE

The torque delivered by the crankshaft to the load is called the *crankshaft torque* and it is the negative of the moment of the couple formed by the forces $\mathbf{F}_{41}$ and $\mathbf{F}_{21}^y$. Therefore, it is obtained from the equation

$$\mathbf{T}_{21} = -F_{41}x\hat{\mathbf{k}} = [(m_{3B} + m_4)\ddot{x} + P]x \tan \phi \,\hat{\mathbf{k}} \tag{14-44}$$

14-9 ENGINE SHAKING FORCES

The inertia force due to the reciprocating masses is shown acting in the positive direction in Fig. 14-21*a*. In Fig. 14-21*b* the forces acting upon the engine block due to these inertia forces are shown. The resultant forces are $\mathbf{F}_{21}$, the force exerted by the crankshaft on the main bearings, and a positive couple formed by the forces $\mathbf{F}_{41}$ and $\mathbf{F}_{21}^y$. The force $\mathbf{F}_{21}^x = -m_B\mathbf{A}_B$ is frequently termed a *shaking force*, and the couple $T = xF_{41}$ a *shaking couple.* As indicated by Eqs. (14-27) and (14-30), the magnitude and direction of this force and couple change with ωt; consequently, the shaking force induces linear vibration of the block in the x direction, and the shaking couple a torsional vibration of the block about the crank center.

A graphical representation of the inertia force is possible if Eq. (14-27) is rearranged as

$$F = m_B r\omega^2 \cos \omega t + m_B r\omega^2 \frac{r}{l} \cos 2\omega t \tag{14-45}$$

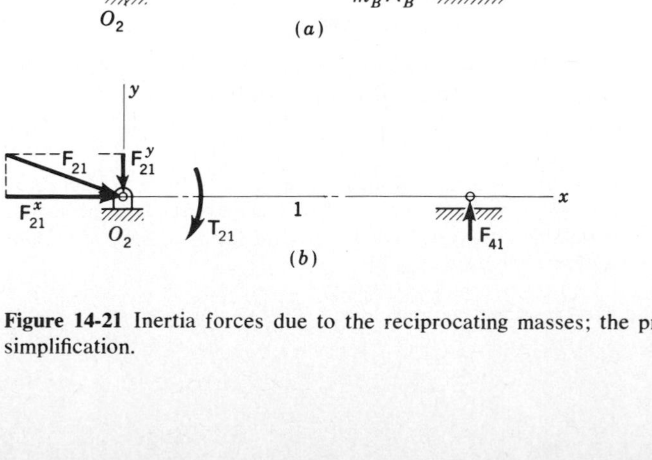

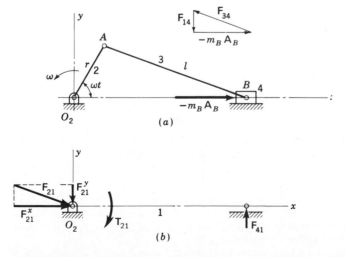

Figure 14-21 Inertia forces due to the reciprocating masses; the primes have been omitted for simplification.

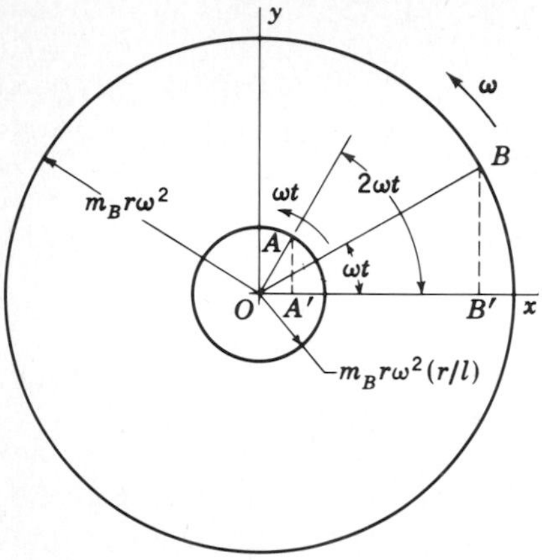

Figure 14-22 Circle diagram for finding inertia forces. Total inertia force is $OA' + OB'$.

where $F = F_{21}^x$ for simplicity of notation. The first term of Eq. (14-45) is represented by the x projection of a vector $m_Br\omega^2$ in length rotating at ω rad/s. This is the primary part of the inertia force. The second term is similarly represented by the x projection of a vector $m_Br\omega^2(r/l)$ in length rotating at 2ω rad/s; this is the secondary part. Such a diagram is shown in Fig. 14-22 for $r/l = \frac{1}{4}$. The total inertia or shaking force is the algebraic sum of the horizontal projections of the two vectors.

14-10 MACHINE-COMPUTATION HINTS

This section contains suggestions for using computers and programmable calculators in solving the dynamics of engine mechanisms. Many of the ideas, however, will be useful for readers using nonprogrammable machines as well as for checking purposes.

Indicator diagrams It would be very convenient if a subprogram for computing the gas forces could be devised and the results used directly in a main program to compute all the resultant bearing forces and crankshaft torques. Unfortunately the theoretical indicator diagram must be manipulated by hand in order to obtain a reasonable approximation to the experimental data. This manipulation can be done graphically or with a computer having a graphical display. The procedure is illustrated by the following example.

Example 14-1 Determine the pressure-vs.-piston-displacement relation for a six cylinder engine having a displacement of 140 in^3, a compression ratio of 8, and a brake horsepower of 57 at 2400 rpm. Use a mechanical efficiency of 75 percent, a card factor of 0.85, a suction pressure of 14.7 psi, and a polytropic exponent of 1.30.

SOLUTION Rearranging Eq. (14-2), we find the brake mean effective pressure as follows:

$$p_b = \frac{(33\,000)(12)(\text{bhp})}{lan} = \frac{(33\,000)(12)(57/6)}{(140/6)(2400/2)} = 135 \text{ psi}$$

Then, from Eq. (14-3), the indicated mean effective pressure is

$$p_i = \frac{p_b}{e_m} = \frac{135}{0.75} = 180 \text{ psi}$$

We must now determine p_4 on the theoretical diagram of Fig. 14-12. Employing Eq. (14-7), we find

$$p_4 = (k-1)\frac{r-1}{r^k - r}\frac{p_i}{f_c} + p_1$$

$$= (1.3-1)\frac{8-1}{8^{1.3}-8}\frac{180}{0.85} + 14.7 = 78.2 \text{ psi}$$

The volume difference $v_1 - v_2$ in Fig. 14-12 is the volume swept out by the piston. Therefore

$$v_1 - v_2 = la = \frac{140}{6} = 23.3 \text{ in}^3$$

Then, from Eq. (*b*) of Sec. 14-2, we have

$$v_1 = \frac{r(v_1 - v_2)}{r-1} = \frac{8(23.3)}{8-1} = 26.6 \text{ in}^3$$

Therefore $\quad v_2 = 26.6 - 23.3 = 3.3 \text{ in}^3$

Then the percentage clearance C is

$$C = \frac{3.3(100)}{23.3} = 14.2\%$$

Expressing volumes as percentages of the displacement volume enables us to write Eq. (14-1) in the form

$$p_x(X+C)^k = p_1(100+C)^k$$

where X is the percentage of piston travel measured from the head end of the stroke. Thus the formula

$$p_{xc} = p_1\left(\frac{100+C}{X+C}\right)^k = 14.7\left(\frac{100+14.2}{X+14.2}\right)^{1.3} \tag{1}$$

is used to compute the pressure during the compression stroke for any piston position between $X = 0$ and $X = 100$ percent. For the expansion stroke Eq. (14-1) becomes

$$p_{xe} = p_4\left(\frac{100+C}{X+C}\right)^k = 78.2\left(\frac{100+14.2}{X+14.2}\right)^{1.3} \tag{2}$$

Equations (1) and (2) are easy to program for machine computation. The results should be displayed and recorded, or printed, for graphical use. Alternately the results can be displayed on the CRT for hand manipulation.

Figure 14-23 shows the plotted results of the computation using $\Delta X = 5$ percent increments. Note particularly how the results have been rounded to obtain a smooth indicator diagram. This rounding will, of course, produce results that will not be exactly duplicated in subsequent trials. The greatest differences will occur in the vicinity of point *B*.

Force analysis In a computer analysis the values of the pressure will be read from a diagram like Fig. 14-23. Since most analysts will want to tabulate these data, a table should be constructed with the first column containing values of

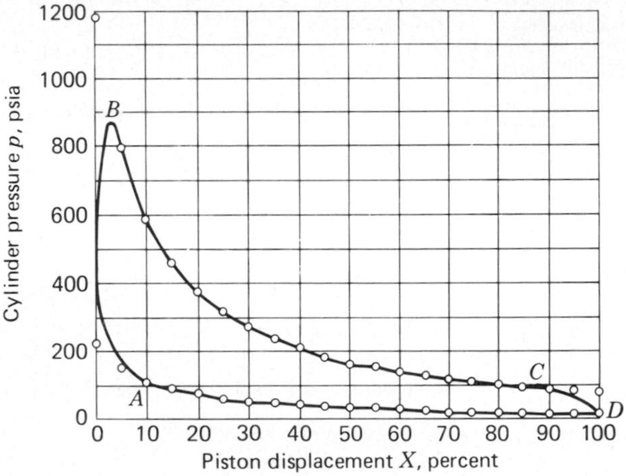

Figure 14-23 Circled points are computer results. Diagram rounded by hand from A to B and from C to D. Point B is about 75 percent of maximum computed pressure at the beginning of the expansion stroke.

the crank angle ωt. For a four-cycle engine values of this angle should be entered from 0 to 720°.

Values of x corresponding to each ωt should be obtained from Eq. (14-9). Then the corresponding piston displacement X in percent is obtained from the equation

$$X = \frac{r + l - x}{2r}(100) \tag{14-46}$$

Some care must be taken in tabulating X and the corresponding pressures. Then the gas forces corresponding to each value of ωt can be computed using the piston area.

The balance of the analysis is perfectly straightforward; use Eqs. (14-11), (14-13), and (14-40) to (14-44) in that order.

PROBLEMS

14-1 A one-cylinder four-cycle engine has a compression ratio of 7.6 and develops 3 bhp at 3000 rpm. Crank length is 0.875 in with a 2.375-in bore. Develop and plot a rounded indicator diagram using a card factor of 0.90, a mechanical efficiency of 72 percent, a suction pressure of 14.7 psi, and a polytropic exponent of 1.30.

14-2 Construct a rounded indicator diagram for a four-cylinder four-cycle gasoline engine having a 3.375-in bore, a 3.5-in stroke, and a compression ratio of 6.25. The operating conditions to be used are 30 hp at 1900 rpm. Use a mechanical efficiency of 72 percent, a card factor of 0.90, and a polytropic exponent of 1.30.

14-3 Construct an indicator diagram for a V6 four-cycle engine having a 100-mm bore, a 90-mm stroke, and a compression ratio of 8.40. This engine develops 150 kW at 4400 rpm. Use a mechanical efficiency of 75 percent, a card factor of 0.88, and 1.30 for the polytropic exponent.

14-4 A single cylinder two-cycle gasoline engine develops 30 kW at 4500 rpm. The engine has an 80-mm bore, a stroke of 70 mm, and a compression ratio of 7.0. Develop a rounded indicator diagram for this engine using a card factor of 0.90, a mechanical efficiency of 65 percent, and a polytropic exponent of 1.30. Use 100 kPa for the suction pressure.

14-5 The engine of Prob. 14-1 has a connecting rod $3\frac{1}{8}$ in long and a weight of 0.214 lb, with the mass center 0.40 in from the crankpin end. Piston weight is 0.393 lb. Find the bearing reactions and the crankshaft torque during the expansion stroke corresponding to a piston displacement of $X = 30$ percent ($\omega t = 60°$). See list of answers for p_e.

14-6 Repeat Prob. 14-5 but do the computations for the compression cycle ($\omega t = 660°$).

14-7 Make a complete force analysis of the engine of Prob. 14-5. Plot a graph of the crankshaft torque vs. crank angle for 720° of crank rotation.

14-8 The engine of Prob. 14-3 uses a connecting rod 350 mm long. The masses are $m_{3A} = 0.80$ kg, $m_{3B} = 0.38$ kg, and $m_4 = 1.64$ kg. Find all the bearing reactions and the crankshaft torque for one cylinder of the engine during the expansion stroke at a piston displacement of $X = 30$ percent ($\omega t = 63.2°$). The pressure should be obtained from the indicator diagram in the list of answers.

14-9 Repeat Prob. 14-8 but do the computations for the same position in the compression cycle ($\omega t = 656.8°$).

14-10 Additional data for the engine of Prob. 14-4 are $l_3 = 110$ mm, $AG_3 = 15$ mm, $m_4 = 0.24$ kg, and $m_3 = 0.13$ kg. Make a complete force analysis of the engine and plot a graph of the crankshaft torque vs. crank angle for 360° of crank rotation.

14-11 The four-cycle engine of Example 14-1 has a stroke of 2.60 in and a connecting rod length of 7.20 in. The weight of the rod is 0.850 lb, and the center of mass is 1.66 in from the crankpin. The piston assembly weighs 1.27 lb. Make a complete force analysis for one cylinder of this engine for 720° of crank rotation. Use 16 psi for the exhaust pressure and 10 psi for the suction pressure. Plot a graph to show the variation of crankshaft torque with crank angle. Use Fig. 14-23 for pressures.

CHAPTER

FIFTEEN

BALANCING

Balancing is the technique of correcting or eliminating unwanted inertia forces and moments. In previous chapters we have seen that the frame forces can vary significantly during a complete cycle of operation. Such forces can cause vibrations which at times may reach dangerous amplitudes. Even if not dangerous, vibrations increase the component stresses and subject bearings to repeated loads which cause parts to fail prematurely by fatigue. Thus it is not sufficient in the design of machinery merely to avoid operation near the critical speeds; we must also eliminate, or at least reduce, the inertia forces which produce these vibrations in the first place.

Production tolerances used in the manufacture of machinery are adjusted as closely as possible without running up the cost of manufacture prohibitively. In general it is more economical to produce parts which are not quite true and then to subject them to a balancing procedure than it is to produce such perfect parts that no correction is needed. Because of this each part produced is an individual case in that no two parts can normally be expected to require the same corrective measures. Thus determining the unbalance and the application of corrections is the principal problem in the study of balancing.

15-1 STATIC UNBALANCE

The arrangement shown in Fig. 15-1*a* consists of a disk-and-shaft combination resting on hard rigid rails so that the shaft, which is assumed to be perfectly straight, can roll without friction. A reference system *xyz* is

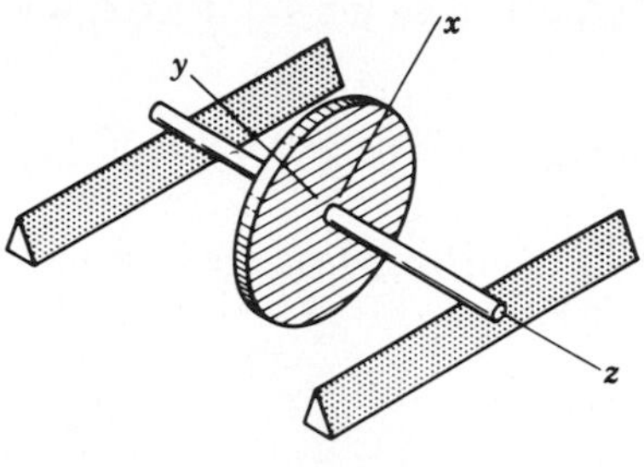

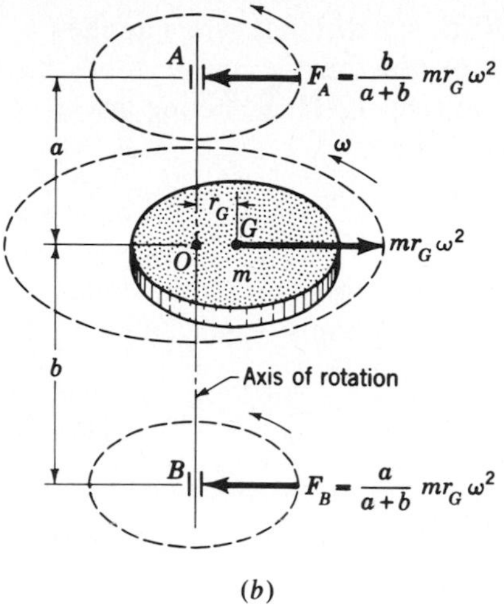

Figure 15-1

attached to the disk and moves with it. Simple experiments to determine whether the disk is statically unbalanced can be conducted as follows. Roll the disk gently by hand and permit it to coast until it comes to rest. Then mark with chalk the lowest point of the periphery of the disk. Repeat four or five times. If the chalk marks are scattered at different places around the periphery, the disk is in static balance. If all the chalk marks are coincident, the disk is statically unbalanced, which means that the axis of the shaft and the center of mass of the disk are not coincident. The position of the chalk marks with respect to the *xy* system indicates the angular location of unbalance but *not* the amount.

It is unlikely that any of the marks will be located 180° from the remaining ones even though it is theoretically possible to obtain static equilibrium with the unbalance above the shaft axis.

If static unbalance is found to exist, it can be corrected by drilling out material at the chalk mark or by adding mass to the periphery 180° from the mark. Since the amount of unbalance is unknown, these corrections must be made by trial and error.

15-2 EQUATION OF MOTION

If an unbalanced disk and shaft is mounted in bearings and caused to rotate, the centrifugal force $mr_G\omega^2$ exists as shown in Fig. 15-1*b*. This force acting upon the shaft produces the rotating bearing reactions shown in the figure.

In order to determine the equation of motion of the system, we specify m as the total mass and m_u as the unbalanced mass. Also, let k be the shaft stiffness, a number that describes the magnitude of a force necessary to bend the shaft a unit distance when applied at O. Thus, k has units of pounds force per inch or newtons per metre. Let c be the coefficient of viscous damping as defined in Sec. 12-11. Selecting any x coordinate normal to the shaft axis, we can now write

$$\sum F_O = -kx - c\dot{x} - m\ddot{x} + m_u r_G \omega^2 \cos \omega t = 0 \tag{a}$$

The solution to this differential equation can be found in any text dealing with differential equations or mechanical vibrations. It is

$$x = \frac{m_u r_G \omega^2 \cos(\omega t - \phi)}{\sqrt{(k - m\omega^2)^2 + c^2\omega^2}} \tag{b}$$

where ϕ is the angle between the force $m_u r_G \omega^2$ and the amplitude X of the shaft vibration; thus ϕ is the *phase angle.* Its value is

$$\phi = \tan^{-1} \frac{c\omega}{k - m\omega^2} \tag{c}$$

Certain simplifications can be made with Eq. (b) to clarify its meaning.

First consider the term $k - m\omega^2$ in the denominator of Eq. (b). If this term were zero, the amplitude of x would be very large because it would be limited only by the damping constant c, which is usually very small. The value of ω that makes the term $k - m\omega^2$ zero is called the *natural angular velocity,* the *critical speed,* and also the *natural circular frequency.* It is designated as ω_n and is seen to be

$$\omega_n = \sqrt{\frac{k}{m}} \tag{15-1}$$

In the study of free or unforced vibrations it is found that a certain value of the viscous factor c will result in no vibration at all. This special value is called the *critical coefficient of viscous damping* and is given by the equation

$$c_c = 2m\omega_n \tag{15-2}$$

The *damping ratio* ζ is the ratio of the actual to the critical and is

$$\zeta = \frac{c}{c_c} = \frac{c}{2m\omega_n} \tag{15-3}$$

For most machine systems in which damping is not deliberately introduced, ζ will be in the approximate range $0.015 \le \zeta \le 0.120$.

Next, note that Eq. (b) can be expressed in the form

$$x = X \cos(\omega t - \phi) \tag{d}$$

If we now divide the numerator and denominator of the amplitude X of Eq. (b) by k, designate the eccentricity as $e = r_G$, and introduce Eqs. (15-1) and (15-3), we obtain the ratio

$$\frac{mX}{m_u e} = \frac{(\omega/\omega_n)^2}{\sqrt{(1-\omega^2/\omega_n^2)^2 + (2\zeta\omega/\omega_n)^2}} \tag{15-4}$$

This is the equation for the amplitude ratio of the vibration of a rotating disk-and-shaft combination. If we neglect damping, let $m = m_u$, and replace e with r_G again, we obtain

$$X = r_G \frac{(\omega/\omega_n)^2}{1-(\omega/\omega_n)^2} \tag{15-5}$$

where r_G is the eccentricity and X is the amplitude of the vibration corresponding to any frequency ratio ω/ω_n. Now if, in Fig. 15-1*b*, we designate O as the center of the shaft at the disk and G as the mass center of the disk, we can draw some interesting conclusions by plotting Eq. (15-5). This is done in Fig. 15-2, where the amplitude is plotted on the vertical axis and the frequency ratio along the abscissa. The natural frequency is ω_n, which corresponds to the critical speed, while ω is the actual speed of the shaft. When rotation is just beginning, ω is much less than ω_n and the graph shows that the amplitude of the vibration is very small. As the shaft speed increases, the amplitude also increases and becomes infinite at the critical speed. As the shaft goes through the critical, the amplitude changes over to a negative value and decreases as the shaft speed increases. The graph shows that the amplitude never returns to zero no matter how much the shaft speed is increased but reaches a limiting value of $-r_G$. Note, in this range, that the disk is rotating about its own center of gravity, which is then coincident with the bearing centerline.

The preceding discussion demonstrates that statically unbalanced rotating systems produce undesirable vibrations and rotating bearing reactions. Using static balancing equipment, the eccentricity r_G can be reduced, but it is

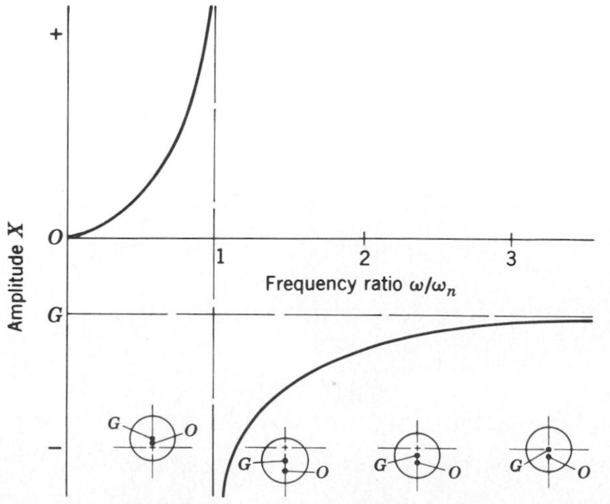

Figure 15-2 The small figures below the graph indicate the relative position of three points for various frequency ratios. The mass center of the disk is at G, the center of the shaft at O, and the axis of rotation at the intersection of the centerlines. Thus, this figure shows both amplitude and phase relationships.

impossible to make it zero. Therefore, no matter how small r_G is made, trouble can always be expected whenever $\omega = \omega_n$. When the operating frequency is higher than the natural frequency, the machine should be designed to pass through the natural frequency as rapidly as possible in order to prevent dangerous vibrations from building up.

15-3 STATIC BALANCING MACHINES

The purpose of a balancing machine is first to indicate whether a part is in balance. If it is out of balance, the machine must measure the unbalance by indicating its *magnitude* and *location.*

Static balancing machines are used only for parts whose axial dimensions are small, such as gears, fans, and impellers, and the machines are often called *single-plane balancers* because the mass must practically lie in a single plane. In the sections to follow we discuss balancing in several planes, but it is important to note here that if several wheels are to be mounted upon a shaft which is to rotate, the parts should be individually statically balanced before mounting. While it is possible to balance the assembly in two planes after the

Figure 15-3 A helicopter-rotor assembly balancer. (*Micro-Poise Engineering and Sales Company, Detroit, Michigan.*)

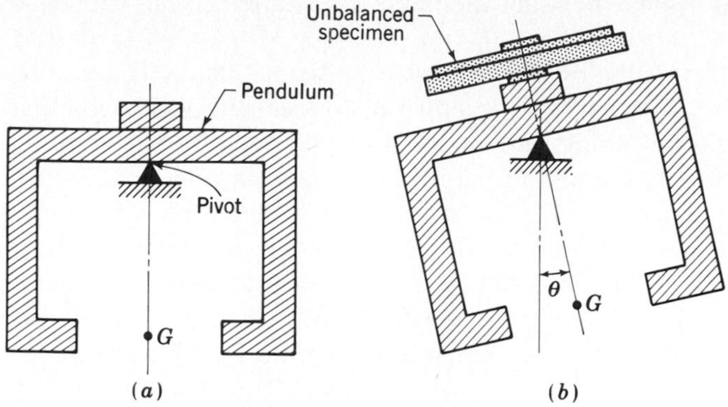

Figure 15-4 Operation of a static balancing machine.

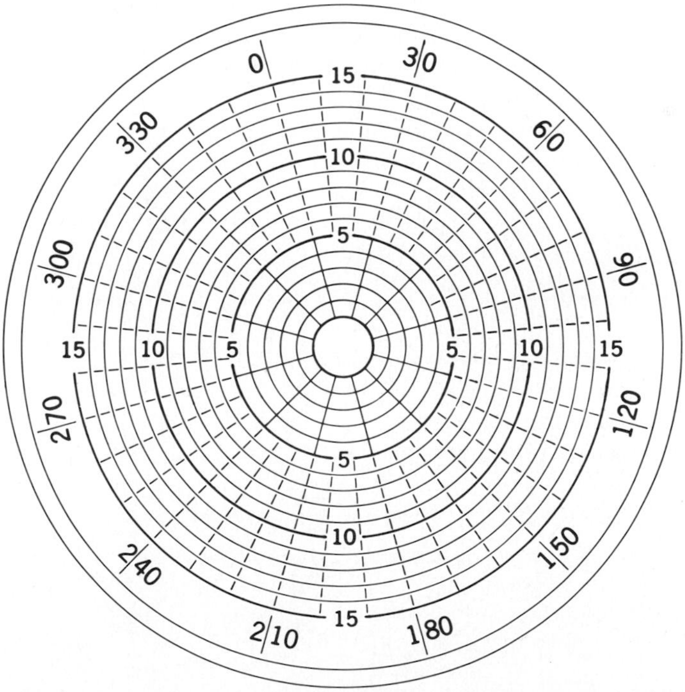

Figure 15-5 Drawing of the universal level used on the Micro-Poise balancer. The numbers on the periphery are degrees; the radial distances are calibrated in units proportional to ounce-inches. The position of the bubble indicates both the location and the magnitude of the unbalance. (*Micro-Poise Engineering and Sales Company, Detroit, Michigan.*)

parts are mounted, additional bending moments inevitably come into existence when this is done.

Static balancing is essentially a weighing process in which the part is acted upon by either a gravity force or a centrifugal force. We have seen that the disk and shaft of the preceding section could be balanced by placing it on two parallel rails, rocking it, and permitting it to seek equilibrium. In this case the location of the unbalance is found through the aid of the force of gravity. Another method of balancing the disk would be to rotate it at a predetermined speed. Then the bearing reactions could be measured and their magnitudes used to indicate the amount of unbalance. Since the part is rotating while the measurements are taken, a stroboscope is used to indicate the location of the required correction.

When machine parts are manufactured in large quantities, a balancer is required which will measure both the amount and location of the unbalance and give the correction directly and quickly. Time can also be saved if it is not necessary to rotate the part. Such a balancing machine is shown in Fig. 15-3. This machine is essentially a pendulum which can tilt in any direction, as illustrated by the schematic drawing of Fig. 15-4*a*. When an unbalanced specimen is mounted on the platform of the machine, the pendulum tilts. The direction of the tilt gives the location of the unbalance, while the angle θ (Fig. 15-4*b*) indicates the magnitude of the unbalance. Some damping is employed to eliminate oscillations of the pendulum. Figure 15-5 shows a universal level which is mounted on the platform of the balancer. A bubble, shown at the center, moves and shows both the location and the magnitude of the correction.

15-4 DYNAMIC UNBALANCE

Figure 15-6 shows a long rotor which is to be mounted in bearings at *A* and *B*. We might suppose that two equal masses m_1 and m_2 are placed at opposite ends of the rotor and at equal distances r_1 and r_2 from the axis of rotation. Since the masses are equal and on opposite sides of the rotational axis, the

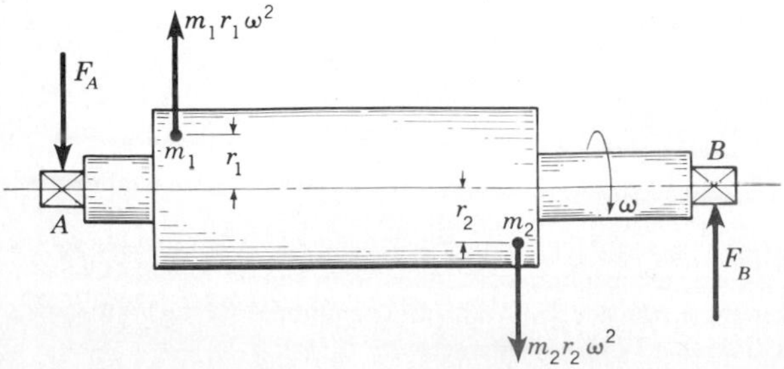

Figure 15-6 The rotor is statically balanced if $m_1 = m_2$ and $r_1 = r_2$ but dynamically unbalanced.

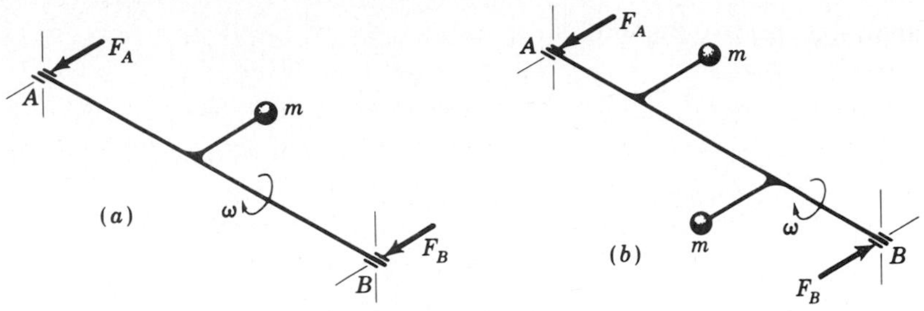

Figure 15-7 (*a*) Static unbalance; when the shaft rotates, both bearing reactions are in the same plane and in the same direction. (*b*) Dynamic unbalance; when the shaft rotates, the unbalance creates a couple tending to turn the shaft end over end. The shaft is in equilibrium because of the opposite couple formed by the bearing reactions. Note that the bearing reactions are still in the same plane but opposite in direction.

rotor can be placed on rails as described earlier to show that it is statically balanced in all angular positions.

If the rotor of Fig. 15-6 is placed in bearings and caused to rotate at an angular velocity ω rad/s, the centrifugal forces $m_1 r_1 \omega^2$ and $m_2 r_2 \omega^2$ act, respectively, at m_1 and m_2 on the rotor ends. These centrifugal forces produce the unequal bearing reactions $\mathbf{F}_A$ and $\mathbf{F}_B$, and the entire system of forces rotates with the rotor at the angular velocity ω. Thus a part may be statically balanced and at the same time dynamically unbalanced (Fig. 15-7).

In the general case distribution of the mass along the axis of the part depends upon the configuration of the part, but errors occur in machining and also in casting and forging. Other errors or unbalance may be caused by improper boring, by keys, and by assembly. It is the designer's responsibility to design so that a line joining all mass centers will be a straight line coinciding with the axis of rotation. However, perfect parts and perfect assembly are seldom attained, and consequently a line from one end of the part to the other, joining all mass centers, will usually be a space curve which may occasionally cross or coincide with the axis of rotation. An unbalanced part, therefore, will usually be out of balance both statically and dynamically. This is the most general kind of unbalance, and if the part is supported by two bearings, one can then expect the magnitudes as well as the directions of these rotating bearing reactions to be different.

15-5 ANALYSIS OF UNBALANCE

In this section we show how to analyze any unbalanced rotating system and determine the proper corrections using graphical methods, vector methods, and computer or calculator programming.

Graphical analysis The two equations

$$\sum \mathbf{F} = 0 \qquad \text{and} \qquad \sum \mathbf{M} = 0 \tag{a}$$

are used to determine the amount and location of the corrections. We begin by noting that the centrifugal force is proportional to the product mr of a rotating eccentric mass. Thus vector quantities, proportional to the centrifugal force of each of the three masses $m_1\mathbf{R}_1$, $m_2\mathbf{R}_2$, and $m_3\mathbf{R}_3$ of Fig. 15-8*a*, will act in radial directions as shown. The first of Eqs. (*a*) is applied by constructing a force polygon (Fig. 15-8*b*). Since this polygon requires another vector $m_c\mathbf{R}_c$ for closure, the magnitude of the correction is $m_c\mathbf{R}_c$ and its direction is parallel to $\mathbf{R}_c$. The three masses of Fig. 15-8 are assumed to rotate in a single plane, and so this is a case of static unbalance.

When the rotating masses are in different planes, both Eqs. (*a*) must be used. Figure 15-9*a* is an end view of a shaft having mounted upon it the three masses m_1, m_2, and m_3 at radial distances R_1, R_2, and R_3, respectively. Figure 15-9*b* is a side view of the same shaft showing left and right correction planes and the distances to the three masses. We want to find the magnitude and angular location of the corrections for each plane.

The first step in the solution is to take a summation of the moments of the centrifugal forces, including the corrections, about some point. We choose to take this summation about A in the left correction plane in order to eliminate the moment of the left correction mass. Thus, applying the second of Eqs. (*a*) gives

$$\sum \mathbf{M}_A = m_1 l_1 \mathbf{R}_1 + m_2 l_2 \mathbf{R}_2 + m_3 l_3 \mathbf{R}_3 + m_R l_R \mathbf{R}_R = 0 \tag{b}$$

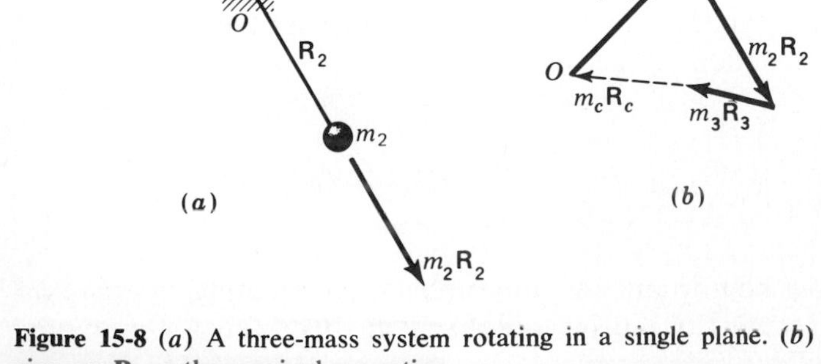

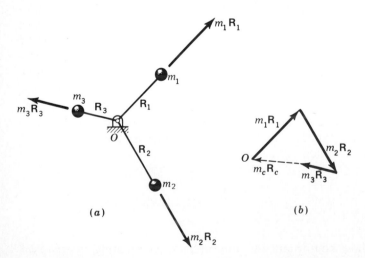

Figure 15-8 (*a*) A three-mass system rotating in a single plane. (*b*) Centrifugal-force polygon gives $m_c\mathbf{R}_c$ as the required correction.

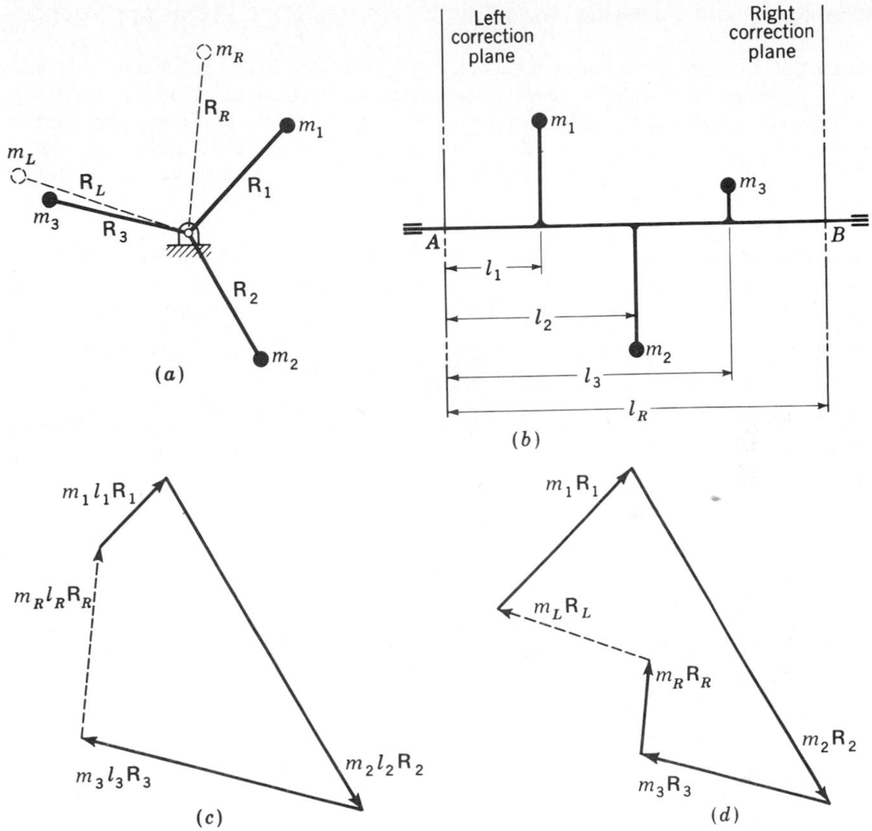

Figure 15-9 Graphical analysis of unbalance.

This is a vector equation in which the directions of the vectors are parallel, respectively, to the vectors $\mathbf{R}_N$ in Fig. 15-9*a*. Consequently the moment polygon of Fig. 15-9*c* can be constructed. The closing vector $m_R l_R \mathbf{R}_R$ gives the magnitude and direction of the correction required for the right-hand plane. The quantities m_R and $\mathbf{R}_R$ can now be found because the magnitude of $\mathbf{R}_R$ is ordinarily given in the problem. Therefore the equation

$$\sum \mathbf{F} = m_1\mathbf{R}_1 + m_2\mathbf{R}_2 + m_3\mathbf{R}_3 + m_R\mathbf{R}_R + m_L\mathbf{R}_L = 0 \qquad (c)$$

can be written. The magnitude of $\mathbf{R}_L$ being given, this equation is solved for the left-hand correction $m_L\mathbf{R}_L$ by constructing the force polygon of Fig. 15-9*d*.

Though Fig. 15-9*c* has been called a moment polygon, it is worth noting that the vectors making up this polygon consist of the moment magnitude and the position vector directions. A true moment polygon would be obtained by rotating the polygon 90° cw since a moment vector is equal to $\mathbf{R} \times \mathbf{F}$.

Vector analysis The following two examples illustrate the vector approach.

Example 15-1 Figure 15-10 represents a rotating system that has been idealized for illustrative purposes. A weightless shaft is supported in bearings at A and B rotates at $\boldsymbol{\omega} = 100\hat{\mathbf{i}}$ rad/s. When U.S. customary units are used, the unbalances are described in ounces. The three weights w_1, w_2, and w_3 are connected to the shaft and rotate with it, causing an unbalance. Determine the bearing reactions at A and B for the particular position shown.

SOLUTION We begin by calculating the centrifugal force due to each rotating weight:

$$m_1r_1\omega^2 = \frac{2(3)(100)^2}{386(16)} = 9.72 \text{ lb} \qquad m_2r_2\omega^2 = \frac{1(2)(100)^2}{386(16)} = 3.24 \text{ lb}$$

$$m_3r_3\omega^2 = \frac{1.5(2.5)(100)^2}{386(16)} = 6.07 \text{ lb}$$

These three forces are parallel to the yz plane, and we can write them in vector form by inspection

$$\mathbf{F}_1 = m_1r_1\omega^2\underline{/\theta_1} = 9.72\underline{/0^\circ} = 9.72\hat{\mathbf{j}}$$
$$\mathbf{F}_2 = m_2r_2\omega^2\underline{/\theta_2} = 3.24\underline{/120^\circ} = -1.62\hat{\mathbf{j}} + 2.81\hat{\mathbf{k}}$$
$$\mathbf{F}_3 = m_3r_3\omega^2\underline{/\theta_3} = 6.07\underline{/195^\circ} = -5.86\hat{\mathbf{j}} - 1.57\hat{\mathbf{k}}$$

where θ, in this example, is measured counterclockwise from y when viewed from the positive end of x. The moments of these forces taken about the bearing at A must be balanced by the moment of the bearing reaction at B. Therefore

$$\sum \mathbf{M}_A = 1\hat{\mathbf{i}} \times 9.72\hat{\mathbf{j}} + 3\hat{\mathbf{i}} \times (-1.62\hat{\mathbf{j}} + 2.81\hat{\mathbf{k}}) + 4\hat{\mathbf{i}} \times (-5.86\hat{\mathbf{j}} - 1.57\hat{\mathbf{k}}) + 6\hat{\mathbf{i}} \times \mathbf{F}_B = 0$$

Solving gives the bearing reaction at B as

$$\mathbf{F}_B = 3.10\hat{\mathbf{j}} - 0.358\hat{\mathbf{k}} \text{ lb}$$

To find the reaction at A we repeat the analysis. Taking moments about B gives

$$\sum \mathbf{M}_B = -2\hat{\mathbf{i}} \times (-5.86\hat{\mathbf{j}} - 1.57\hat{\mathbf{k}}) + (-3\hat{\mathbf{i}}) \times (-1.62\hat{\mathbf{j}} + 2.81\hat{\mathbf{k}}) + (-5\hat{\mathbf{i}}) \times 9.72\hat{\mathbf{j}} + (-6\hat{\mathbf{i}}) \times \mathbf{F}_A = 0$$

Solving again gives

$$\mathbf{F}_A = -5.34\hat{\mathbf{j}} - 0.882\hat{\mathbf{k}} \text{ lb}$$

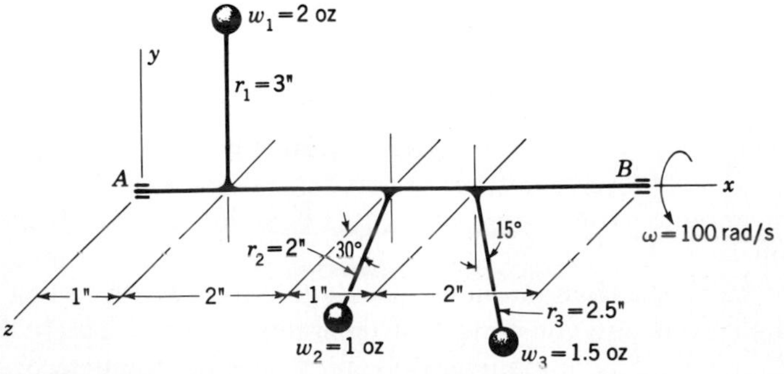

Figure 15-10

The magnitude of the two reactions are found to be $F_A = 5.41$ lb and $F_B = 3.12$ lb. Note that these are rotating reactions and that the static or stationary components due to the gravity force are not included.

Example 15-2 (*a*) What are the bearing reactions for the system shown in Fig. 15-11 if the speed is 750 rpm?

(*b*) Determine the location and magnitude of a balancing mass if it is to be placed at a radius of 0.25 m.

SOLUTION (*a*) The angular velocity of this system is $\omega = 2\pi n/60 = 2\pi(750)/60 = 78.5$ rad/s. The centrifugal forces due to the masses are

$$F_1 = m_1 r_1 \omega^2 = 12(0.2)(78.5)^2(10)^{-3} = 14.8 \text{ kN}$$
$$F_2 = m_2 r_2 \omega^2 = 3(0.3)(78.5)^2(10)^{-3} = 5.55 \text{ kN}$$
$$F_3 = m_3 r_3 \omega^2 = 10(0.15)(78.5)^2(10)^{-3} = 9.24 \text{ kN}$$

In vector form these forces are

$$\mathbf{F}_1 = 14.8\underline{/0^\circ} = 14.8\hat{\mathbf{i}} \qquad \mathbf{F}_2 = 5.55\underline{/135^\circ} = -3.92\hat{\mathbf{i}} + 3.92\hat{\mathbf{j}}$$
$$\mathbf{F}_3 = 9.24\underline{/-150^\circ} = -8.00\hat{\mathbf{i}} - 4.62\hat{\mathbf{j}}$$

To find the bearing reaction at B we take moments about the bearing at A. This equation is written

$$\sum \mathbf{M}_A = 0.3\hat{\mathbf{k}} \times [(14.8\hat{\mathbf{i}}) + (-3.92\hat{\mathbf{i}} + 3.92\hat{\mathbf{j}}) + (-8.00\hat{\mathbf{i}} - 4.62\hat{\mathbf{j}})] + 0.5\hat{\mathbf{k}} \times \mathbf{F}_B = 0$$

Taking the cross products and rearranging gives

$$0.5\hat{\mathbf{k}} \times \mathbf{F}_B = -0.21\hat{\mathbf{i}} - 0.864\hat{\mathbf{j}}$$

When this equation is solved for $\mathbf{F}_B$, we obtain

$$\mathbf{F}_B = 1.73\hat{\mathbf{i}} + 0.42\hat{\mathbf{j}} \qquad \text{and} \qquad F_B = 1.78 \text{ kN} \qquad \textit{Ans.}$$

We can find the reaction at A by summing forces. Thus

$$\begin{aligned} \mathbf{F}_A &= -\mathbf{F}_1 - \mathbf{F}_2 - \mathbf{F}_3 - \mathbf{F}_B \\ &= -14.8\hat{\mathbf{i}} - (-3.92\hat{\mathbf{i}} + 3.92\hat{\mathbf{j}}) - (-8.00\hat{\mathbf{i}} - 4.62\hat{\mathbf{j}}) - (1.73\hat{\mathbf{i}} + 0.42\hat{\mathbf{j}}) \\ &= -4.61\hat{\mathbf{i}} + 0.28\hat{\mathbf{j}} \end{aligned}$$

and $\qquad F_A = 4.62$ kN $\qquad$ *Ans.*

(*b*) Let F_C be the correcting force. Then for zero bearing reactions

$$\sum \mathbf{F} = \mathbf{F}_1 + \mathbf{F}_2 + \mathbf{F}_3 + \mathbf{F}_C = 0$$

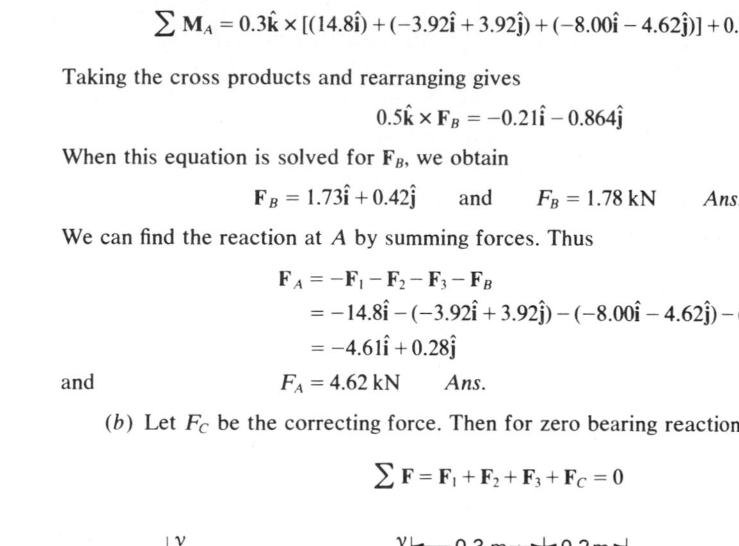

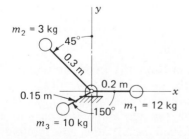

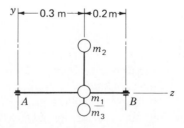

Figure 15-11

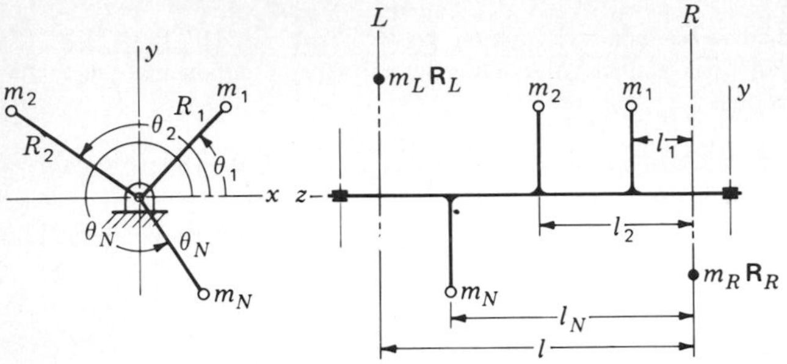

Figure 15-12 Notation for computer solution; corrections not shown on end view.

Thus $$\mathbf{F}_C = -14.8\hat{\mathbf{i}} - (-3.92\hat{\mathbf{i}} + 3.92\hat{\mathbf{j}}) - (-8.00\hat{\mathbf{i}} - 4.62\hat{\mathbf{j}})$$
$$= -2.88\hat{\mathbf{i}} + 0.7\hat{\mathbf{j}} = 2.96\underline{/166^\circ}\text{ kN}$$

and so $$m_C = \frac{F_C}{r_C\omega^2} = \frac{2.96(10)^3}{0.25(78.5)^2} = 1.92\text{ kg} \qquad \textit{Ans.}$$

Computer solution For a computer analysis it is convenient to choose the xy plane as the plane of rotation with z as the axis of rotation, as shown in Fig. 15-12. In this manner the unbalance vectors $m_i\mathbf{R}_i$ and the two correction vectors, $m_L\mathbf{R}_L$ in the left plane and $m_R\mathbf{R}_R$ in the right plane, can be expressed in the two-dimensional polar notation $m\mathbf{R} = mR\underline{/\theta}$. This makes it easy to use the polar-rectangular conversion feature and its inverse, found on programmable calculators.

Note that Fig. 15-12 has $m_1, m_2, \ldots, m_N$ unbalances. By solving Eqs. (b) and (c) for the corrections, we get

$$m_L\mathbf{R}_L = -\sum_{i=1}^{N} \frac{m_i l_i}{l}\mathbf{R}_i \tag{15-6}$$

$$m_R\mathbf{R}_R = -m_L\mathbf{R}_L - \sum_{i=1}^{N} m_i\mathbf{R}_i \tag{15-7}$$

These two equations can easily be programmed for computer solution. If the programmable calculator is used, it is suggested that the summation key be employed with each term of the summation entered using a user-defined key.

15-6 DYNAMIC BALANCING

The units in which unbalance is measured have customarily been the ounce-inch (oz · in), the gram-centimetre (g · cm), and the bastard unit of gram-inch (g · in). If correct practice is followed in the use of SI units, the most appropriate unit of unbalance in SI is the milligram-metre (mg · m) because prefixes in multiples of 1000 are preferred in SI; thus, the prefix *centi*-is not recommended. Furthermore, not more than one prefix should be used in a compound unit and, preferably, the first-named quantity should be prefixed. Thus neither the gram-centimetre nor the kilogram-millimetre, both acceptable in size, should be used.

In this book we use the ounce–inch (oz · in) and the milligram–metre (mg · m) for units of unbalance.

We have seen that static balancing is sufficient for rotating disks, wheels, gears, and the like when the mass can be assumed to exist in a single rotating plane. In the case of longer machine elements, such as turbine rotors or motor armatures, the unbalanced centrifugal forces result in couples whose effect is to tend to cause the rotor to turn end over end. The purpose of balancing is to measure the unbalanced couple and to add a new couple in the opposite direction and of the same magnitude. The new couple is introduced by the addition of masses on two preselected correction planes or by subtracting masses (drilling out) of these two planes. A rotor to be balanced will usually have both static and dynamic unbalance, and consequently the correction masses, their radial location, or both will not be the same for the two correction planes. This also means that the angular separation of the correction masses on the two planes will usually not be 180°. Thus, to balance a rotor, one must measure the magnitude and angular location of the correction mass for each of the two correction planes.

Three methods of measuring the corrections for two planes are in general use, the *pivoted-cradle*, the *nodal-point*, and the *mechanical-compensation methods.*

Figure 15-13 shows a specimen to be balanced mounted on half bearings or rollers attached to a cradle. The right end of the specimen is connected to a drive motor through a universal joint. The cradle can be rocked about either of two points which are adjusted to coincide with the correction planes on the

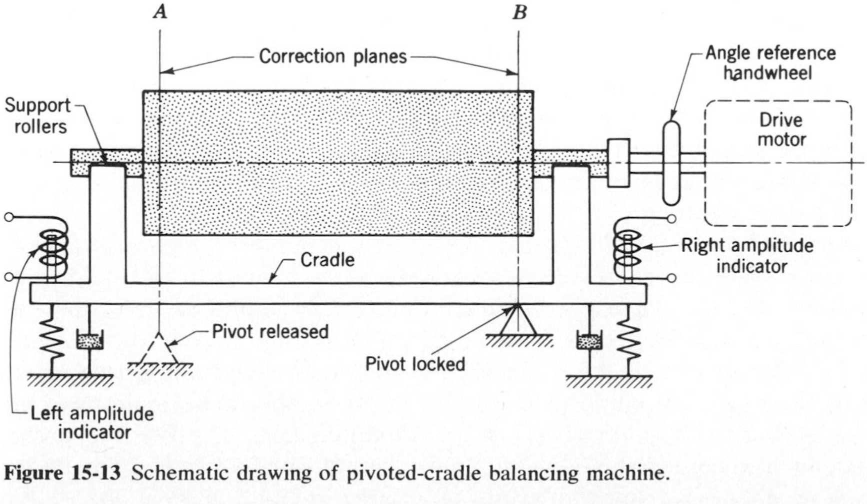

Figure 15-13 Schematic drawing of pivoted-cradle balancing machine.

specimen to be balanced. In the figure the left pivot is shown in the released position and the cradle and specimen are free to rock or oscillate about the right pivot, which is shown in the locked position. Springs and dashpots are secured at each end of the cradle to provide a single-degree-of-freedom vibrating system. Often they are made adjustable so that the natural frequency can be tuned to the motor speed. Also shown are amplitude indicators at each end of the cradle. These transducers are differential transformers, as described in Sec. 17-4, or they may consist of a permanent magnet mounted on the cradle which moves relative to a stationary coil to generate a voltage proportional to the unbalance.

With the pivots located in the two correction planes, one can lock either pivot and take readings of the amount and angle of location of the correction. The readings obtained will be completely independent of the measurements taken in the other correction plane because an unbalance in the plane of the locked pivot will have no moment about that pivot. With the right-hand pivot locked, an unbalance correctable in the left correction plane will cause vibration whose amplitude is measured by the left amplitude indicator. When this correction is made (or measured), the right-hand pivot is released, the left pivot locked, and another set of measurements made for the right-hand correction plane using the right-hand amplitude indicator.

The relation between the amount of unbalance and the measured amplitude is given by Eq. (15-4). Rearranging and substituting r for e gives

$$X = \frac{m_u r(\omega/\omega_n)^2}{m\sqrt{(1-\omega^2/\omega_n^2)^2 + (2\zeta\omega/\omega_n)^2}} \tag{15-8}$$

where $m_u r$ = unbalance
m = mass of cradle and specimen
X = amplitude

This equation shows that the amplitude of the motion X is directly proportional to the unbalance $m_u r$. Figure 15-14*a* shows a plot of it for a particular damping ratio ζ. The figure shows that the machine will be most sensitive near resonance ($\omega = \omega_n$), since in this region the greatest amplitude is recorded for a given unbalance. Damping is deliberately introduced in balancing machines to filter noise and other vibrations that might affect the results. Damping also helps to maintain calibration against effects of temperature and other environmental conditions.

Not shown in Fig. 15-13 is a sine-wave signal generator which is attached to the drive shaft. If the resulting sine wave is compared on a dual-beam oscilloscope with the wave generated by one of the amplitude indicators, a phase difference will be found. This angular phase difference is the angular location of the unbalance. In a balancing machine an electronic phasemeter measures the phase angle and gives the result on another meter calibrated in degrees. To locate the correction on the specimen (Fig. 15-13) the angular reference handwheel is turned by hand until the indicated angle is in line with

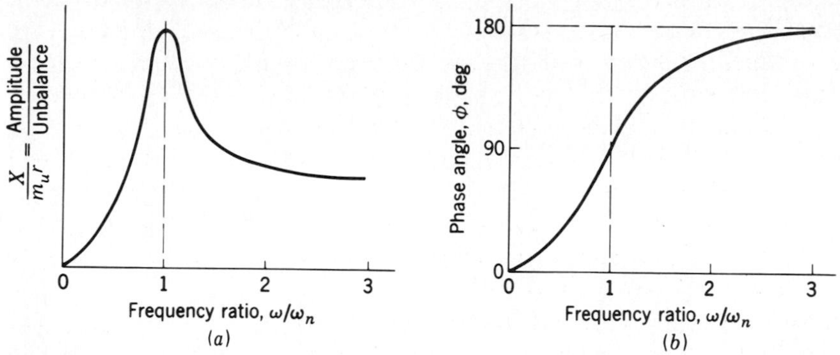

Figure 15-14

a reference pointer. This places the heavy side of the specimen in any preselected position and permits the correction to be made.

By manipulating Eq. (c) of Sec. 15-2 we obtain the equation for the phase angle in parameter form. Thus

$$\phi = \tan^{-1} \frac{2\zeta\omega/\omega_n}{1 - \omega^2/\omega_n^2} \tag{15-9}$$

A plot of this equation for a single damping ratio and for various frequency ratios is shown in Fig. 15-14b. This curve shows that at resonance, when the speed ω of the shaft and the natural frequency ω_n of the shaft system are the same, the displacement lags the unbalance by the angle $\phi = 90°$. If the top of the specimen is turning away from the operator, the unbalance will be horizontal and directly in front of the operator when the displacement is maximum downward. The figure also shows that the angular location approaches 180° as the shaft speed ω is increased above resonance.

15-7 BALANCING MACHINES

A pivoted-cradle balancing machine for high-speed production is illustrated in Fig. 15-15. The shaft-mounted signal generator can be seen at the extreme left.

Nodal-point balancing Plane separation using a point of zero or minimum vibration is called the *nodal-point method* of balancing. To see how this method is used examine Fig. 15-16. Here the specimen to be balanced is shown mounted on bearings which are fastened to a nodal bar. We assume that the specimen is already balanced in the left-hand correction plane and that an unbalance still exists in the right-hand plane, as shown. Because of this unbalance a vibration of the entire assembly takes place, causing the

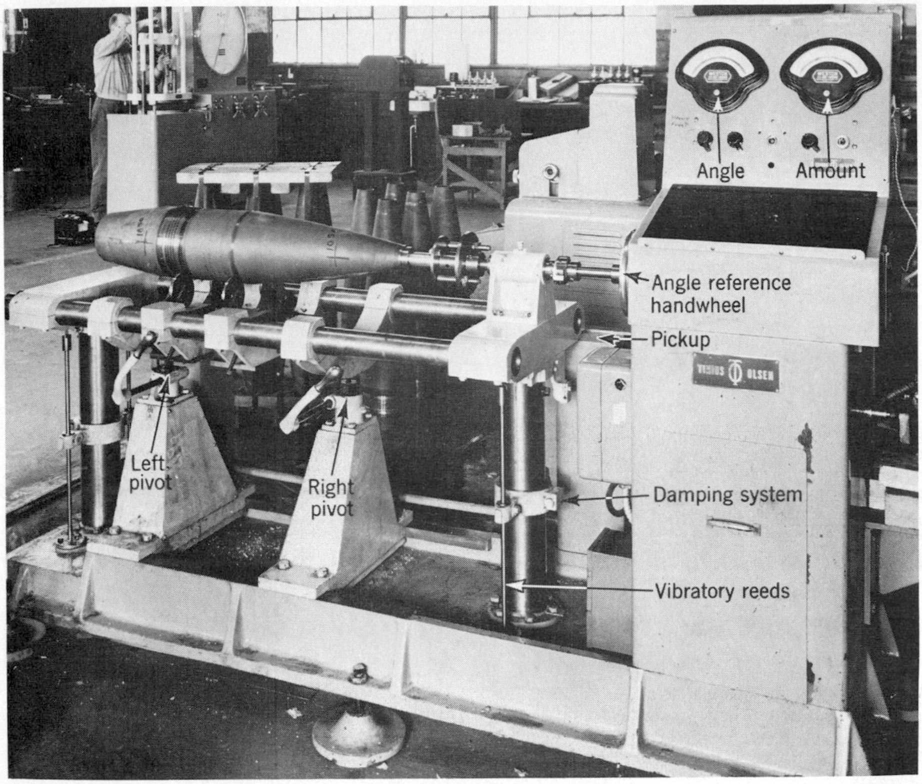

Figure 15-15 A Tinius Olsen static-dynamic pivoted-cradle balancing machine with specimen mounted for balancing. (*Tinius Olsen Testing Machine Company, Willow Grove, Pennsylvania.*)

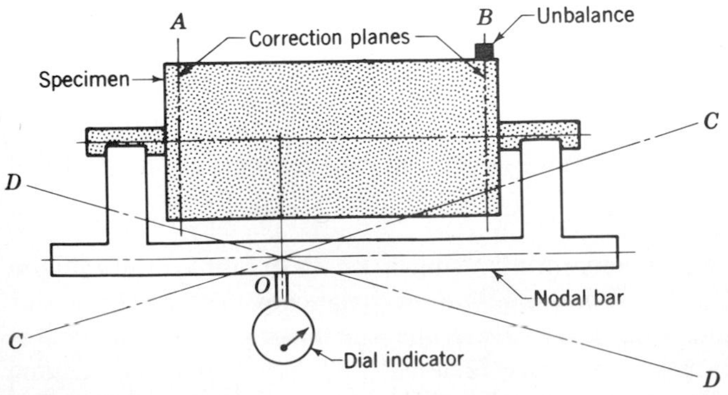

Figure 15-16 Plane separation by the nodal-point method. The nodal bar experiences the same vibration as the specimen.

nodal bar to oscillate about some point O, occupying first position CC and then DD. Point O is easily located by sliding a dial indicator along the nodal bar; a point of zero motion or minimum motion is then readily found. This is the null or nodal point. Its location is the center of oscillation for a center of percussion in the right-hand correction plane.

We assumed, at the beginning of this discussion, that no unbalance existed in the left-hand correction plane. However, if unbalance is present, its magnitude will be given by the dial indicator located at the nodal point just found. Thus, by locating the dial indicator at this nodal point, we measure the unbalance in the left-hand plane without any interference from that in the right-hand plane. In a similar manner another nodal point can be found which will measure only the unbalance in the right-hand correction plane without any interference from that in the left-hand plane.

In commercial balancing machines employing the nodal-point principle the plane separation is accomplished in electrical networks. Typical of these is the Micro Dynamic Balancer; a schematic is shown in Fig. 15–17. On this machine a switching knob selects either correction plane and displays the unbalance on a voltmeter, which is calibrated in appropriate unbalance units.

The computer of Fig. 15-17 contains a filter which eliminates bearing noise and other frequencies not related to the unbalance. A multiplying network is used to give any sensitivity desired and to cause the meter to read in preselected balancing units. The strobe light is driven by an oscillator which is synchronized to the rotor speed.

The rotor is driven at a speed which is much greater than the natural frequency of the system, and since the damping is quite small, Fig. 15-14*b* shows that the phase angle will be approximately 180°. Marked on the right-hand end of the rotor are degrees or numbers which are readable and stationary under the strobe light during rotation of the rotor. Thus it is only necessary to observe the particular station number or degree marking under the strobe light to locate the heavy spot. When the switch is shifted to the other correction plane, the meter again reads the amount and the strobe light illuminates the station. Sometimes as few as five station numbers distributed uniformly around the periphery are adequate for balancing.

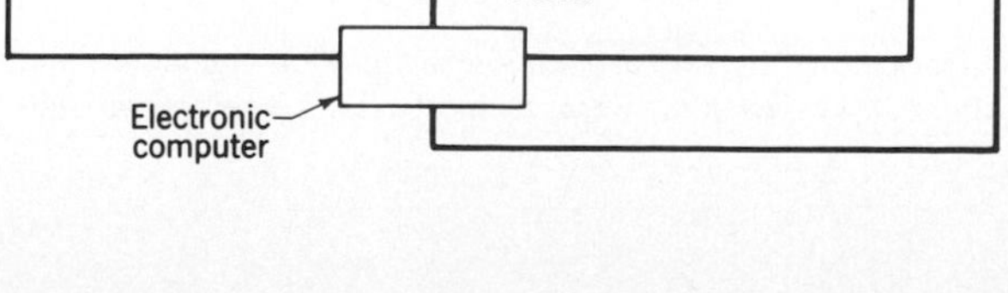

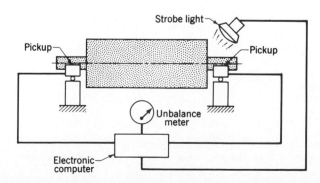

Figure 15-17 Diagram of the electrical circuit in a Micro Dynamic Balancer. (*Micro-Balancing, Inc., Garden City Park, New York.*)

The direction of the vibration is horizontal, and the phase angle is nearly 180°. Thus, rotation such that the top of the rotor moves away from the operator will cause the heavy spot to be in a horizontal plane and on the near side of the axis when illuminated by the strobe lamp. A pointer is usually placed here to indicate its location. If, during production balancing, it is found that the phase angle is less than 180°, the pointer can be shifted slightly to indicate the proper position to observe.

Mechanical compensation An unbalanced rotating rotor located in a balancing machine develops a vibration. One can introduce in the balancing machine counterforces in each correction plane which exactly balance the forces causing the vibration. The result of introducing these is a smooth-running motor. Upon stopping, the location and amount of the counterforce are measured to give the exact correction required. This is called *mechanical compensation.*

When mechanical compensation is used, the speed of the rotor during balancing is not important because the equipment will be in calibration for all speeds. The rotor may be driven by a belt, from a universal joint, or it may be self-driven if, for example, it is a gasoline engine. The electronic equipment is simple, no built-in damping is necessary, and the machine is easy to operate because the unbalance in both correction planes is measured simultaneously and the magnitude and location read directly.

We can understand how mechanical compensation is applied by examining Fig. 15-18*a*. Looking at the end of the rotor, we see one of the correction planes with the unbalance to be corrected represented by wr. Two compensator weights are also shown in the figure. All three of these weights are to rotate with the same angular velocity ω, but the position of the compensator weights relative to one another and their position relative to the unbalanced

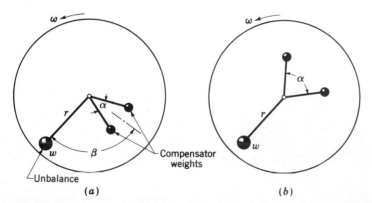

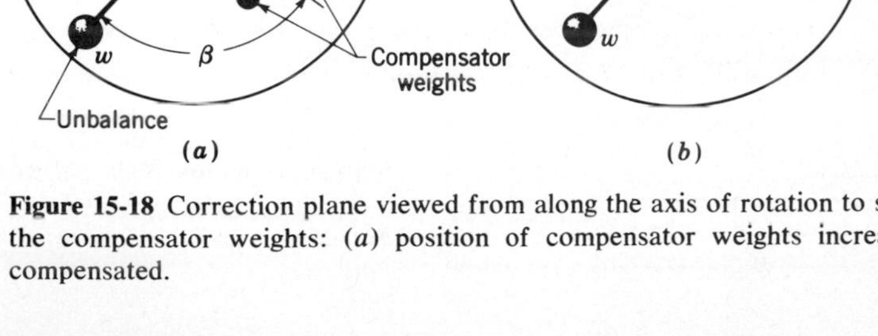

Figure 15-18 Correction plane viewed from along the axis of rotation to show the unbalance and the compensator weights: (*a*) position of compensator weights increases the vibration; (*b*) compensated.

weight can be varied by two controls. One of these controls changes the angle α, that is, the angle between the compensator weights. The other control changes the angular position of the compensator weights relative to the unbalance, i.e., the angle β. The knob which changes the angle β is the *location* control, and when the rotor is compensated (balanced) in this plane, a pointer on the knob indicates the exact angular location of the unbalance. The knob which changes the angle α is the *amount* control, and it also gives a direct reading when the rotor unbalance is compensated. The magnitude of the vibration is measured electrically and displayed on a voltmeter. Thus compensation is secured when the controls are manipulated to make the voltmeter read zero.

15-8 FIELD BALANCING WITH THE PROGRAMMABLE CALCULATOR†

It is possible to balance a machine in the field by balancing a single plane at a time. But cross effects and correction-plane interference often require balancing each end of a rotor two or three times to obtain satisfactory results. Some machines may require as much as an hour to bring them up to full speed, resulting in even more delays in the balancing procedure.

Field balancing is necessary for very large rotors for which balancing machines are impractical. And even though high-speed rotors are balanced in the shop during manufacture, it is frequently necessary to rebalance them in the field because of slight deformations brought on by shipping, by creep, or by high operating temperatures.

Both Rathbone and Thearle‡ have developed methods of two-plane field balancing which can now be expressed in complex-number notation and solved using a programmable calculator. The time saved by using a programmable calculator is several hours when compared with graphical methods or analysis with complex numbers using an ordinary scientific calculator.

In the analysis that follows, boldface letters will be used to represent complex numbers:

$$\mathbf{R} = R\underline{/\theta} = Re^{j\theta} = x + jy$$

In Fig. 15-19 unknown unbalances $\mathbf{M}_L$ and $\mathbf{M}_R$ are assumed to exist in the left- and right-hand corrections planes, respectively. The magnitudes of these unbalances are M_L and M_R, and they are located at angles ϕ_L and ϕ_R from the rotating reference. When these unbalances have been found, their negatives are located in the left and right planes to achieve balance.

† The authors are grateful to W. B. Fagerstrom, of E. I. du Pont de Nemours, Wilmington, Delaware, for contributing some of the ideas for this section.

‡ T. C. Rathbone, Turbine Vibration and Balancing, *Trans. ASME*, 1929, p. 267; E. L. Thearle, Dynamic Balancing in the Field, *Trans. ASME*, 1934, p. 745.

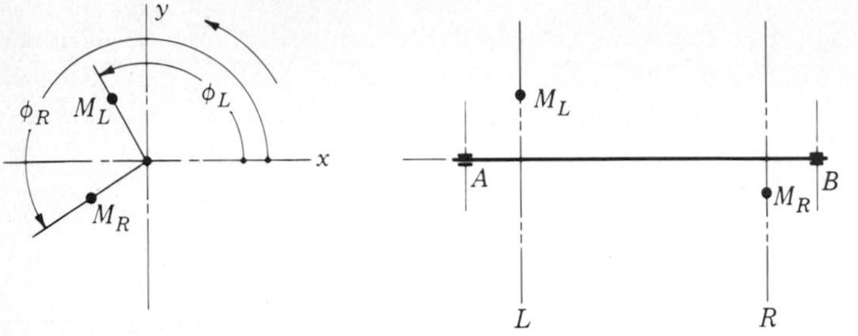

Figure 15-19 Notation for two-plane field balancing. The xy system is the rotating reference.

The rotating unbalances $\mathbf{M}_L$ and $\mathbf{M}_R$ produce disturbances at bearings A and B. Using commercial field-balancing equipment, it is possible to measure the amplitudes and the angular locations of these disturbances. The notation $\mathbf{X} = X\underline{/\phi}$, with appropriate subscripts, will be used to designate these amplitudes.

In field balancing, three runs or tests are made, as follows:

First run. Measure amplitude $\mathbf{X}_A = X_A\underline{/\phi_A}$ at bearing A and amplitude $\mathbf{X}_B = X_B\underline{/\phi_B}$ at bearing B due only to the original unbalances $\mathbf{M}_L = M_L\underline{/\phi_L}$ and $\mathbf{M}_R = M_R\underline{/\phi_R}$.

Second run. Add trial mass $\mathbf{m}_L = m_L\underline{/\theta_L}$ to the left correction plane and measure the amplitudes $\mathbf{X}_{AL} = X_{AL}\underline{/\phi_{AL}}$ and $\mathbf{X}_{BL} = X_{BL}\underline{/\phi_{BL}}$ at the left and right bearings (A and B), respectively.

Third run. Remove trial mass $\mathbf{m}_L = m_L\underline{/\theta_L}$. Add trial mass $\mathbf{m}_R = m_R\underline{/\theta_R}$ to the right-hand correction plane and again measure the bearing amplitudes. These results are designated $\mathbf{X}_{AR} = X_{AR}\underline{/\phi_{AR}}$ for bearing A and $\mathbf{X}_{BR} = X_{BR}\underline{/\phi_{BR}}$ for bearing B.

Note in the above runs that the term "trial mass" means the same as a trial unbalance provided a unit distance from the axis of rotation is used.

To develop the equations for the unbalance to be found we first define the *complex stiffness*, by which we mean the amplitude that would result at either bearing due to a unit unbalance located at the intersection of the rotating reference mark and one of the correction planes. Thus we need to find the complex stiffnesses $\mathbf{A}_L$ and $\mathbf{B}_L$ due to a unit unbalance located at the intersection of the rotating reference mark and plane L. And we require the complex stiffnesses $\mathbf{A}_R$ and $\mathbf{B}_R$ due to a unit unbalance located at the intersection of the rotating reference mark and plane R.

If these stiffnesses were known, we could write the following sets of complex equations:

$$\mathbf{X}_{AL} = \mathbf{X}_A + \mathbf{A}_L\mathbf{m}_L \qquad \mathbf{X}_{BL} = \mathbf{X}_B + \mathbf{B}_L\mathbf{m}_L \tag{a}$$

$$\mathbf{X}_{AR} = \mathbf{X}_A + \mathbf{A}_R\mathbf{m}_R \qquad \mathbf{X}_{BR} = \mathbf{X}_B + \mathbf{B}_R\mathbf{m}_R \tag{b}$$

After the three runs are made, the stiffnesses will be the only unknowns in these equations. Therefore

$$\mathbf{A}_L = \frac{\mathbf{X}_{AL} - \mathbf{X}_A}{\mathbf{m}_L} \qquad \mathbf{B}_L = \frac{\mathbf{X}_{BL} - \mathbf{X}_B}{\mathbf{m}_L}$$

$$\mathbf{A}_R = \frac{\mathbf{X}_{AR} - \mathbf{X}_A}{\mathbf{m}_R} \qquad \mathbf{B}_R = \frac{\mathbf{X}_{BR} - \mathbf{X}_B}{\mathbf{m}_R} \tag{15-10}$$

Then, from the definition of stiffness, we have from the first run

$$\mathbf{X}_A = \mathbf{A}_L\mathbf{M}_L + \mathbf{A}_R\mathbf{M}_R \qquad \mathbf{X}_B = \mathbf{B}_L\mathbf{M}_L + \mathbf{B}_R\mathbf{M}_R \tag{c}$$

Solving this pair of equations simultaneously gives

$$\mathbf{M}_L = \frac{\mathbf{X}_A\mathbf{B}_R - \mathbf{X}_B\mathbf{A}_R}{\mathbf{A}_L\mathbf{B}_R - \mathbf{A}_R\mathbf{B}_L} \qquad \mathbf{M}_R = \frac{\mathbf{X}_B\mathbf{A}_L - \mathbf{X}_A\mathbf{B}_L}{\mathbf{A}_L\mathbf{B}_R - \mathbf{A}_R\mathbf{B}_L} \tag{15-11}$$

These equations can be programmed either in complex polar form or in complex rectangular form. The suggestions that follow were formed assuming a complex rectangular form for the solution.

Since the original data are formulated in polar coordinates, a subroutine should be written to transform the data into rectangular coordinates before storage.

The equations reveal that complex subtraction, division, and multiplication are used often. These operations can be set up as subroutines to be called from the main program. If $\mathbf{A} = a + \hat{\mathbf{j}}b$ and $\mathbf{B} = c + \hat{\mathbf{j}}d$, the formula for complex subtraction is

$$\mathbf{A} - \mathbf{B} = (a - c) + \hat{\mathbf{j}}(b - d) \tag{15-12}$$

For complex multiplication, we have

$$\mathbf{A} \cdot \mathbf{B} = (ac - bd) + \hat{\mathbf{j}}(bc + ad) \tag{15-13}$$

and for complex division the formula is

$$\frac{\mathbf{A}}{\mathbf{B}} = \frac{(ac + bd) + \hat{\mathbf{j}}(bc - ad)}{c^2 + d^2} \tag{15-14}$$

With these subroutines it is an easy matter to program Eqs. (15-10) and (15-11). Indirect addressing may save space.

As a check on your programming, use the following data: $\mathbf{X}_A = 8.6\underline{/63^\circ}$, $\mathbf{X}_B = 6.5\underline{/206^\circ}$, $\mathbf{m}_L = 10\underline{/270^\circ}$, $\mathbf{m}_R = 12\underline{/180^\circ}$, $\mathbf{X}_{AL} = 5.9\underline{/123^\circ}$, $\mathbf{X}_{BL} = 4.5\underline{/228^\circ}$, $\mathbf{X}_{AR} = 6.2\underline{/36^\circ}$, $\mathbf{X}_{BR} = 10.4\underline{/162^\circ}$. The answers are $\mathbf{M}_L = 10.76\underline{/146.6^\circ}$ and $\mathbf{M}_R = 6.20\underline{/245.4^\circ}$.

According to Fagerstrom the vibration angles used can be expressed in two different systems. The first is the *rotating-protractor–stationary-mark system* (RPSM). This is the system used in the preceding analysis and is the one a theoretician would prefer. In actual practice, it is usually easier to have the protractor stationary and use a rotating mark like a key or keyway. This is called the *rotating-mark–stationary-protractor system* (RMSP). The only difference between the two systems is in the sign of the vibration angle, but there is no sign change on the trial or correction masses.

15-9 BALANCING THE SINGLE-CYLINDER ENGINE

The rotating masses in a single-cylinder engine can be balanced using the methods already discussed in this chapter. The reciprocating masses, however, cannot be balanced at all, and so our studies in this section are really concerned with unbalance.

Though the reciprocating masses cannot be balanced using a simple counterweight, it is possible to modify the shaking forces (see Sec. 14-9) by unbalancing the rotating masses. As an example of this let us add a counterweight opposite the crankpin whose mass exceeds the rotating mass by one-half of the reciprocating mass (from one-half to two-thirds of the reciprocating mass is usually added to the counterweight to alter the balance characteristics in single-cylinder engines). We designate the mass of the counterweight by m_C, substitute this mass in Eq. (14-24), and use a negative sign because the counterweight is opposite the crankpin; then the inertia force due to this counterweight is

$$\mathbf{F}_C = -m_C r\omega^2 \cos \omega t\ \hat{\mathbf{i}} - m_C r\omega^2 \sin \omega t\ \hat{\mathbf{j}} \tag{a}$$

Note that the balancing mass and the crankpin both have the same radius. Designating m_A and m_B as the masses of the rotating and the reciprocating parts, respectively, as in Chap. 14, we have

$$m_C = m_A + \frac{m_B}{2} \tag{b}$$

according to the supposition above. Equation (a) can now be written

$$\mathbf{F}_C = -\left(m_A + \frac{m_B}{2}\right) r\omega^2 \cos \omega t\ \hat{\mathbf{i}} - \left(m_A + \frac{m_B}{2}\right) r\omega^2 \sin \omega t\ \hat{\mathbf{j}} \tag{c}$$

The inertia force due to the rotating and reciprocating masses is, from Eqs. (14-28) and (14-29),

$$\mathbf{F}_{A,B} = F^x\hat{\mathbf{i}} + F^y\hat{\mathbf{j}} = \left[(m_A + m_B) r\omega^2 \cos \omega t + m_B r\omega^2 \frac{r}{l} \cos 2\omega t\right]\hat{\mathbf{i}} + m_A r\omega^2 \sin \omega t\ \hat{\mathbf{j}} \tag{d}$$

Adding Eqs. (c) and (d) gives the resultant inertia force as

$$\mathbf{F} = \left(\frac{m_B}{2} r\omega^2 \cos \omega t + m_B r\omega^2 \frac{r}{l} \cos 2\omega t\right)\hat{\mathbf{i}} - \frac{m_B}{2} r\omega^2 \sin \omega t\ \hat{\mathbf{j}} \tag{15-15}$$

The vector

$$\frac{m_B}{2} r\omega^2(\cos \omega t\ \hat{\mathbf{i}} - \sin \omega t\ \hat{\mathbf{j}})$$

is called the *primary component* of Eq. (15-15). This component has a magnitude $m_B r\omega^2/2$ and can be represented as a *backward* (clockwise) rotating vector with angular velocity ω. The remaining component in Eq. (15-15) is called the *secondary component*; it is the x projection of a vector of length $m_B r\omega^2(r/l)$ rotating *forward* (counterclockwise) with an angular velocity of 2ω.

The maximum inertia force occurs when $\omega t = 0$ and from Eq. (15-15) is seen to be

$$F_{\max} = m_B r\omega^2\left(\frac{r}{l} + \frac{1}{2}\right) \tag{e}$$

because $\cos \omega t = \cos 2\omega t = 1$ when $\omega t = 0$. Before the extra counterweight was added, the maximum inertia force was

$$F_{\max} = m_B r\omega^2\left(\frac{r}{l} + 1\right) \tag{f}$$

Thus in this instance the effect of the additional counterweight is to reduce the maximum shaking force by 50 percent of the primary component and to add vertical inertia forces where formerly none existed. Equation (15-15) is plotted as a polar diagram in Fig. 15-20 for an r/l value of $\frac{1}{4}$. Here, the vector OA rotates counterclockwise at 2ω angular velocity. The horizontal projection of this vector OA' is the secondary component. The vector OB, the primary component, rotates clockwise at ω angular velocity. The total shaking

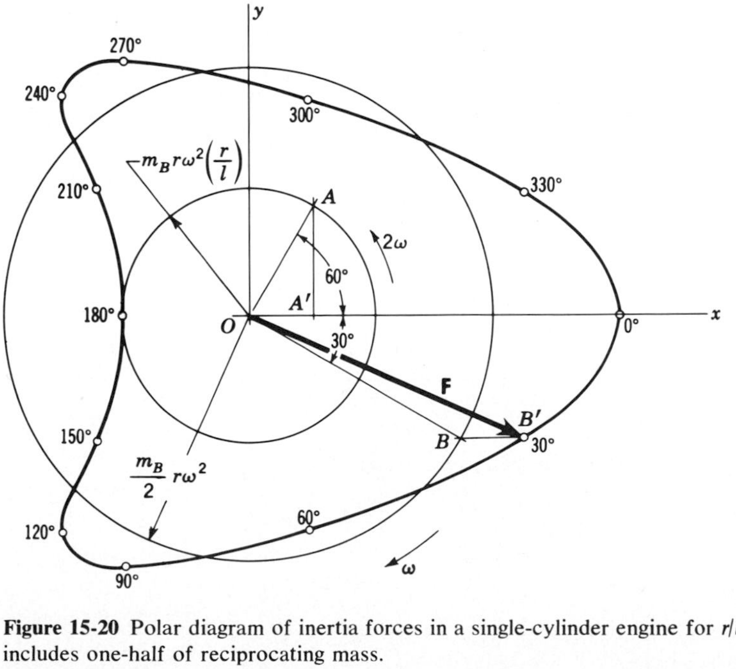

Figure 15-20 Polar diagram of inertia forces in a single-cylinder engine for $r/l = \frac{1}{4}$. Counterweight includes one-half of reciprocating mass.

force F is shown for the 30° position and is the sum of the vectors OB and $BB' = OA'$.

Imaginary-mass approach Stevensen has refined and extended a method of engine balancing which is here called the *imaginary-mass approach.*† It is possible that the method is known in some circles as the *virtual-rotor approach* because it uses what might be called a virtual rotor which counter-rotates to accommodate part of the piston effect in a reciprocating engine.

Before going into details, it is necessary to explain a change in the method of viewing the crank circle of an engine. In developing the imaginary-mass approach in this section and the one to follow, we use the coordinate system of Fig. 15-21*a*. This appears to be a left-handed system because the y axis is located clockwise from x and because positive rotation is shown as clockwise. We adopt this notation because it has been so long used by the automotive industries.‡ If you like, you can think of this system as a right-handed three-dimensional system viewed from the negative z axis.

The imaginary-mass approach uses two fictitious masses, each equal to half the equivalent reciprocating mass at the particular harmonic studied. The

† The presentation here is from Edward N. Stevensen, University of Hartford, class notes, with his permission. While some changes have been made to conform to the notation of this book, the material is all Stevensen's. He refers to Maleev and Lichty [V. L. Maleev, "Internal Combustion Engines," McGraw-Hill, New York, 1933; and L. C. Lichty, "Internal Combustion Engines," 5th Ed., McGraw-Hill, New York, 1939] and states that the method first came to his attention in both the Maleev and Lichty books.

‡ Readers who are antique car buffs will understand this convention because it is the direction in which an antique engine is hand-cranked.

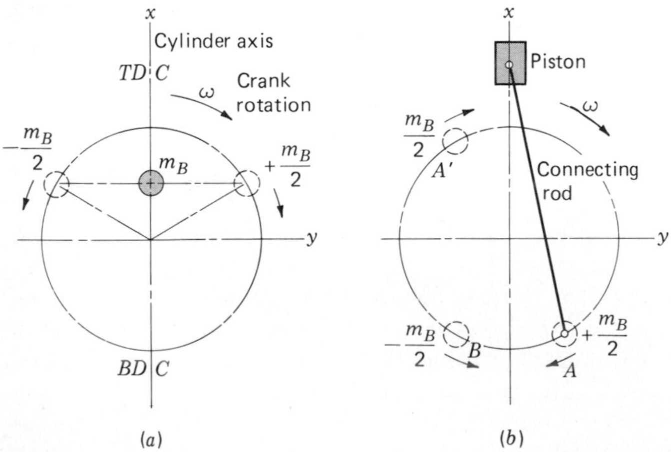

Figure 15-21 Note that axes are right hand but the view of the crank circle is from the negative z axis.

purpose of these fictitious masses is to replace the effects of the reciprocating mass. These imaginary masses rotate about the crank center in opposite directions and with equal velocities. They are arranged so that they come together at both the *top dead center* (TDC) and *bottom dead center* (BDC), as shown in Fig. 15-21*a*. The mass $+m_B/2$ rotates *with* the crank motion; the other mass $-m_B/2$ rotates *opposite* to the crank motion. The mass rotating with the crank is designated on the figure by a plus sign and the mass rotating in the opposite direction by a minus sign. The center of mass of these two rotating masses always lies on the cylinder axis. The imaginary-mass approach was conceived because the piston motion and the resulting inertia force can always be represented by a Fourier series. Such a series has an infinite number of terms, each term representing a simple harmonic motion having a known frequency and amplitude. It turns out that the higher-frequency amplitudes are so small that they can be neglected and hence only a small number of the lower-frequency amplitudes are needed. Also, the odd harmonics (third, fifth, etc.) are not present because of the symmetry of the piston motion.

Each harmonic, the first, second, fourth, etc., is represented by a pair of imaginary masses. The angular velocities of these masses are $\pm\omega$ for the first harmonic, $\pm 2\omega$ for the second, $\pm 4\omega$ for the fourth, and so on. It is rarely necessary to consider the sixth or higher harmonics.

Stevensen gives the following rule for placing the imaginary masses:

> *For any given position of the cranks the positions of the imaginary masses are found, first, by determining the angles of travel of each crank from its top dead center and, second, by moving their imaginary masses, one clockwise and the other counterclockwise, by angles equal to the crank angle times the number of the harmonic.*

All of these angles must be measured from the same dead-center crank position.

Let us apply this approach to the single-cylinder engine, considering only the first harmonic. In Fig. 15-21*b* the mass $+m_B/2$ at A rotates at the velocity ω with the crank while mass $-m_B/2$ at B rotates at the velocity $-\omega$ opposite to crank rotation. The imaginary mass at A can be balanced by adding an equal mass at A' to rotate with the crankshaft. However, the mass at B can be balanced neither by the addition nor by the subtraction of mass from any part of the crankshaft because it is rotating in the opposite direction. When half the mass of the reciprocating parts is balanced in this manner, i.e., by adding the mass at A', the unbalanced part of the first harmonic, due to the mass at B, causes the engine to vibrate in the plane of rotation equally in all directions like a true unbalanced rotating mass.

It is interesting to know that in one-cylinder motorcycle engines a fore-and-aft unbalance is less objectionable than an up-and-down unbalance. For this reason such engines are *overbalanced* by using a counterweight whose mass is more than half the reciprocating mass.

Table 15-1 One-cylinder-engine inertia forces

Type	Equivalent mass	Radius	With cylinder axis (x)	Across cylinder axis (y)
Centrifugal	m_A	r	$m_A r\omega^2 \cos \omega t$	$m_A r\omega^2 \sin \omega t$
	m_{AC}	r_C	$m_{AC} r_C \omega^2 \cos (\omega t + \pi)$	$m_{AC} r_C \omega^2 \sin (\omega t + \pi)$
Reciprocating First harmonic	m_B	r	$m_B r\omega^2 \cos \omega t$	0
	m_{BC}	r_C	$m_{BC} r_C \omega^2 \cos (\omega t + \pi)$	$m_{BC} r_C \omega^2 \sin (\omega t + \pi)$
Second harmonic	$\frac{m_B r}{4l}$	r	$\frac{m_B r}{4l}(r)(2\omega)^2 \cos 2\omega t$	0

It is impossible to balance the second and higher harmonics with masses rotating at crankshaft speeds, since the frequency of the unbalance is higher than that of crankshaft rotation. Balancing of second harmonics has been accomplished by using shafts geared to run at twice the engine crankshaft speed, as in the case of the 1976 Plymouth Arrow engine, but at the cost of complication. It is not usually done.

For ready reference we now summarize the inertia forces in the single-cylinder engine, with balancing masses, in Table 15-1. The equations shown have been obtained from Eqs. (14-24) and (14-27), rewritten so that the effect of the second harmonic is presented as a mass equal to $m_B r/4l$ reciprocating at a speed of 2ω. Note that the subscript C is used to designate the counterweights (balancing masses) and their radii. Since the centrifugal balancing masses will be selected and placed to counterbalance the centrifugal forces, the only unbalance that results along the cylinder axis will be the sum of the last three entries. Similarly, the only unbalance across the cylinder axis will be the value in the fourth row. The maximum values of the unbalance in these two directions can be predetermined in any desired ratio to each other, as indicated previously, and a solution obtained for m_{BC} at the radius r_C.

If this approach is used to include the effect of the fourth harmonic, there will result an additional mass of $m_B r^3/16l^3$ reciprocating at the speed of 2ω and a mass of $-m_B r^3/64l^3$ reciprocating at a speed of 4ω, illustrating the decreasing significance of the higher harmonics.

15-10 BALANCE OF MULTICYLINDER ENGINES

To obtain a basic understanding of the balancing problem in multicylinder engines, let us consider a two-cylinder in-line engine having cranks 180° apart and rotating parts already balanced by counterweights. Such an engine is shown in Fig. 15-22. Applying the imaginary-mass approach for the first harmonics results in the diagram of Fig. 15-22a. This figure shows that masses +1 and +2, rotating clockwise, balance each other, as do masses −1 and −2,

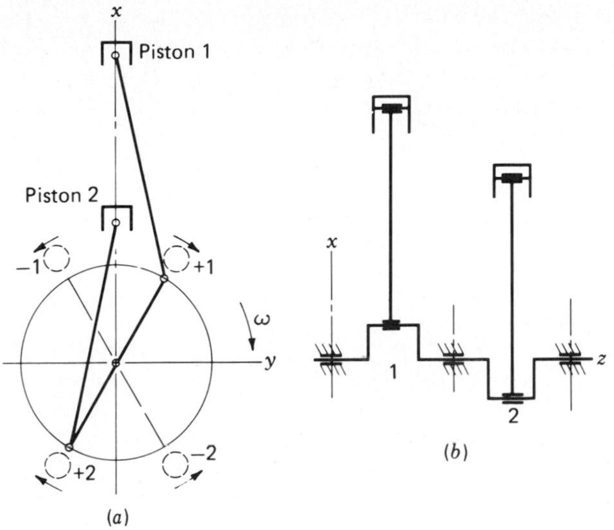

Figure 15-22 (*a*) First harmonics; (*b*) a two-throw crankshaft with three main bearings.

rotating counterclockwise. Thus, the first harmonic forces are inherently balanced for this crank arrangement. Figure 15-22*b* shows, however, that these forces are not in the same plane. For this reason unbalanced couples will be set up which tend to rotate the engine about the y axis. The values of these couples can be determined using the force expressions in Table 15-1 together with the coupling distance because the equations can be applied to each cylinder separately. It is possible to balance the couple due to the real rotating masses as well as the imaginary half-masses that rotate with the engine; however, the couple due to the half-mass of the first harmonic that is counterrotating cannot be balanced.

Figure 15-23*a* shows the location of the imaginary masses for the second harmonic using Stevensen's rule. This diagram shows that the second-harmonic forces are not balanced. Since the greatest unbalance occurs at the dead center positions, the diagrams are usually drawn for this extreme position, with crank 1 at TDC as in Fig. 15-23*b*. This unbalance causes a vibration in the xz plane having the frequency 2ω.

The diagram for the fourth harmonics, not shown, is the same as Fig. 15-23*b* but, of course, the speed is 4ω.

Four-cylinder engine A four-cylinder in-line engine with cranks spaced 180° apart is shown in Fig. 15-24*c*. This engine can be treated as two two-cylinder engines back to back. Thus the first harmonic forces still balance and, in addition, from Fig. 15-24*a* and *c*, the first-harmonic couples also balance. These couples will tend, however, to deflect the center bearing of a three-bearing crankshaft up and down and to bend the center of a two-bearing shaft in the same manner.

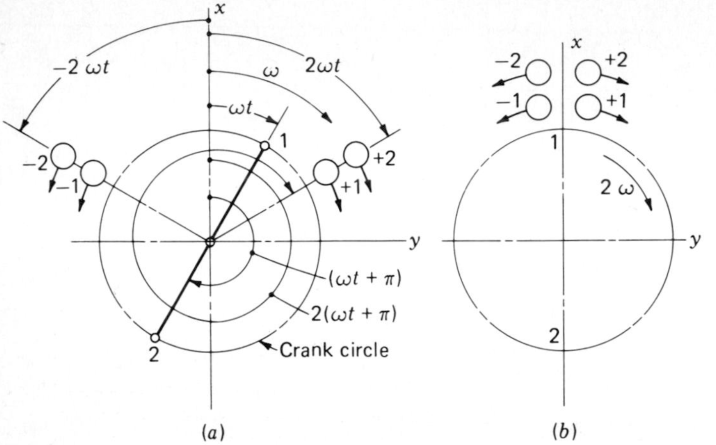

Figure 15-23 Second-harmonic positions of the imaginary masses; (*a*) positions for same crank angle as in Fig. 15-22*a*; (*b*) extreme, or dead-center, positions.

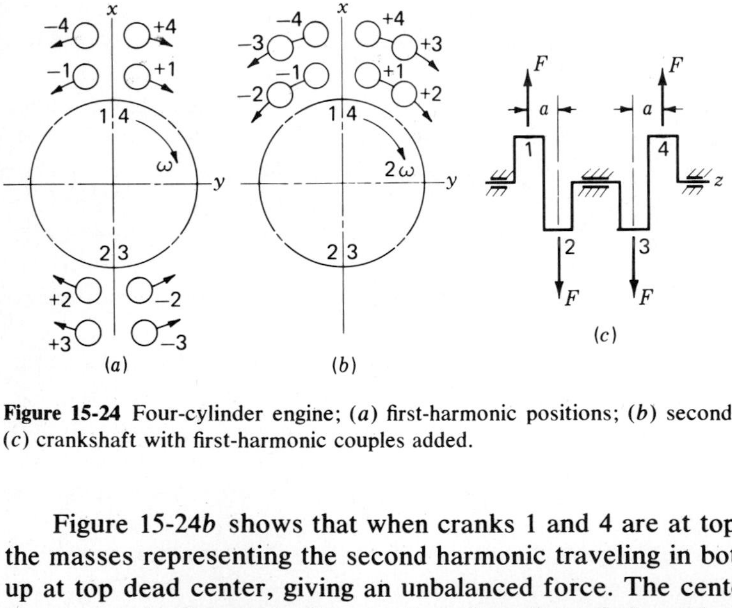

Figure 15-24 Four-cylinder engine; (*a*) first-harmonic positions; (*b*) second-harmonic positions; (*c*) crankshaft with first-harmonic couples added.

Figure 15-24*b* shows that when cranks 1 and 4 are at top dead center, all the masses representing the second harmonic traveling in both directions pile up at top dead center, giving an unbalanced force. The center of mass of all the masses is always on the x axis, and so the unbalanced second harmonics cause a vertical vibration with a frequency of twice the engine speed. This characteristic is typical of all four-cylinder engines with this crank arrangement. Since the masses and forces all act in the same direction, there is no coupling action.

A diagram of the fourth harmonics would be identical to that of Fig. 15-24*b*, and the effects are the same, but they do have a higher frequency and exert less force.

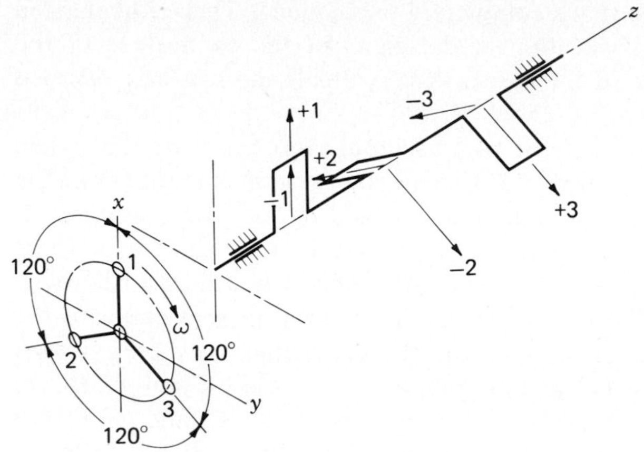

Figure 15-25 Crank arrangement of three-cylinder engine; first-harmonic forces shown.

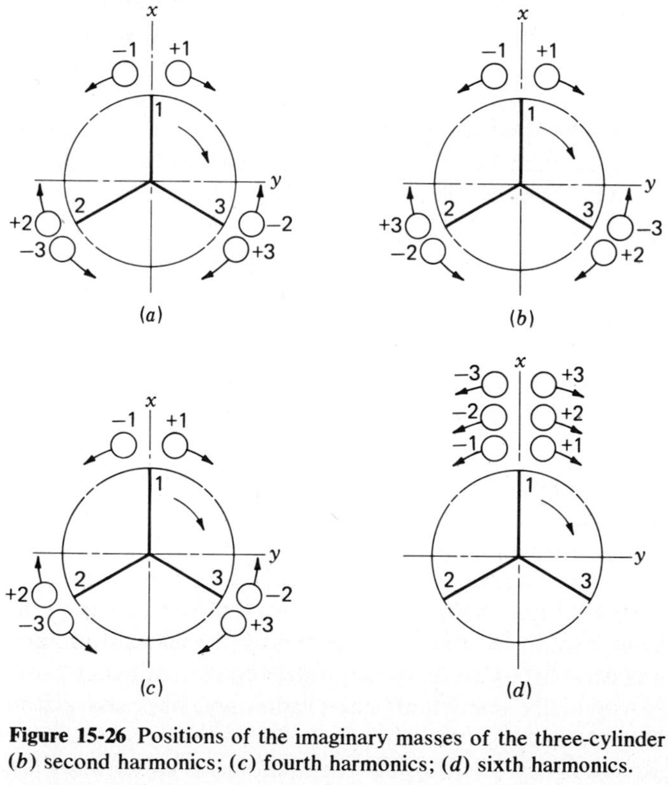

Figure 15-26 Positions of the imaginary masses of the three-cylinder engine: (*a*) first harmonics; (*b*) second harmonics; (*c*) fourth harmonics; (*d*) sixth harmonics.

Three-cylinder engine A three-cylinder in-line engine with cranks spaced 120° apart is shown in Fig. 15-25. Note that the cylinders are numbered according to the order in which they arrive at top dead center. Figure 15-26 shows that the first, second, and fourth harmonic forces are completely balanced, and only the sixth harmonic forces are completely unbalanced. These unbalanced forces will tend to create a vibration in the plane of the centerlines of the cylinders, but the magnitude of the forces is very small and can be neglected as far as vibration is concerned.

An analysis of the couples of the first harmonic forces shows that when crank 1 is on top dead center (Fig. 15-25), there is a vertical component of the forces on cranks 2 and 3 equal in magnitude to half of the force on crank 1. The resultant of these two downward components is equivalent to a force downward, equal in magnitude to the force on crank 1 and located halfway between cranks 2 and 3. Thus a couple is set up with an arm equal to the distance between the center of crank 1 and the centerline between cranks 2 and 3. At the same time, the horizontal components of the +2 and −2 forces cancel each other, as do the horizontal components of the +3 and −3 forces (Fig. 15-25). Therefore, no horizontal couple exists. Similar couples are found for both the second and fourth harmonics. Thus a three-cylinder engine, although inherently balanced for forces in the first, second, and fourth harmonics, still is not free from vibration due to the presence of couples at these harmonics.

Six-cylinder engine If a six-cylinder in-line engine is conceived as a combination of two three-cylinder engines back to back with parallel cylinders, it will have the same inherent balance of the first, second, and fourth harmonics. And, by virtue of symmetry, the couples of each three-cylinder engine will act in opposite directions and will balance each other. These couples, although perfectly balanced, tend to bend the crankshaft and crankcase and necessitate the use of rigid construction for high-speed operation. As before, the sixth harmonics are completely unbalanced and tend to create a vibration in the vertical plane with a frequency of 6ω. The magnitude of these forces, however, is very small and practically negligible as a source of vibration.

Other engines Taking into consideration the cylinder arrangement and crank spacing permits a great many configurations. For any combination, the balancing situation can be investigated to any harmonic desired by the methods outlined in this section. Particular attention must be paid in analyzing to that part of Stevensen's rule which calls for determining the angle of travel from the top dead center of the cylinder under consideration and moving the imaginary masses through the appropriate angles from that same top dead center. This is especially important when radial and opposed-piston engines are investigated.

As practice problems you may wish to use these methods to confirm the following facts:

1. In a three-cylinder radial engine with one crank and three connecting rods having the same crankpin, the negative masses are inherently balanced for the first harmonic forces while the positive masses are always located at

the crankpin. These two findings are inherently true for all radial engines. Also, since the radial engine has its cylinders in a single plane, unbalanced couples do not occur. The three-cylinder engine will have unbalanced forces in the second and higher harmonics.

2. A two-cylinder opposed-piston engine with a crank spacing of 180° is balanced for forces in the first, second, and fourth harmonics but unbalanced for couples.
3. A four-cylinder in-line engine with cranks at 90° is balanced for forces in the first harmonic but unbalanced for couples. In the second harmonic it is balanced for both forces and couples.
4. An eight-cylinder in-line engine with the cranks at 90° is inherently balanced for both forces and couples in the first and second harmonics but unbalanced in the fourth harmonic.
5. An eight-cylinder V engine with cranks at 90° is inherently balanced for forces in the first and second harmonics and for couples in the second. The unbalanced couples in the first harmonic can be balanced by counterweights that introduce an equal and opposite couple. Such an engine is unbalanced for forces in the fourth harmonic.

15-11 BALANCING LINKAGES†

The two problems that arise in balancing linkages are balancing the shaking force and balancing the shaking moment.

In force-balancing a linkage we must concern ourselves with the position of the total center of mass. If a way can be found to cause this total center of mass to remain stationary, the vector sum of all the frame forces will always be zero. Lowen and Berkof‡ have catalogued five methods of force balancing:

1. The method of static balancing, in which concentrated link masses are replaced by systems of masses that are statically equivalent
2. The method of principal vectors, in which an analytical expression is obtained for the center of mass and then manipulated to learn how its trajectory can be influenced
3. The method of linearly independent vectors, in which the center of mass of a mechanism is made stationary, causing the coefficients of the time-dependent terms of the equation describing the trajectory of the total center of mass to vanish
4. The use of cam-driven masses to keep the total center of mass stationary

† Those who wish to investigate this topic in detail should begin with the following reference, in which an entire issue is devoted to the subject of linkage balancing: G. G. Lowen and R. S. Berkof, Survey of Investigations into the Balancing of Linkages, *J. Mech.*, vol. 3, no. 4, p. 221, 1968. This issue contains 11 translations on the subject from the German and Russian literature.

‡ Ibid.

5. The addition of an axially symmetric duplicate mechanism by which the new combined total center of mass is made stationary.

Lowen and Berkof state that very few studies have been reported on the problem of balancing the shaking moment. This problem is discussed further in Sec. 15-12.

Here we present only the Berkof-Lowen method,† which employs the method of linearly independent vectors. The method will be developed completely for the four-bar linkage but only the final results given for a typical six-bar linkage. Here is the procedure. First, find the equation that describes the trajectory of the total center of mass of the linkage. This equation will contain certain terms whose coefficients are time-dependent. Then the total center of mass is made stationary by changing the position of the individual link masses so that the coefficients of all the time-dependent terms vanish. In order to accomplish this, it is necessary to write the equation in such a form that the time-dependent unit vectors contained in the equation are linearly independent.

In Fig. 15-27 a general four-bar linkage is shown having link masses m_2 located at G_2, m_3 located at G_3, and m_4 located at G_4. The coordinates a_i, ϕ_i, describe the positions of these points within each link. We begin by defining the position of the total center of mass of the linkage by the vector $\mathbf{r}_s$:

$$\mathbf{r}_s = \frac{1}{\mathcal{M}}(m_2\mathbf{r}_{s2} + m_3\mathbf{r}_{s3} + m_4\mathbf{r}_{s4}) \qquad (a)$$

where $\mathbf{r}_{s2}$, $\mathbf{r}_{s3}$, and $\mathbf{r}_{s4}$ are the vectors that describe the positions of m_2, m_3, and

† R. S. Berkof and G. G. Lowen, A New Method for Completely Force Balancing Simple Linkages, *J. Eng. Ind. Trans. ASME*, ser. B, vol. 91, no. 1, pp. 21–26, February 1969.

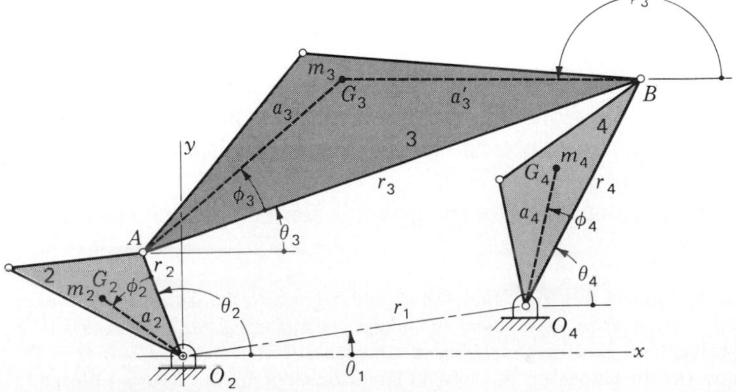

Figure 15-27 Four-bar linkage showing arbitrary positions of the link masses.

m_4, respectively, in the xy coordinate system. Thus, from Fig. 15-27,

$$\mathbf{r}_{s2} = a_2 e^{j(\theta_2+\phi_2)}$$
$$\mathbf{r}_{s3} = r_2 e^{j\theta_2} + a_3 e^{j(\theta_3+\phi_3)} \tag{b}$$
$$\mathbf{r}_{s4} = r_1 e^{j\theta_1} + a_4 e^{j(\theta_4+\phi_4)}$$

The total mass of the mechanism $\mathcal{M}$ is

$$\mathcal{M} = m_2 + m_3 + m_4 \tag{c}$$

Substituting Eq. (b) into (a) gives

$$\mathcal{M}\mathbf{r}_s = (m_2 a_2 e^{j\phi_2} + m_3 r_2)e^{j\theta_2} + (m_3 a_3 e^{j\phi_3})e^{j\theta_3} + (m_4 a_4 e^{j\phi_4})e^{j\theta_4} + m_4 r_1 e^{j\theta_1} \tag{d}$$

where we have used the identity $e^{j(\alpha+\beta)} = e^{j\alpha}e^{j\beta}$. For a four-bar linkage the vector loop-closure equation is

$$r_2 e^{j\theta_2} + r_3 e^{j\theta_3} - r_4 e^{j\theta_4} - r_1 e^{j\theta_1} = 0 \tag{e}$$

Thus, the time-dependent terms $e^{j\theta_2}$, $e^{j\theta_3}$, and $e^{j\theta_4}$ in Eq. (d) are not linearly independent. To make them so, solve Eq. (e) for one of the unit vectors, say $e^{j\theta_3}$, and substitute the result into Eq. (d). Thus

$$e^{j\theta_3} = \frac{1}{r_3}(r_1 e^{j\theta_1} - r_2 e^{j\theta_2} + r_4 e^{j\theta_4}) \tag{f}$$

Equation (d) now becomes

$$\mathcal{M}\mathbf{r}_s = \left(m_2 a_2 e^{j\phi_2} + m_3 r_2 - m_3 a_3 \frac{r_2}{r_3} e^{j\phi_3}\right)e^{j\theta_2} + \left(m_4 a_4 e^{j\phi_4} + m_3 a_3 \frac{r_4}{r_3} e^{j\phi_3}\right)e^{j\theta_4}$$
$$+ \left(m_4 r_1 + m_3 a_3 \frac{r_1}{r_3} e^{j\phi_3}\right)e^{j\theta_1} \tag{g}$$

Equation (g) shows that the center of mass can be made stationary at the position

$$\mathbf{r}_s = \frac{r_1}{r_3 \mathcal{M}}(m_4 r_3 + m_3 a_3 e^{j\phi_3})e^{j\theta_1} \tag{15-16}$$

if the following coefficients of the time-dependent terms vanish

$$m_2 a_2 e^{j\phi_2} + m_3 r_2 - m_3 a_3 \frac{r_2}{r_3} e^{j\phi_3} = 0 \tag{h}$$

$$m_4 a_4 e^{j\phi_4} + m_3 a_3 \frac{r_4}{r_3} e^{j\phi_3} = 0 \tag{i}$$

But Eq. (h) can be simplified by locating G_3 from point B instead of point A (Fig. 15-27). Thus

$$a_3 e^{j\phi_3} = r_3 + a_3' e^{j\phi_3'}$$

With this substitution Eq. (h) becomes

$$m_2 a_2 e^{j\phi_2} - m_3 a_3' \frac{r_2}{r_3} e^{j\phi_3'} = 0 \tag{j}$$

Equations (*i*) and (*j*) must be satisfied to obtain total force balance. These equations yield the two sets of conditions:

$$m_2a_2 = m_3a_3'\frac{r_2}{r_3} \quad \text{and} \quad \phi_2 = \phi_3'$$
$$m_4a_4 = m_3a_3\frac{r_4}{r_3} \quad \text{and} \quad \phi_4 = \phi_3 + \pi \tag{15-17}$$

A study of these conditions will show that the mass and its location can be specified in advance for any single link; and then full balance can be obtained by rearranging the mass of the other two links.

The usual problem in balancing a four-bar linkage is that link lengths r_i are specified in advance because of the function performed. For this situation counterweights can be added to the input and output links in order to redistribute their masses, while the geometry of the third moving link is undisturbed.

When adding counterweights, the following relations must be satisfied:

$$m_ia_i\underline{/\phi_i} = m_i^0a_i^0\underline{/\phi_i^0} + m_i^*a_i^*\underline{/\phi_i^*} \tag{15-18}$$

where m_i^0, a_i^0, ϕ_i^0 are the parameters of the unbalanced linkage, m_i^*, a_i^*, ϕ_i^* are the parameters of the counterweight, and m_i, a_i, ϕ_i are the parameters that result from Eqs. (15-17). A second condition that must generally be satisfied is

$$m_i = m_i^0 + m_i^* \tag{15-19}$$

If the solution to a balancing problem can remain as the mass-distance product $m_i^*a_i^*$, Eq. (15-19) need not be used and Eq. (15-18) can be solved to yield

$$m_i^*a_i^* = \sqrt{(m_ia_i)^2 + (m_i^0a_i^0)^2 - 2(m_ia_i)m_i^0a_i^0)[\cos(\phi_i - \phi_i^0)]} \tag{15-20}$$

$$\phi_i^* = \tan^{-1}\frac{m_ia_i \sin\phi_i - m_i^0a_i^0 \sin\phi_i^0}{m_ia_i \cos\phi_i - m_i^0a_i^0 \cos\phi_i^0} \tag{15-21}$$

Figure 15-28 shows a typical six-bar linkage and the notation. For this, the Berkof-Lowen conditions for total balance are

$$m_2\frac{a_2}{r_2}e^{j\phi_2} = m_5\frac{a_5'b_2}{r_5\,r_2}e^{j(\phi_5'+\alpha_2)} + m_3\frac{a_3'}{r_3}e^{j\phi_3'} \tag{15-22a}$$

$$m_4\frac{a_4}{r_4}e^{j\phi_4} = m_6\frac{a_6'b_4}{r_6\,r_4}e^{j\phi_6'} - m_3\frac{a_3}{r_3}e^{j(\phi_3+\alpha_4)} \tag{15-22b}$$

$$m_5\frac{a_5}{r_5}e^{j\phi_5} = -m_6\frac{a_6}{r_6}e^{j\phi_6} \tag{15-22c}$$

Similar relations can be devised for other six-bar linkages. For total balance Eqs. (15-22) shows that a certain mass-geometry relation between links 5 and 6 must be satisfied, after which the masses of any two links and their locations can be specified. Balance is then achieved by a redistribution of the masses of the remaining three movable links.

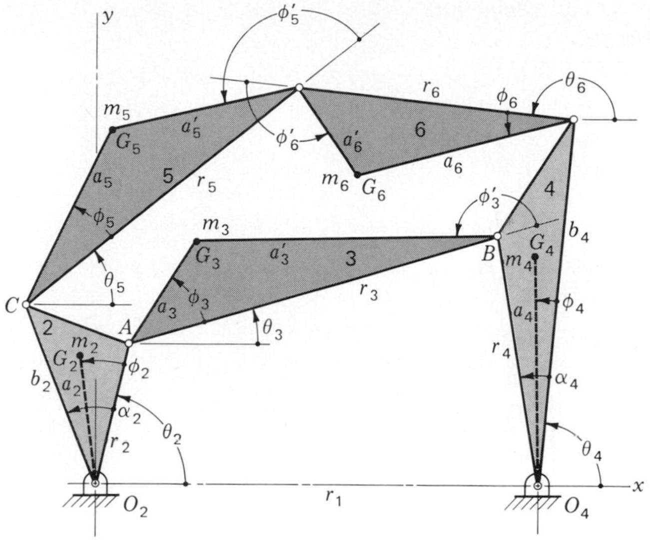

Figure 15-28 Notation for a six-bar linkage.

It is important to note that the addition of counterweights to balance the shaking forces will probably increase the internal bearing forces as well as the shaking moment. Thus only a partial balance may represent the best compromise between these three effects.

Example 15-3 Table 15-2 is a tabulation of the dimensions, masses, and the locations of the mass centers of a four-bar mechanism having link 2 as the input and link 4 as the output. Complete force balancing is desired by adding counterweights to the input and output links. Find the mass-distance values and the angular locations of each counterweight.

SOLUTION From Eqs. (15-17) we first find

$$m_2a_2 = m_3^0 a_3'^0 \frac{r_2}{r_3} = (0.125)(75.6)\frac{50}{150} = 3.15\ \text{g}\cdot\text{m}$$

$$\phi_2 = \phi_3'^0 = 164.1°$$

$$m_4a_4 = m_3^0 a_3^0 \frac{r_4}{r_3} = (0.125)(80)\frac{75}{150} = 5.0\ \text{g}\cdot\text{m}$$

$$\phi_4 = \phi_3^0 + 180 = 15 + 180 = 195°$$

Note that m_2a_2 and m_4a_4 are the mass-distance values *after* the counterweights have been added. Also, note that the link 3 parameters will not be altered.

We next compute

$$m_2^0a_2^0 = (0.046)(25) = 1.5\ \text{g}\cdot\text{m} \qquad m_4^0a_4^0 = (0.054)(40) = 2.16\ \text{g}\cdot\text{m}$$

Using Eq. (15-20), we compute the mass-distance values for link 2 counterweight as

$$\begin{aligned} m_2^*a_2^* &= \sqrt{(m_2a_2)^2 + (m_2^0a_2^0)^2 - 2(m_2a_2)(m_2^0a_2^0)\cos(\phi_2 - \phi_2^0)} \\ &= \sqrt{(3.15)^2 + (1.15)^2 - 2(3.15)(1.15)\cos(164.1° - 0°)} \\ &= 4.27\ \text{g}\cdot\text{m} \end{aligned}$$

Table 15-2 Parameters of an unbalanced four-bar linkage

Link i	1	2	3	4
r_i, mm	140	50	150	75
a_i^0, mm	—	25	80	40
ϕ_i^0	—	0°	15°	0°
a'^0, mm	—	—	75.6	—
$\phi_i'^0$	—	—	164.1°	—
m_i^0, kg	—	0.046	0.125	0.054

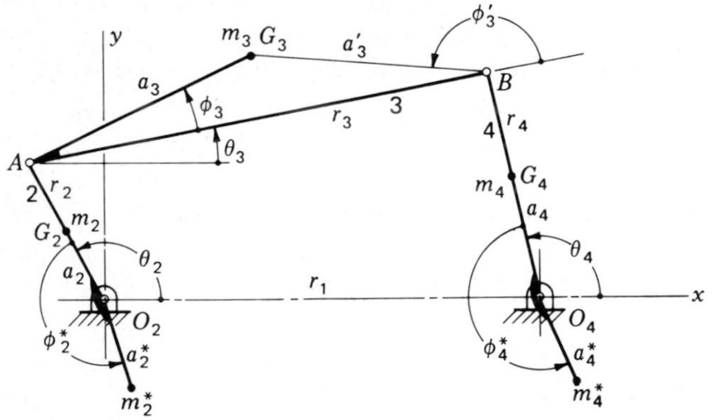

Figure 15-29 Four-bar crank-and-rocker linkage showing counterweights added to input and output links to achieve complete force balance.

From Eq. (15-21) we find the location of this counterweight to be

$$\phi_2^* = \tan^{-1}\frac{m_2a_2 \sin \phi_2 - m_2^0a_2^0 \sin \phi_2^0}{m_2a_2 \cos \phi_2 - m_2^0a_2^0 \cos \phi_2^0}$$

$$= \tan^{-1}\frac{3.15 \sin 164.1° - 1.15 \sin 0°}{3.15 \cos 164.1° - 1.15 \cos 0°} = 168.3°$$

Using the same procedure for link 4 yields

$$m_4^*a_4^* = 7.11 \text{ g} \cdot \text{m} \qquad \text{at } \phi_4^* = 190.5°$$

Figure 15-29 is a scale drawing of the complete linkage with the two counterweights added.

15-12 BALANCING OF MACHINES†

In the previous section we learned how to force-balance a simple linkage by using two or more counterweights, depending upon the number of links

† The material for this section is from E. N. Stevensen, Jr., Balancing of Machines, *J. Eng. Ind., Trans. ASME*, ser. B, vol. 95, pp. 650–656, May 1973. It is included with the advice and consent of Professor Stevensen.

composing the linkage. Unfortunately, this does not balance the shaking moments and, in fact, may even make them worse because of the addition of the counterweights. If a machine is imagined to be composed of several mechanisms, one might consider balancing the machine by balancing each mechanism separately. But this may not result in the best balance for the machine because the addition of a large number of counterweights may cause the inertia torque to be completely unacceptable. Furthermore, unbalance of one mechanism may counteract the unbalance in another, eliminating the need for some of the counterweights in the first place.

Stevensen shows that any single harmonic of unbalanced forces, moments of forces, and torques in a machine can be balanced by the addition of six counterweights. They are arranged on three shafts, two per shaft, driven at the constant speed of the harmonic, and have axes parallel, respectively, to each of the three mutually perpendicular axes through the center of mass of the machine. The method is too complex to be included in this book, but it will be worthwhile to look at the overall approach.

Using the methods of this book together with current computing facilities the linear and angular accelerations of each of the moving mass centers of a machine are computed for points throughout a cycle of motion. The masses and mass moments of inertia of the machine must also be computed or determined experimentally. Then the inertia forces, inertia torques, and the moments of the forces are computed with reference to the three mutually perpendicular coordinate axes through the center of mass of the machine. When these are summed for each point in the cycle, six functions of time will result, three for the forces and three for the moments. With the digital computer it is then possible to use numerical harmonic analysis to define the component harmonics of the unbalanced forces parallel to the three axes, and of the unbalanced moments about these axes.

To balance a single harmonic each component of the unbalance of the machine is now represented in the form $A \cos \omega t + B \sin \omega t$ with appropriate subscripts. Six equations of equilibrium are then written which include the unbalances as well as the effects of the six unknown counterweights. These equations are arranged so that each of the $\sin \omega t$ and $\cos \omega t$ terms is multiplied by parenthetical coefficients. Balance is then achieved by setting the parenthetical terms in each equation equal to zero, much in the manner of the preceding section. This results in 12 equations, all linear, in 12 unknowns. With the locations for the balancing counterweights on the three shafts specified, the 12 equations can be solved for the six mr products and the six phase angles needed for the six balancing weights. Stevensen goes on to show that when less than the necessary three shafts are available, it becomes necessary to optimize some effect of the unbalance, such as the motion of a point on the machine.

PROBLEMS

15-1 Determine the bearing reactions at *A* and *B* for the system shown in the figure if the speed is 300 rpm. Determine the magnitude and angular location of the balancing mass if it is located at a radius of 50 mm.

15-2 The figure shows three weights connected to a shaft that rotates in bearings at *A* and *B*. Determine the magnitude of the bearing reactions if the shaft speed is 300 rpm. A counterweight is to be located at a radius of 10 in. Find the value of the weight and its angular location.

15-3 The figure shows two weights connected to a rotating shaft and mounted outboard of bearings *A* and *B*. If the shaft rotates at 120 rpm, what are the magnitudes of the bearing reactions at *A* and *B*? Suppose the system is to be balanced by subtracting a weight at a radius of 5 in. Determine the amount and angular location of the weight to be removed.

15-4 For a speed of 220 rpm calculate the magnitude and relative angular direction of the bearing reactions at *A* and *B* for the two-mass system shown.

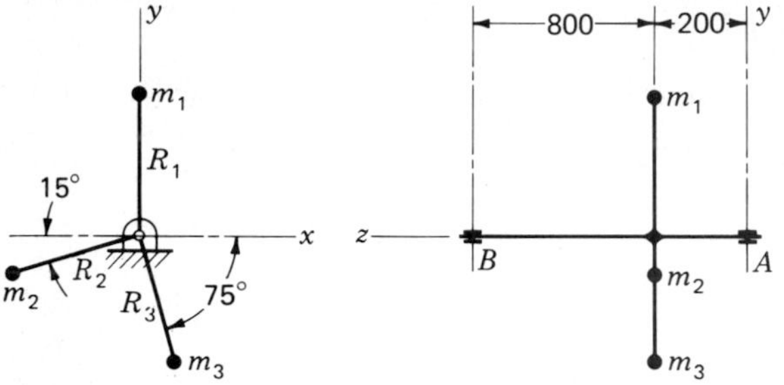

Problem 15-1 Dimensions in millimetres; $R_1 = 25$, $R_2 = 35$, $R_3 = 40$; $m_1 = 2$ kg, $m_2 = 1.5$ kg, $m_3 = 3$ kg.

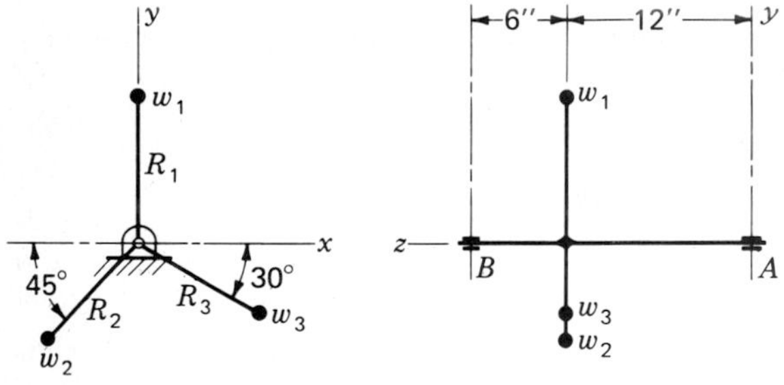

Problem 15-2 $R_1 = 8$ in, $R_2 = 12$ in, $R_3 = 6$ in, $w_1 = 2$ oz, $w_2 = 1.5$ oz, $w_3 = 3$ oz.

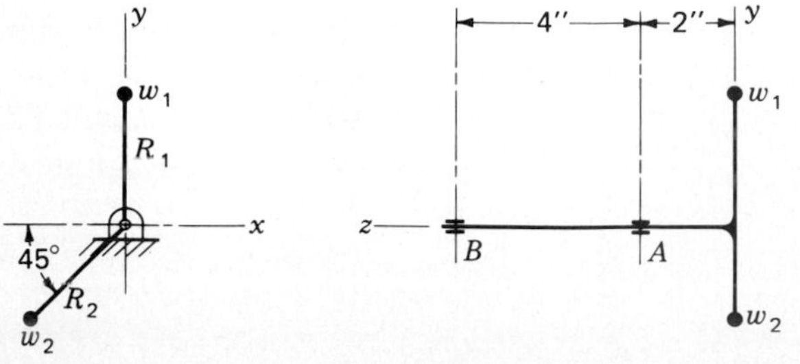

Problem 15-3 $R_1 = 4$ in, $R_2 = 6$ in, $w_1 = 4$ lb, $w_2 = 3$ lb.

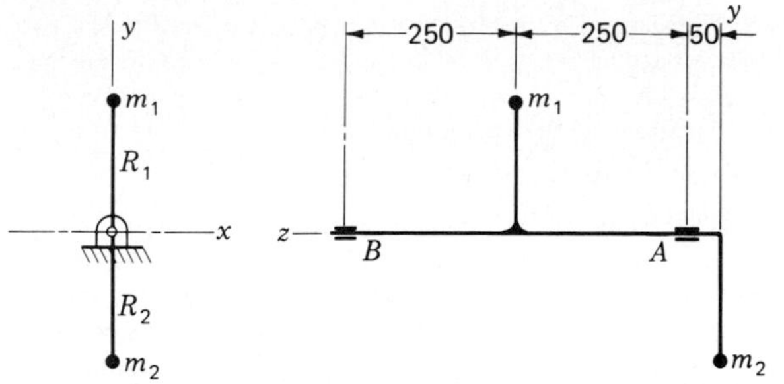

Problem 15-4 Dimensions in millimetres; $R_1 = 60$, $R_2 = 40$, $m_1 = 2$ kg, $m_2 = 1.5$ kg.

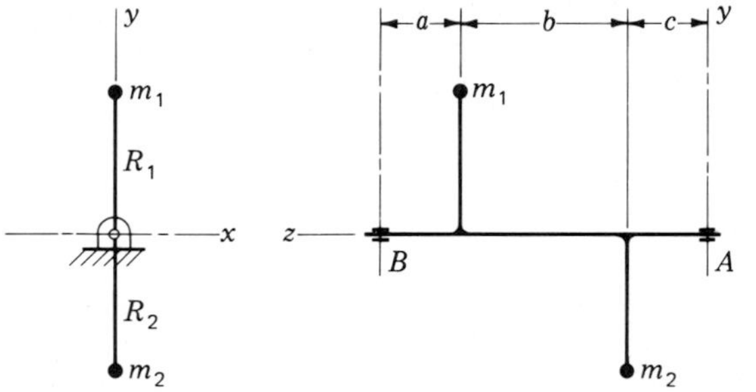

Problems 15-5 and 15-6

15-5 The rotating system shown in the figure has $R_1 = R_2 = 60$ mm, $a = c = 300$ mm, $b = 600$ mm, $m_1 = 1$ kg, and $m_2 = 3$ kg. Find the bearing reactions at A and B and their angular locations measured from a rotating reference mark if the shaft speed is 100 rpm.

15-6 The rotating shaft shown in the figure supports two masses m_1 and m_2, whose weights are 4 and 6 lb, respectively. The dimensions are $R_1 = 4$ in, $R_2 = 3$ in, $a = 2$ in, $b = 8$ in, and $c = 3$ in. Find the magnitude of the rotating bearing reactions at A and B and their angular locations from a rotating reference mark if the shaft rotates at 360 rpm.

15-7 The shaft shown in the figure is to be balanced by placing correction masses in the corrections planes L and R. The weights of the three masses m_1, m_2, and m_3 are, respectively, 4, 3, and 4 oz. The dimensions in inches are $R_1 = 5$, $R_2 = 4$, $R_3 = 5$, $a = 1$, $b = e = 8$, $c = 10$, and $d = 9$. Calculate the magnitude of the corrections in ounce-inches and their angular locations.

15-8 The shaft of Prob. 15-7 is to be balanced by removing weight from the two correction planes. Determine the corrections to be subtracted in ounce-inches and their angular locations.

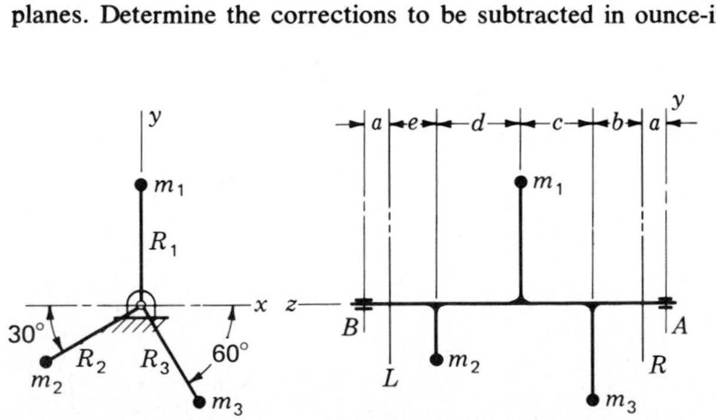

Problems 15-7 to 15-10

15-9 The shaft shown in the figure is to be balanced by subtracting correction masses in the two correction planes L and R. The three masses are $m_1 = 6$ g, $m_2 = 7$ g, and $m_3 = 5$ g. The dimensions in millimetres are $R_1 = 125$, $R_2 = 150$, $R_3 = 100$, $a = 25$, $b = 300$, $c = 600$, $d = 150$, and $e = 75$. Calculate the magnitude and the angular locations of the corrections.

15-10 Repeat Prob. 15-9 if correction masses are to be added to the two planes.

15-11 Solve the two-plane balancing problem as stated in Sec. 15-8.

15-12 A rotor to be balanced in the field yielded an amplitude of 5 at an angle of 142° at the left-hand bearing and an amplitude of 3 at an angle of −22° at the right-hand bearing due to unbalance. To correct this, a trial mass of 12 was added to the left-hand correction plane at an angle of 210° from the rotating reference. A second run then gave left- and right-hand responses of $8\underline{/160^\circ}$ and $4\underline{/260^\circ}$, respectively. The first trial mass was then removed and a second mass of 6 added to the right-hand correction plane at an angle of −70°. The responses to this were $2\underline{/74^\circ}$ and $4.5\underline{/-81^\circ}$ for the left- and right-hand bearings, respectively. Determine the original unbalances.

CHAPTER

SIXTEEN

CAM DYNAMICS

16-1 RIGID- AND ELASTIC-BODY CAM SYSTEMS

Figure 16-1*a* is a cross-sectional view showing the overhead valve arrangement in an automotive engine. In analyzing the dynamics of this or any cam system, we would expect to determine the contact force at the cam surface, the spring force, and the cam-shaft torque, all for a complete rotation of the cam. In one method of analysis the entire cam-follower train, consisting of the push rod, the rocker arm, and the valve stem together with the cam shaft, are considered rigid. If the members are in fact fairly rigid, and if the speed is moderate, such an analysis will usually produce quite satisfactory results. In any event such rigid-body analysis should always be made first.

Sometimes the speeds are so high or the members are so elastic (perhaps because of extreme length) that an elastic-body analysis must be used. This fact is usually discovered when troubles are encountered with the cam system. Such troubles will usually be evidenced by noise, chatter, unusual wear, poor product quality, or perhaps fatigue failure of some of the parts. In other cases laboratory investigation of the performance of a prototype cam system may reveal substantial differences between the theoretical and the observed performance.

Figure 16-1*b* is a mathematical model of an elastic-body cam system. Here m_3 is the mass of the cam and a portion of the cam shaft. The motion machined into the cam is the coordinate y, a function of the cam-shaft angle θ. The bending stiffness of the cam shaft is designated as k_4. The follower retaining spring is k_1. The masses m_1 and m_2 and the stiffnesses k_2 and k_3 are lumped characteristics of the follower train. The dashpots c_i are inserted to

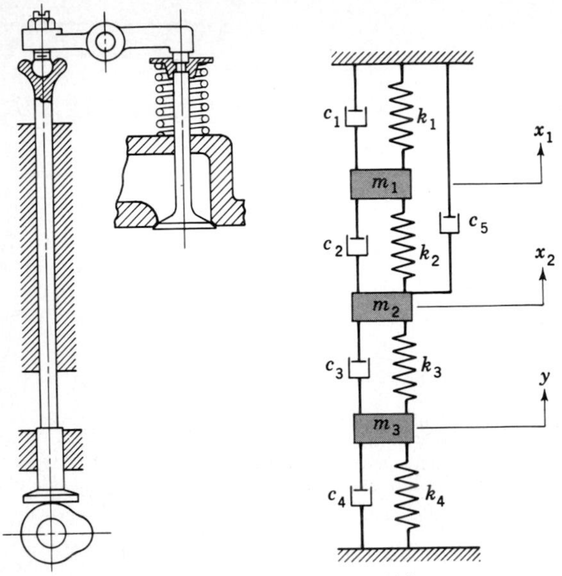

Figure 16-1

represent friction, which, in the analysis, may indicate either viscous or sliding friction or any combination of the two. The system of Fig. 16-1*b* is a rather sophisticated one requiring the solution of three simultaneous differential equations. We will deal with simpler systems in this book.

16-2 ANALYSIS OF AN ECCENTRIC CAM

An eccentric plate cam is a circular disk with the cam-shaft hole drilled off-center. The distance e between the center of the disk and the center of the shaft is called the *eccentricity*. Figure 16-2*a* shows a simple reciprocating-follower eccentric-cam system. It consists of a plate cam, a flat-face follower mass, and a retaining spring of stiffness k. The coordinate y designates the motion of the follower as long as cam contact is made. We arbitrarily select $y = 0$ at the bottom of the stroke. Then the kinematic quantities of interest are

$$y = e - e\cos\omega t \qquad \dot{y} = e\omega\sin\omega t \qquad \ddot{y} = e\omega^2\cos\omega t \tag{16-1}$$

where ωt is the same as the cam angle θ.

To make a rigid-body analysis, we assume no friction and construct a free-body diagram of the follower (Fig. 16-2*b*). In this figure F_{23} is the cam contact force and F_S is the spring force. In general F_{23} and F_S do not have the same line of action, and so a pair of frame forces $F_{13,A}$ and $F_{13,B}$ act at bearings A and B.

Before writing the equation of motion, let us investigate the spring force in more detail. By *spring stiffness k*, also called *spring rate*, we mean the

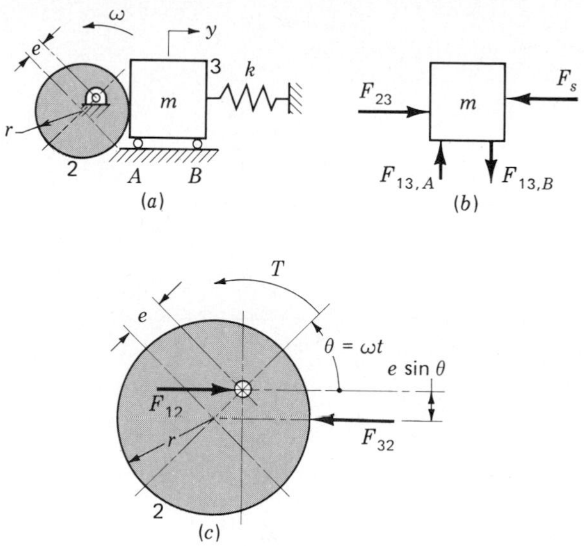

Figure 16-2 (*a*) Eccentric plate cam and flat-face follower; (*b*) free-body diagram of follower; (*c*) free-body diagram of cam.

amount of force necessary to deform the spring a unit length. Thus the units of k will usually be in newtons per metre or in pounds per inch. The purpose of the spring is to keep or retain the follower in contact with the cam. Thus, the spring should exert some force even at the bottom of the stroke where it is extended the most. This force, called the *preload P*, is the force exerted by the spring when $y = 0$. Thus $P = k\delta$, where δ is the distance the spring must be compressed in order to assemble it.

Summing forces on the follower mass in the y direction gives

$$\sum F^y = F_{23} - k(y+\delta) - m\ddot{y} = 0 \qquad (a)$$

Note particularly that F_{23} can only have positive values. Solving Eq. (*a*) for the contact force and substituting the first and third of Eqs. (16-1) gives

$$F_{23} = (ke + P) + (m\omega^2 - k)e \cos \omega t \qquad (16\text{-}2)$$

Figure 16-2*c* is a free-body diagram of the cam. The torque T, applied by the shaft to the cam, is

$$T = F_{23}e \sin \omega t = [(ke + P) + (m\omega^2 - k)e \cos \omega t]e \sin \omega t$$

$$= e(ke + P)\sin \omega t + \frac{e^2}{2}(m\omega^2 - k)\sin 2\omega t \qquad (16\text{-}3)$$

Equation (16-2) and Fig. 16-3*a* show that the contact force F_{23} consists of a constant term $ke + P$ with a cosine wave superimposed on it. The maximum occurs at $\theta = 0°$ and the minimum at $\theta = 180°$. The cosine or variable component has an amplitude that depends upon the square of the cam-shaft velocity. Thus, as the velocity increases, this term increases at a greater rate.

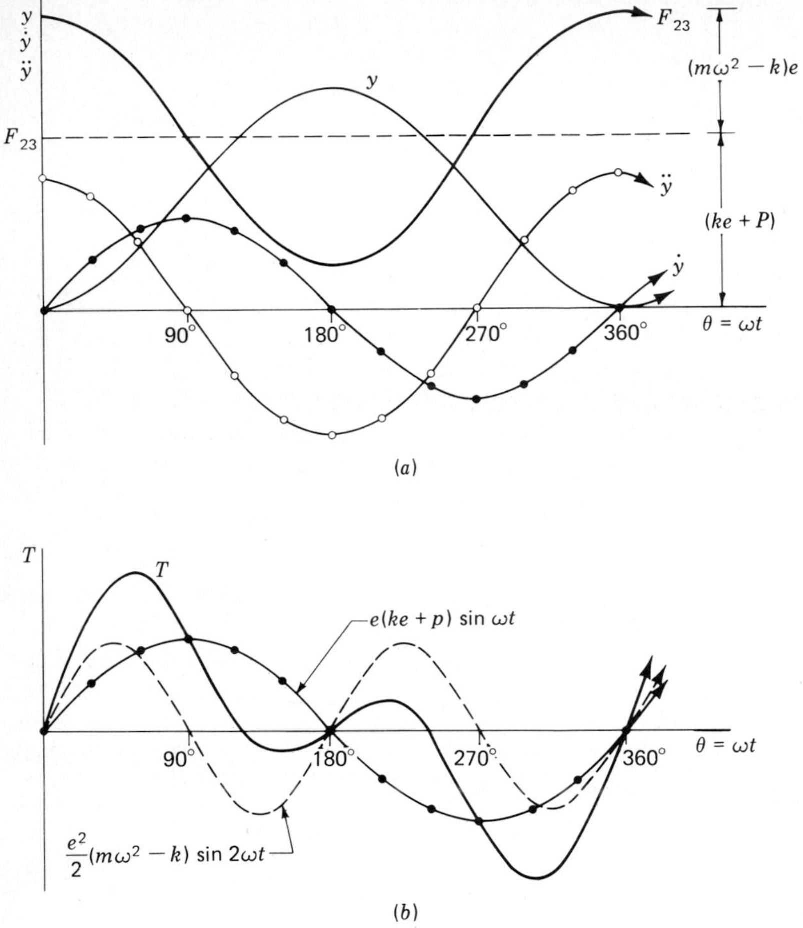

Figure 16-3 (*a*) Plot of displacement, velocity, acceleration, and contact force for an eccentric-cam system; (*b*) graph of torque components and total cam-shaft torque.

At a certain speed the contact force could become zero at $\theta = 180°$. When this happens, there is usually some impact between the cam and follower, resulting in clicking, rattling, or very noisy operation. In effect the sluggishness or inertia of the follower prevents it from following the cam. The result is often called *jump* or *float*. The noise occurs when contact is reestablished. Of course, the purpose of the retaining spring is to prevent this. Since the contact force consists of a cosine wave superimposed on a constant term, all we need do to prevent jump is to move or elevate the cosine wave away from the zero position. To do this, increase the term $ke + P$, either by increasing the preload P or the spring rate k or both.

Having learned that jump begins at $\cos \omega t = -1$ with $F_{23} = 0$, we can solve

Eq. (16-2) for the jump speed. The result is

$$\omega = \sqrt{\frac{2ke + P}{me}} \tag{16-4}$$

Using the same procedure, we find that jump will not occur if

$$P > e(m\omega^2 - 2k) \tag{16-5}$$

Figure 16-3*b* is a plot of Eq. (16-3). Note that the torque consists of a double-frequency component, whose amplitude is a function of the cam velocity squared, superimposed on a single-frequency component whose amplitude is independent of velocity. In this example the area of the torque-displacement diagram in the positive T direction is the same as in the negative T direction. This means that the energy required to drive the follower in the forward direction is recovered when the follower returns. A flywheel or inertia on the cam shaft can be used to handle this fluctuating energy requirement. Of course, if an external load is connected in some manner to the follower system, the energy required to drive this load would lift the torque curve in the positive direction and increase the area in the positive loop of the T curve.

Example 16-1 A cam-and-follower mechanism similar to Fig. 16-2*a* has the cam machined so that it will move the follower to the right through a distance of 40 mm with parabolic motion in 120° of cam rotation, dwell for 30°, then return with parabolic motion to the starting position in the remaining cam angle. The spring rate is 5 kN/m, and the mechanism is assembled with a 35-N preload. The follower mass is 18 kg. Assume no friction. (*a*) Without computing numerical values, sketch approximate graphs of the displacement motion, the acceleration, and the cam-contact force, all vs. cam angle for the entire cycle of events from $\theta = 0$ to $\theta = 360°$ of cam rotation. On this graph show where jump or liftoff is most likely to begin. (*b*) At what cam speed would jump begin? Show calculations.

SOLUTION (*a*) Solving Eq. (*a*) of Sec. 16-2 for the contact force gives

$$F = ky + P + m\ddot{y} \tag{1}$$

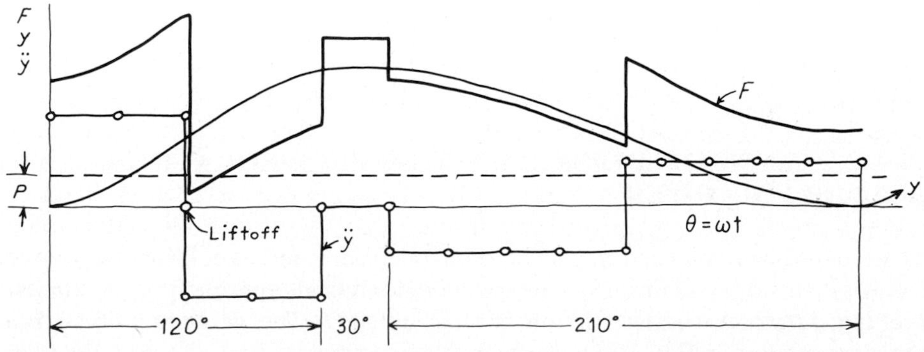

Figure 16-4

which is composed of a constant term P, a term ky that varies like the displacement, and a term $m\ddot{y}$ that varies like the acceleration. Figure 16-4 shows the displacement diagram, the acceleration $\ddot{y}$, and the cam contact force F. Note that jump would first occur at $\omega t = 60°$ since this is the closest approach of F to the zero position.

(*b*) Liftoff would occur at the half-point of rise where $\theta = \beta/2 = 60°$ when the acceleration goes negative. The terms for Eq. (1) are $y = 20$ mm and $ky = 5(20) = 100$ N. The acceleration is

$$\ddot{y} = -\frac{4L\omega^2}{\beta^2} = -\frac{4(0.040)\omega^2}{(120\pi/180)^2} = -0.0365\omega^2 \text{ m/s}^2$$

Substituting these values into Eq. (1) with $F = 0$ gives

$$F = ky + P + m\ddot{y} = 100 + 35 + 18(-0.036\,5\omega^2) = 0$$

or

$$\omega = \sqrt{\frac{100+35}{18(0.365)}} = 14.3 \text{ rad/s}$$

or

$$n = \frac{\omega}{2\pi}(60) = \frac{14.3(60)}{2\pi} = 137 \text{ rpm} \qquad \textit{Ans.}$$

16-3 EFFECT OF SLIDING FRICTION

Let F_μ be the force of sliding friction as defined by Eq. (12-23). Since friction force is always opposite in direction to velocity, let us define a sign function as follows:

$$\text{sign } \dot{y} = \begin{cases} +1 & \dot{y} \geq 0 \\ -1 & \dot{y} < 0 \end{cases} \tag{16-6}$$

With this notation, Eq. (*a*) of Sec. 16-2 can be written

$$\sum F^y = F_{23} - F_\mu \text{ sign } \dot{y} - k(y + \delta) - m\ddot{y} = 0 \tag{a}$$

or

$$F_{23} = F_\mu \text{ sign } \dot{y} + k(y+\delta) + m\ddot{y} \tag{16-7}$$

This equation is plotted for simple harmonic motion with no dwells in Fig. 16-5. Study both parts of this diagram; note that F_μ is positive when $\dot{y}$ is positive and see how F_{23} is obtained by summing the four component curves graphically.

16-4 ANALYSIS OF A DISK CAM WITH RECIPROCATING ROLLER FOLLOWER

In Chap. 12 we analyzed a cam system incorporating a reciprocating roller follower. In this section we present an analytical approach to a similar problem in which sliding friction is also included. The geometry of such a system is shown in Fig. 16-6*a*. In the analysis to follow, the effect of follower weight on bearings B and C is neglected.

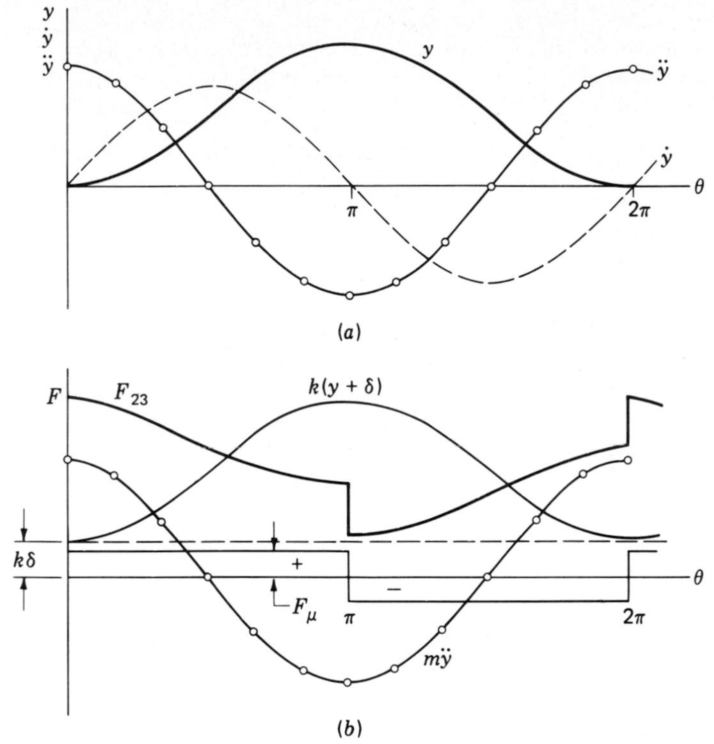

Figure 16-5 Effect of sliding friction on a cam system with harmonic motion: (*a*) graph of displacement, velocity, and acceleration for one motion cycle; (*b*) force diagram showing graph of components F_μ, $k\delta$, $k(y+\delta)$, $m\ddot{y}$, and resultant contact force F_{23}.

Figure 16-6*b* is a free-body diagram of the follower and roller. If y is any motion machined into the cam and $\theta = \omega t$ is the cam angle, at $y = 0$ the follower is at the bottom of its stroke and so $O_2A = R + r$. Therefore

$$a = R + r + y \tag{16-8}$$

In Fig. 16-6*b* the roller contact force forms the angle ϕ, the pressure angle, with the x axis. Since the direction of F_{23} is the same as the normal to the contacting surfaces, the intersection of this line with the x axis is the common instant center of the cam and follower. This means that the velocity of this point is the same no matter whether it is considered as a point on the follower or a point on the cam. Therefore

$$\dot{y} = a\omega \tan \phi$$

and so

$$\tan \phi = \frac{\dot{y}}{a\omega} \tag{16-9}$$

In the analysis to follow, the two bearing reactions are N_B and N_C, the

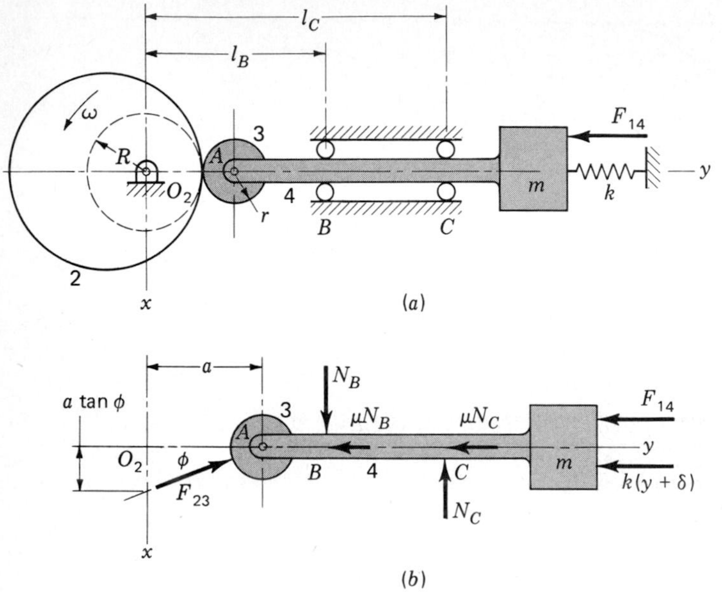

Figure 16-6 (a) Plate cam driving a reciprocating roller-follower system. Base-circle radius is R. (b) Free-body diagram of follower system.

coefficient of sliding friction is μ, and δ is the precompression of the retaining spring. Summing forces in the x and y directions gives

$$\sum F_{3,4}^x = F_{23}^x - N_B + N_C = 0 \qquad (a)$$

$$\sum F_{3,4}^y = F_{23}^y - \mu \text{ sign } \dot{y}\,(N_B + N_C) - F_{14} - k(y+\delta) - m\ddot{y} = 0 \qquad (b)$$

A third equation is obtained by taking moments about A

$$\sum M_A = -N_B(l_B - a) + N_C(l_C - a) = 0 \qquad (c)$$

With the help of Eq. (16-9) these three equations can be solved for the unknowns F_{23}, N_B, and N_C.

First, solve Eq. (c) for N_C. This gives

$$N_C = N_B \frac{l_B - a}{l_C - a} \qquad (d)$$

Now substitute Eq. (d) into Eq. (a) and solve for N_B. The result is

$$N_B = \frac{l_C - a}{l_C - l_B} F_{23}^x \qquad (e)$$

But since $F_{23}^x = F_{23}^y \tan \phi$,

$$N_B = \frac{F_{23}^y(l_C - a)\tan \phi}{l_C - l_B} \qquad (16\text{-}10)$$

Next, substitute Eqs. (*d*) and (16-10) into the friction term of Eq. (*b*)

$$\mu \text{ sign } \dot{y}\ (N_B + N_C) = \mu F_{23}^{y} \tan\phi \text{ sign } \dot{y}\ \frac{l_C + l_B - 2a}{l_C - l_B} \tag{f}$$

Substituting this result back into Eq. (*b*) and solving for F_{23}^{y} gives

$$F_{23}^{y} = \frac{F_{14} + k(y+\delta) + m\ddot{y}}{1 - \mu \tan\phi \text{ sign } \dot{y}\ [(l_C + l_B - 2a)/(l_C - l_B)]} \tag{16-11}$$

For computer or calculator solution, a simple solution to the sign function is

$$\text{sign } \dot{y} = \frac{\dot{y}}{|\dot{y}|} \tag{16-12}$$

Finally, the cam-shaft torque is

$$T = -aF_{23}^{y} \tan\phi \tag{16-13}$$

The equations of this section require the kinematic relations for the appropriate rise and return motions, developed in Chap. 6.

16-5 PROGRAMMING FOR COMPUTER OR CALCULATOR SOLUTION

Throughout this book, particular care has been taken to develop both graphical and analytical solutions to problems before presenting computer or programmable-calculator techniques. Every good engineer knows that such routines must be rigorously checked out; graphical solutions are almost always the most satisfactory means of checking. It is especially important to check computer routines when they are to be turned over to a subordinate for solution; many people blindly accept computer solutions as infallible and so may fail to detect even "obvious" errors.

Programming is apparently a rather personal thing because of the variety of approaches that can be observed in the solution of a single problem when it is presented to many people. To encourage a variety of programming approaches, we present here only the elements or parts of cam programs. You can assemble these parts in the manner best suiting your programming preferences and the facilities available to you.

If we restrict our attention to a reciprocating rise-dwell-return-dwell follower motion, it will be convenient to designate the angles for each of these events as β_1, β_2, β_3, β_4, respectively. For a one-lobed cam these angles would sum to 360°. We can then denote θ_1 as the cam angle during rise. Thus θ_1 has the range $0 \le \theta_1 \le \beta_1$. Using $\theta_3 = \theta - \beta_1 - \beta_2$ as the cam angle for the return motion, we note that its range will be $0 \le \theta_3 \le \beta_3$. If a programmable calculator is to be used, it should be placed in the radian mode.

The kinematic relations will have to be solved first. For illustrative purposes we present here only the basic cam motions—parabolic, simple harmonic, and cycloidal. You will probably want to employ subroutines for these, one for each rise and another for the return. The parabolic motion would require four subroutines, however, two for the rise motion and two for the return. The relations for these three basic motions are as follows.

Parabolic motion The first half of rise is the range $0 \le y \le L/2$. The equations are Eq. (1), $y = 2L(\theta_1/\beta_1)^2$; Eq. (2), $\dot{y} = 4L\omega\theta_1/\beta_1^2$; Eq. (3), $\ddot{y} = 4L\omega^2/\beta_1^2$. The second half of rise corresponds to the range $L/2 < y \le L$; the equations are Eq. (4), $y = L\{1 - 2[1 - (\theta_1/\beta_1)]^2\}$; Eq. (5), $\dot{y} = 4L\omega[1 - (\theta_1/\beta_1)]/\beta_1$; Eq. (6), $\ddot{y} = -4L\omega^2/\beta_1^2$. The first half of the return motion is in the range $L \ge y \ge L/2$; the equations are Eq. (7), $y = L[1 - 2(\theta_3/\beta_3)^2]$; Eq. (8), $\dot{y} = -4L\omega\theta_3/\beta_3^2$; Eq. (9), $\ddot{y} = -4L\omega^2/\beta_3^2$. The range for the second half of the return motion is $L/2 > y \ge 0$, and the equations are Eq. (10), $y = 2L[1 - (\theta_3/\beta_3)]^2$; Eq. (11), $\dot{y} = 4L\omega[(\theta_3/\beta_3) - 1]/\beta_3$; Eq. (12), $\ddot{y} = 4L\omega^2/\beta_3^2$.

Simple harmonic motion The equations for the rise motion are Eq. (13), $y = L[1 - \cos(\pi\theta_1/\beta_1)]/2$; Eq. (14), $\dot{y} = \pi L\omega[\sin(\pi\theta_1/\beta_1)]/2\beta_1$; Eq. (15), $\ddot{y} = L[(\pi\omega/\beta_1)^2 \cos(\pi\theta_1/\beta_1)]/2$. For the return, the equations are Eq. (16), $y = L[1 + \cos(\pi\theta_3/\beta_3)]/2$; Eq. (17), $\dot{y} = -\pi L\omega[\sin(\pi\theta_3/\beta_3)]/2\beta_3$; Eq. (18), $\ddot{y} = -L(\pi\omega/\beta_3)^2[\cos(\pi\theta_3/\beta_3)]/2$.

Cycloidal motion For the rise Eq. (19), $y = L[(\theta_1/\beta_1) - (1/2\pi)\sin(2\pi\theta_1/\beta_1)]$; Eq. (20), $\dot{y} = L\omega[1 - \cos(2\pi\theta_1/\beta_1)]/\beta_1$; Eq. (21), $\ddot{y} = 2\pi L(\omega/\beta_1)^2 \sin(2\pi\theta_1/\beta_1)$. For the return Eq. (22), $y = L[1 - (\theta_3/\beta_3) + (1/2\pi)\sin(2\pi\theta_3/\beta_3)]$; Eq. (23), $\dot{y} = L\omega[\cos(2\pi\theta_3/\beta_3) - 1]/\beta_3$; Eq. (24), $\ddot{y} = -2\pi L(\omega/\beta_3)^2 \sin(2\pi\theta_3/\beta_3)$.

Dynamics program The final step in programming is to work in the dynamics subroutine. If we use the solution in the previous section, the order in which the equations could be solved is Eqs. (16-12), (16-8), (16-9), (16-11), (16-13), F_{23}^x, (16-10), (*d*).

Discontinuities Special precautions must be taken when the motion has a discontinuity. A dwell preceding simple harmonic motion, for example, causes an acceleration discontinuity at the beginning of rise. Discontinuities occur at the beginning and end of parabolic motion and also at the midpoint of rise. One method of getting around such problems is to make the computations at an infinitesimal angle removed from the discontinuity.

Note, too, that the equations of this section do not apply to the dwell periods of the motion.

16-6 ANALYSIS OF ELASTIC CAM SYSTEMS

Figure 16-7 illustrates the effect of follower elasticity upon the displacement and velocity characteristics of a follower system driven by a cycloidal cam.

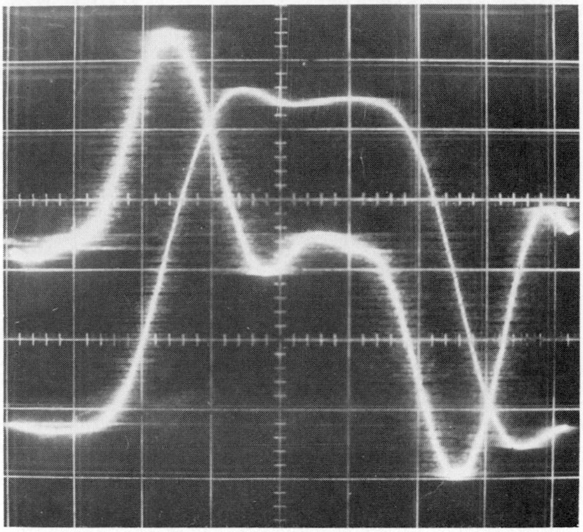

Figure 16-7 Photograph of the oscilloscope traces of the displacement and velocity characteristics of a dwell-rise-dwell-return cam and follower system machined for cycloidal motion. The zero axis of the displacement diagram has been translated downward to obtain a larger diagram in the space available.

To see what has happened, you should compare these diagrams with the theoretical ones in Chap. 6. Though the effect of elasticity is most pronounced for the velocity characteristic, it is usually the modification of the displacement characteristic, especially at the top of rise, that causes the most trouble in practical situations. These troubles are usually evidenced by poor or unreliable product quality when the systems are used in manufacturing or assembly lines, noise, unusual wear, and fatigue failure.

A complete analysis of elastic cam systems requires a good background in vibration studies. To avoid the necessity for this background while still developing a basic understanding we will use an extremely simplified cam system using a linear-motion cam. It must be observed, however, that such a cam system should never be used for high-speed applications.

In Fig. 16-8*a*, k_1 is the retaining spring, m is the lumped mass of the follower, and k_2 represents the stiffness of the follower. Since the follower is usually a rod or a lever, k_2 is many times greater than k_1.

Spring k_1 is assembled with a preload. The coordinate x of the follower motion is chosen at the equilibrium position of the mass after spring k_1 is assembled. Thus k_1 and k_2 will exert equal and opposite preload forces on the mass. Assuming no friction, the free-body diagram of the mass is as shown in Fig. 16-8*b*. To determine the direction of the forces the coordinate x, representing the motion of the follower, has been assumed to be larger than the coordinate y, representing the motion machined into the cam. However, the same result will be obtained if y is assumed larger than x.

Using Fig. 16-8*b*, we find the equation of motion to be

$$\sum F = -k_1 x - k_2(x - y) - m\ddot{x} = 0 \qquad (a)$$

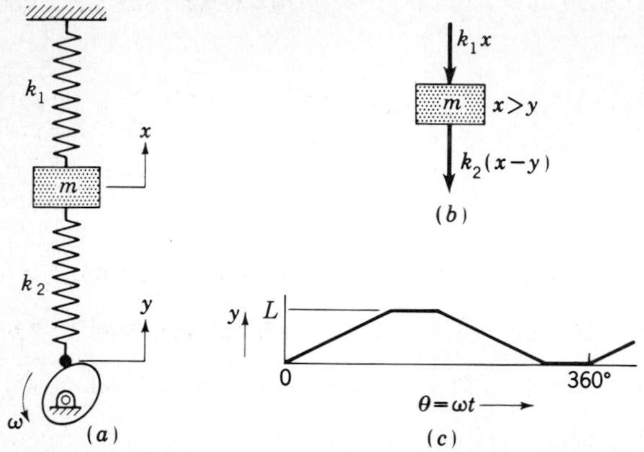

Figure 16-8 (*a*) Undamped model of a cam mechanism. (*b*) Free-body diagram of the mass. (*c*) Displacement diagram.

or

$$\ddot{x} + \frac{k_1 + k_2}{m} x = \frac{k_2}{m} y \tag{16-14}$$

This is the differential equation for the motion of the follower. It can be solved using vibration theory when the function y is specified. This equation must be solved piecewise for each cam event; i.e., the ending conditions for one event, or era of motion, must be used as the beginning or starting conditions for the next era.

Let us analyze the first era of motion using uniform motion as illustrated in Fig. 16-8*c*. First we use the notation

$$\omega_n = \sqrt{\frac{k_1 + k_2}{m}} \tag{16-15}$$

Do not confuse ω_n with the angular cam velocity ω. The quantity ω_n here is called the *natural undamped circular frequency* in vibration theory. The units of ω_n are reciprocal seconds and are usually written as radians per second. This is implied by the circular nature of the quantity.

Equation (16-14) can now be written

$$\ddot{x} + \omega_n^2 x = \frac{k_2 y}{m} \tag{16-16}$$

The solution to this equation is

$$x = A \cos \omega_n t + B \sin \omega_n t + \frac{k_2 y}{m\omega_n^2} \tag{b}$$

where

$$y = \frac{L}{\beta}\theta = \frac{L\omega t}{\beta} \tag{16-17}$$

Of course Eq. (16-17) is valid only during the rise era. You can verify Eq. (*b*) as the solution by substituting it and its second derivative into Eq. (16-16).

The first derivative of Eq. (*b*) is

$$\dot{x} = -A\omega_n \sin \omega_n t + B\omega_n \cos \omega_n t + \frac{k_2 \dot{y}}{m\omega_n^2} \qquad (c)$$

Using $t = 0$ at the beginning of rise with $x = \dot{x} = 0$, we find from Eqs. (b) and (c) that

$$A = 0 \qquad B = -\frac{k_2 \dot{y}}{m\omega_n^3}$$

Thus Eq. (b) becomes

$$x = \frac{k_2}{m\omega_n^2}\left(y - \frac{\dot{y}}{\omega_n} \sin \omega_n t\right) \qquad (16\text{-}18)$$

This equation is plotted in Fig. 16-9. Note that the motion consists of a negative sine term superimposed upon a ramp representing the uniform rise. Because of the additional compression of k_2 during rise, the ramp term $k_2 y/m\omega_n^2$, called the follower command in the figure, becomes less than the rise motion y.

At the end of rise Eqs. (16-16) to (16-18) are no longer valid, and a second era of motion begins. The follower response for this era is shown in Fig. 16-9, but we will not solve for it.†

Equation (16-18) shows that the vibration amplitude $\dot{y}/\omega_n$ can be reduced by making ω_n large, and Eq. (16-15) shows that this can be done by increasing k_2, which means that a very rigid follower should be used.

† This problem can easily be solved using a graphical approach called the *phase-plane method.* A numerical example can be found in Joseph E. Shigley, "Dynamic Analysis of Machines," McGraw-Hill, New York, 1961, p. 583.

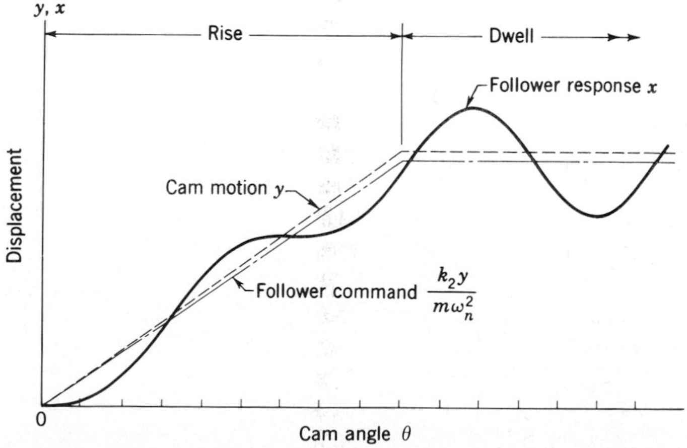

Figure 16-9 Displacement diagram of a uniform-motion cam mechanism showing the follower response.

16-7 UNBALANCE, SPRING SURGE, AND WINDUP

As shown in Fig. 16-10*a*, a disk cam produces unbalance because its mass is not symmetrical with the axis of rotation. This means that two sets of vibratory forces exist, one due to the eccentric cam mass and the other due to the reaction of the follower against the cam. By keeping these effects in mind during design, the engineer can do much to guard against difficulties during operation.

Figure 16-10*b* and *c* shows that the face and cylindrical cams have good balance characteristics. For this reason these are good choices when high-speed operation is involved.

Spring surge It is shown in texts on spring design that helical springs may themselves vibrate when subjected to rapidly varying forces. For example, poorly designed automotive valve springs operating near the critical frequency range permit the valve to open for short intervals during the period the valve is supposed to be closed. Such conditions result in very poor operation of the engine and rapid fatigue failure of the springs themselves. This vibration of the retaining spring, called *spring surge*, has been photographed with high-speed motion-picture cameras and the results exhibited in slow motion. When serious vibrations exist, a clear wave motion can be seen traveling up and down the valve spring.

Windup Figure 16-3*b* is a plot of cam-shaft torque showing that the shaft exerts torque on the cam during a portion of the cycle and that the cam exerts torque on the shaft during another portion of the cycle. This varying torque requirement may cause the shaft to twist, or wind up, as the torque increases during follower rise. Also, during this period, the angular cam velocity is slowed and so is the follower velocity. Near the end of rise the energy stored in the shaft by the windup is released, causing both the follower velocity and

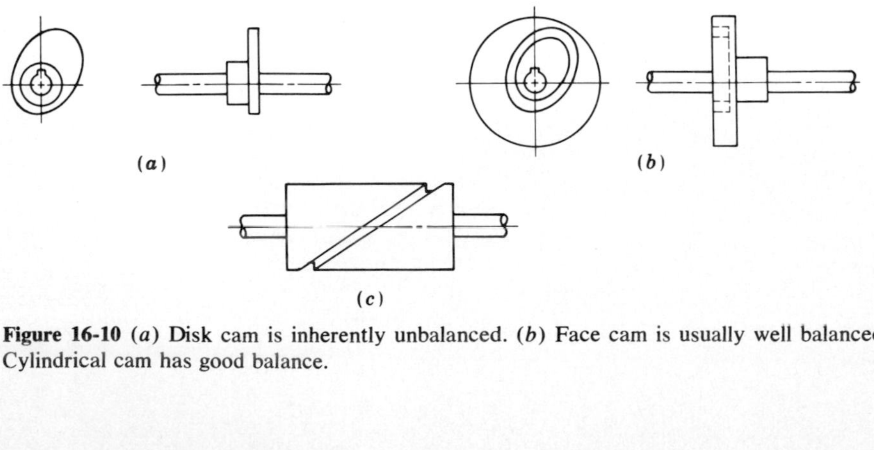

Figure 16-10 (*a*) Disk cam is inherently unbalanced. (*b*) Face cam is usually well balanced. (*c*) Cylindrical cam has good balance.

acceleration to rise above normal values. The resulting kick may produce follower jump or impact.

This effect is most pronounced when heavy loads are being moved by the follower, when the follower moves at a high speed, and when the shaft is flexible.

In most cases a flywheel must be employed in cam systems to provide for the varying torque requirement (see Sec. 17-1). Cam-shaft windup can be prevented to a large extent by mounting the flywheel as close as possible to the cam. Mounting it a long distance from the cam may actually worsen matters.

PROBLEMS

16-1 In part (*a*) of the figure the mass is constrained to move only in the vertical direction. The eccentric cam has an eccentricity of 2 in and a speed of 20 rad/s, and the weight of the mass is 8 lb. Neglecting friction, find the angle $\theta = \omega t$ at the instant jump begins.

16-2 In part (*a*) of the figure the mass m is driven up and down by the eccentric cam and it has a weight of 10 lb. The cam eccentricity is 1 in. Assume no friction.

(*a*) Derive the equation for the contact force.

(*b*) Find the cam velocity ω corresponding to the beginning of jump.

16-3 In part (*a*) of the figure the slider has a mass of 2.5 kg. The cam is a simple eccentric and causes the slider to rise 25 mm with no friction. At what cam speed in revolutions per minute will the slider first lose contact with the cam? Sketch a graph of the contact force at this speed for 360° of cam rotation.

16-4 The cam-and-follower system shown in part (*b*) of the figure has $k = 1$ kN/m, $m = 0.90$ kg, $y = 15 - 15 \cos \omega t$ mm, and $\omega = 60$ rad/s. The retaining spring is assembled with a preload of 2.5 N.

(*a*) Compute the maximum and minimum values of the contact force.

(*b*) If the follower is found to jump off the cam, compute the angle ωt corresponding to the very beginning of jump.

16-5 Part (*b*) of the figure shows the mathematical model of a cam-and-follower system. The motion machined into the cam is to move the mass to the right through a distance of 2 in with parabolic motion in 150° of cam rotation, dwell for 30°, return to the starting position with simple

Problems 16-1 to 16-6

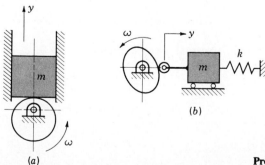

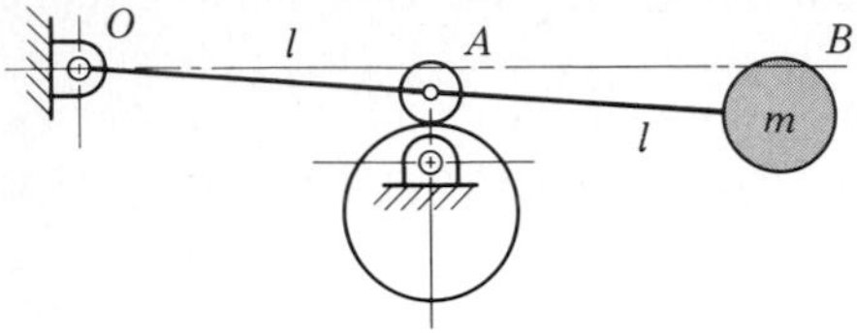

Problem 16-7

harmonic motion in 150°, and dwell for the remaining 30° of cam angle. There is no friction or damping. The spring rate is 40 lb/in, and the spring preload is 6 lb, corresponding to the $y = 0$ position. The weight of the mass is 36 lb.

(*a*) Sketch a displacement diagram showing the follower motion for the entire 360° of cam rotation. Without computing numerical values, superimpose graphs of the acceleration and cam contact force onto the same axes. Show where jump is most likely to begin.

(*b*) At what cam speed in revolutions per minute would jump begin?

16-6 A cam-and-follower mechanism is shown in abstract form in part (*b*) of the figure. The cam is cut so that it causes the mass to move to the right a distance of 25 mm with harmonic motion in 150° of cam rotation, dwell for 30°, then return to the starting position in the remaining 180° of cam rotation, also with harmonic motion. The spring is assembled with a 22-N preload and it has a rate of 4.4 kN/m. The follower mass is 17.5 kg. Compute the cam speed in revolutions per minute at which jump would begin.

16-7 The figure shows a lever *OAB* driven by a cam cut to give the roller a rise of 1 in with parabolic motion and a parabolic return with no dwells. The lever and roller are to be assumed as weightless, and there is no friction. Calculate the jump speed if $l = 5$ in and the mass weighs 5 lb.

16-8 A cam-and-follower system similar to the one of Fig. 16-6 uses a plate cam driven at a speed of 600 rpm and employs simple harmonic rise and parabolic return motions. The events are rise in 150°, dwell for 30°, and return in 180°. The retaining spring has a rate $k = 14$ kN/m with a precompression of 12.5 mm. The follower has a mass of 1.6 kg. The external load is related to the follower motion y by the relation $F = 0.325 - 10.75y$, where y is in metres and F is in kilonewtons. Dimensions in millimetres corresponding to Fig. 16-6 are $R = 20$, $r = 5$, $l_B = 60$, and $l_C = 90$. Using a rise of $L = 20$ mm and assuming no friction, plot the displacement, cam-shaft torque, and radial component of the cam force for one complete revolution of the cam.

16-9 Repeat Prob. 16-8 if the speed is 900 rpm, $F = 0.110 + 10.75y$ kN where y is in metres, and the coefficient of sliding friction is $\mu = 0.025$.

16-10 A plate cam drives a reciprocating roller follower through the distance $L = 1.25$ in with parabolic motion in 120° of cam rotation, dwells for 30°, and returns with cycloidal motion in 120°, then dwells for the remaining cam angle. The external load on the follower is $F_{14} = 36$ lb during rise and zero during the dwells and the return. In the notation of Fig. 16-6, $R = 3$ in, $r = 1$ in, $l_B = 6$ in, $l_C = 8$ in, and $k = 150$ lb/in. The spring is assembled with a preload of 37.5 lb when the follower is at the bottom of its stroke. The weight of the follower is 1.8 lb, and the cam velocity is 140 rad/s. Assuming no friction, plot the displacement, the torque exerted on the cam by the shaft, and the radial component of the contact force exerted by the roller against the cam surface for one complete cycle of motion.

16-11 Repeat Prob. 16-10 if friction exists with $\mu = 0.04$ and the cycloidal return takes place in 180°.

CHAPTER

SEVENTEEN

MACHINE DYNAMICS

17-1 FLYWHEELS

A flywheel is an energy-storage device. It absorbs mechanical energy by increasing its angular velocity and delivers energy by decreasing its velocity. Commonly, the flywheel is used to smooth out the flow of energy between a power source and its load. If the load happens to be a punch press, the actual punching operation requires energy only for a fraction of its motion cycle. If the power source happens to be a two-cylinder four-cycle engine, the engine delivers energy during only about half of its motion cycle. Newer uses being investigated involve using a flywheel to absorb braking energy and deliver accelerating energy for an automobile and to act as energy-smoothing devices for electric utilities as well as solar and wind-power generating facilities. Electric railways have long used regenerative braking by feeding braking energy back into the power lines, but newer and stronger materials now make the flywheel more feasible for such purposes.

Figure 17-1 is a mathematical representation of a flywheel. The flywheel, whose motion is measured by the angular coordinate θ, has a mass moment of inertia I. An input torque T_i, corresponding to a coordinate θ_i, will cause the flywheel speed to increase. And a load or output torque T_o, with corresponding coordinate θ_o, will absorb energy from the flywheel and cause it to slow down. If T_i is considered positive and T_o negative, the equation of motion of the flywheel is

$$\sum M = T_i(\theta_i, \dot{\theta}_i) - T_o(\theta_o, \dot{\theta}_o) - I\ddot{\theta} = 0$$

or

$$I\alpha = T_i(\theta_i, \omega_i) - T_o(\theta_o, \omega_o) \qquad (a)$$

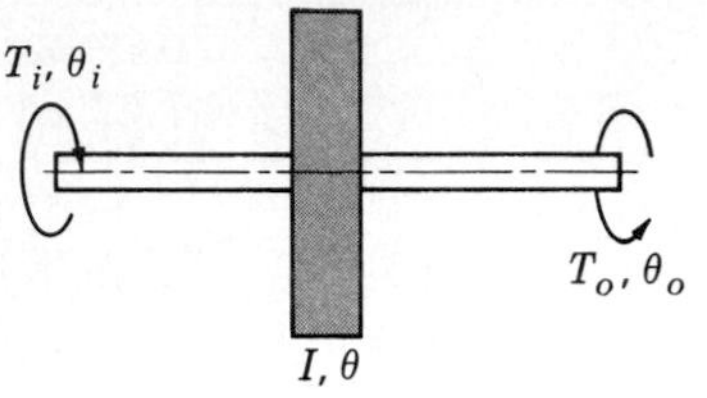

Figure 17-1 Mathematical representation of a flywheel.

Note that both T_i and T_o may depend for their values on the angular displacements θ_i and θ_o as well as their angular velocities ω_i and ω_o. Typically the torque characteristic depends upon only one of these. Thus the torque delivered by an induction motor depends on the speed of the motor. In fact electric-motor manufacturers publish charts detailing the torque-speed characteristics of their various motors.

When the input and output torque functions are given, Eq. (*a*) can be solved for the motion of the flywheel using well-known techniques for solving linear and nonlinear differential equations. Since such methods are beyond the scope of this book we will assume a rigid shaft, giving $\theta_i = \theta = \theta_o$. Thus, Eq. (*a*) becomes

$$I\alpha = T_i(\theta, \omega) - T_o(\theta, \omega) \qquad (b)$$

When the two torque functions are known and the starting values of the displacement θ and velocity ω are given, Eq. (*b*) can be solved for ω and α as functions of time. However, we are not really interested in the instantaneous values of the kinematic quantities. Primarily, we want to know the overall performance of the flywheel. What should its moment of inertia be? How do we match the power source to the load to get an optimum motor or engine? Finally, what are the resulting performance characteristics of the system?

To gain insight into the problem, a hypothetical situation is diagrammed in Fig. 17-2. An input power source subjects a flywheel to a constant torque T_i while the shaft rotates from θ_1 to θ_2. This is a positive torque and is plotted upward. Equation (*b*) indicates that a positive acceleration α will be the result, and so the shaft velocity increases from ω_1 to ω_2. As shown, the shaft now rotates from θ_2 to θ_3 with zero torque and hence, from Eq. (*b*), with zero acceleration. Therefore $\omega_3 = \omega_2$. From θ_3 to θ_4 a load, or output torque, of

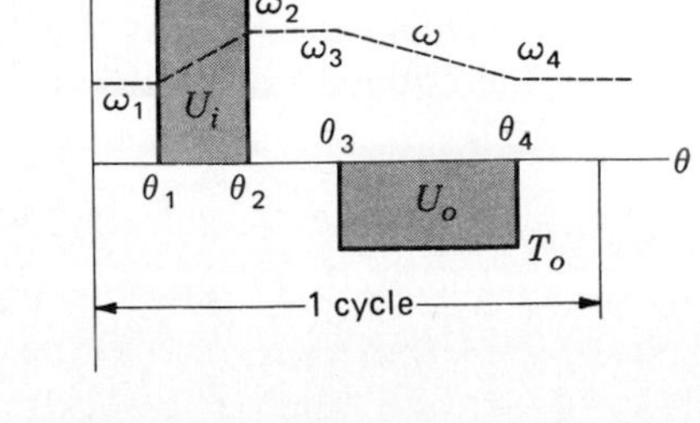

Figure 17-2

constant magnitude is applied, causing the shaft to slow down from ω_3 to ω_4. Note that the output torque is plotted in the negative direction in accordance with Eq. (*b*).

The work input to the flywheel is the area of the rectangle between θ_1 and θ_2, or

$$U_i = T_i(\theta_2 - \theta_1) \tag{c}$$

The work output of the flywheel is the area of the rectangle from θ_3 to θ_4, or

$$U_o = T_o(\theta_4 - \theta_3) \tag{d}$$

If U_o is greater than U_i, the load uses more energy than has been delivered to the flywheel and so ω_4 will be less than ω_1. If $U_o = U_i$, ω_4 will be equal to ω_1 because the gain and losses are equal; we are assuming no friction losses. And finally ω_4 will be greater than ω_1 if $U_i > U_o$.

We can also write these relations in terms of kinetic energy. At $\theta = \theta_1$ the flywheel has a velocity of ω_1 rad/s, and so its kinetic energy is

$$U_1 = \tfrac{1}{2}I\omega_1^2 \tag{e}$$

At $\theta = \theta_2$ the velocity is ω_2, and so

$$U_2 = \tfrac{1}{2}I\omega_2^2 \tag{f}$$

Thus the change in kinetic energy is

$$U_2 - U_1 = \tfrac{1}{2}I(\omega_2^2 - \omega_1^2) \tag{17-1}$$

Many of the torque-displacement functions encountered in practical engineering situations are so complicated that they must be integrated by approximate methods. Figure 17-3, for example, is a plot of the engine torque from Prob. 14-7 for one cycle of motion of a single-cylinder engine. Since a part of the torque curve is negative, the flywheel must return part of the energy back to the engine. Approximate integration of this curve for a cycle of 4π rad yields a mean torque T_m available to drive a load.

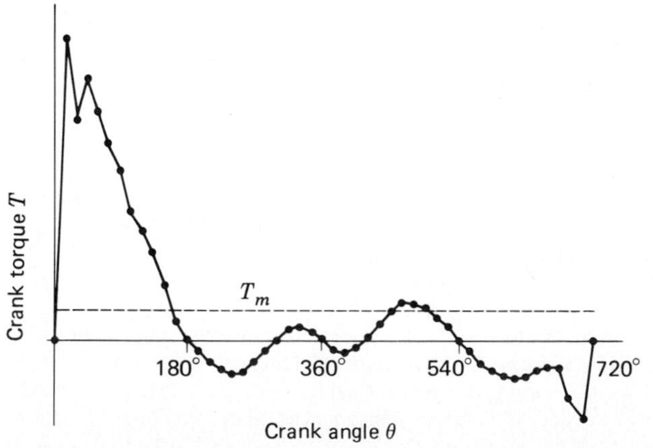

Figure 17-3 Relation between torque and crank angle for a one-cylinder four-cycle internal combustion engine.

The simplest integration routine is Simpson's rule; this approximation can be handled on any computer and is short enough to use on the smallest programmable calculators. In fact this routine is usually found as a part of the library for most calculators and minicomputers. The equation used is

$$\int_{x_0}^{x_n} f(x)\,dx = \frac{h}{3}(f_0 + 4f_1 + 2f_2 + 4f_3 + 2f_4 + \cdots + 2f_{n-2} + 4f_{n-1} + f_n) \qquad (17\text{-}2)$$

where
$$h = \frac{x_n - x_0}{n} \qquad x_n > x_0$$

and n is the number of subintervals used 2, 4, 6, · · ·. If memory is limited, solve Eq. (17-2) in two or more steps, say from 0 to $n/2$ and then from $n/2$ to n.

It is convenient to define a *coefficient of speed fluctuation* as

$$C_s = \frac{\omega_2 - \omega_1}{\omega} \qquad (17\text{-}3)$$

where ω is the nominal angular velocity, given by

$$\omega = \frac{\omega_2 + \omega_1}{2} \qquad (17\text{-}4)$$

Equation (17-1) can be factored to give

$$U_2 - U_1 = \frac{I}{2}(\omega_2 - \omega_1)(\omega_2 + \omega_1)$$

Since $\omega_2 - \omega_1 = C_s\omega$ and $\omega_2 + \omega_1 = 2\omega$, we have

$$U_2 - U_1 = C_s I \omega^2 \qquad (17\text{-}5)$$

Equation (17-5) can be used to obtain an appropriate flywheel inertia corresponding to the energy change $U_2 - U_1$.

Example 17-1 Table 17-1 lists values of the torque used to plot Fig. 17-3. The nominal speed of the engine is to be 250 rad/s. (*a*) Integrate the torque-displacement function for one cycle and find the energy that can be delivered to a load during the cycle. (*b*) Determine the mean torque T_m (see Fig. 17-3). (*c*) The greatest energy fluctuation will occur approximately between $\theta = 15°$ and $\theta = 150°$ on the T_i-T_o diagram; see Fig. 17-3 and note that $T_o = -T_m$. Using a coefficient of speed fluctuation C_s of 0.1, find a suitable value for the flywheel inertia. (*d*) Find ω_2 and ω_1.

SOLUTION (*a*) Using $n = 48$ and $h = 4\pi/48$, we enter the data of Table 17-1 into a computer program and get $U = 3490$ lb · in. This is the energy that can be delivered to the load.

(*b*)
$$T_m = \frac{3490}{4\pi} = 278 \text{ lb} \cdot \text{in} \qquad \textit{Ans.}$$

(*c*) The largest positive loop on the torque-displacement diagram occurs between $\theta = 0$ and $\theta = 180°$. We select this loop as yielding the largest speed change. Subtracting 278 lb · in from the values in Table 17-1 for this loop gives, respectively, −278, 2522, 1812, 2152, 1882, 1562, 1312, 932, 788, 525, 254, −94, and −278 lb · in. Entering the Simpson's approximation

Table 17-1

θ deg	T lb · in	θ deg	T lb · in	θ deg	T lb · in	θ deg	T lb · in	θ deg	T lb · in
0	0	150	532	300	−8	450	242	600	−355
15	2800	165	184	315	89	465	310	615	−371
30	2090	180	0	330	125	480	323	630	−362
45	2430	195	−107	345	85	495	280	645	−312
60	2160	210	−206	360	0	510	206	660	−272
75	1840	225	−280	375	−85	525	107	675	−274
90	1590	240	−323	390	−125	540	0	690	−548
105	1210	255	−310	405	−89	555	−107	705	−760
120	1066	270	−242	420	8	570	−206		
135	803	285	−126	435	126	585	−292		

again, using $n = 12$ and $h = 4\pi/48$, gives $U_2 - U_1 = 3660$ lb · in. We now solve Eq. (17-5) for I and substitute. This gives

$$I = \frac{U_2 - U_1}{C_s\omega^2} = \frac{3660}{0.1(250)^2} = 0.586 \text{ lb} \cdot \text{s}^2 \cdot \text{in} \qquad \textit{Ans.}$$

(*d*) Equations (17-3) and (17-4) can be solved simultaneously for ω_2 and ω_1. Substituting the appropriate values in these two equations yields

$$\omega_2 = \frac{\omega}{2}(2 + C_s) = \frac{250}{2}(2 + 0.1) = 262.5 \text{ rad/s} \qquad \textit{Ans.}$$

$$\omega_1 = 2\omega - \omega_2 = 2(250) - 262.5 = 237.5 \text{ rad/s} \qquad \textit{Ans.}$$

These two speeds occur at $\theta = 180°$ and $\theta = 0$, respectively.

17-2 GYROSCOPES

The gyroscope of Fig. 17-4 is an instrument which has fascinated students of mechanics and applied mathematics for many years. In fact, once the rotor is set spinning, it appears to act like a device possessing intelligence. If we attempt to move some of its parts, it seems not only to resist this motion but even to evade it. We shall even see that it apparently fails to conform to the laws of static equilibrium and of gravitation.

The uses of the gyroscope as turn-and-bank indicators, artificial horizons, and automatic pilots in aircraft and missiles are well known, as is its use in the gyrocompass. For many years it has served as a stabilizer in ships and torpedoes. One also becomes concerned with gyroscopic effects in the design of machines—although not always intentionally. Such effects are present when a motorcycle or bicycle is being ridden; they are also present, owing to the rotating masses, when an airplane or automobile is making a turn. Sometimes these effects are desirable, but more often they are undesirable and designers must account for them in their selection of bearings and

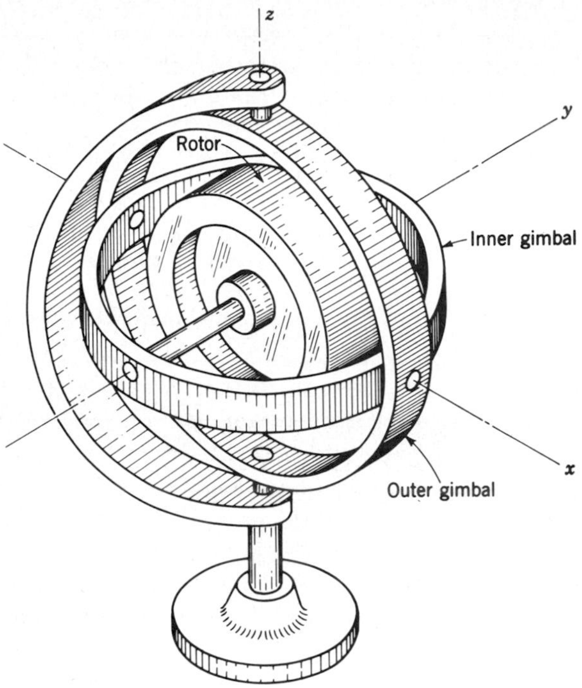

Figure 17-4 A laboratory gyroscope.

rotating parts. It is certainly true that as machine speeds increase to higher and higher values and as factors of safety decrease, we must stop neglecting gyroscopic forces in our machine designs because their values will be more significant. The general equations for the motion of a gyroscope are, indeed, not simple. Fortunately, in designing machines, only a few simple and approximate solutions are necessary.

The rotor of the gyroscope of Fig. 17-4 has a heavy rim and is fastened to a shaft which rotates in bearings in the inner gimbal. The inner gimbal is mounted on pivots so that it is free to rotate about an axis which is perpendicular to the rotational axis of the rotor. It is pivoted to an outer gimbal which can turn about a vertical axis through the frame perpendicular to the plane of the rotor and inner-gimbal axes for the position shown in the figure. Thus the rotor can rotate only about the y axis or, together with the inner gimbal, about the x axis or, with both gimbals, about the z axis. In fact the rotor can simultaneously have these three kinds of rotation. It will be convenient to designate the rotor axis, or y axis, as the *axis of spin.*

To provide a vehicle for the explanation of the simpler motions of a gyroscope it is expedient to perform a series of experiments with the one of Fig. 17-4. In the following we assume that the rotor is spinning and that the pivot friction can be neglected.

1. If the z axis is kept in the vertical position, the gyroscope can be moved

anywhere about a table or a room without altering the direction of the axis of spin. This is a consequence of the law of conservation of moment of momentum. If the axis of spin is to change its direction, the moment-of-momentum vector must also change its direction, but this requires an external torque, which in this experiment we have not supplied. While the rotor is still spinning, we might lift the inner gimbal out of its bearings and move it about. We then find that it can be translated anywhere but that we meet with definite resistance when we attempt to rotate the axis of spin.

2. With the inner gimbal back in the bearings, suppose that pressure is applied, say by a pencil, to the inner gimbal to make it turn about the x axis. Not only do we meet with resistance to the pressure of the pencil but the outer gimbal begins to rotate slowly about the vertical z axis and it continues this rotation until the pressure is released. The pressure of the pencil constitutes a torque on the inner gimbal with the parallel and opposite force of the couple coming from the pivots in the gimbal. In order to study these effects carefully we might cause the rotor to spin in the positive direction, i.e., with the angular-velocity vector pointing in the positive y direction. Then if we apply a positive torque to the inner gimbal (torque vector pointing in the positive x direction), the rotation of the outer gimbal is found to be in the negative z direction. You should note that these effects occur in a right-hand coordinate system. Either a negative spin velocity or a negative torque will cause the gimbal to rotate in the positive z direction for the set of axes shown. The rotation of the spin axis about an axis perpendicular to that of a torque applied to it is called *precession*, and so the application of a torque to the spinning rotor causes it to *precess*. In this example the z axis is called the *axis of precession*.
3. As a third experiment we might apply a torque to the outer gimbal in an attempt to cause it to rotate about the z axis. Such an attempt meets with resistance and causes the inner gimbal with the spin axis to rotate. When the spin axis is in the vertical position, the gyroscope is in stable equilibrium though, and the outer gimbal can then be turned quite freely. Note in this as well as in the previous example that the moment-of-momentum vector is changing its direction because of an application of external torque.

In Fig. 17-5*a* suppose the rotor is spinning about its spin axis with an angular velocity $\boldsymbol{\omega}_s$ while at the same time the spin axis precesses with an angular velocity $\boldsymbol{\omega}_p$. Let the moment of inertia of the rotor about the spin axis be I_s and designate I as the moment of inertia about x and about z, since they are equal. Because the axes of the rotor are the principal axes of inertia, the component of the moment-of-momentum vector† along the spin axis is $\mathbf{H}_s = I_s\boldsymbol{\omega}_s$ and along the precession axis is $\mathbf{H}_p = I\boldsymbol{\omega}_p$. After a small period of

† For a definition of this vector see any applied mechanics text, or, say, F. P. Beer and E. R. Johnston, Jr., "Mechanics for Engineers", 3d ed., chap. 18, McGraw-Hill, New York, 1976.

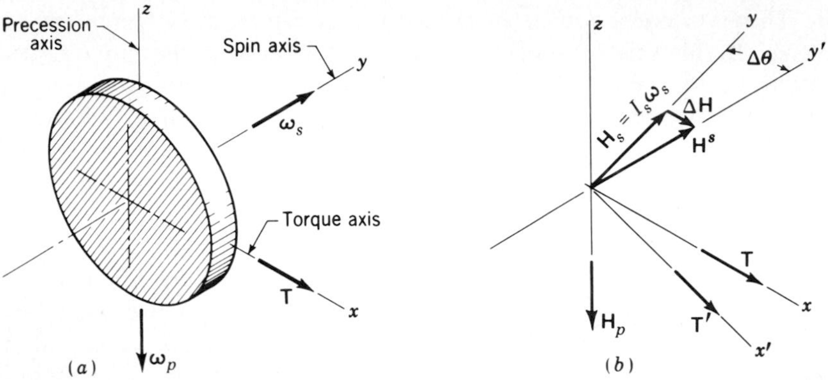

Figure 17-5

time Δt the spin axis has rotated through the angle $\Delta\theta$ to a new position indicated as y' in Fig. 17-5*b*. Thus the component of the moment of momentum along the spin axis is continuously changing its direction during recession. Any vector, such as $\mathbf{H}_s$, rotating with a constant angular velocity $\boldsymbol{\omega}_p$ has a rate of change

$$\dot{\mathbf{H}}_s = \boldsymbol{\omega}_p \times \mathbf{H}_s$$

Since the rate of change of moment of momentum is equal to the external torque acting on the system, we have

$$\mathbf{T} = \dot{\mathbf{H}}_s = \boldsymbol{\omega}_p \times (I_s\boldsymbol{\omega}_s) = I_s\boldsymbol{\omega}_p \times \boldsymbol{\omega}_s \tag{17-6}$$

The direction of the torque required to *maintain* the precession is shown in Fig. 17-5*a*. Figure 17-5*b* shows that the direction of the applied torque must continue to change in order to maintain precession. It also shows that the torque *does not* vary the precessional component of the moment of momentum. It *does* show that the change in the moment of momentum is in the *same* direction as the applied torque. We note further that Eq. (17-6) applies *only* to the maintenance of an existing motion and not to the beginning or ending of a precession. It might be noted, though not demonstrated here, that the beginning or ending of precession is accompanied by vibrations, which usually are damped out quite rapidly by friction.

Example 17-2 Figure 17-6 illustrates a hypothetical problem typical of the situations occurring in the design or analysis of machine systems in which gyroscopic forces must be considered. A round plate designated as 2 rotates about the z' axis with angular velocity ω_2. Mounted on this revolving plate are two bearings A and B, which retain a shaft, and mass 3 rotating at the vector angular velocity $\boldsymbol{\omega}_3$. An xyz system is selected fixed to the shaft and mass and therefore rotating with it. The mass center G defines the origin of this system, and the x axis is coincident with the axis of the shaft rotation. The angular velocity $\boldsymbol{\omega}_3$ is that which an observer stationed on the rotating plate would report the shaft as having. Let the weight of the mass be $W = 10$ lb, its radius of gyration be $k = 2$ in, and its angular velocity

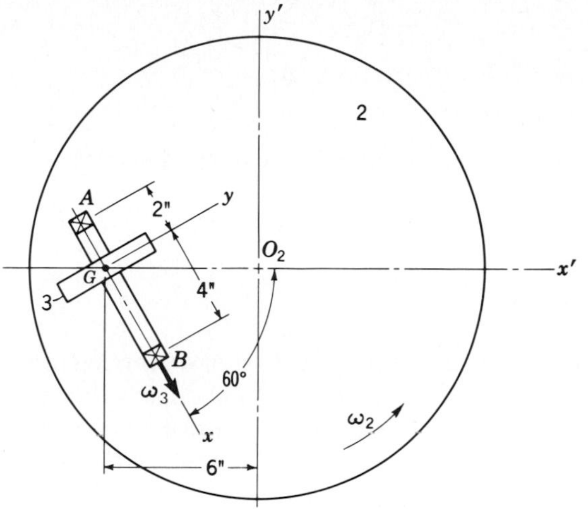

Figure 17-6

be $\boldsymbol{\omega}_3 = 350\hat{\mathbf{i}}$ rad/s. Using $\omega_2 = 5$ rad/s in the direction shown, find the bearing reactions. Assume that the weight of the shaft is negligible and that bearing B is to take only radial load.

SOLUTION Since we are dealing with forces and neglecting friction, superposition can be used. Therefore the bearing reactions at A and B will be calculated first considering that $\boldsymbol{\omega}_3$ is zero. Then to these components we shall add those due to gyroscopic action.

When $\boldsymbol{\omega}_3$ is zero, the methods of Chap. 13 apply. The results are

At A: $$\mathbf{F}_{23} = 1.94\hat{\mathbf{i}} + 2.24\hat{\mathbf{j}} + 6.67\hat{\mathbf{k}} \text{ lb} \tag{1}$$

At B: $$\mathbf{F}_{23} = 1.12\hat{\mathbf{j}} + 3.33\hat{\mathbf{k}} \text{ lb} \tag{2}$$

where the vectors are referred to the xyz system.

The forces due to gyroscope action are found as follows. The x axis is the axis of spin, and the moment of inertia relative to this axis is

$$I_x = mk^2 = \frac{10}{386}(2)^2 = 0.1038 \text{ lb} \cdot \text{s}^2 \cdot \text{in}$$

The precessional velocity is $\boldsymbol{\omega}_2 = 5\hat{\mathbf{k}}' = 5\hat{\mathbf{k}}$ rad/s because an angular-velocity vector is always a free vector. Equation (17-6) now applies where $I_s = I_x$, $\boldsymbol{\omega}_p = \boldsymbol{\omega}_2$, and $\boldsymbol{\omega}_s = \boldsymbol{\omega}_3$. Thus,

$$\mathbf{T} = I_s \boldsymbol{\omega}_2 \times \boldsymbol{\omega}_3 = (0.1038)(5\hat{\mathbf{k}} \times 350\hat{\mathbf{i}}) = 181.5\hat{\mathbf{j}} \text{ lb} \cdot \text{in}$$

The position of B relative to A is $\mathbf{R}_{BA} = 6\hat{\mathbf{i}}$. Taking moments about A, we get

$$\sum \mathbf{M}_A = \mathbf{T} + \mathbf{R}_{BA} \times \mathbf{F}_B = 181.5\hat{\mathbf{j}} + 6\hat{\mathbf{i}} \times \mathbf{F}_B = 0$$

or $$\mathbf{F}_B = 30.2\hat{\mathbf{k}} \text{ lb} \tag{3}$$

Taking moments about B gives

$$\sum \mathbf{M}_B = \mathbf{T} + \mathbf{R}_{AB} \times \mathbf{F}_A = 181.5\hat{\mathbf{j}} + (-6\hat{\mathbf{i}}) \times \mathbf{F}_A = 0$$

or $$\mathbf{F}_A = -30.2\hat{\mathbf{k}} \text{ lb} \tag{4}$$

Adding Eqs. (1) and (4) gives the total reaction at A:

At A: $$\mathbf{F}_{23} = 1.94\hat{\mathbf{i}} + 2.24\hat{\mathbf{j}} - 23.53\hat{\mathbf{k}}\text{ lb} \qquad \textit{Ans.}$$

Similarly, Eqs. (2) and (3) are summed to give the reaction at B:

At B: $$\mathbf{F}_{23} = 1.12\hat{\mathbf{j}} + 33.53\hat{\mathbf{k}}\text{ lb} \qquad \textit{Ans.}$$

We note that the effect of the gyroscope couple is to lift the rear bearing off the plate and to push the front bearing against the plate.

17-3 GOVERNORS

The automatic regulating device known as the *governor* is an example from a large and growing class of mechanical and electromechanical control systems. The *flyball governor* is an example of an all-mechanical control system once widely used to control the speed of steam engines. The availability today of a wide variety of low-priced solid-state electronic devices and transducers makes it possible to regulate mechanical systems to a finer degree and at less cost than with the older all-mechanical devices.

Many mechanical control systems are represented in *block notation* as in Fig. 17-7. Here, θ_i and θ_o represent any set of input and output functions such as angular or linear displacement, or velocity, for example. The system is called a *closed-loop* or *feedback system* because the output θ_o is fed back to the detector at the input so as to measure the error $\mathscr{E}$, which is the difference between the input and the output. The purpose of the controller is to cause this error to become close to zero or even zero. The mechanical characteristics of the system, e.g., mechanical clearances, friction, inertias, and stiffnesses, sometimes cause the output to differ somewhat from the input, and so it is the designer's responsibility to examine these mechanical effects in an effort to minimize the error for all operating conditions.

A closed-loop control system in which the output is directly proportional to the error is called a *proportional-error system.* Figure 17-8 shows the response of such a system to a sudden jump or step change to the input θ_i. The factor ζ is called the *damping factor ratio*; it is a dimensionless number

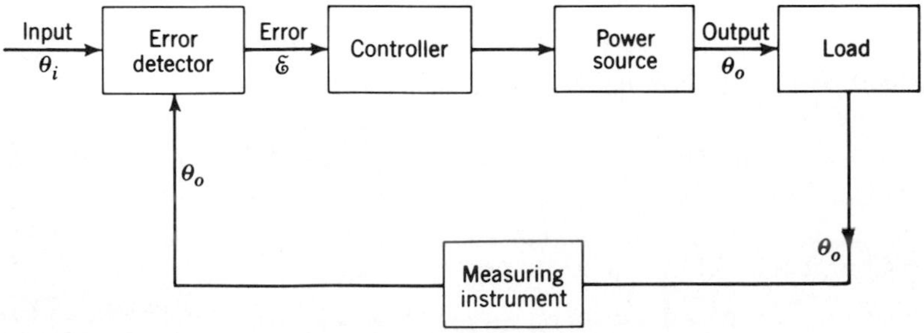

Figure 17-7 Block diagram of a closed-loop system.

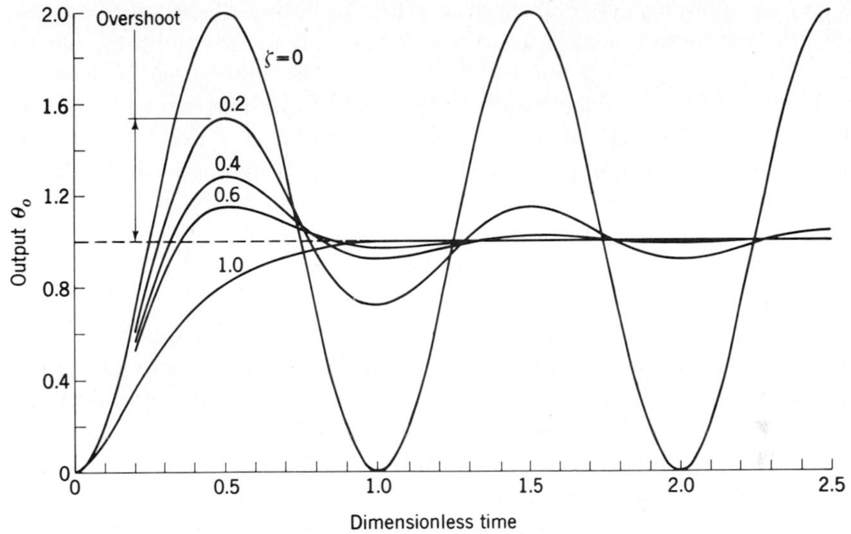

Figure 17-8 Response to a unit step input.

that designates the amount of viscous friction present in the system. As shown, a large value of ζ produces the least overshoot.

The widely used automotive cruise-control system is an excellent example of an electromechanical governor. A transducer is attached to the speedometer cable, and the electrical output of this transducer is the signal θ_o fed to the error detector of Fig. 17-7. In some cases magnets are mounted on the driveshaft of the car to actuate the transducer. In the cruise-control system, the error detector is an electronic regulator, usually mounted under the dash. This regulator is turned on by an engagement switch under or near the steering wheel. A power unit is connected by a chain to the carburetor throttle linkage; the power unit is controlled by the regulator and gets its power from a vacuum port on the engine. Such systems have one or two brake-release switches as well as the engagement switch. The accelerator pedal can also be used to override the system.

17-4 MEASUREMENT OF DYNAMIC RESPONSE

We have now nearly completed our studies of the kinematic and dynamic analysis of a rather large variety of machines and machine systems. Throughout these studies we have found it necessary to make various assumptions regarding friction, rigidity, mass concentration, and moment of inertia. In other cases we have neglected certain items as being of no consequence in the analysis and have assumed almost perfect geometries. However, Murphy's law applies to engineering analysis just as much as it does to everyday living. Machine parts often turn out to be crooked and eccentric; their geometric shapes are often very complicated; and they may fit mating parts with too much or too little clearance. The analyst may produce a brilliant solution to his problem using sophisticated mathematical techniques and the very latest computer-driven graphic displays, but if the real machine does not behave in a similar manner the solution predicts nothing. Good engineering practice requires that engineers verify their analyses using some form of reliable experimentation. With such verification analytical procedures are transformed into reliable methods for improving and optimizing the original design.

This is not a book on engineering experimentation. In the short space we have available we can mention only a few of the most reliable and most used tools and techniques for determining the dynamic behavior of machine systems.

Strain gauges The electrical-resistance strain gauge is generally a printed metallic foil or semiconductor mounted on a thin film backing. Several of these gauges are usually cemented to the mechanical element at the location for which the strain is to be measured. When the mechanical part is subjected to a tensile strain, the resistance of the gauge increases; when the part is subject to a compressive strain, the gauge resistance decreases. Thus strain can be measured merely by measuring the voltage drop across the gauge as it is strained. The basic strain-gauge formula is

$$\frac{\Delta R}{R} = f\frac{\Delta l}{l} = f\epsilon \tag{17-7}$$

where R = gauge resistance
ΔR = change in gauge resistance
l = gauge length
Δl = change in gauge length
f = gauge sensitivity factor
ϵ = unit strain

Note that the gauge sensitivity factor f is merely the constant of proportionality that relates the unit change in gauge resistance to the unit strain.

Foil gauges are made in sizes ranging from about 2 mm square to about 20 mm long; so they can be used in a variety of places, depending upon the distribution of the strain or the geometry of the part.

Since strain can be related to stress and to force or moment, the gauges can be calibrated to record any of these quantities. The gauges should be connected into a bridge circuit or bridge amplifier and the output fed to an oscilloscope for dynamic measurements. Figure 17-9 is a photograph of an oscilloscope recording of the bending force in the lever of a cam-driven oscillating follower for one revolution of the cam. The fuzziness of the trace results from the high amplification needed in this particular application to record a rather small force.

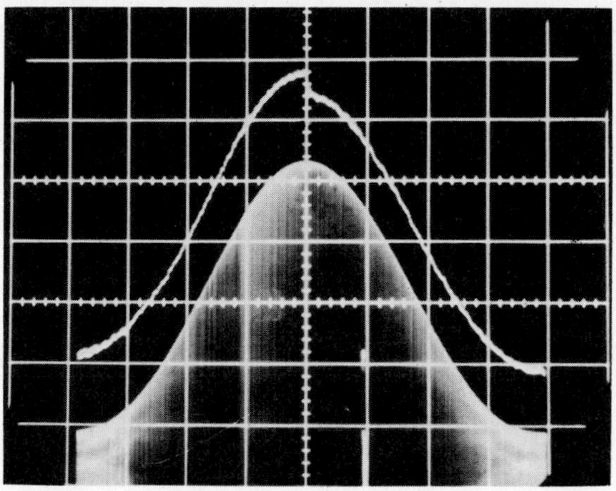

Figure 17-9 Oscilloscope traces of displacement and contact force of an eccentric cam driving a rocking roller follower. The upper trace is the contact force measured by strain gauges aligned to measure bending of the follower lever. For this picture the cam was driven at a slow speed, and so the contact force is nearly the same type of curve as the displacement. The jump at the top of rise is caused by sliding friction, which is reversed in sign when the velocity changes direction. The displacement diagram in the lower part of the picture was generated by a rotary differential transformer connected to the rocker shaft. A rotary potentiometer connected to the cam shaft was used to generate the horizontal sweep signal. (*Courtesy of Professor F. E. Fisher, Mechanical Analysis Laboratory, The University of Michigan.*)

Potentiometers Figure 17-10 shows the schematic diagrams of two displacement transducers called *linear* and *angular* wire-wound *potentiometers*. These are exceedingly useful for the measurement of linear and angular displacements, even at high speeds. The linear potentiometer consists of a coil of resistance wire wound uniformly around a straight insulator. When the output e_o of the circuit is fed into a voltmeter or oscilloscope, the voltage change is directly proportional to the displacement of the movable contact. The contact arms can be connected to any mechanical device and calibrated to measure displacement.

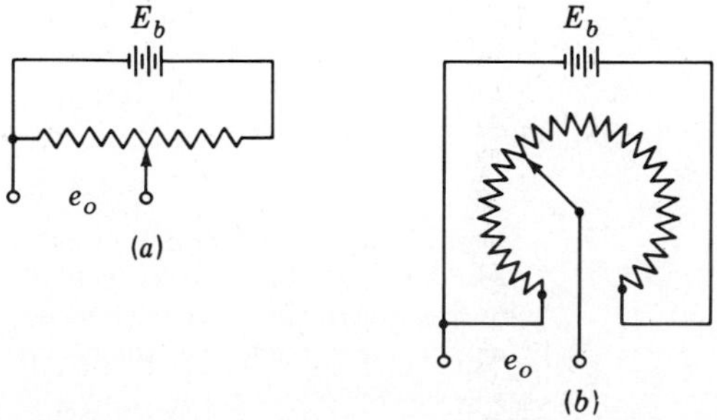

Figure 17-10 (*a*) Reciprocating linear potentiometer; (*b*) rotary or angular potentiometer.

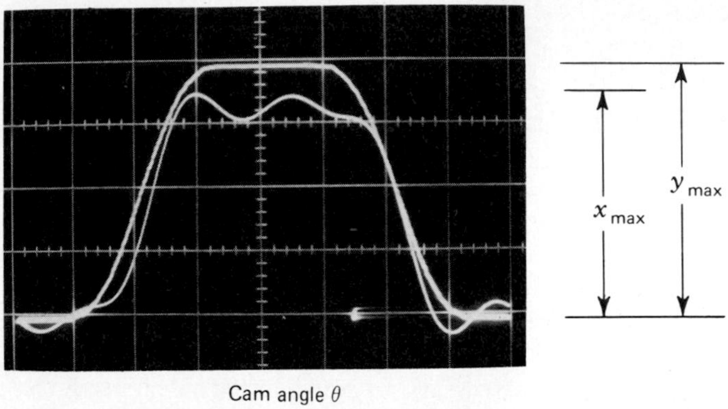

Figure 17-11 Cam-displacement diagrams generated by reciprocating potentiometers. The cam was machined to generate the motion y; output motion of follower is denoted as x. The photograph was made on a dual-beam oscilloscope. The horizontal sweep was generated by a rotary potentiometer connected to the cam shaft.

Figure 17-11 is a photograph obtained from a dual-beam oscilloscope showing the cam-input and lever-output displacements obtained from two reciprocating potentiometers. The cam was machined with a dwell, shown as y_{max}. Due to bending of the lever, the follower output x is not the same as the input y. Note also the phase lag between y and x.

Figure 17-10b shows the schematic of the rotary potentiometer, widely used for dynamic measurements. The horizontal sweep for the photograph of Fig. 17-9 was generated by a rotary potentiometer connected directly to the cam shaft.

Differential transformers Another type of displacement transducer is the differential transformer; linear and angular models are illustrated schematically in Fig. 17-12. When the primary coils are excited by an alternating voltage, voltages are induced in the secondary coils by the magnetic coupling provided by the movable iron cores. For each transformer, the two secondary coils are connected so that the induced voltages oppose each other. Thus the net secondary voltage is zero when the core is centered.

In use, the core is attached to the mechanical element whose motion is to be measured. The primary winding is excited from an audio-frequency source.

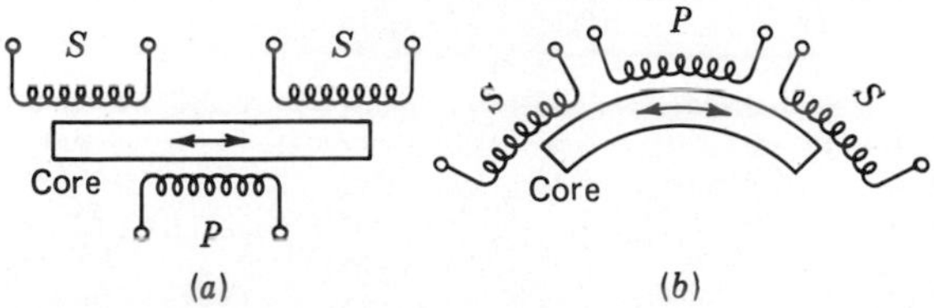

Figure 17-12 Differential transformers; P is the primary coil, S the secondary coil; (a) reciprocating, or linear, model; (b) rotary, or angular, model.

The output of the secondaries is fed to an oscilloscope, which then displays an envelope containing a modulated output at the exciting frequency. The cam-displacement diagram shown in Fig. 17-9 was generated using a rotary differential transformer connected to the shaft of an oscillating follower lever.

Solar cells When the dynamics of very small or lightweight parts must be measured, one must be extremely careful to ensure that the transducer inertia or mass does not affect the motion of the part. *Solar cells* are silicon diodes that measure either displacement or velocity, depending upon the setup, using a beam of light. Nothing need be fastened to the moving part, and so its mass or inertia is not changed. The setup can be arranged so that the solar cell is mounted in a fixed position and the light beam directed in such a manner that the moving part casts a shadow on the solar cell. The trick in the setup is in the diode biasing circuit required to cause the voltage drop across the solar cell to have a linear relationship with the amount of shading produced by the moving part. Fisher† has worked out the details of this procedure using such simple equipment that measurements could be made in almost any home workshop. Figure 17-13*a* is an oscilloscope trace obtained from a solar cell mounted behind a rapidly moving lever as it impacts a rubber stop. Figure 17-13*b* shows two blips obtained as the lever passes two very small solar

† F. E. Fisher and H. H. Alvord, "Instrumentation for Mechanical Analysis," The University of Michigan Summer Conferences, Ann Arbor, Michigan, 1977, pp. 44–58.

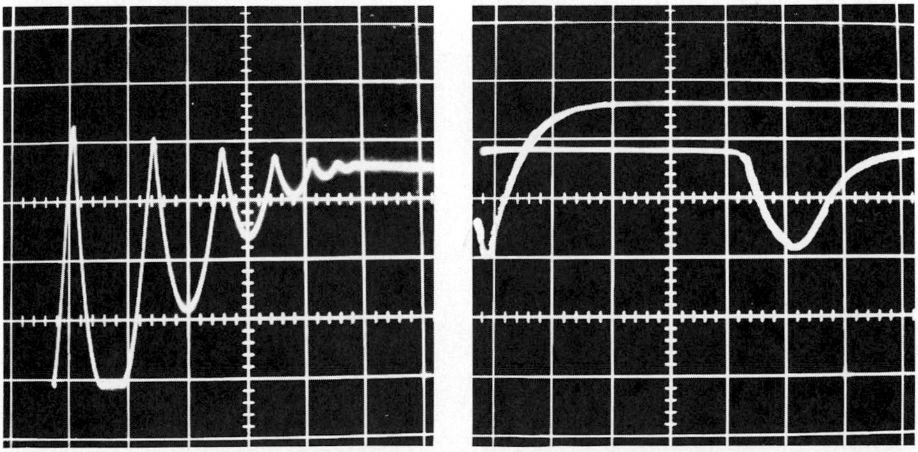

(*a*) (*b*)

Figure 17-13 Examples of motion and velocity measurement using solar cells: (*a*) oscilloscope trace showing the motion of a pendulum mass as it impacts and bounces against a rubber stop; (*b*) pendulum crossing two solar cells causes a blip on each oscilloscope trace. The velocity of the pendulum is computed by dividing the distance between the solar cells by the oscilloscope time between blips. (*Courtesy of Professor F. E. Fisher, Mechanical Analysis Laboratory, The University of Michigan.*)

cells. Observation of the oscilloscope sweep speed together with a measurement of the distance between the two cells provides sufficient information to enable one to compute the velocity of the lever as it passes the cells.

Reflecting transducers The reflecting transducer is another linear-displacement measurement device, but it is used to measure very small motions, about 2 mm or less. In use, a light beam from the transducer is directed onto a small mirror or reflecting surface on a stationary part. Motion of the moving part shades the beam, so that part of the beam is reflected back into the transducer to a photocell and amplified and the resulting voltage is displayed as an oscilloscope trace. The setup is calibrated so that the area of the shaded portion of the beam is proportional to the motion of the part.

Generators Direct-current electric generators are made for use as velocity transducers. They may be obtained in linear form for use in reciprocating motion or in rotary form for use to measure angular velocity. In each case the voltage generated is proportional to the velocity of the part to which the device is connected.

Other instrumentation In this section we have mentioned only a few general-purpose measurement devices. Our aim has been to show what can be done in the area of dynamic measurement rather than to describe all the means and devices that can be used. A search of manufacturers literature will reveal many other techniques and instruments. However, in addition to a collection of transducers, a well-equipped laboratory will have various items of recording and associated instrumentation, such as oscilloscopes and oscillographs, bridge amplifiers, various electrical meters and amplifiers, and strobo-type instruments.

17-5 MACHINE FOUNDATIONS

Large machines, such as engine-generators and presses, including both the driving and driven members, should usually be mounted on a single frame and fixed to a foundation. The foundation should be of reinforced concrete and formed as a thick plate or slab resting on piles driven into the soil or as a huge block resting on the soil.

The bearing layers of the natural soil should be laboratory-tested to determine the permissible bearing pressure before the foundation is designed. The machine foundation should be isolated from the building structure and floor in order to avoid the transmission of vibrations and noise due to unbalanced inertia forces and couples. Particular care should be taken to align the centers of gravity of the machine and foundation block in a vertical direction. Any eccentricity may eventually cause unequal settlement of the foundation and lead to trouble.

It may also be desirable to use vibration absorbers, spring dampers, or other elastic materials between the machine frame and the foundation. They can be selected after a study of the transmissibility characteristics of the machine has been completed.

PROBLEMS

17-1 Table 17-2 lists the output torque for a one-cylinder engine running at 4600 rpm.

(*a*) Find the mean output torque.

(*b*) Determine the mass moment of inertia of an appropriate flywheel using $C_s = 0.025$.

Table 17-2

θ deg	T N · m	θ deg	T N · m	θ deg	T N · m	θ deg	T N · m
0	0	180	0	360	0	540	0
10	17	190	−344	370	−145	550	−344
20	812	200	−540	380	−150	560	−540
30	963	210	−576	390	−7	570	−577
40	1016	220	−570	400	164	580	−572
50	937	230	−638	410	235	590	−643
60	774	240	−785	420	203	600	−793
70	641	250	−879	430	190	610	−893
80	697	260	−814	440	324	620	−836
90	849	270	−571	450	571	630	−605
100	1031	280	−324	460	814	640	−379
110	1027	290	−190	470	879	650	−264
120	902	300	−203	480	785	660	−300
130	712	310	−235	490	638	670	−368
140	607	320	−164	500	570	680	−334
150	594	330	7	510	576	690	−198
160	544	340	150	520	540	700	−56
170	345	350	145	530	344	710	2

Table 17-3

θ deg	T lb · in	θ deg	T lb · in	θ deg	T lb · in	θ deg	T lb · in
0	857	90	7888	180	1801	270	857
10	857	100	8317	190	1629	280	857
20	857	110	8488	200	1458	290	857
30	857	120	8574	210	1372	300	857
40	857	130	8403	220	1115	310	857
50	1287	140	7717	230	1029	320	857
60	2572	150	3515	240	943	330	857
70	5144	160	2144	250	857	340	857
80	6859	170	1972	260	857	350	857

17-2 Using the data of Table 17-2, determine the moment of inertia for a flywheel for a two-cylinder 90° V engine having a single crank. Use $C_s = 0.0125$ and a nominal speed of 4600 rpm. If a cylindrical or disk-type flywheel is to be used, what should be the thickness if it is made of steel and has an outside diameter of 400 mm? Use $\rho = 7.8\ \text{Mg/m}^3$ as the density of steel.

17-3 Using the data of Table 17-1, find the mean output torque and flywheel inertia required for a three-cylinder in-line engine corresponding to a nominal speed of 2400 rpm. Use $C_s = 0.03$.

17-4 The load torque required by a 200-ton punch press is displayed in Table 17-3 for one revolution of the flywheel. The flywheel is to have a nominal velocity of 240 rpm and be designed for a coefficient of speed fluctuation of 0.075.

(*a*) Determine the mean motor torque required at the flywheel shaft and the motor horsepower needed, assuming a constant torque-speed characteristic for the motor.

(*b*) Find the moment of inertia needed for the flywheel.

ANSWERS TO SELECTED PROBLEMS

1-3 $\gamma_{min} = 53°$; $\gamma_{max} = 98°$; at $\theta = 40°$, $\gamma = 59°$; at $\theta = 229°$, $\gamma = 91°$

1-5 (*a*) $m = 1$; (*b*) $m = 1$; (*c*) $m = 0$ (exception); (d) $m = 1$

1-7 $Q = 1.10$

2-1 Spiral

2-3 $\mathbf{R}_{QP} = -7\hat{\mathbf{i}} - 14\hat{\mathbf{j}}$

2-5 $\Delta\mathbf{R}_A = -4.5a\hat{\mathbf{i}}$

2-7 Clockwise; $\mathbf{R} = 4\angle 0°$; $t = 20$; $\Delta\mathbf{R} = 400\angle 0°$

2-9 $\Delta\mathbf{R}_{P_3} = -2.12\hat{\mathbf{i}}_1 + 3.88\hat{\mathbf{j}}_1$; $\Delta\mathbf{R}_{P_{3/2}} = 3\hat{\mathbf{i}}_2$

2-11 $\Delta\mathbf{R}_Q = 1.90\hat{\mathbf{i}} + 1.10\hat{\mathbf{j}} = 2.20\angle 30°$

2-13 $R_C^x = 2.50\cos\theta_2 + \sqrt{36.75 + 6.25\cos^2\theta_2}$

3-1 $\dot{\mathbf{R}} = 314\angle 162°$ in/s

3-3 $\mathbf{V}_{BA} = \mathbf{V}_{B_{3/2}} = 82.6$ mi/h N25°E

3-5 (*a*) $d = 1400$ mm; (*b*) $V_{AB} = -60$m/s; $V_{BA} = 60$ m/s; $\omega_2 = 200$ rad/s cw

3-7 (*a*) Straight line at N48°E; (*b*) no change

3-9 $\omega_3 = 1.43$ rad/s ccw; $\omega_4 = 15.4$ rad/s ccw

3-11 $\mathbf{V}_C = 22.5\angle 284°$ ft/s; $\omega_3 = 0.60$ rad/s ccw

3-13 $\mathbf{V}_C = 0.402\angle 151°$ m/s; $\mathbf{V}_D = 0.290\angle 249°$ m/s

3-15 $\mathbf{V}_B = 4.77\angle 96°$ m/s; $\omega_3 = \omega_4 = 22$ rad/s ccw

3-17 $\omega_6 = 4$ rad/s ccw; $\mathbf{V}_B = 0.963\angle 180°$ ft/s; $\mathbf{V}_C = 2.02\angle 208°$ ft/s; $\mathbf{V}_D = 2.01\angle 205°$ ft/s

3-19 $\omega_3 = 3.23$ rad/s ccw; $\mathbf{V}_B = 16.9\angle -56°$ ft/s

3-21 $\mathbf{V}_C = 9.03\angle 138°$ m/s

3.23 $\mathbf{V}_B = 35.5\angle 240°$ ft/s; $\mathbf{V}_C = 40.9\angle 267°$ ft/s; $\mathbf{V}_D = 31.6\angle -60°$ ft/s

3-25 $\mathbf{V}_B = 1.04\angle -23°$ ft/s

3-27 $\omega_3 = \omega_4 = 14.4$ rad/s ccw; $\omega_5 = \omega_6 = 9.76$ rad/s cw; $\mathbf{V}_E = 77.4\angle -100°$ in/s

3-29 $\omega_3 = 1.61$ rad/s cw

3-31 $\mathbf{V}_E = 10.0\angle 221^\circ$ in/s; $\mathbf{V}_G = 11.6\angle -57^\circ$ in/s; $\omega_3 = \omega_4 = 3.30$ rad/s ccw; $\omega_5 = 25.0$ rad/s cw; $\omega_6 =$ 3.69 rad/s cw

3-33 $\omega_3 = 30$ rad/s cw

4-1 $-4\hat{\mathbf{i}}$ in/s^2

4-3 $\hat{\boldsymbol{\tau}} = 0.300\hat{\mathbf{i}} - 0.954\hat{\mathbf{j}}$; $A^n = 0.0437$ m/s^2; $A^t = 0.0126$ m/s^2; $\rho = 405$ mm

4-5 $\mathbf{A}_A = -7200\hat{\mathbf{i}} + 2400\hat{\mathbf{j}}$ m/s^2

4-7 $\mathbf{V}_B = 12\angle 270^\circ$ ft/s; $\mathbf{V}_C = 8.40\angle 12^\circ$ ft/s; $\mathbf{A}_B = 392\angle 165^\circ$ ft/s^2; $\mathbf{A}_C = 210\angle 240^\circ$ ft/s^2

4-9 $\omega_2 = 386$ rad/s ccw; $\alpha_2 = 557\,000$ rad/s^{2}ccw

4-11 $\alpha_3 = 563$ rad/s^2 ccw; $\alpha_4 = 124$ rad/s^2 ccw

4-13 $\mathbf{A}_C = 3056\angle 113^\circ$ ft/s^2; $\alpha_3 = 1741$ rad/s^2 ccw; $\alpha_4 = 3055$ rad/s^2 ccw

4-15 $\mathbf{A}_C = 2610\angle -69^\circ$ ft/s^2; $\alpha_4 = 1494$ rad/s^2 ccw.

4-17 $\mathbf{A}_B = 16.7\angle 0^\circ$ ft/s^2; $\alpha_3 = 17.5$ rad/s^2 ccw; $\alpha_6 = 10.8$ rad/s^2 cw

4-19 $\alpha_2 = 4180$ rad/s^2 cw

4-21 $\mathbf{A}_C = 450\angle -104^\circ$ m/s^2; $\alpha_3 = 74.1$ rad/s^2 cw

4-23 $\mathbf{A}_B = 2440\angle 240^\circ$ ft/s^2; $\mathbf{A}_D = 4030\angle 120^\circ$ ft/s^2

4-25 $\theta_3 = 15.5^\circ$; $\theta_4 = -8.99^\circ$; $\omega_3 = 47.6$ rad/s ccw; $\omega_4 = 70.5$ rad/s ccw; $\alpha_3 = 3330$ rad/s^2 ccw; $\alpha_4 = 3200$ rad/s^2 ccw

4-27 $\theta_3 = 28.3^\circ$; $\theta_4 = 55.9^\circ$; $\omega_3 = 0.633$ rad/s cw; $\omega_4 = 2.16$ rad/s cw; $\alpha_3 = 7.82$ rad/s^2 ccw; $\alpha_4 = 6.70$ rad/s^2 ccw

4-29 $\theta_3 = 38.4^\circ$; $\theta_4 = 156^\circ$; $\omega_3 = 6.85$ rad/s cw; $\omega_4 = 1.24$ rad/s cw; $\alpha_3 = 62.5$ rad/s^2 ccw; $\alpha_4 = 96.5$ rad/s^2 cw

4-31 $\mathbf{V}_B = 184\angle -19^\circ$ in/s; $\mathbf{A}_B = 2700\angle -172^\circ$ in/s^2; $\omega_4 = 6.57$ rad/s cw; $\alpha_4 = 86.4$ rad/s^2 ccw

4-33 $\mathbf{A}_4 = 29\angle 180^\circ$ m/s^2

4-35 $\mathbf{A}_B = 197\angle -36^\circ$ft/s^2; $\alpha_4 = 45.9$ rad/s^2 ccw

4-37 $\mathbf{A}_{P_4} = 131\angle -65^\circ$ m/s^2

4-39 $\mathbf{A}_{C_4} = 1801\angle 269^\circ$ in/s^2; $\alpha_4 = 6$ rad/s^2 ccw

4-41 $\mathbf{A}_G = 368\angle -64^\circ$ in/s^2; $\alpha_5 = 0$; $\alpha_6 = 106$ rad/s^2 cw

6-3 Face = 150 mm from pivot

6-5 $y'(\beta/2) = \pi L/2\beta$; $y'''(\beta/2) = -\pi^3 L/2\beta^3$; $y''(0) = \pi^2 L/2\beta^2$; $y''(\beta) = -\pi^2 L/2\beta^2$

6-7 *AB*: dwell, $L_1 = 0$, $\beta_1 = 60.0^\circ$; *BC*: full-rise modified harmonic motion, Eq. (6-20), $L_2 = 2.5$ in, $\beta_2 = 61.08^\circ$; *CD*: half-harmonic half-return motion, Eq. (6-26), $L_3 = 0.042$ in, $\beta_3 = 3.96^\circ$; *DE*: uniform motion, $L_4 = 1.0$ in, $\beta_4 = 60.0^\circ$; *EA*: half-cycloidal half-return motion, Eq. (6-31), $L_5 = 1.458$ in, $\beta_5 = 174.96$

6-9 $t_{AB} = 0.025$ s; $\dot{y}_{max} = 200$ in/s; $\dot{y}_{min} = -40$ in/s; $\ddot{y}_{max} = 21\,300$ in/s^2; $\ddot{y}_{min} = -38\,100$ in/s^2

6-11 $\dot{y}_{max} = 41.9$ rad/s; $\ddot{y}_{max} = 7900$ rad/s^2

6-13 Face width = 2.20 in; $\rho_{min} = 2.50$ in

6-15 $R_0 > 19.7$ in; face width > 6.24 in

6-17 $\phi_{max} = 12^\circ$; $R_r < 4.5$ in

6-19 $R_0 > 50$ mm; $\ddot{y}_{max} = 37$ m/s^2

6-21 $R_0 > 65$ mm; $\ddot{y}_{max} = 75$ m/s^2

6-23

$$u = (R_0 + R_c + y)\sin\theta + y'\cos\theta$$
$$v = (R_0 + R_c + y)\cos\theta - y'\sin\theta$$
$$R = \sqrt{(R_0 + R_c + y)^2 + (y')^2}$$
$$\psi = \frac{\pi}{2} - \theta - \tan^{-1}\frac{y'}{R_0 + R_c + y}$$

7-1 160 teeth per inch

7-3 2 mm

7-5 0.8976 teeth per inch, 44.563 in

7-7 12.73 mm, 458.4 mm

7-9 9.19 in

7-11 17 teeth, 51 teeth

7-13 $a = 0.25$ in, $b = 0.3125$ in, $c = 0.0625$ in, $p_c = 0.785$ in, $t = 0.392$ in, $d_{b_2} = 5.64$ in, $d_{b_3} = 8.46$ in, $u_a = 0.62$ in, $u_r = 0.585$ in, $m_c = 1.635$, $p_b = 0.737$ in

7-15 $q_a = 1.07$ in, $q_r = 0.99$ in, $q_t = 2.06$ in, $m_c = 1.64$

7-17 $m_c = 1.56$

7-19 (*a*) $q_a = 1.54$ in, $q_r = 1.52$ in, $q_t = 3.06$ in, $m_c = 1.95$; (*b*) $m_c = 1.55$; no change in pressure angle

7-25 $t_b = 17.14$ mm, $t_a = 6.74$ mm, $\varphi_a = 32.78°$

7-27 $t_b = 1.146$ in

7-29 $t_b = 0.1620$ in, $t_a = 0.0421$ in, $\varphi_a = 35.3°$

7-31 (*a*) $D = 0.1682$ in; (*b*) 9.8268 in

7-33 $m_c = 1.345$

7-35 $m_c = 1.770$

7-37 $a_3 = 1.343$ in

8-1 $p_t = 0.523$ in, $p_n = 0.370$ in, $P_n = 8.48$, $d_2 = 2.5$ in, $d_3 = 4$ in, 42.4, 67.8

8-4 $P_t = 6.93$, $p_t = 0.453$ in, $N_2 = 17$, $N_3 = 31$, $d_2 = 2.45$ in, $d_3 = 4.48$ in

8-7 $m_n = 1.79$, $m_t = 2.87$

8-10 $N_2 = 30$, $N_3 = 60$, $\psi_2 = \psi_3 = 25°$ left-hand, $(d_2 + d_3)/2 = 9.93$ in

8-13 $l = 3.75$ in, $\lambda = 34.37°$, $\psi = 34.37°$, $d_3 = 15.90$ in

8-16 27°, 93°

8-18 $d_2 = 2.125$ in, $d_3 = 3.500$ in, $\gamma_2 = 34.8°$, $\gamma_3 = 70.2°$, $a_2 = 0.1612$ in, $a_3 = 0.0888$ in, $F = 0.559$ in

9-1 $n_8 = 68.2$ rpm cw, $e = -5/88$

9-3 $n_9 = 11.82$ rpm cw

9-5 One solution: $N_3 = 30$ teeth, $N_4 = 25$ teeth, $N_5 = 30$ teeth, $N_6 = 20$ teeth, $N_7 = 25$ teeth, $N_8 = 35$ teeth, $N_{10} = 35$ teeth; output speeds are 200, 214, 322, 385, and 482 rpm

9-7 231 rpm ccw

9-9 645 rpm ccw

9-11 $n_A = -(5/22)n_2$ or opposite in direction; replace gears 4 and 5 with a single gear

9-13 (*a*) 84 teeth, 156 mm; (*b*) $n_A = 6.77$ rpm ccw.

9-15 (*a*) $n_R = 652$ rpm, $n_L = 695$ rpm; (*b*) $n_A = 674$ rpm

10-1 For six points, 0.170, 1.464, 3.706, 6, 6.294, 8.536, and 9.830

10-3 Typical solution: $r_2 = 7.4$ in, $r_3 = 20.9$ in, $e = 8$ in

10-5 Typical solution: $r_1 = 7.63$ ft, $r_2 = 3.22$ ft, $r_3 = 8.48$ ft

10-7 Typical solution: O_2 at $x = -1790$ mm, $y = 320$ mm; $r_2 = 360$ mm, $r_3 = 1990$ mm

10-9 $r_1 = 12$ in, $r_2 = 9$ in, $r_3 = 6$ in, $r_4 = 9$ in; seat locks in open or toggle position which is a 3-4-5 triangle

10-13 and **10-23** $r_2/r_1 = -3.352$, $r_3/r_1 = 0.845$, $r_4/r_1 = 3.485$

10-15 and **10-25** $r_2/r_1 = -2.660$, $r_3/r_1 = 7.430$, $r_4/r_1 = 8.685$

10-17 and **10-27** $r_2/r_1 = -0.385$, $r_3/r_1 = 1.030$, $r_4/r_1 = 0.384$

10-19 and **10-29** $r_2/r_1 = 2.523$, $r_3/r_1 = 3.329$, $r_4/r_1 = -0.556$

10-21 and **10-31** $r_2/r_1 = -1.606$, $r_3/r_1 = 0.925$, $r_4/r_1 = 1.107$

11-1 $m = 2$, including one idle freedom

11-3 $\boldsymbol{\omega}_2 = -2.58\hat{\mathbf{j}}$ rad/s; $\boldsymbol{\omega}_3 = 1.16\hat{\mathbf{i}} - 0.09\hat{\mathbf{j}} + 0.64\hat{\mathbf{k}}$ rad/s; $\mathbf{V}_B = -96\hat{\mathbf{i}} - 50\hat{\mathbf{j}} + 168\hat{\mathbf{k}}$ mm/s

11-5 $\mathbf{R}_{BA} = 5\hat{\mathbf{i}} + 9\hat{\mathbf{j}} - 7\hat{\mathbf{k}}$ in; $\mathbf{V}_A = 180\hat{\mathbf{j}}$ in/s; $\mathbf{V}_B = -231\hat{\mathbf{k}}$ in/s; $\boldsymbol{\omega}_3 = -21.56\hat{\mathbf{i}} + 7.45\hat{\mathbf{j}} - 5.8\hat{\mathbf{k}}$ rad/s; $\boldsymbol{\omega}_4 = -25.7\hat{\mathbf{i}}$ rad/s; $\mathbf{A}_A = 10\,800\hat{\mathbf{i}}$ in/s^2; $\mathbf{A}_B = -5950\hat{\mathbf{i}} - 3087\hat{\mathbf{k}}$ in/s^2; $\boldsymbol{\alpha}_3 = -447\hat{\mathbf{i}} + 588\hat{\mathbf{j}} + 436\hat{\mathbf{k}}$ rad/s^2; $\boldsymbol{\alpha}_4 = -343\hat{\mathbf{i}}$ rad/s^2

11-7 $\Delta\theta_4 = 48°$, time ratio = 1

11-9 $\Delta\theta_4 = 71°$, time ratio = 1.22

11-11 $\mathbf{V}_A = -2.34\hat{\mathbf{i}} - 1.35\hat{\mathbf{j}}$ m/s; $\mathbf{V}_B = 2.33\hat{\mathbf{j}} + 4.58\hat{\mathbf{k}}$ m/s; $\boldsymbol{\omega}_3 = 10.4\hat{\mathbf{i}} - 10.7\hat{\mathbf{j}} + 10.7\hat{\mathbf{j}} + 3.3\hat{\mathbf{k}}$ rad/s; $\boldsymbol{\omega}_4 = 19.5\hat{\mathbf{i}}$ rad/s; $\mathbf{A}_A = 48.6\hat{\mathbf{i}} - 84.2\hat{\mathbf{j}}$ m/s^2; $\mathbf{A}_B = -50\hat{\mathbf{j}} + 122\hat{\mathbf{k}}$ m/s^2; $\boldsymbol{\alpha}_3 = 175\hat{\mathbf{i}} - 111\hat{\mathbf{j}} + 101\hat{\mathbf{k}}$ rad/s^2; $\boldsymbol{\alpha}_4 = 328\hat{\mathbf{i}}$ rad/s^2

11-13 $\mathbf{V}_A = 12\hat{\mathbf{i}} + 20.8\hat{\mathbf{j}} + 41.6\hat{\mathbf{k}}$ in/s; $\mathbf{V}_B = 13.8\hat{\mathbf{i}}$ in/s; $\boldsymbol{\omega}_3 = 2.31\hat{\mathbf{i}} + 6.66\hat{\mathbf{j}} - 3.23\hat{\mathbf{k}}$ rad/s

11-15 (*a*) $m = 1$; (*b*) $\Delta\theta_4 = 90°$; $\Delta R_B = 8$ in; (*c*) $\mathbf{R}_B = 8.32\hat{\mathbf{j}}$ in; $\mathbf{R}_{BA} = 4\hat{\mathbf{i}} + 10.9\hat{\mathbf{j}} - 3.06\hat{\mathbf{k}}$ in

12-3 $P = 1460$ N

12-5 $P = 442$ N

12-7 $M_{12} = -276$ lb · in; $\mathbf{F}_{34} = 338\underline{/26.8°}$ lb; $\mathbf{F}_{14} = 231\underline{/242.1°}$ lb

12-9 $\mathbf{F}_{14} = 318\underline{/-61.7°}$ lb; $\mathbf{F}_{34} = 190\underline{/88.4°}$ lb; $\mathbf{F}_{23} = 228\underline{/56.6°}$ lb; $M_{12} = -761$ lb · in

12-11 Shaft forces: (*a*) $\mathbf{F}_{13} = 2520\underline{/0°}$ lb; (*b*) $\mathbf{F}_{13} = 1049\underline{/225°}$ lb; (*c*) $\mathbf{F}_{13} = 2250\underline{/-45°}$ lb

12-13 $\mathbf{F}_C = 216\underline{/189°}$ lb; $\mathbf{F}_D = 350\underline{/163°}$ lb

12-16 (*a*) F_A(radial) = 570 lb; F_A(thrust) = 85 lb

12-17 $\mathbf{F}_E = 163\hat{\mathbf{i}} - 192\hat{\mathbf{j}} + 355\hat{\mathbf{k}}$ lb; $\mathbf{F}_F = 110\hat{\mathbf{j}} + 145\hat{\mathbf{k}}$ lb

12-19 $\mathbf{F}_{23} = 306\underline{/230.4°}$ kN; $\mathbf{F}_{34} = 387\underline{/239.7°}$ kN; $\mathbf{F}_{14} = 387\underline{/59.7°}$ kN

12-21 $M_{12} = 437$ lb · in

13-1 $I_O = 0.0309$ lb · s^2 · in

13-3 $\mathbf{T}_2 = -190\hat{\mathbf{k}}$ lb · in

13-5 $\mathbf{F}_{14} = 300\underline{/-90°}$ lb; $\mathbf{F}_{34} = 755\underline{/156.6°}$ lb; $\mathbf{F}_{32} = 1535\underline{/7.95°}$ lb; $\mathbf{T}_2 = 2780\hat{\mathbf{k}}$ lb · in

13-7 $\mathbf{T}_2 = -2950\hat{\mathbf{k}}$ N · m; $\mathbf{F}_{14} = 11.7\underline{/205°}$ kN; $\mathbf{F}_{34} = 11.0\underline{/14.8°}$ kN; $\mathbf{F}_{12} = \mathbf{F}_{23} = 9.98\underline{/-20°}$ kN

13-9 $\mathbf{T}_2 = 674\hat{\mathbf{k}}$ N · m; $\mathbf{F}_{14} = 6.98\underline{/-84.9°}$ kN; $\mathbf{F}_{34} = 4.37\underline{/196°}$ kN; $\mathbf{F}_{23} = 2.59\underline{/230°}$ kN

13-11 $\mathbf{T}_2 = 4400\hat{\mathbf{k}}$ N · m; $\mathbf{F}_{14} = 7.57\underline{/282.7°}$ kN; $\mathbf{F}_{34} = 14.4\underline{/56.7°}$ kN; $\mathbf{F}_{23} = 20.9\underline{/45.4°}$ kN

13-13 $\boldsymbol{\alpha}_3 = 200$ rad/s^2 cw; $\mathbf{T}_2 = 11.1\hat{\mathbf{k}}$ N · m; $\mathbf{F}_{14} = 689\underline{/47.2°}$ N

13-15 $\mathbf{T}_2 = -241\hat{\mathbf{k}}$ N · m; $F^y_{14} = 646$ N; $\mathbf{F}_{23} = 3190\hat{\mathbf{i}} - 705\hat{\mathbf{j}}$ N

13-17 $\mathbf{F}_{14} = 546\hat{\mathbf{i}} + 397\hat{\mathbf{j}}$ N; $\mathbf{F}_{23} = -372\hat{\mathbf{i}} - 476\hat{\mathbf{j}}$ N

13-19 $\mathbf{T}_2 = 9750\hat{\mathbf{k}}$ N · m; $F^y_{14} = 22.1$ kN; $\mathbf{F}_{23} = 8.42\hat{\mathbf{i}} + 47\hat{\mathbf{j}}$ kN

13-21 $\mathbf{T}_4 = +22.4\hat{\mathbf{k}}$ lb · in; $\mathbf{F}_{23} = 6.80\underline{/240°}$ lb; $\mathbf{F}_{43} = 45.6\underline{/78.2°}$ lb

13-23 At $\theta_2 = 0°$, $\theta_3 = 120°$, $\theta_4 = 141.8°$, $\boldsymbol{\omega}_3 = 6.67$ rad/s cw, $\boldsymbol{\omega}_4 = 6.67$ rad/s cw, $\boldsymbol{\alpha}_3 = 141$ rad/s^2 cw, $\boldsymbol{\alpha}_4 = 64.1$ rad/s^2 cw, $\mathbf{T}_2 = 7468\hat{\mathbf{k}}$ lb · ft, $\mathbf{F}_{21} = 6734\underline{/-56°}$ lb, and $\mathbf{F}_{41} = 7883\underline{/142.8°}$ lb

14-1 At $X = 30\%$, $p_e = 251$ psi, $p_c = 50$ psi

14-3 See figure

14-5 $F_{41} = -230$ lb, $F_{34} = 935$ lb, $\mathbf{F}_{32} = 941\underline{/164°}$ lb, $T_{21} = 800$ lb · in

14-9 $F_{41} = -0.52$ kN, $\mathbf{F}_{34} = 3.19\underline{/189°}$ kN, $\mathbf{F}_{32} = 10.2\underline{/-38.4°}$ kN, $T_{21} = 191$ N · m

14-11 At $\omega t = 60°$, $X = 28.4\%$, $P = 2600$ lb, $\ddot{x} = -33.6(10)^3$ in/s^2, $F_{41} = -392$ lb, $\mathbf{F}_{34} = 2520\underline{/-8.9°}$ lb, $\mathbf{F}_{32} = 2460\underline{/168°}$ lb, $T_{21} = 3040$ lb · in

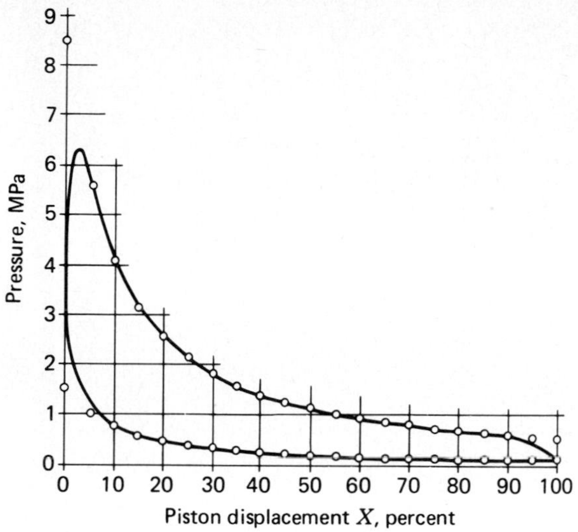

Answer to Problem 14-3

15-1 $\mathbf{F}_A = 64.7\underline{/76.1^\circ}$ kN, $\mathbf{F}_B = 16.2\underline{/76.1^\circ}$ kN, $m_C = 1.64$ kg

15-3 $\mathbf{F}_A = 8.06\underline{/-14.4^\circ}$ lb, $\mathbf{F}_B = 2.68\underline{/165.5^\circ}$ lb, $W_C = 2.63$ lb at $\theta_C = -14.4^\circ$

15-5 $\mathbf{F}_A = 13.15\underline{/90^\circ}$ kN, $\mathbf{F}_B = 0$

15-7 $m_L\mathbf{R}_L = 5.98\underline{/-16.5^\circ}$ oz · in, $m_R\mathbf{R}_R = 7.33\underline{/136.8^\circ}$ oz · in

15-9 Remove $m_l\mathbf{R}_L = 782.1\underline{|180.4^\circ}$ mg · m and $m_R\mathbf{R}_R = 236.8\underline{|301.2^\circ}$ mg · m

15-11 See Sec. 15-8 for answers

16-1 119°

16-3 189 rpm

16-5 Jump would begin at $\theta = 75^\circ$ when $\ddot{y}$ goes negative; $n = 242$ rpm

16-7 21.8 rad/s

16-9 At $\theta = 120^\circ$, $F_{32}^y = -572$ N, $T = 4.04$ N · m; at $\theta = 225^\circ$, $F_{32}^y = -608$ N, $T = -3.87$ N · m

16-11 $\theta = 59.99^\circ$, $F_{32}^y = -278$ lb, $T = 332$ lb · in; $\theta = 255^\circ$, $F_{32}^y = -139$ lb, $T = -104$ lb · in

APPENDIX

Table 1 Standard SI prefixes†,‡

Name	Symbol	Factor
exa	E	1 000 000 000 000 000 000 = 10^{18}
peta	P	1 000 000 000 000 000 = 10^{15}
tera	T	1 000 000 000 000 = 10^{12}
giga	G	1 000 000 000 = 10^{9}
mega	M	1 000 000 = 10^{6}
kilo	k	1 000 = 10^{3}
hecto§	h	100 = 10^{2}
deka§	da	10 = 10^{1}
deci§	d	0.1 = 10^{-1}
centi§	c	0.01 = 10^{-2}
milli	m	0.001 = 10^{-3}
micro	μ	0.000 001 = 10^{-6}
nano	n	0.000 000 001 = 10^{-9}
pico	p	0.000 000 000 001 = 10^{-12}
femto	f	0.000 000 000 000 001 = 10^{-15}
atto	a	0.000 000 000 000 000 001 = 10^{-18}

†If possible, use multiple and submultiple prefixes in steps of 1000. For example, specify length in millimetres, metres, or kilometres, say. In a combination unit, use prefixes only in the numerator. For example, use meganewton per square metre (MN/m^2) but not newton per square centimetre (N/cm^2) or newton per square millimetre (N/mm^2).

‡Spaces are used in SI instead of commas to group numbers to avoid confusion with the practice in some European countries of using commas for decimal points.

§Not recommended but sometimes encountered.

Table 2 Conversion from U.S. customary units to SI units

To convert from	To	Multiply by	
		Accurate†	Common
Foot (ft)	Metre (m)	3.048 000 E − 01*	0.305
Horsepower (hp)	Watt (W)	7.456 999 E + 02	746
Inch (in)	Metre (m)	2.540 000 E − 02*	0.025 4
Mile, U.S. statute (mi)	Metre (m)	1.609 344 E + 03*	1610
Pound force (lb)	Newton (N)	4.448 222 E + 00	4.45
Pound mass (lbm)	Kilogram (kg)	4.535 924 E − 01	0.454
Poundal (lbm · ft/s^2)	Newton (N)	1.382 550 E − 01	0.138
Pound-foot (lb · ft)	Newton-metre (N · m)	1.355 818 E + 00	1.35
	Joule (J)	1.355 818 E + 00	1.35
Pound-foot/second (lb · ft/s)	Watt (W)	1.355 818 E + 00	1.35
Pound-inch (lb · in)	Newton-metre (N · m)	1.128 182 E − 01	0.113
	Joule (J)	1.128 182 E − 01	0.113
Pound-inch/second (lb · in/s)	Watt (W)	1.128 182 E − 01	0.113
Pound/foot^2 (lb/ft^2)	Pascal (Pa)	4.788 026 E + 01	47.9
Pound/inch^2 (lb/in^2), (psi)	Pascal (Pa)	6.894 757 E + 03	6890
Revolutions/minute (rpm)	Radian/second (rad/s)	1.047 198 E − 01	0.105
Slug	Kilogram (kg)	1.459 390 E + 01	14.6
Ton, short (2000 lbm)	Kilogram (kg)	9.071 847 E + 02	907

† An asterisk indicates that the conversion factor is exact.

Table 3 Conversion from SI units to U.S. customary units

To convert from	To	Multiply by	
		Accurate	Common
Joule (J)	Pound-foot (lb · ft)	7.375 620 E − 01	0.737
	Pound-inch (lb · in)	8.850 744 E + 00	8.85
Kilogram (kg)	Pound mass (lbm)	2.204 622 E + 00	2.20
	Slug	6.852 178 E − 02	0.0685
	Ton, short (2000 lbm)	1.102 311 E − 03	0.001 10
Metre (m)	Foot (ft)	3.280 840 E + 00	3.28
	Inch (in)	3.937 008 E + 01	39.4
	Mile (mi)	6.213 712 E + 02	621
Newton (N)	Pound (lb)	2.248 089 E − 01	0.225
	Poundal (lbm · ft/s^2)	7.233 012 E + 00	7.23
Newton-metre (N · m)	Pound-foot (lb · ft)	7.375 620 E − 01	0.737
	Pound-inch (lb · in)	8.850 744 E + 00	8.85
Newton-metre/second (N · m/s)	Horsepower (hp)	1.341 022 E − 03	0.001 34
Pascal (Pa)	Pound/foot^2 (lb/ft^2)	2.088 543 E − 02	0.0209
	Pound/inch^2 (lb/in^2), (psi)	1.450 370 E − 04	0.000 145
Radian/second (rad/s)	Revolutions/minute (rpm)	9.549 297 E + 00	9.55
Watt (W)	Horsepower (hp)	1.341 022 E − 03	0.001 34
	Pound-foot/second (lb · ft/s)	7.375 620 E − 01	0.737
	Pound-inch/second (lb · in/s)	8.850 744 E + 00	8.85

Table 4 Properties of areas

A = area
I = area moment of inertia
J = polar area moment of inertia
k = radius of gyration
$\bar{y}$ = centroidal distance

Rectangle

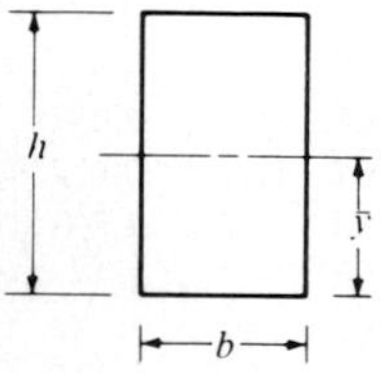

$$A = bh \qquad k = 0.289h$$

$$I = \frac{bh^3}{12} \qquad y = \frac{h}{2}$$

Triangle

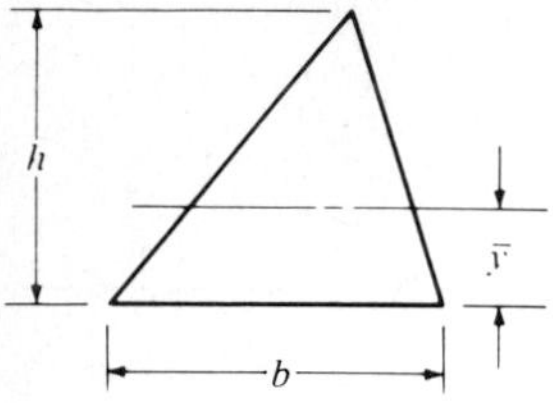

$$A = \frac{bh}{2} \qquad k = 0.236h$$

$$I = \frac{bh^3}{36} \qquad \bar{y} = \frac{h}{3}$$

Circle

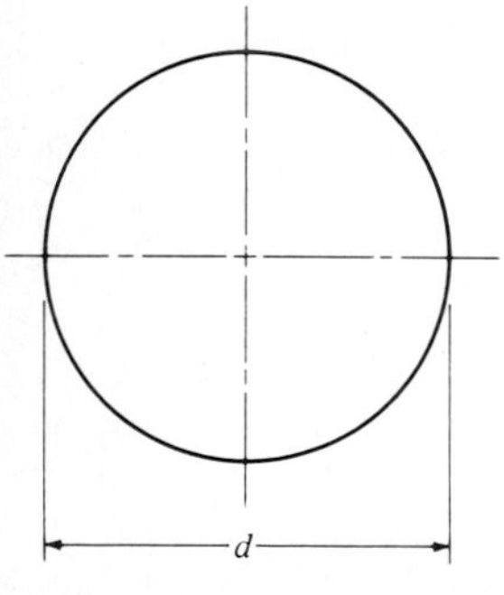

$$A = \frac{\pi d^2}{4} \qquad k = \frac{d}{4}$$

$$I = \frac{\pi d^4}{64} \qquad \bar{y} = \frac{d}{2}$$

$$J = \frac{\pi d^4}{32}$$

Hollow circle

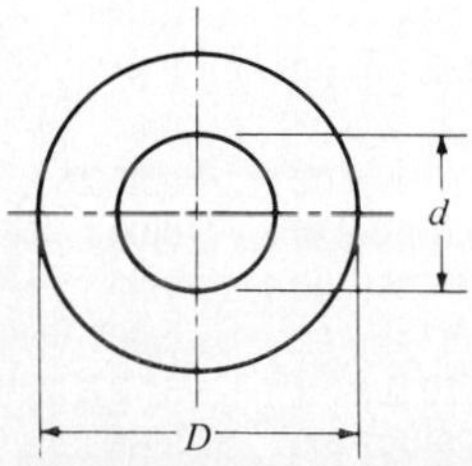

$$A = \frac{\pi}{4}(D^2 - d^2) \qquad k = \frac{1}{4}\sqrt{D^2 + d^2}$$

$$I = \frac{\pi}{64}(D^4 - d^4) \qquad \bar{y} = \frac{D}{2}$$

$$J = \frac{\pi}{32}(D^4 - d^4)$$

Table 5 Mass moments of inertia

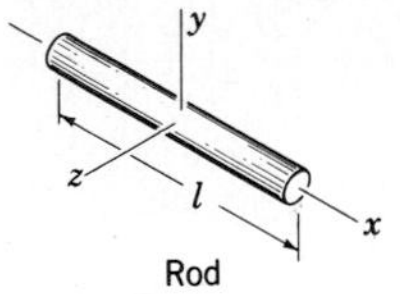

Rod

$$i_y = i_z = \frac{ml^2}{12}$$

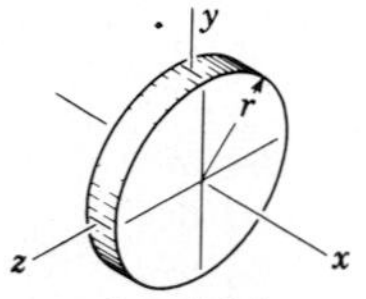

Round disk

$$i_x = \frac{mr^2}{2}$$

$$i_y = i_z = \frac{mr^2}{4}$$

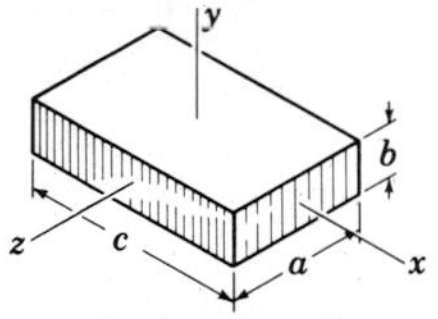

Rectangular prism

$$i_x = \frac{m(a^2 + b^2)}{12}$$

$$i_y = \frac{m(a^2 + c^2)}{12}$$

$$i_z = \frac{m(b^2 + c^2)}{12}$$

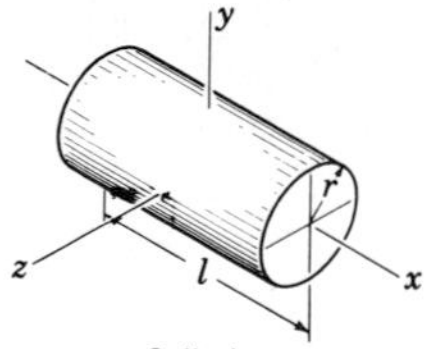

Cylinder

$$i_x = \frac{mr^2}{2}$$

$$i_y = i_z = \frac{m(3r^2 + l^2)}{12}$$

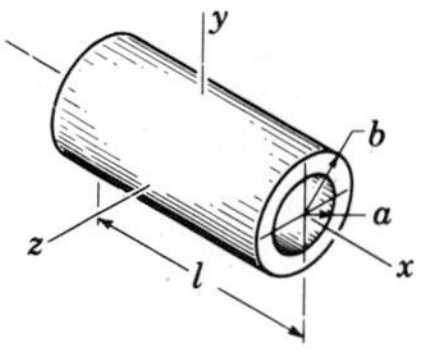

Hollow cylinder

$$i_x = \frac{m(a^2 + b^2)}{2}$$

$$i_y = i_z = \frac{m(3a^2 + 3b^2 + l^2)}{12}$$

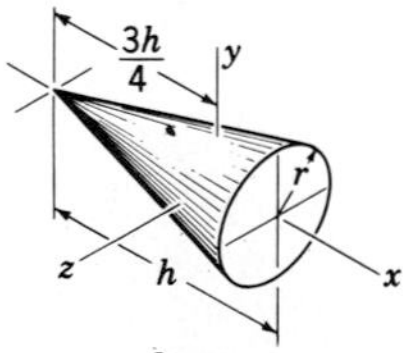

Cone

$$i_x = \frac{3mr^2}{10}$$

$$i_y = i_z = \frac{m(12r^2 + 3h^2)}{80}$$

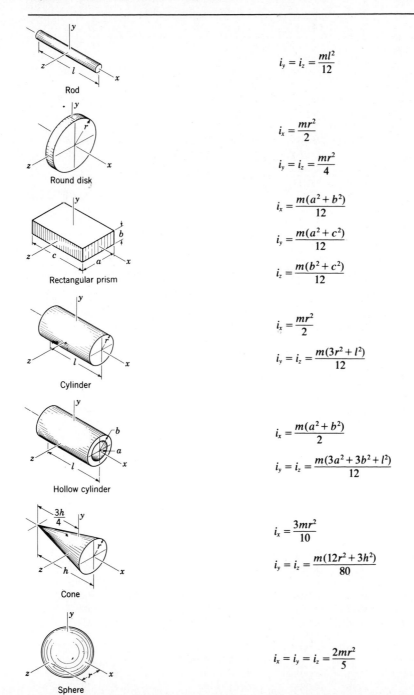

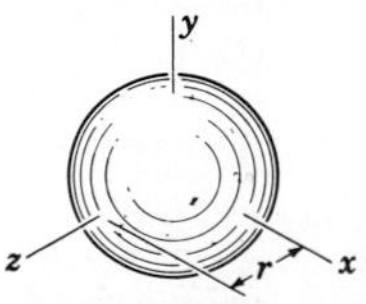

Sphere

$$i_x = i_y = i_z = \frac{2mr^2}{5}$$

Table 6 Involute functions

Deg	*Inv φ*	*Deg*	*Inv φ*	*Deg*	*Inv φ*	*Deg*	*Inv φ*
00.0	.000000						
00.1	.000000	03.1	.000053	06.1	.000404	09.1	.001349
00.2	.000000	03.2	.000058	06.2	.000424	09.2	.001394
00.3	.000000	03.3	.000064	06.3	.000445	09.3	.001440
00.4	.000000	03.4	.000070	06.4	.000467	09.4	.001488
00.5	.000000	03.5	.000076	06.5	.000489	09.5	.001536
00.6	.000000	03.6	.000083	06.6	.000512	09.6	.001586
00.7	.000000	03.7	.000090	06.7	.000536	09.7	.001636
00.8	.000000	03.8	.000097	06.8	.000560	09.8	.001688
00.9	.000001	03.9	.000105	06.9	.000586	09.9	.001740
01.0	.000002	04.0	.000114	07.0	.000612	10.0	.001794
01.1	.000002	04.1	.000122	07.1	.000638	10.1	.001849
01.2	.000003	04.2	.000132	07.2	.000666	10.2	.001905
01.3	.000004	04.3	.000141	07.3	.000694	10.3	.001962
01.4	.000005	04.4	.000151	07.4	.000723	10.4	.002020
01.5	.000006	04.5	.000162	07.5	.000753	10.5	.002079
01.6	.000007	04.6	.000173	07.6	.000783	10.6	.002140
01.7	.000009	04.7	.000184	07.7	.000815	10.7	.002202
01.8	.000010	04.8	.000197	07.8	.000847	10.8	.002265
01.9	.000012	04.9	.000209	07.9	.000880	10.9	.002329
02.0	.000014	05.0	.000222	08.0	.000914	11.0	.002394
02.1	.000016	05.1	.000236	08.1	.000949	11.1	.002461
02.2	.000019	05.2	.000250	08.2	.000985	11.2	.002528
02.3	.000022	05.3	.000265	08.3	.001022	11.3	.002598
02.4	.000025	05.4	.000280	08.4	.001059	11.4	.002668
02.5	.000028	05.5	.000296	08.5	.001098	11.5	.002739
02.6	.000031	05.6	.000312	08.6	.001137	11.6	.002812
02.7	.000035	05.7	.000329	08.7	.001178	11.7	.002894
02.8	.000039	05.8	.000347	08.8	.001219	11.8	.002962
02.9	.000043	05.9	.000366	08.9	.001262	11.9	.003039
03.0	.000048	06.0	.000384	09.0	.001305	12.0	.003117
12.1	.003197	16.3	.007932	20.6	.016337	24.8	.029223
12.2	.003277	16.4	.008082	20.7	.016585	24.9	.029598
12.3	.003360	16.5	.008234	20.8	.016836	25.0	.029975
12.4	.003443			20.9	.017089		
12.5	.003529	16.6	.008388	21.0	.017345	25.1	.030357
		16.7	.008544			25.2	.030741
12.6	.003615	16.8	.008702	21.1	.017603	25.3	.031130
12.7	.003712	16.9	.008863	21.2	.017865	25.4	.031521
12.8	.003792	17.0	.009025	21.3	.018129	25.5	.031917
12.9	.003883			21.4	.018395		
13.0	.003975	17.1	.009189	21.5	.018665	25.6	.032315
		17.2	.009355			25.7	.032718

Table 6 (continued)

Deg	*Inv φ*	*Deg*	*Inv φ*	*Deg*	*Inv φ*	*Deg*	*Inv φ*
13.1	.004069	17.3	.009523	21.6	.018937	25.8	.033124
13.2	.004164	17.4	.009694	21.7	.019212	25.9	.033534
13.3	.004261	17.5	.009866	21.8	.019490	26.0	.033947
13.4	.004359	17.6	.010041	21.9	.019770	26.1	.034364
13.5	.004459	17.7	.010217	22.0	.020054	26.2	.034785
13.6	.004561	17.8	.010396	22.1	.020340	26.3	.035209
13.7	.004664	17.9	.010577	22.2	.020630	26.4	.035637
13.8	.004768	18.0	.010760	22.3	.020921	26.5	.036069
13.9	.004874	18.1	.010946	22.4	.021216	26.6	.036505
14.0	.004982	18.2	.011133	22.5	.021514	26.7	.036945
14.1	.005091	18.3	.011323	22.6	.021815	26.8	.037388
14.2	.005202	18.4	.011515	22.7	.022119	26.9	.037835
14.3	.005315	18.5	.011709	22.8	.022426	27.0	.038287
14.4	.005429	18.6	.011906	22.9	.022736	27.1	.038696
14.5	.005545	18.7	.012105	23.0	.023049	27.2	.039201
14.6	.005662	18.8	.012306	23.1	.023365	27.3	.039664
14.7	.005782	18.9	.012509	23.2	.023684	27.4	.040131
14.8	.005903	19.0	.012715	23.3	.024006	27.5	.040602
14.9	.006025	19.1	.012923	23.4	.024332	27.6	.041076
15.0	.006150	19.2	.013134	23.5	.024660	27.7	.041556
15.1	.006276	19.3	.013346	23.6	.024992	27.8	.042039
15.2	.006404	19.4	.013562	23.7	.025326	27.9	.042526
15.3	.006534	19.5	.013779	23.8	.025664	28.0	.043017
15.4	.006665	19.6	.013999	23.9	.026005	28.1	.043513
15.5	.006799	19.7	.014222	24.0	.026350	28.2	.044012
15.6	.006934	19.8	.014447	24.1	.026697	28.3	.044516
15.7	.007071	19.9	.014674	24.2	.027048	28.4	.045024
15.8	.007209	20.0	.014904	24.3	.027402	28.5	.045537
15.9	.007350	20.1	.015137	24.4	.027760	28.6	.046054
16.0	.007493	20.2	.015372	24.5	.028121	28.7	.046575
16.1	.007637	20.3	.015609	24.6	.028485	28.8	.047100
16.2	.007784	20.4	.015850	24.7	.028852	28.9	.047630
		20.5	.016092			29.0	.048164
29.1	.048702	33.1	.074188	37.1	.108777	41.1	.155025
29.2	.049245	33.2	.074932	37.2	.109779	41.2	.156358
29.3	.049792	33.3	.075683	37.3	.110788	41.3	.157700
29.4	.050344	33.4	.076439	37.4	.111805	41.4	.159052
29.5	.050901	33.5	.077200	73.5	.112828	41.5	.160414
29.6	.051462	33.6	.077968	37.6	.113860	41.6	.161785
29.7	.052027	33.7	.078741	37.7	.114899	41.7	.163165
29.8	.052597	33.8	.079520	37.8	.115945	41.8	.164556
29.9	.053172	33.9	.080305	37.9	.116999	41.9	.165956

Table 6 (continued)

Deg	*Inv* ϕ	*Deg*	*Inv* ϕ	*Deg*	*Inv* ϕ	*Deg*	*Inv* ϕ
30.0	.053751	34.0	.081097	38.0	.118060	42.0	.167366
30.1	.054336	34.1	.081974	38.1	.119130	42.1	.168786
30.2	.054924	34.2	.082697	38.2	.120207	42.2	.170216
30.3	.055519	34.3	.083506	38.3	.121291	42.3	.171656
30.4	.056116	34.4	.084321	38.4	.122384	42.4	.173106
30.5	.056720	34.5	.085142	38.5	.123484	42.5	.174566
30.6	.057267	34.6	.085970	38.6	.124592	42.6	.176037
30.7	.057940	34.7	.086804	38.7	.125709	42.7	.177518
30.8	.058558	34.8	.087644	38.8	.126833	42.8	.179009
30.9	.059181	34.9	.088490	38.9	.127965	42.9	.180511
31.0	.059809	35.0	.089342	39.0	.129106	43.0	.182023
31.1	.060441	35.1	.090201	39.1	.130254	43.1	.183546
31.2	.061079	35.2	.091066	39.2	.131411	43.2	.185080
31.3	.061721	35.3	.091938	39.3	.132576	43.3	.186625
31.4	.062369	35.4	.092816	39.4	.133749	43.4	.188180
31.5	.063022	35.5	.093701	39.5	.134931	43.5	.189746
31.6	.063680	35.6	:094592	39.6	.136122	43.6	.191324
31.7	.064343	35.7	.095490	39.7	.137320	43.7	.192912
31.8	.065012	35.8	.096395	39.8	.138528	43.8	.194511
31.9	.065685	35.9	.097306	39.9	.139743	43.9	.196122
32.0	.066364	36.0	.098224	40.0	.140968	44.0	.197744
32.1	.067048	36.1	.099149	40.1	.142201	44.1	.199377
32.2	.067738	36.2	.100080	40.2	.143443	44.2	.201022
32.3	.068432	36.3	.101019	40.3	.144694	44.3	.202678
32.4	.069133	36.4	.101964	40.4	.145954	44.4	.204346
32.5	.069838	36.5	.102916	40.5	.147222	44.5	.206026
32.6	.070549	36.6	.103875	40.6	.148500	44.6	.207717
32.7	.071266	36.7	.104841	40.7	.149787	44.7	.209420
32.8	.071988	36.8	.105814	40.8	.151082	44.8	.211135
32.9	.072716	36.9	.106795	40:9	.152387	44.9	.212863
33.0	073449	37.0	.107782	41.0	.153702	45.0	.214602

INDEX

INDEX